The Measurement and Control of Diesel Particulate Emissions Part 2

(Selected papers through 1981)

Prepared under the auspices of the Fuels and Lubricants Activity

Published by:
Society of Automotive Engineers, Inc.
400 Commonwealth Drive
Warrendale, PA 15096

ISBN 0-89883-113-X
SAE/PT-82/25

Library of Congress Catalog Card Number: 79-67589
Printed in U.S.A.

PREFACE

THE NEW NATIONAL IMPORTANCE given to energy conservation and the environment has also broadened the segment of the technical community having an intense interest in diesels and their particulate emissions. It is this intensified and broadened interest which SAE hopes to serve with this volume of SAE papers. This book is the second volume in the series dealing with the measurement and control of diesel particulate emissions. The first volume was published in 1979 (1)*.

The oil embargo to the United States in late 1973 dramatically emphasized the limited nature of our domestic petroleum supply and the dependence of the U.S. on foreign crude oil. The increased cost of fuels and the passage by Congress of the Energy Policy and Conservation Act of 1975, which set a passenger-car fuel economy standard that climbs to 27.5 miles per gallon in 1985, made the diesel engine a meaningful alternative to traditional spark-ignition engines in cars, light-duty, and medium-duty trucks. Since heavy-duty diesel powered buses and trucks were already logging large annual mileages (2), the potential for significant increases in the number of diesel engines on the road rekindled concern for particulate emissions, which have always been an inherent part of the diffusion-controlled diesel combustion process. The increased number of diesel powered vehicles shows an increased demand for diesel fuel in the 1980's (3).

For regulatory purposes, particulates are defined to include any solid and liquid (excluding water) material that collects on a sample filter which is maintained at or below 125°F in a diluted exhaust stream. Diesel particulates are typically comprised of carbon, condensed hydrocarbon material and sulfate. Early tests with Constant Volume Sampler (CVS) dilution systems showed that diesel-powered vehicles have higher particulate emission levels than vehicles powered by conventional catalyst-equipped spark-ignition engines burning lead-free gasoline (1). Furthermore, the Environmental Protection Agency has shown that certain components of the soluble organic fraction of the particulate are mutagenic and potentially carcinogenic (4).

The Clean Air Act of August 1977 required that EPA set particulate emission standards for the 1981 model year. The standards for cars and light-duty trucks were proposed in the February 1, 1979 Federal Register (5). The particulate standards in the final regulations are 0.6 gram/mile in 1982 and 0.2 gram/mile in 1985 (6). This Act also sets a 1.0 gram/mile NOx standard for light-duty vehicles for the 1981 model year with a provision allowing waiver of this standard up to 1.5 gram/mile for diesel engines through 1984 (7). The 1.0 gram/mile NOx requirement accentuates the task of meeting the particulate standard since modifications made to the combustion process to decrease NOx generally result in increased particulate levels.

The 1981 heavy-duty vehicle standards for particulates are based on the already existing smoke standards, which were first applied to 1970 model engines. EPA has also proposed that new transient cycle gaseous and particulate standards be implemented in the 1983 model year for heavy-duty engines (8). They have also proposed a 0.25 gram/bhp-hr particulate standard for 1986 (9).

As a result of the expanded role of diesel engines in the transportation system and the continuing concern about the effects of the exhaust from these powerplants on the atmosphere and our well-being, diesel particulate emissions have become a subject of increased interest at SAE technical meetings. A collection of SAE papers dealing primarily with particulate measurement, characterization, and engine baseline levels was bound in the first volume on particulate emissions. This second volume contains papers that shed further insight on the origin of diesel particulates, their fundamental character, the trade-offs involved in tailoring the combustion process for particulate control, and the influence of dilution parameters on the particulate material collected. In addition, papers are included which deal with the engineering trade-offs involved in tailoring the combustion process for particulate control and address the difficult job of collecting and disposing of the particulates in the exhaust system.

*Numbers in parentheses refer to references which are listed at the end of the Preface.

Two technical areas pertaining to particulate emissions that are generally not covered in the SAE literature are: 1) transport and transformation processes in the atmosphere and 2) health and welfare effects. Abstracts of important papers on recent studies in these areas are included in this book to make it a complete and convenient reference dealing with particulate emissions. The proceedings from a recent EPA meeting on diesel engines contains a number of papers on the health aspects of diesel particulate emissions (10). The National Academy of Sciences has recently completed a study on the health effects of exposure to diesel exhausts (11).

This book consists of papers dating back to 1979 and includes papers presented through March 1981. The papers reprinted in their entirety here are only those which were determined by the Editorial Advisory Committee to represent a typical and select cross section of available papers. The remaining papers from the diesel particulate area, published by SAE during the March 1979—March 1981 time period are included in the appended bibliography, which includes abstracts of those papers for the convenience of the reader.

John H. Johnson, Chairman
Frank Black
J. Scott MacDonald
Editorial Advisory Committee
July, 1981

REFERENCES

1. "Measurement and Control of Diesel Particulate Emissions," SAE Progress in Technology (PT-17), 1979.

2. Anil B. Jambekar and John H. Johnson, "An Emission and Fuel Usage Computer Model for Trucks and Buses," SAE Paper 780630, 1978.

3. Anil B. Jambekar and John H. Johnson, "The Effect of Truck Dieselization on Fuel Usage," SAE Paper 810022, 1981.

4. D. S. Barth and S. M. Blacker, "The EPA Program to Assess the Public Health Significance of Diesel Emissions," JAPCA, Vol. 28, No. 8, 1978, pp. 769—771.

5. "EPA-Control of Air Pollution from New Motor Vehicle Engines: Certification and Test Procedures — Particulate Regulations for Light-Duty Diesel Vehicles," Federal Register, Vol. 44, No. 23, February 1, 1979, pp. 6651—6671.

6. "Standard for Emissions of Particulate Regulation for Diesel-Fueled Light-Duty Vehicles and Light-Duty Trucks," Federal Register, Vol. 45, No. 45, March 5, 1980, pp. 14496—14525.

7. "EPA-Waiver of 1981 Automobile Emissions Standards — Guidelines for Applications for Waiver of the 1981 NOx Emission Standard, Based Upon the Use of Diesel Engine Technology," Federal Register, Vol. 43, No. 136, July 14, 1978, pp. 30341—30347.

8. "EPA-1983 and Later Model Year Heavy-Duty Engines — Proposed Gaseous Emission Regulation," Federal Register, Vol. 44, No. 31, February 13, 1979, pp. 9464—9539.

9. "Control of Air Pollution From New Motor Vehicles and New Motor Vehicle Engines; Particulate Regulation for Heavy-Duty Diesel Engines," Federal Register, Vol. 46, No. 4, January 7, 1981, pp. 1910—1967.

10. W. E. Pepelko, R. M. Danner, and N. A. Clarke, Editors, "Health Effects of Diesel Engine Emissions," Proceedings of an International Symposium, December 3—5, 1979, EPA-600/9-80-057a, November 1980.

11. "Health Effects of Exposure to Diesel Exhausts," A Panel Report from the Diesel Impact Study Committee, National Research Council, 1980.

EDITORIAL ADVISORY COMMITTEE PREPARING PT-25

TABLE OF CONTENTS

FORMATION STUDIES

MEASUREMENT TECHNIQUES AND FUNDAMENTAL CONSIDERATIONS

CHARACTERIZATION, ENGINE CONTROLS & FUEL EFFECTS

Bonny Kubik

EXHAUST AFTERTREATMENT

FORMATION STUDIES

Some Rudiments of Diesel Particulate Emissions*

C. A. Amann, D. L. Stivender, S. L. Plee, and J. S. MacDonald
General Motors Research Laboratories
Warren, MI

THE AMERICAN PUBLIC HAS BEEN REGISTERING its interest in fuel-efficient private transportation through its new-car purchasing pattern. In addition, the federal government has imposed a passenger-car fuel-economy standard that climbs to 27.5 mi/gal in 1985. If the 27.5 mi/gal standard were to be achieved exclusively from higher engine thermal efficiency, it would require a 37.5% efficiency increase in just five years. In view of this situation, it is no wonder that industry and government alike have been searching for an alternative to today's dominant reciprocating internal-combustion spark-ignition gasoline engine. Experimental automobiles with steam, gas turbine, and Stirling engines have been built and evaluated. Although these alternative engines may possess other advantages, none of them has yet matched the fuel economy of the conventional gasoline engine.

In contrast, the diesel engine has already demonstrated its ability to surpass the fuel economy of the conventional spark-ignition engine. The magnitude of its fuel-economy advantage over the spark-ignition engine depends on both the duty cycle and the performance potential of the vehicles being compared, but in city-suburban driving the diesel-powered car typically travels 25% farther on a given volume of fuel than its equal-performance counterpart powered by a spark-ignition engine. In terms of gain in fuel economy, this is equivalent to decreasing the mass of a full-sized car with a gasoline engine by about 400 kg (900 lb), i.e., to decreasing the mass of a full-sized car to that of a subcompact.

The nature of the diesel combustion process leads to the emission of exhaust particulates. The particulate mass emitted is of the same order as that issuing from a comparable car powered by a spark-ignition engine operating on leaded gasoline. However the particulate output of that same diesel car is at least an order of magnitude greater than that of the car with the spark-ignition engine that has an exhaust catalyst and is operated on unleaded gasoline.

Out of concern for the effect that increased market penetration of diesel cars may have on total suspended particulates in certain metropolitan regions, the Environmental Protection Agency has proposed an exhaust particulate standard for light-duty diesels of 0.6 g/mi on the Federal Urban Driving Schedule in 1981. For regulatory purposes, particulates comprise any solid or liquid (except water) that collects on a filter inserted in a line which samples a diluted exhaust stream. Under EPA proposed regulations, dilution ratio must be sufficient that the filter temperature does not exceed 52°C (125°F).

Early in their development, particulates

*Paper 800251 presented at the Congress and Exposition, Detroit, Michigan, 1980.

ABSTRACT

In the future the fuel-efficient passenger-car diesel faces difficult particulate emission standards. To provide a foundation for understanding this problem, four aspects of particulate production in such an engine are considered -- the combustion environment for soot formation, the phenomenological mechanisms involved, in-cylinder oxidation of soot, and the effects of exhaust dilution on the particulate mass collected on a sampling filter. Where significant experimental results from diesel engines are lacking, appropriate non-engine experience is reviewed to contribute additional insight. In the concluding section some practical implications are drawn for the diesel engine.

exist as individual sub-micron carbonaceous nuclei of basically spherical shape formed during diesel combustion. These particles grow both by collision and resultant coagulation, and by deposition of additional material on their surfaces. Given the proper environment, both carbon and hydrogen oxidize from these soot particles.

Once the soot particles have been exhausted from the engine, diluted in a tunnel with fresh air, and trapped on the filter, the particulate sample contains a soluble hydrocarbon fraction. The hydrocarbons are adsorbed on the soot particles and under some conditions may also condense from the gaseous stream. The soluble fraction typically comprises 20% (by mass) organic solubles, although substantial variations from this level may occur from one engine type to another and from one operating condition to another. While available epidemiological data do not support any health concern over diesel particulates, biological assessments of the effect of the adsorbed fraction are still underway in a number of laboratories.

Because the proposed 1981 particulate standard of 0.6 g/mi stretches existing diesel technology, and because the proposed 1983 standard of 0.2 g/mi has not yet been demonstrated feasible in a production diesel car, announcement of the proposed standards has intensified research on diesel particulates. Much of the initial work focused on characterization of particulate size and composition [1-8]*. Because of the complexities of conducting measurements in the hostile environment of a diesel cylinder, however, studies of the processes responsible for generating diesel particulates have been more limited. In the indirect-injection engine that typifies current passenger-car diesels, soot production has been followed optically [9] and by sampling the ante-chamber contents [10], but neither of these studies addressed the phenomenological aspects of soot production. Soot formation has been studied more extensively in direct-injection diesels, however [11-13].

Pending completion of additional studies conducted in passenger-car diesel engine configurations, the foundation for an understanding of the rudiments of diesel particulate production is provided by existing literature treating observations of relevant phenomena outside the engine environment. In several instances it is possible to gain insight through simple analyses based on these observations. That is primarily the path followed in this paper. Among the sources of input information are chemical-equilibrium relationships, a variety of laboratory bench experiments, fundamental studies of smoke, and experiences from the carbon black industry.

In this paper four aspects of the particulate production process are considered in turn -- the combustion environment responsible for soot formation, the mechanisms involved, oxidation of soot, and exhaust dilution as it occurs in the particulate sampling process. These four sections are followed by one dealing with some practical implications following from the study of the four aspects considered.

*Numbers in brackets designate References at end of paper.

SOOT FORMATION

The formation of soot particles appears to be inherent in the combustion environment of the diesel engine. Such combustion involves the reactions of fuel introduced into the oxidant as a liquid spray. It has been suggested that knowledge of the laws governing the rate of droplet combustion is a prerequisite to the understanding of spray combustion [14]. Much of the past research into droplet combustion has focused on an isolated single droplet. The most elementary theoretical model treats combustion of a stationary single-component spherical droplet of fuel immersed in a uniform space-field of oxidant and surrounded by a concentric flame zone. Energy from polytropic compression and from the exothermic reaction of fuel and oxygen in the flame zone supplies the necessary latent heat for vaporizing fuel from the droplet surface. The resulting fuel vapor diffuses radially outward into the flame zone at the same time that oxygen from the surrounding oxidant field diffuses radially inward into the flame zone.

As reviewed elsewhere [14], droplet-combustion research has been extended from this elementary level to include consideration of chemical kinetics, nonsteady-state effects, effects of a relative velocity field, multi-component fuels, etc. Of particular interest to the diesel is droplet behavior at supercritical conditions because in the conventional diesel engine the cylinder pressure during fuel injection and combustion typically exceeds the critical pressure of many components of the fuel. It does not necessarily follow that an injected liquid droplet of diesel fuel is immediately transformed into a puff of gas, however, for the critical pressure of a multi-component mixture such as exists in the diesel cylinder may exceed the critical pressure of its individual components [15]. In fact, even in the case of an isolated single-component fuel droplet (n-decane) in air, Faeth et al. found that little evaporation occurred until the droplet reached its critical temperature even though the pressure was more than twice the critical pressure [16]. A high concentration of dissolved gas existing as the supercritical burning condition is approached apparently contributes to this phenomenon [17]. Others have reviewed more generally the existing knowledge of supercritical injection [14,18].

But a fuel spray is not comprised of isolated droplets. In considering several experiments with burning liquid fuel sprays, one reviewer [19] finds no evidence of flames surrounding individual droplets. Rather, most of the droplets evaporate in a region of two-phase mixture near the fuel-nozzle exit, where the temperature is substantially below that of the flame, the mixture is fuel-rich, and mass transfer from the droplet surfaces is impeded by the high concentration of fuel in the gaseous medium surrounding the droplets. This region of two-phase mixture is encompassed by the flame zone. Experiments with spray combustion in a vertical cylindrical furnace [20] suggest that in the flame zone nearest the region of droplet evaporation, some droplets still exist in low concentration and the combustion is essentially rich. In the surrounding flame zone further from the nozzle exit, the local mixture is leaner and oxidation is essentially completed.

Thus it is seen that the spray flame is mixing-controlled, relying on turbulent diffusion to mix the air and the fuel vapor. In one sense it bears some resemblance to the flame surrounding an isolated droplet, in that in both cases the rate of burning depends largely on the rates of diffusion of fuel vapor and air into the flame front from opposite directions.

In the diesel engine the droplets near the fuel nozzle are believed to evaporate fairly rapidly. Although the precise temperature and pressure conditions and the distribution of phases within the diesel spray have not yet been clearly established, it appears that the major portion of the entire spray is in gaseous form [19]. Indeed, some notable success has been achieved in modeling diesel combustion by treating the fuel spray as a gaseous diffusion jet [21,22].

If a detailed understanding of spray combustion in the diesel engine is incomplete, understanding of the chemical mechanism(s) responsible for soot formation is even less satisfactory. There is generally agreement that the rate of formation of soot is kinetically controlled. Beyond that, the precise mechanism by which soot forms in a flame is uncertain. At least eight different theories have been advanced by various investigators [23].

In the face of such uncertainty, those attempting to model soot formation in diesel engines have used an Arrhenius relationship for carbon formation [13,24]. The constants have been selected to fit existing experimental engine data. Such expressions are then used to extrapolate experimental data analytically in order to predict soot formation at new operating conditions.

This is an established and valuable engineering approach, but it is not without its dangers. Since it lacks a firm physical foundation, one can never be certain that extending it beyond the bounds of experience will produce valid results. Moreover, there is no assurance that an expression derived for one engine design will fit a completely different engine design satisfactorily.

This latter concern is particularly troublesome in dealing with the current breed of passenger-car diesels. Nearly all past analyses of soot formation in diesel engines have focused on the direct-injection type of engine. In contrast, all presently commercial passenger-car diesels are of the indirect-injection type.

A cross section through such an engine is shown in Fig. 1. In this engine fuel is sprayed

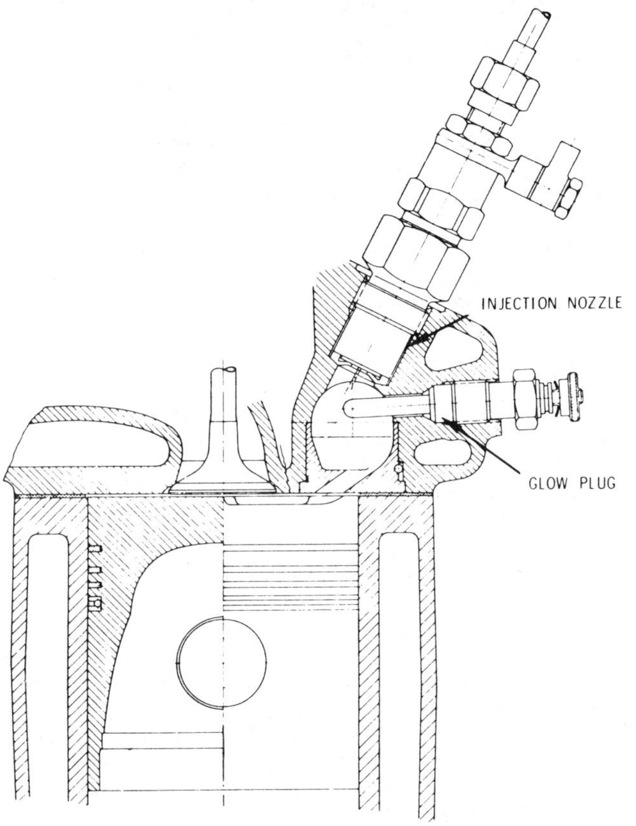

Fig. 1 - Cross section through a typical indirect-injection diesel engine combustion chamber

into an ante-chamber, where it is vigorously mixed with air that has been forced to swirl about the horizontal axis of the chamber as the ascending piston pushes it through the tangential entry passageway. As combustion within the swirl chamber releases energy, and as the piston descends during the expansion stroke to increase the ratio of main-chamber to swirl-chamber volume, burned and unburned mixture is expelled from the swirl chamber into the main chamber. Application to this engine of a heat-release model that uses measured chamber pressures as inputs [25] has indicated that a substantial portion of the energy release during combustion occurs in the main chamber. Clearly, the pattern of combustion and heat release in such an indirect-injection diesel differs markedly from that typical of an engine in which the fuel spray is injected directly into the space above the piston. Although a number of diesel combustion models have been published [13,21,24,26], none of these adequately addresses the complex pattern of combustion in the indirect-injection version of the engine.

EQUILIBRIUM CARBON-PRODUCTION SUBMODEL - In this study the quandary imposed by the absence of an adequate analytical model treating fuel-air mixing and combustion in the indirect-injection diesel is avoided by examining single packets of mixture of arbitrarily specified stoichiometry. No attempt is made to aggregate these packets in a manner that would simulate a total cylinder charge. Thus this study does not deal with an engine combustion model, but rather with a submodel that illustrates the global thermochemical driving forces and the resulting species quantitatively.

The issue of which of the suggested soot formation mechanisms is correct is evaded in this study by considering only the equilibrium compositions toward which the global reactions would be driven. The solid carbon formed through combustion is assumed equivalent to soot. It is not intended that the calculated equilibrium compositions accurately reflect compositions that actually exist for the same packet stoichiometry and temperature in an operating engine. The value of the equilibrium calculation lies rather in whatever insight it is able to provide into the significant factors influencing the formation of soot.

Generally all spontaneous processes act to reduce bound energy and to maximize entropy. Therefore minimization of the Gibbs free energy of the reactants can provide a quantitative assessment of the distribution of reaction species at equilibrium. The computer program used to conduct such a calculation is a modified version of routines attributed to Gordon and McBride [27]. Thermodynamic data for the reaction species are taken from the JANAF tables [28]. Over 50 probable reaction species were considered, mol fractions being calculated for all constituents present in concentrations exceeding 0.1 ppm.

The procedure followed was relatively simple. From a conceptual model of reactant distributions following fuel injection, illustrated in Fig. 2, it was postulated that the spatial distribution in equivalence ratio (ϕ) can encompass all values from zero to infinity. Just as the detailed kinetics are ignored with this approach, the temporal modifications in the distribution of equivalence ratio suggested by Fig. 2 were similarly sidestepped. The logic was that, assuming any equivalence ratio and a reasonable cylinder pressure, somewhere within the heterogeneous mixture there probably exists a packet of reactants that matches those specifications. The temperature of the packet at any time depends upon its history of heating and mixing, and its degree of reaction. A range of reactant temperatures from 500 K to 2700 K was considered in order to cover the entire spectrum of potential product species that might be encountered within the engine cylinder. By dealing only with equilibrium

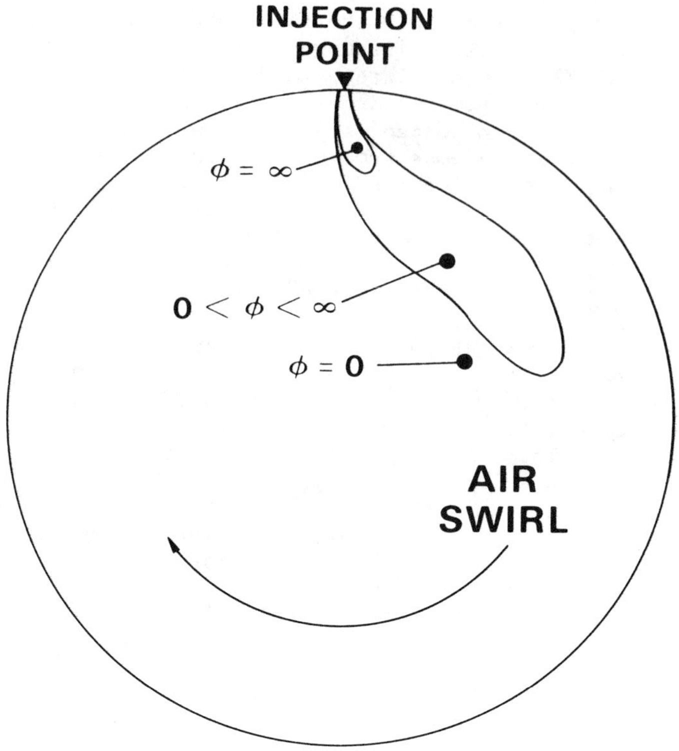

Fig. 2 - Physical representation of a diesel spray in swirling air, showing distribution of reactant equivalence ratio (ϕ)

conditions in the manner described, it becomes unnecessary to know the prior history of the packet -- only the equivalence ratio, pressure and temperature at the point of interest.

A history-dependent mass transfer phenomenon is suggested by Fig. 2 -- distillation and pyrolysis of the fuel followed by a diffusion process that transports the light ends of the fuel and hydrogen away from the original packet. Thus if the original fuel had an average atomic H/C of 2, the possibility exists that such a phenomenon can raise the average H/C ratio of the fuel in the downstream fringe of the vapor cloud to more than 2 while concurrently forcing the H/C ratio of the central spray ligament below 2. To assess the effects of this postulated fuel-fractionation mechanism, reactant H/C ratio was varied from 4 to 0, i.e., from CH_4 to C(s), or β-carbon. From this broad range of equivalence ratios, pressures, temperatures and H/C ratios, those showing strong tendencies toward condensing carbon, or soot, were selected for more detailed consideration.

SPECIES DISTRIBUTION WITH CARBON CONDENSATION - At equilibrium, carbon generally tends to appear precipitously below a certain critical temperature. This temperature was found to be highest for high equivalence ratios and low H/C ratios. For the assumed original fuel H/C ratio of 2 and an equivalence ratio of 3, this temperature is 1500-1600 K and is little affected by pressure. This sudden occurrence of equilibrium solid carbon is evident in Fig. 3, where the equilibrium concentrations of ten constituents are plotted against temperature for a pressure of 100 atm. The dramatic appearance of C(s) below its critical temperature is seen to come primarily at the expense of CO, HCN, C_2H_4 and C_2H_2.

Below 1500 K, where the C(s) mol fraction grows from the ppm to the percentage range, the exchanges occurring among carbon-bearing species illustrate the Boudouard reaction [29]:

$$2CO \rightleftarrows CO_2 + C(s) \mp 170.8 \text{ MJ} \quad (1)$$

and the methane synthesis reaction [29]:

$$CH_4 \rightleftarrows 2H_2 + C(s) \pm 74.9 \text{ MJ} \quad (2)$$

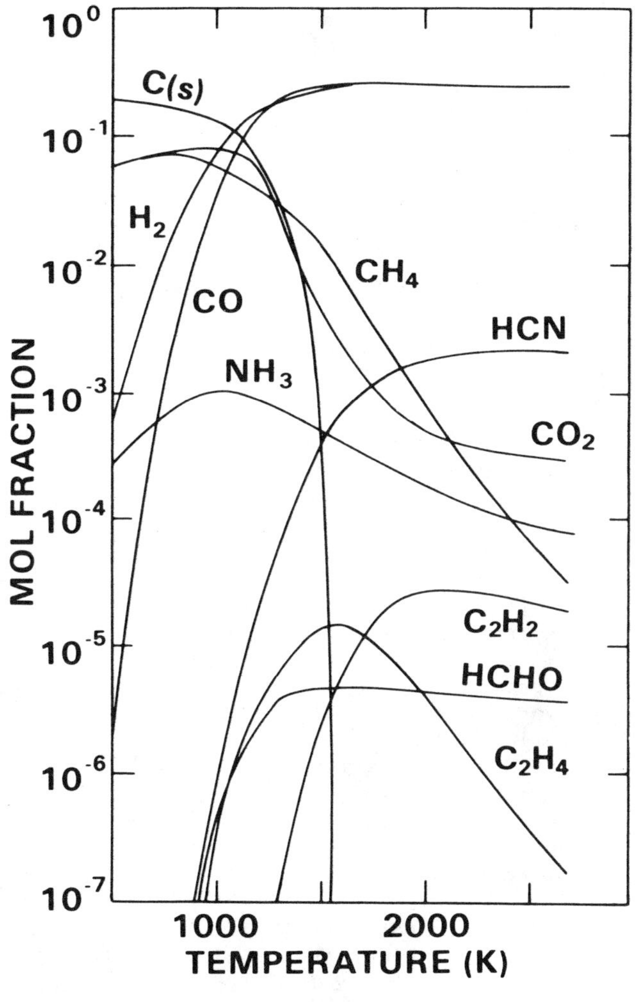

Fig. 3 - Equilibrium composition of combustion products for H/C = 2, ϕ = 3, and P = 100 atm

Due to the changes in volume associated with these reactions, increased pressure tends to increase C(s) in the Boudouard reaction but decrease it in the methane synthesis reaction. This at least partially explains the observation that increased pressure significantly increases the C(s) mol fraction only in the absence of H_2, as discussed subsequently in examining the case of H/C = 0.

The methane synthesis reaction relates to a carbon deposition phenomenon observed when methane is cracked at a carbon surface [30,31]. In this case the deposition of fresh carbon on the previously existing surface has been found to have an adverse effect on subsequent activity of the carbon with air.

EFFECTS OF EQUIVALENCE RATIO AND HYDROGEN/CARBON RATIO - It was previously mentioned that carbon condensation temperature was elevated by increasing equivalence ratio and decreasing H/C ratio. The influence of equivalence ratio on C(s) is shown in Fig. 4 for the

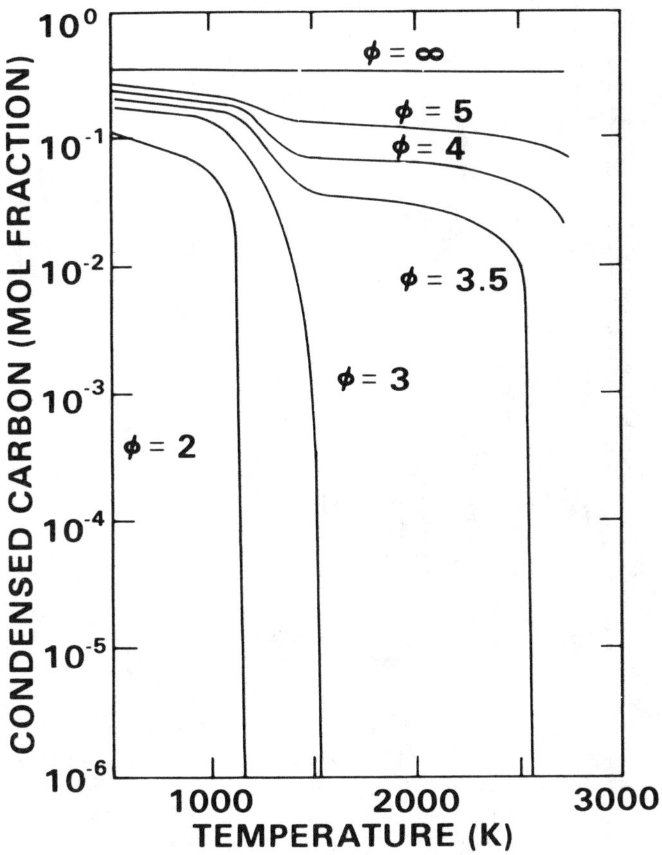

Fig. 4 - Equilibrium carbon concentration for H/C = 2 and P = 100 atm

fuel composition and pressure conditions of Fig. 3, viz., H/C = 2 and pressure = 100 atm. Whereas C(s) appeared in significant quantities only below 1500-1600 K in Fig. 3 (ϕ = 3), it is apparent in Fig. 4 that increasing equivalence ratio only from 3.0 to 3.5 raises the temperature for carbon condensation by 1000 K. This 2500 K formation temperature approximates the adiabatic flame temperature at unity equivalence ratio.

For the same pressure, Fig. 5 and 6 show the equilibrium carbon mol fractions for packets of reactants having H/C ratios of 1.0 and 0, respectively. These concentrations were calculated to assess the ultimate potential influences of fuel distillation, pyrolysis, dehydrogenation and mass transfer, as suggested by Fig. 2.

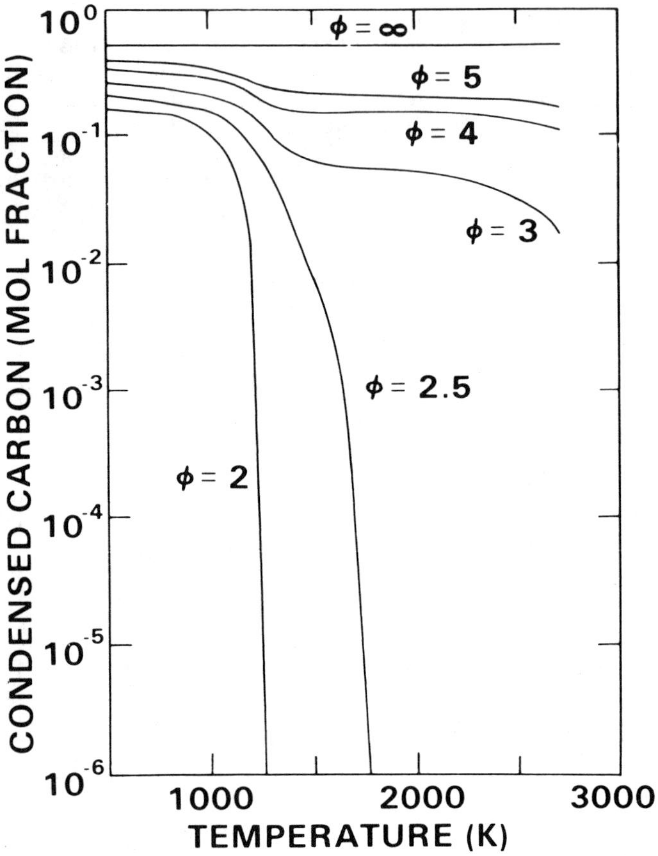

Fig. 5 - Equilibrium carbon concentration for H/C = 1 and P = 100 atm

It is seen from Fig. 5 (H/C = 1) that significantly more equilibrium C(s) is present at ϕ = 3 than for the original fuel (H/C = 2) at the same equivalence ratio. Alternatively, leaner mixtures are needed at H/C = 1 to hold carbon condensation down to the levels calculated for H/C = 2.

The situation of ultimate dehydrogenation (H/C = 0) of the packet is illustrated in Fig. 6. Although this occasion may not be encountered during diesel combustion, it represents a limiting case. It is seen in Fig. 6 that the equivalence ratio must lie below 2 in the temperature range of interest to avoid carbon condensation when H/C = 0.

The offsetting influences of pressure on the Boudouard and methane synthesis reactions were mentioned previously. In the absence of hydrogen (H/C = 0), no CH_4 is possible. Hence the Boudouard reaction, which shifts to the

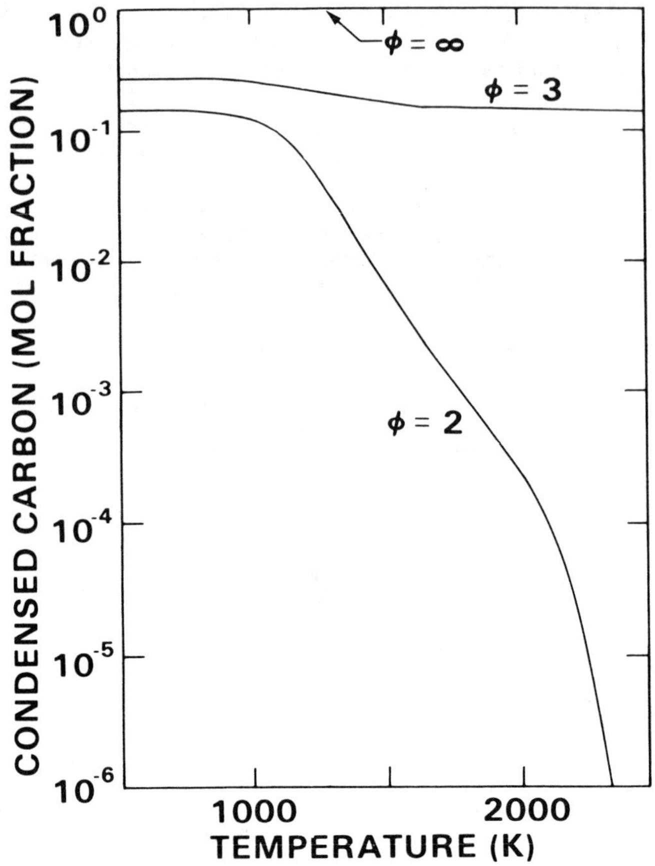

Fig. 6 - Equilibrium carbon concentration for H/C = 0 and P = 100 atm

right with increasing pressure, would be expected to provide lower C(s) at pressures decreased from the 100 atm of Fig. 4 through 6. This is seen to be the case in Fig. 7, where carbon concentration is plotted against temperature for H/C = 0 and ϕ = 2 over two orders of magnitude of pressure. Between 1500 and 2000 K, the equilibrium C(s) mol fraction is approximately proportional to pressure. This large pressure effect was found to decrease as the hydrogen concentration increased toward a H/C ratio of 2 and beyond.

EFFECT OF PACKET PATH - In the limit of chemical equilibrium, paths leading to the condensation of C(s) in diesel combustion can be mapped on plots like Fig. 4 through 7. Any number of paths can be postulated for individual fuel-air packets. As an illustration, isopleths of equivalence ratio from Fig. 4 are duplicated in Fig. 8. Consider the original fuel (H/C = 2) in the core of the injected spray to experience the following three processes:

1. Fuel is injected at $\phi = \infty$, then heated by hot cylinder air and by heat transfer from the neighboring flame.

2. Cylinder air mixes with this fuel at top dead center, decreasing packet equivalence ratio. The contents of the packet consequently react exothermally to increase packet temperature.

3. The packet contents expand as the

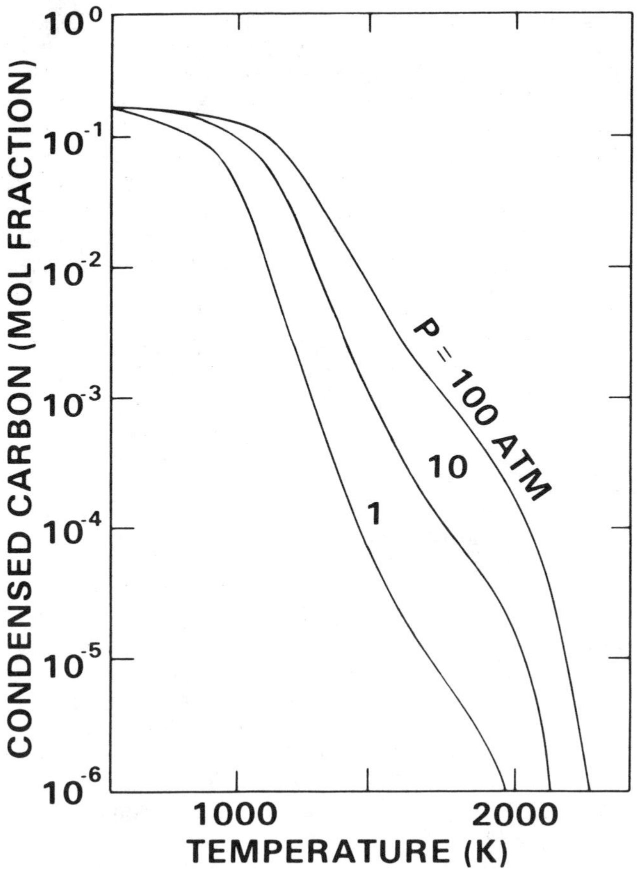

Fig. 7 - Equilibrium carbon concentration for H/C = 0 and ϕ = 2 at three different pressures

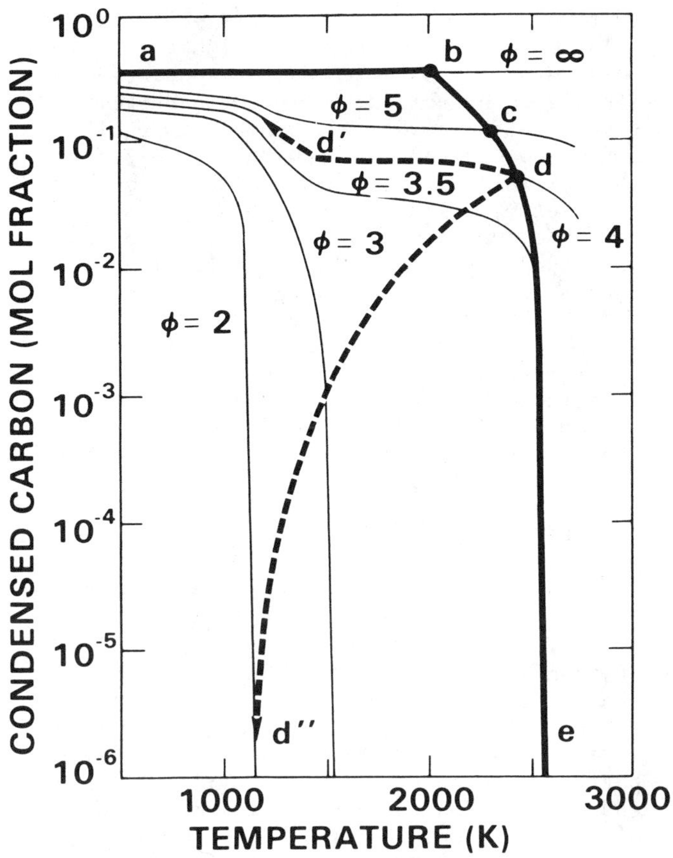

Fig. 8 - Possible paths followed by fuel-air packets for H/C = 2

piston descends on the power stroke. (Because of the comparative insensitivity of equilibrium carbon concentration to pressure at H/C = 2, it is assumed that Fig. 8 is valid for illustrative purposes over the pressure range from 1 to 100 atm.)

The heating of the fuel in the absence of mixing (process 1) is represented by path a-b in Fig. 8. As the fuel begins to mix and react with air (process 2), the resulting packet passes through point c (ϕ = 5) and point d (ϕ = 4).

Suppose that at point d the piston begins to descend (process 3), lowering packet temperature through expansion. Two of the many possible paths are illustrated in Fig. 8. Path d-d' represents expansion at constant equivalence ratio, i.e., without further mixing. In this case additional carbon condenses and the equilibrium carbon content of the packet increases as expansion proceeds.

Path d-d'' represents a possible expansion path wherein simultaneous mixing continues to an equivalence ratio of 2. As illustrated, the cooling effect from expansion is greater than the temperature rise from continued reaction, producing a net decrease in temperature, but also a decrease in packet equilibrium carbon concentration. In fact, for the case illustrated, the equilibrium carbon essentially disappears at ϕ = 2.

Consider a third case, in which mixing at top dead center proceeds beyond ϕ = 4 to ϕ = 3 before expansion commences. This case is represented by path a-b-c-d-e in Fig. 8. As equivalence ratio approaches 3.5, the path becomes tangent to the ϕ = 3.5 isopleth, i.e., at equilibrium the carbon-forming potential disappears when a ϕ of 3.5 is reached. For continued leaning toward ϕ = 3, the path cannot be plotted on Fig. 8 because the temperature of the reacting packet increases beyond 2600 K while the ϕ = 3 isopleth lies well to the left on the plot. During the expansion stroke, if the reasonable assumption is made that equivalence ratio cannot increase as mixing occurs, then the case of no mixing represents the worst case. In this third case condensed carbon is not predicted at equilibrium unless expansion lowers the temperature to about 1550 K (the temperature at which the ϕ = 3 isopleth becomes vertical). Should mixing occur during expansion, an even lower temperature appears to be required for carbon condensation.

These equilibrium examples suggest three approaches to the control of diesel soot formation. First, unsurprisingly, it appears desirable to avoid combustion at extremely rich equivalence ratios. This is apparent in Fig. 8 from the fact that equilibrium carbon concentration is higher at point c ($\phi = 5$) than at point d ($\phi = 4$), where it is higher than at point e ($\phi = 3.5$). Second, it appears desirable to effect combustion at high temperature. This is suggested by the low equilibrium carbon concentration at point e ($\phi = 3.5$). Had the initial temperature been 1000 K lower than at point b, mixing to $\phi = 3.5$ would have moved path b-c-d-e in Fig. 8 to the left and relocated point e on the $\phi = 3.5$ isopleth at a carbon concentration several orders of magnitude higher than that illustrated in Fig. 8. Third, it is desirable to mix the reacting rich packet to a leaner equivalence ratio. This is apparent from a comparison of paths d-d' (no mixing) and d-d'' (mixing). Each of these approaches is considered subsequently under PRACTICAL IMPLICATIONS.

SOOT-FORMATION MECHANISMS

The preceding section has focused on the *environmental* aspect of diesel soot formation, i.e., the conditions existing in the cylinder that influence the production of soot. The *phenomenological* aspect of soot formation is another interesting topic. Early investigators of diesel soot have examined the coagulated particles exhausted from the cylinder [2,12], and coagulation is also the principal subject of this section. Before the coagulation process can occur, however, the individual particles that collide and stick together to form the coagulates must be created. To observe the nucleation and surface growth phenomena responsible for forming these individual particles in diesel combustion would require spatial resolution on the molecular level and temporal resolution on the millisecond level, all in an environment involving temperatures near 2000 K and pressures measured in the tens of atmospheres. For this reason, attention is directed to out-of-engine experiments.

Motivated by an interest in the mechanisms of carbon black formation, Lahaye and Prado have conducted a thorough review of reported studies of the formation of carbon through thermal treatment of hydrocarbon [23]. Commercial carbon black is formed either by heating a hydrocarbon in the absence of oxygen or, alternatively, through partial combustion of a hydrocarbon in the presence of insufficient oxygen. The latter technique appears to be particularly germane to soot formation during diesel combustion.

Lahaye and Prado [23] have concluded that particle growth consists of the simultaneous association of small nuclei and the deposit of large polyaromatic molecules on their surfaces. The polyaromatic macromolecules are produced from the original hydrocarbons, regardless of their composition, by complex reactions that are not yet thoroughly understood. When soot is formed in the absence of oxygen, the macromolecules condense into liquid microdroplets, forming the nuclei, and gas-phase macromolecules then deposit on the nuclei to constitute the main growth step. In the case of flame-formed soot, however, association of particles is much more important in the growth of soot.

At least three theories have been reviewed that purport to explain the formation of the nuclei that serve as the precursors of larger particles [23]. Homann [32] has suggested that nucleation is chemical in nature, resulting from the bonding of large molecules produced by simultaneous polymerization and dehydrogenation of the original hydrocarbons. Tesner et al. [33] have also supported the chemical nucleation hypothesis, but have proposed an alternate chemical route that emphasizes decomposition of the original hydrocarbon rather than polymerization. They have also developed an analytical model based on the concept of chemical nucleation.

The second of these three theories advanced to explain nucleation, supported by the work of Place and Weinberg [34] and by Howard [35], is based on the existence of ions known to be present in the flame region. Howard has developed a qualitative analytical model for nucleation and subsequent surface growth assuming that the positive ions serve as the nuclei. It is suggested that these nuclei undergo growth through surface reaction, establishing crystallites that have the hexagonal-plane structure characteristic of graphite. When these crystallites reach a certain critical size (0.002-0.003 µm), they agglomerate to form the basically spherical individual particles (on the order of 0.01 to 0.05-µm diameter) that subsequently coagulate into clusters and chains. Surface growth continues during this process. Dehydrogenation also continues as the material ages in the flame.

The mechanisms of this model explain the structure of commercial furnace blacks, made by partially combusting liquid hydrocarbons at rich stoichiometry, then quenching the products. According to Heckman [36], furnace blacks contain stacked layer-planes of carbon atoms arranged in hexagonal networks. These crystallites, typically 0.0015 µm thick by 0.002 µm in diameter, are agglomerated to form spherical individual particles. Near the outer surface of a sphere the crystallites are arranged with their layer-planes most nearly parallel to that external surface, like layers of an onion. Near the center the crystallite orientation is more disordered.

These individual particles in turn join to form chainlike coagulates. In furnace blacks few if any discrete spherical particles remain [36]. The individual particles often appear to be fused together at the junctions of adjacent spheres, so that each individual particle tends to lose its distinctive identity and to take on the appearance of a nodule in the coagulate [36,37]. The resulting coagulate then bears some resemblance to a peanut. This fusion of individual particles results from the deposit of new layers of carbon on the surfaces of the growing coagulates and makes for a strong structure [37]. However, the smaller coagulates formed just before quenching are less strongly bonded [37].

The third nucleation theory involves homogeneous nucleation. Originally Parker and Wolfhard [38] explained the appearance of carbon particles in a diffusion flame as being the result of condensation of high-molecular-weight hydrocarbons through pyrolysis of the original fuel until the saturation vapor pressure is exceeded, whereupon condensation into fine droplets occurs. The work of Lahaye and Prado [23] with carbon black formed in the absence of oxygen supports this theory. Ulrich [39] has modeled the theory for flame-formed particles by assuming that macromolecules subject to Brownian motion collide to provide critical nuclei. The model includes allowances for both coagulation and surface growth. Results from this theory also qualitatively support the observations on commercial furnace black mentioned above.

Which, if any, of these three nucleation theories is appropriate for diesel combustion has yet to be proven in an engine. It is even conceivable that different mechanisms dominate in different ranges of temperature and pressure or at different times during the formation of soot. Knowing the precise nucleation mechanism may not even help in the control of particulate formation. Once the nuclei are formed, however, continued chemical potential for surface growth would appear to exist, as indicated by Fig. 4 through 6. An example of such surface growth leading to an apparent fusion of individual particles is provided by the electron micrograph of Fig. 9, taken from a coagulated particle collected from an indirect-injection diesel engine. The fusion of individual particles accounts for the resistance of diesel soot particles to fracture, as was indicated above to be characteristic of carbon black coagulates. Evidence of this strength is provided by the electron micrograph of Fig. 10, in which coagulates that have impacted on a micropore filter are seen to project over and across the pores without losing their integrity.

The preceding discussion has considered nucleation, surface growth and coagulation without treating the relative timing of these three elements of soot production. Some light is shed on this matter by the experiments of Wersborg et al. [40], who studied carbon formation in a low-pressure flat-flame burner consuming premixed acetylene and oxygen at an equivalence ratio of 3. These investigators measured a rate of nucleation that built quickly to a maximum, then just as quickly fell to insignificance. Coagulation began soon after the start of nucleation, but continued long after nucleation ceased. In their particular experiment, surface growth initially accounted for about 80% of the total growth rate of the particle material, but surface growth rate declined to zero about the time that nucleation ceased and during the time that coagulation was significant. Because surface growth occurred during coagulation, the conditions necessary for fusion of individual particles in the coagulate, as discussed above, existed. Coagulates were found to assume roughly spherical shapes during the early stages of their growth, but later to take on a chainlike character. This transformation from spherical to chainlike shape is enhanced by the deterioration of surface growth, since surface growth tends to obscure the identity of individual particles. The growth of chains has been attributed to electrostatic interactions [35], but a computer simulation of the random

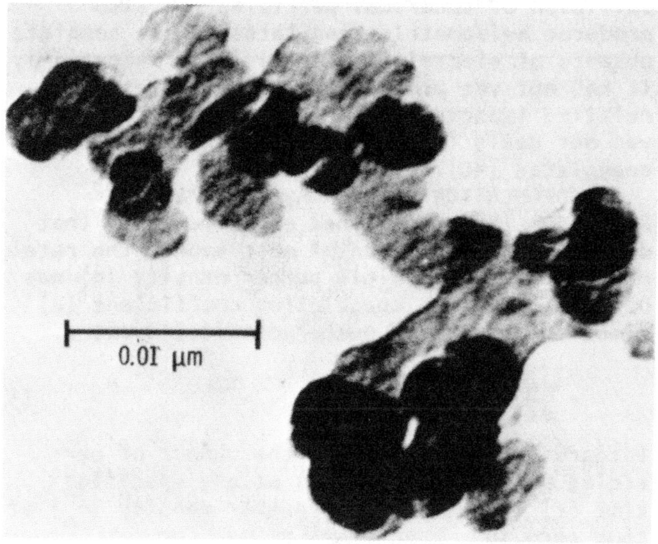

Fig. 9 - Transmission electron micrograph of diesel soot coagulate, showing fusion of original individual particles

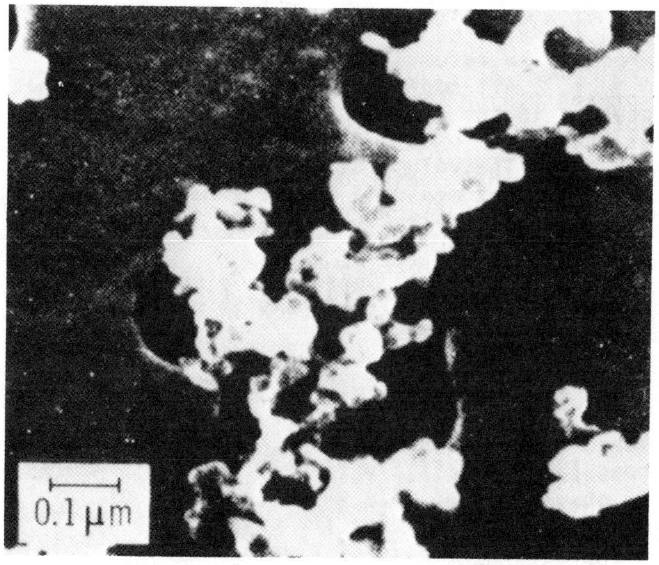

Fig. 10 - Scanning electron micrograph of diesel soot coagulate, demonstrating strength of bonds between individual particles

collision of spherical particles has also produced anisometric coagulates in the complete absence of electrical charges [41]. Apparently it has not yet proven possible to assess the relative importance of electrostatic forces and van der Waals forces in establishing chainlike coagulates [40].

COAGULATION RATE - Whytlaw-Gray and Patterson [42] determined experimentally that during the coagulation of most smokes the rate of decrease of particle number density (n) was proportional to a coagulation coefficient (K) times the square of number density, i.e.,

$$-\frac{dn}{dt} = K\,n^2 \qquad (3)$$

Integrating, the ratio of the number of particles per unit volume (n) at any specified time (t) to the original number density (n_o) at time zero is

$$\frac{n}{n_o} = \frac{1}{1 + (K\,n_o)\,t} \qquad (4)$$

Khan [12] applied the Whytlaw-Gray equation to experimental results from a direct-injection diesel. Recognizing the existing gaps in detailed knowledge of the combustion process, he assumed that:

1. All soot formed during combustion was present in the cylinder at TDC, the beginning of expansion.
2. The volume of soot present at TDC was equal to that measured in the exhaust, i.e., oxidation occurring during expansion was negligible.
3. Coagulate density remained constant and equal to the initial density of the carbon particles.
4. The soot present at TDC was monodisperse, with a diameter of 0.025 µm.
5. Soot was uniformly distributed at TDC.

Based on these assumptions, Khan calculated a coagulation coefficient of 5.93×10^{-2} mm^3/s. He observed that this was five orders of magnitude greater than the coagulation coefficient characteristic of a variety of aerosols under stagnant conditions, attributing the difference to the high air velocities and turbulence within an engine cylinder. Estimating an initial number density of 4.88×10^8 particles/mm^3, Khan calculated that coagulation was essentially completed by 50 deg ATDC over the speed range from 1100 to 2700 r/min.

A shortcoming of Khan's calculation is the assumption of homogeneously distributed soot at TDC. This assumption can be avoided by following a small packet of soot-containing cylinder gas as it expands and mixes with soot-free air from its surroundings during descent of the piston. (The details of this technique are discussed more fully in the subsequent section on SOOT OXIDATION.) This approach precludes calculation of a total coagulate number count in the tailpipe, but it allows an assessment of the effect of mixing rate on coagulation within the packet.

Beginning with a stoichiometric packet at TDC, the variation in number density with crank angle was calculated for the the fast, intermediate and slow mixing rates of Fig. 11.

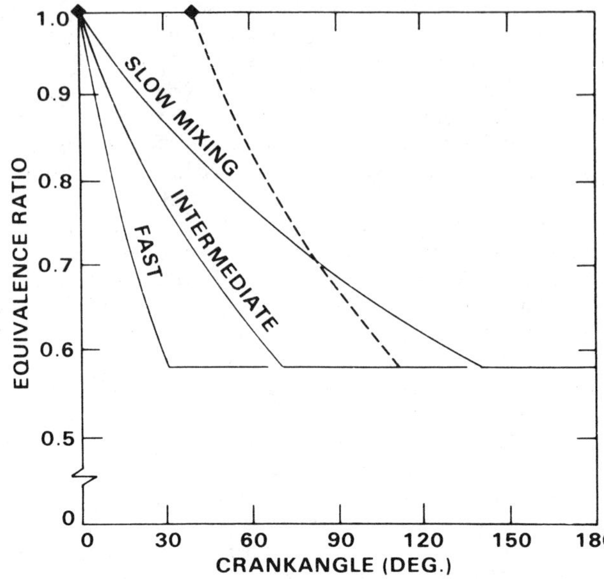

Fig. 11 - Assigned mixing rates for calculation of soot coagulation and oxidation

Following Khan, a nominal coagulation coefficient of 0.05 mm^3/s and an initial concentration of 5×10^8 particles/mm^3 were used. Recognizing the probability of significant variations in the K n_o product from one engine, packet and/or operating condition to another, the calculation was repeated for K n_o products both an order of magnitude higher and lower than the nominal value.

The results for the intermediate mixing rate are given in Fig. 12. Despite the modification to Khan's approach, his conclusion that coagulation of particles existing at TDC is essentially completed at 50 deg ATDC is supported.

Results with the other two mixing rates of Fig. 11 indicated that that variable has a comparatively minor influence on coagulation. This occurs because of the strong inverse dependence of n/n_o on time. Most of the coagulation has already occurred before differences in mixing rates have had an opportunity to establish marked differences in number density from the original common value at TDC.

Extending this logic to particulates formed late during the combustion process, it appears that with time playing a dominant role, whatever particulate coagulation occurs in the engine is essentially completed at the time the exhaust valve opens.

At issue in these calculations for individual packets of diesel combustion products undergoing simultaneous expansion and mixing is the constancy of the coagulation coefficient K. Whytlaw-Gray and Patterson [42] have modified the Smoluchowski equation for coagulation of

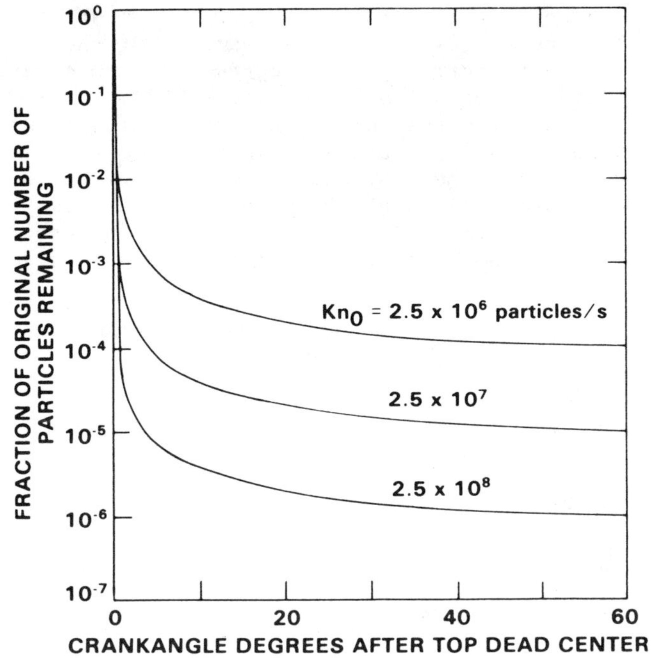

Fig. 12 - Particle coagulation for a medium mixing rate

colloidally suspended particles in a liquid by applying the Cunningham correction to obtain the following expression for coagulation of a monodisperse smoke:

$$\frac{1}{n} - \frac{1}{n_o} = \frac{4}{3}\frac{RT}{\eta N}\left(1 + \frac{B\delta}{r}\right) t \qquad (5)$$

where R = gas constant
T = absolute temperature of gas
η = viscosity of gas
N = Avogadro's number
δ = mean free path of gas molecule
r = radius of particle
B = Millikan's constant $\approx$ 0.9
t = time

The Cunningham correction, in parentheses, accounts for the fact that in an aerosol, unlike a liquid sol, the molecular mean free path is of the same order as the particle size. The resemblance of the coefficient of t in this expression to coagulation coefficient K in the experimentally determined Whytlaw-Gray equation is obvious after a little mathematical manipulation.

Green and Lane [43] review the results of a number of experiments with smoke coagulation in the light of this modified Smoluchowski equation. Immediately obvious is the fact that large monodisperse particles coagulate slower than small ones, other factors remaining constant. This deduction has been confirmed experimentally [43]. Less obvious is the fact that polydisperse smokes coagulate faster than monodisperse smokes, although this typically has only a small effect on the coagulation coefficient [42]. Considering the thermodynamic properties of the gas as the packet of products expands and mixes, the tendency for coagulation rate to decrease as temperature falls is partially counteracted by the associated decrease in viscosity. There is some evidence that chainlike particles coagulate faster than spherical particles, while the effect of electrical charges depends on the distribution of both the charge and the particle size [43]. Experiments demonstrate clearly that mixture motion increases coagulation rate [43]. Weighing the anticipated magnitudes of all these effects, some of which act upon coagulation rate in opposing directions, there appears to be no reason at this time to doubt the conclusion that soot coagulation is essentially completed at the time the exhaust valve opens.

COAGULATE GROWTH - Having considered coagulation rate, attention is now turned to the nature of coagulate growth. A model for the coagulation of polydisperse particles in the environment of an engine cylinder has already been proposed by Dolan [44] and need not be reviewed here. It should be noted, however, that quantitative application of such models to diesel soot within the engine cylinder appears somewhat tenuous unless the coagulation model is coupled to an oxidation model, for both of these processes can occur simultaneously. As the expansion stroke proceeds, a coagulation model that merely traces the collision history of an initial population of soot particles seems bound to be quantitatively inaccurate because it allows for coagulation of individual particles that may actually have disappeared in the engine through oxidation before they coagulate in the model.

The quantitative modeling of soot coagulation within the cylinder of an automotive indirect-injection diesel is also hampered by the absence of a reliable model of diesel combustion that provides a spatially and temporally accurate definition of the distributions of soot within the heterogeneous cylinder gas. Although an attempt to circumvent this problem has provided some encouraging preliminary results [45], it has also served to underscore the magnitude of the relevant unresolved issues.

Faced with this predicament, the course chosen here is to construct an illustrative model for the limited purpose of providing qualitative insight into the nature of the coagulation process. This simplistic model ignores the facts that large particles coagulate slower than small ones and that the probability of collision between particles of unequal size is greater than that between particles of equal size [43]. It also overlooks the previously acknowledged influences of particle shape and electrical charge, neither of which is sufficiently defined to be accommodated analytically. The model begins with a population of monodisperse spherical particules, as indicated on the top line of Fig. 13. After a brief but undefined time increment some fraction f of these spheres is assumed to undergo 1-1 collisions (i.e., collision of two single spheres)

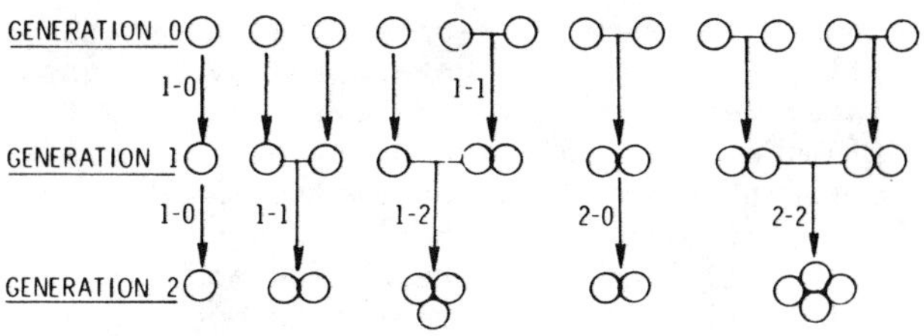

Fig. 13 - Coagulation of individual soot particles

and stick together, forming two-particle coagulates, while (1 - f) of the original particles remain as single spheres. In Fig. 13, f for the zeroth generation is 0.5. The resulting particle population constitutes the "first generation" of coagulation.

In proceeding to the second generation, 1-1, 1-2 and 2-2 collisions become possible. As illustrated in Fig. 13, f for the first generation is 0.75. Fig. 13 is qualitatively suitable for illustrating the development of collision combinations in successive generations, but quantitatively the probability of 1-2 collisions in the first generation is proportional to the numbers of single and two-sphere particles populating the first generation rather than as illustrated.

To put the illustrative model on a quantitative basis, consider an initial population of 6000 identical spherical particles and let f equal 0.5. Then collisions proceed as follows:

Zeroth generation	Collision	First generation
3000 singles	none	3000 singles
3000 singles	1-1	1500 doubles
		TOTAL = 4500 coagulates

In proceeding to the second generation, with f kept equal to 0.5,

First generation	Collision	Second generation
1500 singles	none	1500 singles
1000 singles	1-1	500 doubles
500 singles +500 doubles	1-2	500 triples
750 doubles	none	750 doubles
250 doubles	2-2	125 quadruples
		TOTAL = 3375 coagulates

As succeeding generations are developed with this model, it is evident that the number of particles in each generation is $(1 - f/2)^g$ times the original population, where g is the generation number. It is also seen that the number of single spheres in the largest agglomerate is 2^g. The number of single-sphere particles remaining in any generation, expressed as a fraction of the total number of spheres in the original population, is $(1 - f)^g$. In other words, as coagulation proceeds, the largest coagulates get larger while the original single-sphere particles rapidly approach extinction. This relatively fast disappearance of the smallest particles is in agreement with experimental observations in aerosols [43]. Continuing with the model, this trait of a maturing population is illustrated in Fig. 14, which represents the distribution of particle size, as characterized by the number of individual spherical particles in a coagulate, for arbitrarily chosen third and sixth generations.

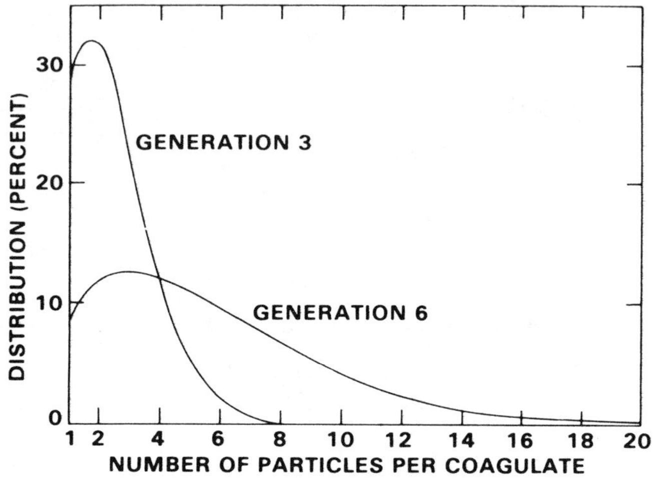

Fig. 14 - Shift in coagulate size distribution as population matures

Measurements of Coagulate Size - Coagulate size is, of course, most often expressed in terms of an effective diameter, not the number of individual particles comprising it. One instrument that has been used widely for this measurement is the Electrical Aerosol Analyzer (EAA) [46]. It measures the mobility of particulate matter in a diluted exhaust gas stream, then relates that mobility to particle size by calibrating against aerosols of known size. Being effective in the particle size range from about 0.01 to 1.0 µm, the EAA deals with nearly all of the particulate mass in diesel exhaust.

Briefly, the EAA works in the following manner. First the particles in the diluted exhaust stream are negatively charged. Then the stream flows down the length of a cylindrical tube containing a centrally mounted axial rod that is positively charged. The voltage on the charged rod controls the size of the particles able to travel the length of the cylindrical annulus without being attracted to the rod. Those particles escaping the rod are collected on a filter at the exit end of the tube and quantified electrically in terms of number density (particles/unit of gas volume).

The particles are graded in increments of diameter, depending on the voltage applied to the charged rod. Number density (n_i) for each diameter (D_i) is transformed to volume concentration (V_i) by

$$V_i = \frac{\pi}{6} D_i^3 n_i \qquad (6)$$

Volume concentration is converted to mass concentration by multiplying by the average particle density.

An alternate method of sizing particles has been employed by Vuk, Jones and Johnson [2]. They used an inertial impactor to sample undiluted exhaust from a direct-injection diesel engine.

In the impactor the exhaust gas flows through a stack of parallel but spatially separated perforated plates oriented normal to the main direction of flow. The holes in each plate are arranged to force impaction of the throughflowing gas against the surface of the downstream plate. The gas stream deflects radially from the point of impingement, depositing on the downstream plate those particles that are too heavy to follow the deflected streamlines. The deflected gas carries the non-impacted particles through holes in this downstream plate and on to new impaction sites on the next plate. Each successive plate contains smaller holes, so each successive downstream plate collects smaller particles. The collected particles are fractionized according to size by measuring the mass increase of each plate. The size range assigned each plate comes from a calibration done with spherical particles of known diameter.

A prominent disadvantage of the inertial impactor is its high particle size range, from 0.4 to 20 μm. This allows most of the particulate mass in diesel exhaust to escape the last plate of the impactor and remain with the exhaust gas stream. An advantage of the inertia impactor is that it facilitates collection of particulate samples of graded size for microscopic examination.

Application of Experimental Measurements in the Model - Vuk et al. [2] capitalized on this advantage to assess the relationship between the number of individual particulates in a coagulate, as viewed through a microscope, and the aerodynamic size of that coagulate, as determined by the inertial impactor. In an effort to minimize the error inherent in counting individual particles from a basically two-dimensional photomicrograph of a coagulate that has a third dimension normal to the plane of view, counting was limited to particles at the small end of the available range, viz., from 0.4 to 4.0 μm.

To correlate the number of individual particles per coagulate with coagulate diameter requires attributing a diameter to a coagulate that is typically not spherical. This task was addressed by Vuk et al. [2] by assigning two diameters to each coagulate. Their "largest sphere" method assigned to the coagulate the diameter of the smallest circle that would circumscribe the entire microscopic image of the coagulate. Their "smallest sphere" method assigned the diameter of the smallest circle through which the image of the coagulate could be passed when optimally oriented to the plane of the circle.

Vuk et al. [2] fit the number of individual particles per coagulate to both of these diameters with straight lines when plotted on logarithmic coordinates (their Fig. 58). Working from their plot, the number of individual particles per coagulate (N) is related to coagulate diameter (D) in microns by

$$N = 40.48\ D_{\ell}^{1.035} \quad \text{(largest-sphere method)} \tag{7}$$

$$N = 97.19\ D_{s}^{1.280} \quad \text{(smallest-sphere method)} \tag{8}$$

Solving these expressions for the size of a single-particle coagulate (i.e., N = 1) gives a diameter of 0.028 μm. As pointed out by Vuk et al. [2], this diameter is in reasonable agreement with the 0.026-μm mean diameter of an individual spherical particle determined for their engine.

Although the foregoing discussion has led to a relationship between the number of individual particles per coagulate and the physical size of the coagulate, both the inertial impactor and the EAA deal with some form of aerodynamic equivalent diameter rather than physical diameter. Consequently Vuk et al. [2] carried their analysis one step farther to correlate the physical coagulate size with its aerodynamic size. Using the smallest-sphere approach, this analysis (their Fig. 59) correlated physical size linearly with aerodynamic size for lower, mean and upper size limits. Their mean- and upper-limit correlations show physical diameters of 2 and 5.5 μm, respectively, when the aerodynamic size is zero. This sort of anomaly can arise if an individual coagulate on a given plate is actually the result of the successive impaction of one or more coagulates on top of or adjacent to the coagulate initially deposited at the point of observation. Because this anomaly was absent from the lower-limit correlation, that correlation was chosen here for further use. It showed that aerodynamic diameter was 25% greater than the physical diameter determined by the smallest-sphere method. Combining this information with the expression above for the smallest-sphere method, the aerodynamic diameter (D_a) of a coagulate from the experiments of Vuk et al. [2] is given in microns by

$$D_a = 0.035\ N^{0.78} \tag{9}$$

This expression is applied to the sixth-generation curve of Fig. 14 to convert the number of particles per coagulate to aerodynamic diameter. At the same time it is assumed that the mass of a coagulate is directly proportional to the number of individual spheres comprising it. The result is the mass-distribution curve of Fig. 15.

Also shown shaded in Fig. 15 is the distribution of volumetric concentration of particulates emitted from the tailpipe of an idling diesel car powered by a 2.1-L engine. These measurements were made in cooperation with Kittelson and Dolan of the University of

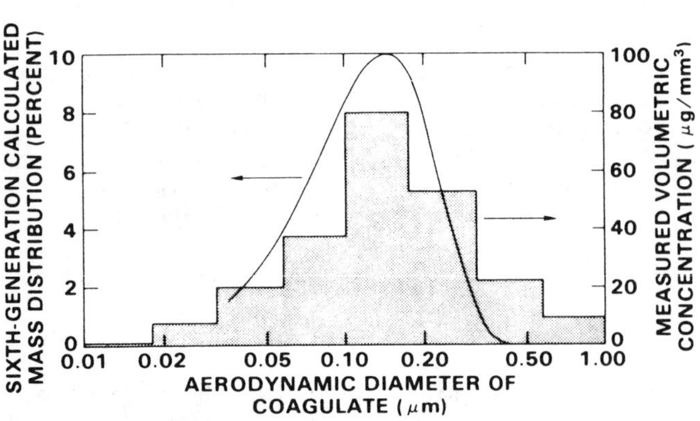

Fig. 15 - Particle size distributions (a) calculated for sixth generation and (b) measured from an idling diesel car

Minnesota at a measuring station 600 mm from the tailpipe exit using an EAA.

In principle, if one chooses a particulate density in order to convert the experimentally determined volumetric concentration to the same mass basis as the calculated sixth-generation curve, the two distributions of Fig. 15 might be expected to share a resemblance. In view of the simplicity of the model, however, it is perhaps surprising that they do.

One obvious difference is that the experimental data extend down to smaller diameters. This possibility was eliminated in the illustrative model by assuming an initially monodisperse population. This has the effect of limiting minimum aerodynamic size in the modeled results to 0.035 μm. Vuk et al. [2] have found the individual particle size in undiluted exhaust to follow a lognormal distribution. This leads to the expectation of a polydisperse initial population, an expectation that would explain the small particulates found in the experimental measurements. Even without initial polydispersion, though, it should be recognized that because the experimental sample was exposed to oxidation in the engine cylinder, smaller-sized individual particles would appear in the exhaust stream even in an initially monodisperse population.

Another potential contributor to particles in the smaller measured ranges is sulfates. Sulfate particles typically have a diameter of about 0.02 μm.

A second obvious difference between the distributions of Fig. 15 is that the experimental distribution has more volume (mass) concentrated in the large size range. A review of Fig. 14 indicates that the model is headed in that direction as further generations are developed.

Of course the illustrative model was never intended to duplicate experimental results, but it has served to provide an appreciation for the manner in which a range of coagulate sizes can develop from even an initially monodisperse particle population.

SOOT OXIDATION

The previous two sections have dealt with the *production* of soot in the cylinder. In addition, in-cylinder *destruction* of soot may occur through oxidation. In general, the rate of such heterogeneous reactions as the oxidation of soot depends both on diffusion of the reactants to and the products from the reaction site, and on the kinetics of the reaction itself. For particles of less than 1-μm diameter, diffusional resistance is minimal and soot oxidation is kinetically controlled [47]. Typically, over 90% of the particulate matter in the exhaust of a diesel engine is smaller than 1 μm [2]. Consequently in applying data on carbon oxidation to diesel soot, caution must be exercised to eliminate experiments in which diffusion plays a significant role.

The experiments of Nagle and Strickland-Constable [48] satisfy this criterion. Beginning from the measured rate of surface recession of a heated rod of pyrolytic graphite upon which a high-velocity jet of oxygen-bearing gas was impinged, they developed the following expression for the oxidation of carbon over the range of 1000-2000°C:

$$\Omega = 12 \left[\left(\frac{K_a P_{02}}{1 + K_z P_{02}}\right) \chi + K_b P_{02} (1 - \chi) \right] \qquad (10)$$

where

Ω = carbon surface recession rate, g/cm^{2}s

$\chi = [1 + K_t/(K_b P_{02})]^{-1}$

K_a = 20 exp (-30 000/RT) g/cm^{2}s atm

K_b = 4.46 x 10^{-3} exp (-15 200/RT) g/cm^{2}s atm

K_t = 1.51 x 10^5 exp (-97 000/RT) g/cm^{2}s

K_z = 21.3 exp (4100/RT) atm^{-1}

In these expressions,

P_{02} = partial pressure of oxygen (atm)

R = gas constant

T = temperature (K)

This relationship is plotted in Fig. 16. The experimental temperature range was subsequently extended to 2400°C by Walls and Strickland-Constable [49], confirming the falloff in recession rate at low oxygen partial pressure that is indicated by Eq. 10.

The structure of soot is highly anisotropic, and the reversal in recession rate at high temperature depicted in Fig. 16 is attributed to the coexistence of "more reactive" and "less reactive" sites [49]. As temperature increases, the fraction of the total number of reaction sites that are of the more reactive type (χ) decreases. At a given temperature, χ depends on the rate of oxidation of less re-

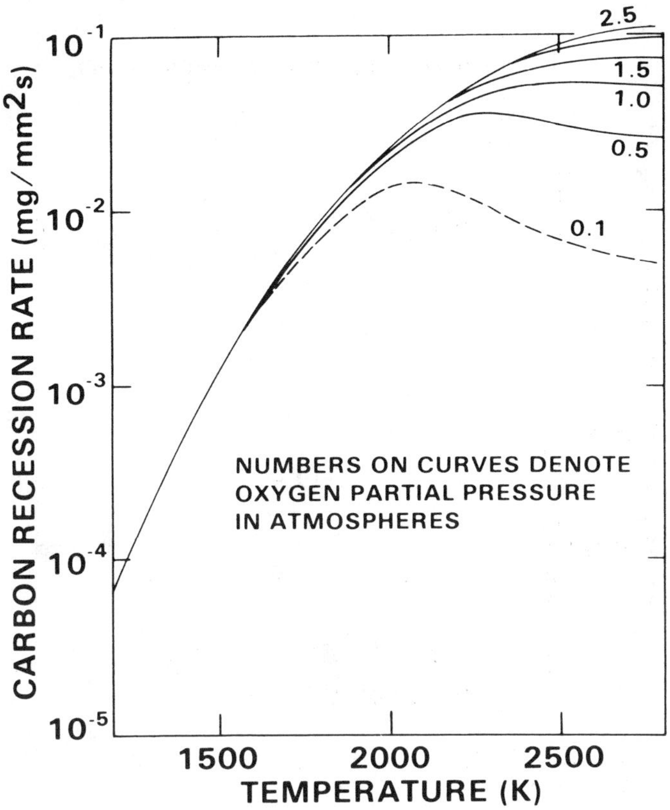

Fig. 16 - Carbon recession rate [48]

active sites and the rate of thermal annealing that converts more reactive sites to less reactive sites.

Recognizing the structural similarity of pyrolytic graphite (which is prepared by thermally cracking hydrocarbon vapor on a high-temperature solid substrate) to that reported for soot, Radcliffe and Appleton [50] undertook a comparison of Eq. 10 to the results of other relevant investigations. Among the principal contributions was the experiment of Lee, Thring and Beer [51] with a laminar diffusion flame of gaseous hydrocarbon in oxygen-enriched air. Another was the experiment of Fenimore and Cooper [52], who created a sooty gas by burning ethylene fuel-rich, then combusting the cooled products of that reaction with additional oxidant in a second burner. They attributed the oxidation of soot in the second reaction to hydroxyl radicals, but Radcliffe and Appleton [50] found the fuel-lean data to fit Eq. 10 quite satisfactorily. Park and Appleton [53] subsequently extended the range of experimental data to 4000 K by using a shock tube to oxidize soot dispersed in a test gas. Reviewing all these data, they concluded that Eq. 10 is generally valid for soot oxidation within a factor of two, a respectable accuracy among kineticists. Roper and Smith [54] recently suggested a resolution of the disagreement over whether the hydroxyl radical was responsible for soot oxidation [52] or whether it was attributable to oxygen [50,53] by explaining that oxidation by OH might be expected to dominate when soot concentrations were small or when a large nonequilibrium source of OH existed close to the soot zone, whereas oxidation by oxygen would be expected to dominate for moderately large soot concentrations.

ENGINE CYLINDER CONDITIONS - In this study Eq. 10 is applied to a single packet of burned mixture submerged within the total cylinder contents. The packet is assumed initially to contain homogeneously distributed combustion products having mixed to stoichiometric proportions, dissociation being neglected. The packet is also assumed to contain a limited mass of solid carbon (soot), present in such a small proportion that its oxidation has an insignificant effect on either the composition or the sensible energy content of the packet gases. Since the packet gases are assumed initially to include no diatomic oxygen, fresh air from the unburned portion of the cylinder contents is allowed to mix into the packet at arbitrarily prescribed rates in order to effect soot oxidation. Thus the total mass contained within the packet increases as the piston descends.

Examination of Eq. 10 reveals that two properties of the gas within the packet are needed to calculate oxidation rate, viz., oxygen partial pressure and packet temperature. Oxygen partial pressure is, of course, the product of cylinder pressure and the mol fraction of oxygen within the packet.

The variation of cylinder pressure with time is taken from instantaneous cylinder pressure data measured in an engine having the combustion chamber illustrated in Fig. 1 [25]. The resulting pressure-volume diagram, shown in Fig. 17, was obtained at an engine speed of 1600 r/min with an overall cylinder equivalence ratio of 0.58.

The mol fraction of oxygen within the packet depends on the rate at which air is mixed into the packet. Three linear rates of air addition are assumed. These rates are

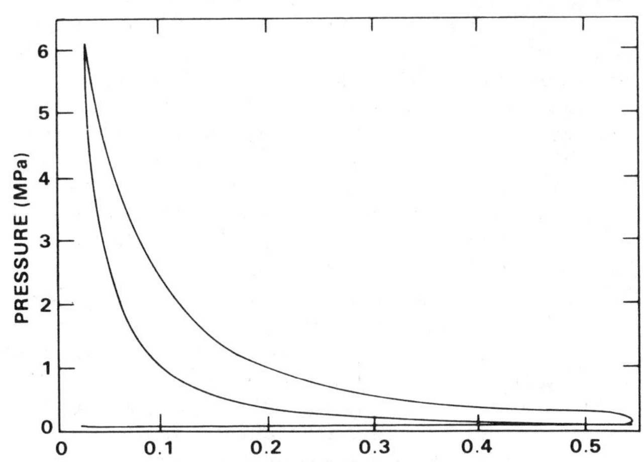

Fig. 17 - Measured pressure-volume diagram for an indirect-injection diesel engine

represented by the three curves of equivalence ratio versus crankangle originating at top dead center in Fig. 11. When a packet equivalence ratio of 0.58 is reached, the packet has attained the average stoichiometry of the total cylinder contents and no further mixing occurs.

Application of a heat-release model to the diagram of Fig. 17 has indicated that at this load, combustion is still occurring 40° after top dead center, which is 48° after the beginning of injection and 18° after the end of injection [25]. To gain some appreciation of the comparative soot oxidation potential in this late-burning portion of the cylinder charge, a case of intermediate mixing rate for a stoichiometric packet originating 40° after top dead center is also considered, as illustrated in Fig. 11.

Packet temperature is at its maximum value at the beginning of the soot oxidation calculation, combustion and mixing to unity equivalence ratio having just been completed. Assigned initial packet temperatures are 2542 and 2436 K at 0° and 40° crankangle, respectively. Packet temperature decreases polytropically (exponent = 4/3) as the cylinder pressure falls in accordance with the expansion stroke defined by Fig. 17. At the same time the air outside of the packet cools polytropically from the compression temperature of 889 K (694 K at 40° crankangle). The expansion process is divided into successive small increments of crankangle. At the end of each increment the appropriate mass of air is mixed into the packet to satisfy the schedule of Fig. 11. In view of the approximate nature of this calculation, the difference in specific heats between the packet gas and the surrounding air is neglected.

COMPARISON OF EARLY AND LATE BURNING - To gain an indication of the difference in oxidation potential between soot formed during early combustion, say at top dead center, and that formed during late combustion, say 40° after top dead center, soot oxidation calculations were completed as described above while holding the mixing rate constant at the intermediate level of Fig. 11. The resulting carbon recession rates are compared in Fig. 18. In both cases the rate rises from zero as air (oxygen) is mixed into the packets, initially devoid of oxygen. As crankangle increases, a peak recession rate is achieved, beyond which the rate rapidly decays to a negligible level.

The nature of these curves is the result of an interplay among three factors. As mixing proceeds from the initial condition, the mol fraction of oxygen increases continuously until the contents of the packet are diluted to the overall cylinder equivalence ratio. At the same time the cylinder total pressure decreases continuously in accordance with Fig. 17. The result of these opposing influences on oxygen partial pressure is to produce a maximum partial pressure at some crankangle during cylinder expansion. Meanwhile packet temperature falls continuously with advancing crankangle, slowing the oxidation rate to an increasing degree as the piston descends. The carbon recession rate is much less for the late-burning packet because both cylinder pressure (hence oxygen partial pressure) and cylinder temperatures are lower in the last part of the mixture to burn.

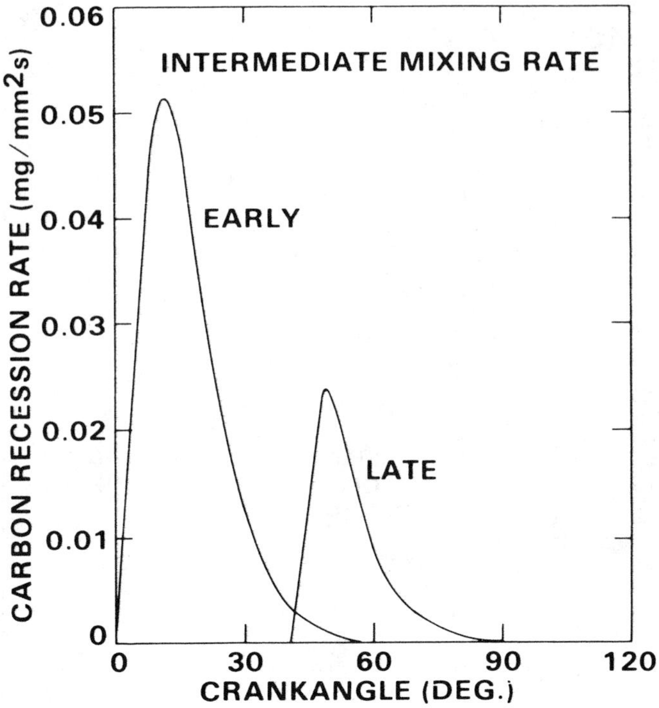

Fig. 18 - Carbon recession rates in early- and late-burned fuel-air packets

The rates of Fig. 18 are transformed into mass oxidized per unit of soot particle surface area in Fig. 19 by integrating the areas under the rate curves. Given equal surface areas,

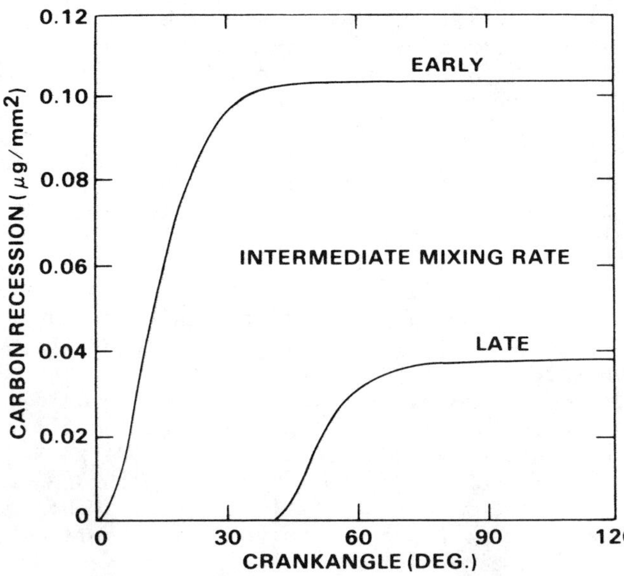

Fig. 19 - Carbon mass oxidized in early- and late-burned fuel-air packets

for the conditions prescribed the mass oxidized is 2.5 times as great in the soot formed early during combustion as in that formed late during combustion. The reason is seen in Fig. 18 not to be the longer in-cylinder time available for oxidation in the early burning mixture, but rather the greater recession rate achieved.

EFFECT OF MIXING RATE - To provide insight into the influence of mixing rate, packets are followed from top dead center through the expansion stroke while air is mixed in at the fast and slow rates of Fig. 11, respectively. The resulting recession rates are plotted in Fig. 20. The initial rate is greatest with fast mixing, even though this results in a lower packet temperature, because that lower temperature is more than offset by a higher oxygen partial pressure. As the piston continues to descend, however, the influence of the lower temperature predominates and the slow mixing rate yields a higher recession rate.

The result of integrating these intersecting rate curves is shown in Fig. 21. For the conditions analyzed, the slow mixing rate is seen to oxidize a greater mass of soot than the fast rate, given equal soot surface areas.

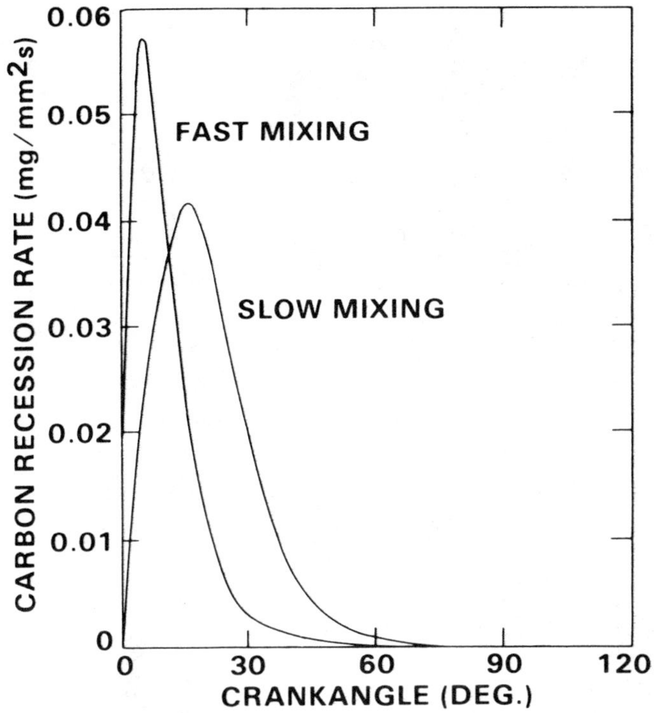

Fig. 20 - Carbon recession rates for two different mixing rates

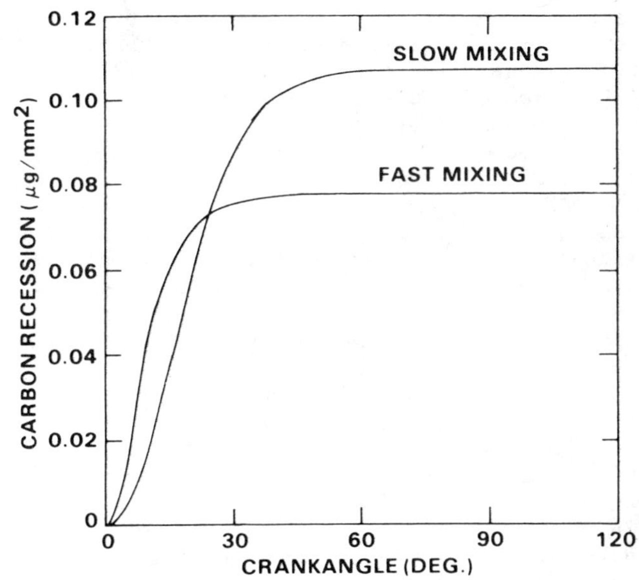

Fig. 21 - Carbon mass oxidized for two different mixing rates

EXHAUST DILUTION

Up to this point the discussion has focused on soot production within the cylinder, but diesel particulate emissions are not measured in the cylinder. The procedure that has emerged as the standard technique for measuring the mass of particulate matter exhausted from a light-duty diesel engine involves weighing a sample from the exhaust stream after the exhaust gas has been diluted with filtered ambient air in a CVS (Constant Volume Sampling) dilution tunnel. In addition to the combustion-generated carbonaceous soot, the filter sample contains a soluble organic fraction typically comprising 10-40% of the total dry sample mass. These soluble hydrocarbons are of particular interest in biological testing. A question exists as to whether the soluble fraction is acquired within the cylinder, downstream in the dilution tunnel, or on the soot collection filter.

The dilution ratio of the tunnel (Λ) is defined here as the ratio of the diluted-stream flow rate to the engine flow rate, both expressed on a molar basis. Assuming adiabatic dilution and constant specific heats, and applying the energy equation, the sample temperature becomes

$$T_s = \frac{T_e + (\Lambda - 1)\, T_a}{\Lambda} \qquad (11)$$

where T_a = absolute temperature of dilution air

T_e = absolute temperature of engine exhaust

Sample temperature is plotted against dilution ratio in Fig. 22 for two different exhaust gas temperatures. The temperature of the diluted stream is seen to fall rapidly during the initial stages of dilution. The EPA has specified a maximum allowable sample temperature of 325 K (125 F). From Fig. 22 this dictates minimum tunnel dilution ratios of 3.5 and 10 for the illustrated exhaust temperatures

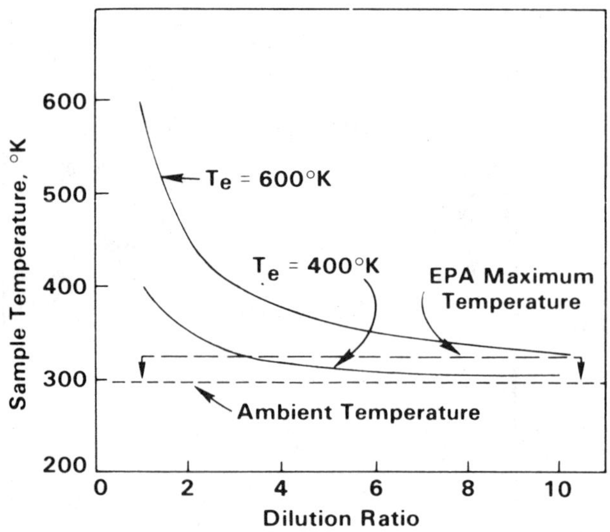

Fig. 22 - Effect of dilution ratio on sample temperature for two different engine exhaust gas temperatures

of 400 and 600 K, respectively.

The coupling that exists between sample temperature and dilution ratio in a conventional CVS tunnel is evident from Eq. 11 and Fig. 22. Using a specially constructed dilution mini-tunnel with provisions for electrically heating the dilution air, these two factors have been experimentally decoupled by varying each one, in turn, independent of the other [55]. Data were accumulated at one typical operating point (1500 r/min, 620 kPa IMEP) from a single-cylinder version of a 2.1-L passenger-car diesel engine. The exhaust hydrocarbon concentration at this point was 13 ppm C_6.

Using the data acquired from this experiment as a base, consideration was given to the mechanism(s) whereby soluble hydrocarbons are associated with the particulate sample collected on the sample filter [56]. Three mechanisms were examined -- condensation/evaporation, diffusion, and adsorption/desorption.

Condensation occurs when the vapor pressure of the gaseous hydrocarbon exceeds its saturated vapor pressure. The vapor pressures of the gaseous hydrocarbons in the dilution tunnel vary directly with their concentration in the engine exhaust stream and inversely with dilution ratio. The saturation vapor pressure is related to the sample temperature and the boiling point of the hydrocarbon. Thus condensation is favored by low dilution ratios, low filter temperatures (which results from a high dilution ratio), high exhaust hydrocarbon concentrations, and high boiling points. For the conditions of the experiment [55] it was found that condensation of exhaust hydrocarbons on the filter was likely to be significant only if all of the hydrocarbons originated from the less volatile half of the lubricating oil, or perhaps if the unburned fuel had been transformed into higher-boiling-point hydrocarbons within the engine cylinder [56]. The boiling points of the fuel components as they entered the engine were all too low for them to condense under the conditions of this experiment. Although most of the material trapped on the particulate filter originates from the fuel, recent experiments with radioactive tracers indicate that a significant fraction can come from the lubricating oil [57]. Weighing the magnitudes of the factors involved, however, it is unlikely that condensation in the dilution tunnel is a major factor accounting for the occurrence of soluble hydrocarbons on the particulate filter for this engine and this operating point. Given an engine or an operating point with a much higher concentration of unburned hydrocarbons in the exhaust, however, condensation could play a more significant role.

The second mechanism examined [56] was mass diffusion. This involves both transport of the hydrocarbons to the surfaces of the soot particles and absorption of the hydrocarbons into the particles. At fixed dilution ratio, diffusion was predicted to increase the soluble hydrocarbon fraction with increasing filter temperature, but the experimental results [55] showed the opposite trend. Consequently diffusion was eliminated as a significant mechanism.

Adsorption/desorption, which involves the adherance of hydrocarbon molecules to the surfaces of the soot particles by either chemical and/ or physical (van der Waals) forces, was the third mechanism considered [56]. By assuming equilibrium between the rates of adsorption and desorption, the particulate data collected in the experiment were fit well by a Langmuir equation relating the fraction of surface occupied by the adsorbate to the partial pressure of the adsorbing hydrocarbon gas and the temperature. Consequently adsorption was accepted as the dominant mechanism responsible for the soluble hydrocarbon fraction appearing on the filter in this test [55]. The occupied surface fraction (θ) was shown to be of the form [56]

$$\theta = \frac{1}{1 + \frac{C\Lambda}{X_{HC}} \exp(-E/RT)} \qquad (12)$$

where C = constant including effects of atmospheric pressure and Arrhenius pre-exponential factors for both desorption and adsorption

Λ = dilution ratio

E = difference between apparent activation energies for desorption and adsorption

R = gas constant

T = local absolute temperature

X_{HC} = gaseous hydrocarbon mol fraction prior to dilution

Unfortunately, not enough is known to determine C and E a priori, but representative values can be deduced [56] from available

Bonny Kubik

experimental data [55]. Beginning with measured particulate concentration at the filter for known engine and tunnel conditions, it is then possible to calculate the expected concentration for other dilution ratios.

The results of such calculations are shown in Fig. 23 for the two engine exhaust temperatures of Fig. 22. The absolute concentrations

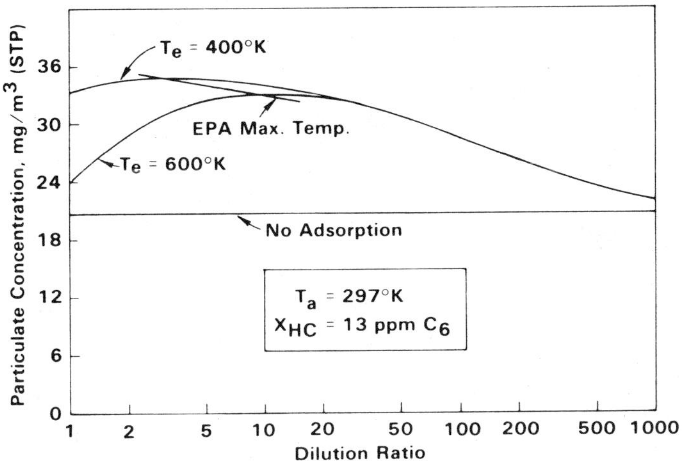

Fig. 23 - Effect of dilution ratio on particulate mass for two different engine exhaust gas temperatures

shown may be valid only for one particular engine and operating condition [55], since the adsorption constants are probably dependent on both engine type and load. However, the relative trends should be applicable to other diesel engines with low hydrocarbon emissions. The horizontal line in Fig. 23 marks the concentration of insoluble particulates (corrected to standard temperature and pressure). The difference between this level and the upper curves represents the adsorption-retained soluble mass.

The competition between dilution ratio and temperature in influencing the fraction of the soot surface occupied by adsorbed hydrocarbons determines the trends of Fig. 23. As dilution ratio is increased, the temperature of the gas stream falls, as shown by Fig. 22. The effect on the exponential term in Eq. 12 is such as to increase θ. At the same time, however, increasing dilution ratio decreases the partial pressure of the gaseous hydrocarbons, decreasing θ in Eq. 12 through the direct effect of Λ. Beginning from a dilution ratio of unity, at low dilution ratios the temperature effect dominates and the adsorbed fraction in Fig. 23 rises with increased dilution. As is evident from Fig. 22, however, at high dilution ratios the temperature is relatively insensitive to further dilution, so the effect of Λ in Eq. 12 becomes dominant and the adsorbed fraction falls.

The 325 K maximum filter temperature limit is cross-plotted in Fig. 23. Dilution tunnels intended for evaluating the particulate emissions from diesel cars would operate somewhat to the right of this limit. Near the prescribed temperature limit the total-concentration curves are reasonably flat. This suggests a general insensitivity of measured particulate mass to tunnel dilution ratio in this region, a characteristic observed experimentally by others [5,58,59].

Measurements made in collaboration with D. B. Kittelson and D. F. Dolan of the University of Minnesota behind the tailpipe of a diesel car powered by a 5.7-L engine indicated a rapid increase in dilution of the exhaust plume once it entered the atmosphere. At idle and low vehicle speed the dilution ratio ranged from 200 at a distance of 2.5 m behind the tailpipe to 1000 at a distance of 8 m. In view of this, the decrease in particulate concentration at high dilution ratios projected by Fig. 23 raises the question of whether the sample collected in the dilution tunnel is representative of diesel-generated particulates as they exist in the atmosphere. In this connection it should be appreciated that the calculations of Fig. 23 do not include the influence of ambient hydrocarbons that may also adsorb on the particulates. Although not significant at the modest dilution ratios typical of a tunnel, they can exert a significant influence at the high dilution ratios typical of diesel exhaust in the atmosphere.

To illustrate, the calculation of Fig. 23 was repeated assuming an ambient hydrocarbon concentration of 0.8 ppm C_6 (approximately 5 ppm C). The extent to which these ambient hydrocarbons affect the measured particulate level depends on their adsorption potentials. The range of possibilities is illustrated by the shaded area in Fig. 24. Based on this result and the probability of other complex reactions occurring in the atmosphere, it is clearly inappropriate to extrapolate this model of adsorption within a dilution tunnel to the global situation that exists in the atmosphere.

Another question is to what degree the

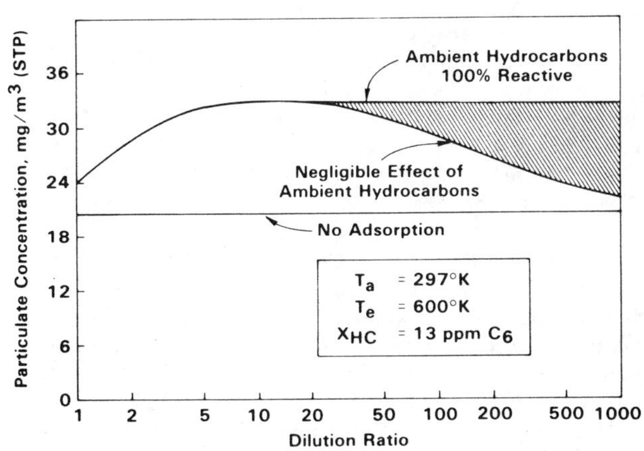

Fig. 24 - Possible effect of background ambient hydrocarbons on particulate mass as dilution occurs in the atmosphere soluble fraction is modified as gaseous hydrocarbons are continually drawn past soot particles trapped on the filter. This contrasts with adsorption occurring upstream in the tunnel, where soot particles and gaseous hydrocarbons flow together toward the filter during a comparatively shorter period of association. (It should be recognized that some hydrocarbons are probably also associated with the soot before it leaves the engine cylinder, of course.) Unfortunately the foregoing equilibrium analysis is unable to resolve this question because time is not one of its dimensions. However, in a separate analysis combining the tunnel temperature with the activation energy of Eq. 12 that was derived from the experiment [55], it was calculated that the time required to reach equilibrium was orders of magnitude less than the residence time of the gas in the tunnel [56]. This suggests that the increase in particulate mass arising from hydrocarbon adsorption during the dilution and sampling process occurred principally ahead of the filter.

PRACTICAL IMPLICATIONS

EFFECT OF LOCALLY LEAN COMBUSTION ON FORMATION - One of the routes suggested by Fig. 8 for avoiding soot formation was to avoid locally rich combustion. The virtues of burning lean to avoid soot formation are widely appreciated by the combustion community. In the limit it seems conceptually reasonable that soot formation could be eliminated by redistributing the fuel in the combustion chamber to provide a premixed lean mixture, but in the compression-ignited diesel engine this introduces a new set of problems.

To illustrate, consider light-load operation at an overall equivalence ratio of 0.2. With a mixture of this equivalence ratio homogeneously distributed throughout the cylinder volume, autoignition within the time available in an operating engine is unlikely. Even if the mixture were to be ignited successfully, it is questionable whether the combustion reaction in so lean a mixture could be completed prior to quenching by cylinder expansion.

A conceptually attractive variation of this approach is to collect all of the fuel that is involved into a smaller richer premixed cloud that is isolated from the chamber walls to avoid wall quenching, maintains a discontinuous stoichiometry boundary at its perimeter to avoid bulk quenching, is rich enough to autoignite within the time available, yet is lean enough to avoid soot formation. A centrally located homogeneous cloud of fuel and air premixed in nearly stoichiometric proportions would satisfy the criteria of this concept. Implementing the concept is another matter, however. With known technology, the time required to mix the rich core of a fuel spray with the cylinder air to achieve the desired homogeneous mixture exceeds typical ignition delay times, so that the desired degree of homogeneity cannot be established before compression ignition occurs.

Even if the homogeneous cloud could be established at an appropriate equivalence ratio, the rate of energy release associated with spontaneous ignition of the complete charge of fuel would be expected to introduce new problems. To quantify this, an engine combustion model was run for a compression ratio of 22, a speed of 1600 r/min, and an overall equivalence ratio of 0.6, with combustion timed always to start at top dead center. The independent variable was combustion duration, which was shortened from the value of 53 crankangle degrees measured for these conditions in an indirect-injection diesel. The progressive shortening of combustion duration reflects the effect of decreasing the degree of diffusion-limited combustion to avoid locally rich mixtures and attendant soot production. The results of this calculation are summarized by Fig. 25, which shows both the increase in peak cylinder pressure (normalized to the peak pressure for a conventional diesel engine) and the decreased time (crankangle) to reach that pressure. The substantial increase in peak pressure accompanying soot control by this means poses engine structural problems. In combination with the large increase in rate of pressure rise promoted by the shortened time to reach peak pressure, it also suggests the potential for objectionable engine combustion noise. The potential for increased NOx production in a nearly stoichiometric mixture undergoing almost instantaneous combustion at diesel compression ratios constitutes an additional concern.

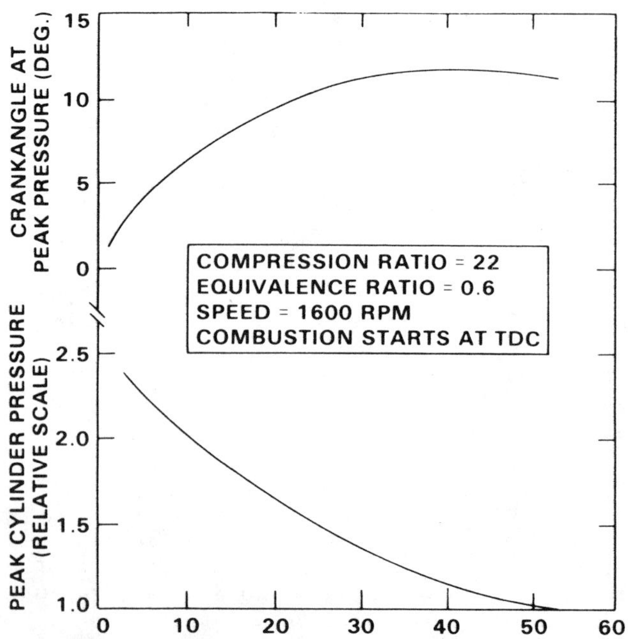

Fig. 25 - Effect of abbreviated combustion duration on peak cylinder pressure and the time to attain it

EFFECT OF HIGH-TEMPERATURE COMBUSTION ON FORMATION - A second way suggested by Fig. 8 to limit soot formation is to burn at high temperature. Reactant temperature can be increased either by heating the inlet air or by preheating the fuel. Contrary to expectations from the equilibrium calculations, Khan has found that heating the inlet air in a direct-injection diesel increases exhaust smoke [11]. There are three possible explanations for this discrepancy.

First, as the inlet air temperature, and with it the compression temperature, is raised, the ignition delay is shortened. Consequently a smaller fraction of the fuel charge is in the chamber at ignition and a greater fraction is injected during the resulting longer diffusion-limited combustion period. The fuel entering the chamber during the ignition delay has the opportunity to vaporize and premix with the cylinder air more effectively than that entering after combustion has started. Thus with the shorter ignition delay effected by heating the intake air, the opportunity exists for a greater fraction of the fuel to react locally at undesirably rich equivalence ratios.

Second, the increased cylinder temperature accompanying inlet air heating increases the opportunity for pyrolysis of the fuel. If dehydrogenation and the mass transfer suggested by Fig. 2 are involved, then the H/C ratio of the fuel in the core of the spray is likely decreased. It is evident from Fig. 4 through 6 that this increases the likelihood for carbon condensation, other factors remaining constant. Experimentally, McFarlane et al. [60] observed an increase in soot formation with temperature in premixed combustion which they suggested could be attributable to pyrolysis, pointing out that in a rich mixture the first fuel to burn consumes the available oxygen, associating the remaining fuel with a composition of gases atypical of equilibrium.

Third, the calculated equilibrium compositions must be carefully interpreted when applied to the rapid process of combustion in an engine cylinder. These compositions represent the limits toward which the global kinetics are driven. They are the result of opposing soot formation and destruction mechanisms, each of which may be characterized by its own rate constants. Thus the kinetics of the two opposing processes can distort the equilibrium representation during the time period required to establish equilibrium.

As for preheating the fuel, Temple-Pediani [61] has demonstrated a general decrease in soot, as reflected by exhaust smoke, when the fuel temperature is raised. In some ranges of load and speed, however, soot production rose with increased fuel temperature. Heated fuel should experience an abbreviated ignition delay, with an increase in soot, as discussed above. However, preheating the fuel should promote faster vaporization of the fuel spray and mixing with air, hence less extensive rich combustion and less soot formation. The occurrence of pyrolysis and dehydrogenation could explain the increased smoke sometimes experienced, especially at high loads.

EFFECT OF MIXING ON FORMATION - The third method suggested by Fig. 8 for minimizing soot formation is by controlling the mixing rate during the combustion reaction. From equilibrium considerations, mixing a rich mixture toward the stoichiometric ratio appears beneficial both during reaction at top dead center (Fig. 8, path d-e) and during expansion (Fig. 8, path d-d''). These two paths suggest that for any given equivalence ratio there is a critical temperature above which significant carbon should not appear, or conversely, at any given temperature a critical equivalence ratio below which significant carbon should not appear. This equilibrium incipent-soot equivalence ratio is a function of H/C ratio and is plotted versus reaction temperature as the solid curves in Fig. 26 for a pressure of 100 atm. (For this purpose, the critical isopleths are those producing only 1 ppm of C(s) at equilibrium.) In support of the previously advanced hypothesis that pyrolysis and dehydrogenation would contribute to soot formation, it is seen that the lower the H/C ratio at a given temperature, the leaner the local mixture must be to avoid incipient carbon formation.

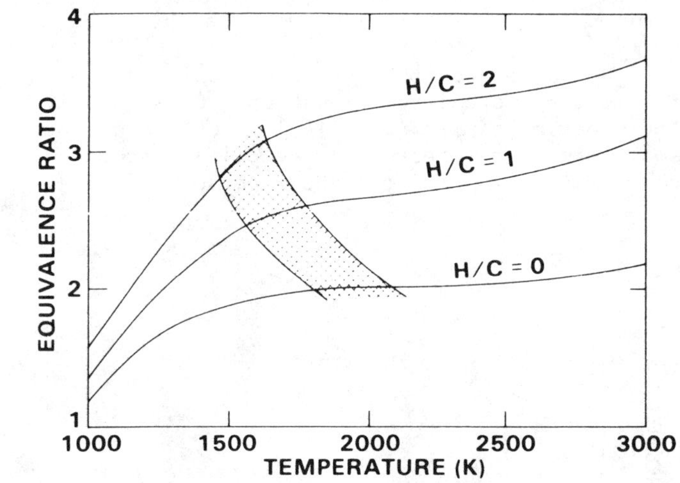

Fig. 26 - Critical equivalence ratio below which carbon condensation does not occur at equilibrium

The band cross-plotted in Fig. 26 indicates a region of equilibrium diabatic flame temperatures calculated for representative fuels having H/C ratios between zero (graphite) and 2.25 (octane). Intersections with the original critical isopleths occur at equivalence ratios of approximately 2, 2.5 and 3 for H/C ratios of 0, 1, and 2, respectively. This set of equilibrium solutions yields a single overall reaction equation for incipient soot formation (letting y represent the H/C ratio):

$$CH_y + \frac{1}{2} O_2 \rightarrow CO + \frac{y}{2} H_2 \qquad (13)$$

This expression suggests that at equilibrium, incipient soot formation occurs when the mixture is enriched to a molar O/C ratio of unity. Then the equivalence ratio for appearance of equilibrium soot is

$$\phi_{is-eq} = 2 + \frac{[H/C]}{2} \qquad (14)$$

As has been noted a number of times previously, chemical equilibrium provides a convenient and well defined starting point for studying soot formation, but soot formation is a kinetically controlled process, not an equilibrium process. Since the kinetics involved are not well defined, and in fact depend on the details of the combustion process of interest, one must turn to experiment. Street and Thomas [62] observed the threshold equivalence ratio for soot formation for premixed mixtures of air and a variety of hydrocarbon fuels, including alkanes, alkenes and aromatics. Their data points are presented on Fig. 27. For the conditions of their experiment, soot formation is avoided by employing premixed stoichiometries leaner than

$$\phi_{is-pm} = \frac{2 + [H/C]}{3} \qquad (15)$$

Both this premixed limit and the equilibrium limit of Eq. 14 are plotted as straight lines on Fig. 27.

But diesel combustion is diffusion-controlled, not premixed combustion. Consequently the experiments of Wright [63] to determine the soot formation threshold in a well stirred reactor are of interest. He used some of the

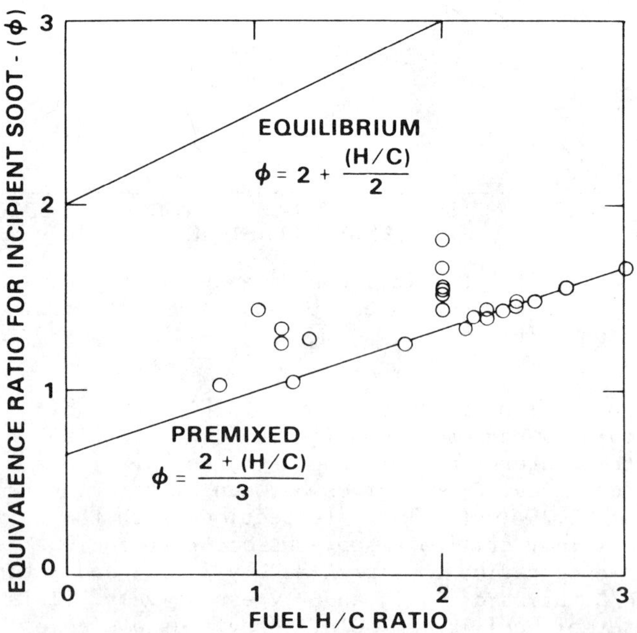

Fig. 27 - Limiting equivalence ratio for soot formation both from equilibrium calculations and from experiments [62] in a premixed fuel-air mixture

same fuels as Street and Thomas [62]. Although his data are omitted from Fig. 27 to avoid confusion, Wright found that the backmixing in his reactor generally allowed a soot-free equivalence ratio from 10 to 40% richer for a given fuel than found for the premixed experiment. In diesel combustion, then, one might expect the onset of soot to occur at some local equivalence ratio between the limits for equilibrium and premixed combustion shown in Fig. 27.

EFFECT OF CYLINDER PRESSURE - As previously discussed in connection with Fig. 7 and Eq. 1 and 2, the equilibrium analysis suggests an increasing propensity for the production of solid carbon during combustion at high cylinder pressures as the H/C ratio of the fuel approaches unity. The carbon oxidation expression of Eq. 10, as plotted in Fig. 16, suggests that for a given molar distribution of species within a packet, high cylinder pressure favors oxidation at temperatures above 2000 K but has little effect at lower temperatures. From these findings at such a fundamental level, it seems unreasonable to expect a simple relationship between diesel soot and cylinder pressure.

A dearth of out-of-engine experiments exists on the effect of diesel-cylinder pressure levels on soot production. In fact, a number of laboratory flame experiments have been conducted at sub-atmospheric pressures. However, Marfarlane et al. [60] studied soot formation in premixed hydrocarbon flames up to 20 atm and found the fraction of fuel carbon converted to soot at fixed equivalence ratio to increase significantly with pressure level. A pressure exponent in the range of 2.5 to 3 was found. They showed how this trend agreed *directionally* with experience in contemporary aero gas turbines. For their measured experimental trend to hold *quantitatively* when extrapolated to ever increasing pressure levels is unreasonable, however, because eventually the fuel runs out of sufficient carbon to satisfy this relationship.

Burning single drops of both n-octane and n-dodecane in compressed air, Kadota et al. [64] found soot production to increase approximately linearly with pressure up to about 10 atm, beyond which it tended to level off and stay reasonably constant to 30 atm. This sudden change in the effect of pressure on soot production coincided with the transition from a laminar to a turbulent flame, so it appears that the nature of the flame structure is of equal importance to the pressure in this experiment.

In fitting an Arrhenius-type relationship to experimental soot data from a direct-injection diesel, Khan et al. [13] found a first-power dependence of the mass rate of soot formation on the partial pressure of the unburned fuel. Given similar mol fractions of chemical species within a packet, this implies a direct proportionality between soot formation rate and cylinder pressure.

Two principal means for controlling cylinder pressure during combustion are compression ratio and rate of heat release. Lowering compression ratio significantly in a passenger-car diesel is hardly a free choice, however, because the compression ratio must be high enough to facilitate cold-engine starting and idling. Stretching out the heat release may be done anyhow to control maximum cylinder pressure, but that involves a tradeoff with engine cycle efficiency as the cycle deviates increasingly from the constant-volume combustion of the Otto cycle at the selected compression ratio.

EFFECT OF QUENCHING ON OXIDATION - The decrease in temperature experienced by a soot-containing packet of diesel combustion products was previously indicated to consist of two components -- cooling through mixing with the surrounding unreacted air and cooling through polytropic expansion of the cylinder contents. This latter component, which occurs because of the motion of the piston in producing useful work, distinguishes the diesel from engines employing continuous combustion, in which the production of work is not so closely coupled to the combustion process.

To evaluate the significance of oxidation-reaction quenching attributable to piston motion, carbon recession was calculated for soot-containing packets initially at top dead center that experienced both the intermediate and slow mixing rates of Fig. 11, but in the absence of piston motion. The results are shown in Fig. 28. (Without piston motion, the abscissa becomes equivalent to a scale of time.) The solid curves are repeated from Fig. 19 and 21 for comparative purposes. The dashed curves indicate from 1.5 to 3 times as much oxidation if work were not extracted from the cylinder gas. The improvement with slow mixing is particularly striking because of the prolonged period during which the products are kept at high temperature in the presence of abundant oxygen.

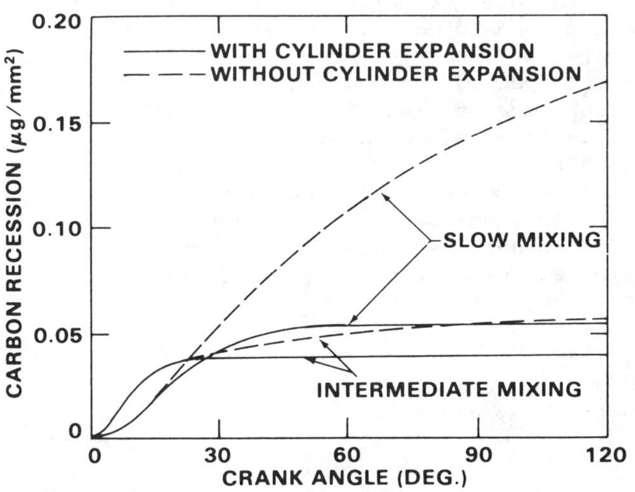

Fig. 28 - Effect of quenching associated with piston motion on carbon recession

OPTIMIZING MIXING RATE FOR OXIDATION - The existence of an optimum mixing rate for maximizing soot oxidation is suggested by Fig. 20, where fast mixing was initially better than slow mixing, but later during expansion proved inferior. To investigate the possibility of an optimum mixing rate, in Fig. 29 the recession rates of Fig. 20 for fast and slow mixing are replotted on the coordinates of Fig. 16. Portions of the oxygen partial pressure isobars from Fig. 16 are also included. For both mixing rates the arch-shaped carbon recession rate trajectories rise abruptly from the top dead center position at the right, then pass through peaks and fall again as they leave the field of the graph at the left. Reference points are marked on each curve at crankangles of 2, 12 and 30°.

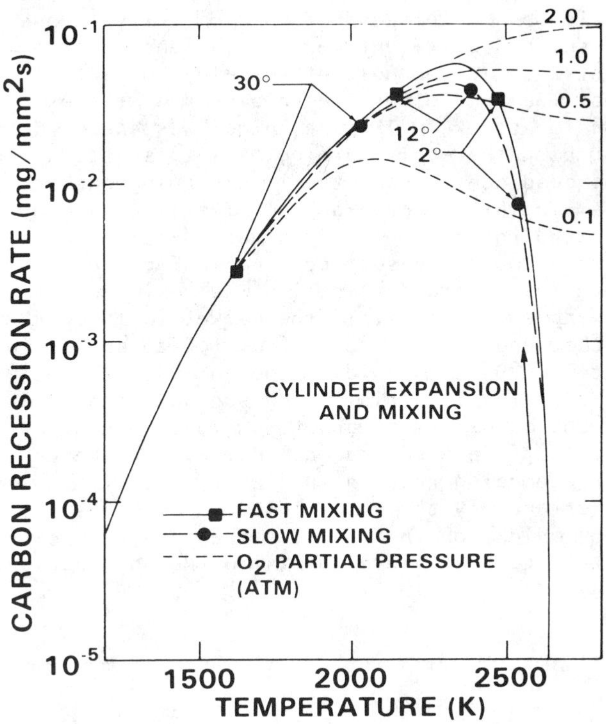

Fig. 29 - Recession-rate trajectories for two different mixing rates

At 2° crankangle the recession rate with fast mixing rate is seen to exceed that with slow mixing by a factor of about five. This is seen to occur despite the lower temperature of the fast-mixed packet because of the overwhelming significance of its higher oxygen partial pressure.

At 12° crankangle both mixing rates provide essentially equal recession rates. Although the fast rate of mixing retains the advantage of a higher oxygen partial pressure, that margin is no longer adequate to overcome the disadvantage of the substantially lower packet temperature it provides.

At the lower packet temperature level

existing 30° after top dead center, the influence of oxygen partial pressure on recession rate is greatly diminished and the lower temperature resulting from fast mixing produces a recession rate lower by a factor of ten than that of the slow mixing rate.

A study of the recession points discussed in connection with Fig. 29 leads to the conclusion that at first, fast mixing is desirable in order to raise the oxygen partial pressure in the packet. Very soon, however, the associated decrease in packet temperature becomes a dominant liability. This suggests that, rather than the linear rate of air addition to the packet represented by Fig. 11, a mixing schedule be used in which air is quickly introduced to the initially stoichiometric packet to provide oxygen, but then no further air is added in order to avoid excessive cooling of the packet.

To test this hypothesis, packet equivalence ratio was decreased in turn from unity to one of several different lean equivalence ratios in just 2° of crankangle, then held constant at that lean ratio for the remainder of the expansion stroke. This technique was executed for both the early-burning packet, starting at top dead center, and the late-burning packet, starting at 40° crankangle. The results are plotted in terms of carbon mass oxidized per unit of particle surface area in Fig. 30. In both cases carbon oxidation was maximized by mixing enough air into the packet to lower the packet equivalence ratio from 1.0 to approximately 0.9, then holding the equivalence ratio at that level. This mixing schedule resulted in the oxidation of about 50% more carbon per unit surface area than calculated for the best corresponding results in Fig. 19 and 21. Unfortunately the stoichiometry found best for soot oxidation in Fig. 30 corresponds approximately to that yielding maximum NOx production.

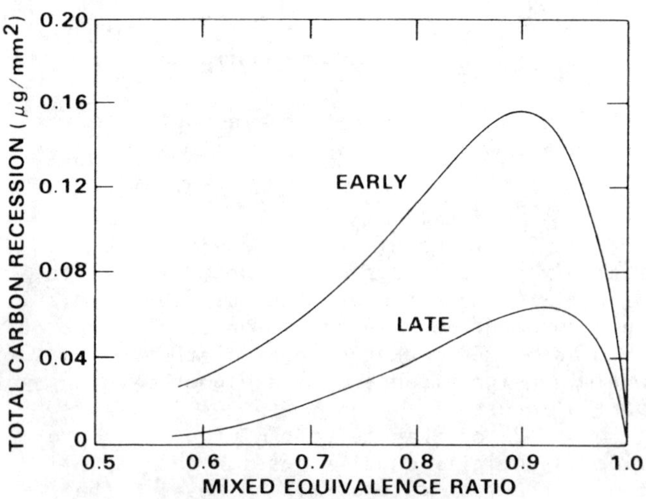

Fig. 30 - Effect of mixed equivalence ratio on total carbon oxidized for early (0° crankangle)- and late (40° crankangle)-burning packets

EFFECT OF COAGULATION ON OXIDATION - Although coagulation by itself has no effect on particulate mass, it has an indirect effect through its influence on oxidation. Diesel-exhaust coagulates revealed by photomicrographs can be broadly classified into two types -- clusters of basically spherical individual particles and string-like chains of the same type of individual particles. Two geometries are hypothesized for assessing the potential effect of these coagulate shapes on soot oxidation in the cylinder.

The first of these is a compacted sphere, i.e., a single solid sphere having the same mass (volume) as the sum of the masses (volumes) of the individual spherical particles comprising it. The diameter of the compacted sphere (D) having the same mass as N individual particles of diameter d is

$$D = d\ (N)^{1/3} \tag{16}$$

The second geometry is the compacted cylinder that would result if a string of tangent individual spherical particles were squeezed together so severely that their shared circular faces became equal in diameter to that of the original individual spheres (d). The length of such a cylinder is

$$L = \frac{2}{3} N\ d \tag{17}$$

In the following analysis L/d is assumed so great that as soot is oxidized from the surfaces of the cylinder, that removed from the end faces is negligibly small relative to that removed from the cylindrical surface.

The two compacted geometries proposed are "worst-case" configurations for a cluster and a chain coagulate, respectively. Mass oxidation rate is enhanced by increased surface area. The compacted sphere has a smooth surface, whereas a closely packed ball of unyielding individual spheres would enjoy greater surface roughness. Moreover, cluster anisometry would offer greater coagulate surface/volume ratio. By the same token, the compacted cylinder is a severe extreme for a chain coagulate. A longer, less compacted chain would be expected for a given number of individual particles, giving the chain a roughened outer surface with more surface area.

The soot recession rates (Ω) calculated previously for several different packets of cylinder gas were expressed in units of mass oxidized per unit time per unit of surface area. By definition

$$\Omega = \frac{1}{S}\frac{dm}{dt} = \frac{\rho}{S}\frac{dV}{dt} = \frac{\rho}{S} \cdot S\frac{dR}{dt} = \rho\frac{dR}{dt} \tag{18}$$

where
m = soot mass
R = radius (of compacted sphere or compacted cylinder)
S = surface area of coagulate
t = time
V = volume of coagulate
ρ = soot density

Thus for either a sphere or a cylinder of high L/d, Ω is a direct measure of the rate of recession of radius. Given the initial geometry of the coagulate and Ω, the rate of soot mass removal through oxidation is then easily calculated.

As an example, consider 20 individual spheres (N = 20) of diameter d. If compacted into a spherical coagulate of the same volume, that larger sphere would have a diameter of

$$N^{1/3}\ d = 2.714\ d \tag{19}$$

If the same 20 individual spheres were compacted into a cylinder of the same diameter as the individual spheres (d) instead, its length would be

$$\frac{2}{3}\ N\ d = 13.33\ d \tag{20}$$

Applying to these three configurations the soot recession rate from Fig. 11 for the intermediate mixing rate beginning at TDC, the results of Fig. 31 are obtained for an initial d of 0.1 μm. The 20 uncoagulated spheres disappear by 30 deg ATDC. So does the compacted cylinder, but if oxidation is arrested at some earlier crank angle, the mass oxidized in the simulated chain coagulate is less than that oxidized if coagulation had not occurred. The worst case is the compacted sphere. In this instance the coagulate persists into the exhaust stream.

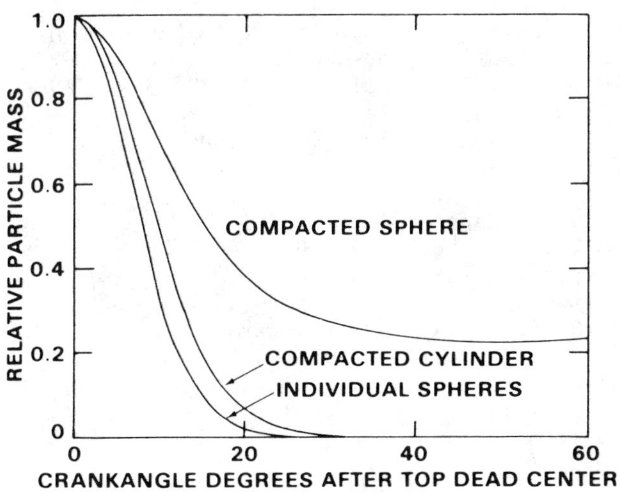

Fig. 31 - Decay of coagulate mass through in-cylinder oxidation

CONCLUSIONS

1. Although the diesel engine is a desirable automotive powerplant because of its proven fuel economy, the production of some soot appears to be an inherent aspect of its diffusion-controlled combustion process.

2. In the absence of both a universally accepted kinetic mechanism for soot formation and a suitable model of combustion in the divided-chamber diesel engine, equilibrium combustion calculations indicate that to minimize the thermochemical potential for carbon condensation, it is desirable to avoid locally rich equivalence ratios, fuels with low hydrogen/carbon ratios, and low reaction temperatures. In the case of reaction temperature, however, experimental evidence indicates a more complex dependence than equilibrium calculations would suggest.

3. Investigators of carbon formation in flames have proposed a variety of detailed mechanisms for the production of carbon, but it is unclear at this time which, if any, is appropriate to diesel combustion. It is generally acknowledged, however, that carbon production involves particle nucleation, followed by growth through both surface deposition and coagulation.

4. Application of established expressions for the coagulation of smoke particles suggests that coagulation in the diesel cylinder is essentially completed prior to opening of the exhaust valve.

5. A simple illustrative model of particle collision illustrates how an initially monodisperse particle population may be transformed into a population having a size distribution typical of that measured in the exhaust of a diesel engine.

6. Application of a published, experimentally based relation for soot oxidation illustrates the conflict between temperature and oxygen partial pressure within a mixture packet in the diesel cylinder as the piston descends. Increased partial pressure promoted by mixing air into the packet favors soot oxidation. However that mixing lowers packet temperature, which, together with the additional gas temperature reduction accompanying cylinder expansion, retards oxidation.

7. Conditions are more favorable for oxidizing soot formed and mixed early than for oxidizing that formed and mixed late during the combustion period.

8. A mixing strategy devised to maximize soot oxidation, based on application of a published, experimentally based soot-oxidation expression, led to a stoichiometry that would be expected to promote production of oxides of nitrogen.

9. Coagulation generally inhibits oxidation by decreasing the exposed surface area of a given mass of soot. This inhibition is least significant if the coagulated individual particles are strung out in a chain and most significant if the same number of individual particles are compacted into a sphere.

10. An increase in particulate mass occurring between the engine exhaust port and the particulate trap in a dilution tunnel, caused by an increase in the soluble particulate fraction, appears attributable primarily to adsorption of hydrocarbons by the soot particles. Under conditions of low filter

temperature and high exhaust hydrocarbon concentration, condensation of high-boiling-point hydrocarbons may also contribute to the soluble fraction to some degree.

ACKNOWLEDGMENTS

The authors gratefully acknowledge the contributions of many colleagues in assembling material for this paper. However, special thanks are due D. C. Siegla, R. W. Talder, J. H. Lienesch, R. W. Amann (Delco-Remy Division), D. L. Hilden, F. R. Faxvog and R. Stevenson.

REFERENCES

1. D. F. Dolan, D. B. Kittelson and K. T. Whitby, "Measurement of Diesel Exhaust Particle Size Distribution." ASME Paper 75-WA/APC-5 (1975).

2. C. T. Vuk, M. A. Jones and J. H. Johnson, "The Measurement and Analysis of the Physical Character of Diesel Particulate Emissions." SAE Transactions, Vol. 85, pp. 556-597, Paper 760131 (1976).

3. D. B. Kittelson, D. F. Dolan and J. A. Verrant, "Investigation of a Diesel Exhaust Aerosol." SAE Paper 780109 (1978).

4. D. F. Dolan and D. B. Kittelson, "Diesel Exhaust Aerosol Particle Size Distributions - Comparison of Theory and Experiment." SAE Transactions, Vol. 87, pp. 462-468, Paper 780110 (1978).

5. F. Black and L. High, "Diesel Hydrocarbon Emissions, Particulate and Gas Phase." Presented at Symposium on Diesel Particulate Emissions Measurement and Characterization, Ann Arbor, Michigan, May 17-18, 1978.

6. R. M. Schreck, J. J. McGrath, S. J. Swarin, W. E. Herin, P. J. Groblicki and J. S. MacDonald, "Characterization of Diesel Exhaust Particulate for Mutagenic Testing." Paper 78-33.5, presented to Air Pollution Control Association, Houston, Texas, June 25, 1978.

7. P. J. Groblicki and C. R. Begeman, "Particle Size Variation in Diesel Car Exhaust." SAE Paper 790421 (1979).

8. R. L. Williams and C. R. Begeman, "Characterization of Exhaust Particulate Matter from Diesel Automobiles." Presented at Seventeenth Annual Purdue Air Quality Conference, Purdue University, West Lafayette, Indiana, May 7, 1979.

9. G. Greeves and J. O. Meehan, "Measurement of Instantaneous Soot Concentration in a Diesel Combustion Chamber." I. Mech. E. Paper C88/75 (1975).

10. V. K. Duggal, T. Priede and I. M. Khan, "A Study of Pollutant Formation Within the Space of a Diesel Engine." SAE Transactions, Vol. 78, pp. 987-1000, Paper 780227 (1978).

11. I. M. Khan, "Formation and Combustion of Carbon in a Diesel Engine." Proceedings of the I. Mech. E., Vol. 184, Part 3J (1970).

12. I. M. Khan, C. H. T. Wang and B. E. Langridge, "Coagulation and Combustion of Soot Particles in Diesel Engines." Combustion and Flame, Vol. 17, pp. 409-419 (1971).

13. I. M. Khan, G. Greeves and C. H. T. Wang, "Factors Affecting Smoke and Gaseous Emissions from Direct Injection Engines and a Method of Calculation." SAE Transactions, Vol. 82, pp. 687-709, Paper 730169 (1973).

14. A. Williams, "Combustion of Droplets of Liquid Fuel: A Review." Combustion and Flame, Vol. 21, pp. 1-31 (1973).

15. J. S. Hsieh, *Principles of Thermodynamics*. McGraw-Hill Book Co. (1975).

16. G. M. Faeth, D. P. Dominicis, J. F. Tulpinsky and D. R. Olson, "Supercritical Bipropellant Droplet Combustion." Twelfth Symposium (International) on Combustion, pp. 9-18 (1969).

17. R. S. Lazar and G. M. Faeth, "Bipropellant Droplet Combustion in the Vicinity of the Critical Point." Thirteenth Symposium (International) on Combustion, pp. 801-811 (1971).

18. G. M. Faeth, "Current Status of Droplet and Liquid Combustion." Progress in Energy and Combustion Science, Vol. 3, pp. 191-224 (1977).

19. N. A. Chigier, "The Atomization and Burning of Liquid Fuel Sprays." Progress in Energy and Combustion Science, Vol. 2, pp. 97-114 (1976).

20. Y. Onuma and M. Ogasawra, "Studies on the Structure of a Spray Combustion Flame." Fifteenth Symposium (International) on Combustion, pp. 453-465 (1975).

21. S. M. Shahed, W. S. Chiu, and V. S. Yumlu, "A Preliminary Model for Formation of Nitric Oxide in Direct Injection Diesel Engines and its Application in Parameteric Studies." SAE Transactions, Vol. 82, pp. 338-351, Paper 730083 (1973).

22. W. S. Chiu, S. M. Shahed and W. T. Lyn, "A Transient Spray Mixing Model for Diesel Combustion." SAE Transactions, Vol. 85, pp. 502-512, Paper 760128 (1976).

23. J. Lahaye and G. Prado, "Mechanisms of Carbon Black Formation." *Chemistry and Physics of Carbon*, Ed. by P. L. Walker, Jr., and P. A. Thrower, Marcel Dekker, Inc. (1978).

24. H. Hiroyasu and T. Kadota, "Models for Combustion and Formation of Nitric Oxide and Soot in Direct Injection Diesel Engines." SAE Transactions, Vol. 85, pp. 502-512, Paper 760129 (1976).

25. R. W. Talder, J. D. Fleming, D. C. Siegla and C. A. Amann, "Dynamometer-Based Evaluation of Low Oxides of Nitrogen, Advanced Concept Diesel Engine for a Passenger Car." SAE Transactions, Vol. 87, pp. 1633-1652, Paper 780343 (1978).

26. E. K. Bastress, K. M. Chng and D. M. Dix, "Models of Combustion and Nitrogen Oxygen Formation in Direct and Indirect Injection Compression-Ignition Engines." Intersociety Energy Conversion Engineering Conference Paper 719053 (1971).

27. S. Gordon and B. J. McBride, "Computer Program for Calculation of Complex Chemi-

cal Equilibrium Composition, Rocket Performance, Incident and Reflected Shocks, and Chapman-Jouguet Detonations." NASA Lewis Research Center, NASA SP-273 (1971).

28. D. R. Stull and H. Prophet, "JANAF Thermochemical Tables." Dow Chemical Co., Midland, Michigan (1970).

29. K. Littlewood, "Gasification: Theory and Application." Progress in Energy and Combustion Science, Vol. 3, pp. 35-71 (1971).

30. M. Kamashita, O. P. Mahajan and P. L. Walker, Jr., "Effect of Carbon Deposition on Porosity and Reactivity of a Lignite Char." Fuel, Vol. 56, pp. 444-450 (1977).

31. A. L. Solano, O. P. Mahajan and P. L. Walker, Jr., "Concurrent Carbon Gasification and Carbon Deposition in Chars." Fuel, Vol. 56, pp. 451-452 (1977).

32. K. H. Homann, "Carbon Formation in Premixed Flames." Combustion and Flame, Vol. 11, pp. 265-287 (1967).

33. P. A. Tesner, T. D. Snegiriovia and V. G. Knorre, "Kinetics of Dispersed Carbon Formation." Combustion and Flame, Vol. 17, pp. 253-260 (1971).

34. E. R. Place and F. J. Weinberg, "The Nucleation of Flame Carbon by Ions and the Effect of Electric Fields." Eleventh Symposium (International) on Combustion, pp. 245-255 (1967).

35. J. Howard, "On the Mechanism of Carbon Formation in Flames." Twelfth Symposium (International) on Combustion, pp. 877-887 (1969).

36. F. A. Heckman, "Microstructure of Carbon Black." Rubber Chemistry and Technology, Vol. 37, pp. 1245-1298 (1964).

37. T. D. Bolt, E. M. Dannenburg, R. E. Dobbin and R. P. Rossman, "Carbon Black Structure Effects in Snythetic Rubbers." Rubber and Plastics Age, December 1960, pp. 1520-1526.

38. W. G. Parker and H. G. Wolfhard, "Carbon Formation in Flames." Journal of the Chemical Society, pp. 2038-2049 (1950).

39. G. D. Ulrich, "Theory of Particle Formation and Growth in Oxide Synthesis Flames." Combustion Science and Technology, Vol. 4, pp. 47-57 (1971).

40. B. L. Wersborg, J. H. Howard and G. C. Williams, "Physical Mechanisms in Carbon Formation in Flames." Fourteenth Symposium (International) on Combustion, pp. 929-940 (1973).

41. A. I. Medalia, "Morphology of Aggregates - I. Calculation of Shape and Bulkiness Factors; Application to Computer-Simulated Random Flocs." Journal of Colloid and Interface Science, Vol. 24, pp. 393-404 (1967).

42. R. Whytlaw-Gray and H. S. Patterson, *Smoke*. Edward Arnold and Co. (1932).

43. H. L. Green and W. R. Lane, *Particulate Clouds: Dusts, Smokes and Mists*. D. Van Nostrand Co., Inc., Second Edition (1945).

44. D. F. Dolan, "Experimental and Theoretical Investigation of Diesel Exhaust Particulate Matter." Ph.D. Thesis, University of Minnesota (1977).

45. D. B. Kittelson, D. F. Dolan and M. C. Beaver, "Exhaust Particle Size Distributions Produced by Diesel and Spark Ignition Engines: Comparison of Theory and Experiment." Paper CSS/CI-79-09, presented at Central States Section Meeting, The Combustion Institute, Columbus, Indiana, April 9-10, 1979.

46. B. Y. H. Liu and D. Y. H. Pui, "On the Performance of the Electrical Aerosol Analyzer." Journal of Aerosol Science, Vol. 6, pp. 249-264 (1975).

47. N. M. Laurendeau, "Heterogeneous Kinetics of Coal Char Gasification and Combustion." Progress in Energy and Combustion Science, Vol. 4, pp. 221-270 (1978).

48. J. Nagle and R. F. Strickland-Constable, "Oxidation of Carbon Between 1000-2000°C." Proceedings of the Fifth Carbon Conference, Vol. 1, pp. 154-164 (1962).

49. J. R. Walls and R. F. Strickland-Constable, "Oxidation of Carbon Between 1000-2400°C." Carbon, Vol. 1, pp. 333-338, Pergamon Press Ltd. (1964).

50. S. W. Radcliffe and J. P. Appleton, "Soot Oxidation Rates in Gas Turbine Engines." Combustion Science and Technology, Vol. 4, pp. 171-175 (1971).

51. K. B. Lee, M. W. Thring and J. M. Beer, "On the Rate of Combustion of Soot in a Laminar Soot Flame." Combustion and Flame, Vol. 6, pp. 137-145 (1962).

52. C. P. Fenimore and G. W. Jones, "Oxidation of Soot by Hydroxyl Radicals." Journal of Physical Chemistry, Vol. 71, No. 2, pp. 593-597 (1967).

53. C. Park and J. P. Appleton, "Shock-Tube Measurements of Soot Oxidation Rates." Combustion and Flame, Vol. 20, pp. 369-379 (1973).

54. F. G. Roper and C. Smith, "Soot Escape from Laminar Air-Starved Hydrocarbon Flames." Combustion and Flame, Vol. 36, pp. 125-138 (1979).

55. J. S. MacDonald, S. L. Plee, J. B. D'arcy and R. M. Schreck, "Experimental Measurements of the Independent Effects of Dilution Ratio and Filter Temperature on Diesel Exhaust Particulate Samples." SAE Paper 800185 (1980).

56. S. L. Plee and J. S. MacDonald, "Some Mechanisms Affecting the Mass of Diesel Exhaust Particulate Collected Following a Dilution Process." SAE Paper 800186 (1980).

57. W. J. Mayer, D. L. Hilden and D. C. Lechman, "The Contribution of Engine Oil to Diesel Exhaust Particulate Emissions." SAE Paper 800256 (1980).

58. P. M. Najt, "Light-Duty Particulate Measurements Using a CVS Tunnel." Presented at Symposium on Diesel Particulate Emissions Measurements and Characterization, Ann Arbor, Michigan, May 17-18, 1978.

59. F. Black and L. High, "Methodology for Determining Particulate and Gaseous Diesel Hydrocarbon Emissions." SAE Paper 790422 (1979).

60. J. J. Macfarlane, F. H. Holderness and F. S. E. Whitcher, "Soot Formation Rates in Premixed C_5 and C_6 Hydrocarbon-Air Flames at Pressures up to 20 Atmospheres." Combustion and Flame, Vol. 8, pp. 215-229 (1964).

61. R. W. Temple-Pediani, "Effect of Pre-injection Fuel Temperature upon Diesel Engine Ignition Delay and Soot Emissions." Proceedings of the I. Mech. E., Vol. 187, pp. 395-404 (1973).

62. J. C. Street and A. Thomas, "Carbon Formation in Pre-mixed Flames." Fuel, Vol. 34, pp. 4-36 (1955).

63. F. J. Wright, "The Formation of Carbon under Well-stirred Conditions." Twelfth Symposium (International) on Combustion, pp. 867-875 (1969).

64. T. Kadota, H. Hiroyasu and A. Farazandehmehr, "Soot Formation by Combustion of a Fuel Droplet in High Pressure Gaseous Environments." Combustion and Flame, Vol. 29, pp. 67-75 (1977).

Soot Formation and Oxidation in Diesel Engines**

H. Hiroyasu, M. Arai, and K. Nakanishi
Dept. of Mechanical Engrg.
Univ. of Hiroshima
(Hiroshima/Japan)

DIESEL ENGINES BECAUSE OF better fuel economy are rapidly gaining a larger share of the market for light load automotive application in addition to their more traditional use for heavy duty trucks. However, the use of the diesel engine in light duty vehicles, if it cannot meet the expected future particulate emission standards might be ruled out. Most of the previous work concerning particulates in diesel engines has been limited to the measurement of their total mass, size distribution and chemical structure after they are emitted in the exhaust, (1 to 14)*. A summary of the general effects of the design and operating parameters on diesel exhaust particulates has been given by Khan (15). In the above cited literature, there is a severe lack of information on particulate formation different stages of during the combustion process. The cause of this lack of information has been due to the complexity of the diesel combustion process and the difficulty in instrumenting the engine to obtain data that would be meaningful to study the particulate formation on temporal and spacial basis in the combustion chamber. (16 to 19) The combustion in both the spray open flame and in a diesel engine are probably essentially diffusion flames. Hydrocarbon flame from a premixed mixture is blue-green in colour, while on the other hand, from a diffusion case the flame is luminous, i.e. yellow due to the radiation from incandescent carbon particles. Radiation from premixed hydrocarbon and air is mainly banded emission spectra due to simple molecules, with some weak continuous radiation, due to chemical processes or a recombination of ions. But the typical emission from a diffusion flame is due to carbon which gives a yellowish luminous flames. (20,21)

From a open spray flame, it is usually possible by using a quenching probe to obtain soot in the luminous zone, but soot generally does not collect on the quenching probe at the top of flame in the case of turbulent diffusion flame.

The photographs of spray combustion in a direct injection engine show that a spray flame

*Numbers in parentheses designate References at end of paper

**Paper 800252 presented at the Congress and Exposition, Detroit, Michigan, 1980.

ABSTRACT

In order to obtain a fundamental understanding of how soot forms and burns up during combustion in diesel engines, the relation between exhaust smoke level and combustion duration in direct injection diesel engine was measured. The effect of different injection timings, engine speeds and loads on combustion duration are presented and discussed. Combustion experiments in the less complex environment of a continuous spray open flame were performed from which soot samples were collected at various locations in the flame. By use of a special technique were obtained information on the mechanism of soot formation and oxidation could not be obtained from engine experiments. The similarity of soot formation and oxidation within diesel combustion chamber and continuous spray open flame are discussed. An analysis of these results shows that the exhaust smoke level is mainly determined by the concentration of soot at the time of exhaust valve opening, i.e. combustion duration is very important. The net soot release in a diesel engine cylinder, for a given fuel, is controlled by the amount of soot oxidation and the rate of combustion.

in a combustion chamber represents quite complicated features, but is fundamentally similar to the diffusion spray open flame. The results of measuring soot on both temporal and spacial basis using sampling probe and other techniques show that a fair amount of soot is obtained during the middle of the combustion period, but this soot is burnt up during the latter part of the combustion period, which explains why we can not obtain soot at the end of combustion period under some engine operating conditions.

It seems that the mechanism of soot formation is important but combustion of soot is a more important factor, since it determines the exhaust smoke level in diesel engines.

In this study, we have chosen as our object the obtaining a fundamental understanding on how soot forms and burns up during combustion period in diesel engines. To this end we have performed combustion experiments in the less complex environment of a continuous spray open flame. Soot samples collected at various locations using a nitrogen cooled quenching probe provides information on soot formation locations and oxidation locations indicating the burn-up of soot that could not be obtained from engine experiments. The relationship between soot release from the exhaust and combustion duration in direct injection diesel engine is discussed. Also the similarity of soot formation and oxidation in diesel combustion and continuous spray open flame will be discussed.

EXPERIMENTAL DETAILS

A four cycle single cylinder direct-injection diesel engine with 135mmϕ×130mm (5.3inϕ×5.1in) with a toroidal combustion chamber was used. Some basic specifications of this engine are listed in table 1. For the measurement of combustion duration in a cylinder, we made a flame luminosity detector with a photo-transistor. The details of this flame luminosity detector and its position on the cylinder head are shown in Fig. 1.

This detector was set up in the observed window with a diameter of 10mm (0.4in) which makes an inclination angle of 45 degrees, and the visual angle of the detector was about 13 degrees. In order to protect the detector from destruction by the combustion pressure and temperature, a glass window was installed in front of the photo-transistor. The sensible wavelength of this photo-transistor is from 0.3μm to 1.1μm, and the most sensible wavelength is at 1.0μm. The wavelength of the visible light ranges from 0.4μm to 0.7μm and the light from the luminous flame of the sprayed combustion ranges from visible to infra-red. Therefore, this detector could detect with accuracy the flame luminosity during combustion. The plane of delineation of the visual direction of the flame luminosity detector and of the fuel spray injected into the combustion chamber are shown schematically in Fig. 2. It is clearly

Table 1 - The Engine Specifications

Bore × stroke (mm)	135×130 (5.3×5.1 in)
Displacement (cu cm)	1861 (113.6 cu in)
Compression ratio	15.5:1
Connecting rod length (mm)	230
Clearance volume (cu cm)	128.3
Piston bowl diameter (mm)	78
Intake valve opening (CA)	7 BTDC
closing (CA)	35 ABDC
Exhaust valve opening (CA)	35 BBDC
closing (CA)	5 ATDC
Injection pump	Bosch AP
Plunger diameter (mm)	10
Diameter of nozzle hole (mm)	0.35
Number of nozzle holes	4
Nozzle opening pressure (MPa)	22

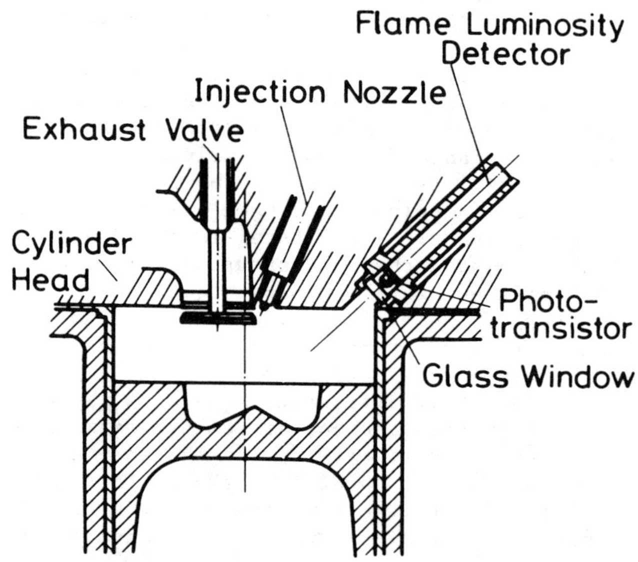

Fig. 1 - The flame luminosity detector location in cylinder head

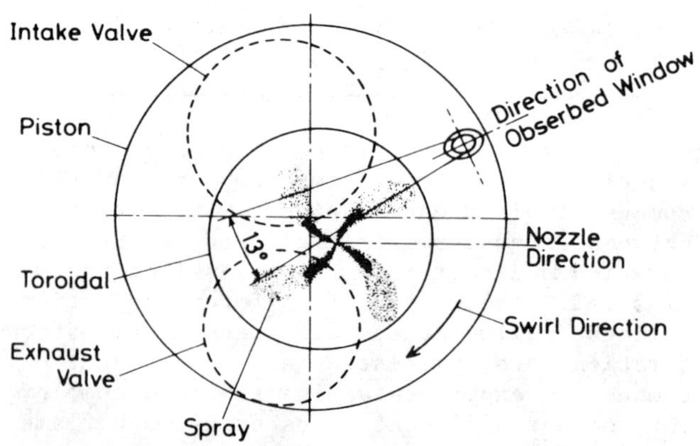

Fig. 2 - Plane delineation of visual direction of flame luminosity detector and injected fuel spray

seen from Figs.1 and 2 that the flame luminosity detector could detect the flame luminosity from the start of ignition in the injected spray to the end of combustion period. Experiments were performed with a number of different injection timings, loads and engine speeds. During these experiments soot was measured by using a Bosch smoke meter at an exhaust mixing chamber 100cm from the exhaust valve.

RESULTS AND DISCUSSIONS

Figures 3, 4 and 5 show the diagram of measured cylinder pressure, needle lift of nozzle, and flame luminosity and also slow the result of the calculated rate of heat release. Figure 3 shows the case of low load when the brake mean effective pressure (BMEP) was 0.35MPa and Figs,4 and 5 show the cases of 0.51 MPa and 0.73 MPa. It is seen from these figures that the maximum pressure was getting higher and the end of combustion determined by the rate of heat release was getting longer with an increase of BMEP. The end of combustion determined by the rate of heat release was 40~60 degrees ATDC. The wave signal of flame luminosity fluctuated greatly. However, the end of flame luminosity detected by the photo-transistor was longer than the end of the heat release. A strong relationship was observed between the end of flame luminosity and BMEP, i.e. the end of flame luminosity changed from 50 to 200 degrees ATDC with an increase of BMEP. In the cases where BMEP were 0.51 MPa and 0.73 MPa, which were middle and high load, the flame luminosity was still detected after the exhaust valve opening. This phenomenon indicated the flame, which involved the last stage of the combustion with very little heat release, existed after the exhaust valve opening. At all speeds, the effect of injection timing on smoke is similar, i.e. the smoke level increased both with an increase of the load and with retarding the injection timing as shown in Figs. 6, 7 and 8.

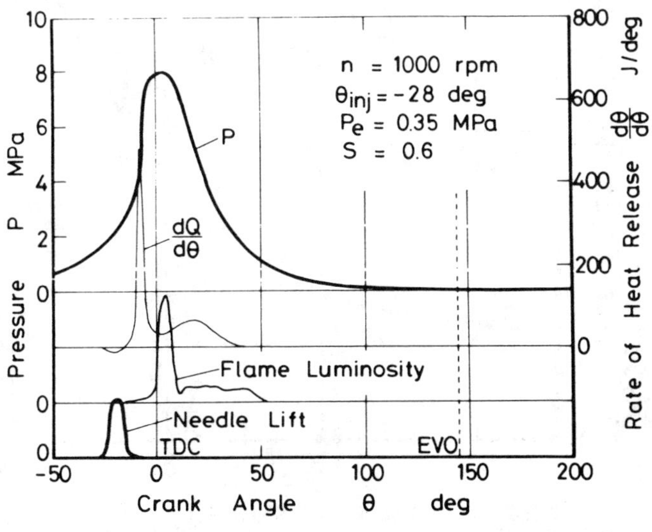

Fig. 3 - Cylinder pressure, needle lift of nozzle, flame luminosity and rate of heat release (BMEP=0.35MPa)

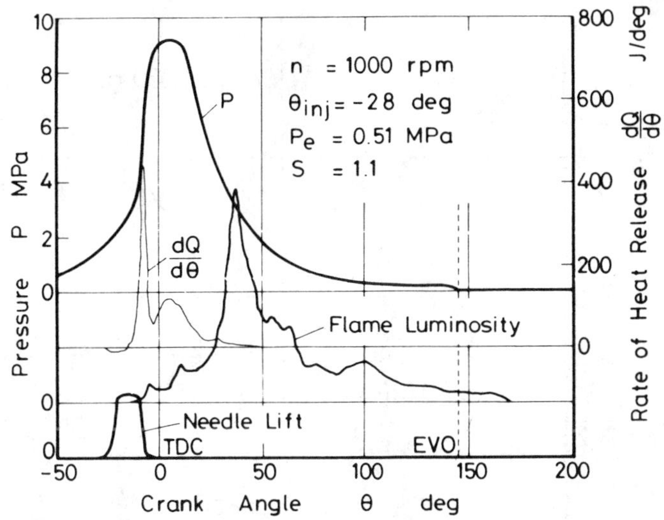

Fig. 4 - Cylinder pressure, needle lift of nozzle, flame luminosity and rate of heat release (BMEP=0.51MPa)

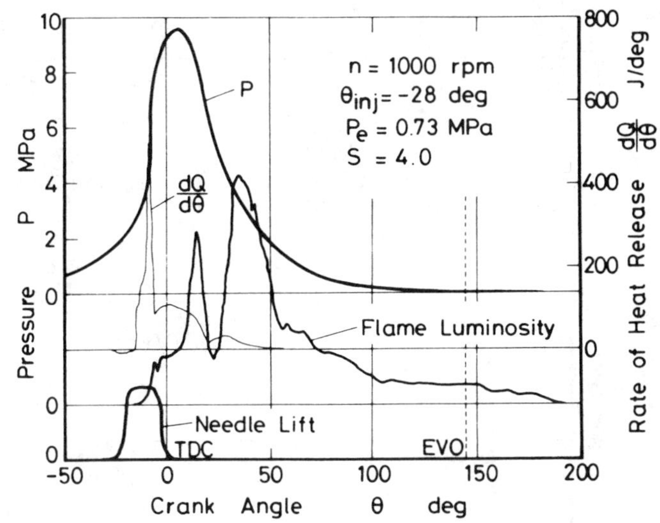

Fig. 5 - Cylinder pressure, needle lift of nozzle, flame luminosity and rate of heat release (BMEP=0.73MPa)

The relationship between the end of flame luminosity and exhaust smoke level are shown in Figs. 9 and 10. From these figures it can be seen that the change in luminosity duration from 50 to 150 degrees ATDC caused an increase in the Bosch smoke number of approximately 1 while the change in duration from 150 to 200 degrees ATDC caused a change in the Bosch smoke number of at least 2. As the duration of the flame luminosity extends to the exhaust valve opening the soot release increases rapidly. Especially, there was a very high value of smoke level when the end of flame luminosity exceeded the opening timing of the exhaust valve.

Figure 11 shows the relation between the

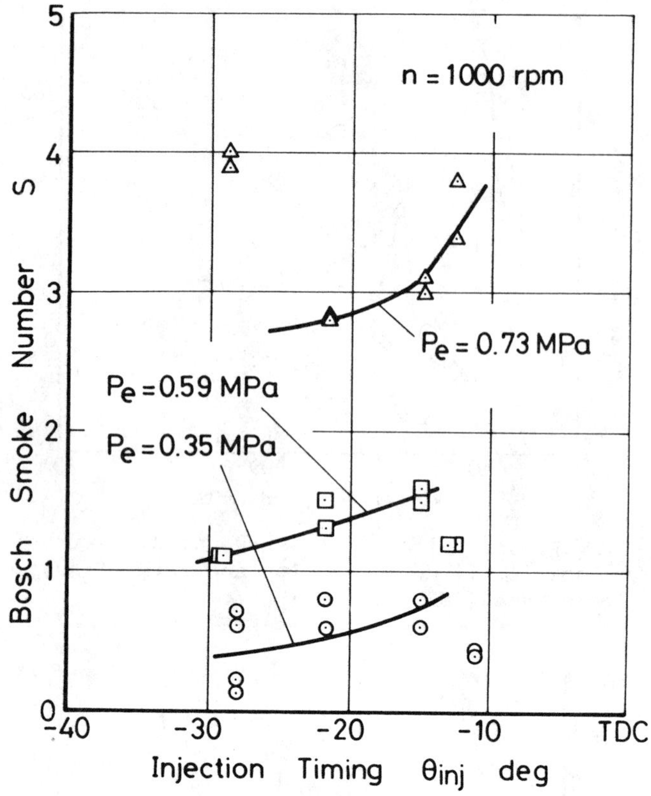

Fig. 6 - Effects of injection timing and load on exhaust smoke (n=1000rpm)

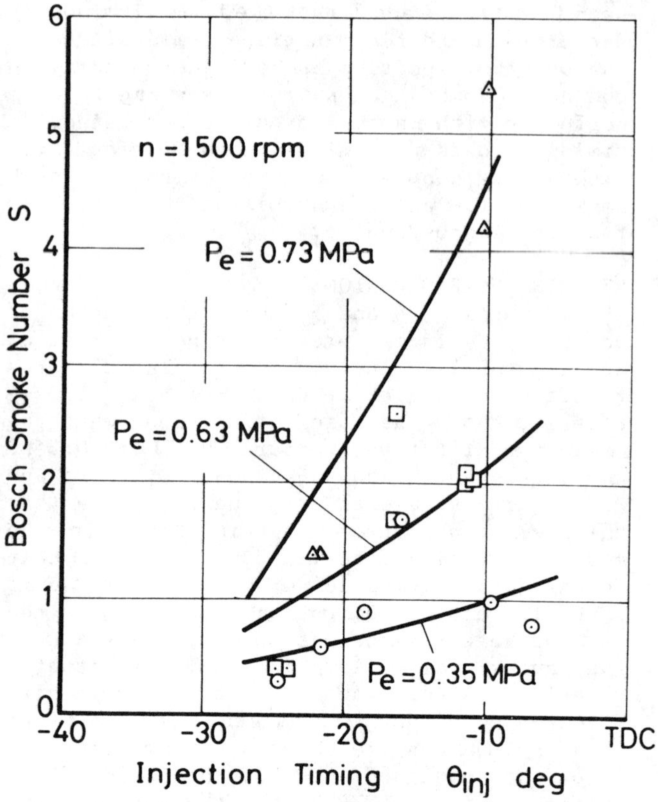

Fig. 8 - Effects of injection timing and load on exhaust smoke (n=1500rpm)

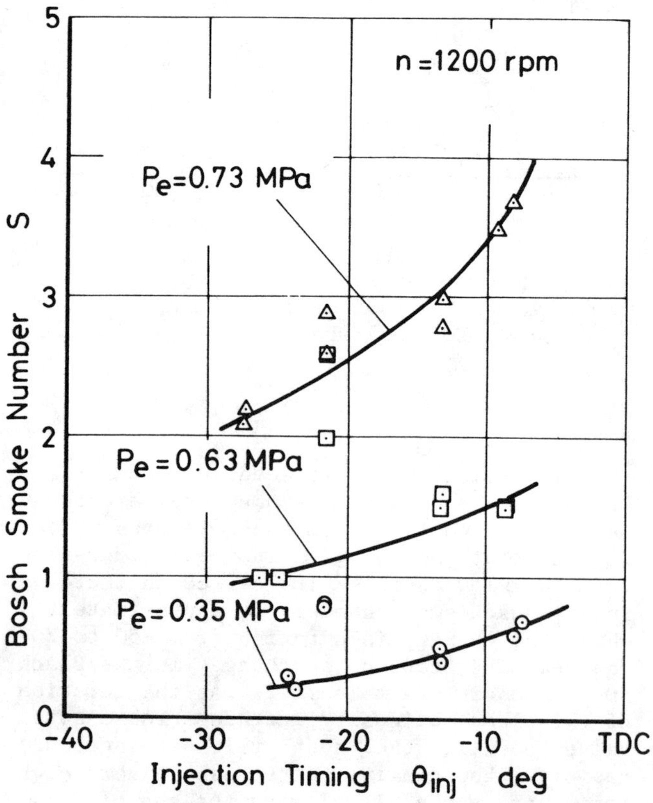

Fig. 7 - Effects of injection timing and load on exhaust smoke (n=1200rpm)

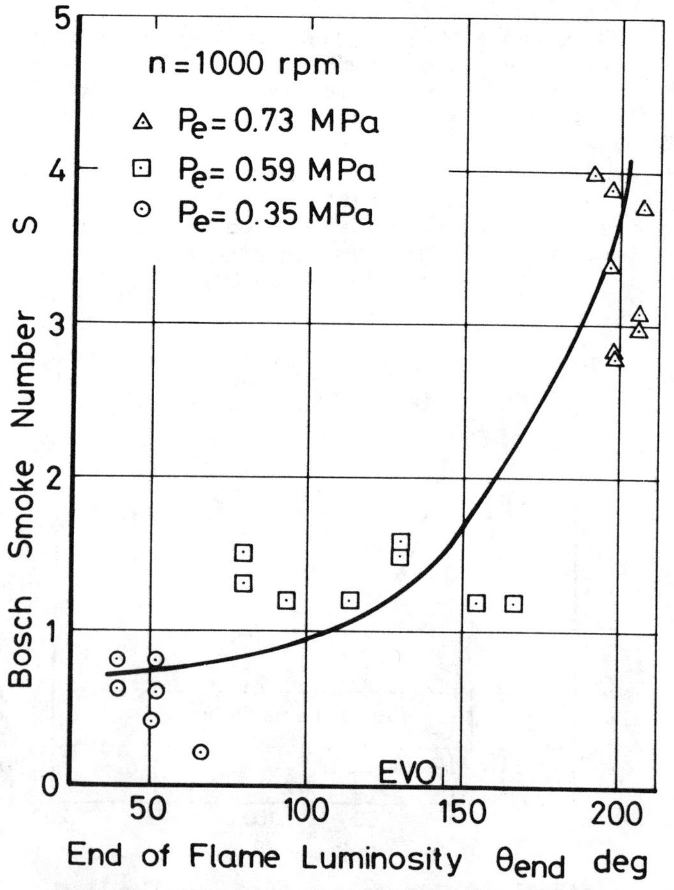

Fig. 9 - Relation between end of flame luminosity and exhaust smoke (n=1000rpm)

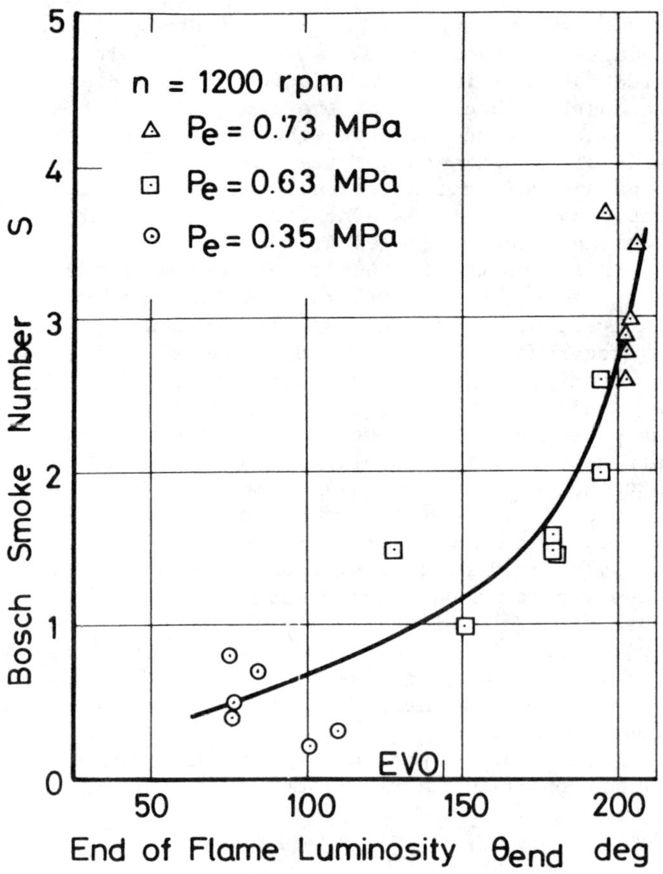

Fig. 10 - Relation between end of flame luminosity and exhaust smoke (n=1200rpm)

BMEP and the combustion duration determined by the flame luminosity detector at various injection timings and engine speeds. The combustion duration increased with BMEP. It is already known that the ignition delay changes with injection timing, i.e. the ignition delay is reduced by retarded injection timing. This phenomenon was observed also in our engine test and the start of the flame luminosity detected with photo-transistor was almost constant.

Figure 12 shows the relation between the exhaust smoke level and the end of flame luminosity with various injection timings and engine speeds. The effects of injection timing and engine speed on the exhaust smoke level are important, but it can be deduced from Fig. 12 that the exhaust smoke level change with the end of flame luminosity is a better representation than changes with injection timing and engine speed.

In the range where the end of flame luminosity did not exceed the opening timing of exhaust valve, that is, 145 degrees ATDC, it can be seen from Fig. 12 that the Bosch smoke number was less than 2 and the effect of injection timing on the exhaust smoke level was minimal. As the end of flame luminosity exceeded this critical point, the exhaust smoke level increased strikingly and the effect of injection timing began to have a strong effect. In the case that the injection timing and the end of flame luminosity were both late, i.e.

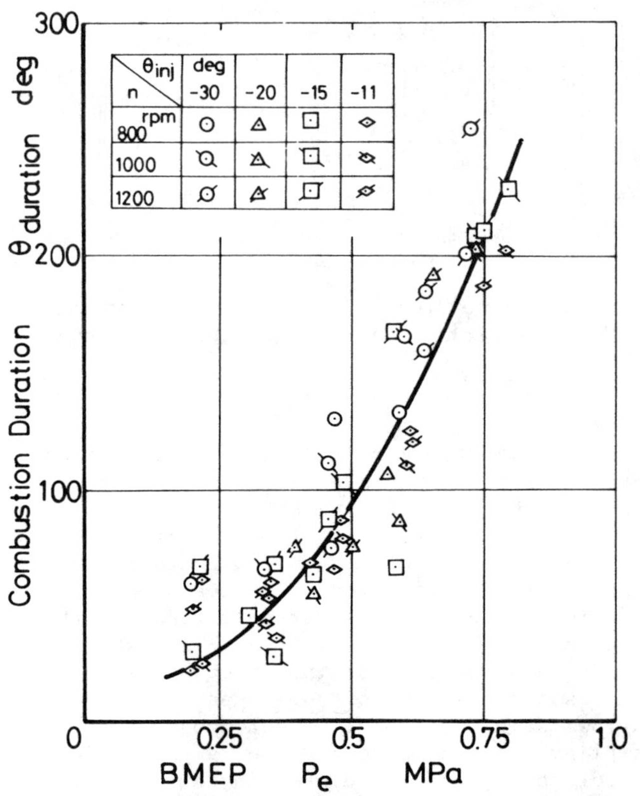

Fig. 11 - Relation between BMEP and combustion duration at various kinds of injection timing and engine speed

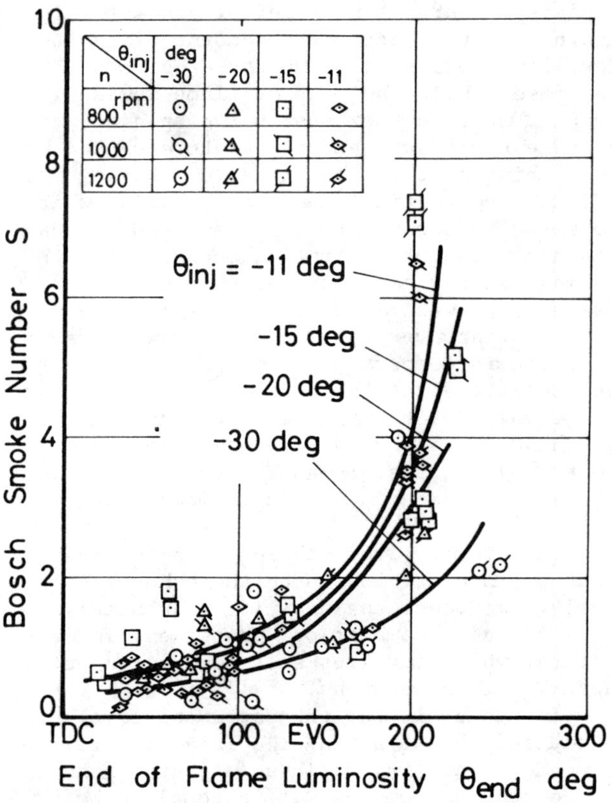

Fig. 12 - Relation between end of flame luminosity and exhaust smoke at various injection timings and engine speeds

θinj=11 degrees BTDC and θend=200 degrees ATDC, the exhaust smoke level was fairly high, i.e. Bosch smoke number 6. In the case where the injection timing was early, the exhaust smoke level increased only to the range that experienced when the end of flame luminosity exceeded the opening timing of the exhaust valve, i.e. 145 degrees ATDC, but the increase of smoke was small compared with the case of reterded injection timing. For example when θinj=30 degrees BTDC and θend=200 degrees ATDC, the Bosch smoke number was only 2. The explanation of the experimental phenomenon where the exhaust smoke level strikingly increased under the operating conditions where the end of flame luminosity exceeded the opening timing of exhaust valve is as follows: The exhaust smoke level from diesel combustion was caused by soot formation and soot oxidation during combustion period. Soot, which was also called pre-soot in flame, was formed mainly from the early to middle stages of combustion duration and burned mainly from middle to final stages of combustion. Therefore, it can be considered that the exhaust smoke level was the remainder of formed soot from soot oxidation. The flame luminosity, which was measured in this experimental study, was the detection of luminous or yellow flame, that is to say, the detection of soot combustion. In the case that the end of flame luminosity did not exceed the timing of exhaust valve opening, the luminous flame, that is soot combustion, did not exist at the time of exhaust valve opening, therefore, the soot combustion was already complete before the time of exhaust valve opening. After that, the amount of exhaust smoke level was small. On the other hand, in the case that the end of flame luminosity exceeded the time of exhaust valve opening, the decrease of pressure and temperature caused by the opening of the exhaust valve, during the period of the continuing soot combustion, which froze the soot combustion process. Additional, quenching of soot combustion occured when the flaming exhaust gases flowed through the narrow clearance between the exhaust valve and the valve seat at a very high speed causing a rapid reduction in temperature and pressure due to the expansion process and due to heat transfer to the valve seat and valve. The result of both mechanisms freezing in the combustion chamber and the heat transfer and expansion during exhaust, the amount of exhaust smoke level was high.

Therefore, from these experimental findings, in order to reduce the exhaust smoke level from diesel engines, it is necessary to complete all combustion processes in the cylinder which involves soot combustion before the exhaust valve opens.

In general, the exhaust smoke level can be reduced by advancing the injection timing or by adding air swirl. These effects, enhance the combustion so that it is complete before the exhaust valve opens.

SOOT CONCENTRATION IN CONTINUOUS SPRAY OPEN FLAME

There are so many factors affecting diesel combustion, therefore it is a very difficult to elucidate how individual factors effect diesel combustion because of the complexity of the combustion process. For these reasons, the special experimental systems, for example, the constant volume combustion vessel (22) or the rapid compression machine (23) which limits some of the variables are used, so that the analysis of the combustion phenomenon can be made some what simpler. Physical factors influencing soot release in fundamental hydrocarbon flame have been studied such as a continuous diffusion flame (24, 25) and a single droplet flame (26,27). These studies have been performed so that a better understanding of soot formation mechanism could be obtained. Now, if diesel combustion can be simulated with continuous spray diffusion flame, then the knowledge gained with continuous spray diffusion flame studies can be applied to diesel combustion. This makes it easier to interprete the diesel combustion process. In spite of the difference of ambient pressure and temperature, an element of diesel spray combustion is identical to the combustion of continuous spray.

Measuring apparatus for soot concentration in the continuous spray flame is shown in Fig.13.

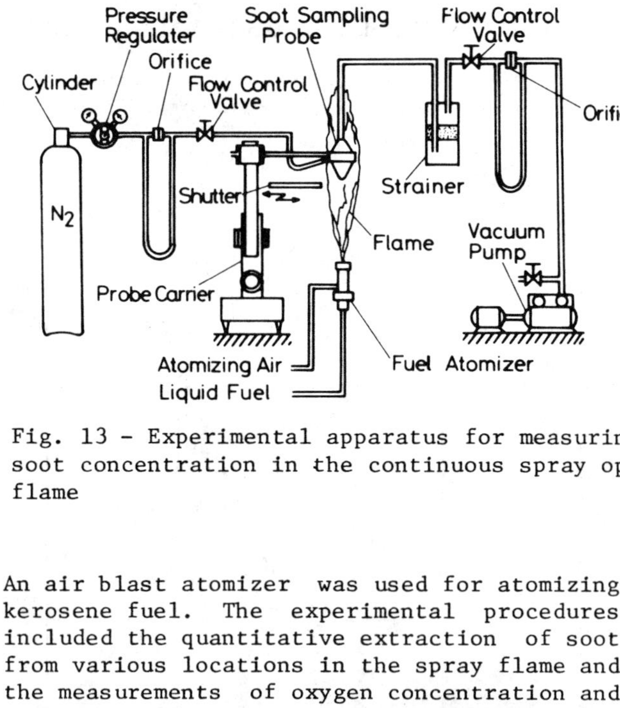

Fig. 13 - Experimental apparatus for measuring soot concentration in the continuous spray open flame

An air blast atomizer was used for atomizing kerosene fuel. The experimental procedures included the quantitative extraction of soot from various locations in the spray flame and the measurements of oxygen concentration and temperature distribution. The measurement of soot in the spray flame was performed by the sampling probe as shown in Fig. 14. The burning gas which contained soot was sampled and was quenched by nitrogen gas through the narrow sampling port. Soot contained in sampling gas was collected on a glass fiber filter paper which had been installed in the sampling probe. The mass of soot was derived from the difference between the mass of filter before and after sampling of soot. The flow

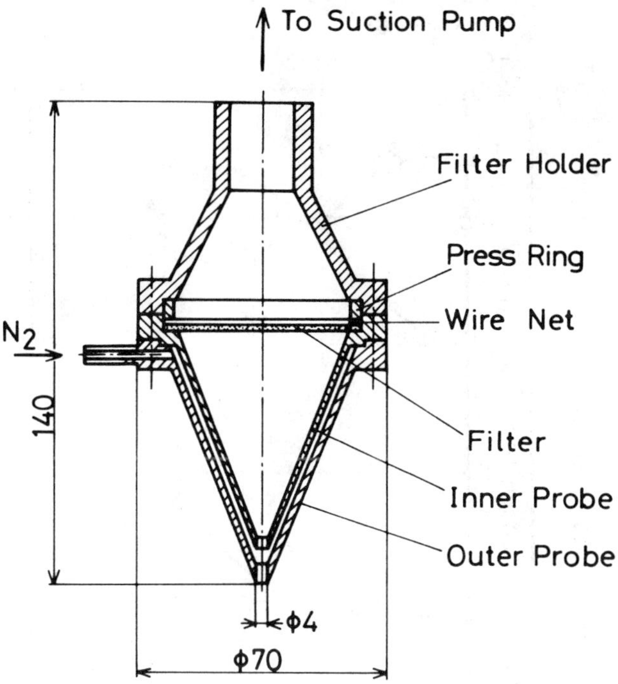

Fig. 14 - Soot sampling probe

pattern around the location of the sampling port was affected only slightly by this method of sampling since the sampling probe was made as an inverse cone shape and the sampling was performed under nearly iso-kinetic conditions. This method of sampling soot had enough sensivity to measure the soot distribution in flames. Soot concentration C_s was defined by the following equation,

$$C_s = \frac{m_s}{(Q_s - Q_{N_2})t_s} \qquad (1)$$

where, m_s is mass of sampled soot, Q_s is sampling gas flow rate containing nitrogen gas, Q_{N_2} is nitrogen gas flow rate and t_s is a sampling time.

Figure 15 shows the distribution of soot concentration along a flame axis. The axis of the abscissa shows the axial distance from the atomizer, and the ordinate shows the soot concentration C_s, and in this case the flame length was about 500mm. It can be understood from these results that there was a considerable amount of soot in the flame in spite of little soot observed at the flame end. The distribution of the soot concentration had the maximum value at the location of 200mm from the atomizer and the oxygen distribution had the minimum value at the same location. From the atomizer to this location, the oxygen concentration decreased and the soot concentration increased. In contrast, from that location to the flame end, the oxygen concentration increased by air entrainment owing to

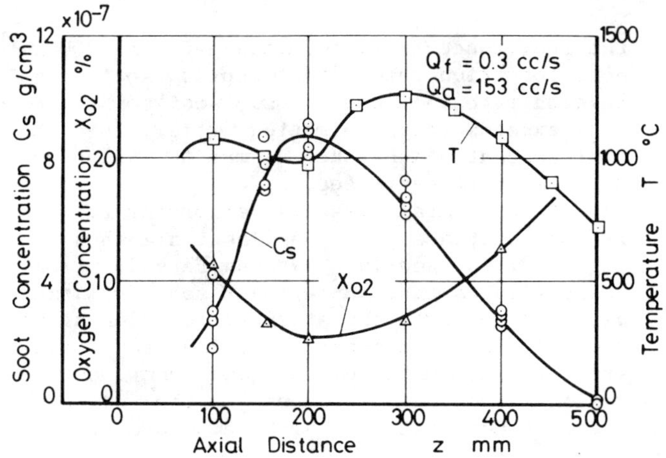

Fig. 15 - Axial distributions of soot, oxygen and temperature

the turbulent diffusion of the flame while the soot concentration decreased. The location of maximum temperature did not correspond to that of maximum soot concentration but corresponded to the location of soot decrease.

Figure 16 shows the relation between the soot concentration and density of fuel droplets measured by the impression method using MgO powder. From this figure, liquid fuel almost evaporated within the region where the soot concentration increased. So, it can be considered that soot was formed from vapour fuel at the early stage of combustion period. The reason that maximum soot concentration appeared in the flame can be considered as follows. In the same way, as the diesel combustion mentioned before, soot concentration in flame was determined by the balance of soot formation rate and soot combustion rate. In the spray flame which was measured in this study, the soot formation rate is greater than soot combustion rate in the range from the atomizer to 200mm, therefore

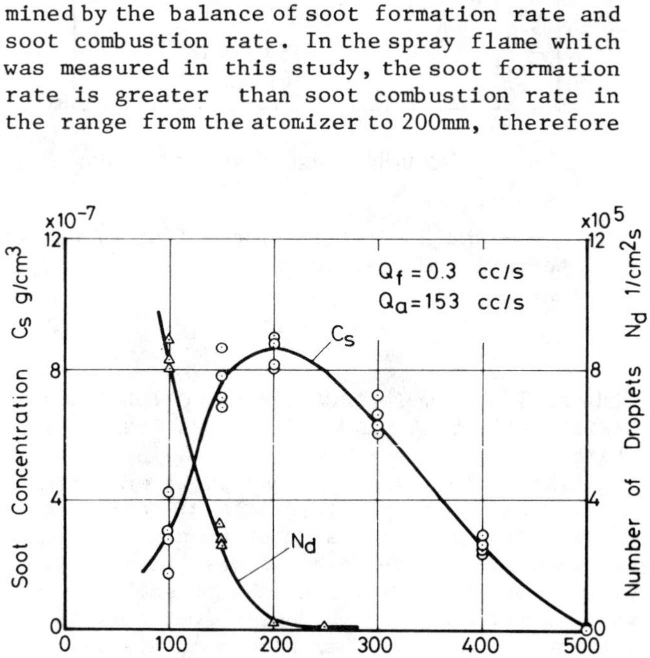

Fig. 16 - Axial distributions of soot and fuel droplets

the soot concentration increases. At 200mm, soot formation rate almost equals soot combustion rate, and over 200mm, soot combustion rate exceeds soot formation rate. Then the soot concentration takes the maximum value and after that it decreases.

Experimental results according to the radial distance of the flame, are shown in Figs. 17, 18 and 19. It can be said from these figures that the region of soot formation was limited from the atomizer to 200mm along the axis and in comparatively low temperature areas inside the turbulent flame front, and the region of soot combustion was, in the following area, of comparatively high temperature.

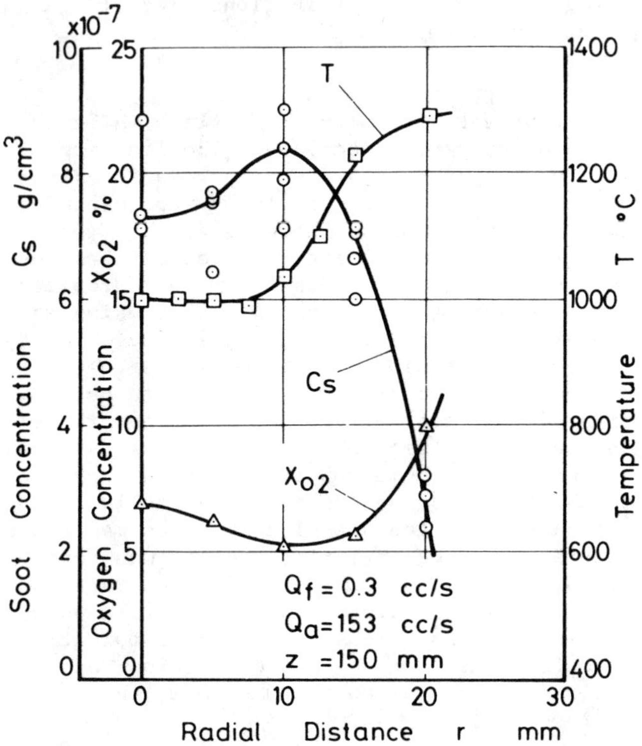

Fig. 17 - Radial distributions of soot, oxygen and temperature (z=150mm)

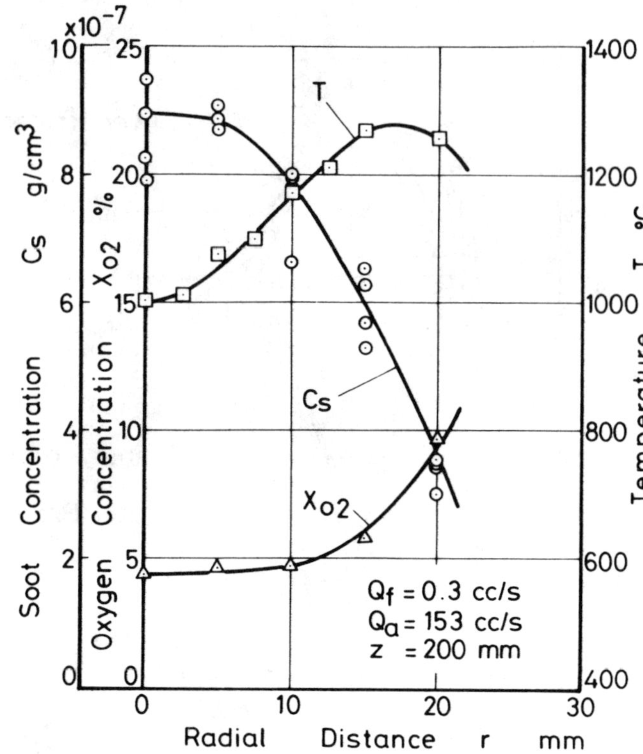

Fig. 18 - Radial distributions of soot, oxygen and temperature (z=200mm)

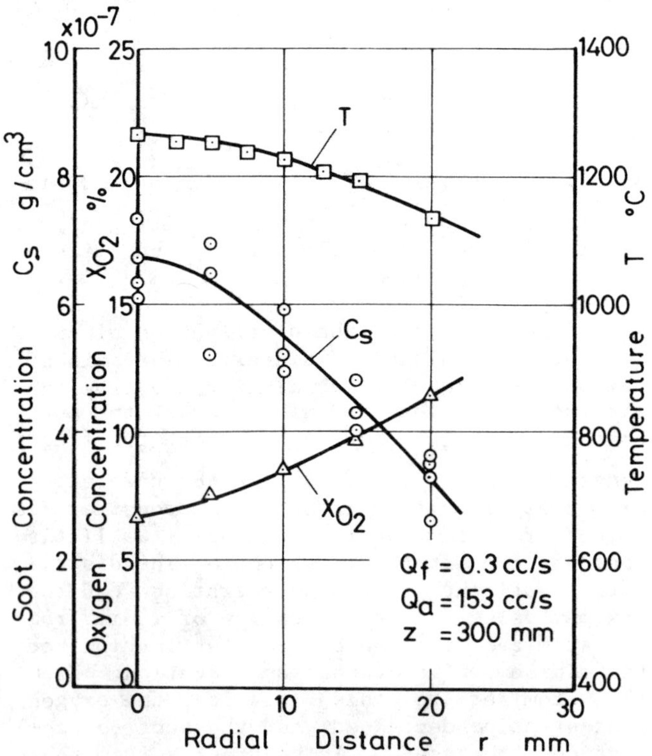

Fig. 19 - Radial distributions of soot, oxygen and temperature (z=300mm)

SIMILARITY OF SOOT FORMATION AND OXIDATION IN DIESEL COMBUSTION AND CONTINUOUS SPRAY OPEN FLAME

The mechanism of exhaust smoke was considered based on the experimental results of diesel and continuous spray open flame combustion. This consideration was based on the general opinion that the exhaust smoke level was determined by the balance of soot formation and soot oxidation. Khan stated in his report (15) that the soot formation is very important and the soot oxidation effect is relatively unimportant. But we considered that, conversely, the soot oxidation is very important. This consideration is based on the experimental results of the continuous spray open flame as shown in Fig.16 and the following assumptions. The mass of formed soot was determined in the early and the middle stages of combustion period and it did not change with the condition of the final stage of combustion period as long as the fuel spray burned with luminous flame. The condition of the final stage of combustion period only effected the soot oxidation. Then, the amount of exhaust smoke level was changed according to the amount of unburned soot. Figure 20(a) gives a schematic illustration of the expected variation of soot formation and soot oxidation in the diesel combustion chamber with crank angle, and its relationship to flame luminosity, heat release and injection rate. Figure 20(b) shows the measured soot concentration in spray open flame along a flame axis, and also give the expected variation of soot formation and soot oxidation. The case which is shown in Fig. 20 is the example of a low exhaust smoke level. The combustion duration of the diesel combustion corresponds to the flame length of the spray open flame in this figure. In this case, the end of flame luminosity does not exceed the opening time of the exhaust valve, and the formed soot almost burns up before the exhaust valve opens. This condition corresponds to the situation where soot is sampled at the end of the continuous spray flame. Figure 21 shows the example that exhaust smoke level is high. The flame luminosity exists after the opening time of the exhaust valve in diesel combustion, in other words, soot oxidation still continues during the exhaust process. The quenching effect due to the exhaust valve opening, the soot, which ought to burn, essentially is exhausted from the exhaust pipe as net soot release. Therefore the exhaust smoke level increases in comparison to the situation of Fig. 20. This case of diesel combustion corresponds to the situation that soot is sampled from the inside of the continuous spray open flame.

If the diesel combustion is similar to the continuous spray open flame by this method, the effects of many complex parameters on the exhaust smoke level from diesel engines by using spray open flame can easily be predicted. For example, the combustion duration in a diesel combustion chamber corresponds to the flame

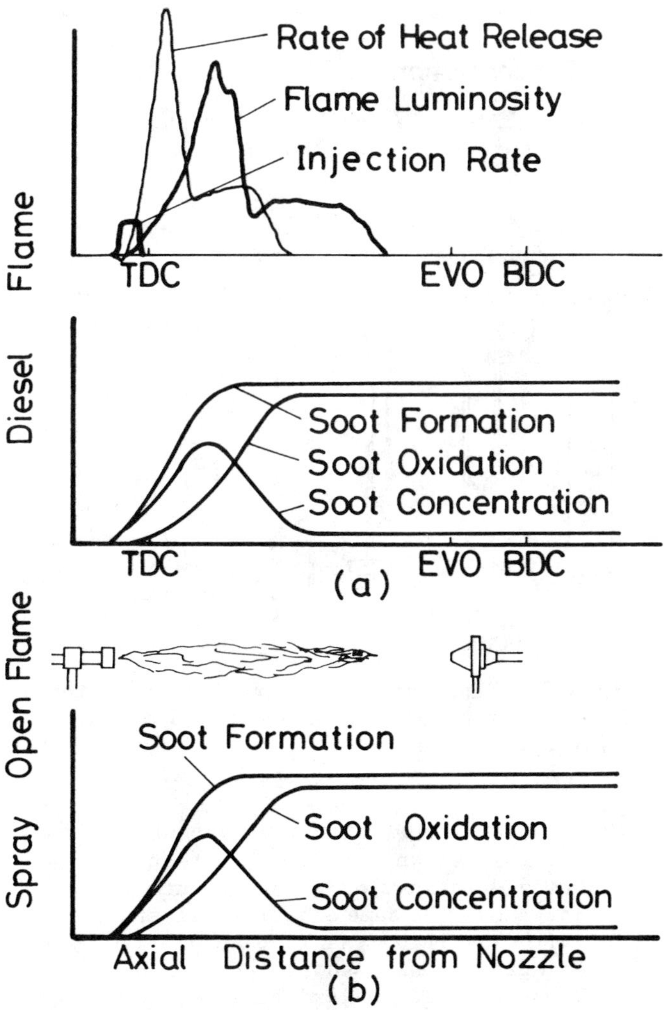

Fig. 20 - Illustrations of soot formation and soot oxidation in the diesel flame and spray open flame (At low exhaust smoke level)

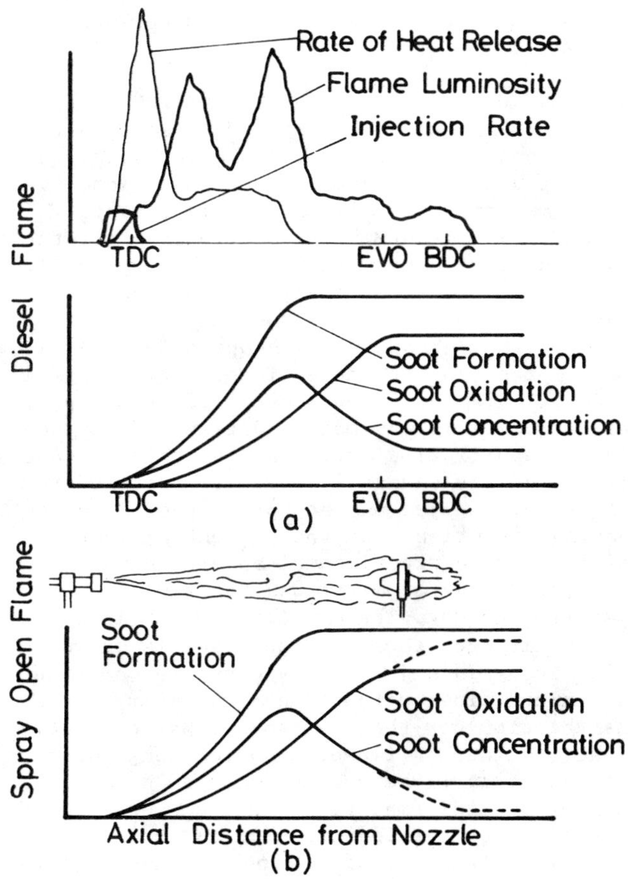

Fig. 21 - Illustrations of soot formation and soot oxidation in the diesel flame and spray open flame (At high exhaust smoke level)

length of spray open flame. The effect of air swirl corresponds to the effects of the air entrainment and so on. The effect of exhaust gas recirculation (EGR) is easily simulated by adding burned gas into the ambient air of spray flame. The effect of injection duration is simulated by the alteration of the flame length of spray flame. And the change of injection timing corresponds to the change of the location of soot sampling from the nozzle in the spray flame.

It is already known that in diesel combustion the effect of EGR caused the increase of the exhaust smoke level. But the reason why is not yet clear. By adding carbon dioxide gas or nitrogen gas into the ambient air of spray diffusion flame, the effect of EGR was tested. It is seen that the flame length increased with an increase in the additional gas concentration of diluent as shown in Fig. 22. Based on the similarity between diesel combustion and continuous spray open flame as mentioned before, the tendency for the smoke to increase is due to the delay of flame luminosity ending in the diesel combustion chamber, as was predicted. In other words, when the combustion duration was extended by the introduction of EGR, so that at the time the exhaust valve opened, soot combustion was still in progress, which caused the soot combustion prosess to stop for reasons presented previously. The result in the exhaust smoke level is increased.

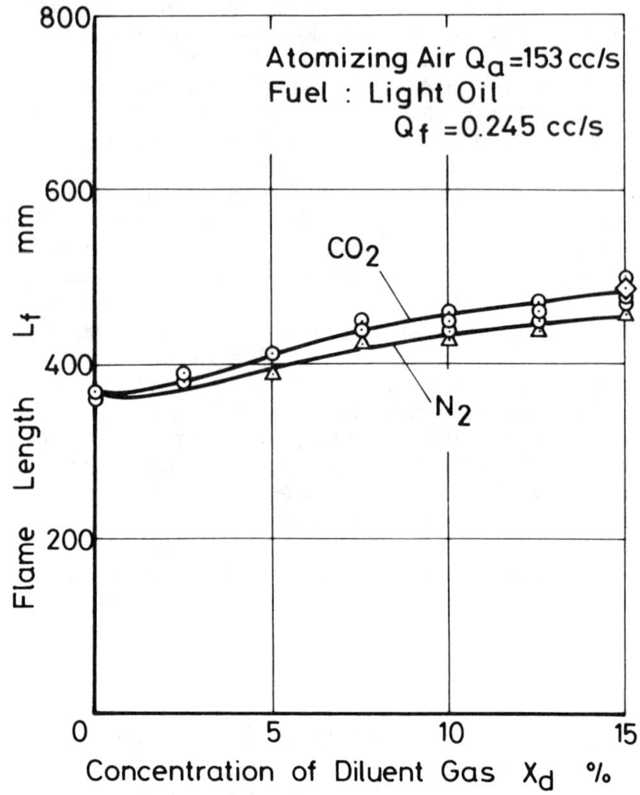

Fig. 22 - Relation between diluent gas concentration and flame length

METHODS FOR REDUCTION OF THE COMBUSTION DURATION

It was shown in the preceeding section that variations in the exhaust smoke level largely reflected changes in the soot combustion, i.e. the combustion duration is very important. The oxidation of the soot still continued after the end of heat release with sufficient oxygen. It must be necessary to control the combustion duration as to finish all the combustion period before the exhaust valve opens in order to reduce the exhaust smoke level from diesel combustion. The flame length of the spray diffusion flame was mainly determined by the fuel vaporization rate, air entrainment and mixing of air and fuel vapours. We tested the effect of air jet on the spray open flame by paying attention to the air entrainment and mixing. The experimental apparatus is shown in Fig.23. The atomizer was the same one that was used before. The jet nozzle which injected air or nitrogen gas into the flame is shown in Fig. 24. This jet nozzle consisted of eight stainless pipes with an inner diameter of 1.7mm. As the tip of the jet nozzle was completly inserted into the flame, the outer configuration of the spray flame was not changed on a large scale and the ambient air did not enter into the flame by the injection of air.

The effect of air jet on the flame length was measured for the various air injection rates and injection locations. The flame length was shortened with an increase of the air jet and

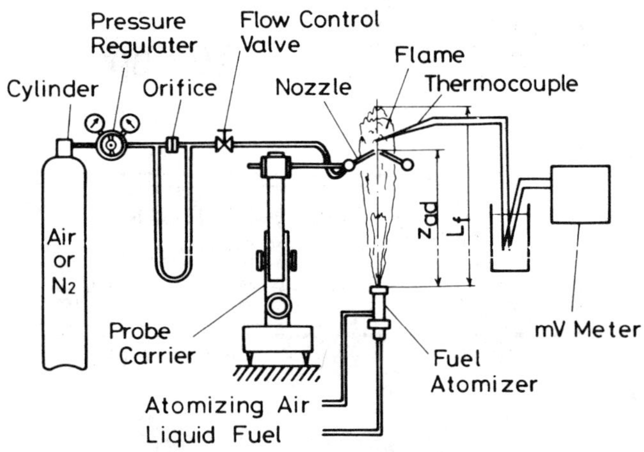

Fig. 23 - Experimental apparatus for tested the effects of air entrainment and mixing in spray open flame

with a closing of the injection distance to the atomizer as shown in Fig. 25, that is, the combustion duration of each element of spray was shortened. The same experimental study was made with nitrogen gas instead of air. The flame length was also shortened as shown in Fig. 26 by a nitrogen gas jet. It showed the same tendency as the air jet but the decreasing ratio was not as large as with air.

The difference of the shortened flame

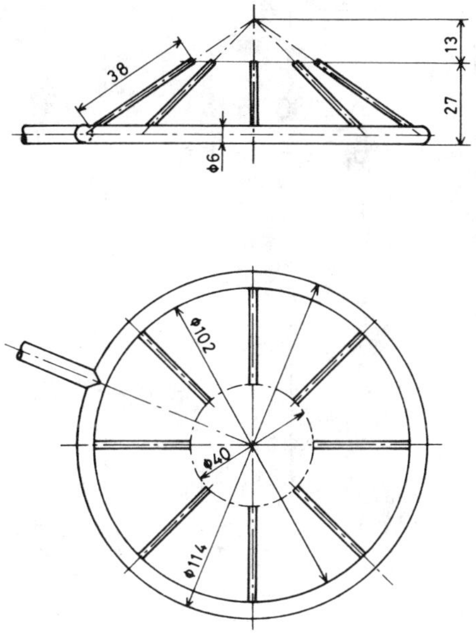

Fig. 24 - The jet nozzle

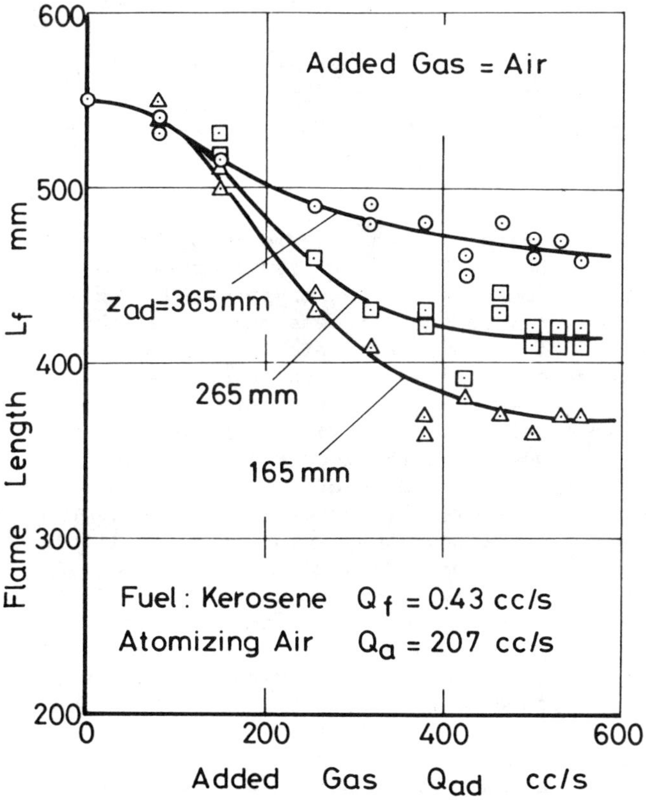

Fig. 25 - Relation between added gas flow rate and flame length (Air)

length between the air jet and the nitrogen gas jet showed the effect of air entrainment of injected air itself. The reduction of the flame length appearing for both gases showed the effect of the promotion of the turbulent mixing. The temperature distribution along the

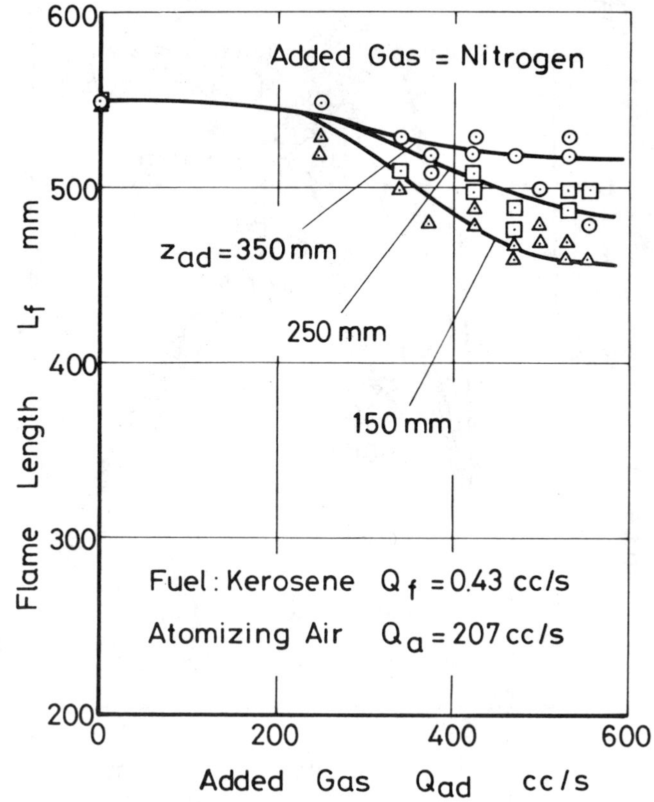

Fig. 26 - Relation between added gas flow rate and flame length (Nitrogen)

flame axis affected by the air jet is shown in Fig.27. Though the flame temperature dropped just after the location of the air injection due to low temperature of the air, but after that, it rose up the higher value than the original flame temperature. This phenomenon showed an increase of the combustion rate of spray by the air jet. Therefore the reduction of flame length by the air jet was due to the increase of the combustion rate and was not due to the quenching of the flame. The air swirl in the cylinder is one of the effective methods of applying the reduction of flame length in the spray open flame to diesel combustion.

A shrouded valve was used to measure the effect of induction swirl on the combustion duration and the exhaust smoke level. The relationship between a swirl ratio and the combustion duration which determined with the end of flame luminosity was shown in Fig. 28. In the case where the load was high,i.e. BMEP was 0.76 MPa, the combustion duration changed little with the swirl ratio, but in the cases where the loads were middle and low,i.e. BMEP were 0.63 MPa and 0.48 MPa, the increase of swirl ratio results in the reduction of the combustion duration. Figure 29 shows the relationship between the swirl ratio and the exhaust smoke level. In the case where the load was high, which the end of flame luminosity timing exceeded the timing of the exhaust valve

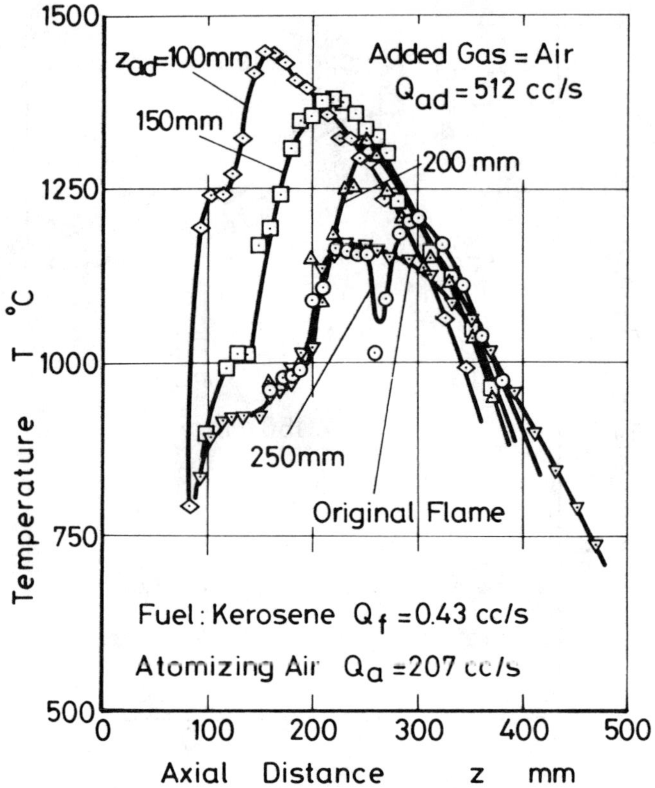

Fig. 27 - Temperature distribution along the flame axis affected by air jet

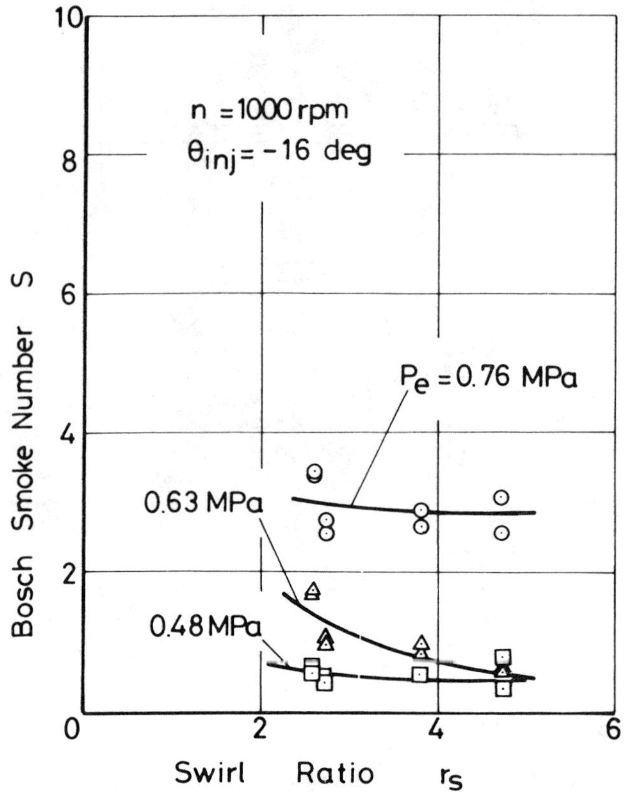

Fig. 29 - Effect of swirl ratio on exhaust smoke

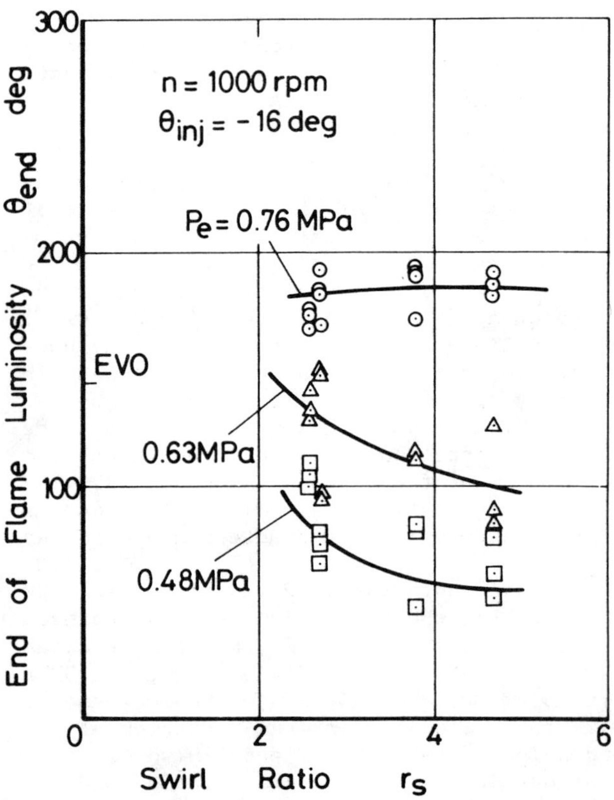

Fig. 28 - Relation between swirl ratio and the end of flame luminosity

open in spite of the increase of the swirl ratio, the effect of the swirl ratio on the exhaust smoke level was not appeared. But in the case where the load was middle, which the end of flame luminosity timing decreased and crossed over the exhaust valve opening timing with increase of the swirl ratio, the exhaust smoke level decreased with the swirl ratio. In the case where the load was low, which the combustion involve the soot oxidation was completed before the exhaust valve open, the exhaust smoke level was low and changed little with increase of the swirl ratio. In diesel combustion, air entrainment or turbulent mixing in the combustion chamber is promoted by air swirl, and as the number of related reports were shown, the exhaust smoke level is decreased by air swirl. But, another method of promoting air entrainment or turbulent mixing in combustion chambers, for example, by air injection by force during the middle stage of combustion which should result in a further reduction of soot by reducing the period of flame combustion duration.

CONCLUSIONS

Combustion duration was measured by using a flame luminosity detector with photo-transistor in a direct injection diesel cylinder. Close relation was observed between the combustion duration and the exhaust smoke level. The combustion duration detected by the flame luminosity detector is quite a bit longer than the one that detected by the rate of heat release. In the case of high load, the wave signal of flame luminosity is continues after the exhaust valve opens. It was found that the longer the end of flame luminosity resulted in a higher level of exhaust smoke.

In order to apply the useful knowledge concerned with continuous spray diffusion flame to diesel combustion, the concentration of soot by using a nitrogen gas quenching probe in the space of continuous spray open flame was measured and used. The experimental data show that maximum soot value and minimum oxygen concentration is in the same position, and the location of maximum temperature did not correspond to that of maximum soot concentration but corresponded to the location of soot decrease.

It has been shown by the schematic illustration given that the expected variation of mass of soot formation and oxidation in the diesel combustion chamber with crank angle is similar to the soot formation and combustion in a spray open flame along a flame axis.

The effects of exhaust gas recirculation, injection duration or load and air swirl on soot release in diesel combustion can be explained by the experimental results obtained from spray open flame.

It must be necessary to control the combustion duration to end before the exhaust valve opens in order to reduce the exhaust smoke level in diesel engines.

In order to reduce the flame length of spray open flame, that is, to shorten the combustion duration, the air jet forced into the flame is effective. Considering these results, it can be predicted that the air jet forced into the diesel combustion flame during the middle stage of combustion period would be effective to reduce the exhaust smoke level in diesel engines.

ACKNOWLEDGEMENT

The authors wish to thank National Science Research Foundation of Japan for their financial support of this research. We wish to give special thanks to graduate students K. Atarashiya, and Y. Oda who assisted in taking the data.

REFERENCES

1. I.M. Khan, G. Greeves and D.M. Probert, "Prediction of Soot and Nitric Oxide Concentrations in Diesel Engine Exhaust", Proc.IME, C142, p.205, 1971.

2. I.M. Khan, C.H.T. Wang and B.E.Langridge, "Coagulation and Combustion of Soot Particles in Diesel Engines", Combustion and Flame, Vol.17, p.409, 1971.

3. D. Broome and I.M. Khan, "The Mechanism of Soot Release from Combustion of Hydrocarbon Fuels with Particular Reference to the Diesel Engine", from "Air Pollution Control in Transport Engine", Institution of Mech. Eng., London, p.185, 1971.

4. R.J. Harmes, D.F. Merrion and H.S. Ford, "Some Effects of Fuel Injection System Parameters on Diesel Exhaust Emissions", SAE Paper NO.710671, August 1971.

5. R.E. Lee, Jr., R.K. Patterson, W.L.Crider and J. Wagman, "Concentration and Particle Size Distribution of Particulate Emissions in Automobile Exhaust", Atomospheric Environ., Vol.5, p.225, 1971.

6. K. Habbi, "Characterization of Particulate Matter in Vehicle Exhaust", Environ. Sci. and Tech., Vol.7, p.223, 1973.

7. R.C. Bascom, W.S. Chiu and R.J. Padd, "Measurement and Evaluation of Diesel Smoke", SAE Paper NO.730212, January 1973.

8. J.T. Ganley and G.S. Springer, "Physical and Chemical Characteristics of Particulates in Spark Ignition Exhaust", Environ.Sci.and Tech., Vol.8, p.341, 1974.

9. C.T. Vuk, M.A. Jones and J.H. Johnson, "The Measurement and Analysis of the Physical Character of Diesel Particulate Emissions", SAE Paper NO.760131, February 1976.

10. W. Bryzik and C.O. Smith, "Relationships between Exhaust Smoke Emissions and Operating Variable in Diesel Engines", SAE Paper NO.770718, September 1977.

11. A. Laresgoiti, A.G. Loos and G.S. Springer, "Particulate and Smoke Emission from a Light Duty Diesel Engine", Environ.Sci. and Tech., Vol.11, p.973, 1977.

12. W.H. Lipkea, J.H. Johnson and C.T.Vuk, "The Physical and Chemical Character of Diesel Particulate Emissions - Measurement Techniques and Fundamental Considerations", SAE Paper NO.780108, 1978.

13. D.F. Dolan and D.B. Kittelson, "Diesel Exhaust Aerosol Particle Size Distributions - Comparison of Theory and Experiment", SAE Paper NO.780110, 1978.

14. N.J. Khatri, J.H.Johnson and D.G.Leddy, "The Characteristic of the Hydrocarbon and Sulfate Fractions of Diesel Exhaust Particulate Matter", SAE Paper NO.780111, Feb.1978.

15. I.M. Khan, "Formation and Combustion of Carbon in a Diesel Engine", from "Diesel Engine Combustion", Proceeding-Int. of Mech. Engineers, Vol.184, Part 3J, p.36, 1969-70.

16. G. Greeves, and J.O. Meelian, "Measurement of Instantaneous Soot Concentration in a Diesel Combustion Chamber", Paper of IME, 1975.

17. V.K. Duggal, T. Prided and I.M. Khan, "A Study of Pollutant Formation within the Combustion Space of a Diesel Engine", SAE Paper NO.780227, 1978.

18. K. Miwa, Y. Nakano and M. Ikegami, "Formation of Carbon and Soot in Swirl Chamber Diesel Engine", Paper No.102, JSME and JSAE Joint Meeting in Tokyo on Aug. 2, 1979.

19. Y. Aoyagi, T. Kamimoto, Y. Matsui and S. Matsuoka, "Study on NO and Soot Formation

in Direct Injection Diesel Engine by Using Gas Sampling Method", Paper NO.103, JSME and JSAE Joint Meeting in Tokyo on Aug. 2, 1979.

20. A.G. Gaydon and H.G. Wolfhard, "Flames" 3rd ed., Chapman and Hall, p.175, 1970.

21. R. Echigo, R. Nishiwaki, and N. Hirata, "A Study on the Radiation of Luminous Flames" 11th Symposium (International) on Combustion, p.381, 1967.

22. K. Atarashiya, Y. Wajiki, M. Arai, T. Kadota and H. Hiroyasu, "The Effect of Air Swirl on the Diesel Combustion by Constant Volume Combustion Vessel", JSME Paper NO.790-16, 1979.

23. H. Kamikubo, "Experimental Study on Analysis the Diesel Combustion with Rapid Compression Machine", Master Thesis of the University of Hiroshima, 1979.

24. Y. Tadokoro, M. Arai, T. Kadota and H. Hiroyasu, "Soot Formation from Liquid Fuel Atomizing Spray Free Flames", Paper presented at the 16th Symposium (Japan) on Combustion, p.137, 1978.

25. T. Kadota and H. Hiroyasu, "Soot Formation from Gaseous Fuel Diffusion Flame", JSME Paper NO.790-16, 1979.

26. T. Kadota, H. Hiroyasu and A. Farazandehmehr, "Soot Formation by Combustion of a Fuel Droplet in High Pressure Gaseous Environments", Combustion and Flame, Vol.29, p.67, 1977.

27. T. Kadota, K. Nakanishi and H.Hiroyasu, "The Effects of Oxygen Concentration and Temperature on the Soot Formation by Combustion of a Fuel Droplet", Paper presented at the 16th Symposium (Japan) on Combustion, p.126, 1978.

A Gas Sampling Study on the Formation Processes of Soot and NO in a DI Diesel Engine**

Yuzo Aoyagi, Takeyuki Kamimoto, Yukio Matsui, and Shin Matsuoka
Tokyo Institute of Technology

SINCE EMISSION GAS AND EXHAUST SMOKE from diesel engines have been an important social problem, the elucidation of the formation mechanism of nitric oxide and soot in the combustion chamber have been receiving mounting interest. Many theoretical models for predicting the exhaust pollutants have hitherto been reported (1, 2, 3, 4, 5).* In the last few years many experimental studies using the gas sampling method have been performed for the purpose of obtaining information based on actual data (6, 7, 8, 9, 10, 11). There are still many unclarified points in the formation mechanism of NO and soot in diesel combustion. However few measurements of the concentration of soot in the combustion chamber that will continue to be a problem, have been made (10).

In this study a method of sampling and measuring soot is devised, and employing this method along with the conventional gas sampling technique the spatial distributions and time histories of the soot, NO and other combustion products concentration are investigated in the combustion chamber of a direct injection engine. At the same time measurements of the local equivalence ratio and flame temperature are carried out. To interpret the temporal and spatial changes of the concentration of each combustion product there is a need to observe the flame motion, therefore high speed photographs are also taken.

In the first part of this report, the formation and exhaust processes of NO and soot are investigated in detail under a basic condition. The next part deals with the effects of injection timing, swirl ratio and combustion chamber shape. New information is obtained on the formation and oxidation of soot, suggesting a method to reduce it.

APPARATUS AND METHOD OF EXPERIMENTS

SAMPLING METHOD OF GAS AND SOOT - The sampl-

---

* Numbers in brackets designate References at end of paper.

**Paper 800254 presented at the Congress and Exposition, Detroit, Michigan, 1980.

ABSTRACT

The concentrations of soot, NO and the other combustion products were measured by incylinder gas sampling in a DI diesel engine. The effects of injection timing, swirl ratio, and combustion chamber geometry on the formation and emission processes of soot and NO were studied. The following results were obtained: (1) Soot is promptly formed in the flame during the early combustion period where the equivalence ratio in the flame is high over 1.0. Thereafter almost all the formed soot is swiftly burnd up by oxidation during the middle combustion period. This process mainly determines the exhaust soot concentration. (2) NO is formed in the flame during the early and middle combustion period where the flame temperature is high over 2000 K. The highest NO concentration is observed at the flame tip swept by the air swirl. Though the concentration of the formed NO decreases by dilusion it nearly constant during the later combustion period. For this reason the concentration of exhaust NO is almost completely governed by the formation process.

ing gas lines from the combustion chamber to the analytical apparatus and the electric power supply to move electro-magnetic sampling valve are shown in Fig. 1. The flow paths and vessels were maintained at a constant temperature of 120°C. The gas withdrawn from the sampling valve ① was passed through a filter ② for removing dust and collected in a stainless steel syringe ③ and glass vessel ④ respectively. The gas in ③ and ④ was for analysis of primary combustion products and of NO. The samples of unburnt hydrocarbons were analyzed in the same way but with the flow paths and vessel ③ heated to 200°C. Before sampling, each vessel and flow path were evacuated and the gas were collected from a number of cycles until the pressure in each vessel reached a pressure of 1.2 absolute atmosphere. The measurement of soot concentration was made by a soot sampler ⑤ . Sampling of

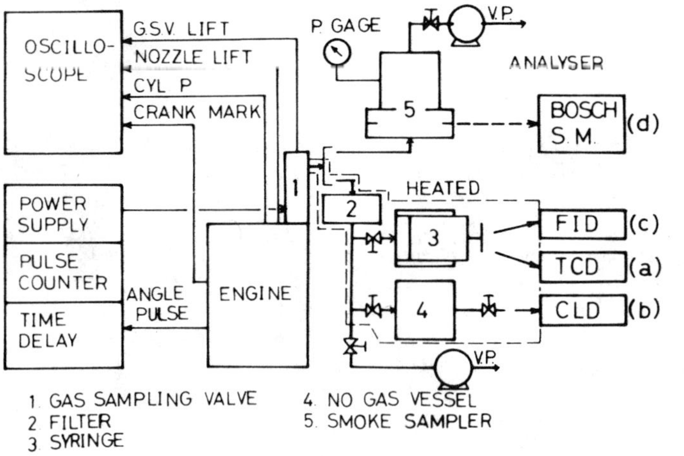

Fig. 1 - Experimental set-up of the engine, sampling valve, and emissions equipment

soot and gas were made separately at the rate of once every 4 cycles. The number of gas sampling at a given crank angle was from 30 to 300 depending on the cylinder pressure at that time so that the total sampling amount would be constant. Fig. 2 shows the way the stainless needle type magnetic sampling valve (Tsukasa Instruments) was attached to the engine. There were two kinds of tips, type 1 and type 2, as shown in Fig. 3 and when the tip protruded into the cylinder, type 2 was employed. The gas sampling duration was 1 ms which was equivalent to about 8°CA at an engine speed of 1250 rpm. The representative sample timing was defined as the middle crank angle of the valve opening duration. The sample volume per one shot in both types is about 0.5 cc at cylinder conditions.

ANALYSES OF GAS COMPONENTS AND SOOT - The analyses completed were for (a) nitrogen (N_2), oxygen (O_2) + Argon (Ar), carbon dioxide (CO_2), carbon monoxide (CO), and methane (CH_4) ; (b) nitric monoxide (NO) and nitric dioxide (NO_2) ; (c) unburnt hydrocarbons and (d) soot. The gases

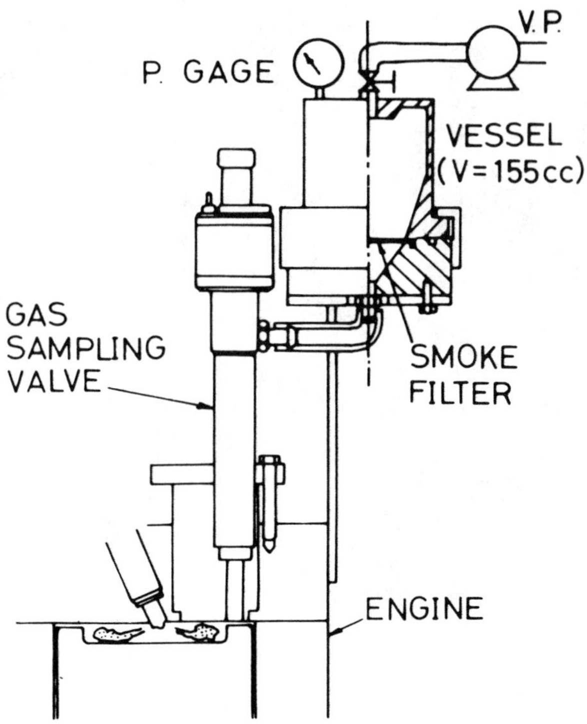

Fig. 2 - Layout of the DI engine and sampling valve

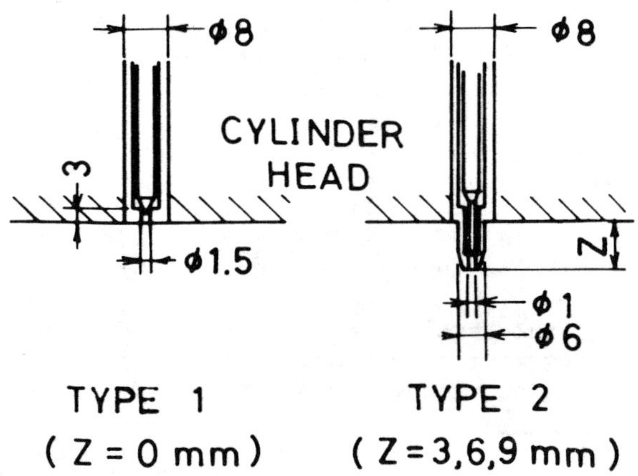

Fig. 3 - Shapes of the two kinds of sampling valve tip

of the (a) group were analyzed with a TCD type gas chromatography and those of the (b) group were analyzed with a chemiluminescence type NO analyser (Yanagimoto ECL-30 model). The unburnt hydrocarbons (c) were analyzed with an FID type detector, and their concentrations were converted to Hexane (C_6H_{14}) from which the concentration of CH_4 obtained from the gas chromatography had been subtracted. The concentrations of each constituent represented in mol fraction and the total mol fraction of the burnt gas constituents except for soot was taken to be 100%. The analyses of the (a) and (b) groups were carried out simultaneously but the others were done on individual engine runs.

Fig. 2 shows the details of the measuring apparatus for soot concentration. A filter paper for the Bosch smoke unit was installed just behind the entrance to the vessel to gather the soot in the collected gas. First the vessel (capacity 155 cc) was evacuated and gas was collected over a number of cycles so that the maximum Bosch number of the filter would not be over about 5. This soiled filter paper was measured by a Bosch smoke meter and converted to gravimetric basis (g/m^3) at standard conditions, considering the amount of collected gas.

Furthermore the fact that the ratio of atoms between carbon and hydrogen in soot is about three (12), using this relationship it is possible to assume that the soot molecule is $CH_{1/3}$, and this is represented in mol fraction in the same way as the gas contituents. This representation facilitates a quantative comparison to the CO_2, CO and CH_4 molecules.

THE FLAME TEMPERATURE was measured by a device of our own design employing the visible two-color method (13). The wavelengths employed were 0.539 and 0.738 μ m. The diameter of the field of detection was about 2 mm. Both the flame temperature and incylinder pressure were statistically averaged by a mini-computer (TOSBAC - 40A).

THE TEST ENGINE was a direct injection type, 4 stroke, water cooled single cylinder engine with a bore x stroke ; 95 mm x 110 mm, compression ratio 14.6. The specification of the injector was 4 holes, with hole diameter 0.20 mm, nozzle opening pressure 170 atm, injection angle 150°. The fuel employed was gas oil with a density of 0.849 (15°C) and cetane number of 57.

LOCAL FUEL-AIR EQUIVALENCE RATIO

Calculation of local fuel-air equivalence ratio (ϕ) is made taking into account the amount of soot in this study using a method (14). The fuel-air mixture is assumed to change into 12 chemical species through combustion.

$$
\begin{aligned}
x_{mix} & \left(C_nH_m + ((n + m/4) / \phi) \cdot (O_2 + 3.7274\, N_2 + 0.0445\, Ar)\right) \\
&= x_1\, CO_2 + x_2\, CO + x_3\, CH_4 + x_4\, C_6H_{14} \\
&+ x_5\, CH_{1/3} + x_6\, H_2 + x_7\, H_2O + x_8\, O_2 \\
&+ x_9\, Ar + x_{10}\, NO + x_{11}\, NO_2 + x_{12}\, N_2 \qquad (1)
\end{aligned}
$$

Where n and m were the number of carbon and hydrogen atoms respectively in one molecule of hydrocarbon fuel, x_i was the mol fraction of the i-th term of constituent in the right side of the equation (1). The total mol fraction of the gas constituents on the right side of the equation excepting x_5 was assumed to be 100%. The left hand side x_{mix} was the mol number of the air fuel mixture required to equal 1 mol of the gas constituents on the right side. The concentrations of H_2O and H_2 in the equation (1) that were not measured were predicted from the water gas reaction (15). The air was assumed to be dry. The ϕ value was calculated by the conservation law of mass about the carbon atoms and oxygen atoms. Thus the fuel-air equivalence ratio is,

$$\phi = 2\left(1 + \frac{m}{4n}\right) \cdot \frac{x_1 + x_2 + x_3 + 6\,x_4 + x_5}{2x_1 + x_2 + x_7 + 2x_8 + x_{10} + 2x_{11}} \qquad (2)$$

Where the mol fraction of the soot x_5 reaches 10% of the total number of carbon atoms, it cannot be ignored. The m/n was taken to be 2.0.

EXPERIMENTAL RESULTS AND DISCUSSION

Injection timing (θ i), swirl ratio (Sr), and combustion chamber shape were chosen as experimental parameters. The engine speed and fueling (overall fuel air equivalence ratio (ϕ m) was 0.5) were kept constant for all sampling tests reported in this study. The standard operating engine condition was fixed, that is, θ i=-15°CA ATDC, Sr=0, shallow dish piston chamber. Fig. 4 shows the two kinds of the piston shapes used in this test. The sampling positions are shown on co-ordinates of the X, Y

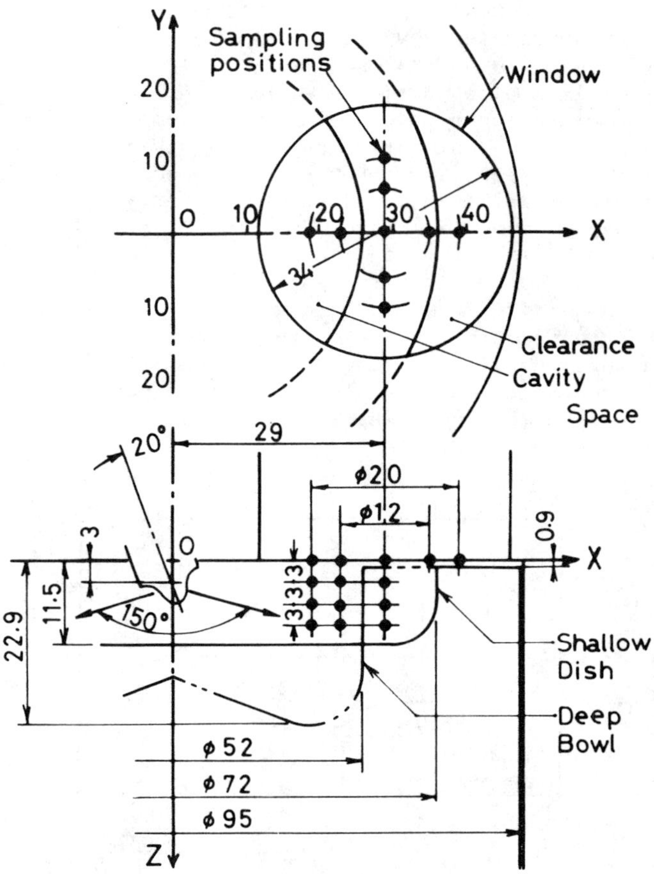

Fig. 4 - Shallow dish and deep bowl combustion chamber and sampling positions

and Z axes in the Figure. The flame in the combustion chamber was observed within a circle of 34 mm in diameter shown by a solid line. The angle of the shroud attached on the inlet valve was adjusted to obtain the arbitrary swirl ratio. Fig. 5 shows the bulk exhaust measurements and performance at each engine condition.

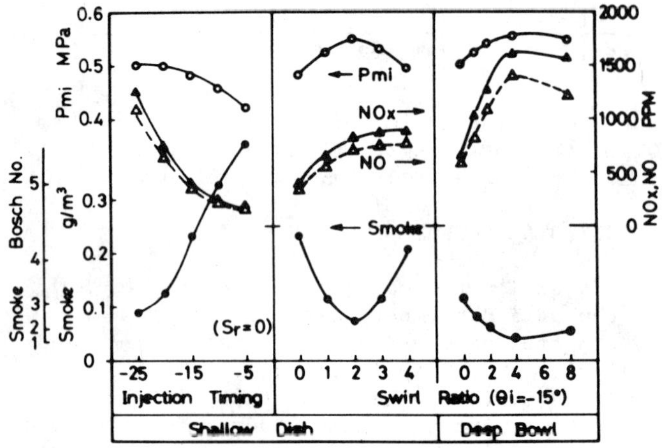

Fig. 5 - Engine test conditions and bulk exhaust measurements at engine speed of 1250 rpm and overall equivalence ratio (ϕm = 0.50)

THE COMBUSTION PROCESS UNDER THE STANDARD ENGINE CONDITION - The cylinder pressure and rato of heat release in this experimental condition are shown in Fig. 6. For convenience of explaining the combustion process, the combustion period is divided into the following three periods according to the rate of heat release. The early combustion period ranges from the start of combustion to the time when the rate of heat release becomes maximum in the diffusion combustion period ; the middle combustion period ranges from the end of prior period to 15° CA after, where the rate of heat release steadily decreases ; and the later combustion period ranges after this, where the rate of heat ralease is low as shown Fig. 6.

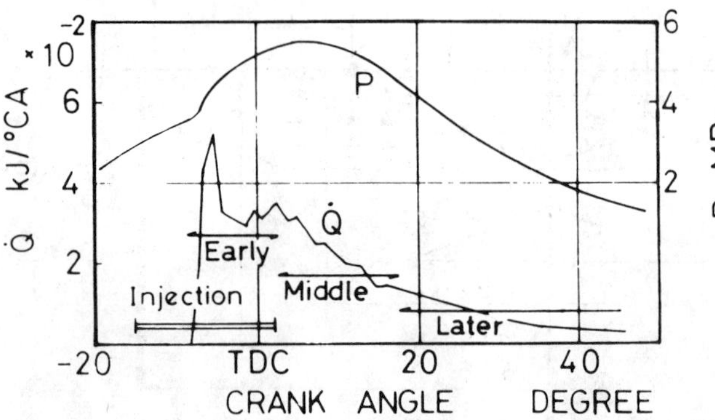

Fig. 6 - Cylinder pressure diagram and rate of heat release at the standard engine condition

The following are the results obtained from representative sampling positions along the spray axis (X axis). Fig. 7 (a) and (b) shows the results of different positions in the piston cavity space. Fig. 7 (c) shows the case in the clearance space. The values labeled Ex. on the bottom line show the exhaust concentration.

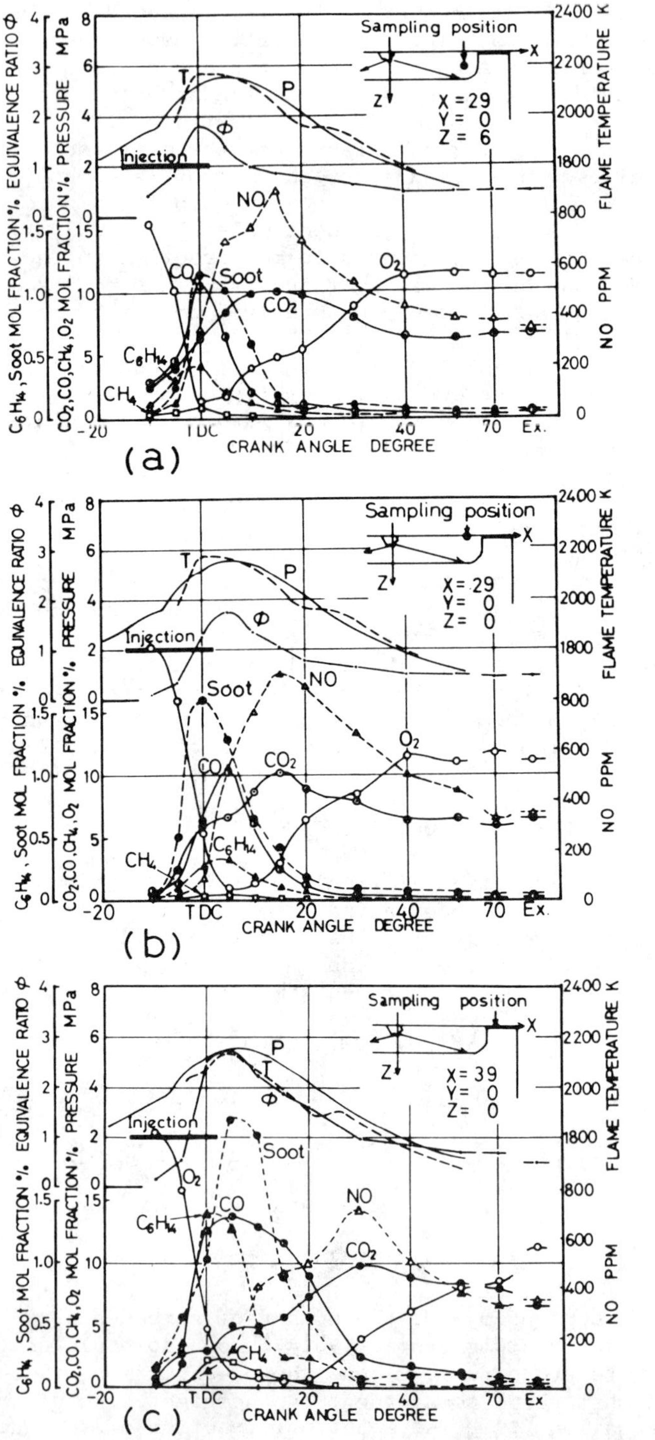

Fig. 7 - Histories of soot, NO and other combustion products with T, P and ϕ under the standard engine condition

The concentration changes of each combustion product at each sampling position are not only the changes due to chemical reaction and turbulent diffusion mixing but include the change due to the movement of the flame. In order to take this phenomenon into account, the combustion photographs under this experimental condition are shown in Fig. 8 (b). The longitudinal flame motion is illustrated in Fig. 9 by observing these photographs.

Histories and Spatial Distribution on Local Equivalence Ratio (ϕ) - In the early combustion period, ϕ rises up rapidly and reaches a peak value. These peak times are different in each sampling position. That is, in Fig. 7 (a) the peak time is TDC, and both in Fig. 7 (b) and 7 (c) it is 5°CA. This difference can be explained from Fig. 8 (b) and Fig. 9. The flame tip after impinging upon the piston surface flows upward passing the sampling position of Fig. 7 (a) at -3°CA ATDC. Furthermore the flame after impinging upon the lower surface of the cylinder head advances along the cylinder head surface in two directions, one towards the cylinder center and the other towards the clearance space, and passes the sampling position of Fig. 7 (b) and Fig. 7 (c) at TDC. As described above the difference in the time when ϕ peaks in each position is due to the time difference when the flame reaches each sampling position. Therefore, the rapid increase of ϕ up until peak is considered to be caused by the rapid movement of the flame. On the other hand the boundary between flame and air region is clear in this period as shown in combustion

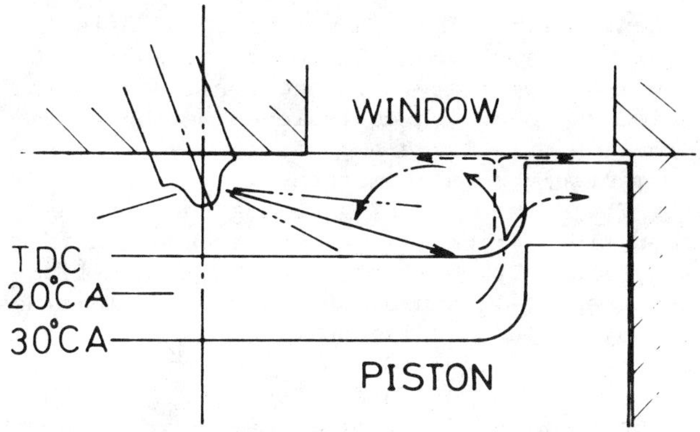

Fig. 9 - Illustration of longitudinal flame motion under the standard engine condition

photographs and the rise of ϕ is expected to be sharp, but actually it is not sharp. This is due to ϕ of the flame tip being low, due to the opening duration of the sampling valve being long in comparison to the velocity of the phenomenon at this period, and due to the variation in the arrival time of the flame tip changing each cycle. From these facts it can be estimated that the equvalence ratio whthin the flame is higher than the measured ϕ in the early combustion period.

The rapid increase in the concentration of CO, soot, unburnt hydrocarbons may be interpreted in the same way as the increase of ϕ, consequently the concentrations of the above mentioned species in the flame are higher than the measured

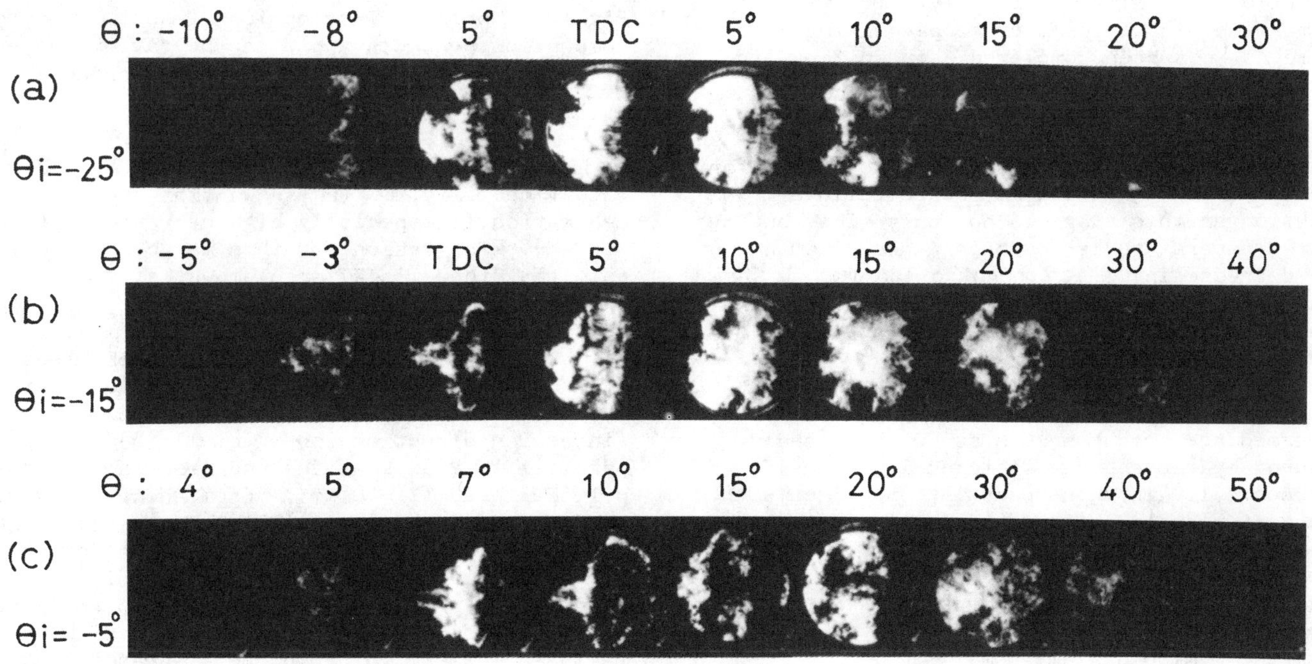

Fig. 8 - Combustion photography in the quiescent state (Sr = 0) under the standard engine condition timing (a) θi = -25°CA, (c) θi = -5°CA

concentration. The maximum value of ϕ is about 2.0 in the cavity space, and it is about 3.0 in the clearance space. In the early combustion period the overrich equivalence ratio causes formation of a large amount of CO and other incompletely burnt components.

The events occuring during the middle combustion period will be described in this paragraph. By observation of the combustion photographs, it can be seen that the flame has already spread widely and moves more slowly as denoted by one dotted chain line shown in Fig. 9 after the end of fuel injection. Therefore it can be considered that the changes of ϕ and the concentration of each constituent are dominated by reaction and diffusion. Now in this period ϕ will steadily decline, and this implies that air is gradually entrained into the flame. CO, soot and unburnt hydrocarbon formed during the prior period consume the entrained air immediately and are converted into final combustion products. The combustion photographs show that the brightness of the flame is great when both the flame temperature and rate of heat release are high. These facts reveal that combustion is the most active in this period. When ϕ declines to 1.0, both CO_2 and NO concentration peak together.

In the later combustion Period the flame motion becomes slower than in the prior period, and ϕ steadily approaches the overall equivalence ratio (ϕm). As the piston descends the flame may be considered to move downwards but the measured ϕ at the lower side of the cylinder (Z = 6 mm) approaches ϕ m, therefore the macroscopic heterogeneity of each constituent along the depth does not seem to be so high.

When the brightness of the flame is observed, it becomes dimmer with only a filament like flame remaining. The microscopic mixing in the flame has not been completed. In spite of the steady increase in the O_2 concentration, the still remaining CO, soot and unburnt hydrocarbons hardly burn, and especially the decrease in soot concentration is small. The reason for the very slow oxidation may be as follows : The dilution at this stage is not only slow, but the flame temperature is low being under 2000 K. The decrease in CO_2 concentration is considered to be caused by the dilution of the combustion gas by air as the increase of O_2 concentration shows.

In the clearance space the maximum value of ϕ is large and this yields a large quantity of intermediate products such as CO, soot and unburnt hydrocarbons. It takes long time for ϕ to become 1.0 in this space in comparison with the cavity space, because the mixing determined by the descent of the piston is slow. These species remain as higher concentrations at this space than at the cavity space in the later combustion period as can be seen in accordance with the luminous flame remaining.

Fig. 10 shows the spatial distribution of ϕ along the Z direction (depth) in the cavity space (X = 29, Y = 0 mm) in time sequence. As gross aspect it can be seen that the heterogenities change to the homogenities along the depth

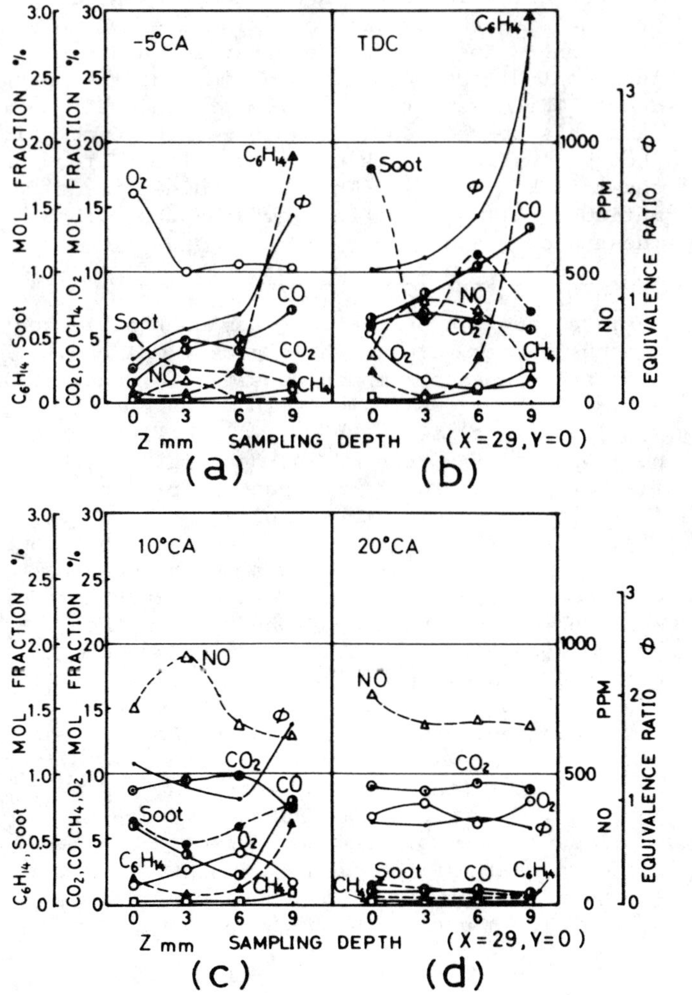

Fig.10 - Spatial distribution of soot, NO, other combustion products and ϕ in the cavity space along the depth under the standard engine condition

with time. Fig. 10 (b) shows that the C_6H_{14} concentration is especially high near the piston surface. The reason for this may be due to the fact that there is a large amount of fuel vapour in the region near the piston surface where fuel spray impinges as shown in Fig. 4.

The histories and spatial distributions of ϕ are summarized in Fig. 11 (a) and (b). Fig. 11 (a) is the distribution along the X axis (injection direction) and Fig. 11 (b) is that parallel to Y axis at the under-surface of the cylinder head (Z = 0 mm). From these diagrams the general trend of ϕ will be described. The ϕ values are maximized at the end of injection in all sampling positions (The hatching in the diagram). In this period spatially ϕ increases as it goes from cylinder center to periphery taking a maximum value in the clearance space. With time the ϕ values at all points gradually decline approaching the overall equivalence ratio (ϕ m). The distribution of ϕ in the X direction during the injection period takes a maximum value at Y = 0 mm. It takes a nearly symmetrical

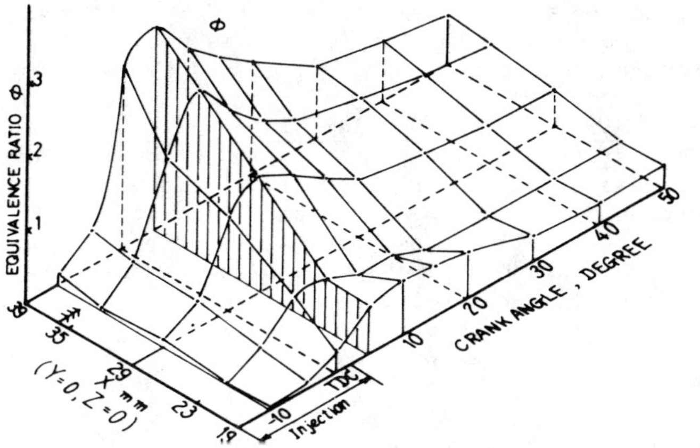

(a) - Along X axis

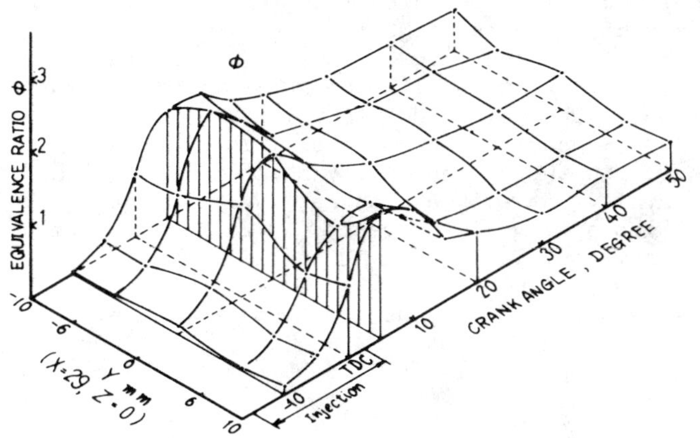

(b) - Along Y axis

Fig.11 - Histories and spatial distributions of local equivalence ratio (ϕ) under the standard engine condition

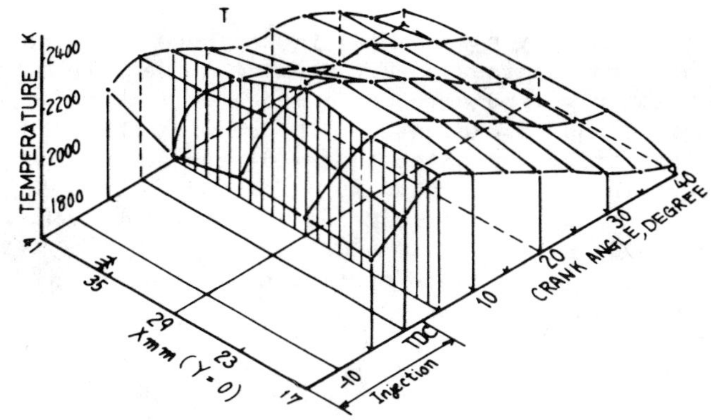

(a) - Along X axis

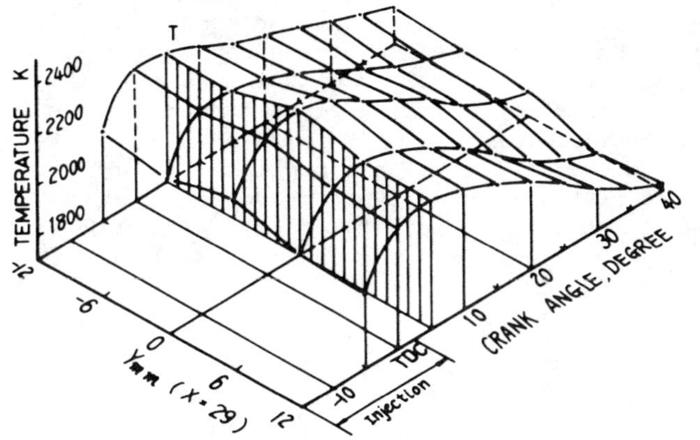

(b) - Along Y axis

Fig.12 - Histories and spatial distributions of flame temperature (T) under the standard engine condition

distribution with respect to the center. With time ϕ declines and the distribution of ϕ becomes even.

The formation of soot and NO depends on the equivalence ratio, flame temperature and pressure. In investigating the formation processes of soot and NO, the temporal and spatial distribution of the flame temperature are summarized and shown in Fig. 12 in the same way as ϕ.

Histories and Spatial Distributions of the Soot Concentration - The detail of the formation and oxidation process of soot obtained from previous data will be discribed in this section. The combustion photographs of Fig. 8 (b) shows that in the early combustion period from TDC to 5°CA ATDC, there is a dark area on the flame surface that has contacted the observation glass window. This is caused by the soot, which is cooled on the glass surface, absorbing the radiation from the flame interior. This phenomenon is observed only in the early combustion period where the soot concentration is high. In this period ϕ is high ranging in value from 2.0 to 3.0 while the temperature is above 2000 K and the pressure is about 5 MPa : thus they are all high through all combustion periods, and the flame is in the most favorable state for soot formation (16).

In the middle combustion period, where ϕ decreases from its maximum value to 1.0 by mixing, the soot in the same way as CO undertakes rapid combustion resulting in the concentration dropping to about 1/40 of its maximum value. In agreement with this phenomenon the soot layer observed in the flame surface steadily disappears. In this period the flame temperature is maintained above 2000 K, and moreover a large amount of air is entrained. These yield a favorable state for oxidation of the soot within the flame.

In the later combustion period where ϕ goes below 1.0, in spite of the increase in the concentration of O_2 the soot concentration hardly shows a decline. This is considered to be caused by the drop in flame temperature resulting in a decrease in the reaction rate. The soot is exhausted through the exhaust system after maintaining the same concentration frozen in this period.

In this experiment the soot formation and oxidation process shows a good resemblance to that of the steady state spray combustion flame

(17).

In the prior works (2) is has been said that in a diesel engine soot it formed in the later combustion period and the formed soot is hardly oxidized and is exhausted. The results obtained in this study indicates that soot formation occurs in the early combustion period and we believe shed new knowledge on the soot formation process.

The histories and spatial distributions of soot concentration are shown in Fig 13. The comparison of the temporal and spatial distributions of ϕ with those of soot concentration indicates that both take a peak value at 5° CA immediately after completion of injection, while in spatial distributions both values are high in the clearance space. Since both factors show temporally and spatially similar tendencies, it can be said that the soot formation and oxidation is principally controlled by ϕ in a diesel combustion chamber where the flame temperature is high over 2000 K.

However there is an exception. In spite of the large ϕ the soot is not present in large quantities in the vicinity of piston surface (X = 29, Y = 0, Z = 9 mm) as shown in Fig. 10 (b).

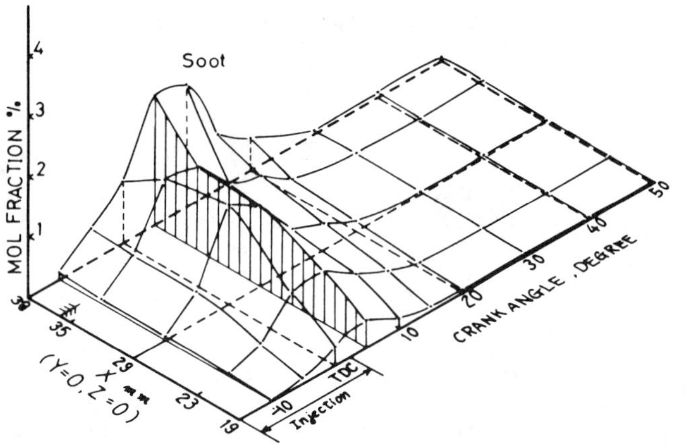

(a) - Along X axis

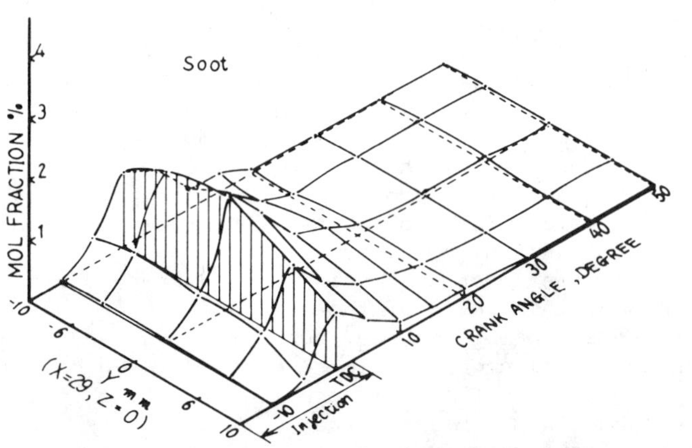

(b) - Along Y axis

Fig.13 - Histories and spatial distributions of soot concentration under the standard engine condition

Since there is a large amount of C_6H_{14}, it can be considered that the temperature is not high enough to decompose the fuel or hydrocarbons at this position.

<u>Histroies and Spatial Distributions of NO Concentration</u> - The concentration of NO increases in the early and middle combustion period, where the pressure and temperature are highest of all the combustion periods, and ϕ is overrich as shown in Fig. 7. NO concentration reaches its peak value when ϕ decreases to about 1.0. This phenomemon has also been observed in the steady state spray combustion flame (18). From the observation of combustion photographs it can be seen that flame spreads widely and is brightest at 15° CA ATDC.

In the later combustion period, where ϕ is below 1.0, NO decreases together with ϕ. The rate of decrease of NO is slightly higher than that of ϕ, concequently some decomposition of NO might exist, but the decrease of NO concentration may be mainly due to dilution by air.

An observation of NO distributions along the depth in the piston cavity space is made from Fig. 10. At -5° CA immediately after the luminous flame is appeared, NO is hardly formed at any position. From TDC to 10° CA the NO concentration becomes high at all positions. Especially the NO Concentration is high at the intermediate depths of Z = 3 and 6 mm, where flame temperature may be high due to the distance from chamber wall.

The histories and spatial distributions of the NO concentration is shown in Fig. 14. A comparison with the distribution of ϕ shows that the NO concentration in both space and time becomes maximum when ϕ decreases to 1.0. The NO concentration becomes maximum spatially in the vicinity of the vertical cavity walls of the piston and becomes maximum temporally from 15° CA to 20° CA. The results of the histories and spatial distribution indicate that NO is steadily fromed in the flame of an overrich mixture when the pressure and temperature are high. The emission concentration of soot is several percent of the maximum concentration while that of NO is about one third of its maximum concentration.

EFFECT OF INJECTION TIMING IN A SHALLOW DISH COMBUSTION CHAMBER - Combustion photographs are shown in Fig. 8 (a), (b) and (c) with injection timing changes of θ i = -25°, -15°, and -5° CA, while the pressure diagrams and the rates of heat release are shown in Fig. 15. Fig. 16 (a) and (b) shows the histories of the flame temperature (T), NO and soot concentration and ϕ summarized in the three injection timings, where the comparative sampling position was X = 29, Y = 0, Z = 0 mm in the cavity space and X = 39, Y = 0, Z = 0 mm in the clearance space.

<u>Histories and Spacial Distribution of Local Equivalence Ratio and Flame Temperature</u> - Fig. 16 (a) and (b) shows that with the retardation of the injection timing the maximum values of ϕ in both the cavity and clearance space increase, and the time which it takes for ϕ to decrease to 1.0 becomes longer. The comparison of the combustion photographs in Fig. 8 shows that the retardation

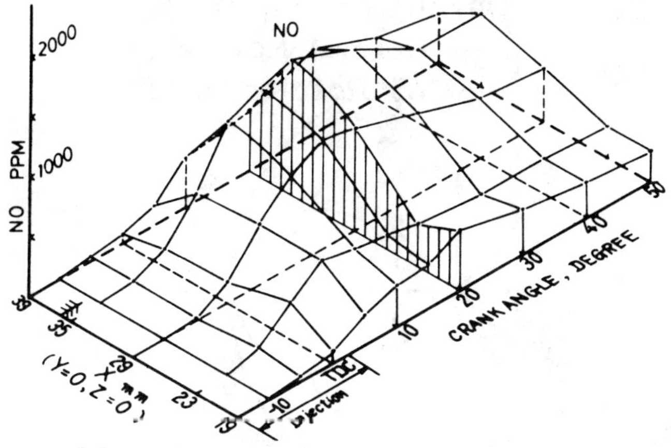

(a) - Along X axis

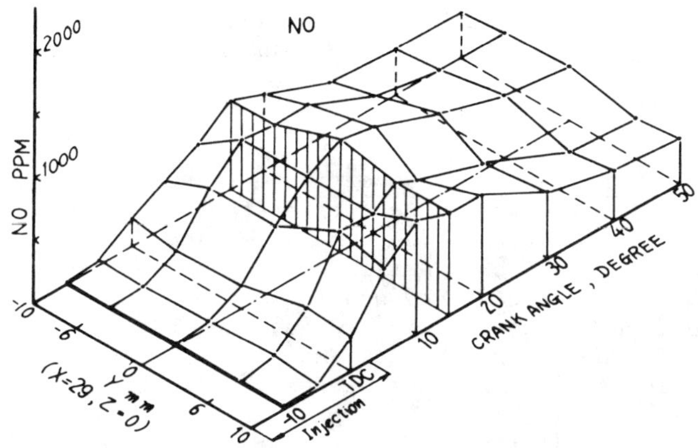

(b) - Along Y axis

Fig.14 - Histories and spatial distributions of NO concentration under the standard engine condition

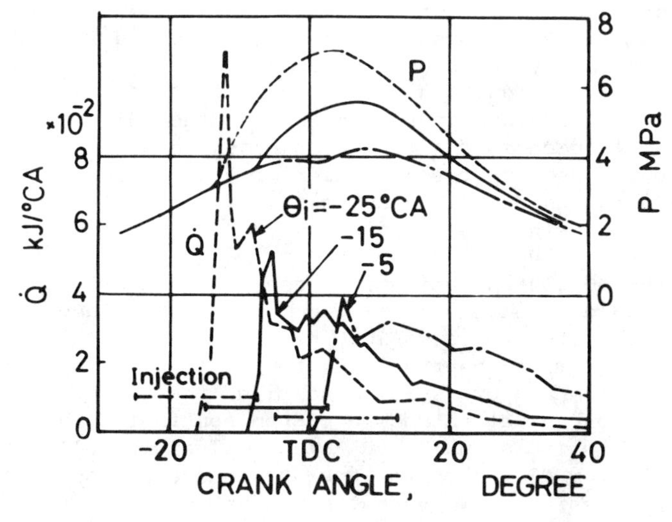

Fig.15 - Cylinder pressure diagrams and rates of heat release with changing injection timing (θ i)

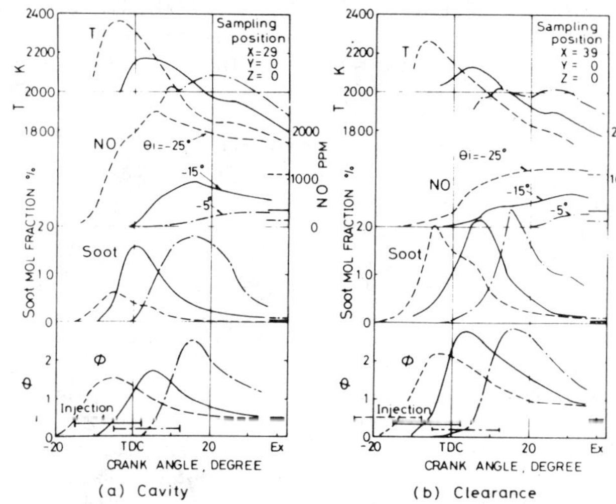

Fig.16 - Variations of NO and soot concentrations, T and ϕ with changing injection timing (θ i) as a parameter

of injection timing prolongs the period from the injection start till the extinction of the luminous flame and reduces the expansion speed of the flame. The comparison of (a) and (b) in Fig. 16 shows that in all injection timings the maximum ϕ at the clearance space is larger than at the cavity space.

As the injection timing is retarded, the maximum value of the flame temperature in both the cavity and clearance space generally becomes lower. This is due to the retardation of the injection timing resulting in combustion proceeding during the expansion stroke and burning in the rich mixture.

The Histories and Spatial Distributions of Soot Concentration - Soot is formed in all injection timings in the early combustion period and reaches maximum concentration at the same time as ϕ becomes maximum. After that with the decrease of ϕ the formed soot decreases, but even more rapidly. This decrease which is caused mainly by oxidation can be seen by the fact that the reduction rate of the soot concentration is considerably larger than that of ϕ. In the later combustion period the soot is nearly constant and it approaches the exhaust concentration. The soot concentration in the clearance space, where ϕ is large, is higher than that in the cavity space.

The retardation of injection timing results in increasing the maximum value of the soot concentration and gives a comparatively high soot concentration in the later combustion period. When this trend is compared to the trend of ϕ, the larger the ϕ value by the retardation of injection timing, the larger both the maximum soot concentration and soot concentration in the later combustion period become.

The duration from injection end to the end of the soot concentration drop is approximately 20° CA, independent of injection timing. In this period : (1) The flame temperature generally

Bonny Kubik

remains above 2000 K, (2) The air entrainment causes the decrease of ϕ, (3) The gas movements within the flame are active. These conditions are all thought to be important factors for the soot oxidation.

Histories and Spatial Destributions of NO Concentration - In all injection timings the NO concentration increases in the cavity space during the early and middle combustion period where the flame temperature is high and ϕ is large through the whole period, and reaches a maximum when ϕ decreases to about 1.0. After this the NO concentration decreases at almost the same reduction rate as ϕ. From this it can be seen that the formed NO is diluted by air before being exhausted. The exhaust NO concentration is from 1/2 to 1/3 of the maximum.

On the other hand in the clearance space in all injection timings ϕ is higher than in the cavity space, and the flame temperature is lower. Due to low flame temperature the NO concentration is lower in the clearance space than in the cavity space.

The change of injection timing influences the histories of ϕ, flame temperature and pressure, and these affect in the amount of NO emission greatly. Thus as the injection timing is advanced, the maximums of both flame temperature and pressure increase while ϕ becomes smaller, and the maximum NO concentration becomes higher.

CHANGING THE SWIRL RATIO IN A SHALLOW DISH COMBUSTION CHAMBER - To investigate the swirl effect both the swirl ratio (Sr = 2) of the optimum swirl condition, when the exhaust soot concentration becomes minimum and output becomes highest, and the swirl ratio (Sr = 4) of the overswirl condition, when both exhaust soot concentration and output deteriorated, were selected as experimental conditions as shown in Fig. 5. In the existence of a swirl the flame masses are swept along by the flow, so these flame masses pass the sampling position one after the other. In this study it is desirous to observe the states and histories of the combustion products within the moving flame masses, so in addition to the normal injection direction as shown in Fig. 17 experiments were made with the injection direction rotated 45°. From the results of these two experiments, the histories of the concentrations in the flame masses were followed continuously.

The combustion photographs of the optimum swirl condition when the injection direction is normal and when rotated 45° are shown in Fig. 18 (a) and (b) respectively. The combustion photographs of the overswirl condition are shown in Fig. 18 (c) and (d). The results of gas sampling, when the distance Z from the under surface of the cylinder head is changed in the cavity space, show that in the existence of a swirl the main part of the flame passes the sampling position (Z = 6 mm), therefore the representative sampling position was selected to be X = 29, Y = 0, Z = 6 mm. The rate of heat release accompanying the changing swirl ratio are summarized in Fig. 19.

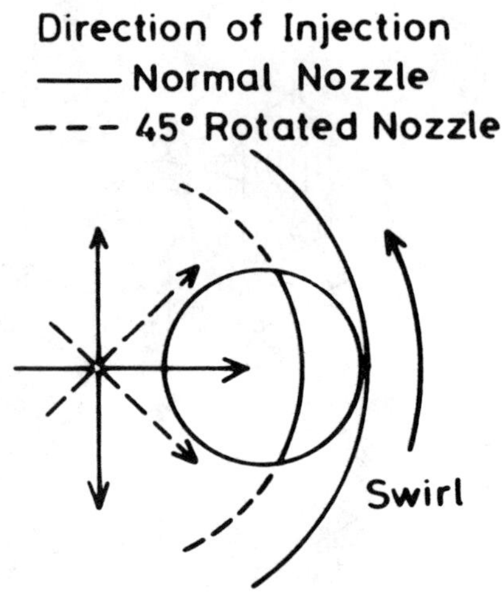

Fig.17 - Normal and 45° rotated direction of injection for tracing the moving flames under swirl conditions

Optimum Swirl Condition - In the optimum swirl the measured results of the flame temperature, NO, CO, soot concentration and ϕ in the cases when the injection directions were both normal (dotted line) and rotated 45° (solid line) are summarized in Fig. 20 (a) and (b).

These results show that the tops appear twice in all curves as already shown by Rhee et al (9). The combustion photographs of Fig. 18 reveal that this is due to the movements of the main part of the flame swept along by the swirl passing the sampling position. The maximum ϕ in both the cavity and clearance space show a reduction to 1/2 in comparison to that of the quiescent state (Sr = 0) and moreover ϕ in the cavity space soon reaches the overall equivalence ratio (ϕ m) at 30° CA ATDC. This shows clearly that in this condition a fuel-air mixing is completed at a relatively early period. The combustion photographs also show that the flame spreads the downstream by the swirl from 10° CA to 15° CA ATDC and is widely distributed in both the cavity and clearance space.

The flame temperature of this condition is around 100 K higher than that in the quiescent state because the mixing is good and combustion proceeds near the stoichiometric equivalence ratio, as described above.

The histories of the soot concentration are qualitatively the same as in the quiescent state. The NO concentration increases in the early and middle combustion periods where the flame temperature exceeds 2000 K.

By the observations of results in both optimum and overswirl conditions the histories of NO concentration show a phenomenon that does not appear in the quiescent state. That is, all the maximum NO concentrations show a sharp peak. The combustion photographs show that the time when NO concentration peaks correspond to the time when the flame tip swept along by the swirl passes the sampling position. The vertical arrow in each figure shows the arrival time of the flame tips.

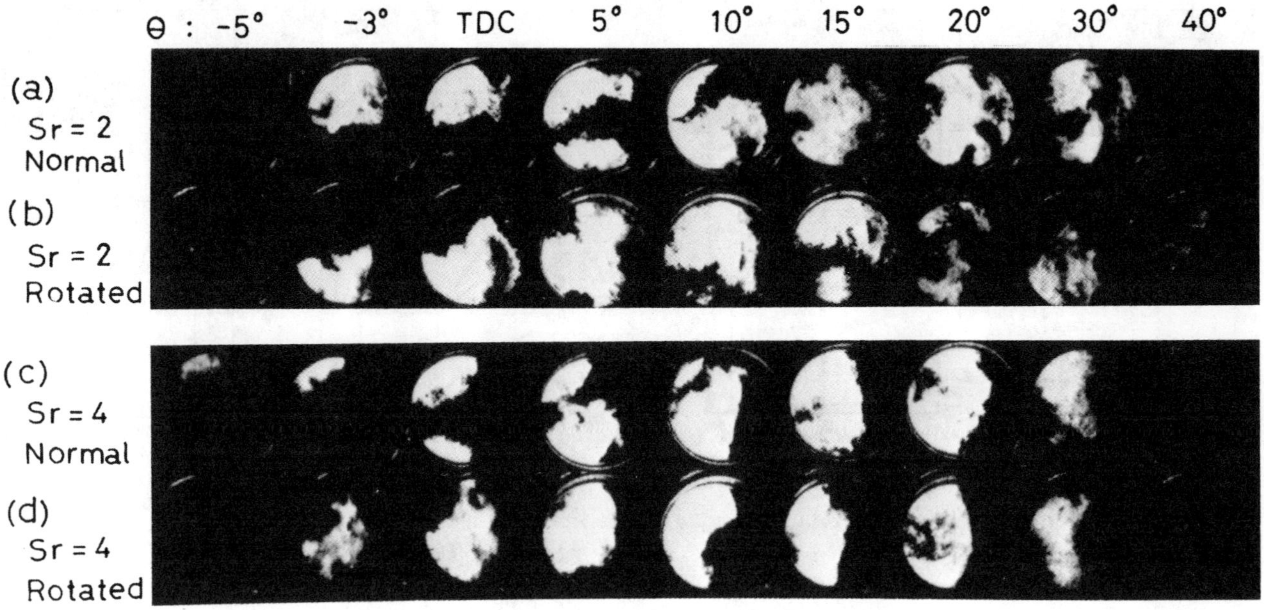

Fig.18 - Combustion photography with changing swirl ratio (Sr) and injection direction (θi) as parameters

The above shows that the highest concentration of NO is formed at the tip of the flame swept by the swirl. This is due to the combustion starting at the tip of the spray in the downstream of the swirl as the combustion photographs show, and due to this part remaining at a high temperature a long time, and moreover due to ϕ being comparatively small. The NO emission becomes higher in the case of a large amount of initial combustion (19) due to this phenomenon.

Overswirl Condition - The results of the cavity and clearance space in the case of overswirl are shown in Fig. 21 (a) and (b) respectively. This swirl strength causes the flame to exist only in the cavity space as the combustion photographs of Fig. 18 (c) and (d) show. This phenomenon is called the thermal pinch effect due to the burned lightweight gases concentrating in the center part of the combustion chamber (20, 21). Then in the cavity space the maximum ϕ is larger than that of the optimum swirl, and accordingly the maximum value of the soot concentration is larger than that of the optimum swirl. The flame in the cavity space continues only stable rotation caused by the strong swirl, and due to the bad mixing the inner burnt gases with the outer air the air entrainment become little. Then the ϕ value after reaching maximum remains roughly constant and even in the later combustion period does not

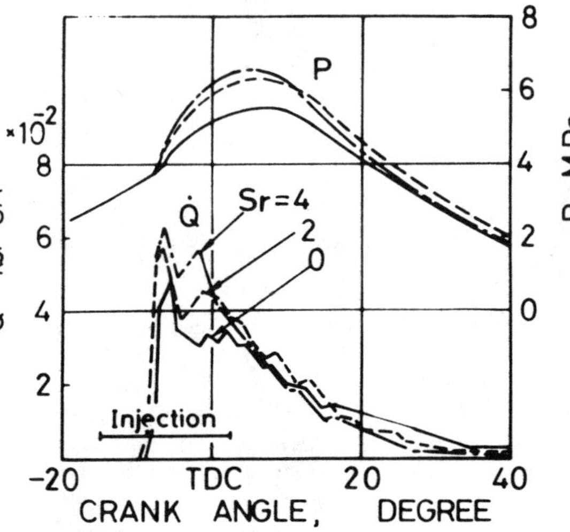

Fig.19 - Cylinder pressure diagrams and rates of heat release with changing the swirl ratio (Sr) as a parameter

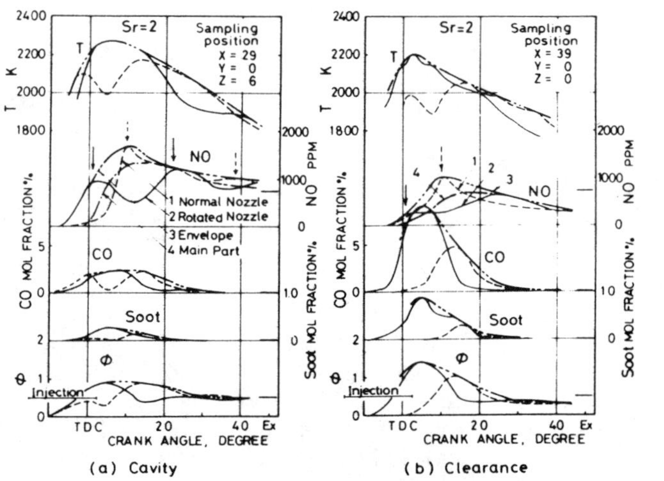

Fig.20 - Soot, CO and NO concentrations, T and ϕ at optimum swirl (Sr = 2)

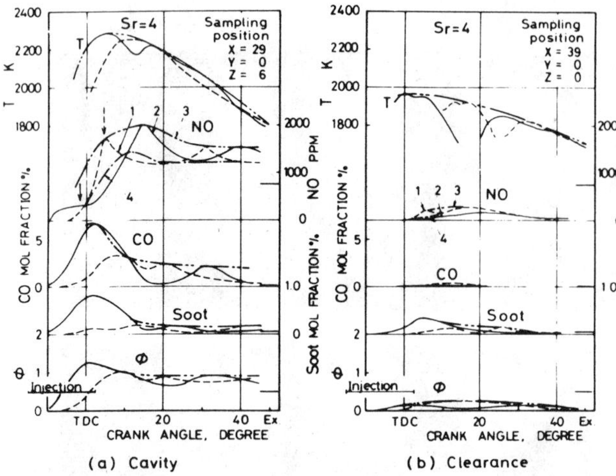

Fig.21 - Soot, CO and NO concentrations, T and ϕ at overswirl (Sr = 4)

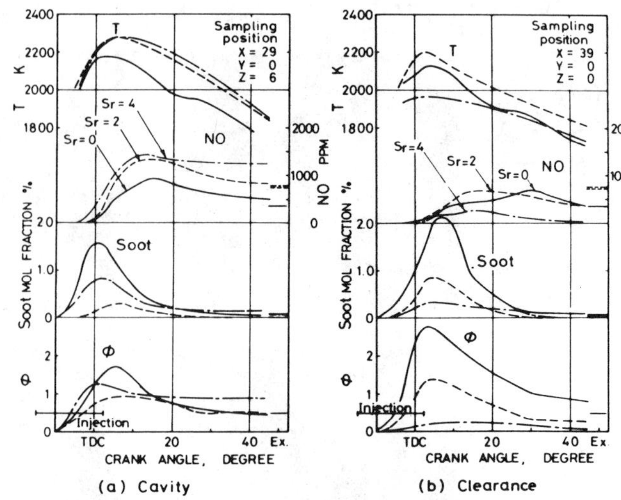

Fig.22 - Variations of soot, CO and NO concentrations, T and ϕ with changing swirl ratio (Sr) as a parameter

reach ϕ m. Because of this the oxidation of soot in the flame is poor and its concentration in the later combustion period does not decrease in the cavity space as much as the decrease in the optimum swirl.

On the other hand in the cavity space ϕ remains at a value near 1.0, consequently the maximum value of the flame temperature is as high in level as that of the optimum swirl, and moreover the flame temperature is maintained at a high value of over 2000 K for a long time. Due to this the maximum NO concentration in the overswirl becomes higher than in the optimum swirl.

In the clearance space, even though the maximum value of ϕ is below 0.3, soot is detected. This does not mean that in spite of ϕ being quite small soot is formed, but it may be inferred that because the ratio between the number of the cycles in which the flames including high soot concentration penetrate the clearance space and the number of sampling cycles is small, the average of a large number of cycles gives the comparatively high soot concentration as the above shown. It is for a similar reason that measurements of flame temperature in the cavity space show a low value.

Effect of Swirl - All the two dotted-chain lines in Fig. 20 and Fig. 21 show the envelope curve of tops in the cases when injection directions were both normal and rotated 45°. When the phases of maximum values of soot, CO and flame temperature correspond with the tops of ϕ, each envelope curve shows the histories in the main part of the flame. The envelope curve of NO as described previously shows the histories of NO concentration in the flame tip on downstream by the swirl. The histories of the NO concentration in the main parts of the flame are shown as one dotted-chain line in each figure. Fig. 22 (a) and (b) shows the history curves of the main part of the flame with changing swirl ratios as a parameter.

Firstly the histories of ϕ will be described. In a quiescent state, ϕ in both the cavity and clearance space are larger compared to the state with a swirl, and moreover ϕ in the clearance space becomes larger than in the cavity space. Thus this may be said to be caused by the condition of over penetration. In the case of the optimum swirl the maximum value of ϕ in both the cavity and clearance space is reduced to around 1/2 of the quiescent state. That is, the injected fuel is divided evenly between the cavity and clearance spaces. As the swirl is intensified further to a overswirl, most of the fuel exists only in the cavity space. For this reason the maximum value of ϕ in the cavity space becomes larger than that of the optimum swirl and ϕ in the later combustion period also is the largest value among all these test runs. From the above observation by intensifying the swirl strength the main part of the flame moves from the periphery to the center of the combustion chamber.

The flame temperature in the cavity space by the presence of the swirl becomes around 100 K high because the combustion with the swirl proceeds near a stoichiometric equivalence ratio. Especially in the case of overswirl the ϕ value in the cavity space is maintained for a long time in the neighborhood of stoichiometry resulting in the flame temperature being maintained for the longest time at a high temperature.

The soot concentration corresponds to the above described changes of ϕ. That is, the soot in a quiescent state remains in the clearance space until the later combustion period while in the overswirl it remains only in the cavity space until the later combustion period. Though the exhaust concentration of the two cases are almost the same, the swirl strength in this way causes a difference in the location where the soot is formed and remained. Under all conditions NO is formed mainly in the cavity space which has a high flame temperature. Though the NO concentration in the cavity space takes the highest value in an overswirl, the exhaust NO concentration takes the same value as that of optimum swirl. This is because the high NO concentration gas in the cavity space is mixed with the low NO concentration gas in the clearance space during the exhaust stroke.

OPTIMUM SWIRL IN DEEP BOWL COMBUSTION CHAMBER - The description will be made of a typical deep bowl combustion chamber which gives a minimum soot emission as shown in Fig. 5 at an optimum swirl (Sr = 4). Fig. 23 shows the combustion photographs in this condition, while Fig. 24 shows the pressure diagram and rate of heat release. The results of gas sampling made at representative sampling positions are shown in Fig. 25 (a), (b) and (c).

The comparison of these results with those of a shallow dish combustion chamber at an optimum swirl shows that the histories of the concentrations of the main combustion products are qualitatively almost the same. The fact that the ϕ value in the cavity space of Fig. 25 (a) reaches ϕ m early at 15° CA ATDC, the fact that the combustion photographs show the flame spreading in both the cavity and clearance space at 10° CA and the fact that the luminous flame disappears at 30 ° CA, all indicate that the fuel-air mixing finishes early in this condition. The combustion completes early in accordance with the change of the rate of heat release.

The maximum soot concentration reaches same value as in the optimum swirl in a shallow dish combustion chamber, while the soot concentration becomes an extremely low value in the later combustion period because of good fuel-air mixing. The exhaust soot concentration also takes the lowest value among all tests reported in this study.

In contrast to the case of the shallow dish type, where NO reaches maximum concentration at about 15° CA, in the deep bowl type the NO concentration increades rapidly from the early combustion period, and continues to increase gradually in the cavity space till about 40° CA. The maximum NO concentration is the highest in the range of the same injection timings in this report, and as a result of this the exhaust level also becomes high. In the cavity space in this condition the peak of NO concentration does not become sharp due to the good mixing. For the same reason the exhaust NO concentration only decreases to 2/3 of the maximum concentration.

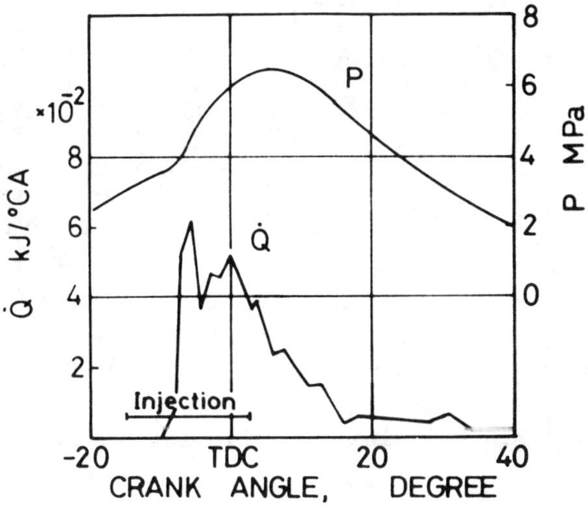

Fig.24 - Cylinder pressure diagram and rate of heat release in a deep bowl combustion chamber

SUMMARY AND CONCLUSIONS

The injection timing, swirl ratio and combustion chamber shape were changed in a direct injection diesel engine, and by gas sampling the formation processes of soot and NO were investigated. The following conclusions can be drawn:

(1) Soot is formed rapidly and reaches maximum concentration in the early combustion period where the equivalence ratio within the flame is large. As the maximum equivalence ratio becomes higher, its maximum concentration becomes higher. After this almost all the formed soot swiftly decreases by oxidation in the same way as CO and hydrocarbons in the middle combustion period where the equivalence ratio is reduced by the air entrained into the flame. Thereafter in the later combustion period the flame temperature falls less than about 2000 K and the flame movements slow down causing difficulties in soot oxidation and consequently soot remains at a constant concentration, and then being exhausted.

(2) NO is formed in the early and middle combustion period where the flame temperature and pressure are high and also the equivalence ratio within the flame is comparatively large, and reaches maximum value when the local equivalence ratio decreases to 1.0. After this NO is diluted by the air entrained into the flame and

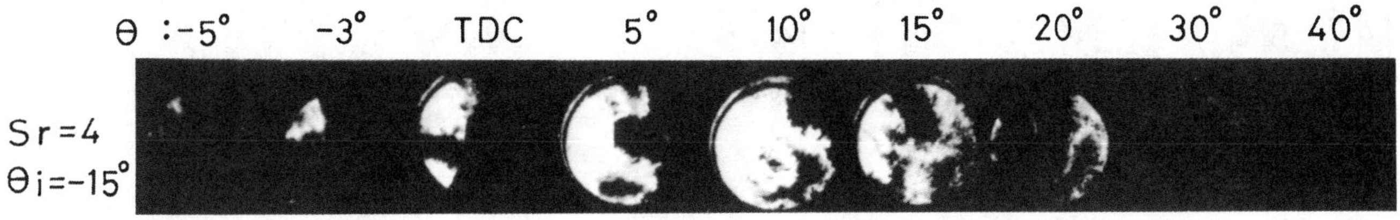

Fig.23 - Combustion photography for deep bowl piston at optimum swirl (Sr = 4)

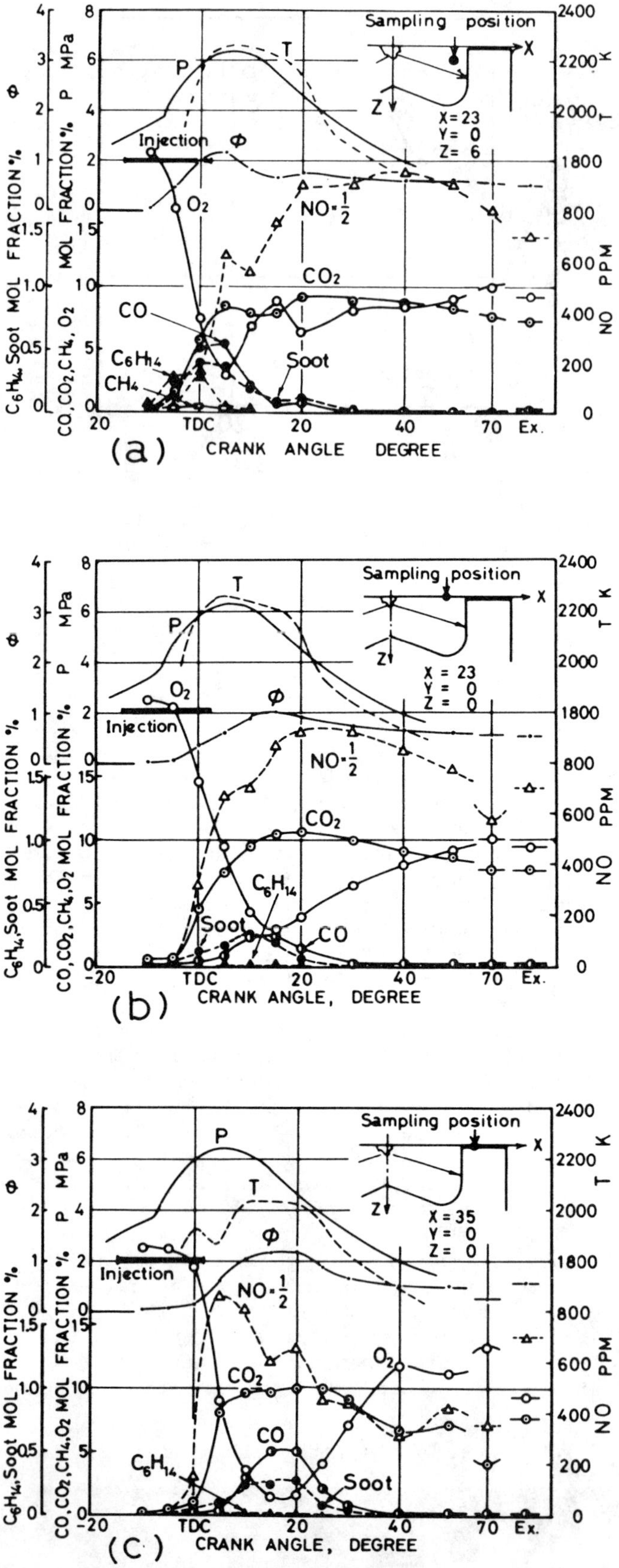

Fig.25 - Histories of soot, NO and other combustion products with T, P and ø in a deep bowl combustion chamber

is exhausted after it has become between 1/3 and 2/3 of the maximum concentration.

(3) The retardation of the injection timing not only causes decreases in pressure and flame temperature, but also an increase in the equivalence ratio within the flame. Dut to this the maximum NO concentration decreases while the maximum soot concentration increases in the combustion chamber.

(4) The optimum swirl condition makes a good fuel distribution in the combustion chamber, and reduces the maximum equivalence ratio. Due to this the maximum concentration of soot becomes low. In a swirl condition the sharp peaks of NO concentration are observed at the flame tip on the downstream side of the swirl.

(5) In a deep bowl combustion chamber the histories of the main combustion products, NO and soot are qualitatively similar to the shallow dish type. The main difference is that the maximum concentration of NO is higher and that of soot is lower because the fuel-air mixing is superior and the local equivalence ratio reaches the overall equivalence ratio earlier in comparison to a shallow dish combustion chamber.

ACKNOWLEDGMENTS

The authors wish to thank Automobile Exhaust Clarification Committee by the Ministry of Education for their financial support of this research.

We would also like to acknowledge the helpful contributions of Hino Motors, Ltd.

Thanks are also due to the graduate students Y. Nakada, K. Hishinuma and Y. Kooguchi who assisted in taking the data.

NOMENCLATURE

P	Cylinder perssure	(MPa)
Pmi	Indicated mean effective pressure	(MPa)
$\dot{Q}$	Rate of heat release	(kJ/°CA)
Sr	Swirl ratio (Angular velocity of air in cavity space at TDC / engine speed)	
T	Flame temperature	(K)
x	Mol fraction of chemical species	
θ	Crank angle from TDC	(°CA ATDC)
θ i	Fuel injection timing measured from the start of needle lift	(°CA ATDC)
ø	Local fuel-air equivalence ratio	
øm	Overall or mean fuel-air equivalence ratio	

REFERENCES

1. E. K. Bastress, K. M. Cheng and D. M. Dix, "Models of Combustion and Nitrogen Oxide Formation in Direct and Indirect Injection Compression Ignition Engines." Paper NO. 719053, IECEC Meeting, Boston, Mass., 1971.

2. I. M. Khan, G. Greeves and C. H. T. Wang, "Factors Affecting Smoke and Gaseous Emissions From Direct Injection Engines and a Method of Calculation." SAE Paper 730169.

3. S. M. Shahed, W. S. Chiu and W. T. Lyn, "A Mathematical Model of Diesel Combustion." Presented at Institution of Mech. Engrs. Conference on "Combustion in Engines", Cranfield Institute of Technolgy, Great Britain, 1975.

4. H. Hiroyasu and T. Kadota. "Models for Combustion and Formation of Nitric Oxide and Soot in Direct Injection Engines." SAE Paper 760129, 1976 SAE Trans.

5. N. D. Whitehouse and N. Baluswamy, "Calculation of Gaseous Products During Combustion in a Diesel Engine Using a Four Zone Model." SAE Paper 770410.

6. D. R. Nightingale, "A Fundamental Investigation into the Problom of NO Formation in Diesel Engines." SAE Trans., Vol. 84, Paper 750848, 1975.

7. J. E. Bennethum, J. N. Mattavi and R. K. Toepe, "Diesel Combustion Chamber Sampling-Hardware, Procedures and Data Interpretation." SAE Trans., Vol. 84, Paper 750849, 1975.

8. N. D. Whitehouse, E. Clough, and P. S. Roberts, "Investigating Diesl Engine Combustion by Means of a Timed Sampling Valve." SAE Paper 770409.

9. K. T. Rhee, P. S. Myers and O. A. Uyehara, "Time-and Space-Resolved Species Determination in Diesel Combustion Using Continuous Flow Gas Sampling." SAE Paper 780226.

10. V. K. Duggal, T. Priede and I. M. Khan, "A Study of Pollutant Formation within the Combustion Space of A Diesel Engine." SAE Paper 780227.

11. I. A. Voiculescu and G. L. Borman, "An Experimental Study of Diesel Engine Cylinder-Averaged NOx Histories." SAE Paper 780228.

12. A. Thomas, "Cabon Formation in Flames" Combustion and Flame, Vol. 6 (1962), 46

13. Y. Matisui, T. Kaminoto and S. Matsuoka, "A Study on the Time and Space Resolved Measurement of Flame and soot concentration in a D. I. Diesel Engine by the Two-Color Method." SAE Paper 790491.

14. D. L. Stivender, "Development of a Fuel-Based Mass Emission Measurement Procedure," SAE Paper 710604.

15. C. Olidara and G. L. Borman, "A Computer Program for Calculating Properties of Equilibrium Combustion Products with Some Applecations to I. C. Engines." SAE Paper 750468

16. J. J. Macfarlane and F. H. Holderness, "Laboratory Studies of Carbon Formation in Fuel-Rich Flame at high Pressures", P. I. M. E., Vol 184 Pt3J.

17. G. P. Prado. "Soot and Hydrocarbon Formation in a Turbulent Diffusion Flame." 16th Symposium (International) on Combustion, PP 561, The Combustion Institute (1976).

18. T. Onuma, M. Ogasawara and T. Inoue, "Further Experiments on the Structure of a Spray Combustion Flame." 16th Symposium (International) on Combustion, PP. 561, The Combustion Institute (1976).

19. T. Suzuki et al, "Controlling Exhaust Emissions from The High Speed Diesel Engine with a Direct Injection Type Combustion Chamber." Instn. Mech. Engrs., Combustion in Engines, C87/75, P. 59, 1975.

20. F. Nagao et al, "A Study on Combustion in Direct-Injection Diesel Engines (1st Report, In the Case of Shallow-Dish Chamber)," Bulletin of the JSME, Vol. 16, No. 93, P. 588, 1973.

21. T. Suzuki and T. Shiozaki, "A New Combustion System for the Diesel Engine and Its Analysis via High Speed Photography," SAE Paper 770674.

The Contribution of Engine Oil to Diesel Exhaust Particulate Emissions**

William J. Mayer and Donald C. Lechman
Analytical Chemistry Dept.
General Motors Research Laboratories
Warren, MI

David L. Hilden
Fuels and Lubricants Dept.
General Motors Research Laboratories
Warren, MI

THE EXCELLENT FUEL ECONOMY OF LIGHT-DUTY DIESEL ENGINES provides a strong incentive to expand their use in the future, but particulate exhaust emissions may be an obstacle. The particulate emissions can appear as black smoke, and there is some concern that they may constitute a health hazard (1)*. Consequently, the Federal government has proposed standards that will limit the amount of particulate matter emitted by automotive engines (2).

Fundamental research on particulate formation during combustion has indicated that particulate matter may be produced by pyrolysis and polymerization of low vapor-pressure liquid hydrocarbon droplets (3-5), such as engine oil. The potential for particulate formation by this process apparently increases as the molecular weight of the hydrocarbon increases. Since the molecular weight of engine oil averages approximately 400 versus about 170 (6) for the fuel, engine oil which finds its way into the combustion chamber might contribute to diesel particulate matter by this same droplet mechanism. Consequently, the contribution of oil to the particulate emissions could be much greater than would be expected based on relative consumption rates of oil and fuel (estimated to be only about 0.25 percent by mass).

There is an additional reason for being interested in the contribution of engine oil to particulate emissions. Ordinarily, a fraction of the particulate material produced by a diesel engine is soluble in organic solvents (7). This soluble (extractable) particulate matter is of special interest since it includes material which, by at least one type of test, is mutagenic to certain bacteria (1). The heaviest portions of the extractable particulate material have properties [including molecular weight distribution (6,8)] very much like those of engine oil. This fact has led to the qualitative conclusion that much of the extractable portion of the particulate emis-

*Numbers in parentheses designate References at end of paper.

**Paper 800256 presented at the Congress and Exposition, Detroit, Michigan, 1980.

ABSTRACT

A radioactive tracer technique was developed to determine the contribution of oil from an engine sump to exhaust particulate matter collected on a filter. The technique was applied to particulate emissions produced by an automotive diesel engine which was operated on an engine dynamometer over a range of steady-state conditions. The results indicated that from 1.5 to 25 mass percent of the particulate matter, depending on speed and load, consisted of material from the engine oil. The oil contribution to the extractable organic portion of the particulate matter ranged between 16 and 80 percent.

sions may be derived from the oil (8).

Although circumstantial evidence suggests that engine oil may have an effect on diesel particulate emissions, the literature does not contain references to direct measurements of this effect. The study described in this report was designed to make such measurements. A radioactive tracer technique was developed to measure not only the contribution of the oil to the total particulate emissions produced by a diesel engine, but also the contribution of the oil to the extractable fraction of the particulate emissions.

EXPERIMENTAL

Since the tracer technique described herein was developed specifically for this study, considerable development and testing were required. Only essential experimental information is presented in this section. Details are in Appendix A.

GENERAL METHOD - The oil contribution experiments described herein are based on a technique called isotope dilution (9). An application of this technique which was used in the present study can be described as follows:

1. A radioactive organic compound containing carbon-14 is added as a tracer to the diesel engine oil.
2. After mixing the tracer and oil thoroughly, the amount of carbon-14 relative to total carbon (primarily carbon-12 and -13) in the oil is measured by methods to be described later.
3. The engine is operated at steady-state, and a portion of the particulate emissions is collected.
4. The amount of carbon-14 in this material (relative to the total carbon) is measured.
5. The oil contribution to the particulate emissions is based on the degree to which the carbon-14 in the particulate material (originally from the oil) is "diluted" by naturally occurring carbon from some other source (e.g., fuel). The development of the equations needed for calculation of the oil contribution to the particulate emissions can be found in Appendix A.

The experiments to determine the engine oil contribution to the particulate emissions proceeded in two steps. First, the oil contribution to the diesel particulate material was determined. Second, the material was extracted with an organic solvent. Both the amount of extractable material and the contribution of oil to this material were measured.

Upon completion of this portion of the study, the oil consumption of the engine was determined with another radioactive tracer technique (10). This was done to determine if the trends in the previous results could be explained by the consumption of oil relative to that of the fuel.

ENGINE - The engine was a 1978, 5.7-L, light-duty diesel. The engine has a V-8 cylinder configuration and compression ratio of 22:1. Fuel injection and compression ignition occur in a swirl-chamber which is connected to the main chamber by a passageway.

Normally, an oil cooler would be used with this engine, but in the present study the oil cooler was eliminated to keep the oil volume to a minimum. (A given reduction in the oil volume allows a similar reduction in the required amount of radioactive tracer.) To keep the oil temperature below the 135°C upper limit recommended by the manufacturer, the engine was cooled by high air flow in the test cell and operated for only short times at those test conditions (high speed and load) which resulted in relatively high oil temperatures.

FUEL AND ENGINE OIL - A commercial No. 2 diesel fuel was used for all of the tests decribed. The fuel analysis is given in Appendix B.

The commercial SE/CD oil used in the study had an SAE viscosity of 15W-40. The oil analysis appears in Appendix C.

TEST SET-UP - The engine was mounted on an electric dynamometer stand in a special test cell designed for handling radioactive material (11). An exhaust crossover allowed the total exhaust to be routed to a plenum through a single pipe. As shown in Figure 1, a portion of the exhaust was pumped from the plenum [via a large-capacity (23.3-L/s) vacuum pump] through a 102-mm diameter filter paper holder which contained a Schleicher and Schuell TE38 filter paper (maximum pore size 5.0 μm). This filter paper was stable and easily handled even when exposed to temperatures as high as 140°C. Air was added to the exhaust (prior to filtration) to keep the temperature at the filter holder at 52°C or below. Cooling the exhaust to this degree is done conventionally to ensure that particulate sampling systems will remove about the same amount of easily condensable hydrocarbon material (12).

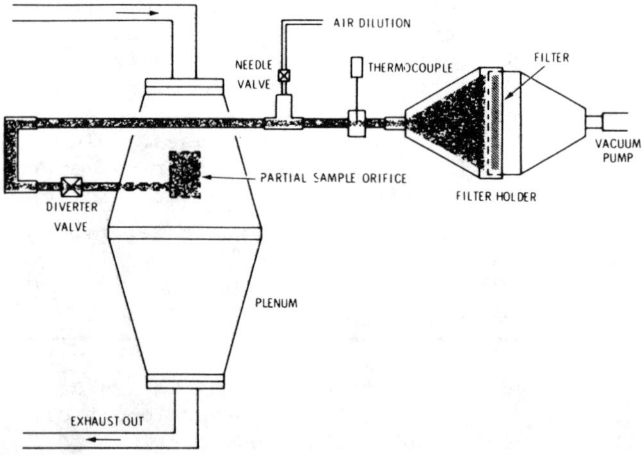

Fig. 1 - Exhaust particulate emissions collection system

TEST CONDITIONS - Data were obtained at steady-state operating conditions from ~30 percent (156 kPa bmep) to ~90 percent (468 kPa bmep) of full load and at engine speeds of 1 000, 2 000, and 3 000 r/min. Limited data were also obtained at 73 kPa bmep, 600 r/min (corresponding to idle). The operating conditions were chosen to cover the broadest possible range of speeds and loads possible with the engine-dynamometer combination. The oil contribution to the particulate emissions was measured at least three times for each test condition.

RADIOACTIVE TRACER ADDED TO THE OIL - The tracer initially chosen for the study was n-octacosane. The radioactive carbon-14 isotope was located at either the 14th or 15th or both the 14th and 15th carbon atom positions in the chain of this 28-carbon atom straight chain paraffin. Pathfinder Laboratories, Inc. of St. Louis supplied the tracer. Since the engine oil was paraffinic and had a median molecular weight of about 400 (6), n-octacosane (MW=395) was considered a good choice to represent the oil.

A critical assumption required for this study was that a proportional amount of tracer accompanied the bulk of the engine oil which went from the engine sump to particulate material collected on a filter. This assumption is not generally true unless the measured oil contribution is essentially independent of the physical and chemical properties of the tracer molecule (assuming the tracer molecule is at least consumed at the same rate as the bulk of the oil). The assumption was tested by repeating the measurements with another tracer which had significantly different properties than those of n-octacosane. The results of this study (Appendix D) generally corroborate those observed with n-octacosane tracer.

MEASUREMENT OF CARBON-14 IN TOTAL PARTICULATE EMISSIONS - To measure the oil contribution to diesel particulate emissions, the amount of carbon-14 relative to total carbon must be determined. The liquid scintillation method was used to measure carbon-14.

To determine the carbon-14 in the individual particulate samples, each sample was combusted in excess oxygen to CO_2 and H_2O. The combustion products were bubbled through a commercial CO_2 absorbing solution (Carbsorb) which, in turn, was mixed with an appropriate liquid scintillator. This general approach was developed prior to the present study (13-16), but its application to the specific problem of diesel particulate emissions is unique. A description of this application appears in Appendix A.

ANALYSIS OF EXTRACTABLE PARTICULATE EMISSIONS - A solution of 80 percent benzene and 20 percent ethanol by volume was used to determine the contribution of the oil to the extractable portion of the particulate emissions. This solvent is polar enough to dissolve partially oxidized organic material as well as hydrocarbons. The choice of solvent and extraction technique was patterned after work by Williams and Begeman (17). Of course, the particular solvent and the way in which it is used to remove the extractable material will determine the amount and type of material extracted.

OIL CONSUMPTION MEASUREMENTS - Oil consumption was measured by a radiometric method (10). The method consists of adding a radioactive tracer (1,2-dibromo-82-octadecane) to the engine oil and measuring the radioactive material that is extracted from the engine exhaust gas at each selected speed-load condition of the engine. The precision of this method based on previous experience is ±10 percent (10).

A portion of the exhaust gas was drawn through a small extraction column (500 mL gas washing bottle equipped with a coarse glass frit) which contained 200 mL of a 1 molar sodium hydroxide solution. The radioactive bromine-containing species in the exhaust were removed with nearly 100 percent efficiency by the column. The amount of exhaust gas flowing through the extraction column was measured with a wet test meter, whereas the total exhaust flow was calculated from air intake and fuel flow measurements. The radioactive bromine content of the extraction solution was determined with a well-type sodium iodide crystal gamma ray counter. The oil consumption rate of the engine was then calculated from the experimentally determined values of sample flow, total exhaust flow, exhaust bromine content, and the radioactive bromine content of the engine oil.

RESULTS AND DISCUSSION

CONTRIBUTION OF ENGINE OIL TO TOTAL PARTICULATE EMISSIONS - Figure 2 shows the percent of total particulate material (the complete sample removed from the exhaust filter) which was derived from the oil as a function of the engine load and speed. Each point in the figure (except for idle) is an average of the data collected for all tests at that condition. The length of each error bar equals twice the standard deviation of the results.

The figure shows an increased contribution of oil to the total particulate emissions when the engine speed is increased. Also, the oil contribution generally decreased as the load increased. The magnitude of the contribution lies between 1.5 percent (for low speed and high load) and 25 percent (for high speed and low load). Note that the maximum effect is approximately 100 times greater than that predicted from the relative oil-fuel consumption rate.

Although the magnitude of the oil contribution proved much greater than the relative oil consumption, it could be possible that the oil contribution would be directly proportional to the oil consumption. This hypothesis was checked by determining the oil and fuel consumption rates for each of the test conditions.

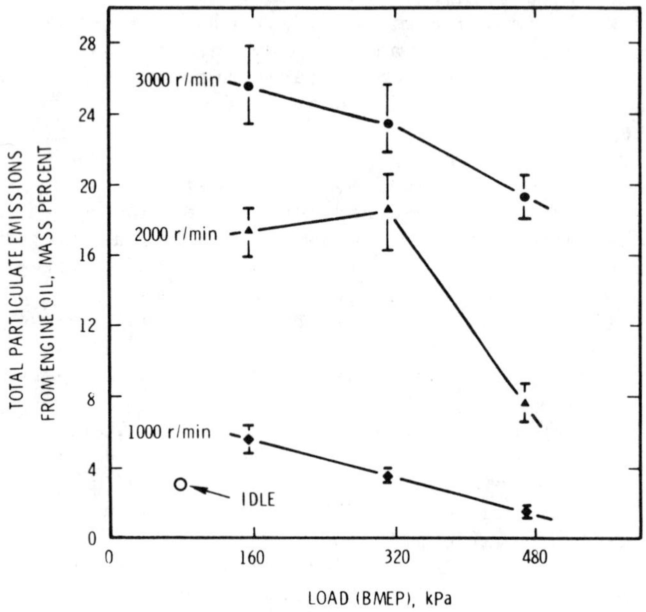

Fig. 2 - Contribution of engine oil to total particulate emissions

These consumption rates are given in Table 1 along with the relative oil consumption (as compared to fuel consumption) and the oil economy. The curves of Figure 2 are repeated on Figure 3 (left hand ordinate with open symbols) along with the corresponding curves for the relative oil consumption (right hand ordinate with solid symbols). The relative oil consumption generally increased with speed as did the percentage of total particulate matter derived from the engine oil. However, it also increased with load at the higher speeds (2 000 and 3 000 r/min) whereas the oil contribution to the particulate matter generally decreased. Only for the 1 000 r/min engine speed did the percentage of oil-derived particulate matter appear to follow the same trend as the relative oil consumption. By dividing the oil contribution to the particulate matter by the corresponding relative oil consumption, it can be shown that a given amount of engine oil supplies from 50 to 280 times (depending on the test conditions) as much material to the particulate emissions as does an equal amount of fuel. Thus, even though the oil consumption is relatively small compared to that of the fuel, the oil can still have a significant effect on particulate emissions.

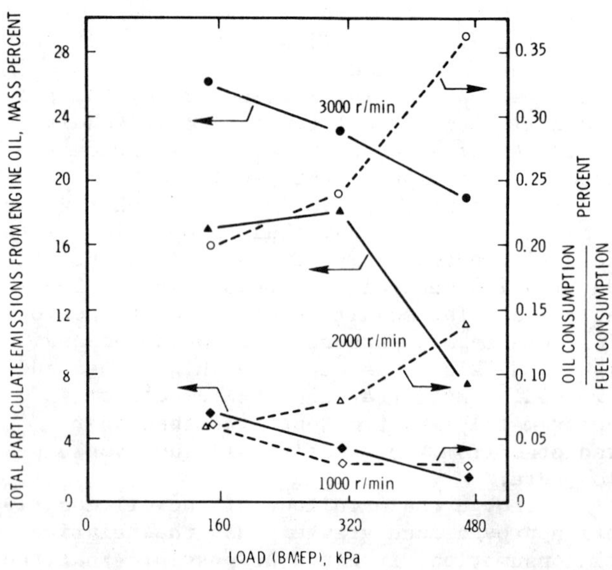

Fig. 3 - Comparison of engine oil contribution to total particulate emissions with relative oil consumption

One possible explanation for the magnitude of the measured contribution of engine oil to the particulate emissions is abnormally high oil consumption (low oil economy). The oil economy measurements given in Table 1 indicate this was not the case for the present study. On the contrary, the oil economy of the test engine proved to be quite good. Similar to typical spark ignition engines (10-11), the oil economy generally decreased as either the speed or load increased.

Table 1 - Oil and Fuel Consumption Rates of the 5.7-L Diesel Engine

Engine Speed, r/min	BMEP, kPa	Fuel Consumption, mg/s	Oil Consumption, mg/s	[Oil/Fuel] x 100, Percent	Oil Economy,* km/L
1 000	156	800	0.47	.06	21 800
	312	1 180	0.38	.03	25 300
	468	1 680	0.50	.03	20 500
2 000	156	1 420	0.9	.06	24 000
	312	2 120	1.8	.08	10 800
	468	3 100	4.5	.14	4 550
3 000	156	2 400	4.7	.20	6 700
	312	3 720	8.8	.24	3 500
	468	4 480	16.3	.36	1 900

*For an N/V of 21.6, N = r/min and V = km/h

Since the percent oil contribution to the particulate emissions was determined in this study rather than the emission rate, mass particulate emission data collected at another laboratory (18) were used with the results shown in Figure 2 to determine the mass of exhaust particulate emissions contributed by the oil. Figure 4 is a bar chart of mass specific total particulate emissions for the operating conditions reported herein. Each bar has been subdivided to show how much of the particulate sample came from the oil. Not only is the percentage oil contribution greatest for the high speed, low load condition, but the total particulate emissions are also greatest for this condition. Assuming the oil contribution could be eliminated, the maximum potential reduction in particulate emissions is

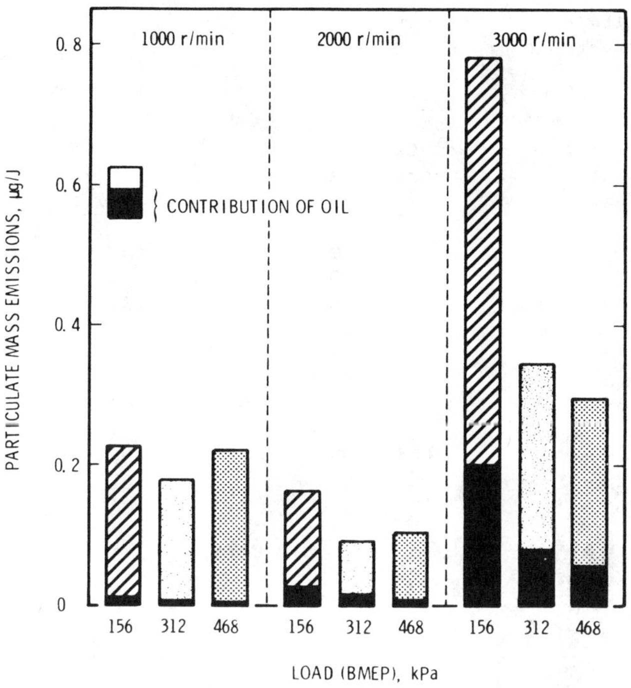

Fig. 4 - Particulate mass emissions from the 5.7-L diesel engine and estimated oil contribution

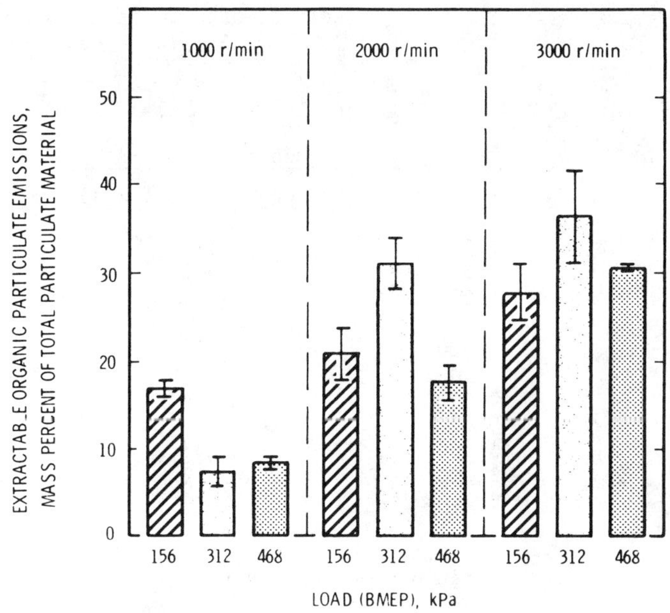

Fig. 5 - Material extracted from total particulate emissions with benzene-ethanol solvent

about 0.2 µg/J at the high speed, low load condition.

The results show that the oil contribution to diesel particulate emissions can be important depending on engine operating conditions. However, since the contribution varies greatly from one speed-load condition to another, the overall effect may be relatively small for cyclic vehicle operation. The contribution of oil to the extractable organic portion of the particulate emissions turns out to be much greater.

CONTRIBUTION OF ENGINE OIL TO EXTRACTABLE ORGANIC PARTICULATE EMISSIONS - Figure 5 shows, for each engine operating condition, the average mass percentage of the total particulate sample which was extractable with the organic solvent. The length of the error bar is twice the standard deviation in each case. The results range from 8 to 36 percent with an overall average of 22 percent. The contribution generally increased with engine speed. Since, as shown in Figure 4, the total particulate emissions also were greatest at the highest engine speed, it follows that the amount of extractable organic particulate matter would be emphasized by increased speed. Although the percentage of extractables appears highest for the mid-load condition at 2 000 and 3 000 r/min, this trend is not observed at 1 000 r/min. Hence, there is no consistent dependence on engine load.

Springer and Baines (7) have determined the organic soluble particulate matter emitted by a vehicle powered by the same model engine tested in the present study. The vehicle was operated over a variety of cyclic modes. Their results (12-19 percent, depending on the operating cycle) are bracketed by the results presented here. Unfortunately, no steady state data for extractable organic particulate emissions are available for comparison.

The results shown in Figure 5 bring up the following question: When organic material is extracted from the total particulate sample, how much of the radioactivity (and, by association, how much of the engine oil contribution) is removed? The results indicate that all (within experimental error) of the radioactivity is extracted. Table 2 shows that the radioactivity initially observed in the total particulate sample that was found in the extract for each engine operating condition varied between 93 and 105 percent. It was, therefore, concluded that all the radioactivity was extracted.

If the radioactive material in the particulate emissions is representative of the oil contribution (as indicated in Appendix D), the extraction results indicate that essentially all of the engine oil contribution should be classified as extractable organic material. Hence, the crosshatched portions of the bars in Figure 4 not only show oil contribution to total particulate mass, but also the approximate oil-derived extractable organic contribution to total particulate mass.

Known values of both the amount of radioactive tracer found in the extractable material and the amount of extractable material can be used to calculate the oil contribution to the

Table 2 - Percentage of Radioactivity From Total Particulate Samples Recovered in Extractable Organic Material

Engine Test Conditions		Radioactivity in Total Particulate Matter*	Radioactivity in Extractable Particulate Matter*	
Speed (r/min)	Load (BMEP in kPa)			Recovery (%)
1 000	156	2 500	2 620	105
1 000	312	1 570	1 540	98
1 000	468	695	708	102
2 000	156	7 380	6 900	93
2 000	312	8 750	8 580	98
2 000	468	3 670	3 730	102
3 000	156	10 200	10 700	105
3 000	312	10 200	10 300	101
3 000	468	8 150	8 480	104

* Disintegrations per second per gram of particulate sample.

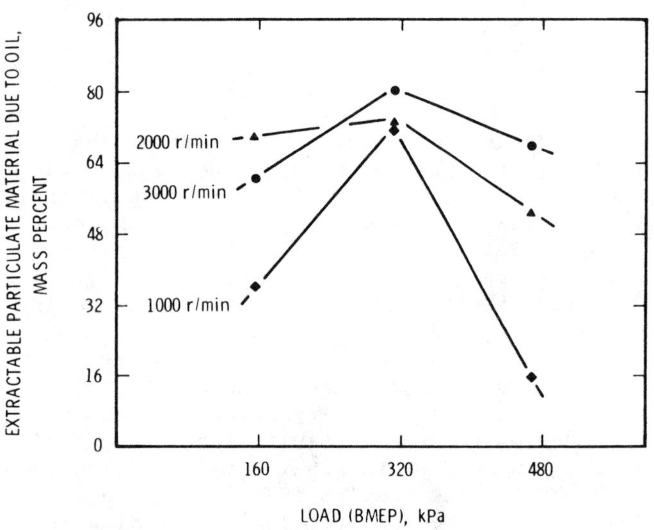

Fig. 6 - Contribution of engine oil to extractable organic particulate emissions

extract (the calculation is shown in Appendix A). These results are shown in Figure 6.

The results are based on the assumption that the oil contribution to the extract has the same composition as the oil from which it was derived. The potential error due to this assumption is small since, as mentioned previously, the oil and extract are both hydrocarbon materials with similar properties. Figure 6 indicates that engine oil can make a major contribution to the extractable organic portion of the diesel particulate emissions. The maximum oil contribution (80 percent) was at the high speed, mid-load condition. On the other hand, when the engine was operated at low speed and high load the oil contribution was minimum (16 percent). A comparison of Figure 4 and Figure 6 indicates that the oil contribution to the total particulate emissions is more likely to end up in the extractable organics than is the fuel contribution. Also, this trend is related to engine speed. For example, at the mid-load, 1 000 r/min operating condition the percentage oil contribution to the extractables is 20 times its contribution (see Figure 3) to total particulate matter. However, when the engine is operated at mid load and 3 000 r/min, the oil contribution to the extract is 3.4 times its contribution to total particulate emissions. This trend holds for the other engine loads, also. That is, at any given load the contribution of oil to the extractable organic material (relative to its contribution to total particulate material) decreases as speed increases. Why this should be the case is unknown.

CONCLUDING REMARKS

This study has shown that engine oil can be a significant source (1.5 to 25 mass percent) of particulate emissions produced by a light-duty diesel engine and is a major source (16 to 80 mass percent) of the extractable organic portion of these emissions. The greatest contribution from the oil was generally observed at the highest test speed.

Previously it was suggested that heated droplets containing relatively high molecular weight species (such as found in engine oil) might pyrolyze and polymerize to form particles. As far as oil consumed by the diesel engine in this study is concerned, this mechanism appears inconsistent with the experimental observation that all of the oil contribution to the particulate material is extractable. The oil contribution has retained the solubility of the lubricating oil from which it was derived (see

Appendix D). Hence, the oil probably contributed to the particulate matter after most of the chemical reactions forming the nonextractable portion of the particulate matter had been completed.

ACKNOWLEDGMENTS

We gratefully acknowledge technical assistance provided by the Oldsmobile Division of General Motors Corporation. We also acknowledge fuel, lubricating oil, and particulate analyses by Analytical Chemistry Department personnel as well as the use of data provided by General Motors Engineering Staff.

REFERENCES

1. R. Zweidinger, S. Tejada, D. Dropkin, J. Huisingh, and L. Claxton, "Characterization of Extractable Organics in Diesel Exhaust Particulates," EPA Symposium on Diesel Particle Emissions Measurement Characterization, Ann Arbor, Michigan, May 17-19, 1978.

2. Environmental Protection Agency, "Control of Air Pollution From New Motor Vehicles and New Motor Vehicle Engines: Certification and Test Procedures, Particulate Regulation For Light-Duty Diesel Vehicles," Federal Register, February 1, 1979.

3. H. Hottel, G. Williams, and H. Simpson, "Combustion of Droplets of Heavy Liquid Fuels," 5th Symposium on Combustion, The Combustion Institute, 101 (1955).

4. E. Masdin and M. Thring, "Combustion of Single Droplets of Liquid Fuel," Journal of the Institute of Fuel 35, 251 (1962).

5. A. Williams, Combustion of Sprays of Liquid Fuels, Elek Science, London (1976).

6. R. Schreck, J. McGrath, S. Swarin, W. Hering, P. Groblicki, and J. MacDonald, "Characterization of Diesel Exhaust Particulate For Mutagenic Testing," Presented to the Air Pollution Control Association, Houston, Texas, June 25, 1978. (Available as General Motors Publication GMR 2755, June 25, 1978).

7. K. Springer and T. Baines, "Emissions From Diesel Versions of Production Passenger Cars," SAE Paper 770818, Presented September 26-30, 1977.

8. F. Black and L. High, "Methodology For Determining Particulate and Gaseous Diesel Hydrocarbon Emissions," SAE Paper 790422, Presented February 26-March 2, 1979.

9. G. Friedlander and J. Kennedy, Nuclear and Radiochemistry, John Wiley and Sons, Inc., New York (1956) p. 324.

10. W. Mayer, D. Krause, C. Moore, and C. Murphy, "Oil Consumption Measurements of Individual Cylinders in Automotive Engines," SAE Paper 700055, January 1970.

11. W. Mayer, C. Moore, D. Krause, W. Lange, C. Murphy and J. Klepaczyk, "Rapid, Precise Measurement of Engine Oil Economy by a Radiometric Method," SAE Paper 660058, January 1966.

12. Coordinating Research Council - Air Pollution Research Advisory Committee diesel exhaust emissions measurement symposium, Chicago, Illinois on April 5-6, 1977.

13. W. Schoniger, Mikrochemica Acta 1955, 123 (1955).

14. W. Schoniger, Fact and Methods for Scientific Research 1, 1 (1960).

15. N. Kaartinen, Packard Technical Bulletin 18, Packard Instrument Co., Downers Grove, Illinois (1969).

16. W. Kisielski and E. Buess, Liquid Scintillation Science and Technology, Academic Press, Inc. New York (1976) p. 299.

17. R. Williams and C. Begeman, "Characterization of Exhaust Particulate Matter From Diesel Automobiles," Presented at the 17th Annual Purdue Air Quality Conference, Purdue University, Lafayette, Indiana, May 7, 1979. (Available as General Motors Research Publication GMR 2970, May 7, 1979).

18. Communication from Donald M. Scott, General Motors Engineering Staff, September 13, 1977.

APPENDIX A

SUPPLEMENTARY EXPERIMENTAL DETAILS

RADIOACTIVE ENGINE OIL TRACER - Based on the approximate fuel and oil consumption rates of the diesel engine used in this study, it was calculated that 132 MBq [a megabecquerel (MBq) is the amount of tracer required to produce one million disintegrations per second] of ^{14}C should be placed in the engine's oil to detect an oil contribution to the particulate emissions of 0.1 mass percent. For an extra margin of experimental sensitivity, 183 MBq (364 mg) of the radioactive ^{14}C tracer were added to the oil.

Prior to addition of the radioactive tracer, the oil used during engine break-in was drained. Fresh oil (4.7-L), in which the tracer had been dissolved, was then added to the sump. The engine was then operated long enough to mix the tracer with any remaining unactivated oil which had not previously drained from the engine. The end of the required mixing period was signified when a constant amount of radiation per unit volume of oil was noted. The amount of ^{14}C in the oil was determined to be 35 200 d/s / mL oil. This concentration was measured periodically during the tests which followed and was found to remain constant. Of course, a change in the activity of the oil with time would indicate that the tracer was not being consumed at the same rate as the oil. In such a case, the radioactive material clearly would not be properly tracing the oil consumption.

ANALYSIS OF TOTAL PARTICULATE EMISSIONS - Diesel particulate samples (typically 20-30 mg) were weighed into a combustion boat and placed in one end of a quartz combustion tube (30.5 cm long, 1.3 cm diameter), Figure A-1. The other

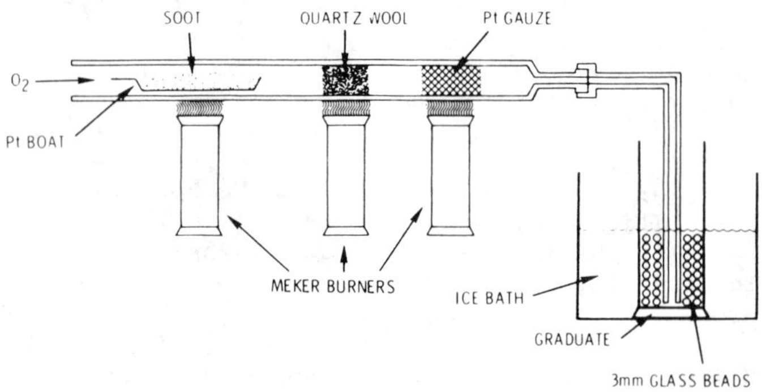

Fig. A-1 - Apparatus for combustion analysis of radioactivity in total particulate emissions

end of the tube was filled with platinum gauze catalyst. Quartz wool was placed between the boat-shaped container and catalyst. Oxygen (2.1 mL/s) was passed into the combustion tube, over the particulate sample and then through the catalyst. The oxygen then passed through a capillary into the bottom of a 25 mL graduated cylinder which was half full of 3 mm diameter glass beads and contained 5 mL of Carbsorb solution.* The graduated cylinder was cooled in an ice bath.

After the apparatus was assembled as described, a Meker burner was placed under one end of the combustion tube to bring the catalyst up to temperature (~850°C). Once the catalyst was heated, another Meker burner was used to heat the quartz wool plug and a third was used to heat the particulate sample. Complete combustion of the sample to CO_2 and H_2O was ensured by the catalyst. When the entire sample had been combusted (i.e., when the container was empty) the Carbsorb solution was quantitatively transferred to a scintillator vial along with 15 mL of Permafluor V* scintillator solution. The vial was capped and shaken and was then ready for counting by a TriCarb Model 3310 Liquid Scintillation Spectrometer.*

Quantitative recovery of carbon-14 by the method just described was determined by adding a given amount of ^{14}C tracer to a particulate sample containing no activity and analyzing for radioactivity as previously described. The results of this analysis were compared to those derived by adding the same amount of ^{14}C directly to the scintillation solution. The comparison indicated that 98 percent of the tracer in the particulate sample was consistently accounted for.

ANALYSIS OF EXTRACT - Particulate samples were extracted with 20 mL of solvent (80 percent benzene and 20 percent ethanol by volume) using a Soxhlet apparatus. The sample was repetitively washed with distilled solvent and then the solvent (containing extractable organic material) was removed by evaporation. Following the extraction, one-fourth of the solvent was used to determine ^{14}C, and the remainder was used to determine the amount of extractable organic material contained in the original particulate sample. The first portion was pipetted into a scintillator vial to which 15 mL of standard scintillator solution was added. This mixture was placed in the liquid scintillation counter and "counted" in the same manner as described in the "Combustion Analysis of Total Particulate Emissions" portion of Appendix A. An internal standard was used to correct the count rate for losses due to photon quenching.

The second portion (three-fourths) of the extraction solvent was transferred to a thin preweighed metal planchet. The solvent was evaporated from this sample by applying low heat and allowing the final evaporation to occur in the atmosphere with no additional heat. The weight of extractable organic particulate material, which remained behind on the planchet, was determined by difference.

DERIVATION OF THE EXPRESSION USED TO CALCULATE THE OIL CONTRIBUTION TO THE PARTICULATE EMISSIONS - The mass of particulate carbon emitted by the engine per unit time (C_p) is given by equation (1):

$$C_p = X\,C_o + Y\,C_f \qquad (1)$$

where C_o = mass of oil carbon consumed by the engine per unit time,

C_f = mass of fuel carbon consumed by the engine per unit time,

X = fraction of oil carbon converted into particulate carbon, and

Y = fraction of fuel carbon converted into particulate carbon.

Equation (2) follows when equation (1) is divided by C_p

$$1 = \frac{X\,C_o}{C_p} + \frac{Y\,C_f}{C_p} = a + b \qquad (2)$$

where a = fraction of particulate carbon derived from oil and

b = fraction of particulate carbon derived from fuel.

When radioactive carbon is added to either the oil or fuel or both it is assumed the additional carbon will contribute the same proportion to the particulate matter as either the oil or fuel. Hence,

$$T_p = X\,T_o + Y\,T_f \qquad (3)$$

*Carbsorb, Permafluor V, and TriCarb are brand names used by Packard Instrument Co. of Downers Grove, Illinois, for their CO_2 absorber, corresponding scintillator solution, and liquid scintillation spectrometer, respectively.

where T_p = mass production rate of radioactive particulate carbon,

T_o = mass consumption rate of radioactive oil carbon, and

T_f = mass consumption rate of radioactive fuel carbon.

The ratio of radioactive particulate matter to all particulate matter (whether radioactive or not) is given by dividing equation (3) by C_p,

$$\frac{T_p}{C_p} = \frac{X\,T_o}{C_p} + \frac{Y\,T_f}{C_p} = \frac{X\,T_o}{C_p}\frac{C_o}{C_o} + \frac{Y\,T_f}{C_p}\frac{C_f}{C_f} =$$

$$a\,\frac{T_o}{C_o} + b\,\frac{T_f}{C_f}\,. \qquad (4)$$

For the experiments described herein, a radioactive tracer was added to the oil only, so it follows that the ratio of T_f to C_f equals zero. Hence,

$$\frac{T_p}{C_p} = a\,\frac{T_o}{C_o} \qquad (5)$$

and

$$a = [T_p/C_p]\;/\;[T_o/C_o] \qquad (6)$$

Both T_p and T_o are proportional to the number of radioactive carbon disintegrations per second. That is,

$$T_p = \sigma(d/s)_p \text{ and } T_o = \sigma(d/s)_o \qquad (7)$$

where σ = proportionality constant and

(d/s) = disintegrations of radioactive tracer per second.

When equation (7) is substituted into equation (6) "a" can be expressed in experimentally measured quantities.

$$a = \frac{\sigma(d/s)_p}{C_p} \Big/ \frac{\sigma(d/s)_o}{C_o}$$

$$a = \frac{\dfrac{(d/s)_p}{(g_p)}\;\dfrac{g_p}{gC_p}}{\dfrac{(d/s)_o}{\text{mL oil}}\;\dfrac{\text{mL oil}}{g_o}\;\dfrac{g_o}{gC_o}} \qquad (8)$$

where

g_p = mass of a particulate sample in grams,

gC_p = mass of carbon in the same particulate sample in grams,

g_o = mass of an oil sample in grams, and

gC_o = mass of carbon in the same oil sample in grams.

APPENDIX B

ANALYSIS OF DIESEL FUEL NO. 2

Specific Gravity at 15.6°C	0.845
Sulfur, Mass %	0.33
HC, Volume % (by F.I.A.)	
Paraffins	66
Olefins	4
Aromatics	30
H/C Atom Ratio	1.80
Distillation (Evaporated Basis)	

Percent of Sample, by Volume	Temperature, °C
IBP	170.5
5	211.0
10	216.5
15	224.5
20	231.5
30	240.5
40	248.5
50	257.5
60	266.5
70	277.5
80	288.5
90	304.5
95	317.5
END	332.0

Recovery, %: 98.5
Residue, %: 1.3
Loss, %: 0.2

APPENDIX C

ANALYSIS OF SE/CD ENGINE OIL

15W-40 SAE VISCOSITY

Specific Gravity at 15.6°C	0.864
Kinematic Viscosity at 38°C (m^2/s)	117.84×10^{-6}
Kinematic Viscosity at 99°C (m^2/s)	12.28×10^{-6}
Total Acid Number	5.06
Total Base Number	5.93
Sulfur, Mass %	0.37
Phosphorous, Mass %	0.15
Calcium, Mass %	0.23
Zinc, Mass %	0.17

APPENDIX D

CONTRIBUTION OF ENGINE OIL TO DIESEL EXHAUST PARTICULATE EMISSIONS AS DETERMINED WITH THE RADIOACTIVE TRACER 1,2-DIBROMO-82-OCTADECANE

The use of any oil tracer implies that the engine oil and the tracer contribute in the same proportion to both the total and the extractable particulate material. In general this is not necessarily true. However, as indicated in the body of this paper, it appears that n-octacosane tracer adds to the particulate material without undergoing significant chemical change. If the oil does likewise, the results of the study should be independent of the tracer which is used as long as the consumption rate of tracer and oil are the same (an observation found to be true for n-octacosane tracer).

To test the hypothesis that the results are independent of the tracer, the data were retaken using the tracer employed for oil consumption measurements, 1,2-dibromo-82-octadecane, rather than n-octacosane. Not only is this a different tracer molecule, it also contains a different radioactive element than the n-octacosane (bromine-82 rather than carbon-14). The bromine emits gamma rays instead of beta rays and consequently the analysis scheme can be greatly simplified from that described before. The use of 1,2-dibromo-82-octadecane provides a test of the entire carbon-14 methodology which was developed for this study.

The sump of the engine was filled with oil containing 1,2-dibromo-82-octadecane to give a specific activity of 1.1 x 10^6 disintegrations per second per milliliter after mixing.

The particulate samples for the speed-load ranges were collected in the same manner as described in the experimental section of this paper. However, after the particulate material was collected, it was removed from the filter paper, and weighed samples (20 to 30 mg) were placed in Soxhlet extraction thimbles. The thimbles were analyzed for radioactivity in a well-type sodium iodide crystal gamma ray counter. The results for the oil contribution with this tracer are shown in Table D-1. (The n-octacosane data are also given for comparison.)

The results are generally in good agreement for the two tracers except for one point (3 000 r/min, 468 kPa). For this one point the observed oil contribution to the particulate material was much less for the 1,2-dibromo-82-octadecane tracer than for the n-octacosane tracer (7.4 percent rather than 19.2 percent). The agreement between the data is sufficient so that the overall conclusion of the study, viz., that the oil contribution to particulate emissions is unaffected by the choice of the tracer, is valid at least for this particular engine.

Table D-1 - Comparison of Oil Contribution to Particulate Emissions as Measured With Two Radioactive Tracers

Engine Speed r/min	BMEP, kPa	Contribution of Oil to Particulate Emissions: n[14,15-^{14}C] octacosane, Mass %	1,2-dibromo-82-octadecane, Mass %
1 000	156	5.6	4.3
	312	3.5	3.2
	468	1.6	1.6
2 000	156	17.4	20.9
	312	18.4	14.3
	468	7.6	6.1
3 000	156	25.5	23.9
	312	23.4	22.4
	468	19.2	7.4

The thimbles were removed from the well-type crystal scintillation detector, placed in the Soxhlet apparatus, and extracted with benzene/ethanol. Again, as in the case of the n-octacosane, virtually all the radioactive material was soluble. This supports the observation made previously that all of the engine oil contribution to the particulate emissions is soluble in the benzene/ethanol solvent.

Another experiment also suggests that engine oil is virtually unchanged in the process of contributing to particulate emissions. Gel permeation chromatography was used to analyze the extractable organic particulate material. The retention time of the radioactive material in the extract was found equal to that of n-octacosane. This result indicates that, qualitatively at least, the tracer is undergoing little, if any, chemical reaction when it is transported from the engine sump to particulate material in the exhaust.

MEASUREMENT TECHNIQUES AND FUNDAMENTAL CONSIDERATIONS

Experimental Measurements of the Independent Effects of Dilution Ratio and Filter Temperature on Diesel Exhaust Particulate Samples**

J. Scott MacDonald and Steven L. Plee
Engine Research Dept.
General Motors Research Laboratories
Warren, MI

James B. D'Arcy and Richard M. Schreck
Biomedical Science Dept.
General Motors Research Laboratories
Warren, MI

INCREASED EMPHASIS ON PASSENGER CAR FUEL ECONOMY has made the light-duty diesel engine attractive as a passenger car powerplant. As a result, the number of vehicles powered by diesel engines is expected to increase significantly in the near future. One characteristic of the diesel engine that causes concern is its tendency to generate exhaust particulate matter. To limit the amount of particulate matter added to the atmosphere by diesel-powered passenger cars, EPA proposed that exhaust particulates be regulated for light-duty vehicles starting in the 1981 model year [1]*.

In collecting samples of diesel particulates, dilution tunnels are used to simulate the atmospheric dilution process [2,3,4,5]. The dilution process is very important since it can influence the total quantity and composition of the particulate material collected. Diesel particulates are comprised of two major components: an insoluble fraction comprised primarily of carbonaceous material and a soluble fraction made up primarily of hydrocarbon material. It is the soluble fraction that is influenced by the dilution and cooling process which occurs when the exhaust gas leaves the exhaust pipe and enters either the atmosphere or a dilution system. Obtaining an accurate measure of the soluble

*Numbers in parentheses designate References at the end of paper.

**Paper 800185 presented at the Congress and Exposition, Detroit, Michigan, 1980.

ABSTRACT

A dilution mini-tunnel is described that allows collection of diesel exhaust particulate samples with independent control of the dilution ratio and the sample filter temperature. This tunnel was used to determine the individual effects of these two tunnel operating variables on the samples collected from a single-cylinder indirect-injection test engine run at constant speed and load. Either increasing the filter temperature at fixed dilution ratio or increasing the dilution ratio at fixed filter temperature resulted in a decrease in total particulate mass. These changes in total mass were attributed to changes in the soluble fraction of the particulate sample.

portion is very important since it is this portion that tends to be mutagenic in Ames bioassay tests and has therefore been suspected by some investigators of having potentially undesirable health effects.

Concern about the validity of the samples collected from these dilution tunnels exists for several reasons. First, as a test vehicle is driven over a transient test procedure, large variations in the dilution ratio and mixture temperature occur in constant-volume-sampling (CVS) dilution systems. Since the total flow through a CVS system is constant, the quantity of dilution air decreases each time the exhaust flow rate increases. This causes large excursions in the dilution ratio. Of course, for a given dilution system flow rate, the absolute magnitudes of the dilution ratios and mixture temperatures which occur during a test depend on the engine exhaust flow rate and exhaust gas temperature.

Second, dilution systems currently used to sample diesel particulates vary widely in both size and configuration. Many of the tunnels were originally designed for measuring lead salts emitted from gasoline powerplants burning leaded fuels or for measuring sulfates from catalyst-equipped automobiles [2,3,5,6,7]. Since the flow rates of these dilution systems are not the same, the range of dilution ratios covered during an FTP test of a given vehicle would be different for each dilution system. The effects of these tunnel-to-tunnel differences and differences in engine displacement on particulate-emission results have not been clearly shown. For regulation purposes, EPA specified that the quantity of dilution air should be sufficient to keep the gas temperature at the sample probe at or below 52°C (125°F) [1]. While reference to this temperature level has been made many times in the past, its significance has not been demonstrated [8,9,10,11]. It is also unclear whether or not such a temperature specification will lead to consistent particulate samples, independent of dilution ratio and sample temperature variations.

Third and perhaps most important is the concern some investigators feel that these tunnels do not truly simulate the atmospheric dilution process [12,13]. Calculations based on simple condensation theory have shown that conditions may occur in dilution tunnels which could yield more condensation of hydrocarbon material than would occur in the atmospheric dilution process [10,12]. Long residence times at moderate dilution ratios may result in volatile particulate matter that not only increases the particulate mass unrealistically, but may also create harmful species that are not found in the atmosphere.

Hence, many questions remain concerning the validity of samples obtained with today's dilution tunnels. To answer some of these questions, a need exists to determine the effects of dilution ratio and sample temperature on the diesel particulate matter sampled.

PREVIOUS EXPERIMENTAL WORK - Some work has been done with CVS dilution tunnels to determine the effect of dilution ratio and filter temperature on the mass of particulate trapped and the percent of the sample which was extractable. Of course with these systems the filter temperature was a function of the dilution ratio for a given engine or vehicle operating condition. Results from these tests have been mixed.

Williams and Begeman [14] measured particulates from an Oldsmobile diesel and varied the total tunnel flow from 18 m^3/min (650 CFM) to 34 m^3/min (1200 CFM). This yielded dilution ratios of approximately 5.8 and 11.0:1 for the low and high fan speeds, respectively. Doubling the dilution ratio in this fashion decreased the filter temperature from 44°C (111°F) to 36°C (95°F) at a steady vehicle speed of 64 km/h (40 mi/h). Although the total particulate mass emission rates were within the measurement repeatability of the sampling system, the extractable portion of the particulate material did change. Simultaneously increasing the dilution ratio and decreasing the filter temperature in this manner increased the extractable percent from 30.5 to 32.5. Similar trends were also measured at 32, 48, 80 and 97 km/h (20, 30, 50 and 60 mi/h) cruise conditions.

Similar results were reported by Frisch et al. [13]. They studied the effects of fuel type and dilution ratio on the particulate emission from a Caterpillar 3208 direct-injection naturally aspirated diesel engine. This steady-state investigation was run on an engine dynamometer test facility. For all of the fuels tested, the particulate mass concentration increased as the dilution ratio was varied from 1:1 (no dilution) up to 25:1. These investigators attributed these increases in particulate mass to increases in the quantity of hydrocarbon material that condensed on the soot at the lower dilution temperatures in the tunnel. The change in particulate mass with dilution ratio was greatest for the least volatile fuel.

Other investigators have reported that varying the dilution ratio does not alter the particulate sample. Najt [15] measured the particulate level of an Opel 2.1-L diesel engine and varied the dilution ratio from 3.5:1 to 12:1. For a steady-state test point which simulated a 64 km/h (40 mi/h) road-load test condition, this increase in dilution ratio decreased the filter temperature from 60°C (140°F) to 32°C (90°F). Neither the mass nor the soluble portion of the particulate sample was changed by these variations in the dilution process.

Black and High [16] also reported data that showed that the particulate sample was not altered by the quantity of dilution. In their work, particulate samples from a Caterpillar 3304 diesel engine operating at a steady-state condition were obtained at dilution ratios of 10:1, 20:1 and 275:1. The extractable portions of the particulate samples were essentially the same for all of the conditions tested.

Hence, it has not been clear from previous work whether or not the quantity of dilution and the resulting sample temperature at the filter influence the particulate sample. A need still exists to determine the independent effects of dilution ratio and sample temperature on the diesel particulate matter sampled.

CURRENT EXPERIMENTAL WORK - This paper discusses a dilution parameter study which differs from previous work in that the dilution ratio and filter temperature effects were examined independently. This was accomplished with a dilution system that was developed for steady-state engine operation and which is different from conventional CVS dilution systems in several respects. This dilution system is described in detail in the Experimental section of this paper.

In the dilution experiments, the dilution ratio and filter temperature were controlled independently to identify their separate effects and to provide insight into the mechanisms which control the deposition of volatile organic matter on the carbon soot. Sufficient particulate sample was obtained at each test condition so that the effects of these parameters on the total particulate mass and on the soluble organic fraction of the particulate material could be determined.

EXPERIMENTAL

DILUTION SYSTEM

Design Requirements - Requirements for a particulate dilution system used for engine experiments in a test cell are different than those for a system used for vehicle testing. Most engine experiments are conducted at steady-state conditions, while most vehicle testing is done using transient driving schedules. Also, engine test cells are generally more confined than vehicle emission laboratories, making size and noise important design considerations.

Steady-state engine operation eliminates the need for a constant volume flow system. Engine particulate emission rates can be determined from the measured concentrations by using the dilution ratio, which can be easily calculated from an exhaust tracer gas rather than from a total dilution system flow rate. Furthermore, since the dilution ratio provides a measure of the relative quantity of exhaust gas entering the dilution system, it is not necessary to dilute 100 percent of the engine exhaust gas. Diluting only a fraction of the total exhaust is logical since only a few percent of the total exhaust flow is generally used for mass concentration measurements. In doing this, however, care must be taken to ensure that a representative sample is drawn into and out of the dilution system. Of course the big advantage of not diluting all of the engine exhaust flow is that the total volume of gas handled is reduced, which results in a significant reduction in the dilution system size.

An added benefit of using a fraction of the total exhaust flow is the flexibility which it provides. Small exhaust-sample flow rates allow a large range of dilution ratios to be considered without requiring enormous flow capacities. The resulting dilution air flows can be easily supplied, filtered, and temperature-controlled, if necessary.

Final Configuration - The dilution system used in this work evolved during an experimental program which was conducted over a period of several months. For reader convenience, the final configuration is discussed in this section of the paper, and the details of the system development are discussed in Appendix A.

A schematic of the particulate dilution system and the associated sampling equipment is shown in Fig. 1. As shown in this figure, the

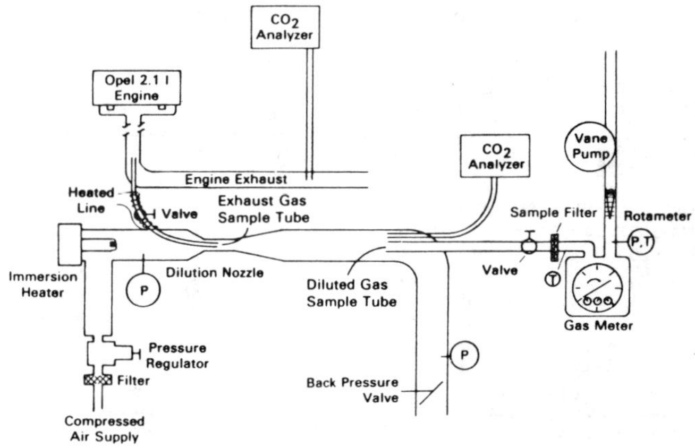

Fig. 1 - Dilution mini-tunnel schematic

heart of this dilution system is the converging-diverging dilution nozzle. During a test, prefiltered air from a compressed air source flows through the converging-diverging nozzle, creating a sub-atmospheric pressure at the nozzle throat. This low pressure draws exhaust gas into the dilution system, where it mixes with the clean dilution air. A portion of the diluted exhaust stream is withdrawn from the dilution tunnel through an absolute filter using a conventional particulate sampling network. The dilution ratio in the tunnel is determined using CO_2 as an exhaust gas tracer.

Use of this approach allowed an available compressed air source to be used to provide air for dilution without the need for a dilution-system pump or blower. Also, diluting only part of the exhaust gas resulted in a relatively small dilution system, as shown in the photograph in Fig. 2.

The dimensions of the dilution-system hardware used in the dilution-parameter study are listed in Table 1. Note that two nozzle throat diameters and several exhaust gas sample tube sizes were used to generate a large range of dilution ratios. The dilution ratios obtained with the various combinations of dilution system hardware are discussed in the

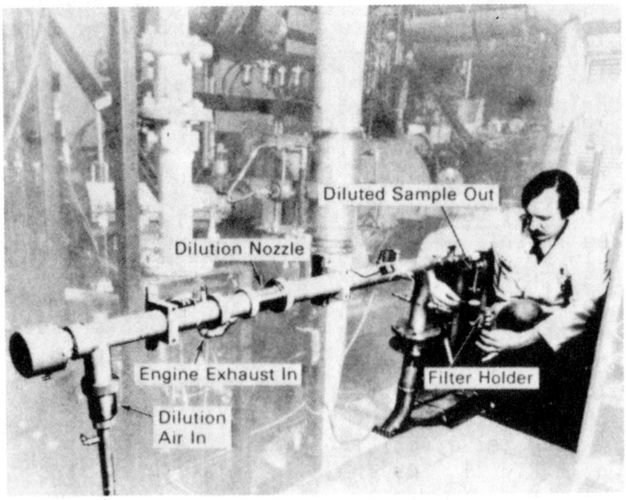

Fig. 2 - Dilution mini-tunnel

Table 1 - Dilution System Dimensions

Diameter Upstream of Dilution Nozzle	60 mm
Diameter Downstream of Dilution Nozzle	60 mm
Dilution Nozzle Throat Diameter	6.35 mm or 4.76 mm
Exhaust Gas Sample Tube O.D.	4.76 mm or 3.18 mm
Exhaust Gas Sample Tube I.D.	3.96 mm, 2.36 mm, 2.03 mm or 0.91 mm
Diluted Gas Sample Tube O.D.	15.9 mm
Diluted Gas Sample Tube I.D.	15.4 mm
Overall Dilution System Length	1.68 m

next section of this paper. A 6.25-mm O.D. tube mounted centrally in the exhaust pipe was used to transfer the exhaust gas from the engine exhaust system to the exhaust gas sample tube in the dilution system. The size of this sample probe was not varied during the dilution experiment.

With the exception of the dilution nozzles, which were made of brass, all of the dilution system components were made of stainless steel.

A thermocouple mounted behind the particulate filter was used to monitor the sample temperature. This temperature was controlled by regulating the quantity of heat supplied to the dilution air by the immersion heater shown in Fig. 1. A constant amount of heat was also supplied to the short sample tube between the engine exhaust system and the dilution tunnel to prevent condensation of water and to minimize the loss of particulate matter. The entire dilution system was insulated during this parametric study in order to minimize the influence of variations in ambient temperature on the diluted sample temperature.

Performance Characteristics - Several tests were conducted to characterize the performance of the dilution system as a function of the independent control parameters (e.g., nozzle size, exhaust gas sample-tube size, and upstream pressure) and to ensure that a representative sample would be obtained with this system. The results of these characterization tests are discussed in this section of the paper.

For a given set of nozzle and exhaust sample tube dimensions and a constant downstream pressure, the dilution ratio for this system depends only on the nozzle upstream pressure. Fig. 3 shows how the dilution

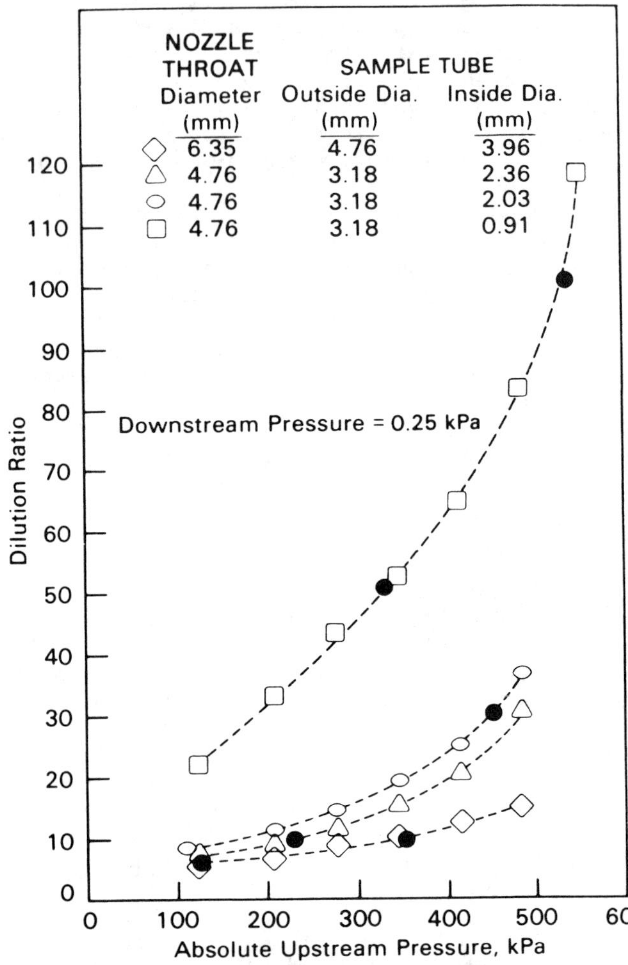

Fig. 3 - Dilution ratio variations with upstream pressure and nozzle dimensions

ratio was varied from 5:1 to 120:1 by varying the upstream pressure for several sample tube-nozzle combinations. Notice that the sensitivity of the dilution ratio to upstream pressure increased as the sample-tube inside diameter was decreased. The highest dilution ratios were obtained with the smallest sample-tube inside diameters, which were achieved by inserting a small orifice plate in the sample tube to restrict the flow of exhaust gas to the dilution tunnel. The solid symbols in Fig. 3 denote the actual nozzle operating conditions used in the dilution-parameter study described in this paper.

Of course varying the upstream pressure to

the dilution nozzle varies the mass flow rate in the system, and consequently the residence time. The mass flow rates for the two nozzle sizes described are shown in Fig. 4. Once the pressure ratio across the nozzle becomes critical, the system behaves like a conventional sonic nozzle (e.g., the mass flow rate is directly proportional to the nozzle upstream pressure and inversely proportional to the square root of the upstream temperature).

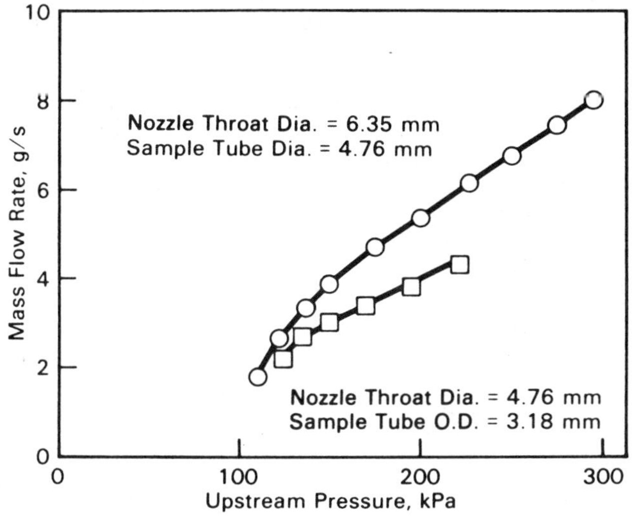

Fig. 4 - Nozzle flow vs. upstream pressure

Although the mass flow data is not required for making routine particulate measurements, it is useful for determining the required dilution air flow rates, the diluted-mixture gas velocity, residence time, and the exhaust sample flow rate into the dilution system.

To ensure that the gases in the dilution system were well mixed, the distribution of CO_2 at the diluted gas sampling point was checked by traversing the dilution system with a small sample probe. Fig. 5 shows that the measured CO_2 concentration was essentially constant across the dilution tunnel downstream cross section. These measurements show that samples taken from the center of the dilution tunnel yield representative measurements of the diluted exhaust stream.

TEST APPARATUS

Engine - A 0.5-L single-cylinder engine, derived from a production 2.1-L diesel engine, was used for this study. Fuel was supplied to the operating cylinder by a production Bosch EP/VE distributor pump and pintle nozzle. The injection timing was set at four crankshaft degrees BTDC at the engine test conditions used. The nozzles for the unused cylinders were mounted in an accumulator which delivered the unused fuel through a heat exchanger back to the supply pump. Fuel consumed by the engine was measured with a burette system. Airflow to the engine was regulated and was measured with a sonic-nozzle metering system.

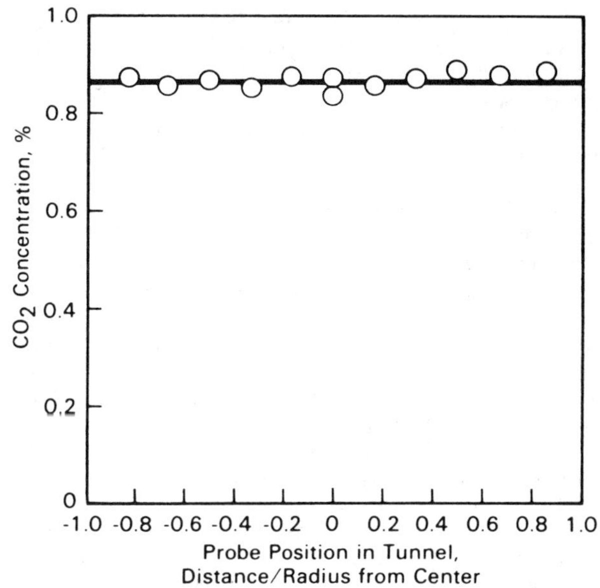

Fig. 5 - CO_2 distribution in the mini-tunnel at the sampling location

The engine combustion chamber was stock except that the glow plug was replaced with a flush-mounted plug. While this change may have altered the magnitude of the particulate output from the engine, it has no influence on the outcome of this experiment.

Instrumentation - Exhaust gas analyzers were used to measure the concentrations of CO, CO_2, O_2, NOx and HC for each data point. These data were used to calculate the exhaust air-fuel ratio as a check on the measured air and fuel flow rates. The levels of the gaseous emissions were also used as indicators of the repeatability of the combustion process from day to day.

A long-cell CO_2 analyzer was used to measure the CO_2 concentration in the dilution air and in the diluted exhaust stream for use in determining the dilution ratio.

Exhaust smoke was measured with both a Celesco in-line opacity meter and a Bosch spot meter.

The volume of gas drawn through a Dexiglass No. 225 glass fiber filter in collecting the particulate sample was measured with a Singer dry-gas meter. In addition, a rotameter was used to set and maintain the sample flow rate through the filter during the sampling process. A Sartorious microbalance was used to measure the masses of the filters before and after the particulate samples were obtained.

Fuel injection timing was obtained from oscilloscope traces of the nozzle opening and closing events. This signal was generated by a specially modified nozzle in which the nozzle seat functioned as a switch in a DC circuit.

TEST CONDITIONS - For all of the tests discussed in this paper, the engine was run at a constant load and speed of 620 kPa IMEP and 1500 r/min, respectively. This test condition required a nominal air-fuel ratio of 26:1 and produced an average hydrocarbon concentration of 13 ppm C_6 as measured with a heated FID hydrocarbon analyzer.

A nominal particulate sample flow rate of 50 L/min was used throughout the dilution-parameter test series. This flow rate provided isokinetic sampling conditions at the sample probe when the maximum dilution tunnel flow rate was used (i.e., at the 100:1 dilution ratio test point). At the other test points, sampling was not isokinetic. This deviation from isokinetic sampling did not cause concern since it has been demonstrated that isokinetic sampling conditions are not required for diesel particulates [5,15].

TEST PROCEDURE - Once the test engine reached equilibrium, preparations were made to obtain a particulate sample. A filter was placed in the filter holder to simulate sampling conditions. The dilution air pressure and temperature upstream of the dilution nozzle were then adjusted to yield the desired dilution ratio and sample temperature. After the dilution system reached equilibrium at the desired conditions, a clean pre-weighed 47-mm Dexiglass No. 225 glass fiber filter was inserted in the filter holder and a particulate sample was started.

During the sampling period, several complete sets of engine performance and emissions data were recorded in addition to the dilution tunnel CO_2 concentration and the pressure drop across the particulate filter. The pressure-drop data were used as a measure of the particulate loading on the filter. Each sample period was terminated when the pressure drop reached 250 mm (10 in.) of mercury. At a flow rate of 50 L/min, this criterion yielded samples of 2 to 4 mg. Hence, to gather the 20 mg sample desired for Soxhlet extraction, between 5 and 10 samples were required for each test condition. Exhaust opacity measurements from the Celesco smoke meter were also monitored throughout the particulate sampling period. Once the particulate sampling period was completed, smoke was also measured with a Bosch spot smoke meter. These smoke measurements were necessary to ensure that the particulate concentration in the exhaust did not vary significantly with time at a fixed operating point.

After each sampling period, the sample tube between the exhaust pipe and the dilution tunnel was back-flushed to minimize particulate build-up on the sample tube walls.

To minimize the effect of day-to-day changes in particulate output from the engine, particulate samples were obtained for the entire dilution ratio spectrum or filter temperature spectrum on the same day. This method was somewhat time consuming since it was necessary to wait for the dilution conditions to reach equilibrium each time one of these independent parameters was varied. However, this technique did prevent day-to-day variations in the exhaust particulate concentration from confounding the study.

The CO_2 concentration in the dilution air was measured several times throughout each test day even though it remained essentially constant for the entire experiment.

In addition to the main test series, tests were run to determine whether or not the dilution and sampling methods generated artificial effects on the particulate samples. These tests, which are discussed in Appendix B of this report, showed that over the range tested, variations in the dilution-tunnel residence time and filtering material did not affect the test results.

ANALYTICAL PROCEDURE

Dilution Ratio - The dilution ratio was defined as the molar ratio of the diluted exhaust to the undiluted exhaust. Hence, the particulate concentration in the sample stream multiplied by the dilution ratio yields the exhaust gas concentration. The dilution ratio was calculated from the CO_2 concentrations using the following equation, which is derived in Appendix C.

$$DR = \frac{[CO_2]_{exh,wet} - [CO_2]_{air,wet}}{[CO_2]_{dil,wet} - [CO_2]_{air,wet}}$$

where $[CO_2]_{exh,wet}$ is the CO_2 concentration in the exhaust gas on a wet basis,

$[CO_2]_{air,wet}$ is the CO_2 concentration in the dilution air on a wet basis,

and $[CO_2]_{dil,wet}$ is the CO_2 concentration in the diluted exhaust gas on a wet basis.

Of course, since the CO_2 analyzer yielded these measurements on a dry basis, the readings were adjusted for the water which was removed in the sampling process. Techniques used to convert the "dry" concentrations to "wet" concentrations have been presented previously [17].

Particulate Mass - The quantity of mass trapped on the filter medium was determined by weighing the filter before and after the sample was obtained. A 12-hour stabilization period in a humidity- and temperature-controlled room was allowed prior to each weighing.

After the particulate mass was determined, the samples were stored in a freezer at -20°C until the Soxhlet extraction could be conducted. This procedure was followed to minimize the chances of losing part of the volatile fraction of the total particulate sample.

Particulate Extraction - The amount of organic soluble material associated with the exhaust particulate was determined by a continuous extraction using a solid-liquid Soxhlet apparatus. Each sample was comprised of between 5 and 11 47-mm filters which contained a total particulate mass of approximately 20 mg. The filters were folded and placed in a pre-weighed 33-mm x 90-mm Pyrex extraction thimble with a sintered glass bottom. The 250-mL solvent flask, containing 125 mL of dichloromethane solvent and several glass beads to aid in boiling, was heated by an electrically heated water bath to avoid excessive glass temperatures. The five-minute cycle time of this set-up was maintained for four hours to yield approximately 50 cycles.*

After the four-hour extraction, the residual solvent in the extraction apparatus was siphoned into the flask and the thimble removed. The thimble containing the filters and the insoluble residue was allowed to air dry. The thimbles were re-weighed on consecutive days until a constant mass measurement was obtained.

The solvent was evaporated from the solution under a stream of nitrogen to a volume of approximately 10 mL. To increase the speed of this step, the samples were maintained at 30°C in a water bath. After the concentrated solution was transferred to a tarred vial, the samples were evaporated to dryness under nitrogen and the mass of the extract was determined. The results of the solvent extractions, reported as mass percents extractable, were determined from the weight losses of the thimbles.

RESULTS AND DISCUSSION

FILTER-TEMPERATURE EFFECTS - Fig. 6 shows the observed relationship between filter temperature and particulate mass concentration at a constant dilution ratio of 10.8:1. Note that each data point represents the average of 5 to 11 individual measurements. The error bands correspond to plus or minus one standard deviation from the average. As shown in Fig. 6, an increase in filter temperature from 35°C to 100°C reduced the average mass concentration by approximately 13 percent.

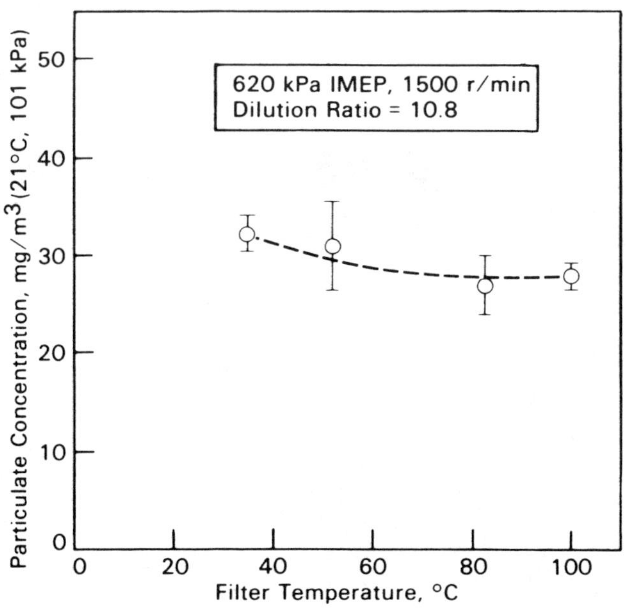

Fig. 6 - Particulate concentration vs. filter temperature

The effect of filter temperature on the soluble portion of the sample obtained using the Soxhlet extraction technique described earlier is shown in Fig. 7. Only one measure of the soluble fraction could be obtained at each filter temperature because of the large amount of mass required (>20 mg) for the analysis. As with the total mass concentration, an increase in filter temperature at constant dilution ratio reduced the soluble fraction from approximately 35 percent at 35°C to 20 percent at the highest filter temperature.

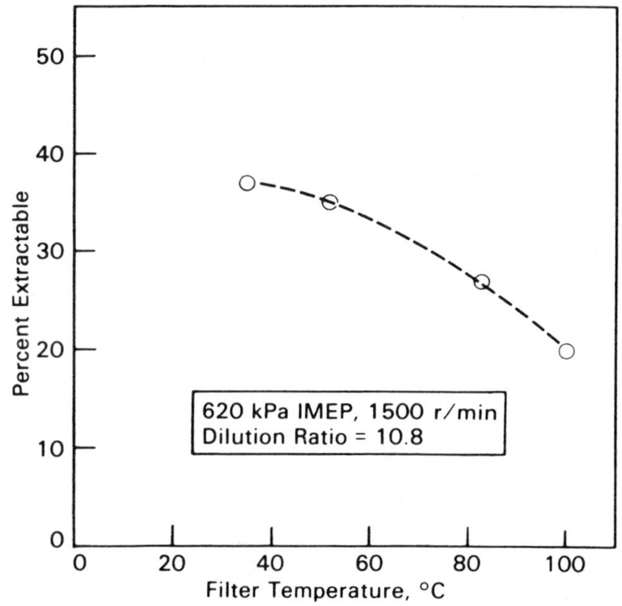

Fig. 7 - Soluble fraction vs. filter temperature

Fig. 8 shows that the entire change in the mass concentration is accounted for by the change in the soluble portion of the particulate sample. This is not surprising. Because the insoluble portion of the particulate material is primarily carbonaceous, it was not expected to change with the variations in temperature studied here.

DILUTION-RATIO EFFECTS - Fig. 9 illustrates the effect of dilution ratio on the total particulate mass concentration at a constant filter temperature of 52°C. Again, the encircled data represent the averages of 5 to 11 individual measurements and the error bands correspond to plus or minus one standard deviation from the averages. These results show a nonlinear decrease in concentration of about 9 mg/m^3, or 26 percent, between dilution ratios

*A cycle is completed each time the solvent is siphoned from the extractor.

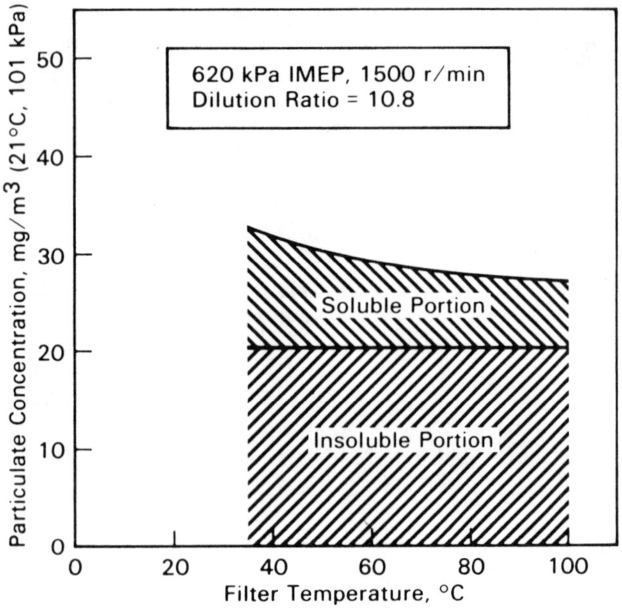

Fig. 8 - Effect of sample temperature on particulate make-up

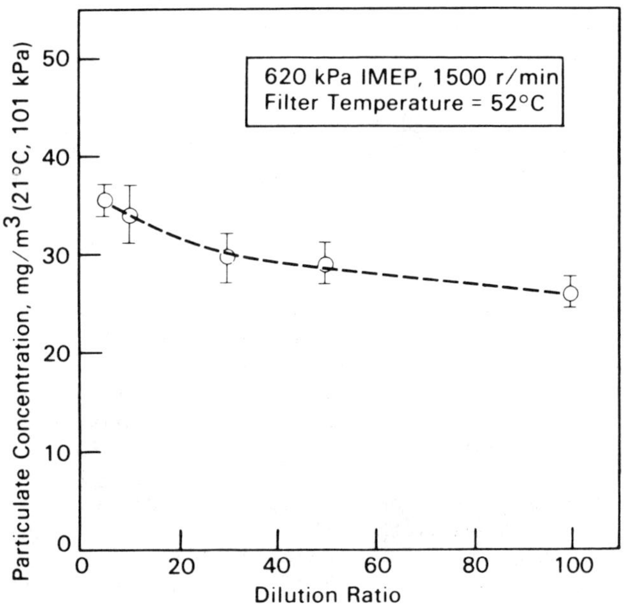

Fig. 9 - Particulate concentration vs. dilution ratio

of 5:1 and 100:1, with a substantial reduction occurring between 5:1 and 30:1. After the initial decrease, the particulate mass concentration declines at a much slower rate between 30:1 and 100:1.

Once again, these changes in the particulate concentration were attributed to changes in the soluble portion of the particulate sample. Unfortunately this could not be shown conclusively by the soluble-fraction measurements. The extractions of the particulate samples obtained at different dilution ratios provided inconsistent results. Although the trend of the soluble fraction data was the same as reported for the total mass data, there was an apparent anomaly in the magnitudes of the measurements.

DATA REPEATABILITY - The effects of varying filter temperature and dilution ratio on the particulate mass concentration were not large, especially when compared to day-to-day variations sometimes observed in diesel-engine particulate concentrations. Both the filter temperature test series and the dilution ratio test series were rerun on several different occasions to ensure that the observed trends were real. The trends did repeat and were therefore considered to be significant. As an example of this repeatability, Fig. 10 shows several sets of particulate mass data plotted against dilution ratio which were obtained on different days. The numbers next to the data

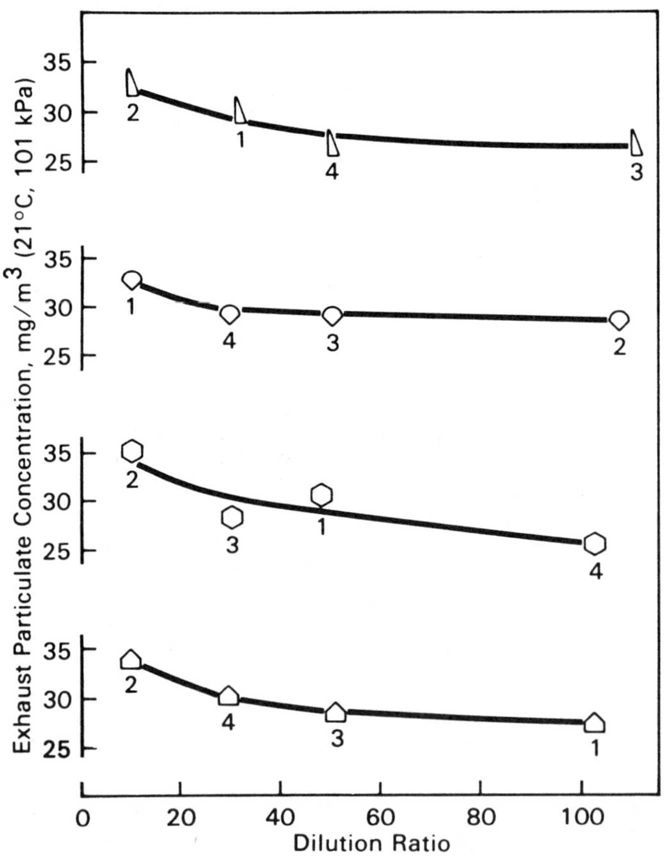

Fig. 10 - Particulate concentration vs. dilution ratio on different days

points indicate the order in which the tests were run.

COMPARISON WITH PREVIOUS WORK - The results of this study are in general agreement with those reported from earlier work with conventional dilution systems [9,14,15,16]. In other dilution systems, the sample-filter temperature decreases as the dilution ratio is increased. This dilution ratio-sample temperature interaction has been calculated for an adibatic dilution process [18], and is shown in Fig. 11 for reference. Because of this interaction,

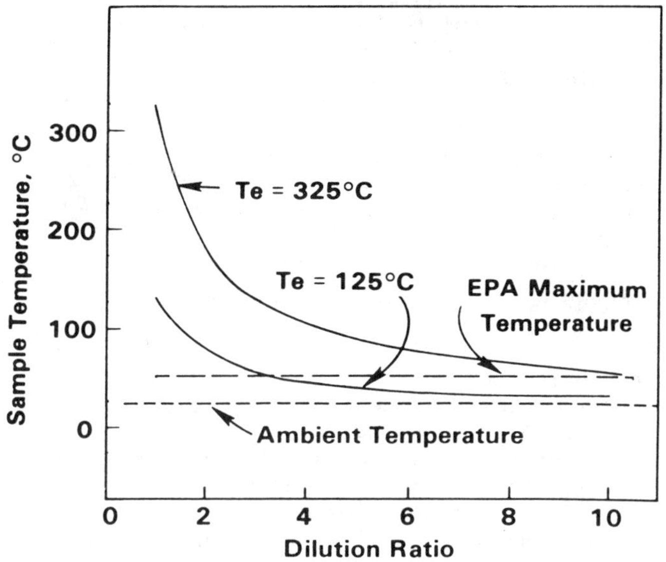

Fig. 11 - Sample temperature vs. dilution ratio for an adiabatic dilution process

the independent effects of these two parameters tend to counteract one another. This cancelling process, plus the fact that the individual effects are small, is probably the reason that results from previous dilution studies have shown little or no effect due to dilution ratio variations [13,14,15,16]. In the earlier work that did show an effect due to varying the dilution ratio, the effect was attributed to the resulting change in filter temperature. This study has shown that both dilution ratio and filter temperature influence the particulate mass, even though the filter temperature effect may dominate in some systems.

DATA LIMITATIONS - The authors feel there are two things a reader should keep in mind while interpreting this work. First, the observed variations in particulate mass with variations in dilution ratio and filter temperature may have occurred either on the sample filter or during the dilution process in the dilution tunnel. The tests reported here were not designed to separate dilution tunnel effects from sampling effects. This is also true of other investigations in which the effect of dilution ratio on particulate mass has been studied. Second, these experiments were limited to one engine type, one engine load and one engine speed. Other diesel engines or other engine operating conditions, which yield different exhaust temperatures and different hydrocarbon concentrations, would probably yield filter temperature and dilution ratio effects which were similar, but different in magnitude.

DATA UTILITY - Having the effects of dilution ratio and filter temperature separated provides information that is helpful in understanding the processes that occur when particulate- and hydrocarbon-laden diesel exhaust is diluted. This is the real utility of the information generated in this experimental work. In a companion paper [19], the data which have been presented in this report are used to identify the primary mechanism that governs the interaction between the gaseous hydrocarbons and the particulate aerosol. This theoretical analysis considers the mechanisms of condensation, diffusion, and adsorption.

SUMMARY

A small dilution system was developed for obtaining particulate mass measurements during steady-state diesel engine tests. This system requires no pumps or blowers and offers a great deal of flexibility in controlling the dilution process. Independent control of both the dilution ratio from 5:1 to 120:1 and the sample temperature from 30°C to 100°C has been demonstrated.

A series of experiments was conducted with this dilution system to establish the individual effects of filter temperature and dilution ratio on particulate samples collected by filtration from the exhaust of an automotive diesel engine. In this work, the exhaust particulate aerosol was generated with a single-cylinder test engine operating at constant load and speed.

Increasing the filter temperature from 35°C to 100°C, while maintaining a constant 10.8:1 dilution ratio, decreased the particulate mass by approximately 13 percent. Increasing the dilution ratio from 5:1 to 100:1 at a constant filter temperature of 52°C decreased the particulate mass approximately 26 percent. These changes in the total particulate mass were attributed to changes in the soluble portion of the particulate sample.

ACKNOWLEDGMENT

The authors gratefully acknowledge Stewart R. McEwen and Louis S. Jerzy, Engine Research Department, for their efforts during the experimental portion of this project. Their perseverance and patience in collecting 20 mg of particulate sample at the high dilution ratios was especially appreciated.

REFERENCES

1. Environmental Protection Agency, "Control of Air Pollution from New Motor Vehicles and New Motor Vehicle Engines; Certification and Test Procedures, Particulate Regulation for Light-Duty Diesel Vehicles," Federal Register, February 1, 1979.

2. C. T. Hare, K. J. Springer and R. L. Bradow, "Fuel and Additive Effects on Diesel Particulate Development and Demonstration of Methodology," SAE paper 760130, presented February 23-27, 1976.

3. C. R. Begeman, M. W. Jackson and G. J. Nebel, "Sulfate Emissions from Catalyst-Equipped Automobiles," SAE paper 741060, presented October 21-25, 1974.

4. D. Collins, R. D. Cuthbertson, R. W. Gawen and R. W. Wheeler, "The Use of Constant Volume Sampler and Dilution Tunnel to Compare the Total Particulates from a Range of Automotive Engines," SAE paper 750904, presented October 13-17, 1975.

5. W. H. Lipkea, J. H. Johnson and C. T. Vuk, "The Physical and Chemical Character of Diesel Particulate Emissions - Measurement Techniques and Fundamental Considerations," SAE paper 780108, SP-430, presented February 27-March 3, 1978.

6. K. Habibi, "Characterization of Particulate Matter in Vehicle Exhaust," Environmental Science and Technology, Vol. 7, No. 3, March 1973, pp. 223-234.

7. J. H. Somers, R. Lawrence, C. E. Fett, T. M. Baines and R. J. Garbe, "Sulfuric Acid Emissions from Light Duty Vehicles," SAE paper 760034, presented February 23-27, 1976.

8. K. J. Springer and R. C. Stahman, "Diesel Car Emissions - Emphasis on Particulate and Sulfate," SAE paper 770254, presented February 28-March 4, 1977.

9. K. J. Springer and R. C. Stahman, "Unregulated Emissions from Diesels Used in Trucks and Buses," SAE paper 770258, presented February 28-March 4, 1977.

10. N. J. Khatri, J. H. Johnson and D. G. Leddy, "The Characterization of the Hydrocarbon and Sulfate Fractions of Diesel Particulate Matter," SAE paper 780111, presented February 27-March 3, 1978.

11. K. J. Springer and T. M. Baines, "Emissions from Diesel Versions of Production Passenger Cars," SAE paper 770818, presented September 26-30, 1977.

12. D. B. Kittelson and D. F. Dolan, "Dynamics of Sampling and Measurement of Diesel Engine Exhaust Aerosols," Paper presented at the Conference on Carbonaceous Particles in the Atmosphere, Berkeley, California, March 20-22, 1978.

13. L. E. Frisch, J. H. Johnson and D. G. Leddy, "Effect of Fuels and Dilution Ratio on Diesel Particulate Emissions," SAE paper 790417, presented February 26-March 2, 1979.

14. R. L. Williams and C. R. Begeman, "Characterization of Exhaust Particulate Matter from Diesel Automobiles," Paper presented at 17th Annual Purdue Air Quality Conference, Purdue University, W. Lafayette, Indiana, May 7, 1979.

15. P. Najt, "Light-Duty Particulate Measurements Using a CVS Tunnel," presented at Symposium on Diesel Particulate Emissions, Measure-ment, and Characterization, Ann Arbor, Michigan, May 17-19, 1978.

16. F. Black and L. High, "Methodology for Determining Particulate and Gaseous Diesel Hydrocarbon Emissions," SAE paper 790422, presented February 26-March 2, 1979.

17. D. L. Stivender, "Development of a Fuel-Based Mass Emission Measurement Procedure," SAE paper 710604, presented June 7-11, 1971.

18. C. A. Amann, D. L. Stivender, S. L. Plee and J. S. MacDonald, "Some Rudiments of Diesel Particulate Emissions," Paper to be presented at 1980 SAE Congress, Detroit, Michigan, February 25-29, 1980.

19. S. L. Plee and J. S. MacDonald, "Some Mechanisms Affecting the Mass of Diesel Exhaust Particulate Collected Following a Dilution Process," Paper to be presented at 1980 SAE Congress, Detroit, Michigan, February 25-29, 1980.

APPENDIX A - DILUTION SYSTEM DEVELOPMENT

Several limitations and compromises were found in the development of the dilution mini-tunnel. Some of these observations are discussed in this appendix for the reader interested in sizing a similar dilution apparatus for his particular application.

The two nozzle designs shown in Fig. A-1 were originally considered for this system.

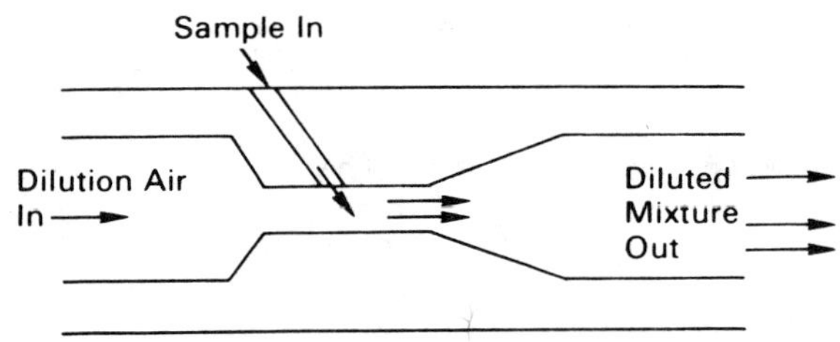

DESIGN I

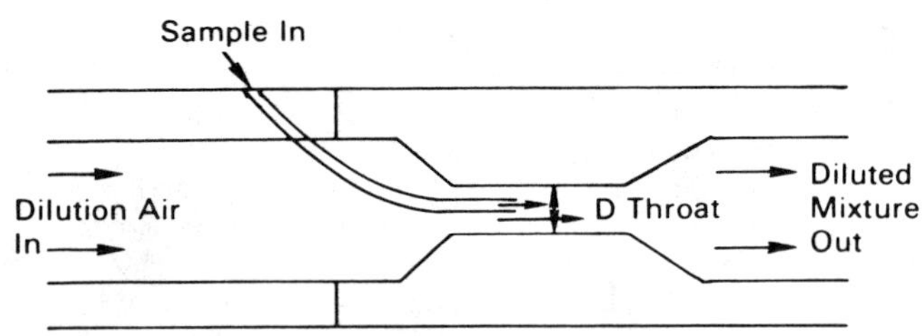

DESIGN II

Fig. A-1 - Two dilution nozzle designs

The Design I version operated well initially, but its performance degraded with use as the nozzle throat and sample outlet port became contaminated. In contrast, the Design II nozzle performance was not altered with use. Consequently, further design development was done with nozzles of the Design II configuration.

In developing a system that would provide

the desired dilution ratios and flow rates, the sample-tube outside and inside diameters and the nozzle throat and downstream diameters were found to be most important.

The nozzle throat diameter and the sample-tube outside diameter determined the flow area at the throat, and hence the vacuum generated at a given dilution air flow rate. The sample-tube inside diameter governed the sample gas flow from the exhaust system to the nozzle throat for a given throat pressure and engine backpressure. A sufficient nozzle downstream diameter was necessary to ensure that the desired downstream pressure could be maintained for all desired system flow rates. A constant, slightly positive backpressure (e.g., 25 mm of water) is desirable to provide stable nozzle performance and to avoid uncontrolled leaks into the dilution system.

In experimenting with different sample tubes and nozzles, two important observations were made. First, very small sample-tube inside diameters resulted in dilution-ratio drift which was caused by particulate contamination in the sample tube. Second, it was desirable to have the total dilution-system flow rate much larger than the filter-sample flow rate which was withdrawn downstream of the dilution nozzle. In cases where the dilute sample flow was a significant part of the total system flow, the pressure increase across the sample filter changed the system back pressure. This in turn changed the nozzle flow and the dilution ratio. While this variation in back pressure can be compensated for by opening the dilution system backpressure valve, it is an added step in the sampling process that can be avoided.

The dilution-system configuration and final dimensions stated in the Experimental section of this paper were found to provide the best compromise of the trade-offs involved.

APPENDIX B - RESIDENCE TIME AND FILTER MATERIAL EFFECTS ON PARTICULATE MASS

A series of tests was run to ensure that the observed effects were not artifacts of the experimental procedure. These tests are described in this Appendix.

One parameter of concern that varied throughout the course of the experiment was the residence time of the exhaust particulate aerosol in the mini-tunnel. As mentioned previously in the discussion of the mini-tunnel operation, the dilution ratio was changed by either varying the upstream pressure to the nozzle or by changing the dimensions of the nozzle flow passages. When the dilution ratio was varied by changing the upstream pressure, the tunnel flow rate, and consequently the residence time, was directly affected.

The first tests were run to determine what effect, if any, these variations in residence time might have on the particulate sample. At a constant dilution ratio of 10:1, the appropriate set of nozzle hardware was selected to provide residence times which bracketed most of the conditions encountered in the dilution-ratio experiment. The mass concentrations obtained at these minimum and maximum residence times agreed within the repeatability of the measurements. Therefore, it was concluded that the variation in residence times that occurred in the dilution-ratio test series did not cause the observed changes in particulate mass concentration.

A second concern was the filter material. As mentioned in a previous section, glass fiber filters were used to collect the particulate material in this work. However, during this program EPA stated that Teflon-coated glass fiber filters were superior to glass fiber filters and would be specified for vehicle certification testing. This opinion later appeared in the Federal Register [1] and in an SAE paper [16].

Although the authors were not necessarily convinced of the superiority of the Teflon-coated filters, the filter-temperature phase of the program was repeated using these filters to ensure that the results were not slanted by the glass fiber filter material. The results of this test series showed that the mass concentrations were not affected by filter type.

Thus, the observed variations in the particulate sample are believed to be due to the controlled variations in the dilution ratio and filter temperature.

APPENDIX C - DILUTION-RATIO CALCULATION

The flow of CO_2 through the dilution system is shown below:

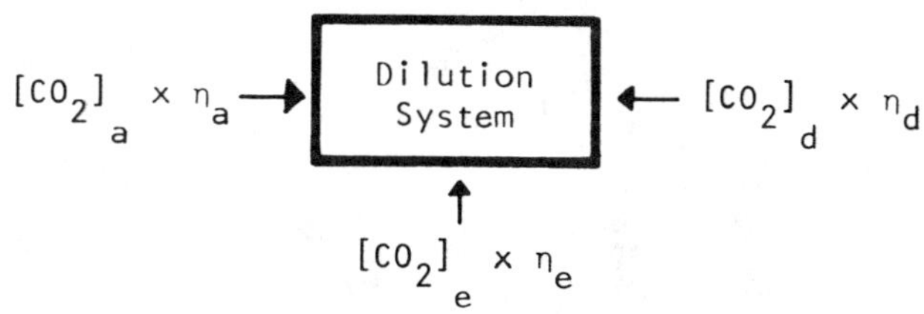

$[CO_2]$ represents the CO_2 concentration in molar percent, η represents the molar flow, and the subscripts a, e, and d represent the dilution air, the engine exhaust and the total flow in the dilution system.

Defining the dilution ratio as $\frac{\eta_d}{\eta_e}$ allows the desired engine particulate-mass concentration to be calculated from the measured dilution system concentration.

That is: $$\frac{\dot{M}_{particulates}}{\eta_e} = \frac{\dot{M}_{particulates}}{\eta_d} \times \frac{\eta_d}{\eta_e}$$

Since the CO_2 into the system must equal the CO_2 out of the system, equation (1) can be written:

$$[CO_2]_a \times \eta_a + [CO_2]_e \times \eta_e = [CO_2]_d \times \eta_d \quad (1)$$

Also, assuming no chemical reactions occur, the moles of gas into and out of the system are equal.

$$\eta_a + \eta_e = \eta_d \qquad (2)$$

Combining equations (1) and (2) and solving for the dilution ratio yields:

$$D.R. = \frac{\eta_d}{\eta_e} = \frac{[CO_2]_{e,wet} - [CO_2]_{a,wet}}{[CO_2]_{d,wet} - [CO_2]_{a,wet}} \qquad (3)$$

Since the concentrations measured with the NDIR CO_2 analyzers are on a dry basis, they must be put on a wet basis before the dilution ratio can be calculated.

The concentration of CO_2 in the engine exhaust on a wet basis can be determined from the measured "dry" concentration using the following equation.

$$[CO_2]_{e,wet} = [CO_2]_{e,dry} \times \frac{1}{1 + [H_2O]_e} \qquad (4)$$

The molar fraction of water, $[H_2O]_e$, was determined using the algorithms in Reference 16.

Since the dilution air has a dew point of ∿ -23°C (-10°F), the ice bath did not remove any moisture from the sample. Therefore, the wet and dry concentrations were equal.

$$[CO_2]_{a,wet} = [CO_2]_{a,dry} \qquad (5)$$

Water in the diluted sample originated in the engine exhaust gas. Therefore,

$$[H_2O]_{d,dry} = [H_2O]_{e,dry} \times \frac{1}{(D.R.)_{dry}} \qquad (6)$$

$$\text{where} \quad (D.R.)_{dry} = \frac{[CO_2]_{e,dry} - [CO_2]_a}{[CO_2]_{d,dry} - [CO_2]_a}$$

The wet CO_2 concentrations from the dilution system can now be determined from the measured concentration.

$$[CO_2]_{d,wet} = [CO_2]_{d,dry} \times \frac{1}{1 + [H_2O]_{d,dry}} \qquad (7)$$

Using equations (4), (6) and (7), the dry concentrations can be corrected for the water removed in the sampling system. Eq. (3) can then be solved for the dilution ratio.

Some Mechanisms Affecting the Mass of Diesel Exhaust Particulate Collected Following a Dilution Process**

Steven L. Plee and J. Scott MacDonald
Engine Research Dept.
General Motors Research Laboratories
Warren, MI

ABSTRACT

Previously measured particulate mass concentrations from a single-cylinder indirect-injection diesel, obtained under conditions of both varying dilution ratio and varying filter temperature, are examined in detail. Considering the mechanisms of condensation, adsorption, and diffusion, the observed variations in total particulate mass are attributed primarily to the adsorption and desorption of exhaust hydrocarbons on the solid particulate matter. A simple Langmuir adsorption model is used to explain qualitatively the observed effects of dilution ratio and sample temperature. Only under conditions of relatively high hydrocarbon emission is the condensation mechanism also shown to be important. The simple adsorption analysis also predicts the trends observed in CVS (Constant Volume Sampling) dilution tunnels in which filter temperature and dilution ratio change simultaneously.

MEASUREMENT OF DIESEL PARTICULATES by the Federally proposed procedure requires the use of a dilution tunnel to simulate the atmospheric dilution process [1].* Several investigators have already considered the effects of dilution parameters on diesel particulate mass measurement [2-5]. However, results to date are conflicting because sample temperature and dilution ratio vary simultaneously with the addition of ambient dilution air.

BACKGROUND - In an earlier study, MacDonald et al. [6] examined the effect of filter temperature and dilution ratio independently using a specially designed dilution tunnel. At a constant dilution ratio of 10.8:1, the total particulate mass concentration, corrected to standard conditions, decreased with increasing sample temperature (Fig. 1). The change was approximately 13 percent for a 65°K rise in filter temperature. Fig. 2 illustrates the effect of dilution ratio on the measured particulate concentration at a constant filter temperature of 325°K. Increasing the dilution ratio from 5:1 to 100:1 produced a nonlinear decrease in the particulate concentration of about 26 percent. In both cases, the observed variations in the total particulate mass were attributed to changes in the soluble portion of the sample [6].

During dilution the hydrocarbons present in the sampling system may affect both the total particulate mass concentration and its soluble fraction through condensation-evaporation, diffusion, or adsorption-desorption mechanisms. The extent that each of these mechanisms participates in the process depends on the partial pressure of the hydrocarbon vapor (dilution ratio), the local temperature, and the composition of the extractable material.

Williams and Begeman [3], Black and High [5], and Schreck et al. [7] have analyzed the soluble fraction obtained from various diesel

*References are designated by numbers in brakets and are listed at the end of this paper.

**Paper 800186 presented at the Congress and Exposition, Detroit, Michigan, 1980.

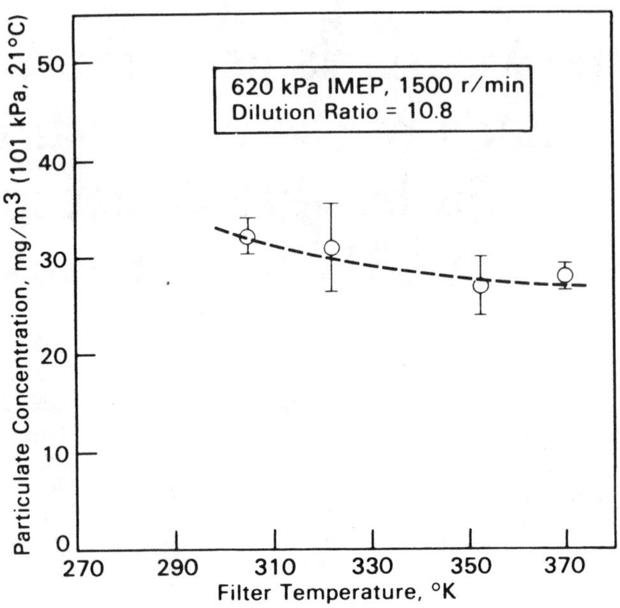

Fig. 1 - Variation in particulate concentration with filter temperature (constant dilution ratio) (from MacDonald et al. [6])

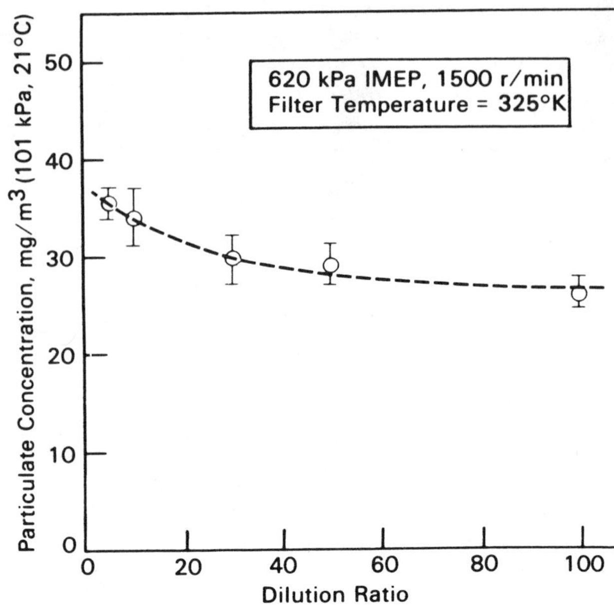

Fig. 2 - Variation in particulate concentration with dilution ratio (constant filter temperature) (from MacDonald et al. [6])

Table 1 - Analysis of Soluble Fraction

Engine	Operating Condition	Carbon No.	Distillation Temp. (°K)	Ref.
2.1-L Peugeot	2400 r/min, 9kW	C_{16} - C_{40} [1] C_{10} - C_{300} [2]	573-733[1]	[7]
5.7-L Olds	1975 FTP	C_{14} - C_{45} [1]	---	[5]
2.0-L Nissan	1975 FTP	C_{12} - C_{39} [1]	---	[5]
1.5-L Rabbit	1975 FTP	C_{14} - C_{38} [1]	---	[5]
5.7-L Olds	idle to 100 km/hr	C_{17} - C_{38} [1] C_{11} - C_{370} [2]	578-778[1]	[3]

1 by gas chromatography

2 by liquid exclusion chromatography

engines. Their results are summarized in Table 1. In general, the solubles are composed of both unburned and partially burned components of the fuel and lubricant [3,5]. As a result, the volatility of the soluble fraction is usually between that of the fuel (carbon number C_{11}-C_{40}, boiling point 431-612°K) and the lubricant (carbon number C_{16}-C_{240}, boiling point 672-811°K), as shown in Fig. 3 [3]. However, both the carbon number and boiling point distribution of the solubles can vary in this range as a function of engine load and particularly engine type [5,7].

PRESENT STUDY - Since the soluble fraction obtained by Soxhlet extraction depends to some extent on both the type of solvent and the composition of the extract, quantitative correlation of Soxhlet data is extremely difficult

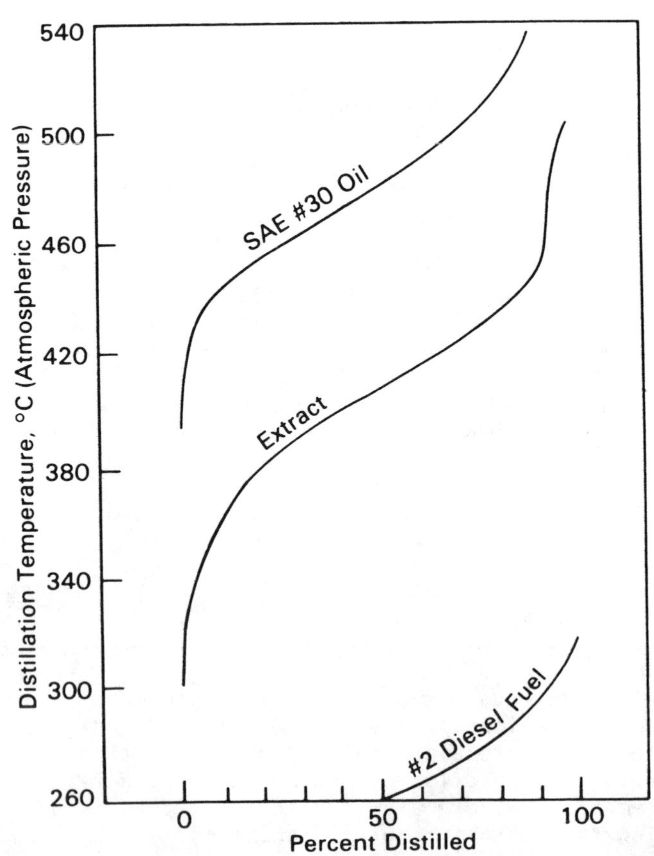

Fig. 3 - Comparison of soluble fraction to fuel and lubricant by GC distillation (from Williams and Begeman [3])

at the present time. Therefore, the following discussion considers only variations in the total particulate mass concentration. To predict changes in the particulate concentration accurately, the entire boiling point distribution should be used to simulate the real system. In this study, a simple analysis using discrete points along the distillation curve is used to approximate the actual boiling points and to assess trends as a function of dilution ratio and filter temperature. Experimental data obtained by MacDonald et al. [6] are examined using simple condensation-evaporation, diffusion, and adsorption-desorption models.

CONDENSATION - EVAPORATION

Kittelson and Dolan [8] and Khatri et al. [9] have suggested that condensation of hydrocarbon vapors may be responsible for the presence of significant organic extractables on the solid particulates. By definition, condensation occurs when the vapor pressure of the gaseous hydrocarbons (P_{HC}) exceeds its saturated vapor pressure (P_{sat}).

DERIVATION - Using the Clausius-Clapeyron equation, the saturated vapor pressure is given by

$$\frac{dP}{P} = \frac{h_{fg}\ dT}{R\ T^2} \tag{1}$$

where h_{fg} = enthalpy of vaporization (cal/mole),

T = local saturation temperature (°K),

and R = gas constant (cal/mole °K).

Assuming a constant enthalpy of vaporization, the integration of Equation (1) yields

$$P_{sat} = P_{atm} \exp\left[\frac{-h_{fg}}{R}\left(\frac{1}{T} - \frac{1}{T_b}\right)\right] \tag{2}$$

The saturation pressure is seen to be a function of the local saturation temperature (T) and the boiling point (T_b) of the hydrocarbons at atmospheric pressure (P_{atm}).

The total hydrocarbon vapor pressure in the dilution tunnel is

$$P_{HC} = \frac{X_{HC}}{DR}\ P_{atm} \tag{3}$$

where P_{atm} ≃ pressure in dilution tunnel (tunnel pressure typically 0.25 kPa above atmospheric),

X_{HC} = total hydrocarbon mole fraction before dilution,

and DR = dilution ratio (molar ratio of diluted exhaust to undiluted exhaust).

Therefore, if the ratio of P_{HC} to P_{sat} exceeds unity, condensation will occur, and

$$\frac{P_{HC}}{P_{sat}} = \frac{X_{HC}}{DR \exp\left[\frac{-h_{fg}}{R}\left(\frac{1}{T} - \frac{1}{T_b}\right)\right]} \tag{4}$$

The implicit assumption of equilibrium between condensation and evaporation appears valid since characteristic times for condensation and evaporation (computed in the manner discussed in Reference [8]) are an order of magnitude lower than estimated residence times in the dilution mini-tunnel and are more than four orders of magnitude lower than filtration times [6].

Equation (4) can also be used to estimate the fraction of hydrocarbon vapor which is present on the particles at a given dilution ratio and filter temperature. This fraction is defined as

$$F = \frac{X_t - X_{gas}}{X_t} \tag{5}$$

where $X_t = \frac{X_{HC}}{DR}$ = total hydrocarbon mole fraction after dilution,

and X_{gas} = maximum hydrocarbon mole fraction possible in the gas phase.

The maximum amount of hydrocarbon vapor present in the gas phase is given by the ratio of the saturated vapor pressure (Equation (2)) to the ambient pressure,

$$X_{gas} = \exp\left[\frac{-h_{fg}}{R}\left(\frac{1}{T} - \frac{1}{T_b}\right)\right] \tag{6}$$

Substituting into Equation 5, the fraction of hydrocarbons on the particulates (F) becomes

$$F = 1 - \frac{DR \exp\left[\frac{-h_{fg}}{R}\left(\frac{1}{T} - \frac{1}{T_b}\right)\right]}{X_{HC}} \tag{7}$$

which is related to the saturation ratio defined by Equation (4).

APPLICATION - Using Equation (4), Kittelson and Dolan [8] have considered the effect of hydrocarbon condensation and evaporation during exhaust dilution in the atmosphere. As illustrated in Fig. 4, a peak occurs in the saturation ratio (P_{HC}/P_{sat}) at moderate dilution ratio because of the competition between partial pressure and temperature effects under simulated roadway dilution. The exact magnitude of the saturation ratio is a function of the exhaust temperature, the boiling point of the hydrocarbon, and the initial hydrocarbon mole fraction. Note that for the results presented in Fig. 4, Kittelson and Dolan [8]

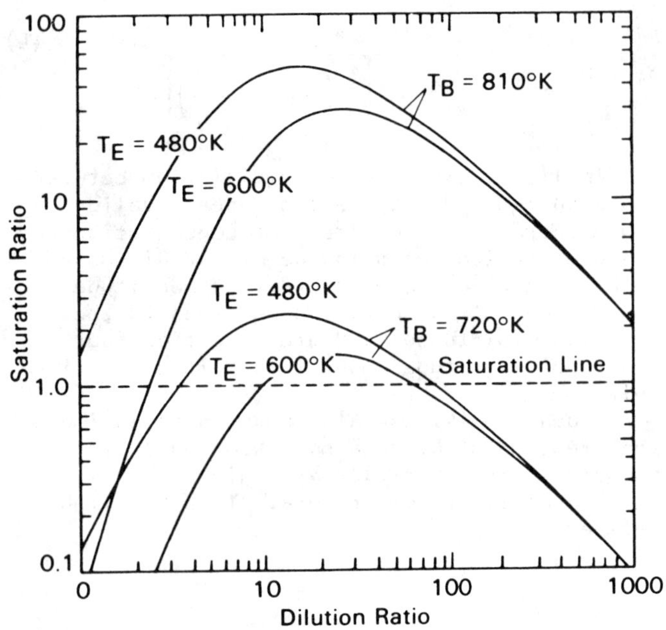

Fig. 4 - Calculated saturation ratios for condensable hydrocarbons in an exhaust stream undergoing adiabatic dilution with ambient air (from Kittelson and Dolan [8])

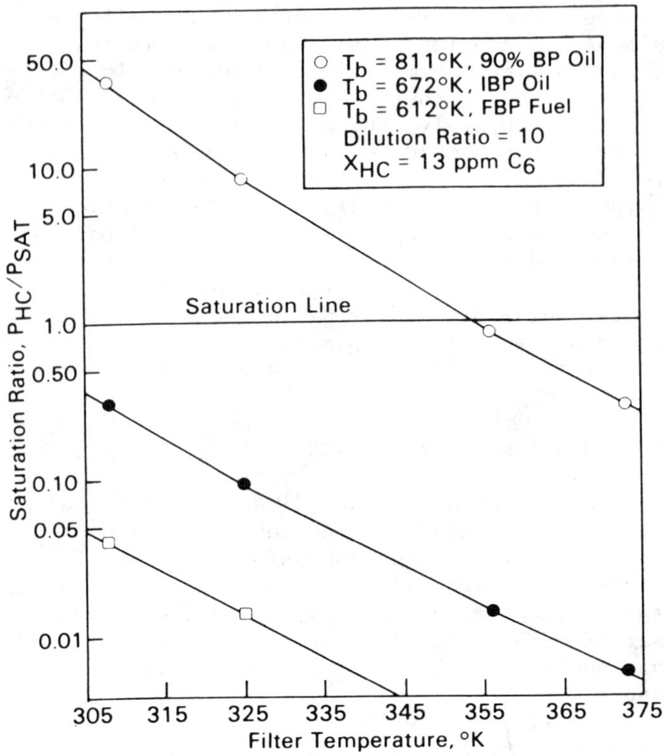

Fig. 5 - Saturation ratio versus filter temperature at various boiling points

assumed adiabatic dilution with ambient air (300°K), constant specific heats, and a total hydrocarbon mole fraction of 40 ppm. The enthalpy of vaporization was calculated from Trouton's rule [10], which relates h_{fg} to the boiling point temperature,

$$\frac{h_{fg}}{R} = 10.5\ T_b \tag{8}$$

In MacDonald et al. [6], dilution ratio and filter temperature were varied independently; thus, one would not expect the saturation ratio curves of Fig. 4. Variations in the saturation ratio with filter temperature (constant dilution ratio) and dilution ratio (constant filter temperature) are shown in Fig. 5 and 6, respectively, for the experimental conditions of MacDonald et al. [6]. For these calculations, the total hydrocarbon mole fraction is assumed to be 13 ppm C_6, based on the heated FID measurements from the test engine [6]. However, note that the total hydrocarbon mole fraction emitted by the engine may have been slightly higher than 13 ppm C_6 because of hydrocarbons trapped on the FID filter paper. Depending on the engine type and operating condition, this trapped HC fraction can be 1-3 percent of the measured FID reading [5]. Although higher hydrocarbon concentrations will raise the saturation ratio curves shown in Fig. 5 and 6 slightly, the results should not be altered significantly.

Since the saturation ratio is a function of the boiling point temperature, various points along the distillation curves of both the fuel

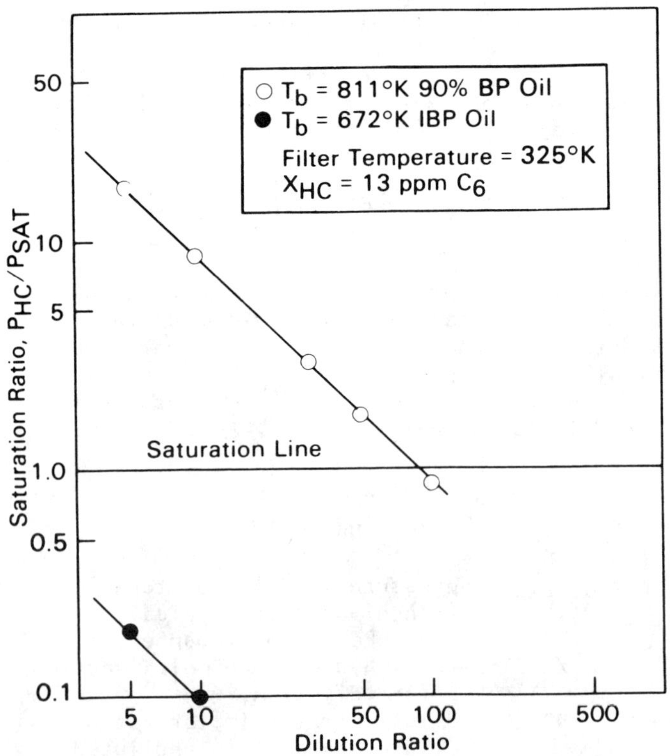

Fig. 6 - Saturation ratio versus dilution ratio at various boiling points

and the lubricant were selected to represent the final (= 90-percent boiling point of the oil = 811°K), 50-percent (= initial boiling point of the oil = 672°K), and initial boiling point (= final boiling point of the fuel = 612°K) of the extract (see Fig. 3). In reality, a distribution of hydrocarbons exists in the dilution tunnel such that only a relatively low concentration is present at the 90-percent point of the lubricant. Thus, the 811°K boiling point calculations at 13 ppm HC represent a worst case. In actual practice, the initial boiling point of the oil (= 50-percent point of extract) is expected to be more representative of the actual hydrocarbon distribution. Thus, although the exact amount of hydrocarbons present on the particles cannot be determined a priori because of the distribution of boiling points and carbon numbers, important trends can be deduced from the equations.

According to Fig. 5, no condensation is expected for hydrocarbons characterized by the 50-percent distillation point of the extract (= initial boiling point of the lubricant). However, condensation of the heavier hydrocarbons is a possibility at filter temperatures less than 355°K.

A similar graph of calculated saturation ratios is shown in Fig. 6 as a function of dilution ratio. At a constant filter temperature of 325°K, no condensation is predicted at dilution ratios beyond 85:1 because of the low hydrocarbon partial pressure. However, condensation may occur at dilution ratios less than 85:1 if, for example, the boiling point temperature is 811°K. For the more realistic case approximated by the initial boiling point of the oil (= 50-percent point of extract) no condensation is expected at any dilution ratio or filter temperature examined by MacDonald et al. [6].

Calculated values of the condensed fraction are shown in Fig. 7 and 8 for different filter temperature and dilution ratio variations, respectively. Here calculations are based on the 90-percent distillation point of the oil. According to Fig. 7, condensation-evaporation theory predicts a sharp reduction in the fraction condensed as the filter temperature is increased. However, the total mass measurements (Fig. 1) do not reflect this filter temperature effect. Furthermore, the non-linear variation of particulate mass with dilution ratio (Fig. 2) does not agree with the linear trend predicted by Equation (7) and illustrated in Fig. 8. Thus, simple condensation-evaporation theory alone does not accurately predict the experimentally observed trends with varying filter temperature and dilution ratio.

In addition, this mechanism incorrectly estimates the magnitude of the fraction condensed. According to Fig. 7 and 8, nearly all of the gaseous hydrocarbons condense at the lower filter temperatures, whereas no hydrocarbons are expected at temperatures and dilution ratios in excess of 355°K and 85:1,

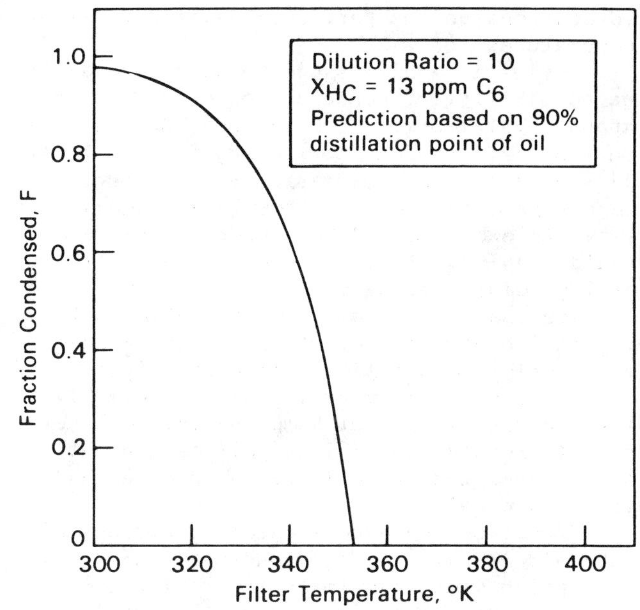

Fig. 7 - Influence of filter temperature on fraction condensed

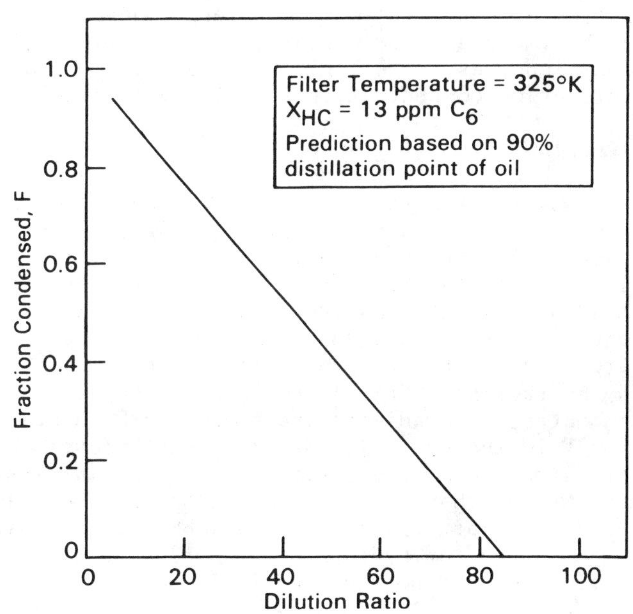

Fig. 8 - Influence of dilution ratio on fraction condensed

respectively. Since this calculation represents a worst case, the use of lower boiling points and of concentrations more typical of the hydrocarbon fraction will produce similar curves but at lower values of F. In fact, for the 50-percent boiling point of the extract (= initial boiling point of the oil), no condensation is predicted for these experimental conditions. However, Soxhlet data indicate the presence of hydrocarbons on the particulates at all filter temperatures [6].

A similar analysis can be used to examine the possibility of condensation occurring in the exhaust system prior to dilution. According to Equation (4), hydrocarbons characterized by the 811°K distillation point should not condense in the raw exhaust (DR=1), since temperatures were always in excess of 400°K. Furthermore, should liquid droplets of 0.1 μm diameter and 811°K boiling point already exist, calculations also indicate that plenty of time is available for evaporation of these droplets in the exhaust system prior to dilution. Although higher boiling point hydrocarbons could conceivably exist in the exhaust or sampling system, they are not expected to be in significant quantity to influence the overall trends observed by MacDonald et al. [6].

Thus, for the experimental conditions of MacDonald et al. [6], the condensation-evaporation mechanism fails to predict the proper trends and magnitudes of the total mass variations with either filter temperature or dilution ratio. This indicates that such other mechanisms as mass diffusion and adsorption-desorption may be primarily responsible for changes in the total particulate mass concentration during dilution. However, condensation may be important for certain other experimental conditions, as for example, if the total gaseous hydrocarbon concentration is relatively high.

DIFFUSION

Mass diffusion involves both the transport of hydrocarbons to the particle surface (interphase diffusion) and the penetration of the carbon particles by hydrocarbon vapors (intraphase diffusion). Unlike condensation, which requires partial pressures in excess of the saturated vapor pressure, diffusion can occur at any hydrocarbon partial pressure as long as concentration gradients are present. Evidence from film theory [11] indicates that interphase diffusion is extremely rapid, orders of magnitude faster than with a catalyst pellet, because the diesel particulate size is submicron. Thus, the diffusion of hydrocarbon vapors into the particle should be the rate-limiting mass transfer process.

DERIVATION - The diffusion mechanism is defined by Fick's Law

$$\frac{\dot{M}_a}{A} = \frac{-D\,\partial C_a}{\partial X} \tag{9}$$

where $\dot{M}_a$ = mass diffusion rate,

A = area,

D = diffusion coefficient,

and $\frac{\partial C_A}{\partial X}$ = concentration gradient in the x-direction.

For non-activated diffusion, the rate of mass flux is proportional to the square root of the temperature [12] and increases with the partial pressure of the diffusing species. For activated diffusion, in which chemical reactions occur on the particle, the rate increases exponentially with temperature and is given by an Arrhenius expression [13]. Under steady-state conditions, the mass diffusion rate is zero since no concentration gradient exists.

APPLICATION - Therefore, with both activated and non-activated diffusion, the quantity of hydrocarbons absorbed increases with temperature. However, MacDonald et al. [6] observed that the particulate mass decreased with increasing filter temperature. Since this is completely opposite the trend predicted by Fick's law, the diffusion of hydrocarbon vapors into the particulates is not expected to be the rate-controlling step. The fact that particle residence time in the dilution tunnel is not an important variable [6] supports this conclusion.

ADSORPTION-DESORPTION

The adsorption process can also occur at saturation ratios less than unity. In fact, physical adsorption and desorption have been observed at saturation ratios as low as 0.01 [12]. During adsorption, gas molecules adhere to solid surfaces as a result of intermolecular or van der Waals attractive forces. In comparison to diffusion, the rate of adsorption and desorption is quite rapid, even at low temperatures [11]. In addition, the physical adsorption process is highly reversible, since the adsorbed gas is desorbed without a change in composition. Physical adsorption is also characterized by a relatively low activation energy [11].

Chemical interaction between the adsorbate and the adsorbing medium is also possible. This process is known as chemisorption. Unlike physical adsorption, chemisorption rates have a higher activation energy and are considered important at elevated temperatures [11]. However, the energy required for the chemisorbed molecules to react with other species is still less than the energy required when the two species react in the gas phase [13]. Because the original adsorbate changes during chemical reaction, the chemisorption process is usually irreversible [11].

Carbon is a particularly difficult substance to characterize because of its porosity [12]. Since little is presently known about the adsorption properties of volatile hydrocarbons on diesel soot, any attempt to separate physical and chemical adsorption is beyond the scope of this study. The primary objective here is to determine how condensation-evaporation, diffusion, or adsorption-desorption influence the total particulate mass with variations in filter temperature and dilution ratio. Therefore, only a simple Langmuir adsorption mechanism is considered in an effort to identify appropriate trends. Note that for extremely porous materials on the order of 1500 m^2/g, Dubinin

[14] recommends the use of capillary condensation theory for the calculation of the particle-bound volatiles. However, since soot particles have specific surface areas on the order of 100 m^2/g [15,16], this capillary condensation approach also represents an approximation.

DERIVATION - The Langmuir adsorption concept is fundamental to the treatment of both physical and chemical adsorption [11,17] and is based on the following assumptions:

1. The solid surface contains a fixed number of adsorption sites, each holding one adsorbed molecule.
2. The heat of adsorption is constant and does not depend on site location or the fraction of sites covered (not entirely valid for solid carbon but probably a good approximation over the narrow temperature range of this study).
3. There is no interaction between molecules on different sites.
4. Equilibrium exists between the adsorption and desorption process.

Using a kinetic approach, Langmuir described the adsorption rate as

$$r_{ads} = k_1 P (1 - \Theta) \quad (10)$$

where k_1 = rate coefficient for adsorption,
P = partial pressure of the adsorbing gas,
and $(1 - \Theta)$ = fraction of the surface which is unoccupied.

The rate of desorption is assumed to be equal to the product of the fraction of sites occupied (Θ) and the rate coefficient for desorption (k_2),

$$r_{des} = k_2 \Theta \quad (11)$$

Note that both rate coefficients are of Arrhenius form,

$$k_1 = Z_1 \exp(-E_a/RT) \quad (12)$$

and

$$k_2 = Z_2 \exp(-E_d/RT) \quad (13)$$

where Z_1, Z_2 = pre-exponential factors,
E_a, E_d = apparent activation energies (constant) for adsorption and desorption, respectively,
R = gas constant,
and T = local temperature.

At equilibrium, the rates of adsorption and desorption are equal

$$k_1 P (1 - \Theta) = k_2 \Theta \quad (14)$$

Therefore, solving for theta yields

$$\Theta = \frac{1}{1 + \frac{k_2}{k_1 P}} \quad (15)$$

Since the hydrocarbon partial pressure is

$$P \sim \frac{X_{HC} P_{atm}}{DR} \quad (16)$$

theta becomes

$$\Theta = \frac{1}{1 + \frac{(K)(DR)}{X_{HC}} \exp(-E/RT)} \quad (17)$$

where K = constant which includes Z_1, Z_2, and P_{atm}
and $E = E_d - E_a$.

The derivation of Equation (17) is similar to the approach used by Natusch and Tomkins [18] to examine the adsorption of polynuclear aromatic hydrocarbons onto fly ash. In their analysis, they indicated that the activation energy for desorption is larger than for adsorption [18].

APPLICATION - Fig. 9 and 10 show the influence of filter temperature and dilution ratio, respectively, on the fraction of sites occupied (Θ) for the range of dilution parameters investigated by MacDonald et al. [6]. Unfortunately, not enough information is available at the present time to determine a priori the constant (K) and the apparent activation energy (E) in Equation (17). Therefore, values of C (equal to K/X_{HC}) and E/R have been chosen arbitrarily to illustrate the effects of partial pressure and temperature on theta. Even without knowledge of the kinetic constants, certain trends are evident in Fig. 9 and 10.

Assuming that changes in Θ are related to variations in the total mass, the adsorption curves of Fig. 9 show the same influence of filter temperature as the total particulate concentration measurements (Fig. 1). In agreement with the experimental data [6], the decrease in total particulates is nonlinear and does not drop off sharply with filter temperature. In addition, adsorption theory predicts a finite value of Θ at high temperatures, unlike simple condensation theory. This is in general agreement with the Soxhlet results, which show that substantial amounts of solubles are present on the particles between filter temperatures of 308-373°K [6].

Calculated values of theta also show the same trends with dilution ratio (Fig. 10) as the total particulate concentration measurements (Fig. 2). Simple adsorption theory predicts a nonlinear decrease in Θ with decreasing partial pressure (increasing dilution ratio), instead of the linear dependence predicted by the condensation model. In addition, adsorption theory indicates that a finite amount of hydrocarbons should be present on the solid particulates at high dilution ratios.

Although both physical and chemical adsorption have been considered in the simple Langmuir model, the weak temperature dependence shown in Fig. 1 indicates that physical adsorption is probably the dominant mechanism. Based on the

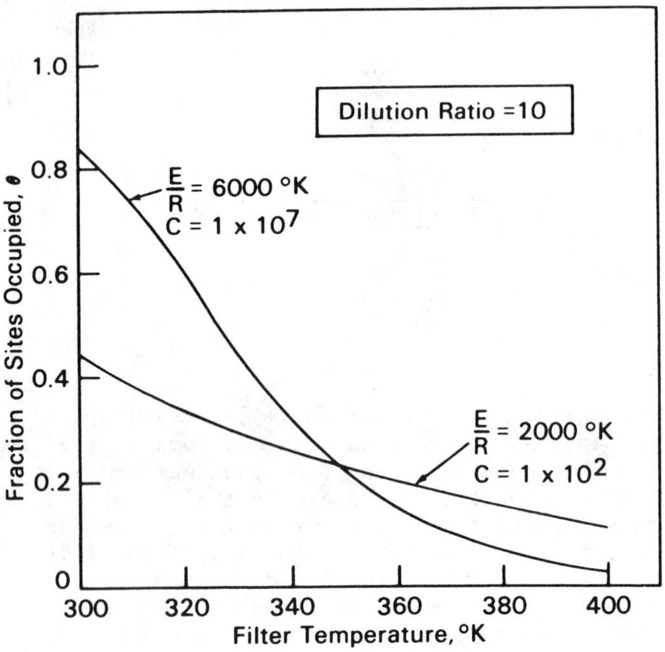

Fig. 9 - Influence of filter temperature on fraction of sites occupied

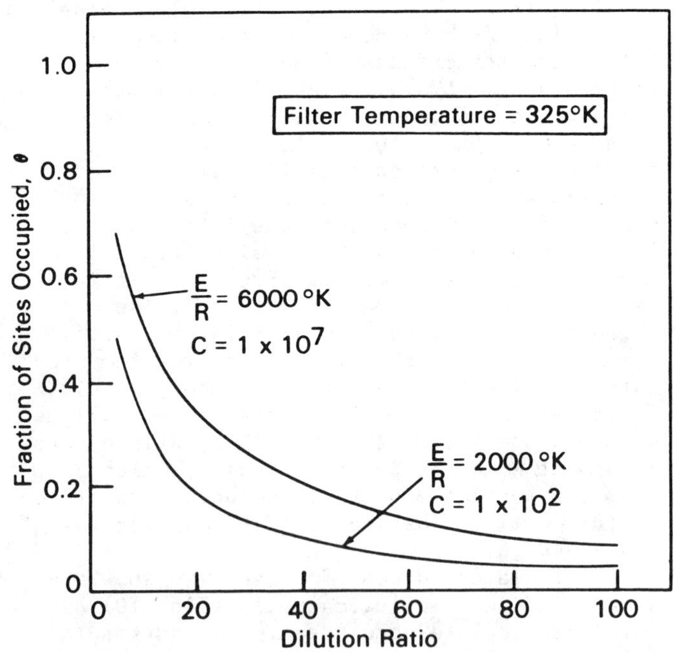

Fig. 10 - Influence of dilution ratio on fraction of sites occupied

experimental data, the apparent activation energy in Equation (17) is approximately 7000 cal/mole. The time for adsorption, defined as the average length of time that elapses before an adsorbed species is desorbed, varies exponentially with temperature and activation energy [19]. According to Thomas and Thomas [19], the adsorption time is orders of magnitude less than the residence time in the dilution tunnel when evaluated at 7000 cal/mole and 300-400°K. Thus, the assumption of equilibrium between adsorption and desorption seems valid for the conditions of this study.

In conventional CVS dilution tunnels, the sample temperature and dilution ratio do not vary independently. Instead, the sample temperature decreases with increasing dilution, as discussed in References [6,20]. Depending on the change in sample temperature with dilution and the apparent activation energy in the adsorption model, no appreciable change in particulate concentration would be predicted over a relatively large range of dilution ratios. The simple adsorption model represents a tool for evaluating these effects under conditions when condensation is unlikely.

Application of the simple adsorption analysis to particulate data obtained from a CVS dilution tunnel is illustrated in Fig. 11, using the experimental results of Najt [4]. In this analysis, it is assumed that changes in particulate mass are directly proportional to Θ (Eq. 17). The constants in the adsorption model were obtained by fitting the correlation to the experimental data of MacDonald et al. [6], in which filter temperature and dilution ratio were varied independently. Values of E equal to 7000

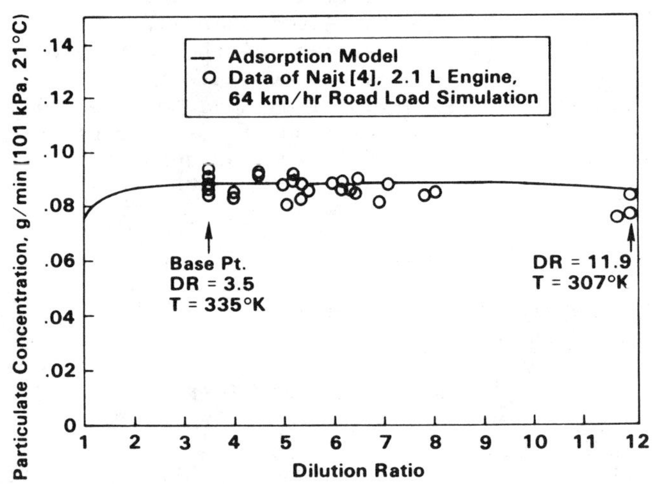

Fig. 11 - Comparison of adsorption model with experimental data of Najt [4]

cal/mole and K equal to 15 400 were found to correlate changes in the particulate mass to within one standard deviation of the experimental average reported in Reference [6]. Since Najt [4] used the same engine as MacDonald et al. [6], the adsorption parameters should also be applicable to the CVS dilution tunnel data. As shown in Fig. 11, calculated changes in the particulate concentration from a prescribed base point (DR=3.5, T=335°K) agree quite well with the experimental data [4]. As predicted by the simple analysis, no significant change occurs in the total particulate mass concentration between dilution ratios of 3:1 to 12:1 for this particular engine and operating condition because of the competition between dilution ratio and temperature effects. The shape of the adsorption curve at various exhaust gas temperatures, hydrocarbon concentrations, and dilution ratios is discussed in more detail in Reference [20].

Black and High [5] also observed that the total particulate mass concentration was unaffected by dilution ratio in the range of 8.7:1 to 18.3:1, while Frisch et al. [2] and Williams and Begeman [3] found only a slight increase in mass with increasing dilution. These results are also in general agreement with the adsorption mechanism. However, since different engines were used in these studies, the adsorption parameters used in Fig. 11 are not expected to be entirely valid. Thus, results can only be compared qualitatively with the model. In addition, hydrocarbon concentrations may be high enough for condensation also to be important in some of these other studies.

SUMMARY

Experimentally observed variations in the measured particulate mass concentration with filter temperature and dilution ratio have been interpreted using models of adsorption-desorption, evaporation-condensation, and diffusion mechanisms. For relatively low gaseous hydrocarbon levels, the analysis showed the following:

1. A simple Langmuir adsorption model explains qualitatively the observed temperature and dilution ratio dependence.
2. Condensation-evaporation and diffusion mechanisms fail to predict the observed trends with the dilution parameters.
3. Therefore, the adsorption-desorption of hydrocarbon vapors on solid particulates appears to be the dominant mechanism. However, both condensation and adsorption may be important under conditions of relatively high hydrocarbon concentrations.
4. The simple adsorption analysis has also been extended to CVS dilution tunnels in which both filter temperature and dilution ratio vary simultaneously. Calculations agree with the experimental data of Najt [4] and show that the effects of dilution ratio and filter temperature can cancel over the range of dilution ratios typical of CVS tunnels.

ACKNOWLEDGMENTS

The authors appreciate the valuable discussions with R. L. Williams (Environmental Science Department, GM Research) and P. M. Najt (GM Engineering Staff) during the course of this investigation.

REFERENCES

1. Environmental Protection Agency, "Control of Air Pollution from New Motor Vehicles and New Motor Vehicle Engines; Certification and Test Procedures, Particulate Regulation for Light-Duty Diesel Vehicles," _Federal Register_, February 1, 1979.

2. L. E. Frisch, J. H. Johnson, and D. G. Leddy, "Effect of Fuels and Dilution Ratio on Diesel Particulate Emissions," SAE paper 790417, presented February 26-March 2, 1979.

3. R. L. Williams and C. R. Begeman, "Characterization of Exhaust Particulate Matter from Diesel Automobiles," Paper presented at 17th Annual Purdue Air Quality Conference, Purdue University, W. Lafayette, Indiana, May 7, 1979.

4. P. Najt, "Light-Duty Particulate Measurements Using a CVS Tunnel," presented at Symposium on Diesel Particulate Emissions, Measurement, and Characterization, Ann Arbor, Michigan, May 17-19, 1978.

5. F. Black and L. High, "Methodology for Determining Particulate and Gaseous Diesel Hydrocarbon Emissions," SAE paper 790422, presented February 26-March 2, 1979.

6. J. S. MacDonald, S. L. Plee, J. B. D'Arcy, and R. M. Schreck, "Experimental Measurements of the Independent Effects of Dilution Ratio and Filter Temperature on Diesel Exhaust Particulate Samples," Paper to be presented at 1980 SAE Congress, Detroit, Michigan, February 25-29, 1980.

7. R. M. Schreck, J. J. McGrath, S. J. Swarin, W. E. Hering, P. J. Groblicki, and J. S. MacDonald, "Characterization of Diesel Exhaust Particulate for Mutagenic Testing," paper No. 78-33.5, presented at the Air Pollution Control Association, Houston, Texas, June 25, 1978.

8. D. B. Kittelson and D. F. Dolan, "Dynamics of Sampling and Measurement of Diesel Engine Exhaust Aerosols," presented at the Conference on Carbonaceous Particles in the Atmosphere, Berkeley, California, March 20-22, 1978.

9. N. J. Khatri, J. H. Johnson, and D. G. Leddy, "The Characterization of the Hydrocarbon and Sulfate Fractions of Diesel Particulate Matter," SAE paper 780111, presented February 27-March 2, 1978.

10. G. W. Castellan, _Physical Chemistry_, 2nd edition, Addison-Wesley, Reading, Massachusetts, 1971.

11. J. J. Carberry, _Chemical and Catalytic Reaction Engineering_, McGraw-Hill, New York, 1976.

12. D. O. Hayward and B. M. Trapnell, _Chemisorption_, 2nd edition, Butterworths, London, 1964.

13. P. L. Walker, Jr., L. G. Austin, and S. P. Nandi, "Activated Diffusion of Gases in Molecular-Sieve Materials," _Chemistry and Physics of Carbon_, Vol. 2, edited by P. L. Walker, Jr., Marcel-Dekker, New York, 1966, pp. 257-371.

14. M. M. Dubinin, "Porous Structure and Adsorption Properties of Active Carbons," _Chemistry and Physics of Carbon_, Vol. 2, edited

by P. L. Walker, Jr., Marcel-Dekker, New York, 1966, pp. 51-120.

15. V. G. Knorre, T. D. Snegireva, T. V. Tekvnova, and A. V. Chulkov, "A Study of the Thermal Decomposition of Acetylene and the Properties of the Soot Formed under the Conditions of a Constant Volume Bomb," Combustion Explosion and Shockwaves, Vol. 8, No. 4, 1972, pp. 437-439.

16. K. B. Lee, M. W. Thring, and J. M. Beér, "On the Rate of Combustion of Soot in a Laminar Soot Flame," Combustion Flame, Vol. 6, 1962, pp. 137-145.

17. W. J. Moore, Physical Chemistry, 4th edition, Prentice-Hall Inc., New Jersey, 1972.

18. D. F. S. Natusch, and B. A. Tomkins, "Theoretical Consideration of the Adsorption of Polynuclear Aromatic Hydrocarbon Vapor onto Fly Ash in a Coal-Fired Power Plant," Carcinogenesis, Vol. 3, Polynuclear Aromatic Hydrocarbons, edited by P. W. Jones and R. I. Freudenthal, Raven Press, New York, 1978, pp. 145-153.

19. J. M. Thomas and W. J. Thomas, Introduction to the Principles of Heterogeneous Catalysis, Academic Press, New York, 1967.

20. C. A. Amann, D. L. Stivender, S. L. Plee, and J. S. MacDonald, "Some Rudiments of Diesel Particulate Emissions," Paper to be presented at 1980 SAE Congress, Detroit, Michigan, February 25-29, 1980.

Analysis of the Physical Characteristics of Diesel Particulate Matter Using Transmission Electron Microscope Techniques**

Kenneth Carpenter and John H. Johnson
Michigan Technological Univ.

ALTHOUGH THE DIESEL ENGINE has been used extensively in heavy-duty truck applications, the traditional power plant for passenger cars and most light and medium-duty truck applications has been the gasoline burning, spark-ignition engine. But now, with increasing fuel costs, sometimes doubtful supplies, and a greater awareness of the potential effects of environmental pollution, there is a growing interest in using the diesel engine in a greater variety of motor vehicle applications.

Despite inherent operating advantages of the diesel engine, the exhaust is subject to two major criticisms - visible smoke and odor. In this regard, the United States Environmental Protection Agency has been directed by the Clean Air Act of August, 1977, to establish particulate matter standards for diesel powered motor vehicles for the 1981 model year. In order to set viable standards, it will be necessary to gather information relevent to the physical and chemical nature of the exhaust particulate.

This paper has been drawn from the thesis of one of the authors (1)*. It is an extension of earlier research work at Michigan Technological University (MTU) involved with the physi-

**Paper 790815 presented at the Off-Highway Vehicle Meeting and Exposition, Milwaukee, Wisconsin, September, 1979.

*Numbers in parentheses designate References at end of paper.

ABSTRACT

An Andersen Impactor was used to collect particulate samples in both the undiluted and diluted exhaust from a Caterpillar 3150 diesel engine operated on the EPA 13-mode cycle. A total of 24 samples were examined using the transmission electron microscope and approximately 300 photomicrographs were taken. The microscope analysis and photomicrographs revealed details concerning the physical characteristics of the particulate and permitted a direct visual comparison of the samples collected.

The photomicrographs were used to obtain diameter measurements of the basic individual spherical particles that comprise the much larger aggregates/agglomerates. Nearly 11,000 basic particles were measured and the observed range of diameters was 70-1200 Å. The mean particle diameters in the undiluted and diluted exhaust samples were 479 Å and 436 Å respectively. These size measurements were grouped, by operating mode, to provide size distributions for the basic spherical particles.

At the lower temperatures of the diluted exhaust, evidence of hydrocarbon condensation is illustrated in many of the photomicrographs. In some instances, the condensed coating is so heavy as to completely obscure all detail of the aggregate/agglomerate on which it is adsorbed and the differences between undiluted and diluted exhaust samples are quickly discerned. Several of the samples showed what appeared to be "droplets" or "pools of liquid".

cal (2,3) and chemical characterization (4,5, 6,7) of diesel particulate matter in both the undiluted (2,6) and diluted exhaust (3,4,5,6,7). It makes use of the undiluted measurement techniques and understanding of the physical characteristics of diesel particles developed in Reference 2. This paper is an extension of the data in Reference 2 in the following areas: 1) diluted exhaust conditions, 2) more extensive engine and impactor plate conditions, and 3) improved particle sizing techniques.

Early in the MTU diesel particulate research program, it was recognized that undiluted diesel particulates above 500°F contained few adsorbed hydrocarbons as compared to those below 500°F (2). Further work showed increased masses of particulates in the diluted exhaust as compared to the undiluted exhaust when operating at light-loads (under 50%) (4). It was also theoretically recognized that the dilution process could cause condensation of the heavy molecular weight organic compounds (4). This paper is directed at showing actual Transmission Electron Microscope (TEM) photomicrographs of both undiluted and diluted exhaust particulate matter samples using the techniques outlined in References 2 and 5. The results show that: 1) liquid hydrocarbons are adsorbed on the solid particles in the dilute exhaust, 2) that liquid hydrocarbon particles are present in the diluted exhaust, and 3) the basic spherical particle sizes are similar to those found in Reference 2 although the mean particle size appears to be larger. This latter finding is felt to be a result of improved sizing techniques used in the present study. Such information is essential in the modeling and control analysis of diesel exhaust particulate matter.

The overall paper is divided into two basic sections. The first section describes the EXPERIMENTAL TEST SET-UP AND ANALYSIS PROCEDURES and includes a brief description of the engine, dilution tunnel, test fuel, the sampling/analysis methods and the engine test conditions. The second section presents the EXPERIMENTAL RESULTS which includes the particle size distributions for individual modes and photomicrographs of typical and atypical particles found in both the undiluted and diluted exhaust.

EXPERIMENTAL TEST SET-UP AND ANALYSIS PROCEDURES

In order to generate diluted diesel exhaust, a dilution tunnel was constructed in the MTU Diesel Dynamometer Test Cell. The stainless steel tunnel, 12 inches in diameter and some 20 feet long, utilizes a constant volume sampler (Roots blower) to draw the air/exhaust mixture through the tunnel at 540 actual cubic feet per minute (1,4,5). The method of exhaust injection to the tunnel is counter-current to the tunnel flow, using a four port injection nozzle. Air/exhaust mixing is natural and no mechanical means are used to increase turbulence. Dilution ratios up to 50:1 are possible with the current equipment and downstream sampling may be performed with a variety of instrumentation (5). Continuous monitoring of tunnel volumetric flow rate, mixture temperature, and NO/NO_x concentrations afford evaluation of tunnel performance. The dilution tunnel and related equipment are shown schematically in Figure 1. Note the sampling points for both diluted and undiluted exhaust.

The collection of particulate matter samples was done using an Andersen Mark III Inertial Impactor coupled to an RAC Stak-sampler. These two pieces of equipment and their general application are shown in Figures 2 and 3. The particulate samples for TEM analysis were collected directly on standard 3.05 mm carbon coated TEM specimen grids which had been cemented to the collection plates of the Andersen Impactor. Using this method, samples could be placed directly into the TEM without further processing. Such analysis permitted the direct visual observation of the collected sample, as well as the making of photomicrographs. The photomicrographs, were used in conjunction with a low power optical microscope to provide basic particle size information for the collected samples as well as insight into the internal structures of the particulates.

In the present study, particulate samples were collected only on impactor plates 4 and 7. This was done so the sampling procedure would be consistent with previous work (2) and, at the same time, reasonably limit the number of samples to be analyzed. Although the combined quantity of particulate collected on these two plates represents only 5-10% of the total particulate collected, including back-up filter, the findings of previous investigations (2) justify the assumption that the basic spherical particles analyzed would be representative of all basic particles collected on all stages of the Andersen Impactor.

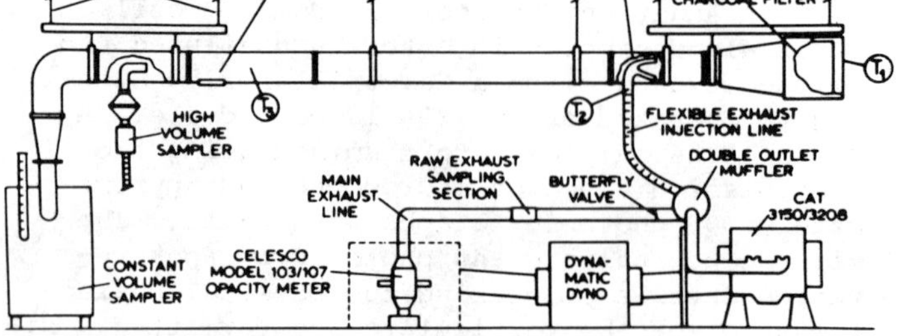

Fig. 1 - Schematic of the dilution tunnel system used at Michigan Technological University. T_1 = inlet air temperature, T_2 = inlet exhaust temperature, T_3 = mixture temperature

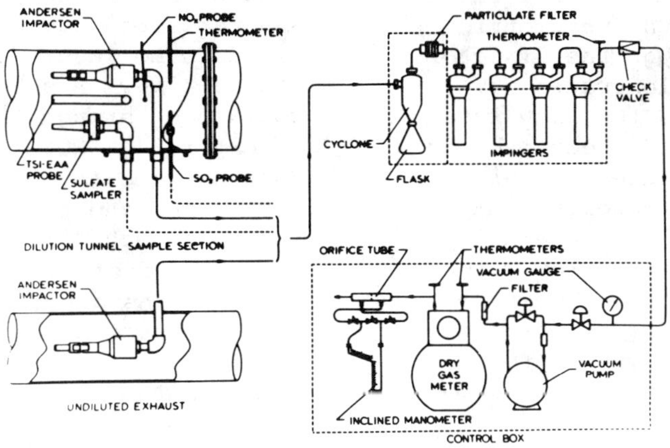

Fig. 2 - General sample collection system and RAC Stak-sampler. Vuk et al (2) and Lipkea et al (5)

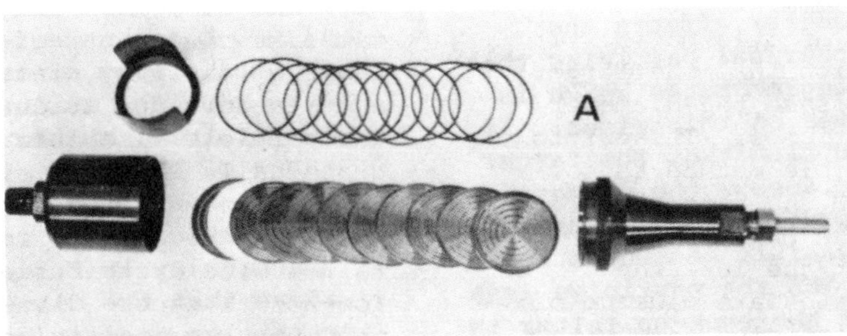

(A) - Andersen Mark III Sampler

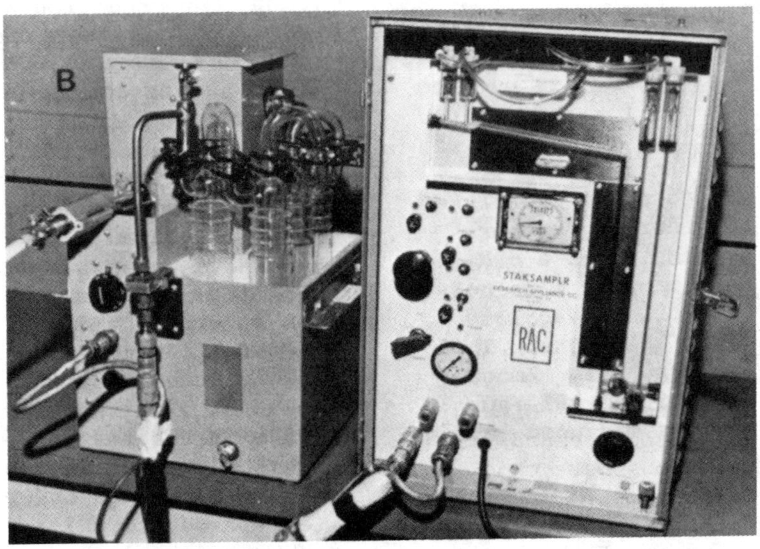

(B) - The RAC Stak-sampler

Fig. 3 - Particulate matter sampling apparatus

The size of the aggregate/agglomerate expected on any given plate is dependent on the cut-off diameter of the plate. This cut-off diameter is a function of the volumetric flow rate and temperature through the impactor. Since measurement of these agglomerates was not an objective of this study, the flow rate during successive samplings was selected based on Figure 15 of Reference 2. Based on this developed undiluted sampling procedure, impactor flow rates were generally maintained around 0.80-0.85 ACFM. Using this flow rate in both the undiluted and diluted exhaust results in the variation in cut-off diameter for Mode 6 - impactor plate 4 from 2.64 μm to 2.02 μm respectively and for Mode 6 - impactor plate 7 from 0.55 μm to .41 μm respectively. This difference progressively decreases for the lighter load modes and becomes practically zero for the idle modes. We do not feel that this variation in cut-off diameter affects the analysis of the basic spherical particles that make up the aggregates/agglomerates which is one of the main objectives of this paper. It is our opinion and assumption that the larger particles can be used to assess the diameters of the basic spherical particles. In addition, we are not aware of a feasible method to collect and analyze in the TEM the particles less than 0.4 μm collected on the back-up filter in the impactor. We feel that the less than 0.4 μm aggregates/agglomerates are made up of the same basic spherical particles only with fewer numbers (2). Complete details of the sampling procedure are presented in References 2 and 5.

A 1973 Caterpillar 3150, naturally-aspirated, direct-injection diesel engine was used to generate the exhaust particles for this study. The engine operating conditions were based on the EPA 13-Mode Cycle as given in Table 1. Note that sampling in this case was restricted to the first six modes in order to further limit the total number of samples to a manageable number. A total of 24 samples were collected for TEM analysis. Mobil Delvac 1330 lubricating oil was used for all tests. This oil is a series 3 additive-type oil. Amoco No. 2 (Premier) diesel fuel of 39° API and 0.29 ± 0.06% by mass of sulfur was used throughout the tests.

The sample analysis was done on an Hitachi HU-11 Transmission Electron Microscope and approximately 300 photomicrographs were made of random portions of the 24 samples. Once the photomicrographs were made, the next task was to perform the actual measurement of the basic particles and develop the size distribution. To measure the individual spherical particles, a low-power optical microscope with a built-in eyepiece grid was used. This grid was calibrated, at a given magnification setting, with a precision scale intended for this purpose. To make particle measurements, the investigator would simply place a photomicrograph on the stage of the microscope and use the calibrated grid to determine particle size. As each of the particles was measured, it was marked with a bodkin to insure that it was not measured a second time. A schematic of what the analyst would see through the microscope is illustrated in Figure 4. Note that this method allows the user to make an accurate measurement in two dimensions and thus determine the average diameter of each particle. The results, grouped into size distribution data and additional TEM observations are discussed in the EXPERIMENTAL RESULTS section of the paper.

EXPERIMENTAL RESULTS

SIZE DISTRIBUTIONS FOR INDIVIDUAL MODES - Using the low power optical microscope with calibrated grid and the photomicrographs, 11,000 basic spherical particles were measured to obtain size distribution data. These cumulative results, including the sample collection temperature, number of particles measured, the range of observed sizes, mean particle size, and standard deviation are presented in Table 2. The particle size information is also shown graphically in Figure 5. Previous work has shown that the diameters of approximately 600 particles need to be measured in order to obtain a size distribution curve when plotted on log-normal paper (2). Note that in some cases, the number of particles measured was less than 600. This was due either to the fact that the collected sample mass was too small, or that the particles were heavily coated with condensed hydrocarbons and were not well defined. This latter condition was most prominent on samples collected in the diluted exhaust. In three instances, all for samples collected in the diluted exhaust, no measurements were obtained because of hydrocarbon coatings.

The highlights of Table 2 are summarized in Table 3. It is interesting to note that particle diameters ranged from 97 Å to 1223 Å in the undiluted exhaust, with the largest being 12.6 times the size of the smallest. In the diluted exhaust, sizes ranged from 63 Å to 1158 Å, with the largest being 18.4 times the size of the smallest. Also of interest is the fact that the smallest particle for each exhaust state was observed in the sample collected on Plate 7 in Mode 4. The largest size particles were also found on Plate 7, but in samples collected at Mode 1-7-13 (idle). The significance of this observation, if any, has not been established. Note that the very large atypical particles are not considered here, but are discussed in the following section. The calculated mean diameters were quite close, being 479 Å for the undiluted exhaust and 436 Å for the diluted. In both cases the standard deviation was 114 Å. Note that the average particle diameter for diluted exhaust samples

Table 1 - Engine Characteristics of the Caterpillar 3150 Diesel Engine Operated on the EPA 13-Mode Cycle

Mode	Speed* (RPM)	Percent of Rated Load	Torque (ft-lb$_f$)	BMEP (lb$_f$/in^2)	Manifold Exhaust Temp., F	Air/Fuel Ratio	Smoke, % Opacity	Hydrocarbon Concentration in Exhaust by FID (2), ppm C	Weighting Factor
1-7-13	600	0	7**	0	213	154	1.8	368	.20
2	1800	2	22**	5.8	295	102	.3	354	.08
3	1800	25	88	23.3	424	65	.3	390	.08
4	1800	50	177	46.6	600	43	.7	570	.08
5	1800	75	266	70.0	801	30	1.9	1080	.08
6	1800	100	354	93.0	1033	23	8.0	960	.08
8	3000	100	308	81.0	1206	22	9.0	630	.08
9	3000	75	231	60.5	936	27	3.1	960	.08
10	3000	50	154	40.5	750	36	2.7	975	.08
11	3000	25	77	19.0	586	54	2.2	525	.08
12	3000	2	56**	14.5	561	62	2.4	570	.08

*Rated Speed = 3000 RPM Intermediate Speed = 1800 RPM Low Idle Speed = 600 RPM
**Fluid Drag Prevented Attainment of Required Load
● Measured with Celesco Model 107 Smokemeter (Green Light Source)
○ Measured with Celesco Model 103 Smokemeter (Red Light Source) and Corrected to Green Light Wavelength (8,9)
◐ Particulate Matter Samples Collected at these Modes

Table 2 - Summary of TEM Analysis Results

Mode	Type of Exhaust Sampled	Volume Dilution Ratio	Mixture/Undiluted Exhaust Sample Temperature (F)	Andersen Stage	Number of Particles Measured	Range of Observed Sizes (Å)	Mean Particle Size (Å)	Standard Deviation (Å)
1-7-13	UD	--	161/161	4	646	129-965	539	125
				7	350	257-1223	705	142
	D	23.8	72/163	4	463	257-1094	632	135
				7	378	225-1158	632	139
2	UD	--	284/284	4	526	193-772	386	105
				7	416	129-965	492	119
	D	7.8	98/280	4	505	129-1094	410	128
				7	NA*	NA	NA	NA
3	UD	--	368/368	4	371	225-1158	723	153
				7	619	161-1158	585	130
	D	11.6	100/369	4	NA	NA	NA	NA
				7	NA	NA	NA	NA
4	UD	--	496/496	4	394	129-772	364	102
				7	513	97-708	338	91
	D	16.0	101/503	4	574	64-1094	422	135
				7	583	63-1094	433	115
5	UD	--	645/645	4	602	129-1030	386	107
				7	601	129-708	391	103
	D	23.0	100/655	4	339	129-644	352	88
				7	368	97-708	339	91
6	UD	--	815/815	4	732	154-925	437	99
				7	666	128-801	405	97
	D	33.8	106/847	4	627	160-897	380	98
				7	733	129-933	327	97

*Data Not Obtainable UD - Undiluted D - Diluted

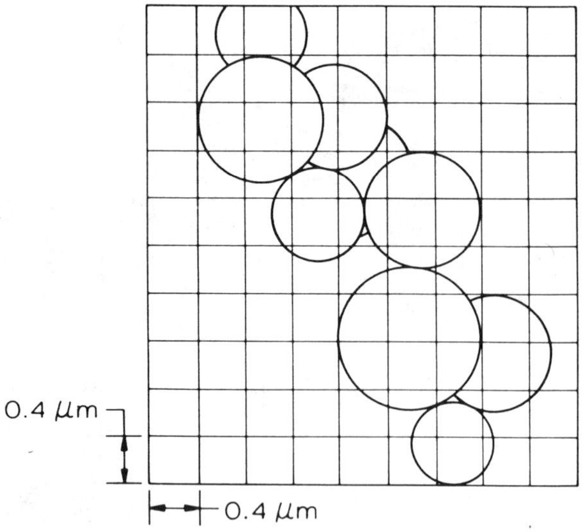

Fig. 4 - Grid technique for sizing individual spherical particles on TEM photomicrographs

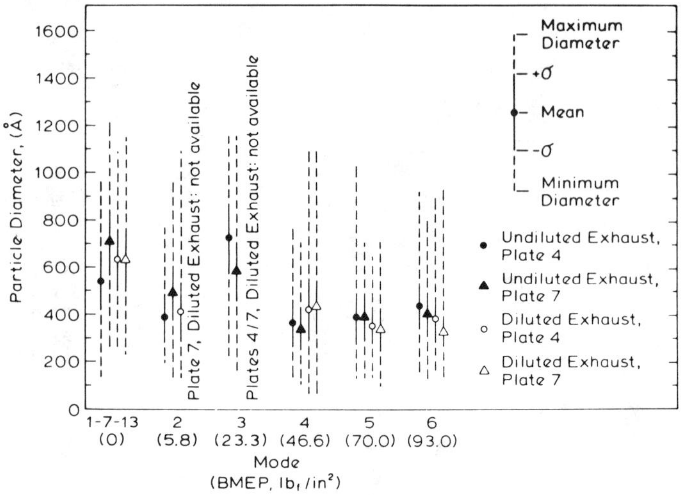

Fig. 5 - Summary of TEM analysis results

Table 3 - Highlights of TEM Analysis

Undiluted Exhaust	
Number of particles measured:	6436
Range of observed sizes*:	
Smallest:	97 Å (Mode 4, Plate 7)
Largest:	1223 Å (Mode 1-7-13, Plate 7)
Mean particle size:	479 Å
Standard deviation:	114 Å
Diluted Exhaust	
Number of particles measured:	4570
Range of observed sizes*:	
Smallest:	63 Å (Mode 4, Plate 7)
Largest:	1158 Å (Mode 1-7-13, Plate 7)
Mean particle size:	436 Å
Standard deviation:	114 Å

*Does not include very large atypical particles

was smaller than the average particle diameter for samples collected in the undiluted exhaust.

Even though all of the information presented thus far is interesting and useful, it does not provide any insight as to the basic spherical particle size distributions. Such distributions may be obtained by plotting the diameter of the individual spherical particle as a cumulative percent of particles less than the given particle diameter versus the particle diameter. The plots are prepared on log-normal paper which is ideally suited for this purpose since they tend to follow the log-normal distribution. Size distributions are shown in Figures 6, 7 and 8. The data shown in these figures are summarized in Table 4. It is important to note carefully the distinction between the average particle diameter (mean) as used above, and the 50% particle diameter (median) obtained from Figures 6, 7 and 8 and Table 4. The nature of diesel exhaust particulate matter is such that the mean diameter will always be quantitatively greater than the median or 50% diameter for a given sample. A comparison of Tables 2 and 4 bears this out. Although a size shift may reasonably be expected when comparing the size distributions of the basic spherical particles for the undiluted and diluted exhausts, the small shift evident in Figures 6, 7 and 8 is not thought to be significant.

The individual distributions for Plates 4 and 7 were compared for samples collected in the undiluted exhaust and the results are shown in Figures 9, 10 and 11. The results are not conclusive, but the distributions for Modes 3, 4 and 6 show that smaller aggregates are made up of smaller size individual particles, while the Mode 2 and Mode 5 distributions are mixed.

CATERPILLAR 3150 DIESEL

Mode	1-7-13	2
Speed, rpm	600	1800
BMEP, lb_f/in^2	--	5.8
Exhaust Temp.,°F	161	284
Mixture Temp.,°F	72	98
Vol. Dilution Ratio	24	8

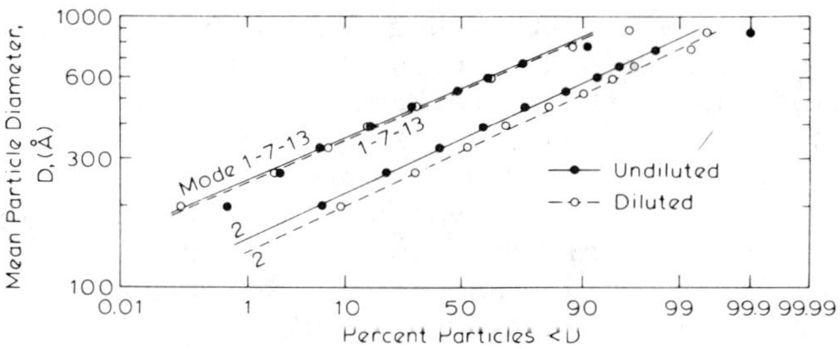

Fig. 6 - Cumulative particle size distributions for Mode 1-7-13 (idle) and Mode 2

CATERPILLAR 3150 DIESEL

Mode	3	4
Speed, rpm	1800	1800
BMEP, lb_f/in^2	23.3	46.6
Exhaust Temp.,°F	368	496
Mixture Temp.,°F	100	101
Vol. Dilution Ratio	12	16

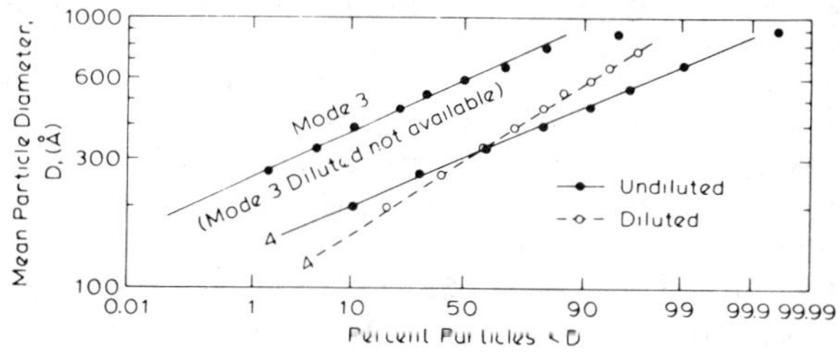

Fig. 7 - Cumulative particle size distributions for Mode 3 and Mode 4

CATERPILLAR 3150 DIESEL

Mode	5	6
Speed, rpm	1800	1800
BMEP, lb_f/in^2	70.0	93.0
Exhaust Temp.,°F	645	815
Mixture Temp.,°F	100	106
Vol. Dilution Ratio	23	34

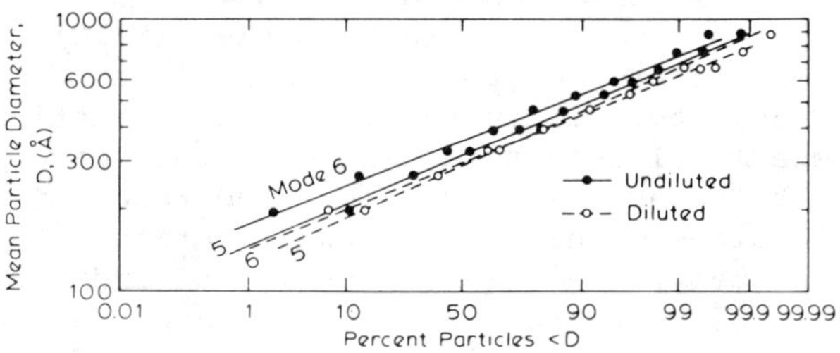

Fig. 8 - Cumulative particle size distributions for Mode 5 and Mode 6

CATERPILLAR 3150 DIESEL

Mode	1-7-13	2
Speed, rpm	600	1800
BMEP, lb_f/in^2	--	5.8
Exhaust Temp.,°F	161	284

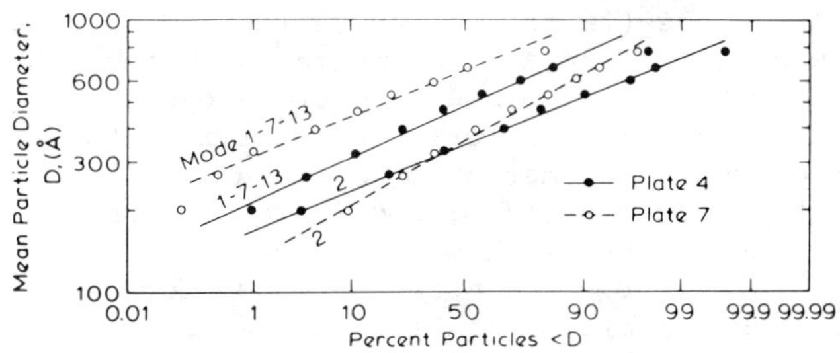

Fig. 9 - Cumulative particle size distributions for Mode 1-7-13 (idle) and Mode 2

CATERPILLAR 3150 DIESEL

Mode	3	4
Speed, rpm	1800	1800
BMEP, lb_f/in^2	23.3	46.6
Exhaust Temp.,°F	368	496

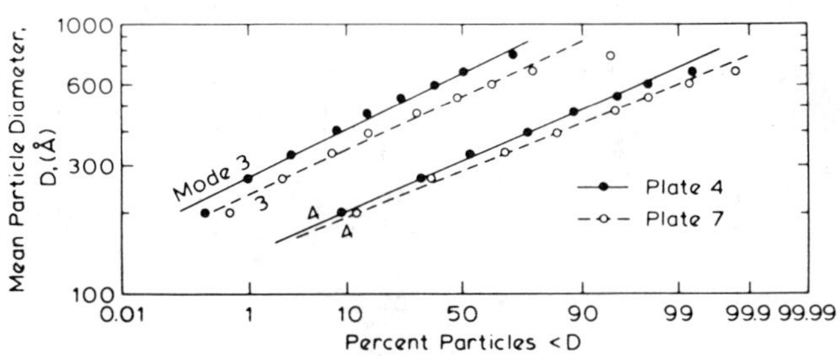

Fig. 10 - Cumulative particle size distributions for Mode 3 and Mode 4

CATERPILLAR 3150 DIESEL

Mode	5	6
Speed, rpm	1800	1800
BMEP, lb_f/in^2	70.0	93.0
Exhaust Temp.,°F	645	815

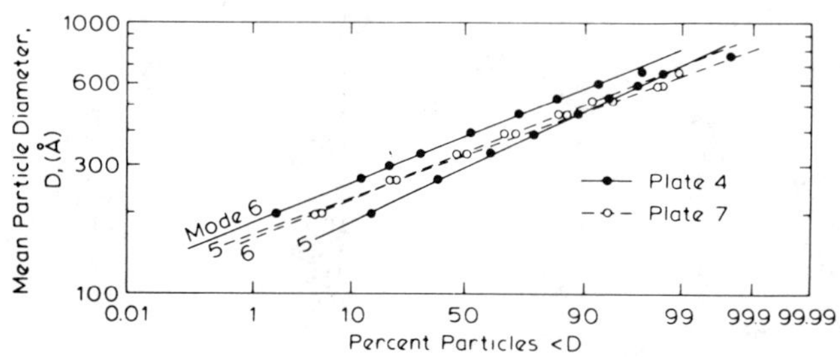

Figure 11 - Cumulative particle size distributions for Mode 5 and Mode 6

Only the results for the idle mode are contradictory to this. The other important result is that Plates 4 and 7 do not show significant differences in size distribution for the same mode. These data help to support the basic assumption that the basic spherical particles that make up the aggregates/agglomerates can be determined by sizing the larger particles on the impactor plates.

In concluding the discussion on the size distributions of the individual spherical particles, it is necessary to comment on the variability between modes, and to compare these findings with the results of earlier work. It is interesting to note that although the size distributions for Modes 2, 4, 5 and 6 are rather closely grouped, the distributions for Modes 1-7-13 and 3 indicate particles of significantly greater relative diameter. This holds true for samples collected in both the undiluted and diluted exhaust. The reason for this is not known since the engine operating parameters were consistent for the mode, and reasonable care was exercised during the sample collection and analysis process to ensure statistically valid results.

Further, when the results as a whole (all modes combined) are compared with the results of earlier work (2) these new results indicate a 60% increase in the median particle diameter. To be specific, an average median diameter (the 50% diameter) of 260 Å was reported by Vuk et al (2), whereas this work has determined the average median diameter to be 417 Å. If the results for Modes 1-7-13 and 3 are discounted, these results still show a 27% increase in the median diameter. One possible cause for these differences could be the subjective nature of the particulate measurements. Experience has shown that the results are quite dependent on both the individual performing the measurements and the technique used. We feel these results are probably more accurate due to the study of more photomicrographs with a better sizing technique.

TEM PHOTOMICROGRAPH OBSERVATIONS - This section is devoted to the discussion and illustration of a number of unusual and interesting observations recorded during the TEM work. For some of these, there are reasonable explanations. For the rest, any explanation would, at best, be vague and uncertain.

One of these observations concerns the wide range of individual spherical particles found within a given sample. Although each sample represents a single engine operating mode, thus isolating the effects of engine speed and load, there still exists a variety of particle sizes as shown in Figure 12. This is characteristic of samples collected at all modes, as noted previously in Table 2.

Examination of the photomicrograph shown in Figure 12 reveals that the individual particles can be grouped by size. That is, the agglomerate can be broken down into smaller segments (aggregates) that are comprised of individual spherical particles of approximately the same diameter. It is thought that these aggregates are made up of basic individual particles that have experienced the same history of formation and thus have similar physical and chemical characteristics (5,10). Differences of temperature, pressure, fuel concentration, etc. within the engine's combustion chamber will result in particles that have different characteristics. Thus the range of particle sizes within a given agglomerate. Unfortunately, the size uniformity of individual particles that make up the aggregates was not closely examined during this project and therefore, no conclusions can be formed.

Table 4 - Spherical Particle Size Data as a Function of Various Engine Parameters

Mode	Type of Exhaust Sampled	Sample Temperature, (F)	Air/Fuel Ratio	BMEP, (lb_f/in^2)	%<500 Å	%<200 Å	50% Size, (Å)	Transit* Time, (sec)
1-7-13	UD	161	154	--	35.0	.08	565	.52
	D	72	154	--	37.0	.10	557	1.34
2	UD	284	102	22	81.7	6.1	356	.17
	D**	98	102	22	87.6	10.0	323	1.34
3	UD	368	65	23.3	30.3	.16	605	.15
	D	100	65	23.3	NA***	NA	NA	1.34
4	UD	496	43	46.6	94.1	10.5	300	.13
	D	101	43	46.6	85.3	22.0	293	1.34
5	UD	645	30	70.0	91.7	8.1	316	.11
	D	100	30	70.0	94.5	13.1	291	1.34
6	UD	815	23	93	85.8	2.7	360	.10
	D	106	23	93	94.8	9.3	302	1.34

*Time Required for Particulate to Travel from Exhaust Port to Point of Sampling
**Includes Plate 4 Only
***Data Not Obtainable

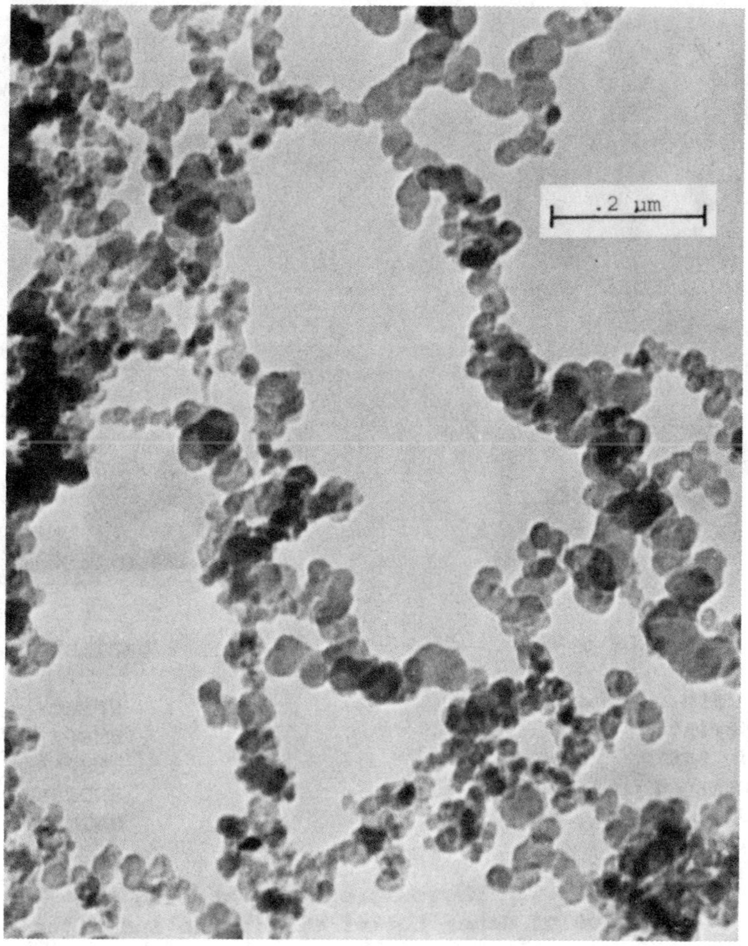

UNDILUTED EXHAUST - MODE 6

SPEED: 1800 RPM
BMEP: 93 lb_f/in^2
TEMPERATURE: 815°F
ANDERSEN PLATE: 7
MAGNIFICATION: 63.4 KX
(ENLARGED 1.8x)

Fig. 12 - TEM photomicrograph showing wide range of individual particle sizes

One of the most frequent observations, particularly in samples collected at the high temperature modes, involves the presence of very dense nuclei within the individual spherical particles. These nuclei, shown in Figure 13 have been observed by others (2) and are believed to be crystalline in nature (5,10). Note that some of the particles exhibit a single nuclei while others contain a large number of relatively smaller nuclei. Although the exact composition of the nuclei has not been definitely established it is thought to be comprised of mostly carbon.

Related to these nuclei, but possibly having achieved a greater degree of development, are the internal structures shown in Figure 14. These photomicrographs show what appears to be a structure that has developed throughout the aggregate. Although not observed as frequently as the basic nuclei, these structures were noted in both the undiluted and diluted exhaust samples collected at Modes 1-7-13 and 3. While a number of different crystalline structures have been observed within particulate matter (10), none seem to correspond to those shown here.

It is well known that certain engine operating conditions, such as Mode 3, exhibit high concentrations of unburned hydrocarbons in the exhaust. It is also known that upon cooling, whether by dilution or other means, these hydrocarbons will condense, either into

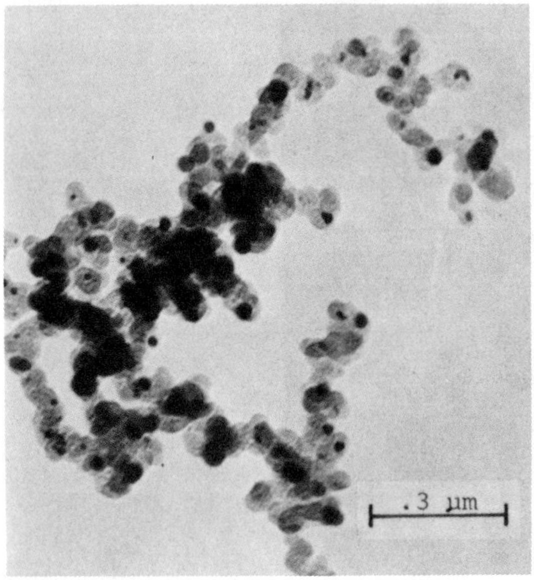

UNDILUTED EXHAUST - MODE 5

SPEED: 1800 RPM
BMEP: 70 lb_f/in^2
TEMPERATURE: 645°F
ANDERSEN PLATE: 4
MAGNIFICATION: 61 KX

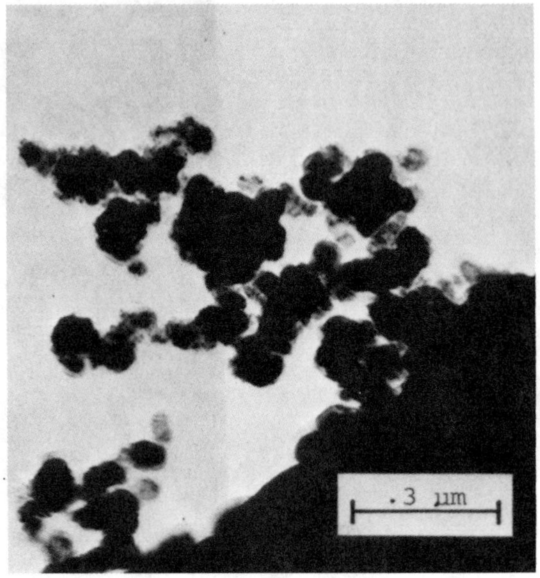

UNDILUTED EXHAUST - MODE 3

SPEED: 1800 RPM
BMEP: 23.3 lb_f/in^2
TEMPERATURE: 368°F
ANDERSEN PLATE: 4
MAGNIFICATION: 63.4 KX

Fig. 13 - TEM photomicrographs showing the presence of dense nuclei within the individual spherical particles

liquid droplets or adsorb onto the surface of the solid particulate matter (4,5,6). The degree of condensation would, of course, be dependent on both temperature and the hydrocarbons' partial pressure and, most condensed hydrocarbons would probably be of high molecular weight and result in a thick liquid.

Evidence that these hydrocarbons do condense is provided by the photomicrographs presented in Figures 15, 16, 17, 18 and 19. The first two figures compare samples collected in the undiluted and diluted exhaust for Modes 2 and 6. Note that with the sample collected in the diluted exhaust for Mode 2, the coating of hydrocarbons almost totally obscures the detail of the individual spherical particles that make up the agglomerate. In contrast, the Mode 6 sample collected in the diluted exhaust exhibits no discernible hydrocarbon coating. It should be pointed out that samples collected in the undiluted exhaust for Mode 1-7-13 frequently had minor coatings of condensed hydrocarbons. This was likely due to the relatively low exhaust temperature at this engine condition. A further comparison of hydrocarbon condensation in the diluted exhaust, for samples collected at Modes 2, 3 and 4, is shown in Figure 17. Note that Mode 3 exhibits the heaviest hydrocarbon coating while Mode 4 shows the least.

Figure 18 shows what appear to be liquid droplets that have been collected on the TEM grid. These samples were collected at Mode 3, in the diluted exhaust, at a mixture temperature of approximately 104°F. This mode appears to have the highest concentrations of hydrocarbons, as shown in Figure 17 and evidenced by the fact that measurement of the individual spherical particles was not possible for samples collected at Mode 3 in the diluted exhaust due to the very heavy hydrocarbon coatings.

Finally, consider the photomicrographs shown in Figure 19. These show what appear to be small "pools" of liquid. Note that these are concentrated around the agglomerates, and are very transparent and irregular in shape. Since the operating vacuum of the electron microscope is in the range of 1.9×10^{-6} to 1.9×10^{-7} lb_f/in^2, the continued presence of these droplets on the sample after placement within the microscope would indicate a strongly bound hydrocarbon compound of high molecular weight. Lighter hydrocarbons would simply vaporize in the microscope's vacuum and their existence could not be demonstrated.

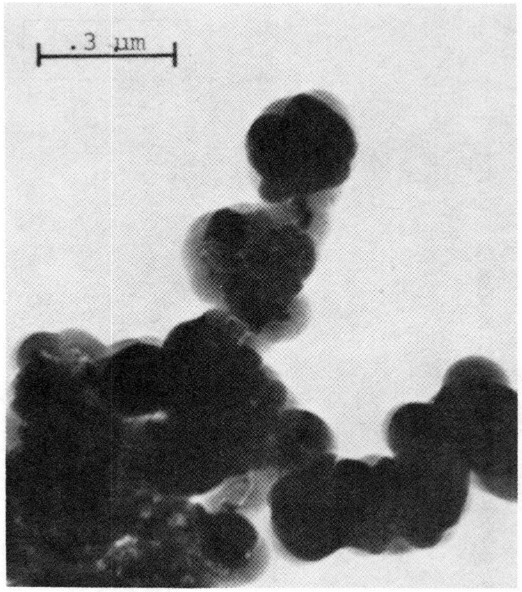

UNDILUTED EXHAUST - MODE 5

SPEED: 1800 RPM
BMEP: 70 lb_f/in^2
TEMPERATURE: 645°F
ANDERSEN PLATE: 4
MAGNIFICATION: 61 KX

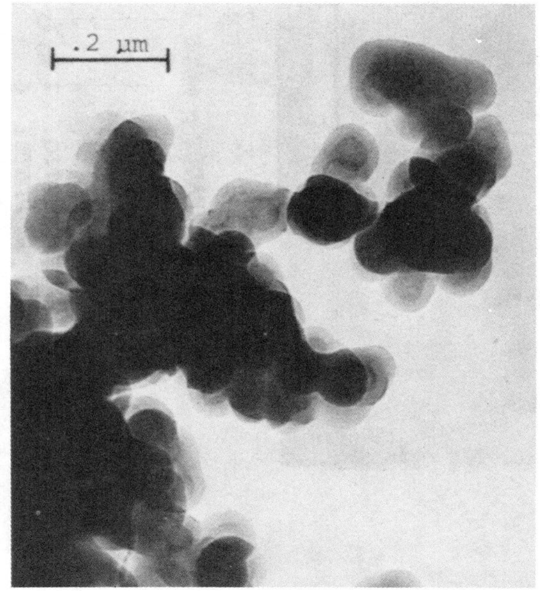

DILUTED EXHAUST - MODE 1-7-13

SPEED: 600 RPM
BMEP: 0 lb_f/in^2
TEMPERATURE: 72°F
DILUTION RATIO: 23.8:1
ANDERSEN PLATE: 7
MAGNIFICATION: 79 KX

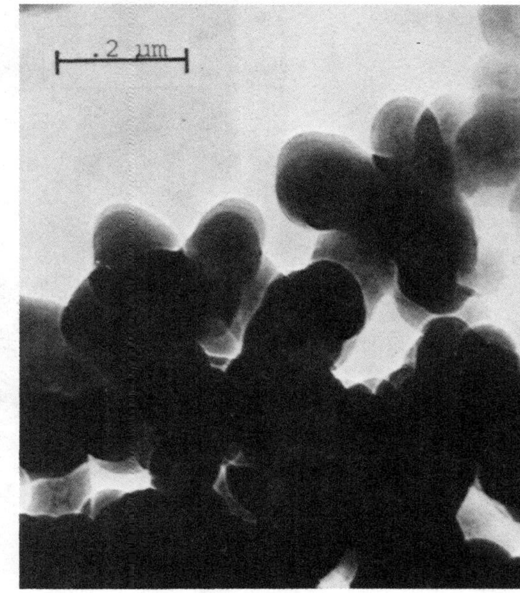

DILUTED EXHAUST - MODE 1-7-13

SPEED: 600 RPM
BMEP: 0 lb_f/in^2
TEMPERATURE: 72°F
DILUTION RATIO: 23.8:1
ANDERSEN PLATE: 7
MAGNIFICATION: 86.5 KX

Fig. 14 - TEM photomicrographs of particulate samples exhibiting apparent internal structure

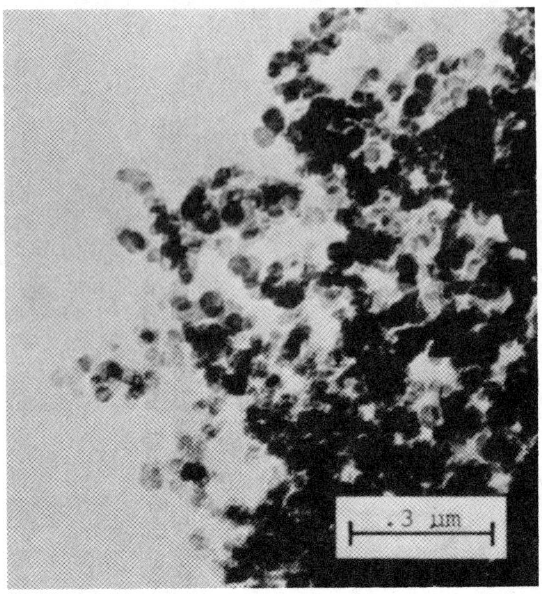

UNDILUTED EXHAUST - MODE 2

SPEED: 1800 RPM
BMEP: 22 lb_f/in^2
TEMPERATURE : 284°F
ANDERSEN PLATE: 7
MAGNIFICATION: 63.4 KX

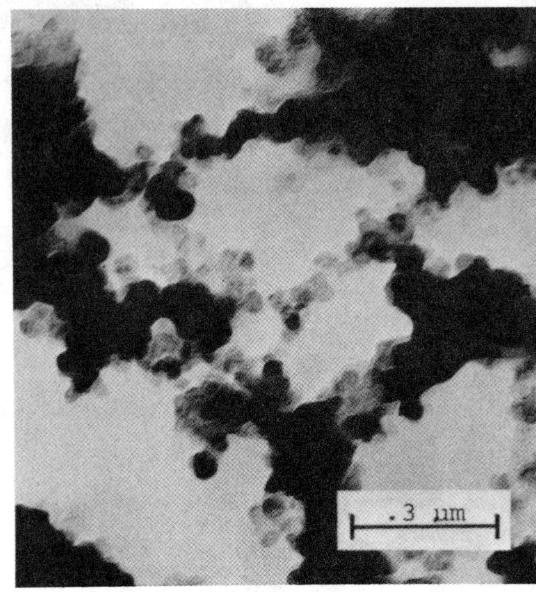

DILUTED EXHAUST - MODE 2

SPEED: 1800 RPM
BMEP: 22 lb_f/in^2
TEMPERATURE: 98°F
DILUTION RATIO: 7.8:1
ANDERSEN PLATE: 4
MAGNIFICATION: 63.4 KX

Fig. 15 - Hydrocarbon condensation at Mode 2

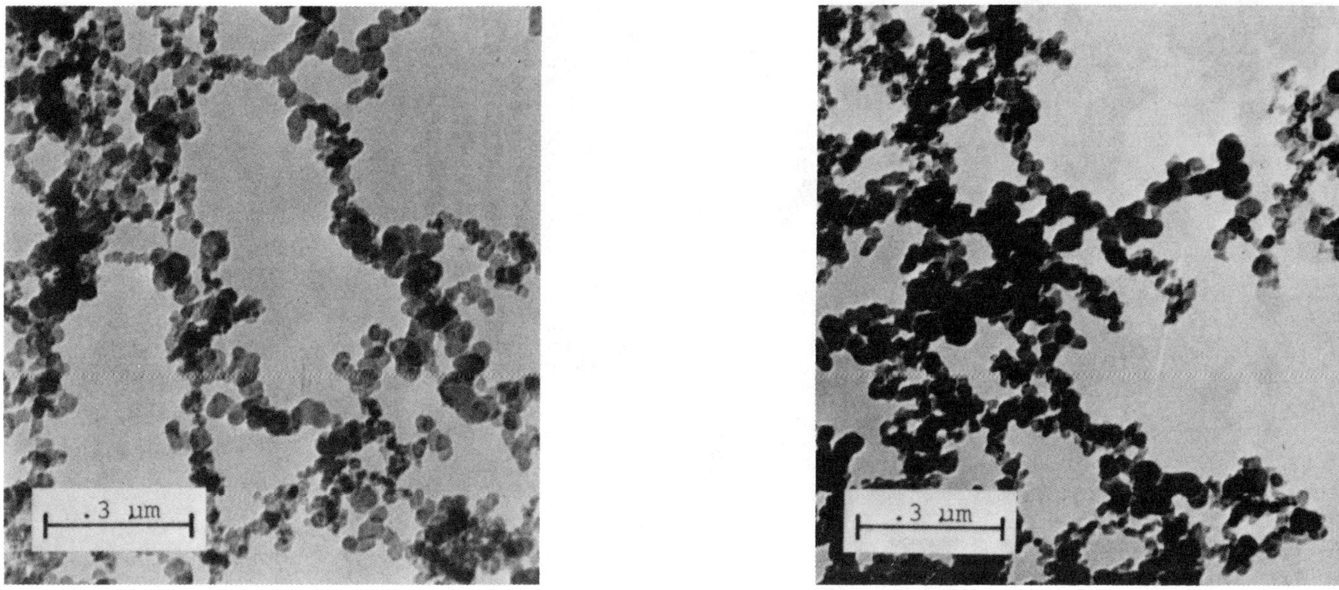

UNDILUTED EXHAUST - MODE 6

SPEED: 1800 RPM
BMEP: 93 lb_f/in^2
TEMPERATURE: 815°F
ANDERSEN PLATE: 7
MAGNIFICATION: 63.4 KX

DILUTED EXHAUST - MODE 6

SPEED: 1800 RPM
BMEP: 93 lb_f/in^2
TEMPERATURE: 106°F
DILUTION RATIO: 34:1
ANDERSEN PLATE: 7
MAGNIFICATION: 63.4 KX

Fig. 16 - Hydrocarbon condensation at Mode 6

The fact that hydrocarbons do condense and adsorb on the particles is acceptable, but many liquid hydrocarbons could not exist in the high vacuum of the TEM while being subjected to the electron beam. Either the vaccum or the electron beam would cause all except the heavy molecular weight liquid hydrocarbons to evaporate. In fact, what appeared to be an evaporation process was noted several times during the TEM analysis. The evaporation took on the form of "rapid boiling" and continued until the "liquid" was gone.

Considered last are the two very large individual spherical particles shown in Figure 20. Large particles such as these, although previously observed (8), are atypical. The larger of the two is approximately 2500 Å in diameter with an inner core diameter of approximately 2000 Å. The other particle is 2/3 the size of the first, while the first is 5.5 times the size of the average individual spherical particle. Although much larger than the typical particle, the form is the same. That is, a more dense center with an outer coating. It is also interesting to note that the smaller particle also exhibits the small nuclei found in many of the other particles. Most definitely, these particles had a unique history of formation.

Many of the observations presented here are atypical, and although their presence is documented, an explanation for their existence is not readily available. Providing these explanations will be the task of future diesel exhaust investigations.

CONCLUSIONS

Based on the results gathered in this paper with the Caterpillar 3150 engine, and the observations made during the sample analysis, the following conclusions have been formed:

1. The individual spherical particles, which make up the larger aggregates and agglomerates, had a range of diameters of 63 Å to 1223 Å. The average size in the undiluted exhaust was 479 Å, while the mean size in the diluted exhaust was 436 Å. A few very large atypical spherical particles, up to 2500 Å in diameter, were observed.

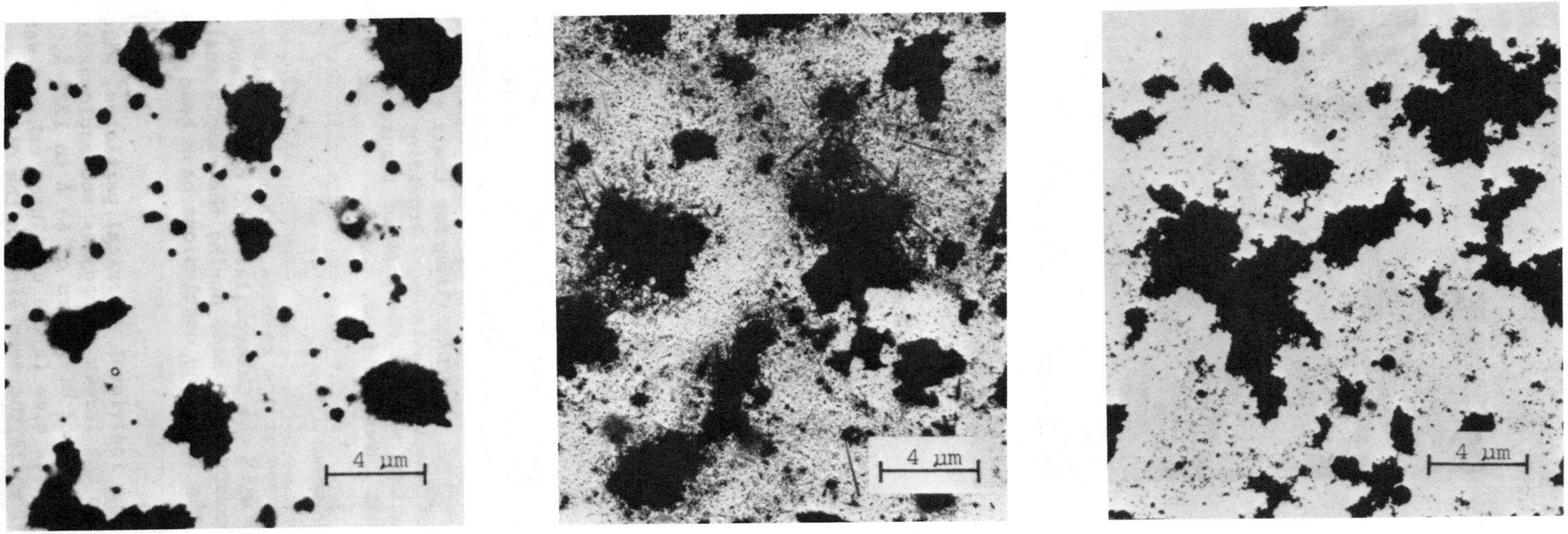

DILUTED EXHAUST

MODE 2
SPEED: 1800 RPM
BMEP: 22 lb_f/in^2
TEMPERATURE: 98°F
DILUTION RATIO: 8:1
ANDERSEN PLATE: 7
MAGNIFICATION: 3.6 KX

MODE 3
SPEED: 1800 RPM
BMEP: 23.3 lb_f/in^2
TEMPERATURE: 100°F
DILUTION RATIO: 12:1
ANDERSEN PLATE: 7
MAGNIFICATION: 3.6 KX

MODE 4
SPEED: 1800 RPM
BMEP: 46.6 lb_f/in^2
TEMPERATURE: 101°F
DILUTION RATIO: 16:1
ANDERSEN PLATE: 7
MAGNIFICATION: 3.6 KX

Fig. 17 - Comparison of hydrocarbon condensation on particulate samples collected in the diluted exhaust at Modes 2, 3 and 4

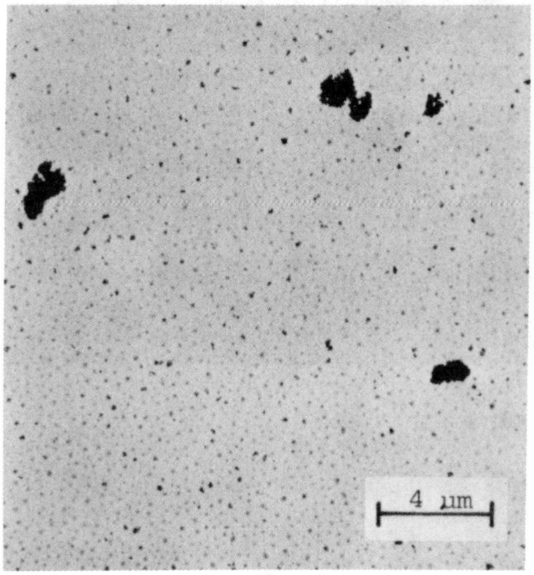

MAGNIFICATION: 3.5 KX

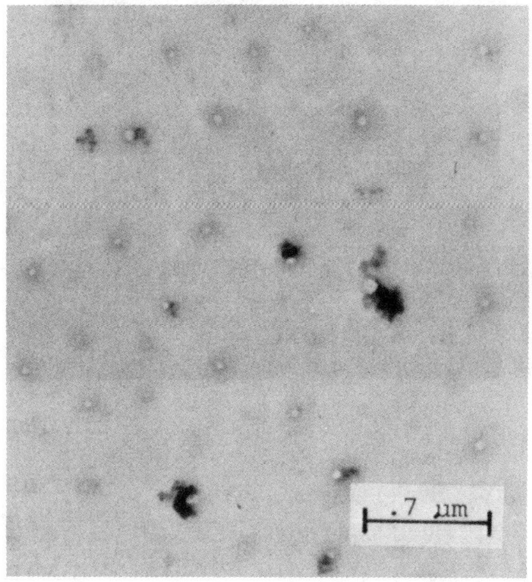

MAGNIFICATION: 23.7 KX

DILUTED EXHAUST

MODE 3
SPEED: 1800 RPM
BMEP: 23.3 lb_f/in^2
TEMPERATURE: 100°F
DILUTION RATIO: 12:1
ANDERSEN PLATE: 4

Fig. 18 - Droplets of condensed hydrocarbons at Mode 3

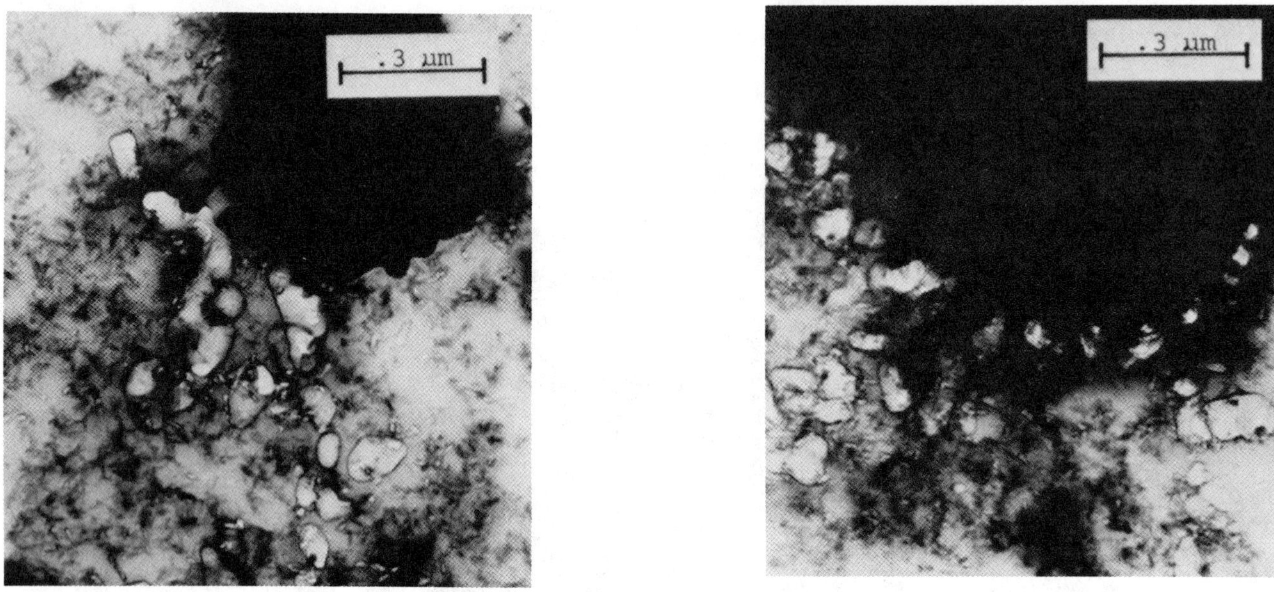

DILUTED EXHAUST - MODE 3

SPEED: 1800 RPM
BMEP: 23.3 lb_f/in^2
TEMPERATURE: 100°F
DILUTION RATIO: 12:1
ANDERSEN PLATE: 7
MAGNIFICATION: 63.4 KX

Fig. 19 - Large droplets of condensed hydrocarbons collected in the diluted exhaust at Mode 3

2. For those modes exhibiting large concentrations of unburned hydrocarbons in the exhaust, there were correspondingly large amounts of condensed hydrocarbons on the particulate samples collected in the diluted exhaust. In some instances, notably Modes 2 and 3, coatings of condensed hydrocarbons were so heavy as to totally obscure the aggregate. Samples showing droplets of condensed hydrocarbons were also observed.

3. Although unidentifiable, several aggregates possessed well developed internal structures. Such formations have been previously reported.

The results in this paper provide better quantitative data as to the physical character of the basic diesel particles that make up the aggregates/agglomerates. In addition, the photomicrographs of particles in the undiluted and diluted exhaust provide experimental evidence that both separate liquid hydrocarbons and liquid hydrocarbons adsorbed on the solid particles exist in diluted diesel exhaust and not in the undiluted exhaust. No significant size differences in the basic solid particles were found between the undiluted and diluted exhaust as would be expected. These results will hopefully add to the information needed for accurately modeling of the physical and chemical processes taking place in the particle formation and coagulation processes (11).

ACKNOWLEDGEMENT

This project was financed with Federal Funds from the Environmental Protection Agency (EPA) under Grant Number R802596020. The authors express our appreciation to the EPA Office of Mobile Source Air Pollution in Ann Arbor, Michigan for making this research possible. The assistance of Tom Williams in the analysis of the photomicrographs is greatly appreciated.

REFERENCES

1. K. Carpenter, "The Design and Development of a Dilution Tunnel for the Physical and Chemical Characterization of Diesel Particulate Matter," M.S. Thesis, Michigan Technological University, 1978.

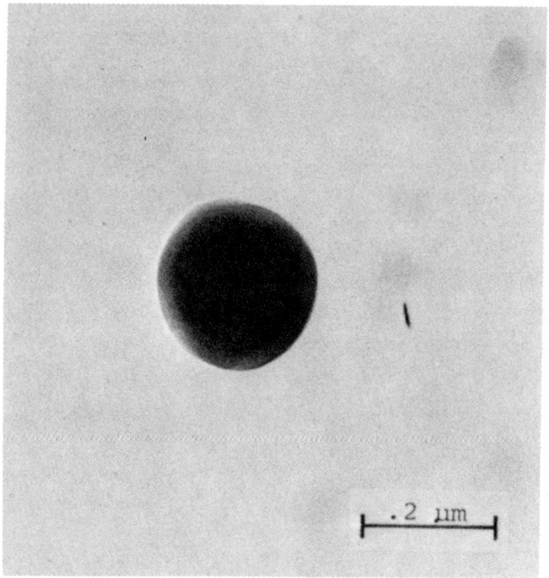

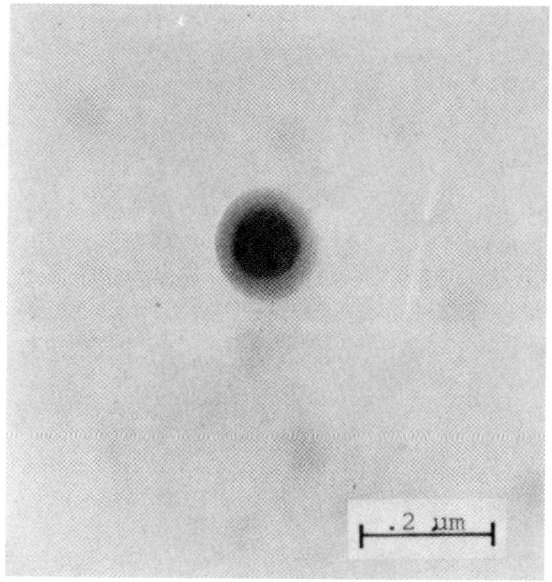

DILUTED EXHAUST - MODE 1-7-13

SPEED: 600 RPM
BMEP: 0 lb_f/in^2
TEMPERATURE: 72°F
DILUTION RATIO: 24:1
ANDERSEN PLATE: 7
MAGNIFICATION: 86.5 KX

Fig. 20 - Atypical large diameter particles

2. C. T. Vuk, M. A. Jones and J. H. Johnson, "The Measurement and Analysis of the Physical Character of Diesel Particulate Emissions," SAE Paper No. 760131, SAE Transactions, 1976.

3. N. J. Khatri, J. H. Johnson, "Physical Size Distribution Characterization of Diesel Particulate Matter and the Study of the Coagulation Process," SAE Paper No. 780788, Presented at the SAE Off-Highway Vehicle Meeting and Exposition, Milwaukee, September 11-14, 1978, SAE Transactions, 1978.

4. N. J. Khatri, J. H. Johnson and D. G. Leddy, "The Characterization of the Hydrocarbon and Sulfate Fraction of Diesel Particulate Matter," SAE Paper No. 780111, SAE Transactions, 1978.

5. W. H. Lipkea, J. H. Johnson and C. T. Vuk, "The Physical and Chemical Character of Diesel Particulate Emissions-Measurement Techniques and Fundamental Considerations," SAE Special Publication 430, 1978, SAE Transactions, 1978.

6. L. E. Frisch, J. H. Johnson and D. G. Leddy, "Effect of Fuels and Dilution Ratio on Diesel Particulate Emissions," SAE Paper No. 790417, Presented at SAE Congress and Exposition, Detroit, February 26 - March 2, 1979.

7. E. F. Funkenbusch, D. G. Leddy and J. H. Johnson, "Characterization of the Organic Soluble Fraction of Diesel Particulate Emissions," SAE Paper No. 790418, Presented at SAE Congress and Exposition, Detroit, February 26 - March 2, 1979.

8. M. A. Jones, "The Effect of Engine Load and Speed on the Physical Characteristics of Diesel Particulate Emissions," M.S. Thesis, Michigan Technological University, 1975.

9. "Smoke Meter Round Robin Results," CAPI-1-64, Coordinating Research Council, Inc., New York, 1976.

10. D. Broome, "The Mechanisms of Soot Release from Combustion of Hydrocarbon Fuels with Particular Reference to the Diesel Engine," Lucus Engineering Review, Vol. 6, p. 46, November, 1973.

11. D. B. Kittelson, D. F. Dolan and M. C. Beaver, "Exhaust Particle Size Distributions Produced by Diesel and Spark Ignition Engines: Comparison of Theory and Experiment," Paper No. CSS/CI-79-09, Presented at Central States Section, The Combustion Institute, April 9-10, 1979.

Light Absorption Measurements of Diesel Particulate Matter**

Huel C. Scherrer, David B. Kittelson, and Daniel F. Dolan
Dept. of Mechanical Engineering
Univ. of Minnesota

LIGHT ABSORPTION and extinction properties as employed by Bosch Spot or smoke opacity measurements have traditionally served as a convenient means of monitoring diesel soot emissions. However, increasing interest in automotive diesel engines makes the optical properties of the diesel particles an emission concern in and of itself. The purpose of this paper is to present absorption and scattering measurement results for the visible spectrum, analyze these results against the varying physical characteristics of the diesel particulates, and reconcile these measurements with electrodynamic theory.

Measurements were made on laboratory generated, dilute diesel exhaust aerosols, with both direct injection and swirl prechamber engines used under varying load conditions. Light absorption was determined by the Lin-Baker-Charlson(1)* (integrating plate) method. The non-volatile mass fraction was determined by vacuum volatilization of filter samples, and the particle size classified by an electrical aerosol analyzer. In addition, light scattering was measured by an integrating nephelometer for comparison of the relative importance of absorption and scattering in determining light extinction.

*Numbers in parentheses designate References at end of paper.

**Paper 810181 presented at the International Congress and Exposition, Detroit, Michigan, February, 1981.

EXPERIMENTAL

Figure 1 shows schematically the exhaust sampling system used. The system has been described in detail by Dolan et.al.(2). Two naturally aspirated four stroke cycle four cylinder engines were used as particle sources: a 1.5 liter swirl chamber engine and a 3.6 liter direct injection engine. Dilute exhaust aerosols were collected on Nuclepore membrane filters of 0.4 μm pore size for measurement of the absorption coefficient. Due to the extremely light filter loadings required for this method, parallel filter samples at higher flow were taken on Pallflex TX40HI20WW teflon coated glass fiber filters for accurate determination of the Nuclepore filter loadings. Aerosol flow rates through both filters were measured by capillary tube flow meters, and the Nuclepore filter loading in particulate mass deposited per unit area was calculated. The Pallflex filter samples were also used for volatile fraction determinations. The samples were vacuum volatilized at an absolute pressure of less than 1 kPa (7.6 mm Hg) at 175°C for 24 hours.

ABSTRACT

Light absorption and scattering coefficients have been measured for particles emitted by two diesel engines; one direct injection and one indirect injection, operating over a range of speeds and loads. Integrating plate absorption measurements yield a specific absorption coefficient of 9.1 square meters per gram of non-volatile particulate matter at 550 nm wavelength. This absorption coefficient is inversely proportional to wavelength and independent of engine operating conditions or type. The scattering coefficient was simultaneously measured as 1.3 square meters per gram of undifferentiated particulate matter. These experimental results are shown to be in the range predicted by theoretical absorption and scattering calculations, which have been made for elongated carbon-void particles.

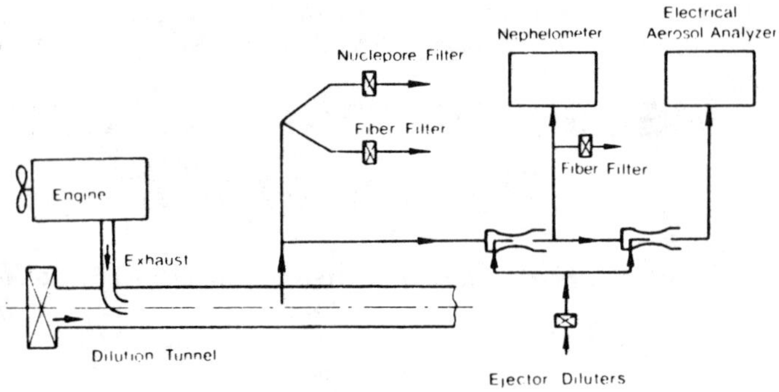

Fig. 1 - Exhaust sampling system for absorption, scattering, and size distribution measurements

The absorption measurements were performed using a Beckman DK-1A spectrophotometer, adapted as shown in figure 2. A monochromatic light beam is alternately directed to reference and sample paths by a semi-circular rotating mirror (a). An unloaded reference filter (b) and opal glass diffuser (c) are placed in the reference beam, and the loaded filter with the particles facing the light source (d) and opal glass diffuser (e) are placed in the sample beam. The opal glass diffuser approximates an integrating sphere, with the light intensity in the forward direction nearly proportional to the light transmitted and forward scattered by the filter and aerosol. Backscatter from the sub-micron diesel particles is negligible. The sample and reference beams are directed to a photomultiplier (g) by a synchronous semi-circular rotating mirror (f). The sample and reference opal glass diffusers are positioned so that an equal solid angle of light is received from each by the photomultiplier. The light intensity of the sample beam is recorded as a percentage of the light intensity of the reference beam, as the wavelength is varied by the monochromator.

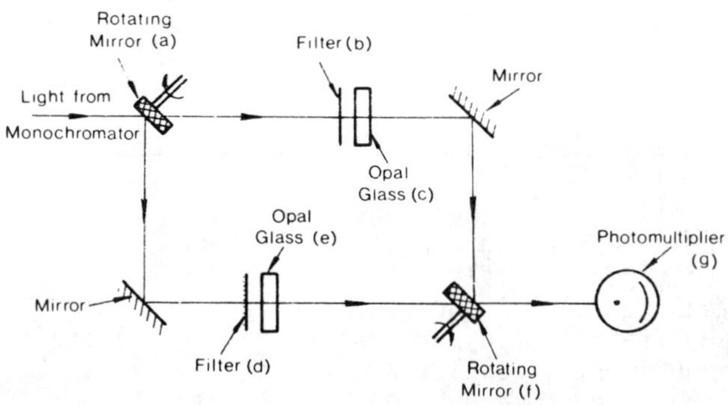

Fig. 2 - Optical system for integrating plate absorption measurements

Light absorption by the aerosol particles deposited on a filter surface follows the Beer-Lambert law:

$$I = I_o e^{-b_{abs} x} \quad (1)$$

where I_o is the intensity of the incident beam, I is the intensity of the transmitted plus scattered light, b_{abs} is the absorption coefficient of the aerosol, and x is the pathlength through the particles. This may be rewritten as:

$$I = I_o e^{-(b_{abs}/\rho)(\rho x)} = I_o e^{-R_{abs} L} \quad (2)$$

where ρ is the particle mass per unit volume of deposit on the filter, L is the filter mass loading per unit area, and R_{abs} is the cross section for absorption per unit mass of aerosol, the quantity to be experimentally determined.

Scattering cross section per unit volume of aerosol at 535 nm is measured directly by the integrating nephelometer. Aerosol mass concentration at the nephelometer inlet is determined by a Pallflex TX40HI20WW filter sample. These two measurements jointly yield scattering cross section per unit mass of diesel particulate matter.

RESULTS

Table 1 summarizes the results of the measurements made on the two engines. Optical absorption and scattering coefficients as well as their sums, the extinction coefficients, are listed. The absorption cross section per unit mass of undifferentiated particulate matter is seen to vary appreciably with engine type and operating conditions. However, the absorption cross section per unit mass of non-volatile particulate matter, which is tabulated in the last column of the table, is essentially independent of engine type and operating conditions.

This result is further supported by figure 3, which is a plot of absorption cross section per unit mass total particulate matter measured at a wavelength of 550 nm against non-volatile mass fraction. The slope of the linear regression line gives a light absorption cross section per unit mass of non-volatile particulate matter of 9.05 ± 1.13 m^2/g at 550 nm. Here the upper and lower bounds include the 95% confidence interval. This result compares well with spectrophone measurements (3,4). The intercept of the regression line, 0.07 ± 0.85 m^2/g, gives a measure of the light absorption coefficient per unit mass of volatile particulate matter. Its small value, which does not differ significantly from zero, suggests that the particles are relatively non-light absorbing.

The absorption measurements also show that the absorption coefficient is, essentially, inversely proportional to wavelenth. In

Table 1 - Summary of light absorption and scattering measurements. Trials 1 - 6 are for a direct injection, naturally aspirated engine with trials 11 - 16 for a swirl prechamber naturally aspirated engine.

Trial #	RPM	BMEP (MPa)	R_{abs}[a] (m^2/g)	R_{scat}[b] (m^2/g)	R_{ext} (m^2/g)	Non-Volatile fraction	Nuclepore loading, L (ug/cm^2)	R abs for non-volatile mass (m^2/g)
1	1495	0.42	4.52	0.98	5.50	.507	9.36	8.91
2	2000	0.62	7.59	1.73	9.32	.880	15.87	8.62
3	1493	0.44	5.46	1.02	6.48	.584	14.23	9.35
5	1196	0.25	4.01	1.64	5.65	.443	15.14	9.05
6	1507	0.43	4.22	0.22 [c]	4.44 [c]	.422	6.38	10.01
11	2050	0.23	8.14	1.53	9.67	.901	16.83	9.04
12	2950	0.47	8.93	0.55 [c]	9.48 [c]	.947	30.26	9.43
13	2020	0.23	8.83	0.98	9.81	.948	8.89	9.32
14	1500	0.01	6.13	1.18	7.31	.726	10.92	8.45
15	2040	0.23	8.37	1.29	9.66	.887	27.86	9.44
16	2500	0.12	4.95 [d]	1.35	6.30 [d]	.847	38.62	5.84 [d]
Mean Values:			6.6	1.30	7.9			9.16e
95% Confidence Band:			±4.5	±0.66	±4.5			±1.02

Superscripts

a - At 550 nm wavelength

b - At 535 nm wavelength

c - Nephelometer response too low for accurate measurement of light scattering. These results not included in calculation of average.

d - Wild point, not included in average calculations

e - Note that this result is slightly different from the result obtained from the slope of the regression line in figure 3. We consider the value obtained from the slope of the regression, line 9.05 ± 1.13 to be more representative.

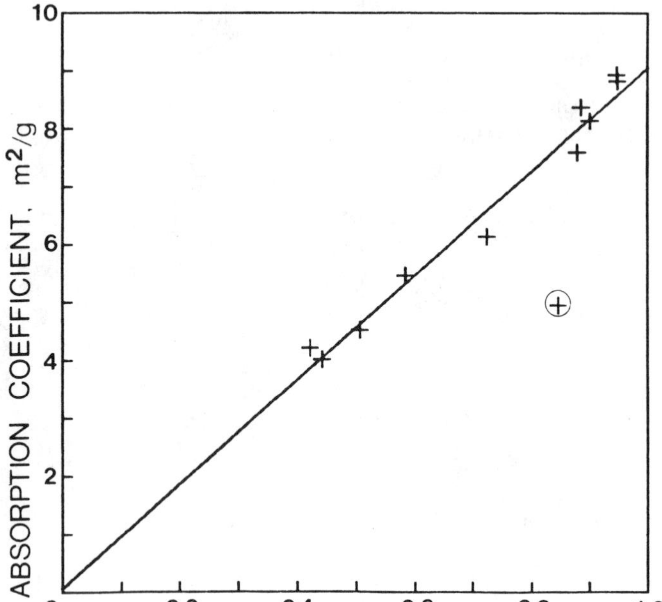

Fig. 3 - Absorption coefficient for 550 nm light versus non-volatile mass fraction by vacuum volatilization. The parameters of the regression line and their 95% confidence intervals are: Intercept = 0.07 ± 0.85 m^2/g
Slope = 9.05 ± 1.13 m^2/g
The circled point was not included in the analysis

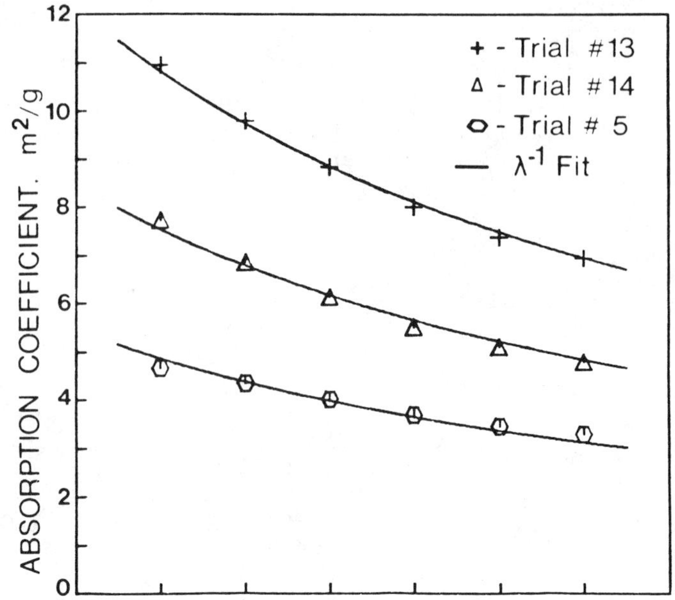

Fig. 4 - Absorption coefficient versus wavelength for 3 trials, with different volatile fractions

Trial	Volatile Fraction (%)
13	5.2
14	27.4
5	55.7

figure 4 the spectral absorption coefficient is plotted against wavelength for three representative samples. The solid lines, which are best fits to a $1/\lambda$ dependence, describe the experimental results very well.

Table 1 also presents the results of the light scattering measurements. These measurements were made with an experimental system that was optimized for absorption and size distribution determinations. Consequently it was necessary to operate the integrating nephelometer near the limits of its sensitivity, which introduces some uncertainty into these results. However, it is clear from these simultaneous measurements that light scattering by these particles is much smaller than light absorption. Within the limitations of these measurements, a dependence of scattering on engine type or operating conditions is not apparent. The average of these measurements gives a scattering coefficient of 1.3 ± 0.3 m^2/g.

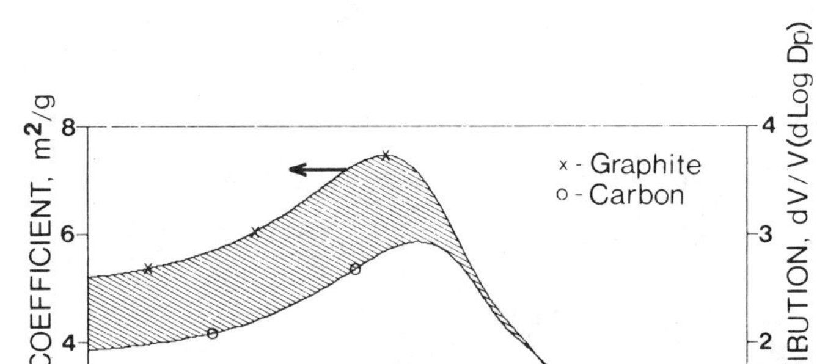

Fig. 5 - Particle volume distribution and Mie scattering coefficient range versus particle diameter

THEORETICAL CONSIDERATIONS

MIE SPHERICAL PARTICLE THEORY - Figures 5 and 6 show the results of Mie* spherical particle theory calculations of light absorption and scattering coefficients, respectively, plotted against particle diameter. The log-normal fit of a particle size distribution measured with the electrical aerosol analyzer in a typical trial, in this case trial 15, is

* Gustav Mie (1868-1957).

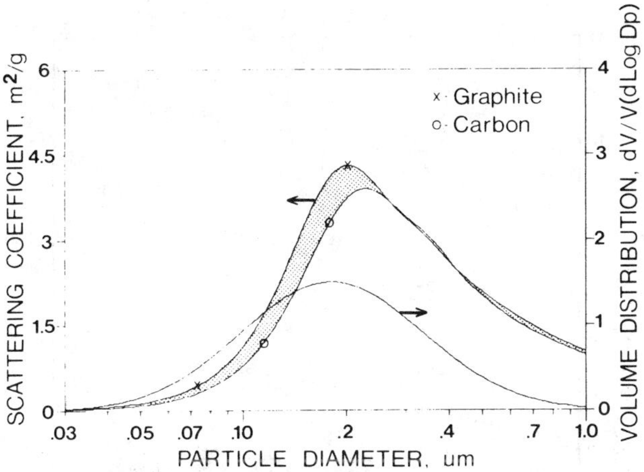

Fig. 6 - Particle volume distribution and Mie scattering coefficient range versus particle diameter

also plotted in both figures. The size dependent absorption and scattering coefficients (5) integrated over this size distribution yield the following theoretical values for trial 15:

$$\bar{R}_{SCAT} = \frac{1}{V}\int R_{SCAT}(D_p)\,\frac{dV}{d(\log D_p)}\,d(\log D_p) \quad (3)$$

$$\bar{R}_{ABS} = \frac{1}{V}\int R_{ABS}(D_p)\,\frac{dV}{d(\log D_p)}\,d(\log D_p) \quad (4)$$

to yield thoretical values:

$$2.4\ m^2/g \le \bar{R}_{SCAT} \le 3.6\ m^2/g \quad (5)$$

$$4.7\ m^2/g \le \bar{R}_{ABS} \le 7.4\ m^2/g \quad (6)$$

Maximum and minimum values given correspond to graphite (refractive index m = 2.0 - 1.0i), and amorphous carbon (m = 1.95 - .66i), respectively. A density of 2.0 g/cm³ is assumed in both cases. Disagreement with experimental scattering and absorption values for trial 15 of 1.29 m²/g and 9.66 m²/g, respectively, is noted. Therefore a more realistic model which includes the effects of a porous, non-spherical particle on theoretical absorption and scattering will be considered.

SHAPE AND DENSITY EFFECTS - The physical structure of a diesel particle is significantly different from the solid sphere assumed for Mie calculations. Density measurements (6) give results as low as 0.7 g/cm³ versus 2.0 g/cm³ for solid carbon, which suggests a very porous structure. Electron micrographs show the particles to have irregular, elongated shapes. Light absorption and scattering effects due to the loosely packed, non-spherical nature of the particles may be considered by examining a simplified model. The porous structure of the particle will result in an effective complex refractive index approximated by the volume mean refractive index (7):

$$\bar{m} = \frac{1}{V}\int m(V)\,dV \quad (7)$$

and:

$$\bar{\rho} = \frac{1}{V}\int \rho(V)\,dV \quad (8)$$

The density parameter $\bar{\rho}$ determines the mean refractive index by specifying the proportions of void and solid carbon which form the particle. Thus:

$$\bar{m} = (m_s - 1)\,(\bar{\rho}/\rho_s) + 1 \quad (9)$$

where the subscript s denotes the solid material.

The elongated particle shape is approximated by a right rectangular prism. The elongation parameter ϵ is given by the height to base ratio of the prism. The equivalent diameter of the prism is defined as the diameter of a sphere of equal volume.

A finite elements solution for scattering and absorption described by Purcell and Pennypacker (8), with improved numerical techniques (9), may be applied to the prism. Three mutually perpendicular orientations of the particle are evaluated and averaged. Results for independently varying density, elongation, and equivalent diameter with 550 nm light are shown in figures 7 - 10. Upper and lower limits given by graphite and carbon are shown in each graph. Figure 7 shows the effect on scattering and absorption for varying density. An equivalent diameter of 0.1 μm is used, with an elongation ϵ = 3. Figures 8 and 9 show the variation of absorption and scattering coefficients with elongation for particles of 0.10 μm equivalent diameter, 1.0 g/cm³ mean density; and of 0.04 μm equivalent diameter, 2.0 g/cm³ mean density, respectively. It is interesting to note that for a small solid particle, absorption and scattering strongly increase with elongation, while for a larger, loosely packed particle, optical interaction is nearly independent of elongation. Figure 10 gives absorption and scattering as a function of particle diameter for ϵ = 3 and $\bar{\rho}$ = 1.0 g/cm³.

These results show that theoretical absorption coefficients for porous, elongated particles are larger than those of solid carbon spheres of the same equivalent diameter, and are consistent with the experimentally observed absorption coefficient of 9.05 m²/g .

The range of size, elongation and density parameters investigated theoretically were chosen to be representative of diesel particles. Within this range the absorption was found to be relatively insensitive to these parameters. This is in keeping with the experimentally observed constancy of absorption for two types of engines under varying operating conditions. A maximum equivalent diameter of 0.2 μm was used in order to limit computation costs which

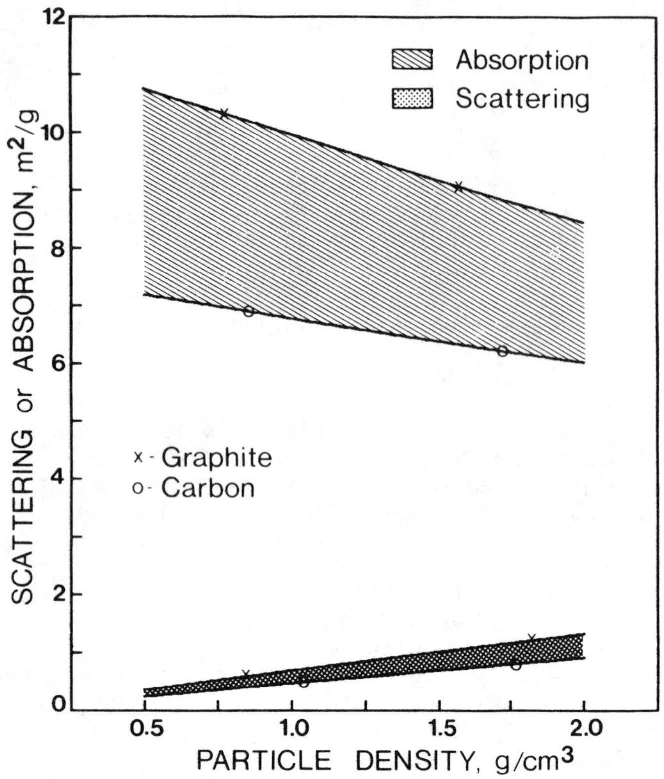

Fig. 7 - Theoretical scattering and absorption for a right rectangular prism of varying density (elongation = 3.0, equivalent diamteter = 0.1 μm)

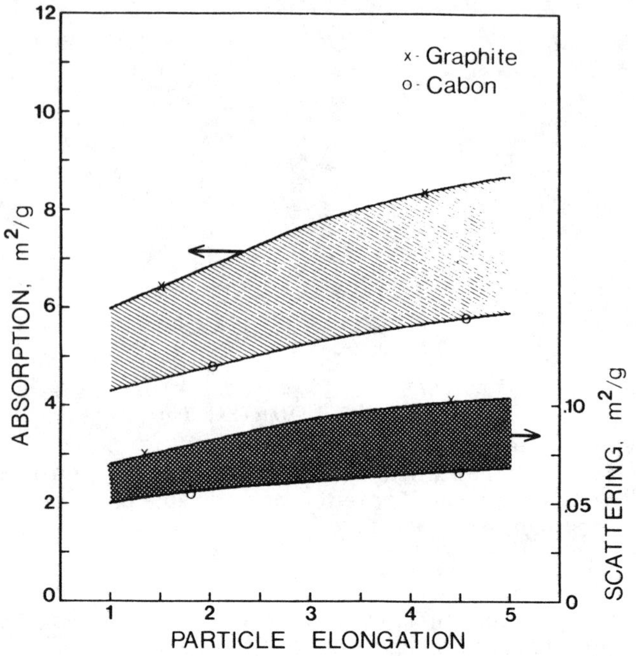

Fig. 9 - Theoretical scattering and absorption for a small, solid prism of varying elongation (density = 2.0 g/cm³, equivalent diameter = 0.04 μm)

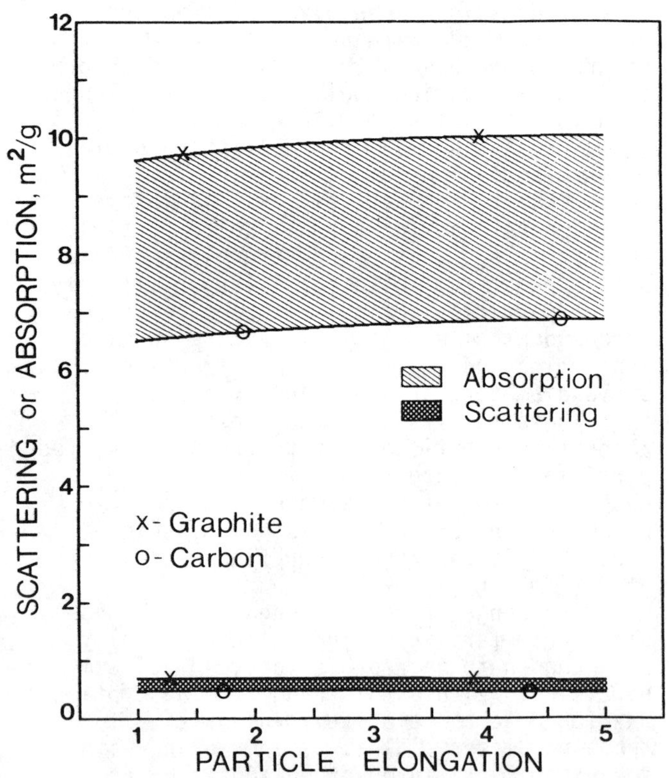

Fig. 8 - Theoretical scattering and absorption for a right rectangular prism of varying elongation (density - 1.0 g/cm³, equivalent diameter = 0.1 μm)

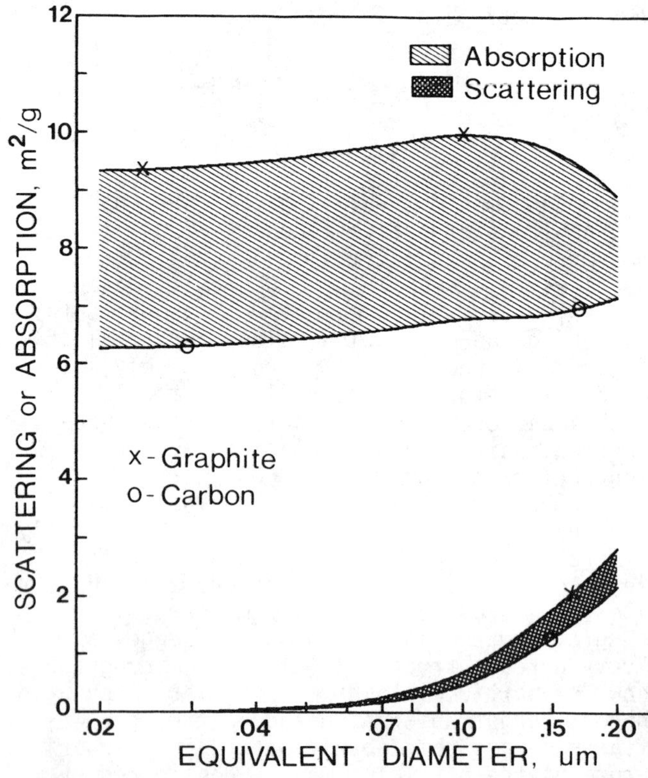

Fig. 10 - Theoretical scattering and absorption for a right rectangular prism of varying equivalent diameter (density = 1.0 g/cm³, elongation = 3.0)

vary with its ninth power. By analogy to the spherical particle results given in figure 5, the absorption coefficient will probably decrease for particles much larger than 0.2 μm equivalent diameter.

Scattering is calculated to be both size and density dependent, with the experimental value of 1.3 m^2/g in general agreement with theoretical values.

CONCLUSIONS

In conclusion, an absorption coefficient of 9.05 m^2/g of non-volatile aerosol mass was observed at 550 nm, independent of both engine type and operating conditions. Further investigation, however, is needed to verify this result for other engines and operating conditions, especially other engine types such as two stroke cycle and turbocharged diesel engines. This result is quantitatively supported by calculations for rectangular prisms of carbon-void structure, which gives values significantly different from those for Mie absorption and scattering by solid spheres. Absorption is experimentally determined to be inversely proportional to wavelength, and found, on average, to account for more than 80% of the light extinction by diesel exhaust aerosols.

ACKNOWLEDGEMENTS

The authors would like to thank Prof. Y. L. Yung for a copy of his finite elements code, and Mr. William Alden for his help in constructing the apparatus. This work has been supported by the Coordinating Research Council and the Engine Manufacturers Association.

REFERENCES

1. C. Lin, M. Baker, and R.J. Charlson, Appl. Opt. 12, 1356 (1973).
2. D.F. Dolan, D.B. Kittelson, and D.Y.H. Pui, SAE paper 800187.
3. Japar, S.M. and Szkarlat, C., "Mass Monitoring of Carbonaceous Aerosols by Spectrophone," Combust. Sci. Tech. 24, 143 (1980).
4. D.M. Roessler and F.R. Faxvog, J. Opt. Soc. Am. 69, 1699 (1979).
5. D.M. Roessler and F.R. Faxvog, Appl. Opt. 17, 2612 (1978).
6. D.B. Kittelson, D.F. Dolan, and J.A. Verrant, SAE paper 780109, p. 6.
7. M. Kerker, "The Scattering of Light," Academic Press, 1969, p. 247.
8. E.M. Purcell and C.R. Pennypacker, Astrophysical J., 186, 705 (1973).
9. Y.L. Yung, Apl. Opt. 17, 3707 (1978).

800336

The Oxidation of Soot Deposits from Diesel Engines*

K. Otto, M. H. Sieg, M. Zinbo, and L. Bartosiewicz
Engineering and Research Staff
Ford Motor Co.
Dearborn, MI

THERE IS CONSIDERABLE INTEREST in the reduction of the amount of particulates from the exhaust of diesel engines. The principal objective of the present work is to define with some precision the conditions which are required to ignite soot deposits in a diesel exhaust system. Proper ignition would then lead to an accelerated combustion caused by the high heat of combustion of carbon.

The present investigation of diesel soot is an extension of our previous work dealing with the reactivity and with gasification kinetics of various forms of carbon, such as graphite and coal chars. Results from those studies suggest that various physical and chemical properties may influence ignition and combustion characteristics of soot. Such properties, including particle size, specific surface area, bulk density, adsorbed species and embedded inorganic contaminants were characterized for various soot samples. Ignition temperatures and oxidation rates as functions of temperature, oxygen pressure and burn-off were then measured.

EXPERIMENTAL SECTION

MATERIALS - Most of the soot deposits used in our study were collected on a pair of cylindrical filters in the exhaust system of an Opel 2.1 l diesel engine. The filters consisted of resin-bonded borosilicate microfibers and measured 33 mm in diameter, 180 mm in height. Each pair of filters was typically exposed to the exhaust gas for four CVS-CH cycles during chassis-dynamometer testing, sometimes with exhaust-gas recirculation. The soot particles were washed off each filter with about 400 ml of acetone and dried by evaporating the acetone in air on a hot plate at 120°C. For most of the experiments, especially for extensive kinetic measurements, the soot was pressed into cylindrical pellets of 9 mm diameter to facilitate handling. The weight of each soot sample was about 300 mg and represented the soot yield of three filters. Because of considerable variation in engine operation during soot formation and collection, uniformity of the chemical and physical structure cannot be expected on a microscopic scale.

For comparison diesel soot samples derived from another Opel, a VW Rabbit, and an Oldsmobile Cutlass engine were prepared in the same fashion.

The acetone treatment produced distinct changes in some soot properties. For example, an increase in bulk density and a decrease in organic adsorbate were observed. These changes, in turn, can influence soot ignition and combustion to some extent, as will be

*Paper 800336 presented at the Congress and Exposition, Detroit, Michigan, February, 1980.

ABSTRACT

Combustion and gasification characteristics of deposits from diesel exhaust were investigated. Properties which may influence soot reactivity, such as morphology, specific surface area, organic adsorbate, and occlusions of inorganic origin were characterized. Effects of temperature, oxygen pressure, and stepwise combustion on the reactivity were determined. Ignition and combustion kinetics of various soot deposits were measured gravimetrically, and requirements for fast soot oxidation in the presence of an exhaust gas were assessed. Reliable completion of soot combustion after ignition depends strongly on oxygen pressure and bulk density of the soot. The ignition temperature can be lowered by suitable catalysts. Organic molecules, adsorbed on the soot surface, may also have a significant influence on the ignition of soot deposits. The results confirm that soot particles suspended in exhaust gas are not oxidized significantly under the conditions prevailing in diesel exhaust.

discussed below.

It was possible to remove a few small samples by mechanical means without solvent treatment from some of the filters. These samples were used to assess the influence of the acetone treatment. Three additional samples of soot, deposited in different types of traps, were used for further comparison.

The possibility of lowering the ignition temperature of soot by an inorganic catalyst was explored by using barium as a prototype. For these experiments soot was impregnated with an aqueous solution of analytical grade barium acetate to yield a loading of 1 atom% Ba. This soot was dried at 100°C in air for two hours.

For a direct comparison of soot reactivities with a well-defined carbon, SP1 graphite (Union Carbide, less than 1 ppm ash) was used.

GRAVIMETRIC MEASUREMENTS - Gasification rates and surface areas were measured gravimetrically with a Cahn R. G. balance, as described earlier (1,2).* Samples were degassed, usually at 400°C, for several hours initially and before each new charge of gas was admitted to the balance system. Gasification rates of soot by steam or carbon dioxide were measured at a reactant pressure of 2.7 kPa (0.026 atm) with 56 kPa (0.55 atm) of Ar as a carrier gas. Gasification by oxygen was measured at various oxygen pressures using ambient air diluted with Ar to yield a total pressure of 100 kPa. The gas mixture was recirculated by a glass pump containing a Teflon gasket (1). The reaction rate was measured as a function of temperature, in a convenient temperature range, and as a function of burn-off down to about 20% of the original sample weight.

Surface areas were measured in situ either by the adsorption of carbon dioxide overnight at the temperature of dry ice and 20 kPa, or the adsorption of Ar for 70 minutes at the temperature of liquid nitrogen and 3.3 kPa. The surface area was evaluated according to the BET method, as previously described (1). The specific surface area of soot was found to be larger than that of graphite by two orders of magnitude.

Another series of experiments was carried out by thermogravimetric analysis (TGA), for which a DuPont 951 unit was employed. Samples of only about 10 mg, used in powder form, were heated from room temperature to 750°C at a rate of 20°C/min in either argon or air at a flow rate of 60 cm^3/min.

SOOT COMBUSTION IN SIMULATED EXHAUST - To assess the influence of various components present in automobile exhaust on combustion characteristics, several soot and graphite samples were gradually heated in simulated automobile exhaust. For this purpose a pulse-flame generator, used in the laboratory testing of thermal and chemical deactivation of automobile exhaust catalysts (3), was modified to produce a continuously burning isooctane flame. The exhaust, at a flow rate of about 2500 cm^3/min, contained typically 1.2% CO, 500 - 1000 ppm unburnt hydrocarbons and a negligible amount of NO in addition to CO_2 and water. The oxygen content was varied by admitting secondary oxygen downstream of the flame. The remainder of the exhaust gas was nitrogen. Samples were placed into the combustor tube either as pellets or as powder in a combustion boat. For direct temperature measurements a chromel-alumel thermocouple in a stainless steel sheath was either attached to the sample pellet by a platinum wire, or placed into the powder-filled combustion boat.

---

* Numbers in parentheses designate references at end of paper

EXPERIMENTAL RESULTS AND DISCUSSION

PARTICLE SIZE - There is agreement in numerous publications that primary soot particles from diesel engines are more or less spherical with an average diameter of about 10 - 30 nm (4 - 9), comprising about 10^6 C-atoms. It has been deduced from theoretical considerations that soot particles measuring less than 10 nm in diameter are not stable and should burn up in a luminous diffusion flame (10). The primary spherical particles coalesce into larger agglomerates of irregular shape, which have a diameter of about 0.1 - 1 μm (5 - 9, 11). There is evidence that the final size of these agglomerates does not depend strongly on engine-operation parameters (11), although stickiness of the surface, the exhaust temperature, and the number of particle collisions are expected to have some effect on the final particle size (4).

We noticed considerable differences in the apparent density of diesel-soot deposits. Thus soot removed from an electrostatic precipitator had a bulk density of only 0.02 g/cm^3, as measured in a graduated cylinder after vigorous tapping. Another sample, shaken out of a monolithic trap, had a bulk density of 0.056 g/cm^3. When acetone was added and evaporated, the soot samples collapsed and the bulk density increased to 0.4 g/cm^3. On the other hand, soot removed from an exhaust-recirculation valve had the same density of 0.4 g/cm^3 without acetone treatment.

The appearance of the soot particles studied here was found to be in agreement with the description given in the literature. When accumulated on a filter the soot structure under magnification has a cauliflower-like appearance, as shown by the micrographs in Fig. 1.

For all practical purposes soot can be considered amorphous, although x-ray diffraction indicates some ordered hexagonal arrays of carbon atoms which are stacked in parallel layers of 2 to 20 units (5, 12), with an interlayer spacing which is about 3 to 5% larger than that in graphite (12). The ordered basal planes measure about 2 - 3 nm across. Thus it appears that, in general, primary soot particles from various sources do not differ very much in their physical structure, but that the final particle size and the bulk density depend on the number of collisions and on other conditions.

SURFACE AREA AND PORE STRUCTURE - The specific surface area of soot can be measured rather

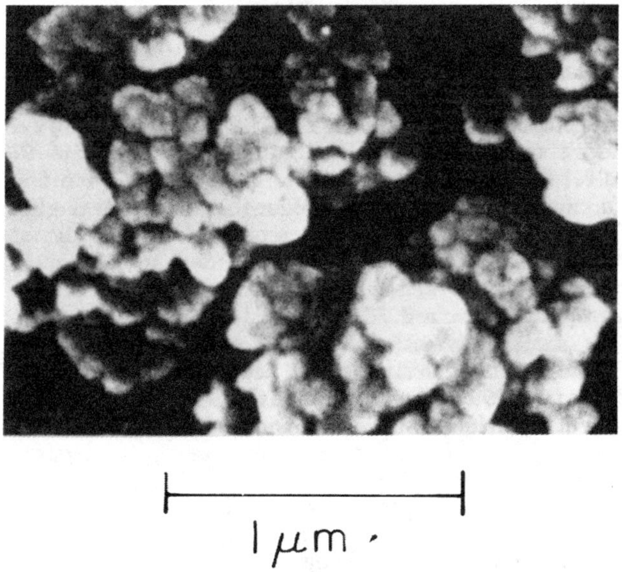

Fig. 1 - Micrographs of diesel soot on filter

easily and reliably by the amount of adsorbate which can be accommodated. However, since soot has a relatively large surface area of about 200 m^2/g, caution has to be exercised to avoid erroneous adsorption data due to extremely narrow pores. For example, BET surface areas derived for coals or fresh coal chars were found to be strongly dependent on the adsorbate used for the measurement. Thus adsorption of argon (1) or nitrogen (13) at the temperature of liquid nitrogen yielded BET areas which were about one order of magnitude smaller than those measured by the adsorption of carbon dioxide at the temperature of dry ice or at 25°C. The discrepancy is explained by micropores in the carbon structure which are comparable in size to the adsorbed species. At low temperatures an adsorbate atom or molecule requires exceedingly long diffusion times to penetrate very narrow pores. The difference between results obtained with different adsorbates becomes gradually smaller as the pores are widened during the gasification process (1). The soot samples examined here did not indicate such a problem. Table 1 shows the BET areas for two samples, as measured by the adsorption of argon at the temperature of liquid nitrogen, and of carbon dioxide at the temperature of dry ice. As the adsorption isotherms were approximated by the adsorbate uptake at one pressure only, the experimental error of the surface-area measurements is judged to be ± 10%. The agreement of the data obtained by the two adsorption methods, shown in Table 1, is satisfactory, even at the stage where the samples have undergone only little gasification. The absence of diffusional hindrance during these adsorption experiments is supported by the experimental fact that for the two adsorbates used here, more than 90% of the total amount was taken up within 20 minutes.

TABLE 1 - Comparison of BET Measurements Based on the Adsorption of Argon and Carbon Dioxide

	Burn-off %	BET area (m^2/g)		ratio CO_2/Ar
		CO_2	Ar	
Sample 1	0	86	104	0.83
	11	164	146	1.12
	18	202	200	1.01
	28	268	296	0.91
Sample 2	17	260	246	1.06
	38	314	316	0.99
	71	327	333	0.98

Since soot gasification is a surface process it is natural to calculate reaction rates per unit surface area rather than per unit weight. This approach is especially important if the specific surface area changes significantly with progressive burn-off. Examination of the soot surfaces disclosed areas as small as 80 m^2/g, for freshly devolatilized soot, and as large as 400 m^2/g, for a sample which had been gasified in air by about 70%. The typical increase in surface area as a function of burn-off in air at a gasification rate of about 10 $\mu g/(m^2\ min)$ is shown in Fig. 2. The development of the surface area is clearly influenced by the initial devolatilization temperature. Thus, as Fig. 2 shows, the specific surface area of several soot samples devolatilized at 700 - 800°C is consistently larger, by about 35%, than those devolatilized in the range from 400 - 500°C. Interestingly the surface development of soot also depends on the gasification agent in the same manner observed in the case of coals (14). For example, at comparable rates, gasification by water or carbon dioxide produces significantly larger surface areas, as shown in Fig. 3, than does

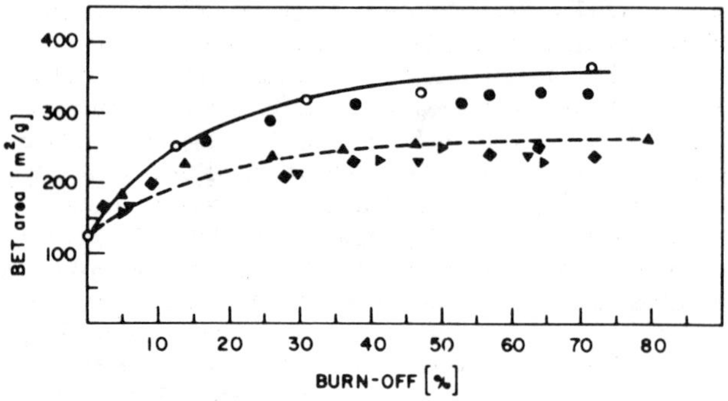

Fig. 2 - Specific surface area of diesel soot as a function of burnoff in air at °C: 410 (▲,▶), 425 (◆), 495 (▼), 700 (●), 800 (o)

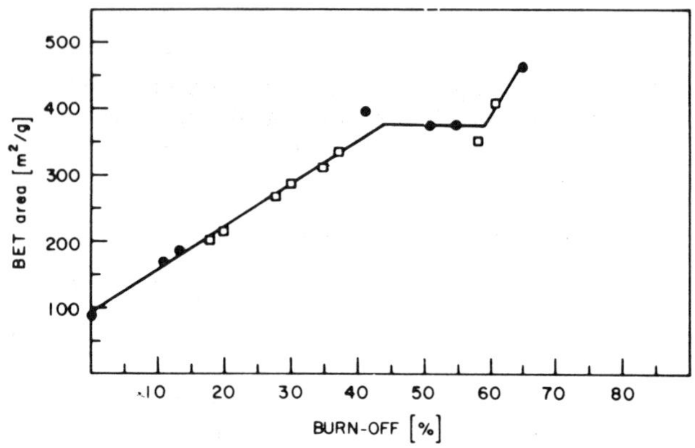

Fig. 3 - Effect of gasification agent on the development of the specific surface area of diesel soot at 400°C in air (●), in H_2O or CO_2 (□)

gasification by oxygen (Fig. 2). This difference in surface development is more pronounced at medium and high levels of burn-off, in agreement with the findings of Tomkow et al. (14) on coals.

A brief calculation can be made to test whether the assumption of smooth and spherical primary soot particles of an average radius of 10 nm is consistent with the adsorption measurements. Assuming a density of 2 g/cm^3 (15) for these spheres the specific surface area as a function of the radius of the sphere, r, is given by

$$a = 1.5 \times 10^{-6}/r \ (m^2/g) \qquad (1).$$

With a radius of 10 nm, a surface area of 150 m^2/g follows, in reasonable agreement with the measured surface area of the soot (see Table 1).

Furthermore the minimum pore radius in a network of uniform spheres of radius r is given by a trigonal opening between three spheres touching each other. The radius of this opening follows as

$$r_p = (2/\sqrt{3} - 1)\, r = 0.15\, r \qquad (2).$$

For a sphere of radius 10 nm the diameter of the pore opening is 3 nm. Measurements of diffusion coefficients in activated carbon (16) led to the conclusion that diameters of micropores larger than 0.7 nm do not cause an increase in the time which is required to reach adsorption equilibrium.

Thus both the total surface area and the absence of diffusional hindrance during physical adsorption are consistent with the assumption of spherical carbon particles of approximately uniform size, as mentioned above. This result, of course, cannot be used to conclude that gasification rates are equally unaffected by diffusion, as mentioned below.

ORGANIC COMPONENTS - Soot carbon, in general, contains an appreciable amount of organic molecules, which have received increasing attention. Because of the high complexity of the engine combustion process and differences in engine conditions, the organic component of diesel soot is as yet poorly defined. The overall weight ratio of hydrogen to soot carbon can be used to characterize the organic component semiquantitatively. This ratio can be measured by simple analysis of the combustion products of the soot and is known to vary in a wide range from about 0.01 - 0.15 (4, 17, 18). For example, a comparison (17) of two different diesel engines, a Caterpillar 3208 (4-stroke) and a Detroit Diesel 6V-71, showed consistently H/C weight ratios of about 0.03 and 0.12, respectively. Similar differences were found in a comparison between a Peugeot 204D diesel engine and various Mercedes diesel engines (18). Analysis of one of our fresh soot samples yielded an H/C weight ratio of 0.027, which decreased to 0.006 after the sample had reached 70% burn-off. The ratio for the fresh sample may have been lowered by a loss of organic molecules during the acetone treatment. The ratio after 70% burn-off indicates that a small portion of hydrogen is chemically bonded to soot carbon and not merely adsorbed as a component of organic molecules on the carbon surface. These hydrogen atoms are presumably left with carbon atoms from the fuel, which have been incompletely stripped during soot formation. A more discriminating analysis of organic components associated with soot requires thermogravimetric analysis, or extensive solvent extraction followed by analyses based on gas chromatography and mass spectrometry.

Most of the soot hydrogen is associated with organic molecules derived from the components of fuel or engine oil (17), which are adsorbed on the large surface area of soot. The adsorption equilibrium depends strongly on the soot temperature. In one investigation (6) soot collected at 130°C at the exhaust manifold, lost 75 wt% in volatile components when heated to 350°C in air. The corresponding loss of a sample collected at 250°C was only 10 wt%. The large difference in the amount of organic material deposited on soot under varying exhaust conditions observed in these devolatilization experiments was confirmed by solvent extraction. Thus Funkenbuscn et al. (19) used an elaborate extraction technique on soot derived from the the diluted exhaust of a 1976 Caterpillar Model 3208 diesel engine and measured an organic fraction in the range 10-90 wt%, depending on fuel and operation mode. In our gravimetric studies the weight loss up to 750°C in an inert atmosphere was usually about 10 - 20%.

Because of known large variations in quality and quantity of the organic deposit, we made no attempt to characterize the adsorbed components in greater detail. However, a brief examination was carried out to evaluate redistribution by acetone of organic compounds associated with soot from one of the fiber filters of the Opel 2.1 l diesel engine. Analyses by gas chromatography and mass spectrometry showed that the organic compounds extracted by acetone amounted to 1.7 wt% of the soot sample. A duplicate sample after an acetone washing, followed by evaporation of the acetone, contained 7.2 wt% in volatile organic material, as measured by thermogravimetric analysis. Thus not more than 24% of the organic component was dissolved during the acetone treatment. No organic compounds were found in acetone used for washing of an unused filter element. Of the acetone-soluble compounds, 63% were n-alkanes distributed between n-tridecane (C_{13}) and n-pentasocane (C_{25}). The rest of the organic compounds soluble in acetone (37%) were branched or oxygenated hydrocarbons and compounds consisting of fused rings. The distribution of the compounds indicates that a substantial amount of the organic molecules was derived from the engine oil. The H/C weight ratio of 0.03, mentioned above, is consistent with the presence of a significant portion of oxygenated compounds.

A few key experiments were also carried out to assess the adsorbate influence and, in particular, the effects of the acetone washing on soot ignition. The results will be discussed below together with the data of the reaction studies. It should be stressed that the main purpose of this investigation is to evaluate the reactivity of soot carbon per se. The possibility of adsorbed species acting as a "lighter fluid" and associated topics require a more detailed study and will be examined separately.

INORGANIC COMPONENTS - The definition of soot reactivity is further complicated by a substantial amount of inorganic matter which is incorporated into the soot. Ash, comprised of a variety of inorganic oxides, accounted for 2 - 6 wt% of our soot samples. Assuming soot formation at a rate of 0.5 g/mile (20), an exhaust trap designed for periodic combustion of soot deposits would accumulate ash residue on the order of 1000 g over 80,000 km (50,000 miles), if the ash were fully retained. On the positive side, ash components have a potential to aid the soot combustion catalytically. Elements, such as K, Ca, Fe and many others, are known to increase the gasification rate of carbon significantly (21). An evaluation of catalysis by inorganic soot contaminants involves a high degree of complexity and demands further extensive experiments. The present report will be essentially limited to a qualitative characterization of the inorganic contaminants and to a demonstration that a catalytic lowering of soot-ignition temperature is possible.

The existence of a large variety of elements in exhaust gas has been observed in numerous studies of automotive catalysts (22). These elements are derived from additives and contaminants in fuel and engine oil, and from engine construction materials. Basically the same contaminants are present in diesel exhaust, and up to 20 trace elements have been identified in samples of diesel soot (9, 23). The quantities of inorganic deposits vary, of course, strongly with their concentrations in fuel and engine oil. Elemental analyses of the soot samples used in our investigation by x-ray diffraction in connection with a scanning electron microscope, revealed Fe, Ba, Ca, Zn, and Ni as major components S, P, Cr and Cu as minor components. It should be stressed that the distribution of the inorganic deposit is uneven on a microscopic scale. Some elements, for example iron, have a strong tendency to concentrate in certain spots, while others, such as barium, are much more evenly distributed. Occasionally isolated areas with substantial deposits of Cr or Cu were detected, while other areas of the same soot sample were completely free of these elements.

From a catalytic viewpoint Ca and Ba are the more interesting elements. It is known that as a fuel additive, Ba can suppress soot formation in diesel engines significantly (24). Moreover both elements were found to increase steam-gasification rates of graphite (25) and coal chars (26) strongly. The catalytic action of other elements, for example iron, depends on their oxidation state. The alkaline earths Ba and Ca were found to constitute up to 0.088 and 0.18 wt%, respectively, of particulates from a Mercedes 240D exhaust (23). The source of the alkaline earths is not entirely clear; it can be either the engine oil, or the fuel, or both. Because of the catalytic potential of Ca and Ba it was of interest to analyze the engine oil and the diesel fuel used in one of our tests by emission spectroscopy. The analysis is given in Table 2. Based on a fuel economy of 32 miles/gallon (13.5 km/l) and an engine-oil consumption of 1 quart per 1000 miles (0.6 l per 1000 km), the amounts of Ca and Ba which passed through the exhaust system per km can be calculated. Assuming that the alkaline-earth species are deposited onto the soot in proportion to their concentrations, the relative amounts shown in Table 2 are calculated with reference to the amount of Ba derived from the engine oil. These relative deposits show that the source of most of the calcium is the oil additive, while barium is carried predominantly by the fuel. The ratio of the amounts carried per mile Ba:Ca = 14:84 is consistent

TABLE 2 - Analysis of Calcium and Barium from Fuel and Engine Oil

Analysis	Calcium	Barium
Diesel fuel	0.2 ppm	1.16 ppm
Engine oil (Shell Rotella)	912 ppm	11.2 ppm
Relative amounts carried per mile		
From fuel	2.2	12.9
From oil	81.4	1.0

with the intensities of Ba and Ca displayed in elemental maps of soot produced during this test; although an accurate quantitative analysis is not available.

GRAVIMETRIC STUDIES OF SOOT IGNITION - After assessing those properties of soot which have a potential influence on its reactivity, soot oxidation per se will be discussed now. Most of the soot samples used in the following experiments were our "standard" samples removed from an exhaust filter of an Opel engine by acetone washing, as described in the experimental section. The use of any of the other soot samples, mentioned above, will be clearly identified in the description of the experiments.

Thermogravimetric analysis (TGA) is a useful method for evaluating devolatilization and ignition characteristics of soot. The typical weight loss of our "standard" diesel soot, is shown in Fig. 4. Heating in

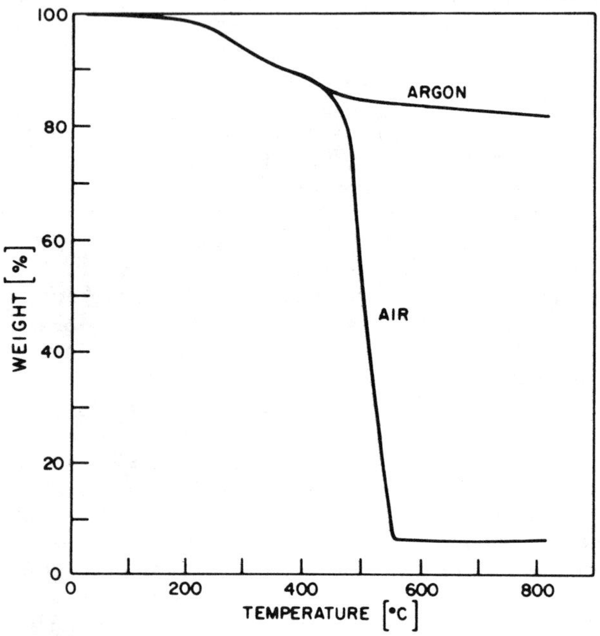

Fig. 4 - TGA trace of diesel soot in argon and in air

argon to 110°C (230°F) at a constant rate of 20°C/min results in a weight loss of about 1 - 2%. This weight change is probably due mostly to the loss of water which was adsorbed from the atmosphere on the soot surface after the acetone treatment. This amount of water accounts for roughly 1/3 of a monolayer coverage on the soot surface. Further heating in Ar removes hydrocarbons amounting to 10 - 20% of the sample weight. When heated in air the weight loss of a duplicate sample coincides with that measured in Ar until about 420°C. At this temperature the weight loss steepens as ignition occurs. Within about five minutes a constant weight of about 5% of the original sample weight is reached, which is attributable to the ash residue.

An ignition temperature was defined from the TGA traces by that point at which the trace measured in air diverged from the trace of a corresponding sample in Ar by bending downward. This temperature is approximately given by the point at which the TGA trace in air starts to deviate from a straight line segment before an abrupt change caused by soot combustion occurs. Since the curvature before onset of the combustion process is not very pronounced, the ignition temperature is known with an accuracy of only about 10°C. Table 3 shows that the ignition temperature range is 350 - 450°C. As initial ignition does not always lead to complete combustion, an ignition temperature defined in this manner has to be considered as a lower limit.

TABLE 3 - Ignition Temperature Measured by TGA

Sample designation	Ignition temperature, °C
(1) Opel 2.1 l	450
(2) Opel 2.1 l*	420
(2) Opel 2.1 l	415
(3) Opel 2.1 l	430
(4) Opel T602	420
(5) Opel T602	420
(6) Oldsmobile Cutlass*	420
(6) Oldsmobile Cutlass	430
(7) Oldsmobile Cutlass*	370
(7) Oldsmobile Cutlass	380
(7) Oldsmobile Cutlass	350
(8) VW Rabbit	420

* Samples not subjected to acetone wash

Ignition was also evaluated for pellets of our standard soot with a Cahn R. G. balance. In this case the soot samples were preheated to 700°C in vacuo, cooled to room temperature and reheated slowly in air until ignition, immediately followed by a large increase in temperature, was observed. This temperature differs from the ignition temperature defined above for the TGA experiments and will be referred to as "light-off" temperature for distinction. The light-off temperature is expected to be somewhat higher than the ignition temperature, as light-off is accompanied by sustained soot combustion at a higher rate and any effect of adsorbed hydrocarbons is diminished because of the preheating. A series of light-off temperatures is given in Table 4. In spite of differences in thermal treatment and heat conductivity there is reasonable agreement between the range of temperatures in Tables 3 and 4. The relatively large interval of 350 - 450°C encompasses most of

TABLE 4 - Light-off Temperature Measured by Cahn Balance

Sample designation	Burn-off wt %	Light-off temperature, °C
(1) Opel 2.1 1	5	400
	20	378
(2) Opel 2.1 1	0	458
(3) Opel 2.1 1	17	420
	26	393
	38	394
	64	437

these temperatures. It is noteworthy that a light-off temperature in air could not be detected for spectroscopically pure SP1 graphite even when heated to 900°C.

These results indicate that ignition is not a well-defined process. Ignition is expected to start at a location of high activity, for example at a catalytic center. To sustain and accelerate the initial oxidation, it is necessary that the reaction be fast enough and that the soot deposit retain a sufficient amount of the heat released during oxidation to cause self-sustained combustion at a higher rate.

It is important to keep in mind that due to our procedure of soot collection most of the samples had to be exposed to acetone. The acetone wash can remove, in part, or can redistribute the adsorbed organic molecules. Either change can influence the ignition temperature. It is therefore essential to compare samples which were removed from our filters by shaking (indicated in Table 3 by an asterisk) with samples subjected to an acetone wash. For further comparison one sample of soot was removed mechanically from a ceramic trap and one portion was subjected to TGA in this form. Another portion of the same sample was analyzed after immersion in acetone and evaporation of the solvent in a water bath. The acetone treatment was accompanied by a large increase in bulk density from 0.056 to 0.39 g/cm^3. The two TGA curves, measured in air, are given in Fig. 5. Both samples contained the same amount of organic material, 4.5 wt%, and the ignition temperatures as well as the combustion temperatures, defined by the steepest slope, are in agreement within about 10°C. It appears that careful washing with acetone does not remove the major part of the adsorbed hydrocarbons. A portion of the dissolved hydrocarbons is probably readsorbed onto the soot surface during acetone evaporation. It should be noted that the amount of volatile material is atypically low in this case.

At the same time acetone can cause distinct changes in a TGA curve, as an instructive example in Fig. 6 shows. The acetone-washed part of a sample of our standard soot, removed from a fiber filter (Opel 2.1 1) shows smooth combustion characteristics (dashed curve). Ignition in air at a flow rate of 60 cm^3/min takes place at 420°C, and combustion is complete at 550°C. In the absence of an acetone

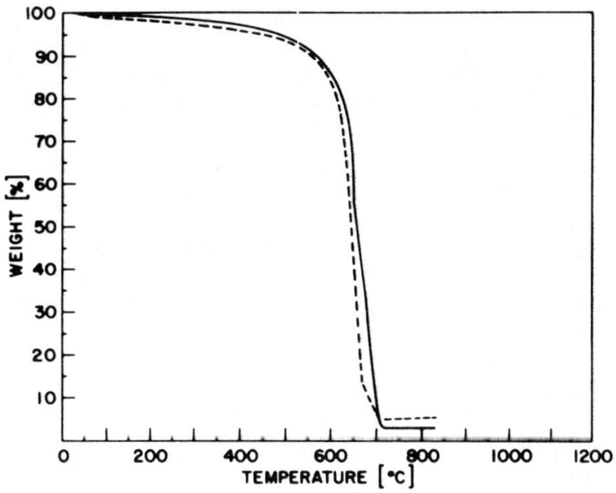

Fig. 5 - TGA traces of soot before (——) and after (----) acetone wash

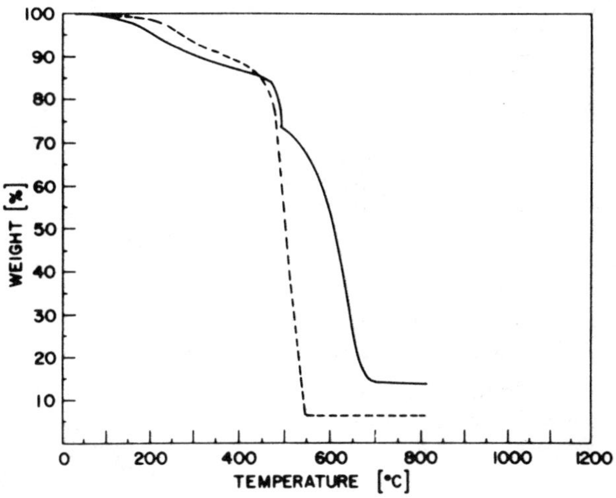

Fig. 6 - TGA trace showing combustion quenching without acetone wash (——), after acetone wash (-----)

treatment the weight loss preceding ignition (solid curve) is somewhat shifted compared to the dashed curve, but the total weight loss due to volatile material is similar. Thus at 800°C, when heated in argon, the weight loss accounted for 18.3 and 19.2 wt% with and without acetone treatment, respectively. Ignition in air occurs in Fig. 6 at the same temperature for both TGA traces. However, without acetone treatment (solid curve) a pronounced combustion delay occurs, starting at 490°C; combustion is completed only at 710°C. The increase in residue associated with the dashed trace originates from filter material which remained with the soot during the mechanical removal procedure. These TGA traces indicate that a substantial portion of the adsorbate has been released into the gas phase by the time ignition occurs. Furthermore the remainder of the hydrocarbons on the soot surface does not appear to play a major part in the combustion characteristics. The break in the solid curve during combustion is probably caused by the poor heat conductivity within the fluffy soot structure. In contrast, the heat conductivity is improved by the acetone treatment because of an increased bulk density, mentioned above, and a smoother combustion trace results (dashed curve). Further evidence for this conclusion will be given below, in connection with experiments carried out with simulated exhaust. Our limited comparisons indicate that soot combustion is usually, but not always, helped rather than impaired by the solvent treatment.

Up to this point there has been no evidence that adsorbed organic molecules have a pronounced influence on light-off or ignition temperature. However, one particular sample, removed from an EGR valve of an Opel 2.3 l diesel engine, showed a behavior different from the other samples. This sample had a relatively high bulk density of 0.4 g/cm^3 even without an acetone treatment. Some of the reaction characteristics are illustrated in Fig. 7. Without acetone treatment, heated in Ar, the soot lost almost twice the amount of volatile material compared to previous samples. When the same sample was cooled to about 150°C and reheated in air, ignition was recorded at about 420°C. A duplicate sample heated in air from the start yielded an ignition temperature of 315°C (Fig. 7). Another sample, taken from the same batch of soot, was wetted with acetone and dried before heating in air. The TGA trace of this sample shows an initial weight loss about 2% smaller than that of the untreated soot sample. This result suggests that the acetone removed some moisture which was not readsorbed. The acetone treatment here seems to increase the ignition temperature by about 30°C, although a comparison is difficult because of the shifted desorption pattern. The point at which combustion is

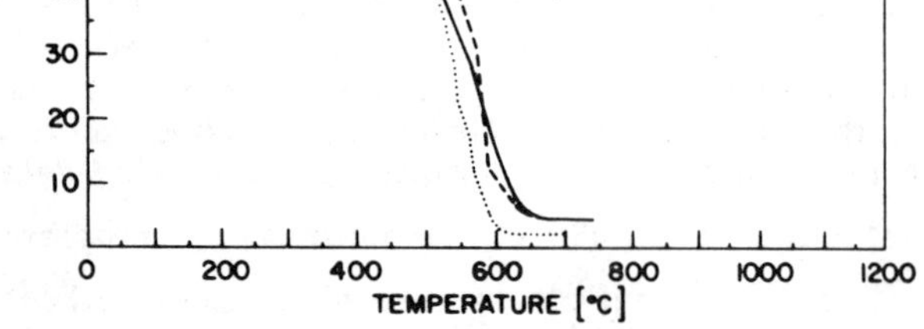

Fig. 7 - TGA traces of soot removed from exhaust mixing valve (Opel diesel 2.3 l); (a) without acetone treatment, heated in air (———), heated in Ar (—. —.), followed by cooling and heating in air (.....); (b) with acetone treatment, heated in air (-----)

completed is not affected by the acetone treatment. These results indicate that organic molecules on soot in the recirculation valve differ significantly from those passing normally through the exhaust system. It is possible that these hydrocarbons, derived either from fuel or engine oil, are partly oxidized and more reactive than those adsorbed on soot trapped on a standard exhaust filter and thus serve as a "lighter fluid". Further experiements are in progress to identify the chemical nature of such reactive surface species, which are as yet hypothetical. It is unsatisfactory to explain the results of the TGA traces in Fig. 7 by catalysis due to inorganic ash components.

OXIDATION RATES VERSUS BURN-OFF AND OXYGEN PRESSURE - To study soot reactivity more systematically, reaction rates of soot pellets weighing about 300 mg were measured gravimetrically under various conditions. Furthermore, a comparison was made with the oxidation rates of spectroscopically pure SP1 graphite, which is known to be a carbon of very low reactivity. Fig. 8 illustrates the difference between reaction rates of graphite and soot in atmospheric air at 70% burn-off. Arrhenius plots, used in this comparison, describe the reaction rate of carbon, v (per unit surface area) as a function of the absolute temperature T

$$v = v_o \exp\left[-E/(RT)\right] \qquad (3),$$

with two kinetic parameters: E, an apparent activation energy, and v_o, a frequency factor, which is proportional to the density of reaction sites; R is the gas constant. The oxidation rate of soot exceeds that of graphite by two orders of magnitude. The apparent activation energy, which is proportional to the slope of the Arrhenius plot, averages 178.2 ± 24.7 kJ/mol for graphite in the burn-off range 0 - 70%. The soot sample, which was also oxidized in several steps at different oxygen pressures, yielded an overall apparent activation energy of 142 ± 21 kJ/mol. These ranges of the activation energy derived for graphite and soot are in agreement with many literature data (27 - 35); although in some instances activation energies for carbon oxidation have been reported in the wide range from 71 to over 420 kJ/mol (34).

While the specific oxidation rate of graphite per unit surface area remains constant with burn-off, the rate for soot declines. This decline is shown in Fig. 9 for soot combustion at 350°C at three oxygen pressures. A considerable degree of scatter illustrates that the rate at a given degree of burn-off is poorly defined for soot. Reproducibility at a given pressure is exemplified in Fig. 9 by two runs measured at 0.097 atm oxygen. Rates below 0.3 $\mu g/(m^2$ min) had to be derived by extrapolation and are inherently less accurate.

Oxidation rates of soot at 350°C and 50% burn-off, taken from Fig. 9, are represented by the solid line in Fig. 10. The line has been derived by a least-squares fit with a heavily weighted origin. The dashed line in the same figure is based on a semi-empirical correlation of soot reactivity and oxygen pressure at 350°C, derived by Lee et al. (35) for a soot from a

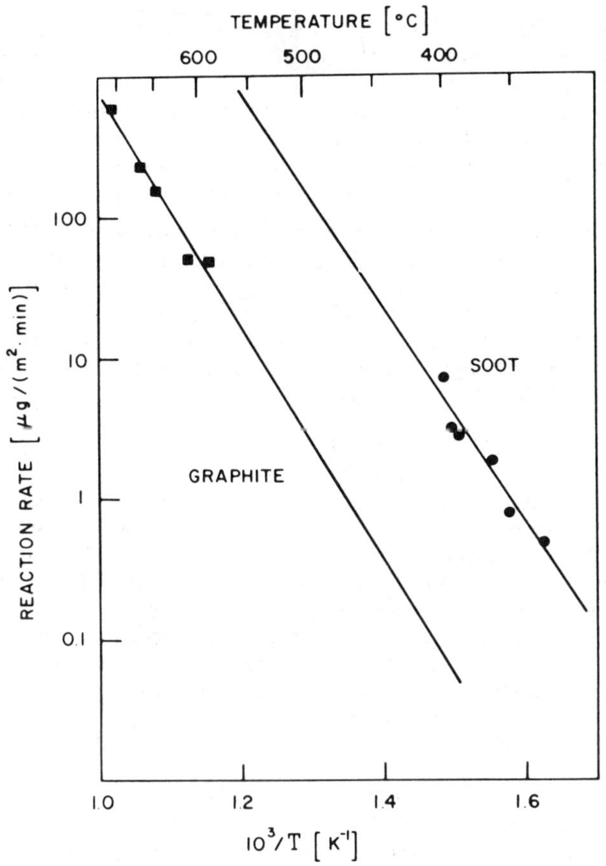

Fig. 8 - Reactivity of graphite and soot in air at 70% burn-off

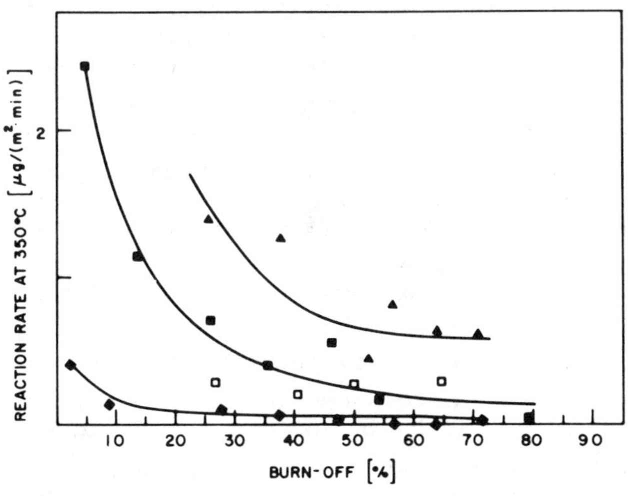

Fig. 9 - Decline of soot reactivity with burn-off at three oxygen pressures: (◆) 0.028, (■,□) 0.097, (▲) 0.21 atm

mixed hydrocarbon flame. The agreement of the two lines in Fig. 10 is reasonable considering the strong inherent fluctuations of soot reactivity. The scatter is emphasized by including in Fig. 10 points from measurements on other soot samples. It is safe to

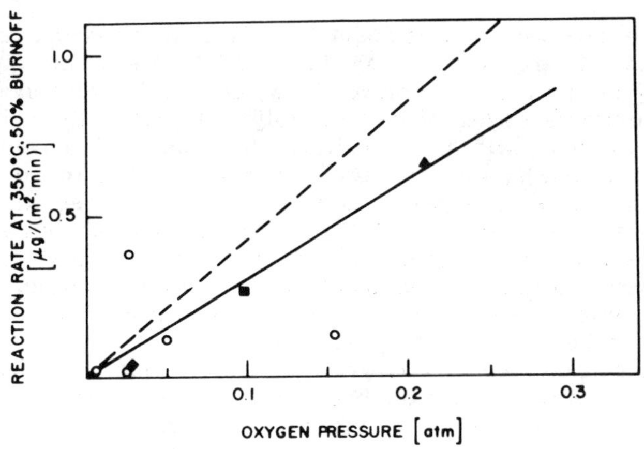

Fig. 10 - Influence of oxygen pressure on soot reactivity data from Fig. 9 (———), from reference (35) (-----)

conclude that the soot reactivity depends strongly on oxygen pressure and that the reaction with oxygen is described satisfactorily by first order kinetics (29, 30, 36).

It is suspected that the poor reproducibility of the reaction rates is caused, in part, by local overheating due to the highly exothermic reaction. To check our basic results, we also employed steam at a pressure of 0.026 atm as a gasification agent. Since the reaction between water and carbon is endothermic, the problem of local overheating is eliminated. Soot gasification by steam, like that by air, is characterized by a significant rate decrease with burn-off, shown for two stages, at 28 and 58%, in Fig. 11. The soot reactivity here is greater than that of graphite by one order of magnitude. The apparent activation energy of

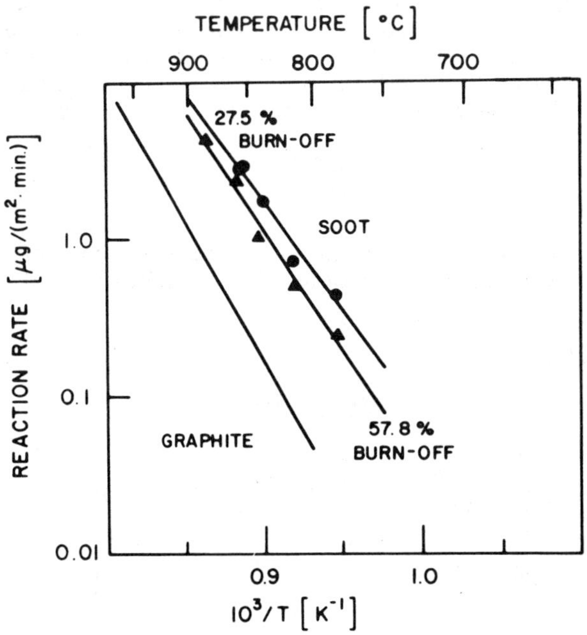

Fig. 11 - Steam gasification of SP1 graphite and diesel soot

the two soot runs is about 40 - 70 kJ/mol lower than that of pure graphite (335 kJ/mol) (2). The relatively small decrease is most likely caused by diffusion hindrance rather than by catalysis (2, 25, 26); it is consistent with the conclusion that soot has a relatively open pore structure. It is interesting to note that the steam-gasification rate of soot agrees with that of a coal char of low reactivity with a similar specific surface area (37). When carbon dioxide was used as a gasification agent in a cursory evaluation, a decline in reactivity with burn-off was also observed. As expected from the literature (34), gasification by carbon dioxide was slower than gasification by steam by roughly a factor of three. The apparent activation energy was the same, within our experimental error, for both reactions.

The empirical rate data of soot oxidation can be used to predict the time required for the combustion of a soot particle at a selected temperature and oxygen pressure. The reaction rate v in terms of particle radius r is

$$v = - d(M/A)/dt = - (\rho/3)\, dr/dt \qquad (4)$$

where M is the particle mass, $\rho = 2.0$ g/cm^3 the soot density (15), A the specific surface area, and t the time. Assuming as a first approximation that v remains constant with burn-off, eq. (4) is integrated to obtain

$$t = (\rho/3v)\,(r_i - r) \qquad (5)$$

as an expression for the time required to reduce a soot sphere from radius r_i to radius r. From our experimental data a standard reaction rate of 0.457 µg/(m^2 min) is deduced at 350°C and 50% burn-off, assuming an oxygen pressure of 0.15 atm, which is typical for diesel exhaust. Using an activation energy of 142 kJ/mol for extrapolation based on the Arrhenius equation, time values for complete combustion of a spherical soot particle with a radius of 10 nm, calculated for several temperatures, are given in Table 5. The data indicate that a temperature exceeding 800°C would be necessary to oxidize soot particles passing through the exhaust system of a diesel engine in a reasonable length of time in the absence of thermal runaway.

TABLE 5 - Calculated Combustion Time of a Spherical Soot Particle with a Radius r = 10 nm

Temperature °C	Rate g/(m^{2}min)	Combustion time min
350	0.5×10^{-6}	1.5×10^{4}
400	3.5×10^{-6}	1.9×10^{3}
600	1.2×10^{-3}	5.6
800	0.05	0.14

$P(O_2)$: 0.15 atm, apparent activation energy: 142 kJ/mol

A comparison was made of combustion time calculated by the method above with the combustion time of a 20 nm particle reported by Lee et al. (35). The comparison required an extrapolation of our data to 1355°C to approximate the conditions employed in run 5 of Lee et al. The combustion time of the spherical particle calculated using eq. (5) is larger than the measured value by 45%. The agreement is good considering the uncertainty of ± 15% in the apparent activation energy which enters exponentially into the extrapolation of our data.

REACTION STUDIES WITH SIMULATED EXHAUST - To demonstrate that the results of our combustion studies in air, discussed above, are valid under actual exhaust conditions, it is necessary to assess the combined effects of all the components that constitute a typical exhaust-gas mixture. For this purpose soot was subjected to an exhaust gas produced by the combustion of isooctane in a laboratory device (3), modified to produce a continuously burning flame. The oxygen content of the exhaust was varied by introducing secondary oxygen between flame and soot sample. The soot, either in powder or pellet form, was exposed to a heating rate of 6°C per minute, except in two runs, where samples were heated at 35°C/min for comparison. Ignition, followed by thermal runaway, was monitored by the increase in temperature, measured by a thermocouple in contact with the soot, and by the recorded decrease in oxygen concentration by about 0.6% and the accompanying increase in CO by about 0.5%, observed downstream of the soot sample. Results are summarized in Table 6.

At a concentration of 2.5% oxygen sustained fast combustion could not be achieved with either soot powder or pellet. Completion of the combustion was evident only at an ambient temperature of 800°C in the case of the powder and at a significantly lower temperature ≤ 630°C in the case of the soot pellet.

At 5% oxygen, again, sustained combustion did not occur in two samples of soot powder; combustion was complete below 610°C. In another instance, under the same experimental conditions, local ignition took place briefly at 437°C, but did not spread: only a small portion of the soot was consumed. A second ignition event was observed at 522°C, which led to complete combustion by the time a temperature of 590°C was reached. The incombustible residue amounted to about 5% of the initial weight. Ignition temperature of soot pellets at 5% oxygen was 513°C on the average, or 590°C at the higher heating rate (see Table 6).

At a concentration of 10% oxygen soot ignited consistently at 460°C, regardless of the heating rate, for powder and pellet samples alike. In comparison, no ignition was observed with SP1 graphite, even when a pellet was heated to 700°C, in agreement with the

TABLE 6 - Soot Combustion in Pulse-Flame Apparatus

Sample	Oxygen atm	Ignition temperature °C
powder	0.025	---
pellet	0.025	---
powder	0.05	---
powder	0.05	437 (522)
pellet	0.05	513**
pellet*	0.05	590
powder	0.10	460
powder*	0.10	455
SP1 graphite	0.05	---
powder with 1 atom% Ba	0.05	311 (387)

* heating rate 35 °C/min, otherwise 6 °C/min

** average of three samples

gravimetric results discussed above.

A cursory study was carried out to evaluate whether a suitable catalyst can lower the ignition temperature of diesel-soot deposits. Catalytic ignition had been indicated in our experiments by the tendency of the soot powder, especially at lower oxygen pressures, to ignite at points of contact with the thermocouple sheath, followed sometimes by quenching of the combustion. Since barium has been recognized recently as a strong steam-gasification catalyst (2, 25, 26, 38, 39), we mixed 1 atom% (11.4 wt%) Ba from an acetate solution with soot. When heated in exhaust gas containing 5% oxygen, partial combustion was observed at 311°C, followed by quenching of the reaction and rekindling at 387°C. The observed smoke suppression by fuel containing Ba which was reported to lower particulates by up to 40% (24, 40) is probably closely related with the observed catalytic increase in soot reactivity. The details of catalytic ignition by various elements are currently being studied.

CONCLUSIONS

To a first approximation primary soot particles produced in diesel engines can be considered spherical with a radius of 10 nm. These particles agglomerate to form deposits which can vary in bulk density by one order of magnitude. Soot structure is affected by substantial amounts of hydrocarbon deposit and embedded inorganic matter. Because of a pronounced heterogeneity of the soot surface on a microscopic level, ignition is expected to occur at a center of higher reactivity, which can be produced, for example, by a catalytic particle on the carbon surface. The combustion studies show that a temperature of at least 460°C is required to ensure reliable combustion, provided the oxygen concentration is sufficiently high. On the other hand a temperature of 350°C or less may be sufficient to trigger ignition, especially if an extraneous gasification catalyst is employed. A comparison of the ignition temperatures of soot samples containing varying amounts of adsorbed organic molecules with other samples that had been subjected to an acetone wash indicates, as shown in Figs. 4-6, that volatile molecules, at least at moderate levels, have only a small effect on soot combustion. In one instance, however, organic adsorbate was apparently effective in lowering the ignition temperature (see Fig. 7). An oxygen pressure of 0.1 atm is sufficient to sustain soot ignition and accelerate combustion. The reliability of soot ignition improves as the soot's bulk density increases. This conclusion follows from ignition studies of soot samples with and without acetone treatment and of soot in pellet and powder form. The minimum concentration of oxygen necessary for rapid soot combustion is readily available in a diesel-exhaust system. Under some driving conditions, for example during a CVS-CH cycle, a temperature of 460°C is not reached in the exhaust system of some diesel cars. However, under somewhat more severe driving conditions temperatures above 500°C were measured (41). It appears at present that a reliable lowering of the ignition temperature by an inorganic catalyst or by organic compounds derived from the fuel or engine oil merits further experimental work.

Ignition and combustion of soot deposits are complicated by the fact that partial combustion produces deposits which become increasingly more resistant toward oxidation. As a result, considerable amounts of carbonaceous material of low reactivity may accumulate. Such a residue may not burn at the temperatures which are available in diesel exhaust systems.

ACKNOWLEDGMENTS

We thank W. R. Wade, J. E. White, E. T. King, and J. J. Florek for providing us with the soot samples and for helpful technical information. The assistance of R. Friedman with microscopic analyses of the soot structure is gratefully acknowledged. We also appreciate suggestions by S. Gratch and L. R. Mahoney after their critical reading of this manuscript.

REFERENCES

1. K. Otto and M. Shelef, "Catalytic Steam Gasification of Carbons: Effect of Ni and K on Specific Rates," presented at the Sixth International Congress on Catalysis, London, July 1976; Proceedings, Vol. 2, pp. 1092-1100, 1977.

2. K. Otto and M. Shelef, "Catalytic Steam Gasification of Graphite: Effects of Intercalated and Externally Added Ru, Rh, Pd and Pt," Carbon, Vol. 15, pp. 317-325, 1977.

3. K. Otto, R. A. Dalla Betta, and H. C. Yao, "A Laboratory Method for the Simulation of Automobile Exhaust and Studies of Catalyst Poisoning," J. Air Pollution Control Association, Vol. 24, pp. 596-600, 1974.

4. C. T. Vuk, M. A. Jones, and J. H. Johnson, "The Measurement and Analysis of the Physical Character of Diesel Particulate Emissions," SAE Paper No. 760131, SAE Transactions, 1976.

5. W. H. Lipkea, J. H. Johnson, and C. T. Vuk, "The Physical and Chemical Character of Diesel Particulate Emissions - Measurement Techniques and Fundamental Considerations," SAE Special Publication 430, 1978.

6. R. M. Schreck, J. J. McGrath, S. J. Swarin, W. E. Hering, P. J. Groblicki, and J. S. MacDonald, "Characterization of Diesel Exhaust Particulate Under Different Engine Load Conditions," Paper presented at the 71st Annual Meeting of the Air Pollution Control Association, Houston, Preprint 78-33.5, June 25-30, 1978.

7. H. Gg. Wagner, "Soot Formation in Combustion," Plenary Lecture presented at the 17th Combustion Symposium, Leeds, October 1978.

8. D. B. Kittleson and D. F. Dolan, "Diesel Exhaust Aerosols," presented at Meeting of American Chemical Society, Honolulu, Hawaii, Preprint, April 1 - 6, 1979.

9. J. W. Frey and M. Corn, "Diesel Exhaust Particulates," Nature, Vol. 216, pp. 615-616, 1967.

10. C. P. Fenimore and G. W. Jones, "Coagulation of Soot to Smoke in Hydrocarbon Flames," Combustion and Flame, Vol. 13, pp. 303-310, 1969.

11. P. J. Groblicki and C. R. Begeman, "Particle Size Variation in Diesel Car Exhaust," SAE Paper No. 790421, presented at SAE Congress and Exposition, Detroit, February 26 - March 2, 1979.

12. K. H. Homan, "Carbon Formation in Premixed Flames," Combustion and Flame, Vol.11, pp. 265-287, 1967.

13. O. P. Mahajan and P. L. Walker, Jr., "Water Adsorption on Coals," Fuel, Vol. 50, pp. 308-317, 1971.

14. K. Tomkow, T. Siemieniewska, F. Czechowski, and A. Jankowska, "Formation of Porous Structures in Activated Brown-Coal Chars Using O_2, CO_2, and H_2O as Activating Agents," Fuel, Vol. 56, pp. 267-270, 1977.

15. Handbook of Chemistry and Physics, 52nd Edition, p. B-79, The Chemical Rubber Co, Cleveland, Ohio, 1971-1972.

16. H. J. Schroter, H. Juntgen, and W. Peters, "The Determination of Diffusion Coefficients in Activated Carbons Made from Bituminous Coals," Carbon, Vol. 11, pp. 93-102, 1973.

17. C. T. Hare and R. L. Bradow, "Characterization of Heavy-Duty Diesel Gaseous and Particulate Emissions, and Effects of Fuel Composition," SAE Paper No. 790490, presented at SAE Congress and Exposition, Detroit, February 26 - March 2, 1979.

18. "Investigation of Diesel-powered Vehicle Emissions: VII. Removal of Exhaust Particulates from Mercedes 300D Diesel Car," Report EPA-460/3-76-034, p. 144, February 1977.

19. E. F. Funkenbusch, D. G. Leddy, and J. H. Johnson, "The Characterization of the Soluble Organic Fraction of Diesel Particulate Matter," SAE Paper No. 790418, presented at SAE Congress and Exposition, Detroit, February 26 - March 2, 1979.

20. C. Bassoli, G. M. Cornetti, C. Biaggini, and A. Di Lorenzo, "Exhaust Emissions from a European Light Duty Turbocharged Diesel," SAE Paper No. 790316, presented at SAE Congress and Exposition, Detroit, February 26 - March 2, 1979.

21. P. L. Walker, Jr., M. Shelef, and R. A. Anderson, "Catalysis of Carbon Gasification," in Chemistry and Physics of Carbon, Ed. P. L. Walker, Jr., Vol. 4, p. 287, Marcel Dekker, New York, 1968.

22. M. Shelef, K. Otto, and N. C. Otto, "Poisoning of Automotive Catalysts," Advances in Catalysis, Vol. 27, pp. 311-365, 1978.

23. C. T. Hare and T. M. Baines, "Characterization of Particulate and Gaseous Emissions from Two Diesel Automobiles as Functions of Fuel and Driving Cycle," SAE Paper No. 790424, presented at SAE Congress and Exposition, Detroit, February 26 - March 2, 1979.

24. N. D. Apostelescu, R. D. Matthew, and R. F. Sawyer, "Effects of Barium-Based Fuel Additive on Particulate Emissions from Diesel Engines," SAE Paper No. 770828, presented at SAE Congress, Detroit, September 26 - 30, 1977.

25. K. Otto, L. Bartosiewicz, and M. Shelef, "Catalytic Steam Gasification of Graphite: Effects of Calcium, Strontium, and Barium With and Without Sulfur," Carbon, Vol. 17, pp. 351-357, 1979.

26. K. Otto, L. Bartosiewicz, and M. Shelef, "Effects of Calcium, Strontium, and Barium as Catalysts and Sulfur Scavengers in the Steam Gasification of Coal Chars," Fuel, Vol. 58, pp. 565-572, 1979.

27. G. Blyholder and H. Eyring, "Kinetics of Graphite Oxidation," J. Phys. Chemistry, Vol. 61, pp. 682-688, 1957.

28. R. H. Essenhigh, R. Froberg, and J. B. Howard, "Combustion Behavior of Small Particles," Ind. Eng. Chemistry, Vol. 57, No. 9, pp. 32-43, 1965.

29. E. A. Gulbransen and K. F. Andrew, "Reactions of Artificial Graphite," Ind. Eng. Chemistry, Vol. 44, pp. 1034-1038, 1952.

30. J D. Lambert, "The Oxidation of Carbon," Trans. Faraday Soc., Vol. 32, pp. 452-462, 1936.

31. W. K. Lewis, E. R. Gilliland, and R. R. Paxton, "Low-Temperature Oxidation of Carbon," Ind. Eng. Chemistry, Vol. 46, pp. 1327-1331, 1954.

32. A. S. Parker and H. C. Hottel, "Combustion Rate of Carbon," Ind. Eng. Chemistry, Vol. 28, pp. 1334-1341, 1936.

33. I. W. Smith and R. J. Tyler, "Internal Burning of Pulverized Semi-Anthracite: the Relation Between Particle Structure and Reactivity," Fuel, Vol. 51, pp. 312-321, 1972.

34. P. L. Walker, Jr., F. Rusinko, Jr., and L. G. Austin, "Gas Reactions of Carbon," Advances in Catalysis, Vol. 11, p. 161, 1959.

35. K. B. Lee, M. W. Thring, and J. M. Beer, "On the Rate of Combustion of Soot in a Laminar Soot Flame," Combustion and Flame, Vol. 6, pp. 137-145, 1962.

36. M. C. Chen, C. J. Christensen, and H. Eyring "Kinetics of Oxidation of Pyrolytic Carbon," J. Phys. Chemistry, Vol. 59, pp. 1146-1148, 1955.

37. K. Otto, L. Bartosiewicz, and M. Shelef, "Catalysis of Carbon-Steam Gasification by Ash Components from Two Lignites," Fuel, Vol.58, pp. 85-91,

1979.

38. K. Otto, L. Bartosiewicz, and M. Shelef, "Steam Gasification of Carbons: Catalytic Effects of Ca, Sr, and Ba," 14th Biennial Conference on Carbon, Extended Abstracts, pp. 157-158, The American Carbon Society and the Pennsylvania State University, June 25-29, 1979.

39. D. W. McKee, "Catalysis of the Graphite-Water Vapor Reaction by Alkaline Earth Salts," 14th Biennial Conference on Carbon, Extended Abstracts, pp. 155-156, The American Carbon Society and the Pennsylvania State University, June 25-29, 1979.

40. Study of the University of California, "Barium Explored as Smoke Suppressant," Automotive Engineering, Vol. 85, No. 11, pp. 53-55, November 1977.

41. "Investigation of Diesel-Powered Vehicle Emissions: VIII. Removal of Exhaust Particulates from Mercedes 300D Diesel Car," Report EPA-460/3-77-007, p. 34, June 1977.

Origins of Diesel Particulate Mass Emission**

G. Greeves and C. H. T. Wang
Lucas CAV Ltd.

THE EMISSION OF PARTICULATES, NO_x and HC are of current concern in the development of diesel engines for use in the USA. Regulations for diesel passenger cars are of the most immediate concern and the predominant combustion system is indirect injection (IDI). Figures 1, 2 and 3 show baseline data for a wide range of different models of diesel powered passenger cars, and also the proposed regulatory requirements for emissions and fuel economy. These data are based partly on those given in (1,2,3,4,5,6)* and partly on results obtained by Lucas CAV Limited. It is clear that the diesel car fuel economy allows all but the larger vehicles to meet future C.A.F.E. fuel economy requirements, but only the smallest vehicles (2000 lbs) are close to meeting the requirement of particulate = 0.2 g/mile and NO_x = 1.0 g/mile. The observed increase in fuel consumption and emissions with vehicle weight is expected, to the extent that the required engine power output and fuel consumed must increase.

Exhaust gas recirculation can be used to give NO_x reductions of up to 50% and particulate traps in the exhaust system are under development (2,7). However the best reported collection efficiency of these traps is about 70%, and Figure 1 suggests that a successful trap design would need to be combined with combustion optimisation to provide an adequate margin for meeting particulates < 0.2 g/mile. There is therefore a need to improve our understanding of in-cylinder particulate formation and control. It is appropriate to consider the effects of various parameters at steady state conditions.

In the following sections, data from a parametric study of an IDI engine are presented and used to derive a correlation between particulates, black smoke and HC emissions. This correlation shows the origins of particulates to be the same as the origins of soot and HC emissions, which are then discussed.

DEFINITIONS - In this paper particulate mass emission is taken to mean all material which is collected from a diesel engine exhaust on a filter paper, when the exhaust gases have been diluted with air to achieve a sample temperature < 50°C. The sample is allowed to dry prior to weighing to a controlled ambient humidity level.

The term HC emission is taken to mean all hydrocarbon species which are detected by a heated hydrogen flame ionization detector generally used for monitoring HC in the diesel exhaust. The constituents vary from methane to the heaviest hydrocarbons contained in the fuel that are in the vapour phase, and pass through the sampling line maintained at about 190°C. It should be noted that some of the very heavy hydrocarbons formed by agglomeration are condensed and these, as well as the solid phase, i.e. soot, are filtered from the exhaust gas sample prior to it being passed into the sample line.

Exhaust smoke in this paper is taken to

* Numbers in paretheses designate References at end of paper.

**Paper 810260 presented at the International Congress and Exposition, Detroit, Michigan February, 1981.

ABSTRACT

Emissions results are presented from an experimental study of an indirect injection diesel engine. The mass measurements of particulate emission are correlated with the measurements of smoke and HC emission. The correlation provides a less expensive way of carrying out preliminary combustion optimisation work. It is concluded that practically all of the particulate mass emission is accounted for by: (a) the black smoke or soot formed in the high temperature fuel-rich regions of the diffusion phase of burning, (b) that fraction (about 50%) of the total HC mass emission which condenses at the particulate sampling filter. The total HC mass emission is itself a function of three distinct sources in the combustion process.

The understanding gained is then used to define three combustion ideals for optimising diesel combustion to minimise fuel consumption and emissions of smoke, NO_x, particulates, HC, CO, odour and noise.

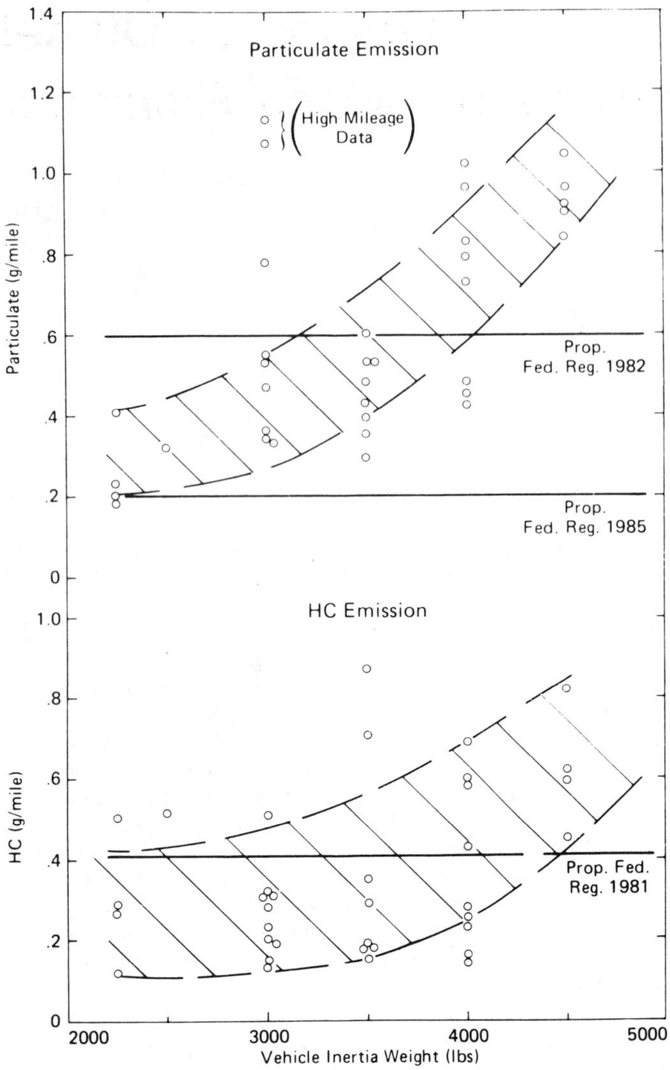

Fig. 1 - Emissions baseline data for diesel powered passenger cars on FTP Urban cycle

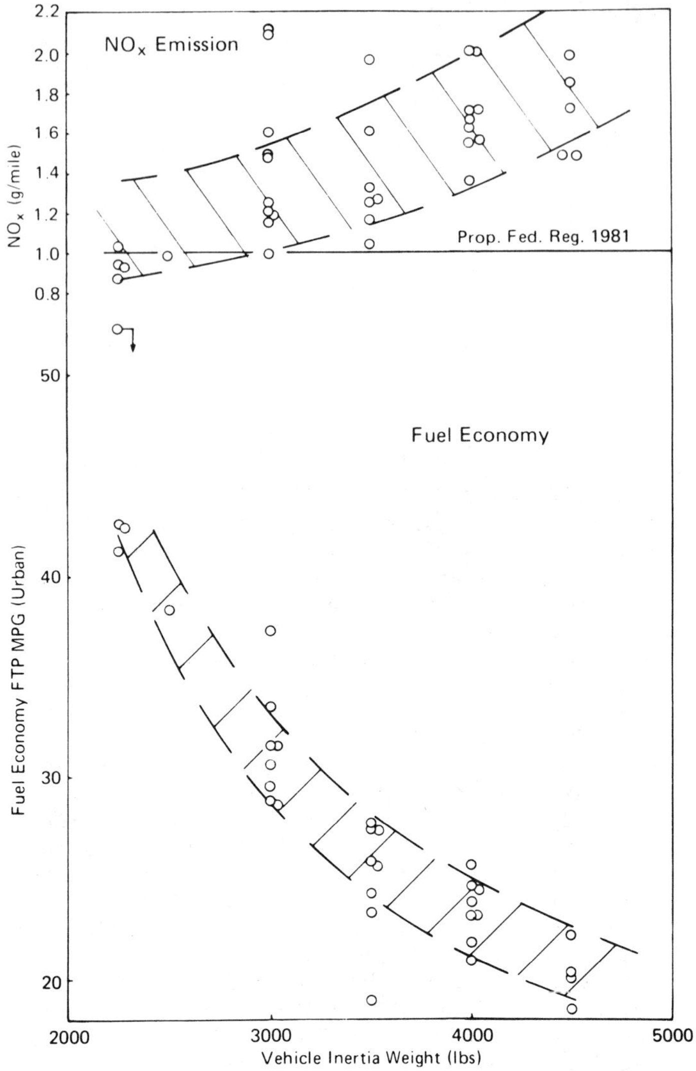

Fig. 2 - NO_x emissions and fuel economy baseline data for diesel powered passenger cars on FTP Urban cycle

mean the black smoke (soot) which is observed in the diesel engine exhaust particularly at the higher load conditions. The method of measurement is by withdrawal of a hot exhaust sample close to the engine manifold and a known volume of exhaust gas is passed through a filter. The filter is then evaluated by optical means to determine the Bosch smoke number.

PARAMETRIC STUDY AND CORRELATION OF PARTICULATE MASS EMISSIONS

An extensive emissions and performance study of the engine detailed in Table 1, has been made at steady state conditions on an engine test bed. Particulate mass emissions were measured in a dilution tunnel according to EPA procedures involving collection on a filter material at a temperature < 50°C. NO_x was measured by chemiluminescent analyser, HC by heated FID, and smoke by a filter method (Bosch smoke meter). All of the measurements of NO_x, HC and smoke were made 'raw and hot', that is close to the engine manifold according to normal procedures for measurements on heavy duty diesel engines. The particulate measurement facility was specially developed for the project and required the use of an exhaust splitter to satisfy the 50°C requirement at the highest engine loads and speeds. The splitting ratio was determined by monitoring NO_x in both raw and diluted states.

Figures 4, 5 and 6 show the results obtained on the effect of varying the dynamic timing of fuel injection. As injection timing is retarded, the NO_x, smoke, and particulates decrease at 100% load conditions, but at 0% load there is little change of exhaust smoke while HC and particulates increase. These results demonstrate that the trends of particulate mass emission sometimes follow that of black smoke, and sometimes that of HC emissions.

It was anticipated that the mass emission of

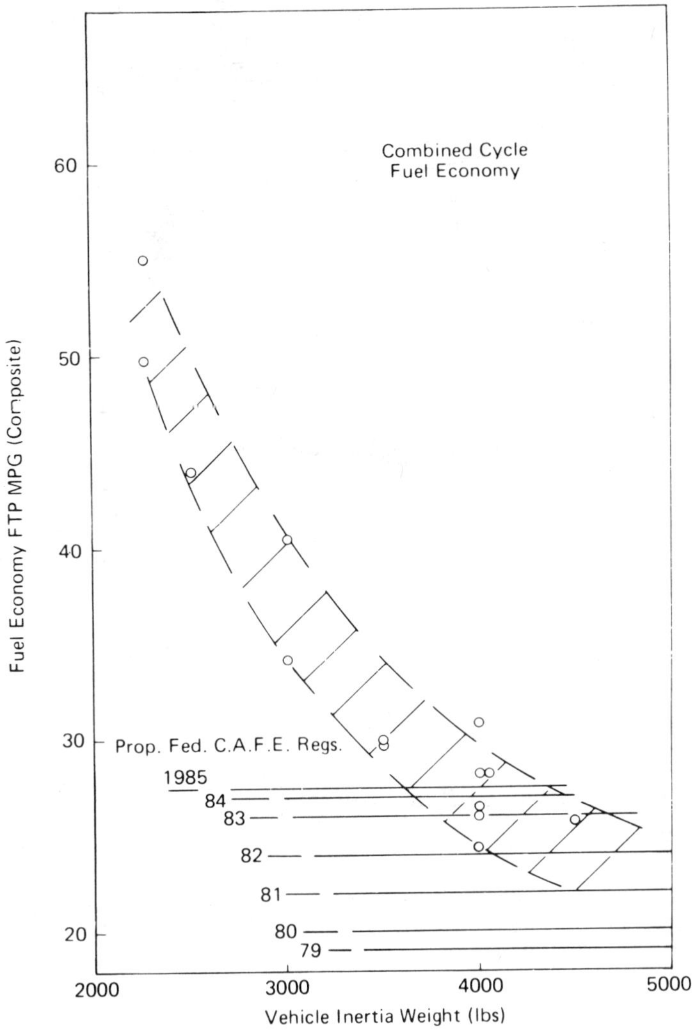

Fig. 3 - Fuel economy on composite cycle for diesel powered passenger cars

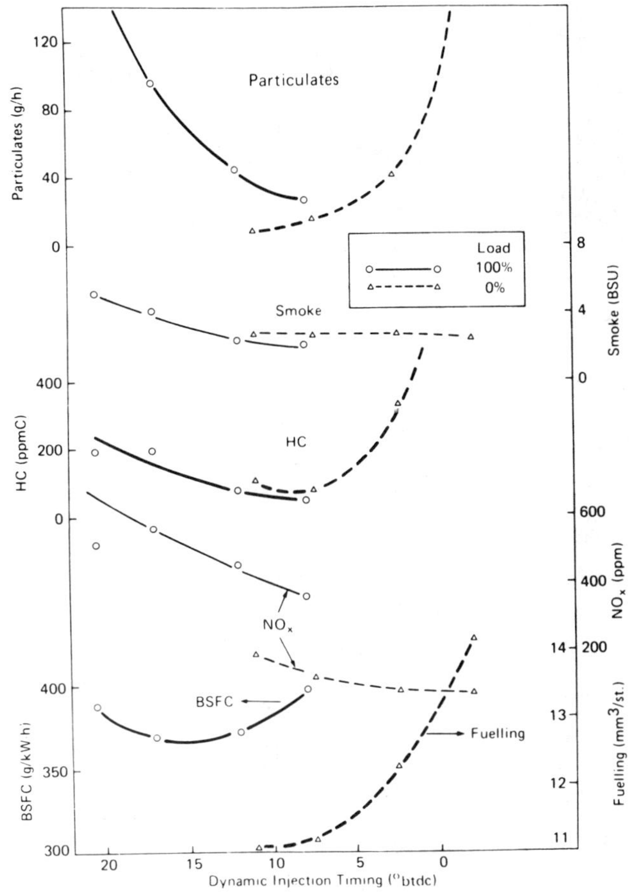

Fig. 4 - Effects of injection timing and load on emissions and fuel economy, engine A, engine speed 4400 rev/min

Table 1 - Details of Engine A

Swept Volume per cylinder	=	520 cm^3
Number of cylinders	=	4
Combustion chamber	=	Comet IDI
Nominal compression ratio	=	22:1
Maximum engine speed	=	4400 rev/min
Vehicle application inertia weight	=	3000 lb

diesel exhaust particulate is equal to the mass emission of soot (black smoke) plus the mass emission of HC which is liquid at the particulate sampling temperature (50°C). It was therefore decided to attempt to fit the engine measurements of particulate, smoke and HC to a simple linear equation of the form:

$$\underset{(g/m^3)}{\text{Particulate}} = A \times \underset{(g/m^3)}{\text{Smoke}} + B \times \underset{(g/m^3)}{\text{HC}} + C \qquad (1)$$

In equation (1) all the measurements are expressed in mass per unit volume of raw exhaust gas, and the equation is to be applied to individual steady-state test conditions. In the case of HC, the ppm volume concentration is converted to a mass assuming a molecular weight of 13.85 (Figure 7), which is consistent with the conversion constants specified in U.S. emissions measurement procedures. In the case of smoke, the correlation given in (8) is used to convert the smoke numbers of the Bosch smokemeter into a solids content (Table 2). A computer programme was available to calculate the values of the constants A, B and C for a least squares fit of equation (1) to a set of data. Also, a set of data was available for a DI engine, by kind permission of the members of a Joint Project detailed in the acknowledgement section of this paper.

Table 3 shows the correlations obtained by computer fitting to sets of data covering a range of engine speeds, loads, and fuel injection timings. The first correlation is for the IDI engine detailed in Table 1, the second correlation is for the data obtained on the DI engine, the third correlation is for the combined

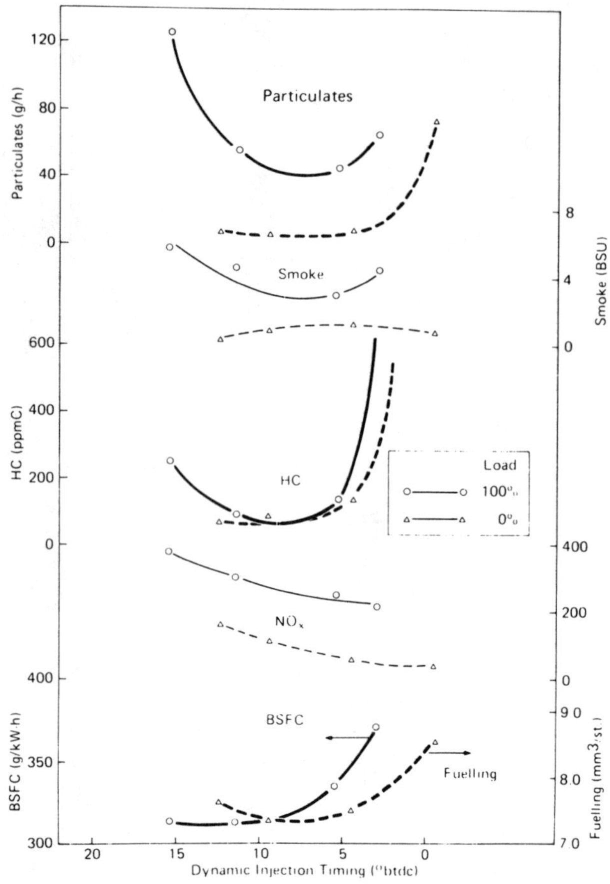

Fig. 5 - Effects of injection timing and load on emissions and fuel economy, engine A, engine speed 2500 rev/min

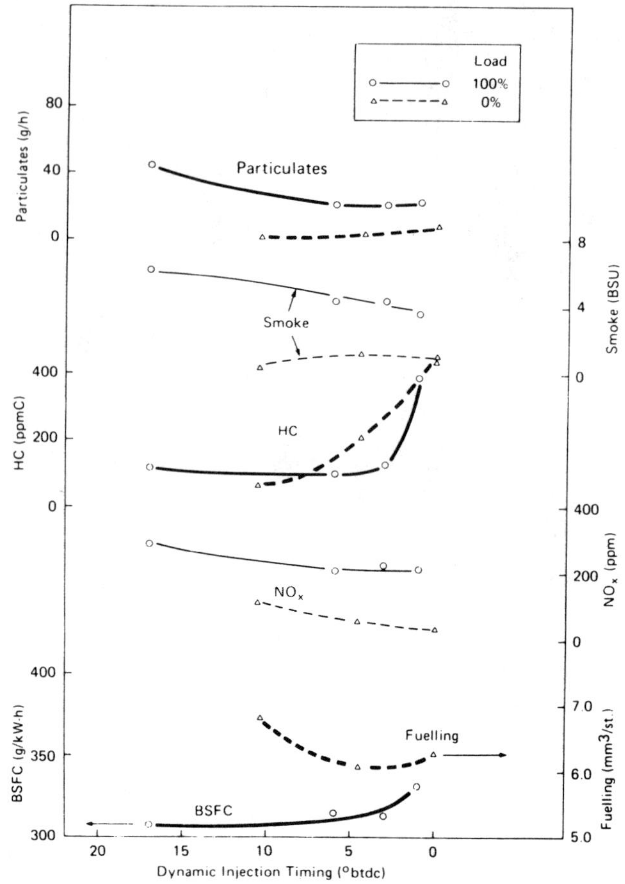

Fig. 6 - Effects of injection timing and load on emissions and fuel economy, engine A, engine speed 1000 rev/min

data of the two engines. From these results it is clear that the value of C is nearly zero and the fourth correlation is for the combined data with C set equal to zero. This correlation is illustrated in Figure 8. The final result is therefore:

$$\underset{(g/m^3)}{\text{Particulate}} = 1.024 \times \underset{(g/m^3)}{\text{Smoke}} + 0.505 \times \underset{(g/m^3)}{\text{HC}} \qquad (2)$$

This correlation is based on 142 data points which cover smoke values ranging from 0.002 to 0.83 g/m^3 and HC values ranging from 0.03 to 0.8 g/m^3. The correlation has a standard error of 0.066 g/m^3. This standard error is greater than desirable, but the correlation is considered to be a significant improvement on previously suggested correlations, which omit the contribution of HC to particulate mass (e.g. Fig. 20 in (9)).

Part of this standard error is undoubtedly due to inaccuracies in the measurement of smoke, HC and particulates, and also in the correlation for converting smoke number to equivalent concentration of solids. It is therefore difficult to assess any small variation in the true values of the constants A and B with change of engine test condition. For example if the mode of

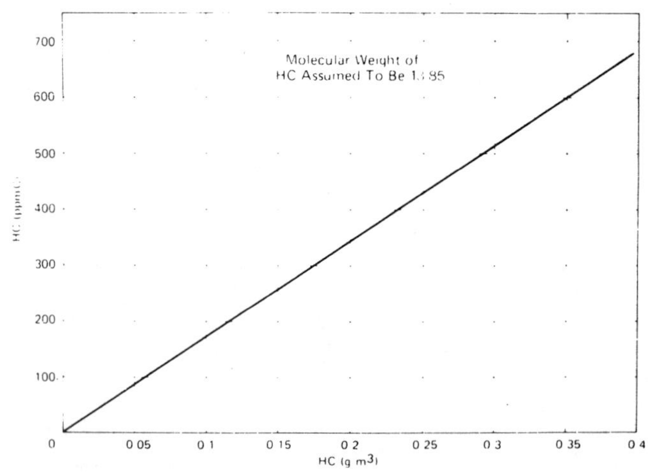

Fig. 7 - Conversion graph for HC (ppm) to HC (g/m^3)

combustion changes, then the boiling range of the HC emission might be expected to change and hence the fraction of HC which condenses on the particulate filter. This would imply a change in the value of the constant B with a change of

TABLE 2 - Conversion of Smoke to Equivalent Solids (see Reference 8)

Gravimetric Calibration of Bosch Smokemeter

Smoke Units	Solids Content gm/m^3	Smoke Units	Solids Content gm/m^3
1.0	0.028	4.0	0.279
1	0.033	1	0.295
2	0.037	2	0.311
3	0.041	3	0.329
4	0.046	4	0.347
5	0.051	5	0.366
6	0.056	6	0.386
7	0.062	7	0.407
8	0.068	8	0.429
9	0.074	9	0.450
2.0	0.079	5.0	0.474
1	0.086	1	0.499
2	0.092	2	0.525
3	0.099	3	0.551
4	0.107	4	0.579
5	0.115	5	0.607
6	0.123	6	0.636
7	0.131	7	0.666
8	0.139	8	0.698
9	0.147	9	0.730
3.0	0.156	6.0	0.763
1	0.166	1	0.796
2	0.175	2	0.832
3	0.186	3	0.869
4	0.197	4	0.906
5	0.209	5	0.943
6	0.222	6	0.980
7	0.236	7	1.019
8	0.249	8	1.058
9	0.264	9	1.097
4.0	0.279	7.0	1.136

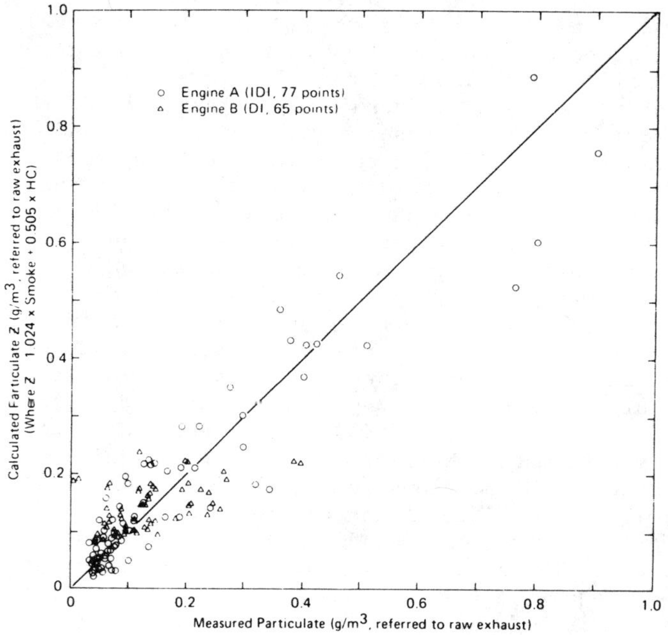

Fig. 8 - Final correlation of measured particulate emission with measured smoke and HC emission, combined data for engines A and B, constant C = 0

engine operating condition. The results obtained suggest that this does not happen to any major extent for the conditions tested.

At this point it is appropriate to discuss the values obtained for the constants A and B in the final correlation given in equation (2) above.

VALUES OF THE CORRELATION CONSTANTS - The constant A = 1.024 is close to unity. This suggests that the mass of soot inferred from the smokemeter filter paper, for the hot raw exhaust, corresponds to the mass of soot collected on the particulate filter paper from the diluted exhaust gas.

Table 3 - Correlations of Particulate Mass Emission with Smoke and HC Emissions Obtained by Computer Fitting of Engine Steady State Test Data

Correlation	Engine Code	Equation: Particulates = A × Smoke + B × (HC) + C, i.e. $z = A \times x + B \times y + C$, where z, x, y in g/m^3	Standard Error g/m^3	Number of Data
1	A (IDI)	$z = 1.022x + 0.579y - 0.012$	0.065	77
2	B (DI)	$z = 1.243x + 0.461y - 0.0004$	0.063	65
3	A + B	$z = 1.046x + 0.537y - 0.010$	0.066	142
4	A + B	$z = 1.024x + 0.505y$ (C set = 0)	0.066	142

The constant B = 0.505 suggests that only 50% of the HC which is sampled from the hot raw exhaust, condenses to liquid on the particulate sampling filter, at < 50°C. Figure 24 in reference (9) gives results on an analysis of the liquid part of a diesel particulate sample and also the distillation curve for the fuel used. It is clear that the distillation curve for the liquid part of the particulate is some 100°C lower than that for the fuel. However these distillation curves omit the HC species which pass through the particulate filter in the gaseous state. Reference (10) gives the mole fraction of the hydrocarbons taken from a diesel exhaust by carbon number. The distribution of hydrocarbon species range from C_1 up to C_{24}, see Figure 1 of reference (10). From these data we have estimated that the fraction of HC, which condenses at 50°C and for a dilution ratio of 10:1, to be 25%, 37% and 43% for raw HC concentrations of 100, 500 and 1000 ppm respectively. These values are lower than the constant B in equation (2). A much more detailed analysis, of which chemical species are included in the 'hot raw' HC measurement and also which species condense or are absorbed by soot on the smoke and particulate filters, would be required to fully explain the value of the constant B.

As discussed earlier, the value of constant C in the correlation is near zero. This implies that the measurements of soot by the smokemeter, plus the measurements of HC from the hot raw exhaust, account for practically all of the solid or liquid matter collected on the particulate sampling filter.

These results were obtained with a current (1980) U.K. gas oil as fuel (cetane number = 53) and for four stroke engines in good mechanical condition. It is possible that the value of the constant B in the correlation may increase with a higher boiling point fuel, or with engines where lubricating oil gives a significant contribution to smoke, HC and particulate emissions.

Figure 9 shows the correlation applied to the results of engine A for two different conditions. Figure 9(a) shows that retarding the injection timing, at 0% load, greatly increases the ignition delay. There is a corresponding increase in the 'lean limit' source of HC emission (11), and both the measured particulate emission and that calculated from the correlation follow the rising HC trend. Smoke is low and relatively constant owing to the low overall equivalence ratio.

Figure 9(b) shows that increase of overall equivalence ratio at high load conditions greatly increases exhaust smoke. Both the measured particulate and that calculated from the correlation follow the rising smoke trend. An injection timing was chosen so that ignition delay is short and HC relatively constant.

The results in Figure 9 demonstrate how the correlation given in equation (2) may be used to estimate the trends of particulate emission from smoke and HC measurements with change of engine test conditions. Also the effect of the standard error, quoted previously for the correlation, is greatly reduced when the correlation is used to estimate an overall emissions cycle result from individual load and speed test data.

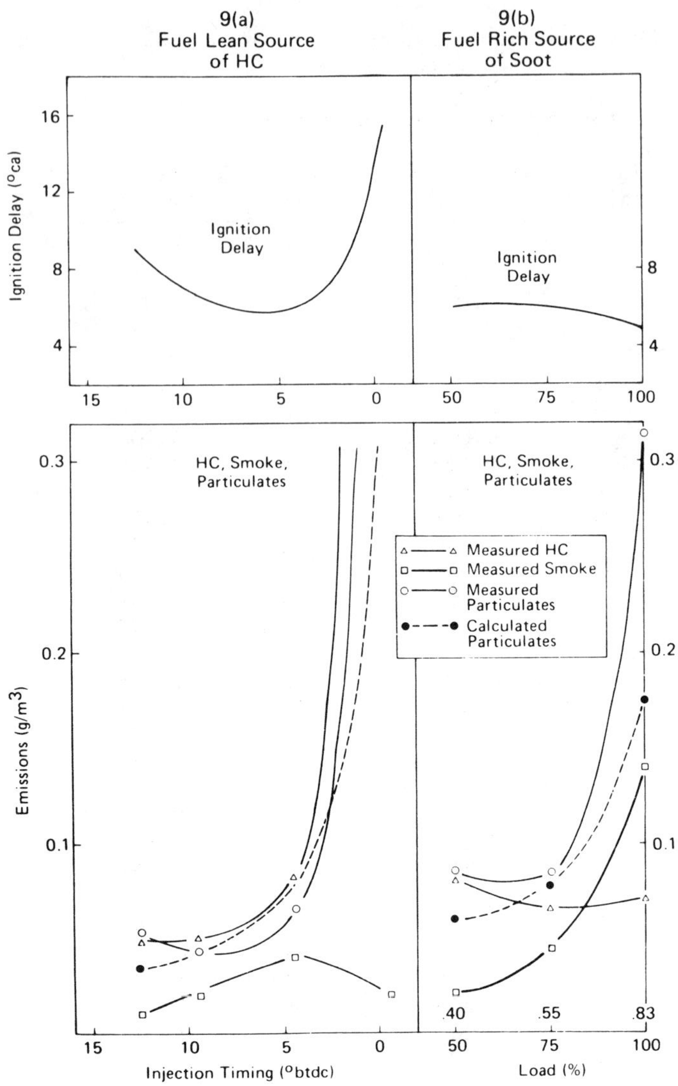

Fig. 9 - Comparison of measured particulate emission with that calculated by equation (2) from measured smoke and HC, engine A, engine speed 2500 rev/min;
(a) effect of injection timing at 0% load (fuel lean source of HC)
(b) effect of equivalence ratio at injection timing of 6° BTDC (fuel rich source of soot)

It has been found that smoke and HC measurements can be taken 3 times more quickly than particulate measurements, owing to the long sampling times required at some engine conditions for reliable particulate mass measurements. The correlation can thus be used to carry out preliminary combustion optimisation work with 3 times as many test points, or at 1/3rd of the cost for the same number of test points. Also there is a greater prospect of measuring HC and smoke during engine transients than measuring particulate mass.

However an equally important aspect of the correlation, is the implication that our previous understanding of smoke and HC formation in diesel combustion can now be directly applied to understanding particulate formation.

ORIGINS OF PARTICULATE FORMATION

The correlation described in the previous section is an essential step in establishing the origins of particulate emission. Whatever the precise values of the constants A and B, the correlation shows that the origins of particulate emission are the origins of HC emission plus the origins of black smoke or soot formation. This means that the previous understanding developed for HC formation (11) and black smoke formation (12) can be directly applied. In this paper we will summarise the origins of HC and black smoke emissions and the reader is referred to the appropriate references for the detailed explanations, supporting data, and combustion modelling work.

ORIGINS OF HC EMISSION - The major part of the injected fuel is consumed by fast combustion reactions at near stoichiometric conditions with practically no HC emission. The emission of HC is related to the equivalence ratio and temperature time histories of a small fraction of the injected fuel. HC emission occurs when the maximum mixture temperature attained is low or there is an absence of oxygen. This occurs when the fuel mixture is too lean or too rich to be completely consumed by fast combustion reactions.

There are three principal sources of HC emission in diesel combustion as follows.

Too-Lean-To-Burn Mixtures - During the ignition delay period, some of the fuel mixes to ultra-lean equivalence ratios. When ignition occurs in a near stoichiometric mixture, the flame front propagates until a lean limit of propagation is reached. This ultra-lean mixture is only partially consumed by relatively slow thermal oxidation reactions.

It is found that the HC emission from this source correlates with the ignition delay, since the latter directly controls the amount of too-lean to-burn mixture which is formed, Figure 10(a). Also there is a critical ignition delay, below which there is no further reduction of HC emission and hence below which the contribution from the lean limit source becomes negligible.

Nozzle Sac Volume Source - After the end of the main injection of fuel, there is a small quantity of fuel left in the nozzle tip of direct injection engines. Some of this fuel is observed to leave the nozzle tip, owing to fuel evaporation at the high temperatures and the high gas flow past the nozzle tip. This fuel is an emissions source because it enters the combustion chamber late in the engine cycle and also because it has a low velocity for mixing. It is found that the HC emission from this source correlates with the volume of the nozzle sac plus the volume of the nozzle spray holes, Figure 10(b).

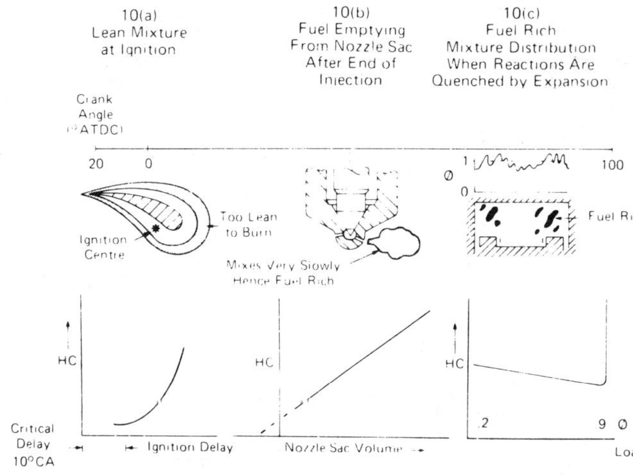

Fig. 10 - Summary of principal sources of HC emission, see reference (11)

Fuel-Rich Mixtures Late in Engine Cycle - Experiments show that when fuel delivered is increased, the HC emission can remain relatively constant until a certain value of delivery is reached. Above this value there is a dramatic increase in HC emissions, Figure 10(c). In a well optimised engine, this limiting overall equivalence ratio is $\emptyset \simeq 0.9$. This represents a considerable over-fuelling of the engine, since smoke limits engine operation to much lower values of $\emptyset$. Some current engine designs benefit from liquid fuel impingement on the combustion chamber wall (13). However in some cases there may be inadequate rates of air swirl and/or wall temperature (e.g. at low engine speeds) and this can give fuel-rich mixtures late in the engine cycle and a contribution to HC emission.

ORIGIN OF SOOT EMISSION - Soot forms when fuel-rich mixtures are subjected to a high temperature. The more fuel-rich the mixture, or the higher the temperature, the greater the yield of soot from the carbon available in the fuel. The emission of soot is therefore concerned with the main part of the injected fuel and in particular with that part which is injected after the ignition delay period. Fuel injected before ignition has time to mix to leaner conditions before the temperature rises.

In the terminology as above for HC, there is essentially a single source of soot emission, that is the temperature and equivalence ratio history of the main part of the injected fuel. Figure 11 illustrates the soot formation and oxidation process. The curves for temperature, equivalence ratio, cumulative soot formation, and cumulative soot release have been taken from a computer model for soot formation (14). These data are for a multizone model of the diesel fuel-air mixing process, but for present purposes we can visualise this as the time history along the axis of a burning fuel spray jet. The equation for soot formation is a semi-empirical equation developed in the

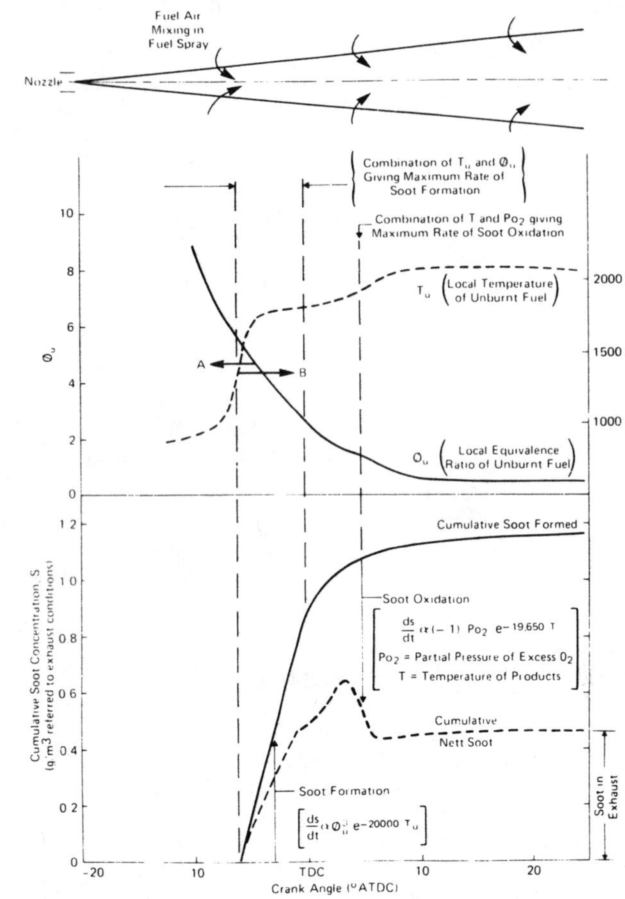

Fig. 11 - Soot formation and oxidation in diesel combustion, see computer model of reference (14)

authors' laboratory. This gives a good qualitative simulation of the effect of a wide range of parameters in diesel combustion for a given type of fuel. The equation used for soot oxidation is that from the work of Lee (15). Thus soot formation has a strong dependence on temperature and on lack of oxygen (high Ø) while soot oxidation has a strong dependence on temperature and needs excess oxygen ($Ø < 1.0$). The highest rates of soot formation occur on the fuel rich side, and the highest rates of soot oxidation occur on the fuel lean side, of the high temperature stoichiometric mixing region. Nearly 62% of the soot formed is oxidised in the example of Figure 11. However it is found that the effect of a change of operating variables on soot emission is explained by the effect on the soot formation equation, rather than the soot oxidation equation, (14). This is not to say that soot oxidation is unimportant, merely that for the conditions prevailing in diesel combustion, soot formation has the dominant affect.

From the above discussion, the important parameters for soot emission are the local temperature and equivalence ratio histories. Computer model calculations reported in (12,14) give detailed insight as to how various design and operating parameters such as fuel injection timing, injection rate, fuelling, air swirl rate and engine speed affect the temperature and equivalence ratio time histories and the resulting soot emission.

Basically any change which advances the equivalence ratio profile in the direction A in Figure 11, and or retards the temperature profile in the direction B, will reduce the soot emission. For example if water is added to the fuel this moves the temperature profile in the direction B due to the additional latent heat, and also the equivalence ratio profile moves in the direction A owing to the increased spray momentum and faster entrainment of air. The result is a larger reduction in soot emission (16). However with water injection the mixture temperatures are also reduced during the ignition delay period and hence the ignition delay lengthens and the lean limit emission of HC increases.

The importance of the temperature and equivalence profiles can be further demonstrated in a laboratory turbulent diffusion flame. Soot formation is considered to occur from the fuel vapour phase during diesel combustion (14). We can consider a simple turbulent flame with a gaseous fuel which is sufficiently reactive at atmospheric conditions to form appreciable quantities of soot. Figure 12 shows a quantitative interpretation of some observations made

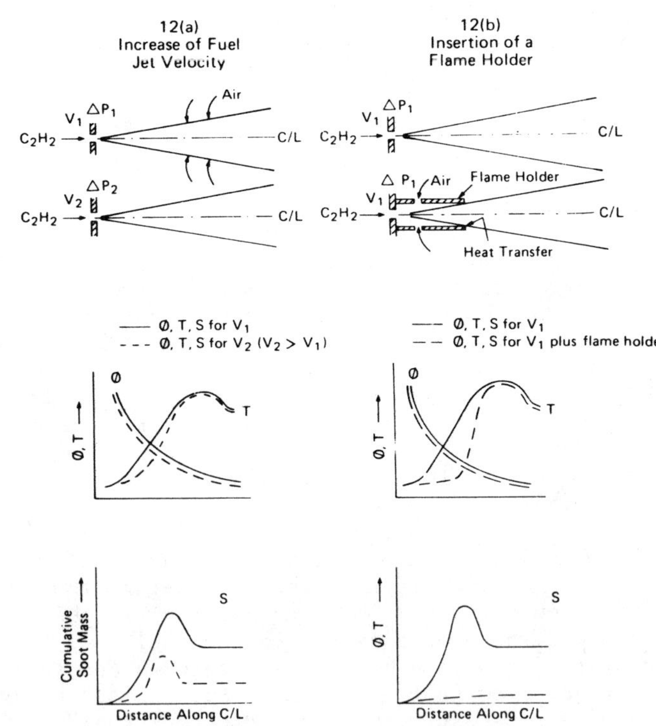

Fig. 12 - Interpretation of observations on an acetylene - air laboratory diffusion flame; (a) effect of increasing fuel jet velocity, (b) effect of suppressing temperature in the fuel rich region by insertion of a flame holder

in the authors' laboratory on an acetylene - air turbulent diffusion flame.

A bunsen burner jet orifice represents a hole type injection nozzle and has been operated with pure acetylene as the fuel gas and with entrainment of ambient air as the oxidant. A fuel pressure can be chosen so that there is an easily seen soot emission from the wake of the flame. Figure 12(a) shows that as the fuel pressure and hence injection velocity is increased from V_1 to V_2 the soot emission is decreased. The effect is to increase the jet velocities, but the rate of the exothermic reactions which cause the temperature rise are controlled by chemical kinetics and occur at a similar rate. In terms of distance from the nozzle, the temperature rise occurs further downstream, and the temperature profile becomes retarded relative to the Ø profile and hence soot formation and emission is reduced. This corresponds to increasing the fuel-air mixing rate by fuel injection rate or air swirl rate in a diesel combustion system.

Figure 12(b) shows what happens when we revert to the injection velocity V_1, which produces a high soot emission, and then install a flame holder tube. The effect of the flame holder is to remove heat from the upstream fuel-rich region, and so prevent 'light back' of the exothermic reactions into the upstream fuel-rich mixtures. This has a dramatic effect in retarding the temperature rise profile and soot emission is reduced to practically zero; this operating condition is generally referred to as a premixed flame. It should be noted that this soot reduction is achieved with no increase in the energy provided for fuel-air mixing and hence with no increase in the rate of fuel-air mixing.

The laboratory tests for Figure 12(b) are for the ambient conditions which do not produce compression ignition; the fuel jet is artificially ignited. If the air temperature is increased, so that there is spontaneous ignition immediately the acetylene is introduced, then the flame will almost certainly 'light back' within the flame holder tube; that is the heat transfer by the flame holder tube will be insufficient to prevent the flame lighting back. The temperatures and soot formation in the fuel-rich region will then rise to similar values to that without the flame tube, with the consequent loss of control of soot emission. In addition to the requirement for compression ignition there is considerable recirculation in the confines of a diesel combustion chamber, which would further increase the tendancy to 'light back'. It should be noted that all experiments with acetylene should be treated with the utmost caution and the experiment suggested above, with raised air temperature, may well prove much too hazardous to perform!

From the above discussion it is concluded that the insertion of such an apparently simple device as a flame holder tube is impractical for diesel combustion systems. Nevertheless the aim of reducing the temperatures in the fuel-rich mixtures, relative to the temperatures which are naturally achieved in an adiabatic diffusion flame, is still valid. For example it is likely that this effect is used to advantage in high swirl direct injection combustion systems where fuel mixes from the combustion chamber wall.

Heat transfer to the combustion chamber wall will help to lower the temperature in the fuel-rich mixture regions close to the wall and hence minimise soot formation. This effect, combined with the high rate of fuel-air mixing, accounts for the substantial increase in smoke limited BMEP which has been observed when swirl ratio, and the amount of fuel mixed from the wall, is increased on high swirl DI engines (13).

If we return now to the main parameters which are of direct interest in diesel combustion, these are illustrated in Figure 13.

Increase of air inlet temperature or increase of compression ratio will increase the end of compression temperature and hence the rate of soot formation. Figure 13(a) shows results for air inlet temperature. The quantity of fuel injected will have an obvious effect on the equivalence ratio profiles of the later parts of the injected fuel as illustrated in Figure 13(b). The effect of increasing the rate of fuel-air mixing, by increase of the injection rate or the air swirl rate, is to reduce soot emission as in Figure 13(c) or increase smoke limited fuelling as in 13(d). This is because the equivalence ratio profile becomes more advanced relative to the temperature profile illustrated in Figure 11, and hence the rate of soot formation is reduced.

In quiescent diesel combustion systems, increase of fuel-air mixing rate by increase of injection rate, has given progressive reductions of black smoke, see Figure 6 of (17). In swirl type combustion systems, increase of injection rate has given a smoke reduction, but there is a maximum useful injection rate (M.U.R.) beyond which there is no further improvement (18). However with swirl type combustion systems, a progressive improvement in smoke limited power has been observed with increase of swirl rate (13). This is achieved when volumetric efficiency is maintained constant, and the injection system is optimised to fully benefit from fuel-air mixing from the combustion chamber wall.

The above discussion of soot formation has referred to earlier modelling and experimental work on soot emission in direct injection engines. The same understanding is applicable to indirect injection engines, with the proviso that we can correctly assess the relevant features of the fuel-air mixing. A computer model has been developed for fuel-air mixing in indirect engines (19), and illustrates several important features of the fuel-air mixing in a two-chamber system. This model, however, has not yet been applied to the prediction of soot emission.

LUBRICATING OIL AS AN EMISSIONS SOURCE - In the above discussion it was assumed that the sources of HC, soot and hence particulate emission are derived from the injected fuel

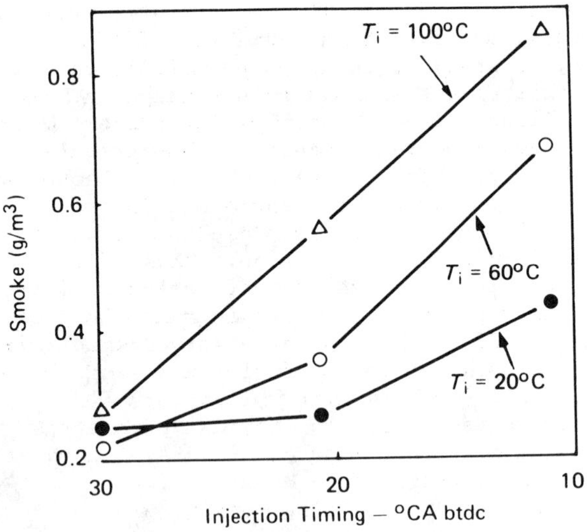

13(a)
Effect of Inlet Air Temperature T_i
on Exhaust Smoke
(See Fig. 3.5 of Ref. 21)

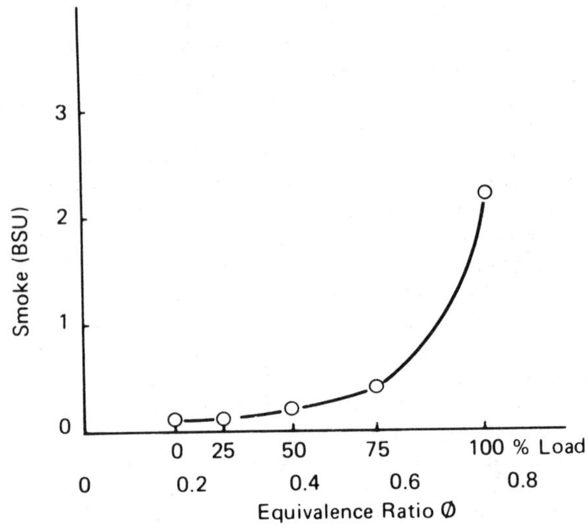

13(b)
Effect of Overall Equivalence

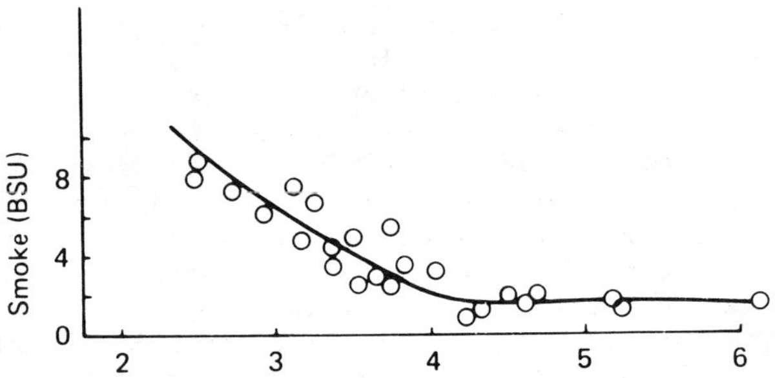

$$\left[\frac{F}{\theta}\,\frac{d_o}{d}\left(\frac{n_o}{n}\right)^{1/3}\right]\left(mm^3/\rho\,CA\right)$$

13(c)
Effect of Fuel-Air Mixing Rate
— Increase of Injection Rate
(See Fig. 4 of Ref. 18)

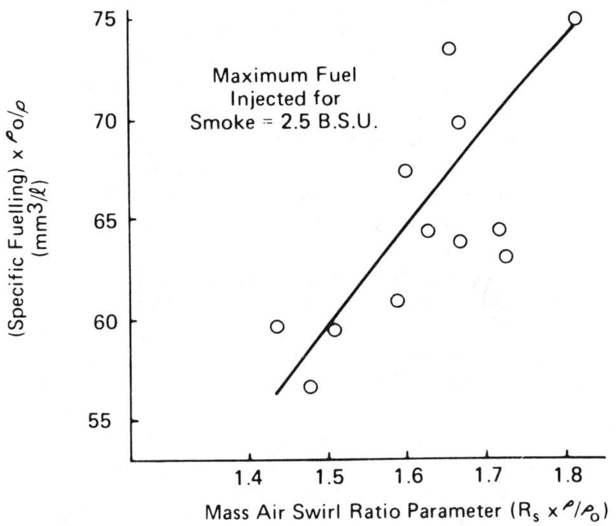

13(d)
Effect of Fuel-Air Mixing Rate
— Increase of Air Swirl Rate
(See Fig. 9 of Ref. 13)

Fig. 13 - Illustration of main parameters which affect soot formation in diesel combustion, engine test bed results

rather than the lubricating oil. The evidence from studies on a number of four-stroke engines in good mechanical condition is that the contribution from the lubricating oil is relatively small. Good mechanical condition excludes the 'running-in' phase of a new engine and also the latter part of the useful life of an engine. We can quote an example here for a vehicle fitted with an IDI engine in good condition and giving 0.33 g/mile of particulates on the LA4 cycle. An engine which had done 70,000 miles was fitted to the same vehicle and gave 1.1 g/mile of particulates. Loss of oil control is suspected in these circumstances.

High levels of EGR may be necessary to meet stringent NO_X emissions targets. However high levels of EGR can accelerate the rate of engine wear owing to the deposition of soot on the cylinder wall during the induction and compression process. The possible adverse effect of high levels of EGR on engine wear, and hence particulates at high mileage, should be borne in mind when setting excessively stringent NO_x emissions targets.

IN-CYLINDER CONTROL OF PARTICULATE EMISSIONS

With any practical control measure, it is necessary to consider any adverse affects on other emissions or performance parameters. Consequently it is necessary to consider the optimisation of diesel combustion for fuel economy, combustion noise and emissions of smoke, NO_x, HC, particulates, CO and odour. The understanding gained by Lucas CAV in this paper, and that of references (11,12,13,14,18,19,20), has been used to construct three ideals for optimising diesel combustion as shown in Figure 14.

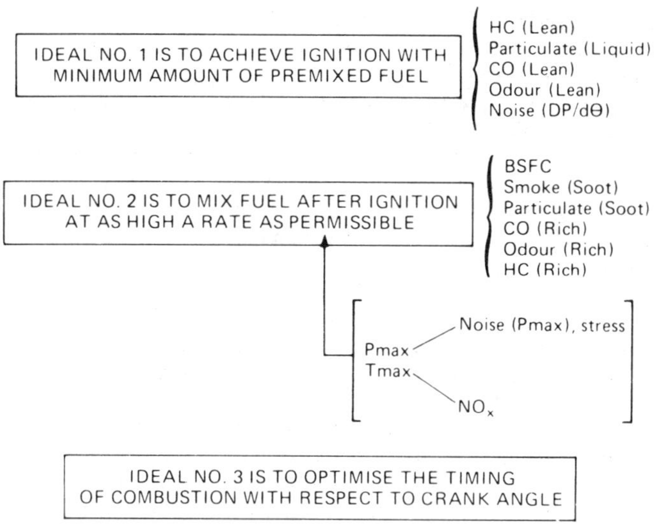

Figure 14 - Optimisation of diesel combustion

OPTIMISATION OF DIESEL COMBUSTION - Ideal No. 1 is concerned with the premixed phase of combustion and is to achieve ignition with the minimum amount of premixed fuel. If this is achieved, then the contribution of the lean limit source in Figure 10(a) to HC, particulates, CO, odour and the rate of cylinder pressure rise (noise) is minimised.

Ideal No. 2 is concerned with the diffusion phase of combustion and is to mix the rest of the fuel, injected after ignition, at as high a rate as "permissible". This will minimise the burn time to give optimum cycle efficiency and hence minimise BSFC. The high temperature fuel rich regions which produce soot will be 'leaned out' more rapidly and hence reduce their contribution to soot and particulate emissions. Finally, the fuel-rich sources of CO, odour and HC will be minimised. However there is a maximum "permissible" rate of fuel-air mixing. This is because the faster the mixing, the higher the maximum cylinder pressure and temperature become, and both noise and NO_x emissions increase.

Ideal No. 3 is simply to time the combustion event defined by ideals Nos. 1 and 2 in an optimum way with respect to crank angle.

The three combustion ideals just defined apply equally to direct and indirect injection diesel engines. However the achievement of these ideals involves many complex interactions in practical combustion systems. Often there are quite different ways of approaching the ideals, for example in the relative contribution which injection velocity and air swirl velocity make to fuel-air mixing. A combination of experimental method and computer modelling has been used to provide a preliminary understanding of combustion systems and the potential development areas. However much remains to be done to fully understand the quantitative details of fuel-air mixing and local temperatures, and how the various performance and emissions aspects interact.

PRACTICAL CONTROL MEASURES - It is clear that optimising diesel combustion is partly the responsibility of the engine maker and partly the responsibility of the fuel injection maker. The particulate mass emissions can be minimised as follows:

(a) Aim to achieve sub-scritical ignition delay at all operating loads and speeds by providing a high end of compression temperature, and optimum injection timing schedules over the load and speed range.

(b) Minimise the sac volume in the nozzles of direct injection engines.

(c) Provide fast rates of fuel-air mixing by optimising the fuel injection and air motion characteristics of both direct and indirect injection engines.

These measures for controlling particulate emissions are essentially those known previously for optimising fuel economy, smoke, and gaseous emissions. Additional measures for controlling NO_x emissions include the retard of fuel injection timing, combined with increase of compression ratio and the use of EGR. Current well optimised IDI engines correspond to the lower data values given in Figure 1 for particulate emissions, and current well optimised swirl type DI engines to the highest smoke limited BMEP values given in (13). The combustion ideals outlined in the previous section show the direction in which in-cylinder optimisation should move, but in practice the gains are unlikely to provide sufficient improvement for the more stringent emissions targets.

If we consider a 3000 lb inertia weight car on the LA4 emissions cycle, then a result of particulates = 0.3 g/mile, HC = 0.2 g/mile and NO_x = 1.2 g/mile (without EGR) already represents a fairly well optimised IDI engine. In these circumstances some form of exhaust aftertreatment, such as a particulate trap, may be necessary to meet significantly lower levels of particulate emission.

CONCLUSION

A correlation between the measured values of particulates, HC and smoke emissions has been demonstrated for a given fuel. This correlation provides a less expensive way of carrying out preliminary engine optimisation work for minimising particulate emissions. The correlation also provides the key towards establishing the in-cylinder origins of particulate formation as the origins of HC plus the origins of soot formation. The origins of HC and soot, and hence particulate formation, are understood in qualitative terms. This has allowed the formulation of three combustion ideals to aim for when optimising the temperature and fuel-air mixing aspects of diesel combustion for fuel economy, emissions and noise.

ACKNOWLEDGEMENTS

The authors wish to thank the directors of Lucas CAV Limited and Mr J Righton, Group Director Technical Policy, Lucas Industries, for permission to publish this paper. Particular thanks are due to Dr B A Jarrett, Technical Director of Lucas CAV Limited and to Mr H C Grigg, Chief Research Engineer, for their support and comments.

The authors would like to acknowledge the work of Mr R Freese and his team in developing the dilution tunnel splitting technique, and carrying out the IDI engine study at Ricardo.

The data points, for the DI engine shown in Figure 8, were obtained as part of a Joint Research Project carried out by SGRD Limited, and sponsored by Perkins Engine Company Limited, Leyland Vehicles Limited, Rolls-Royce Motors Limited, Shell UK Oil Limited, Lucas CAV Limited and the UK Department of Industry. The authors would like to thank the sponsoring Companies for their permission to include these data points.

REFERENCES

1. Eugene Danielson, "Particulate Measurement - Light Duty Diesel Particulate Baseline Test Results." EPA Technical Support Report SDSB 79-03, January 1979.

2. G M Cornetti, "Lightweight Automotive Diesel Data Base." Fiat Contract Number DOT-TSC-1424, Nov. 77 - Sept. 79.

3. F Black and L High, "Methodology for Determining Particulate and Gaseous Diesel Hydrocarbon Emissions." SAE Paper 790422.

4. C H Hare and T M Baines, "Characterisation of Particulate and Gaseous Emissions from Two Diesel Automobiles as Functions of Fuel and Driving Cycle." SAE Paper 790424.

5. R L Williams and S J Swarim, "Benzo (a) Pyrene Emissions from Gasoline and Diesel Automobile." SAE Paper 790419.

6. K J Springer and R C Stahman, "Diesel Car Emissions - Emphasis on Particulate and Sulfate." SAE Paper 770254.

7. Federal Register/Vol. 45, No. 140/ Friday July 18, 1980/Rules and Regulations, Page Nos. 48133-48141.

8. R A C Fosberry and D E Gee, "Some Experiments on the Measurement of Exhaust Smoke Emission from Diesel Engines." MIRA Report 1961/5.

9. C T Vuk, M A Jones, J H Johnson, "The Measurement and Analysis of the Physical Character of Diesel Particulate Emissions." SAE Paper 760131.

10. R W Hurn, "Air Pollution and the Compression-ignition Engine." 12th Symposium (International) on Combustion, the Combustion Institute, 1968.

11. G Greeves, I M Khan, C H T Wang and I Fenne, "Origins of Hydrocarbon Emissions from Diesel Engines. SAE Paper 770259.

12. I M Khan, G Greeves and C H T Wang, "Factors Affecting Smoke and Gaseous Emissions from Direct Injection Engines and a Method of Calculation." SAE Paper 730169.

13. G Greeves, C H T Wang and G A Kyriazis, "Inlet Port Design and Fuel Injection Rate Requirements for Direct Injection Diesel Engines." FISITA XVIII International Congress 5-8 May 1980, Hamburg.

14. I M Khan and G Greeves, "A Method for Calculating the Formation and Combustion of Soot in Diesel Engines". 1973 International Seminar, International Centre for Heat and Mass Transfer, Belgrade, Yugoslavia.

15. K B Lee, M W Thring and J M Beer, Combustion and Flame, Vol. 6, p. 137, 1962.

16. G Greeves, I M Khan and G Onion, "Effect of Water Introduction on Diesel Engine Combustion and Emissions." Paper No. 25, 16th Symposium (International) on Combustion, The Combustion Institute, 1976.

17. S M Shahed, P F Flynn and W T Lyn, "A Model for the Formation of Emissions in a Direct Injection Diesel Engine." G.M. Symposium on Combustion Modelling, Nov. 6-7, 1978.

18. G Greeves, "Response of Diesel Combustion Systems to Increase of Fuel Injection Rate." SAE Paper 790037.

19. G Greeves and I M Partridge, "Experiments and Calculations on Optimising Fuel Economy and Emissions in Passenger Car Diesel Engines." Conference on Passenger Car Power Plant of the Future, I.Mech.E., London, 2-4 October 1979.

20. A R Reading and G Greeves, "Measurement of Diesel Exhaust Odorants and Effect of Engine Variables." SAE Paper 800424.

21. I M Khan, "Formation and Combustion of Carbon in a Diesel Engine." I.Mech.E. Proc., 1969-70, Vol. 184.

The Use of a Thermogravimetric Analyser for the Investigation of Particulates and Hydrocarbons in Diesel Engine Exhaust**

R. D. Cuthbertson, H. C. Stinton, and R. W. Wheeler
Ricardo Consulting Engineers Ltd.
Bridge Works
Shoreham-by-Sea
West Sussex
(England)

IT WAS INEVITABLE THAT following on from the successful application of advanced gas analytical techniques to the control of the Los Angeles smog problem there would be a similar move to control the less severe but more obvious problem of objectionable diesel engine fumes.

Although the problem is much greater in Europe, Victorian attitudes of the inevitability of smoke and smells have restricted attacks on it to lenient and ineffective smoke opacity legislation. In the USA the greater sophistication of public opinion brought about by the smog publicity caused an outcry against the city and school diesel buses operating in residential areas. The observer did not need to be a chemist to relate cause and effect.

Research at this point is divided into three streams:

1) Engine fuel injection system research to reduce the fuel admitted after the correct end-of-injection. This very quickly reduced the problem to manageable proportions (1)*.
2) Diesel odour research to define the exhaust components responsible which effectively characterized them as pyrolysed and oxidized hydrocarbons (2)(3).
3) Application of modified gasoline pollution technology to the regulatory control of the emissions of carbon monoxide (CO), nitrogen oxides (NOx) and hydrocarbons (HC) from automotive diesel engines (4).

These researches revealed that the major problem resided in the diesel hydrocarbons

* Numbers in parentheses designate References at end of paper.

**Paper 790814 presented at the Off-Highway Vehicle Meeting and Exposition, Milwaukee, Wisconsin, September, 1979.

ABSTRACT

The conditions necessary for obtaining samples of diesel particulates in CVS tunnels have been investigated in relation to the simultaneous analysis of the gaseous hydrocarbons which were present in equilibrium with the particulate borne hydrocarbons.

Thermogravimetric analysis has been developed to separate the components of particulates in a simple way which is useful in diesel engine research. A reliable instrument has been interfaced with a heated flame ionization detector which gave additional or corroborative information.

A system has been evolved for total diesel hydrocarbon analysis using thermogravimetry for those hydrocarbons trapped on the particulate filter, and heated flame ionization detection for those passing through the filter. The results are compared with total hydrocarbon analysis as used in the Federal Test Procedure and show a small gain in hydrocarbon recovery with an increase in overall information for the new system.

because of the greater span of physical properties than is seen in the gasoline engine. However, the engine developers needed a rough and ready analysis of the total unburnt HC in the exhaust and applied the technique of heated flame ionization detection (HFID) with such great success that it became the prime candidate for regulatory use (5).

As a result of these researches the truck diesel engine (the only automotive diesel in widespread use in the USA) was regulated by the 13 mode steady state cycle and since smoke was seen to be a transient operation problem the "lug down" cycle was devised and used with an exhaust plume smokemeter.

The advent of the energy crisis, particularly the concept of the corporate average fuel economy (CAFE) for cars, brought threats of an explosion in diesel car production. This prompted a demand by U.S. Congress in August 1977 for a review of diesel engine emissions, a new approach to smoke by particulate control and an examination of their toxicity.

The Environmental Protection Agency (EPA) responded to this demand with researches which soon revealed the complexity of the problem of regulating in a fair manner the HCs and particulates. These contain a baffling array of materials such as smog HCs, odours, irritants, lube oil, carcinogens, mutagens and black opaque carbon with some of these appearing in both the HC fraction measured by HFID and in the particulates collected by the Constant Volume Sampling (CVS) dilution tunnel procedure.

The EPA researchers applied advanced gas analysis techniques (HPLC, GC, GC-MS) to the HC adsorbed on or absorbed in the carbon structure (6). Work in Europe using mass-spectroscopy of the gases evolved from diesel particulates when heated showed how strongly the polycyclic hydrocarbons were bonded to the carbon (C). One group used a thermogravimetric analyser (TGA) to separate carbon and HC in a nitrogen (N_2) atmosphere (7), another group used a temperature controlled probe in the MS inlet (8).

The work described in this paper was aimed at the development of a simple procedure for investigating the large spectrum of HCs existing in their gaseous, liquid and solid states in cooled and diluted diesel exhaust.

THE PROBLEM

At present diesel hydrocarbons are evaluated by HFID in filtered exhaust gas at a temperature of 190°C; under these conditions only a small fraction of the HC is lost in the soot remaining on the filter. Unfortunately it is probable that some biologically active HCs are lost here.

The particulates which are carbon with associated volatile material - HC, sulphuric acid (H_2SO_4), nitric acid, etc. - are evaluated, using CVS dilution tunnel development, and filtration on teflon loaded glass fibre paper. The C/HC ratio may be from ∿ 4:1 to 0.2:1 and in a typical vehicle engine over the Federal Test Procedure (FTP) the particulate bound HC may be from ∿ 10% to ∿ 50% of the total HC by HFID.

When diesel exhaust is regulated using total HC by HFID and total particulates, clearly much information is lost, for example, the fraction of the total HC which takes part in smog aerochemistry and the fraction containing carcinogens. The emission of black opaque carbon also remains unknown.

Several routes to some of these further data exist, e.g. using two HFIDs, one with a hot filter at 190°C, the other with a cold (< 52°C) filter, should give the carbon bound HC by difference. Solvent extraction of the total particulate filter or the evaporation of the HC from the filter in a hot nitrogen stream can also be used to obtain these data.

The first alternative is very demanding technically and expensive. Solvent extraction is excellent but very slow and needs trained laboratory staff. The last method is the clear favourite because the apparatus required already exists in principle - the TGA, in which a sample can be heated through a set programme in an inert or reactive gas flow. It operates gravimetrically giving the weight of carbon and of volatiles in a sample (9). If it is interfaced with an HFID or flame photometric analyser the volatiles can be separated into HC and H_2SO_4. The relationship between the weight loss and the temperature programme could give crude data on the HC boiling ranges.

Experience in the rubber industry (9) has shown the Stanton Redcroft TG750 to be very robust and it was chosen therefore as the basic instrument for modification to the specific requirements of our particulate research programmes.

THERMOGRAVIMETRIC ANALYSIS

PRELIMINARY INVESTIGATION - The general layout of the TG750 is shown in Fig. 1 and details of the furnace are shown in Fig. 2. Points to notice are the very small sample holder capable of holding only a few confetti sized pieces of a particulate filter, the flow of inert gas (nitrogen) through the glass envelope which carries the volatiles down through the furnace, and the water cooled furnace crown and walls where volatiles can condense and be lost. Tests immediately showed that the sample size was totally inadequate since it is mostly fibre glass, that extreme care was needed to avoid loss of carbon by oxygen ingress, and that indeed much HC was collected on the furnace interior.

Furnace Redesign and Addition of HFID - During the early tests it became obvious that

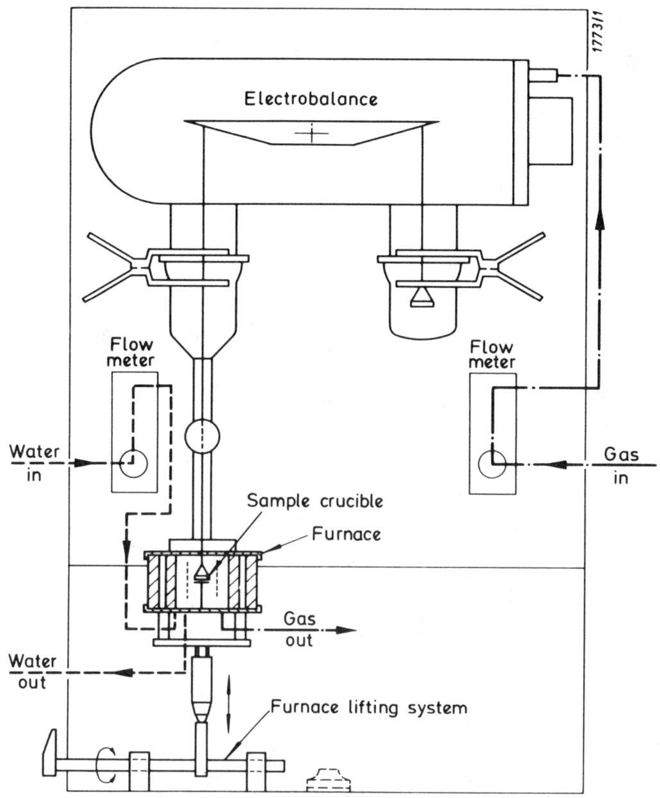

Fig. 1 - Diagram of standard Stanton Redcroft TG750 thermobalance

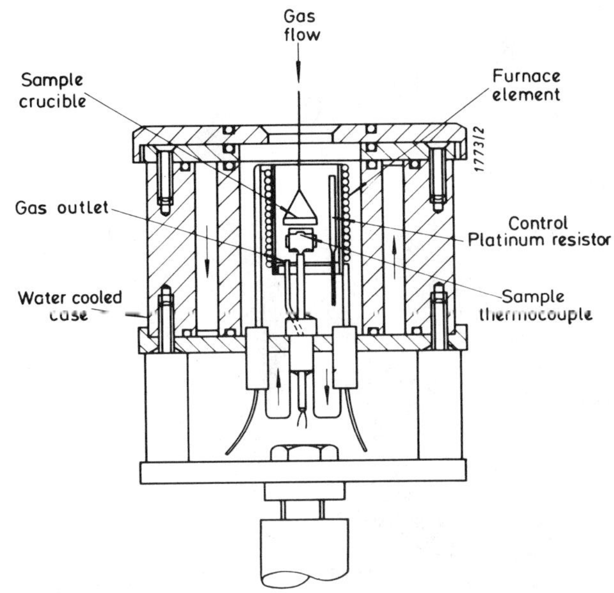

Fig. 2 - Standard micro-furnace as fitted to Stanton Redcroft TG750 thermobalance

the addition of an HFID would give very valuable corroborative evidence and in the case of samples from cars equipped with oxidation catalysts the only method of assessing the true value of HC because of the presence of H_2SO_4. Thus the furnace was rebuilt as shown in Fig. 3 to contain a 9 cm filter paper packaged and wire wrapped to 1.5 cm diameter and 2 cm long. The design was dictated by the need for a short heated (330°C) line to the HFID which had to be fed through and insulated from the water cooled furnace crown.

The new furnace incorporated a tubular heat shield to prevent volatiles reaching the water cooled walls and the nitrogen flows from both top and bottom were arranged to carry gases as rapidly as possible to the heated line inlet.

This design was used throughout the tests with considerable success. The weak points were the 'O' ring gas seal, a small loss of volatiles on the water cooled crown and more seriously, the HFID trace showed random rapid fluctuations which were believed to be due to gas pulsing from the counter flows of nitrogen at the heated line inlet.

A future design which avoids some of these weaknesses is shown compared to the present arrangement in Fig. 3.

The HFID, constructed by one of the team (Cuthbertson), used a Perkin Elmer F11 chromatograph with the detector operating under depression to avoid the requirement of a heated sample pump. The sample was diluted with nitrogen at the detector capillary so that only a small flow of nitrogen was needed through the TGA and thus the balance mechanism was not disturbed aerodynamically.

The complete flow chart is given in Fig. 4.

SYSTEM CHECKOUT AND TROUBLE-SHOOTING - From the beginning, the balance, its readout, and the temperature programmer behaved impeccably. The main problems centred on the gas flows carrying the volatiles through the TGA and the HFID. It was essential to avoid leaks and obtain stability in the output over a useful time and temperature schedule. To this end actual diesel exhaust soot was tested in the quantities expected to be collected on the filters during an FTP (i.e. from 10 to 50 mg). The rate of temperature rise must be adequate to allow time for evaporation, but be as short as possible for test convenience and to prevent trace oxygen leakage into the system with consequent loss of carbon by oxidation. Figure 5 shows TGA and HFID curves superimposed for two heating rates, 50 and 10°C/min, with the chart recorder running 5 x slower at 10°C/min and the HFID gain 5 x higher. The slower temperature rise volatilizes the HC at slightly lower temperatures due to more time being available for diffusion from the filter package. Otherwise the curves are almost identical in shape and area and it was decided to use a temperature rise rate of 50°C/min in the subsequent tests.

Operation of the HFID - Considerable work was done to optimize the HFID flow settings when

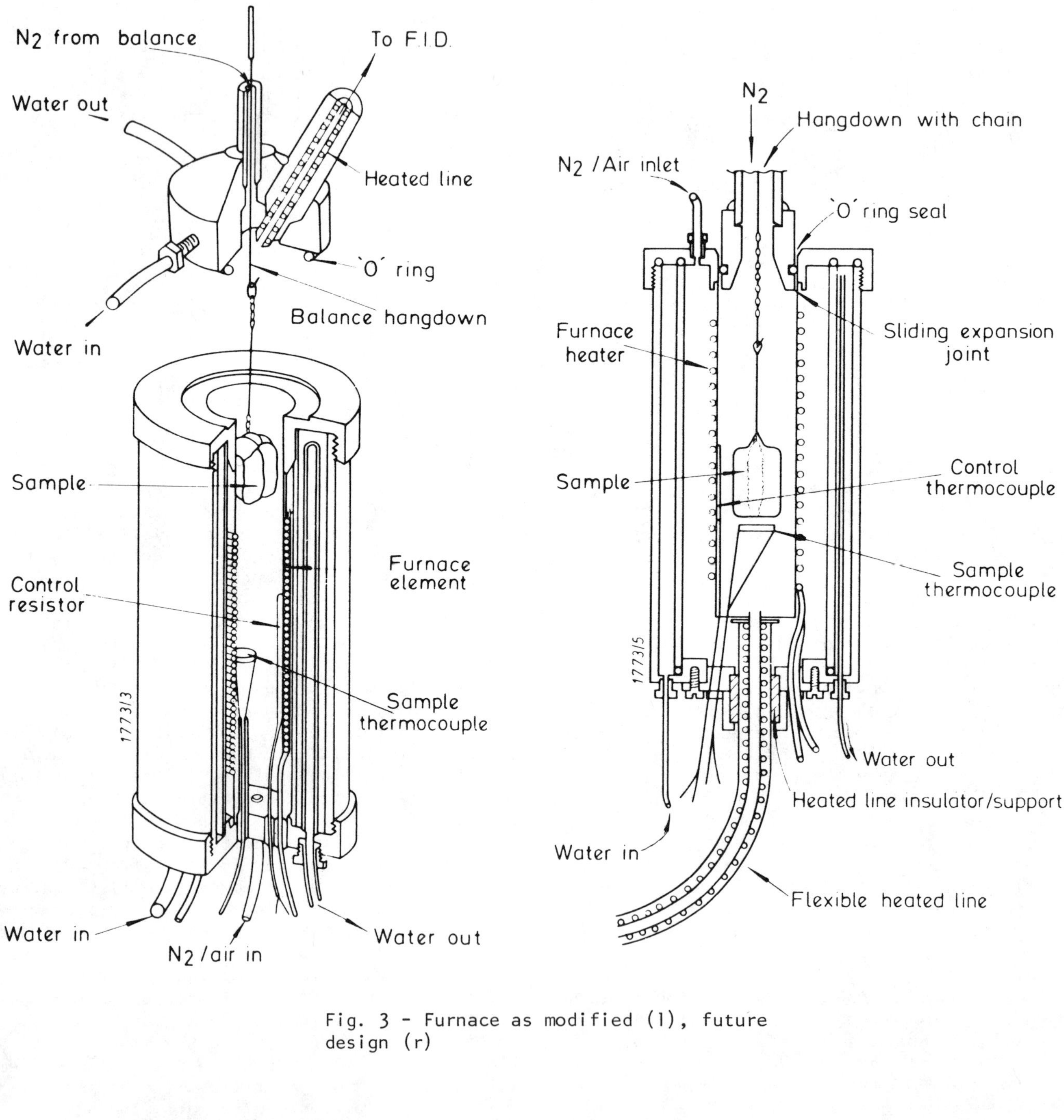

Fig. 3 - Furnace as modified (l), future design (r)

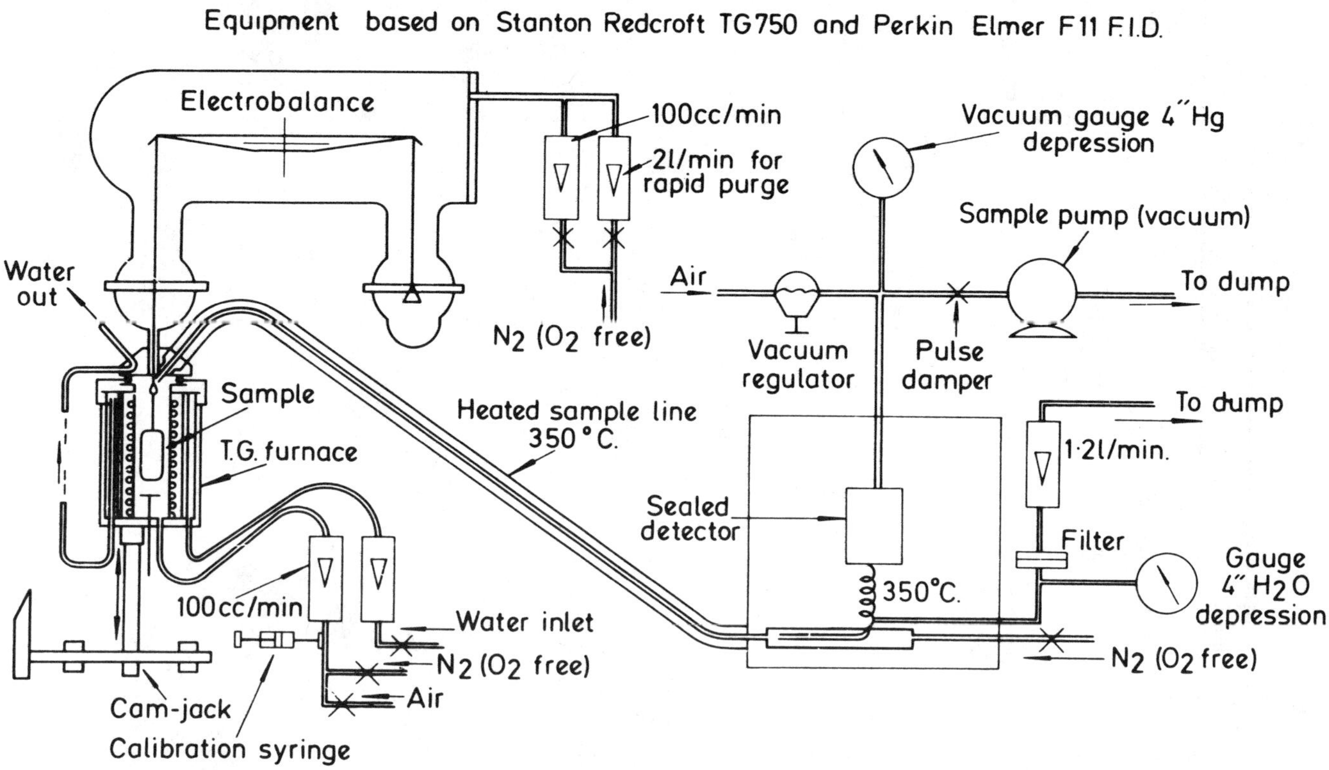

Fig. 4 - Diagram of modified thermogravimetric analyser interfaced with a flame ionization detector

using diesel soot samples; the values shown in Fig. 4 gave HC readings in a stable and linear range at 50°C/min temperature rise. There was no oxygen in the sample and therefore no oxygen response is seen in the HC response. Hydrogen fuel was used since it allowed our burner to operate more reliably under depression; the carbon counting (CC) numbers were estimated experimentally.

Oxygen was excluded from the system eventually by eliminating, where possible, all rubber and plastic tubing. The effect of oxygen leakage is seen in Fig. 6 where the slope of the weight loss curve is too steep at temperatures above 550°C due to carbon dioxide (CO_2) formation.

Figure 6, however, shows typical features of the thermograms:
1) The HFID trace shows the rise of HC evolution clearly and is preferred to the TGA's first derivative of weight loss.
2) The HFID trace has two parts, the later "hump" matches the evolution of high boiling species such as polynuclear aromatic hydrocarbons (PNAH) and lub oil.
3) The long tail of the HFID trace is due to residual "hang-up" in the furnace crown.
4) The carbon burns violently if too much air is allowed to enter above 500°C and causes a "kick" on the temperature curve; an initial introduction of 5% oxygen in the nitrogen was

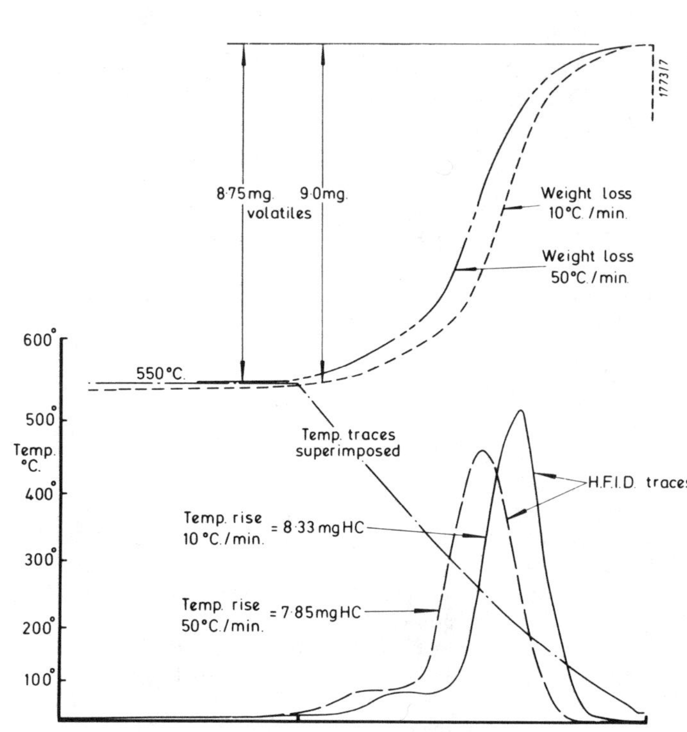

Fig. 5 - Effect of heating rate on the evolution of hydrocarbons in the TGA

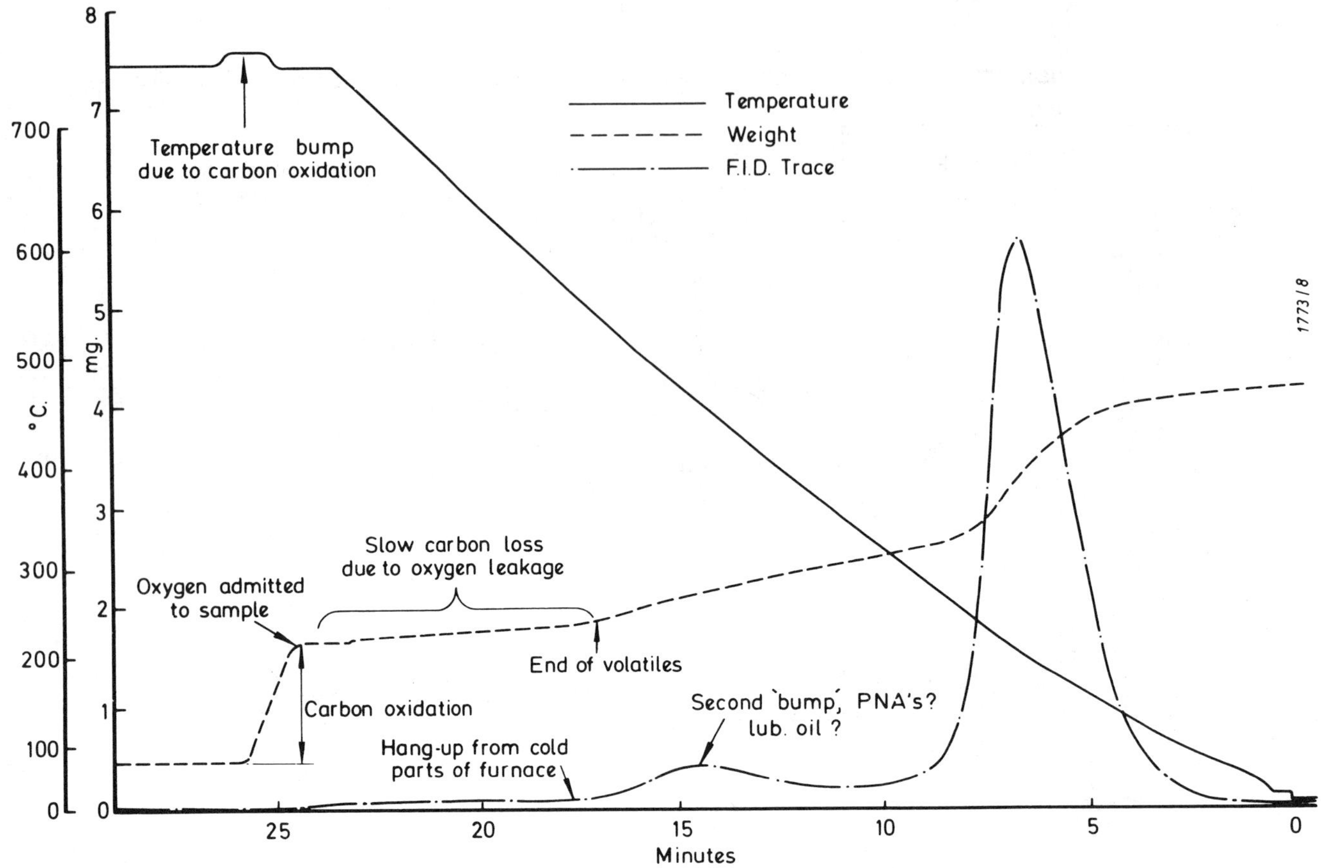

Fig. 6 - TGA interfaced with an HFID - weight loss and HC curves for a soot sample

found to be adequate to prevent this temperature rise.

PRESENTATION OF THE PARTICULATE SAMPLE TO THE TGA - As already suggested the soot is collected in a CVS dilution tunnel by filtration on an inert "paper" and it is usual to obtain from 10-50 mg, fairly evenly spread across the sampling area. The furnace will just contain one 9 cm paper; larger filter papers need to be halved across a diameter or segmented.

Filter types selected for testing included:

1) Pallflex - teflon covered glass fibre.
2) Whatmans type GF/A
3) Whatmans type GF/D
4) Experimental quartz fibre
5) Carbon fibre of high absorptivity. (Obtained from the Chemical Defence Experimental Station, Porton, Wiltshire, UK.)

The Pallflex was unsuitable for TG work as it decomposed and lost weight above 350°C. However, a direct comparison of Pallflex and GF/A filters in a CVS tunnel with duplicate filter holders gave identical particulate loadings from several FTP tests on diesel cars. Therefore Pallflex offered no advantages in this work.

The GF/A proved suitable, except for its low melting point of ∿ 800°C which allowed it to melt in the exothermic carbon burn off. Sulphuric acid also reacted with the GF/A paper and could not all be evaporated off. GF/D on the other hand was too bulky and could not be fitted into the present furnace. Both papers gave a blank weight loss; a 9 cm circle of GF/A weighing 340 mg for example lost 2.5 mg as a typical value. This loss could be reduced to 0.5 mg by preheating to 500°C and followed by very careful storage. For routine work standard packs of GF/A were used together with a running blank check filter paper.

The quartz filter was assessed in an attempt to avoid the blank error; the first trials showed an even higher blank loss value than the GF/A and it was abandoned.

However, the blank losses in no instance gave rise to any measurable HC by the HFID.

The carbon cloth, because of its strong adsorption of water vapour, CO_2, etc., was found to be quite unweighable but proved very effective as a total particulate and reactive HC collector for subsequent solvent extraction.

All further testing was performed using GF/A, although GF/D might prove to be a better filter medium according to A.D. Little (10) but its use would necessitate a larger furnace.

Some final tests on paper folding procedures and carbon burn off temperatures defined the oxygen level (initially < 5%), the temperature (550°C) and the package size (9 cm paper folded to 2 cm length x 1.5 cm diameter) so that the carbon oxidized quietly and did not

Table 1 - Carbon Counting Efficiencies of the Flame Ionization Detector Interfaced with the Thermobalance for a Number of Hydrocarbons

Material	Chain Length (No. of Carbon Atoms)	Carbon Counting Efficiency %
Hexane	6	80
Decane	10	83
Dodecane	12	73
Eicosane	20	94
Docosane	22	88
Tetracosane	24	90
Octacosane	28	89
Dodec-1-ene	12	65
Benzene	6 carbon atom ring	74
Toluene	6 carbon atom ring + 1 side carbon = 7	82

melt the GF/A. If melting occurred some carbon was encapsulated in the melt and lost.

CARBON COUNTING EFFICIENCY OF THE TGA/HFID - The weight loss of volatile matter in a flow of nitrogen up to the plateau at $\sim 500^{\circ}C$ cannot be ascribed in total to HC since water, inorganic acids and ammonium salts will also volatilize. It was therefore important to obtain an accurate value of HC from the HFID reading by submitting known hydrocarbons to thermal analysis (which gives the true weight used) and integrating the HFID trace. The HFID was standardized by injecting into the TGA 5 ml of propane (= 9.15 mg at $20^{\circ}C$) which flowed via the furnace through the heated line into the HFID.

Table 1 shows the carbon counting for a batch of pure hydrocarbons expressed as a percentage of full recovery. It will be seen that the higher molecular weight compounds gave the highest efficiency and some of the more volatile ones gave the lowest. The latter effect was almost certainly due to the difficulty of weighing the samples with evaporation and buoyancy effects intervening.

Clearly only a rough estimate can be made of the weight volatilized in the TGA using these data which suggest an overall recovery of approximately 83%. Many typical diesel HCs must be tested to determine this value with greater accuracy. Using this technique on a typical TGA trace gave the results shown in Fig. 7. The HFID integration needed a carbon counting efficiency of 78% to predict the observed volatile weight loss.

A further test using diesel fuel (British DERV) placed on a new filter paper in the TGA needed a carbon counting efficiency of 74% to predict the observed weight loss.

We thus have 83% for higher paraffins 74% for lower (DERV) and 78% for diesel particulate volatiles so the assumption is made that 78% carbon counting is suitable for our TGA/HFID.

Suitability for Routine Engine Development - Despite the incomplete development of the system it was applied with great success to programmes involving particulate reduction by filters and catalysts. Three typical thermograms are shown for general interest (Figs. 8, 9 and 10). They display the breakdown of the particulates into carbon, volatiles, HC and

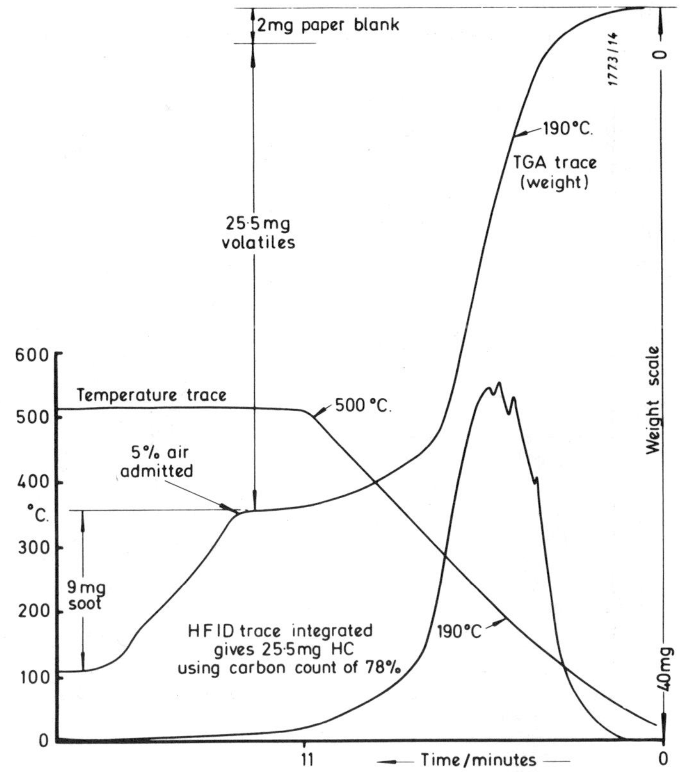

Fig. 7 - TGA/HFID traces showing equality of HC mass using a carbon counting factor of 78%

non-HC volatiles. The ratio C/HC gives vital information on engine build and operation. The HFID trace often shows two peaks, the second of these being probably PNAHs and lub oil fractions.

The carbon count of 78% is vindicated in Fig. 8 but in Figs. 9 and 10 there are clearly non-HC volatiles present, probably H_2SO_4, which are discussed in a later section. Figure 10, showing catalysed particulates, indicates how H_2SO_4 is produced and how the C/HC ratio is greatly increased when the catalyst is sufficiently hot.

At this point it was clear that TGA/HFID could provide valuable information on particulate samples and attention was directed therefore towards the actual sampling procedure in the CVS dilution tunnel.

PARTICULATE SAMPLING IN THE DILUTION TUNNEL

A tunnel was specially prepared for this research using stainless steel sheeting. It is 15" in diameter and 12' in length from exhaust entry to sample probes. It was connected to a standard CVS equipment which had a total throughput of 363 cfm with clean air taken from a hydrocarbon free source external to the laboratory. For the preliminary tests on particulate trapping efficiency four probes carrying external filters were used. They were

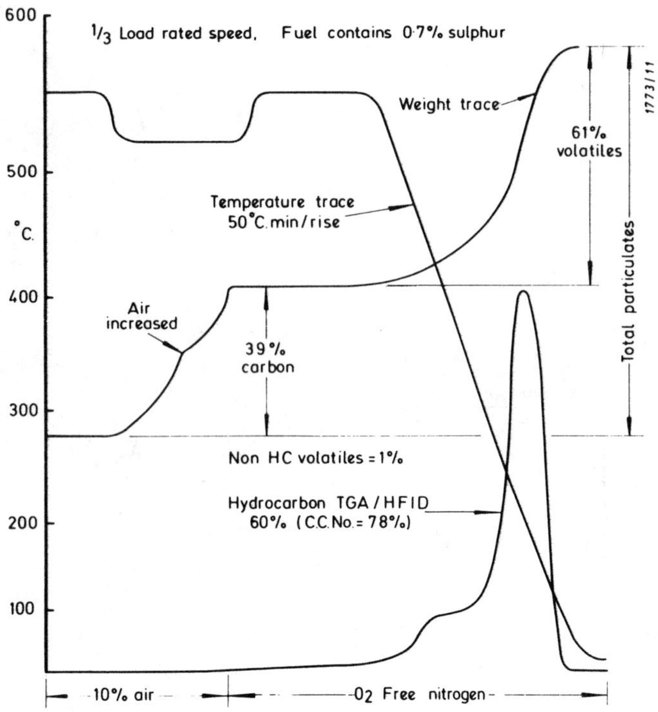

Fig. 8 - TG of particulates from a DI diesel engine - showing that volatile material is HC

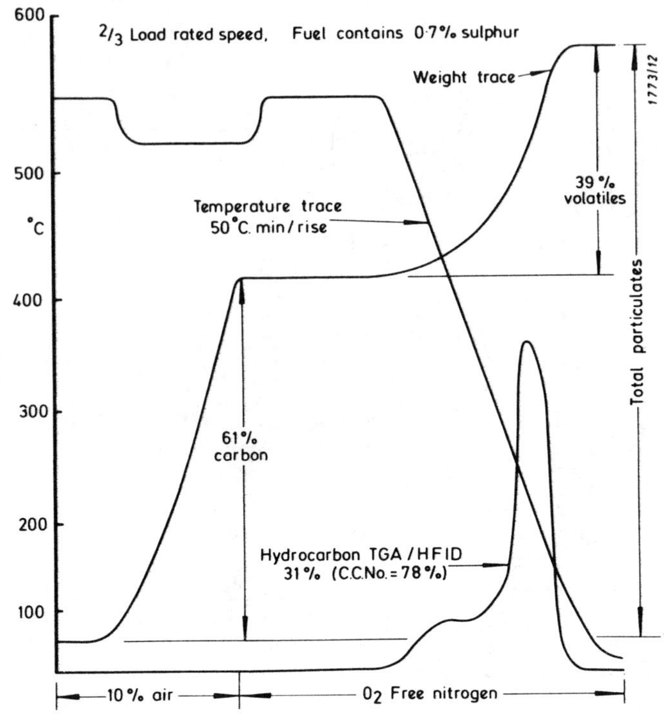

Fig. 9 - TG of particulates from a DI diesel engine - showing some non-HC volatile material

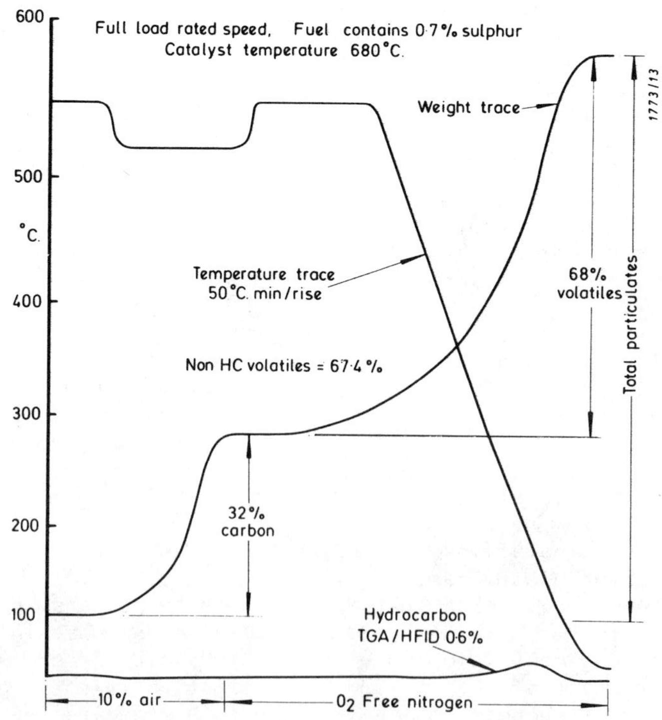

Fig. 10 - TG of particulates from a DI diesel engine fitted with a platinum catalyst - showing a high sulphuric acid presence

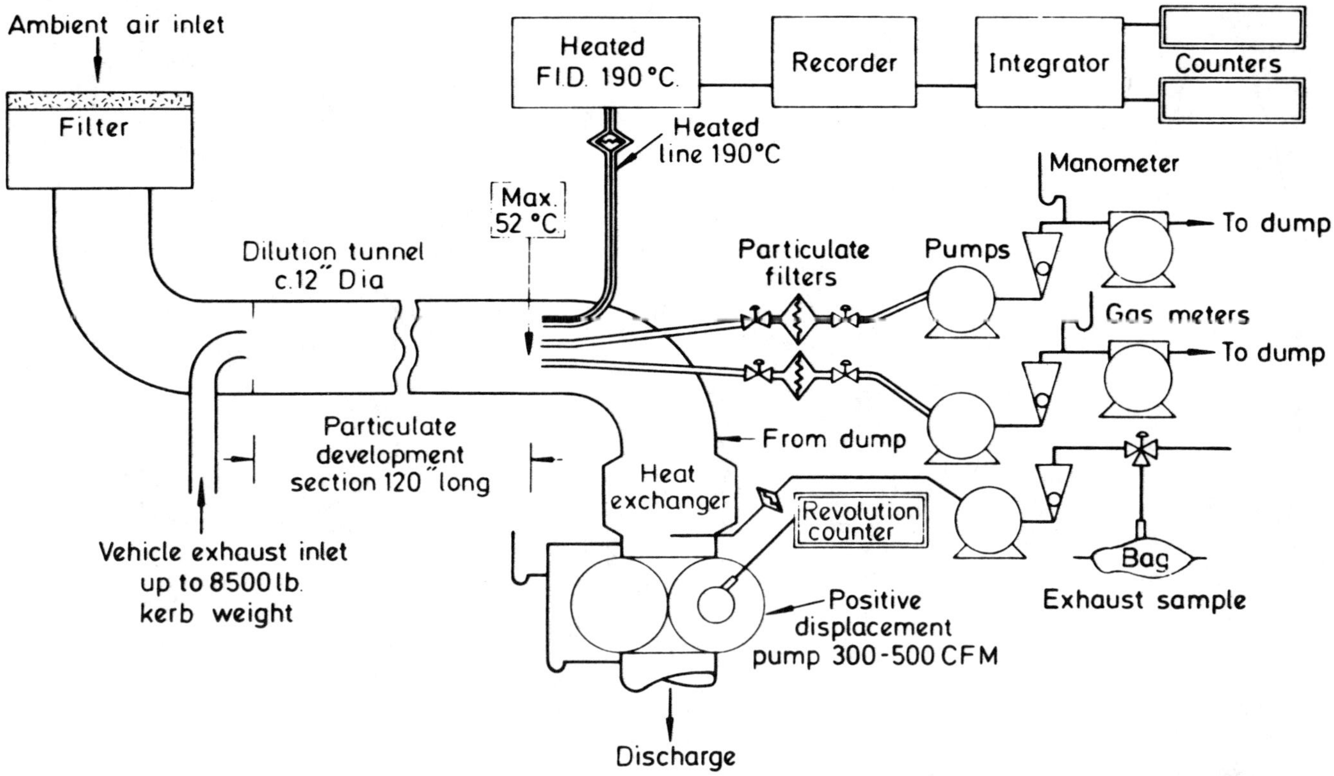

Fig. 11 - Diesel exhaust particulate sampling system

arranged equally spaced at approximately half tunnel radius. The sample flow through each filter was 1 cfm and the filter holders were able to accommodate 14 cm or 9 cm papers.

One of the four probe assemblies was designed such that it could be replaced by a special heated probe and filter for the purpose of obtaining total HC by HFID in the normally accepted manner, whilst another one provided particulate samples. Figure 11 shows the general principles of the tunnel and its accessories.

TUNNEL EFFICIENCY - In order to test the particulate mixing in the tunnel and its collection on the four filters it was necessary to disperse a realistic aerosol of known weight into the tunnel inlet and to assess the mixing efficiency by measuring the increase in weight of each of the filters.

Of course it is impossible to redisperse diesel soot in its original particulate size range; it was decided therefore to disperse dioctylphthalate (DOP) and hexadecane using a motorized syringe and stainless steel hot plate assembly (Fig. 12). The hot plate temperature was set at the minimum required to evaporate the liquids at approximately 0.2 g/min into the full CVS flow.

It proved difficult to carry out these experiments as originally contemplated since the complete evaporation of the fluids was

Table 2 - Distribution of Particulate Matter Between the Four Filters Fitted to the Dilution Tunnel

	Quantity of Test Material Injected as Indicated by Loading on Filter			
Filter Number	No. 1	No. 2	No. 3	No. 4
Test Material				
DOP	1.844	1.859	1.868	1.764
	4.6	4.73	4.6	4.6
Hexadecane	1.63	1.63	1.69	1.71
	1.55	1.58	1.58	1.4

difficult to observe and a slight lacquering of the hot plate occurred. However, Table 2 shows good agreement between the four filters with the exception of two tests with filter No. 4. The differences were considered to be due to slight filter leakage. The best estimate of the percentage recovery gave 98% for both liquids; a value confirmed by later tests using the same technique but with HFID assessment of the gases passing the filter (Table 3).

Direct Measure of Tunnel Loss - These tests did not give a meaningful indication of the possible losses in the tunnel system itself. A direct test was therefore carried out in a similar tunnel which had the facility of an accurate circumferential joint dividing it into two equal parts. This allowed a tenth part of the tunnel length to be cleared by scouring with GF/A filter papers until a TGA revealed

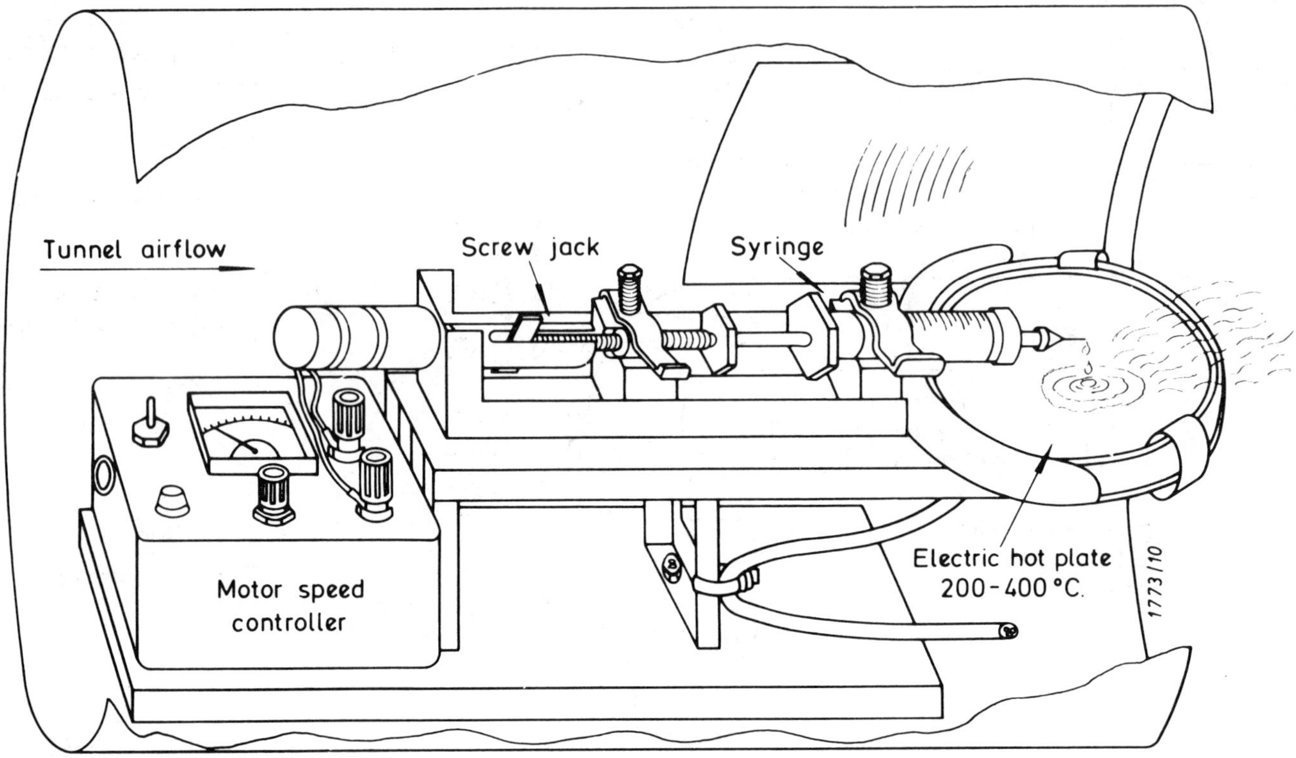

Fig. 12 - Sketch of the "hot spot" evaporation

Table 3 - Results Showing the Distribution of some Pure Hydrocarbons between a Filter and an HFID Located after the Filter and Comparing them with a standard HFID

				% Recovery of Total Amount Injected into Dilution Tunnel		
Material	No. of Carbon Atoms	Boiling Point at 1 bar	HC by HFID	HC on Filter by TGA +	HC by HFID by Cold Filter =	Combined HC
Dioctyl-phthalate (DOP)	24	230	-	98	0.5	98.5
Propane	3	-42	100	0	100	100
Decane	10	174	100	0	100	100
Tetradecane	14	254	46	3	42	45
Hexadecane	16	287	-	98	2	100
Eicosane	20	343	75	99	1	100
Docosane	22	369	54	100	2.5	102.5
Octacosane	28	432	70	100	1	101
Petroleum Fraction	-	200-260	90	1	87	88
Diesel Fuel (DERV)	-	200-350	100	27	74	101
Lubricating Oil (Talpa 20)	-	> 350	53	90	9	99

that the papers showed no further particulate weight loss.

Two complete FTP runs with a two litre diesel car were carried out using this tunnel, after which the test area was recleaned carefully using one 9 cm filter paper. Thermogravimetric analysis showed that the particulate on the whole tunnel (making an assumption that the deposition was even) was 0.5% of the total particulates observed and hence negligible compared to other potential errors.

The ratio of HC to carbon on the material cleaned from the walls was 37:73 which was identical to the ratio for the main particulate filter.

This series of tests relates only to the behaviour of certain particulates in the Ricardo tunnels and gives no information on overall test repeatability taking engine behaviour and drive cycle variability into account. Researches in this area at Ricardo by W.F. Ball (11) have shown the importance of prior operational parameters on the subsequent collection of particulates. Thus a limited steady state emission can have different values depending on just how this "steady state" condition is reached, e.g. whether from a high or from a low load. These factors make it very difficult to obtain good test repeatability for particulate emissions unless complete vehicle/CVS tunnel system preconditioning is carefully controlled.

VOLATILE HYDROCARBONS PASSING THROUGH THE FILTERS - The use of TGA/HFID offers a sensitive analysis for the HC trapped on a filter but further researches were required to decide how best this information could be matched to the amount of gaseous HC as measured by continuous integration from an HFID or by collection in a tedlar bag. To this end two HFIDs, manufactured by Analysis Automation of Oxford, England, were set up (with integration and recording facilities) in order to measure the gaseous HCs. The burner fuel used was 40/60 hydrogen/helium and these HFIDs together with their heated sample lines could be heated over a range of temperatures from 100°C to 190°C.

This equipment allowed for the connection of two HFIDs to any of the four probes and/or filters.

Integrated HFID versus Bag Collection - Initially a simple low cost total HC analysis system was investigated in which the gaseous HC passing through a cold (< 125°F) GF/A filter was analysed by both integrated HFID at 100°C and by standard CVS tedlar bag collection techniques.

The exhaust was produced by a two litre diesel car with an HC emission of 0.6 g/mi It was driven over two hot start FTPs.

As expected the bag showed HC values approximately 15% lower than the levels measured by the HFID system proving that further HC condensation is occurring downstream of a cold filter making the bag collection inadmissable. The HFID temperature of 100°C was chosen for operational convenience but the result suggested that a temperature of 130°C would give a better compromise between HC loss by condensation and by oxidation. Time did not allow a temperature scan of volatile HC downstream of the cold filter at this stage; instead a range of pure hydrocarbons and some oil products were evaporated into the tunnel to examine the tendency of GF/A filters to trap them or to pass them through.

TRAPPING TENDENCY OF GF/A FILTERS - It is almost certain that the carbon structure of the soot is responsible for the retention of the most important HC pollutants on the filter (7). However difficulties arise over preparing simulated HC free diesel soot filters although close approximations to bulk diesel carbon can be obtained from other sources (8) which are essentially HC free.

Desorption of HC from a diesel sample filter has to be effected at up to 550°C at which temperature the carbon may be altered in adsorptivity. Nevertheless a few tests were undertaken using pre-evaporated used diesel soot filters which had been heated at 550°C to constant weight.

Approximately 20 mg of hexadecane, a diesel fuel (DERV), and a straight mineral lubricating oil (Talpa 20), were in turn placed on three similar used (and desorbed) filters. These were taken up to 550°C at 50°/min in nitrogen gas in the TGA and held at 550°C until a constant weight was indicated. Only in the case of Talpa 20 did the filter show a weight gain amounting to 2.5% of the added oil. A check test in which the oil was placed in a platinum cup in the TGA gave 100% evaporation suggesting that the 2.5% gain might be due to either adsorption of heavy ends or their pyrolysis to carbon being catalysed by the glass surface or by the diesel carbon.

In passing it might be noted that TGA has been used in the past for oil evaporation studies (12).

At this stage it was decided to abandon work on carbon adsorption as being outside our terms of reference, and requiring mass-spectrometry, but clearly a subject open to further developments. Instead, work on the trapping efficiency of a range of hydrocarbon materials on GF/A filters was extended to the list shown in Table 3 despite the disadvantage of having no carbon adsorption effects.

TESTING WELL-DEFINED HYDROCARBONS IN THE CVS TUNNEL - The objective of this work was to provide some information on the behaviour of the tunnel/filter/HFID configurations using well-defined materials before moving on to an assessment of the various HC splits possible with actual diesel exhaust.

Particular problem areas were in obtaining a total HC value by HFID (190°C) which would give a fair comparison with the combined value of HC obtained from the cold filter by TGA/HFID (130°C) downstream of the cold filter. In these tests the heated sample line (190°C) passed only 2 litre/min of CVS tunnel gas and was a standard piece of equipment as used for obtaining total diesel HC by the CVS mixing box technique. Some doubts had already been expressed about the design problems of heated lines which have low flow and which project into large cool CVS gas flows (13).

These doubts were confirmed by the results shown in Table 3 when the total HC by HFID (190°C) were compared to the final combined column (cold filter HC by TGA + HC by HFID after cold filter). It may be seen that the comparison was good with the exception of the paraffins (C_{20}, C_{22} and C_{28}) and Talpa 20 all of which might be expected to suffer "hang-up" in the standard HFID sampling system.

The combined value for tetradecane and the petroleum fraction did not reach 100% recovery because of losses on the hot plate injection system; however, in both cases the HFID agreed with the combined value.

These results demonstrate clearly that the total volatile HC was best estimated for hydrocarbons boiling above ∿ 330°C by the combined procedure. Improvements in the standard HFID were needed particularly in the area of the probe inlet and the filter temperature where losses occur. This conclusion is in line with that reached at Ricardo in connection with the total HC analysis in two-stroke cars fuelled by petrol/oil mixtures. Hydrocarbon by CVS bag was identical to that measured by the light duty diesel vehicle CVS/mixing box/HFID method, since the latter failed to record lubricating oil.

The exhaust gas in this case has a "hole" in its hydrocarbon boiling range between ∿ 180°C and ∿ 330°C (13).

Improved HFID Sampling System - The main problem is to ensure that the filter near the probe inlet has an even temperature over its face of 190°C. If the filter has a thick bed it is difficult to ensure an even temperature by conduction from the external heaters. A thin filter paper can be heated evenly if the sample flow is adequate and at 190°C which requires a long (165 cm) section of heated probe before the filter.

The latter choice was made and a heated filter holder was constructed for a 9 cm GF/A filter. It was connected to a stainless steel probe (165 cm x 2 cm diameter) which was heated electrically by a wire coil. The gas flow was 56 litre/min through the filter only 2 litre/min of which was taken through the stainless steel heated sample line to the HFID. This latter line was thus maintained clean and soot free.

Preliminary tests with this system showed that the hot filter paper retained only ∿ 10% as much HC by TGA/HFID as did the parallel cold filter. Some difficulty was experienced in keeping the filter temperature up to and not exceeding 190°C due to the variable tunnel gas temperature.

THE TEMPERATURE SENSITIVITY OF THE SAMPLING SYSTEM - A series of car runs was driven with the filter temperature varying from 25°C to 180°C with the results shown in Fig. 13. The car had a high hydrocarbon emission and was run over an FTP to each (hot start) test but even so some test variability occurred necessitating slight smoothing of the curves.

The most interesting point was the demonstration that 50°C was the temperature above which the particulates begin to fall in agreement with the EPA figure of 125°F (52°C). Since the carbon content by TGA was constant the HC content of the particulates also fell above 50°C. The results obtained at a filter temperature below 50°C showed considerable scatter compared with those at 50°C and above. This seemed to be due to the capture of lighter volatiles on the filter, these can be lost or retained depending on the pre-weighing procedure used. There might be some practical advantage in maintaining the tunnel temperature at 52°C throughout the test by a feedback controlled intake air heater. At the high temperature end the curves demonstrated the advisability of keeping the filter and sampling system temperature above 170°C confirming the choice of 190°C for the HFID.

These tests were sufficiently encouraging for us to proceed to a series of car tests to compare the hot and cold filter procedures for HC and particulates.

Time did not allow the repetition of the tests on specific hydrocarbon materials using the new hot sample system.

FEDERAL TEST PROCEDURE RUNS WITH DIESEL CARS

Four cars were available from normal development schedules as and when their programmes allowed.

They were used in the "as received" condition and run over the FTP from a cold start. Therefore each test should be regarded as quite separate as far as the cars are concerned but the analytical and test equipment were kept standard.

The complete results as measured are shown in Fig. 14 in bar form. The filter particulates are shown as the sum of the lower two blocks in each bar, with that for the second (cold filter) bar being the accepted "total particulates". Each bar segment is additive as representative fractions of additives by different methods. The carbon fraction should of course have been equal for both filters on each test the slight discrepancies were due to a combination of flow differences and possibly filter inefficiencies.

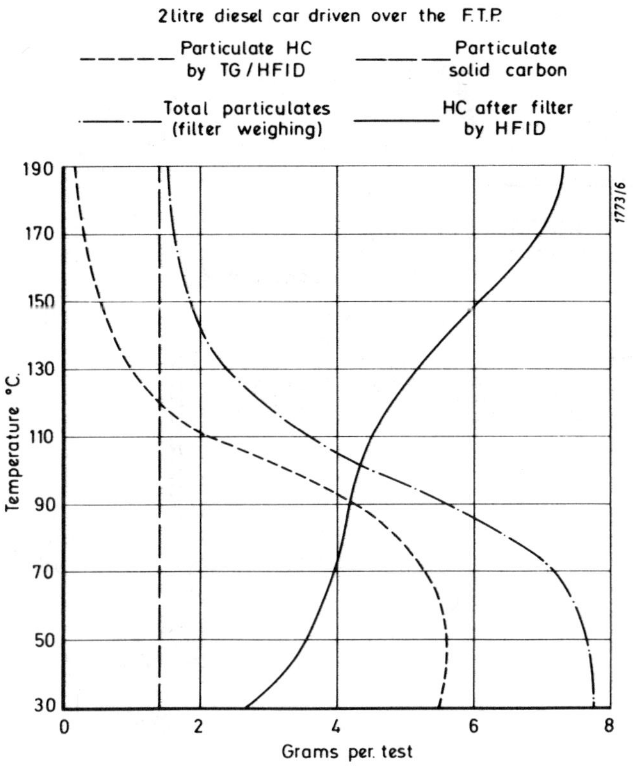

Fig. 13 - Variation of particulates and HC with filter temperature

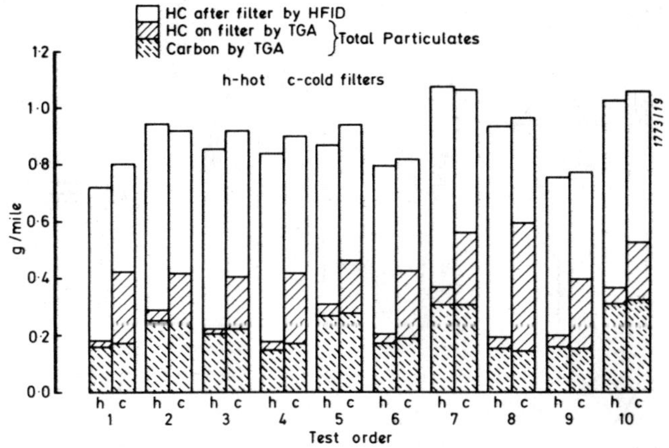

Fig. 14 - Ten FTP tests using four diesel cars for comparing hot filter with cold filter particulates and gaseous HC

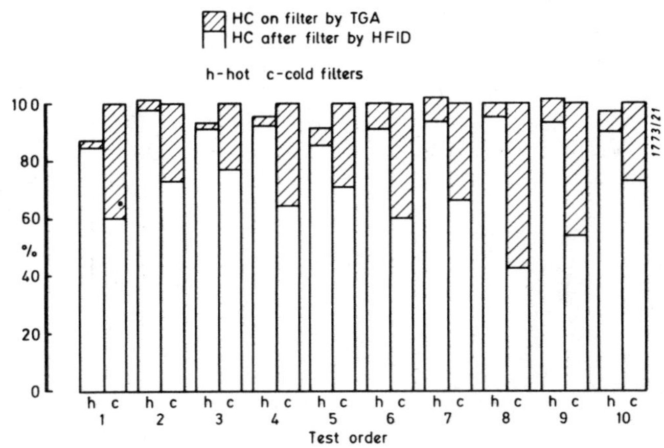

Fig. 15 - Ten FTP runs comparing total HC by hot filter HFID (190°C) with total HC by cold filter HFID (130°C)

In order to show up the differences between the hot and the cold filter HC more clearly, the bar charts were redrawn with the cold filter HC taken as 100%. The results given in Fig. 15 show how little HC remained on the hot filters and how close the hot system HC came to the cold. When averaged, the HC measured by the hot system was shown to be approximately 97% of the value given by the cold system.

SPECIAL PROBLEMS OF THERMOGRAVIMETRY

Whilst these researches were in progress three questions were asked regarding a) the need for a new and expensive TGA, b) the use of an integral HFID, and c) the validity of taking the weight loss in nitrogen, up to a temperature of 550°C, as filter bound HC mass.

A SUBSTITUTE FOR TGA - In fact it is possible to dispense with the TGA in favour of the system shown in Fig. 16 in which only the temperature controlled furnace is used. At the end of the particulate test (in which HC is also integrated through the cold filter by HFID) the filter is transferred to the furnace and heated in nitrogen with the desorbed HC being integrated by the same HFID. The carbon remaining can be estimated by the introduction of oxygen, and the evolved gases passed through a CO oxidation tube to an integrating CO_2 analyser.

This system has been used at Ricardo and gave essentially identical final results to the TGA/HFID. It was only slightly less costly and was less convenient in operation.

SOLVENT EXTRACTION VERSUS TGA - Solvent extraction of the filters has been practised for the analysis of PNAH material but requires a long (∿ 12 hours) Soxhlet extraction and some doubts exist as to the total degree of hydrocarbon desorption effected.

To compare the values of HC by solvent extraction and by TGA four hot start FTP tests were completed using the same car. Each test produced two identical filters one of which was solvent extracted the other was placed in the TGA.

The HC in the Soxhlet extract was also measured by TGA. The 100 ml bulk extract in either dichloromethane (DCM) or 50/50 DCM/acetone was evaporated in a nitrogen stream to

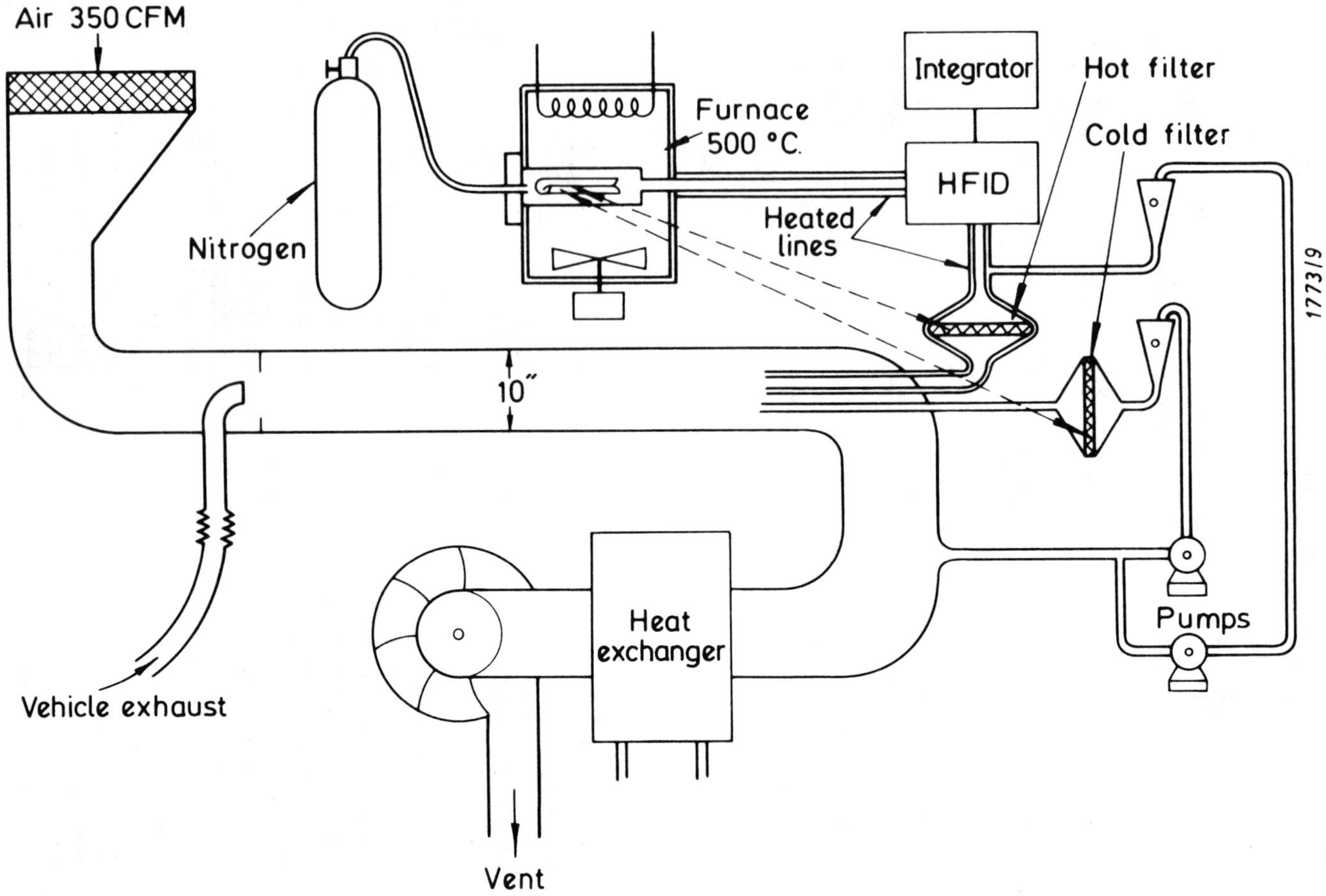

Fig. 16 - Analysis system for gaseous and particulate bound hydrocarbons

a low volume which was transferred to a folded TGA filter paper package (GF/A) and held in the cold TGA in a nitrogen stream to constant weight. The normal TGA heating procedure was then applied.

The extracted filter paper was also submitted to TGA when it was shown that a small evolution of non-extractable HC occurred which could be up to 20% of the extracted HC. Dichloromethane was less efficient than DCM/acetone as a solvent.

The results averaged over the four filters are shown in Fig. 17. They show that TGA is slightly better than solvent extraction if only a mass emission is required. The last bar shows how a very small, almost negligible, amount of the extract is pyrolysed to carbon in the TGA. However it is possible that this derives from the pyrolysis of PNAH and could represent a serious loss of toxic agents.

HOW MUCH OF THE WEIGHT LOSS IS HC? - The second and third questions are related since the HFID equipment was added to answer the last question - assuming the problem of carbon counting could be solved.

A better system for obtaining reliable carbon counting efficiencies is needed. Our figure of 78% is based on too few tests but is believed to be adequate for this stage in the work which has involved only normal diesel particulates from non-additive fuels and with no catalyst in the exhaust. When a platinum oxidation catalyst is fitted the TGA volatiles can be almost completely composed of H_2SO_4 and in this case the TGA/HFID link-up is the only source of HC values. This effect is shown in Fig. 18 and Fig. 10 on particulates from a four litre DI diesel engine running at rated speed and load both with and without a catalyst fitted. The high sulphur (0.7%) fuel and the catalyst gave a sulphate emission more than double the carbonaceous particulate emission. Only approximately 50% of the sulphate appeared as volatile weight loss in the TGA the residue appeared to react with the glass fibre and became fixed as so-called "inert material" in Fig. 18.

Inert material is defined as the difference between the total particulate weight on the filter, and the sum of the volatiles and carbon

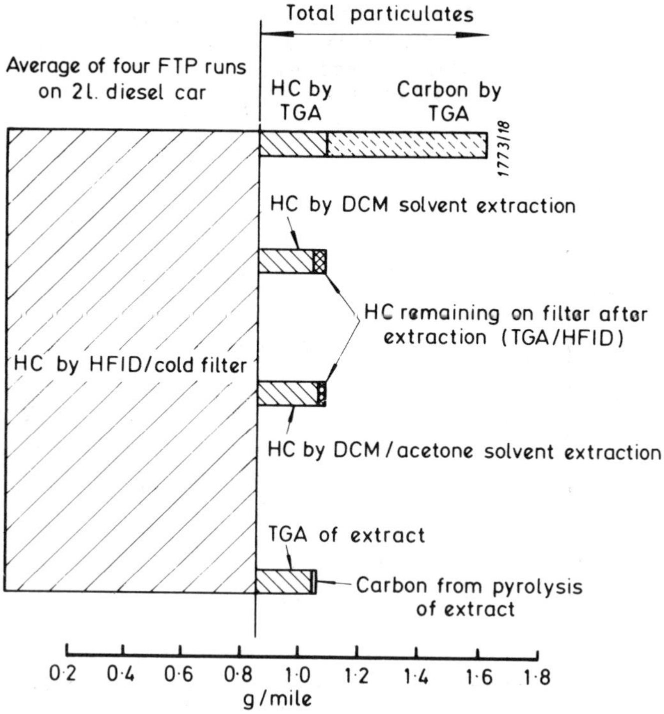

Fig. 17 - Comparison of TGA/HFID and solvent extraction for estimating HC on particulate filters

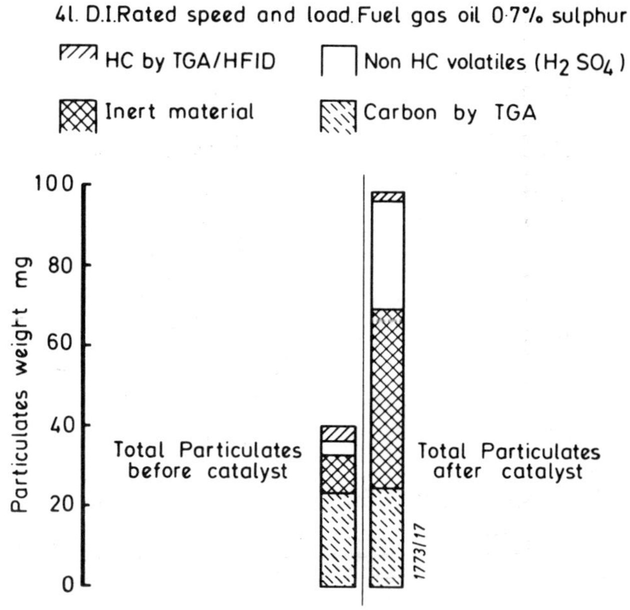

Fig. 18 - Comparison of particulates before and after a platinum catalyst fitted to a DI diesel engine

in the TGA. Usually it is a very small quantity consisting of extraneous "dust", iron oxide, and oil additive ash, but in this case it was very large. The effect could be simulated easily be adding a few milligrams of H_2SO_4 to a fresh filter package in the TGA and heating to 550°C when the weight remained well above that of the original filter when all volatile H_2SO_4 was driven off.

No attempt was made to eliminate this problem by the use of non-reactive fibre filters, or to quantify the effect using total sulphate analysis. Clearly this needs investigating, as does the effect of H_2SO_4 on carbon counting efficiencies.

HYDROCARBON SEPARATIONS IN THE TGA/HFID - Only the crudest separations of hydrocarbons by boiling point could be expected from this apparatus, an expectation borne out by tests on mixtures of pure paraffins. A 50/50 mixture of dodecane (C_{12}) and eicosane (C_{20}) could be partially separated (Fig. 19) but the addition of cetane (hexadecane C_{16}) almost completely removed the two peaks.

The presence of adsorptive carbon however can alter the situation with some strongly sorbed hydrocarbons and as shown already in Fig 8 and Fig. 9 there was a second "hump" in HC desorption starting at ∿ 300°C and ending at ∿ 500°C. It is tempting to ascribe this to strongly adsorbed higher aromatics and certainly this is the region where PNAHs desorb (7)(8).

Lubricating oil may also contribute to this "hump" as is shown in Fig. 20. A typical filter obtained under full load conditions on a DI diesel engine was halved and to one half was added an ether solution of Aeroshell 100 lubricating oil so that the volatile HC was approximately doubled. The superimposed TGs demonstrate how the heavy lub oil restrained the evolution of the lighter diesel hydrocarbons and how the second "hump" was almost quadrupled but remained clearly defined.

DISCUSSION

UTILITY AND ACCURACY OF THE SYSTEM - As stated in the introduction this work was aimed at the development of a simple procedure for the analysis of hydrocarbons and particulates which could help in diesel engine depollution. When the HFID accessory was added this ideal was somewhat departed from, but it must be said that in its operation the whole apparatus has been extraordinarily reliable, so much so, that in the later stages of this work it had also serviced four engine and car development programmes on a routine basis.

It has been shown that there were some losses of hydrocarbons in the TGA/HFID which occurred by condensation, pyrolysis and also possibly in the HFID by inefficient carbon counting. The HC measured by weight loss in the TGA was probably a slight over estimate when a low sulphur fuel and no catalyst was used. The weight losses were slightly higher

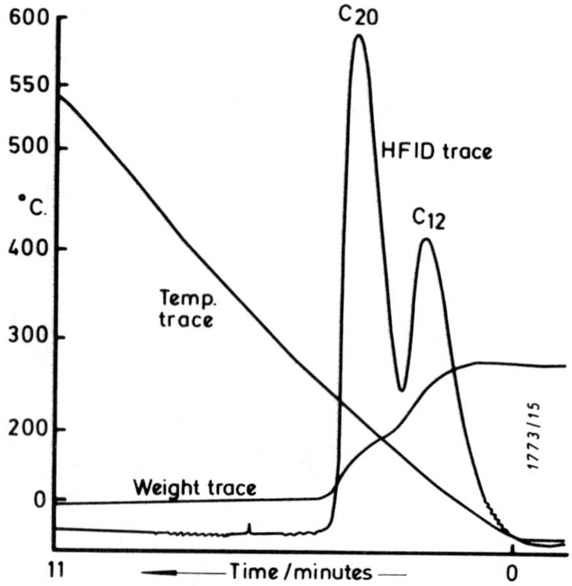

Fig. 19 - TG showing a separation between dodecane and eicosane

by solvent extraction which was seen to be an incomplete process.

A further stage of TGA development as suggested is unlikely to give much more accurate weight information but would be more convenient in use and have a much better interface with the HFID and have better carbon counting ability. It could be a completely integrated instrument.

The Ratio of Carbon to Hydrocarbon in Particulates - The ratio C/HC coupled with their absolute values has been found to be of great use in defining the engine parameters and operational modes which cause particulate emissions. For example, a ratio of 4:1 from a diesel car particulate taken over the FTP at a total level of 0.8 g/ml indicates that the engine is rated too highly. On the other hand a ratio of 1:4 at the same total level of emission indicates poor injection (secondaries, nozzle dribble, etc.). When catalysts are used the ratio can be related to the catalytic activity and the temperature of the matrix.

THE OPTIMUM PARTICULATE AND HYDROCARBON SAMPLING TEMPERATURES - The CVS tunnels which were constructed very much along the lines proposed by the EPA proved to be more efficient than expected. There was no problem in obtaining duplicate filter papers which held convenient weights of particulates (∿ 20 mg) from 0.5-1% of the total CVS flow.

The choice of filter material cannot be said to have been settled in these tests. A ceramic fibre which is more weight stable particularly in the presence of sulphuric acid is badly needed.

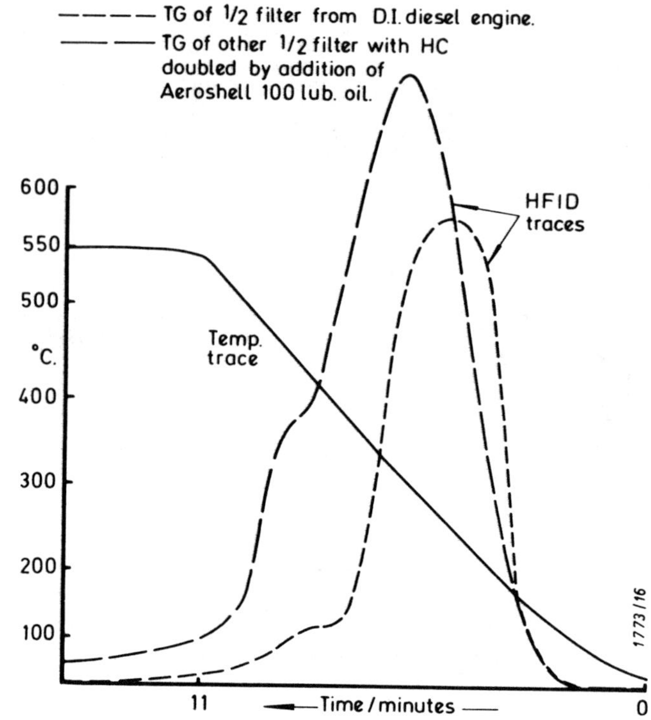

Fig. 20 - Superimposed TGs showing the effect of the addition of lubricating oil to diesel particulates

The tests to determine the optimum filter temperature gave some support to the use of 125°F as the maximum allowable. However, the variable results obtained when the filter was maintained at a lower temperature than this gave cause for disquiet leading to the suggestion that a constant tunnel temperature might be advantageous for good repeatability.

The most difficult problem encountered was to build a heated probe and filter holder that would give a filter with an even temperature of 190°C and holding ∿ 20 mg of particulate. Eventually when such a filter was produced it was found that this total HC by HFID (190°C) system gave considerably higher figures than were given by a good standard system.

Although no statistical testing was employed when using the complete system with hot and cold filters the strong impression was gained that the total HC, measured by TGA on the cold filter particulates added to the HC measured downstream of the filter by HFID (130°C), was much more repeatable than the total HC by HFID (190°C).

The 3% HC bonus for the cold system was expected since at 190°C undoubtedly some HC must oxidize and be lost in the hot system. However, 3% was a rather smaller difference than had been expected perhaps because there is also a loss of HC on the cold filter during the pre-weighing and stabilization period.

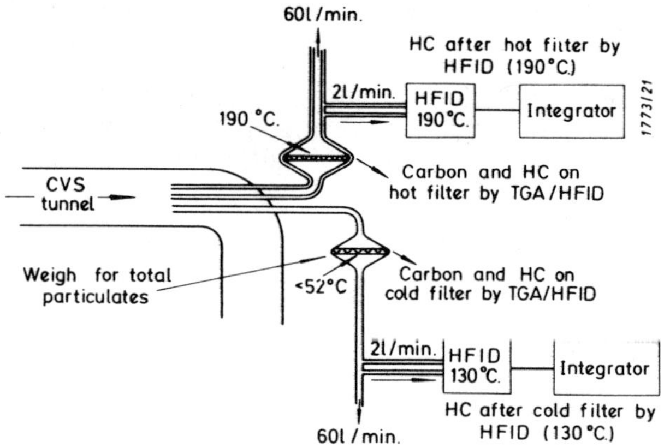

Fig. 21 - Final system as used on ten FTP car tests

Until much more work is done using this system in many more laboratories, its benefits, if any, cannot be assessed. However, it would appear that it will not overturn traditional procedures only provide extra information.

An interesting possibility for further research is that the mutagens and carcinogens may be concentrated in the hot filter HC and that this may relate to the second "hump" in the TGA/HFID traces in some way. The application of PNAH analysis and bio-assay by Ames test to the hot filter material, and also to the residual HC on filters heated to progressively higher temperatures in the TGA/HFID, might be a useful approach.

CONCLUSIONS

Thermogravimetry has been shown to be a useful tool for diesel engine development particularly if it is interfaced with a heated flame ionization analyser. It offers a new way of separating and describing the various groups of hydrocarbons in diesel exhaust in a reliable fashion using emissions laboratory staff.

The CVS tunnel was shown to have good mixing and very low losses. The methods of sampling and analysing for HC and particulates are best summarized diagrammatically in Fig. 21.

The results confirm the validity of the diesel hydrocarbon analysis procedure by heated flame ionization detector (decreed by EPA) as long as the whole system is maintained at the correct temperature (190°C).

It was never the concern of this research to work towards a new regulatory procedure and despite its advantages in diesel particulate research and development any use this work might have in the regulatory field is purely the concern of ambient air quality specialists.

ACKNOWLEDGEMENT

This work was completed under EPA Contract No. 68-02-2751. The authors wish to express their indebtedness to Mrs Carmen O'Leary and Mr Philip Shore for running the TGA/HFID and solvent extractions. The Manager of the Ricardo Pollution Laboratory, Mr R.W. Gawen, provided facilities and much helpful advice.

REFERENCES

1. H.S. Ford, D.F. Merrion and R.J. Hames, "Reducing Hydrocarbons and Odour in Diesel Exhaust by Fuel Injector Design." Society of Automotive Engineers Paper 700734.
2. D.A. Kendall, P.L. Levins and G. Leonardos, "Diesel Exhaust Odour Analysis by Sensory Techniques." Society of Automotive Engineers Paper 740215.
3. P.L. Levins, D.A. Kendall, A.B. Caragay, G. Leonardos and J.E. Oberholtzer, "Chemical Analysis of Diesel Exhaust Odour Species." Society of Automotive Engineers Paper 740216.
4. R.C. Bascom and G.C. Hass, "Status Report on the Development of the 1973 Californian Diesel Emission Standards." Society of Automotive Engineers Paper 700671.
5. H.W. Pearsall, "Measuring the Total Hydrocarbons in Diesel Exhaust." Society of Automotive Engineers Paper 670089.
6. F. Black and L. High, "Methodology for Determining Particulate and Gaseous Diesel Hydrocarbon Emissions." Society of Automotive Engineers Paper 790422.
7. G.M. Cornetti, "Lightweight Automotive Diesel Data Base." DOT, DOE, EPA, NHTSA, "Workshop on Unregulated Diesel Emissions and their Potential Health Effects." Washington, April 1978, Transcript of Proceedings, p. 250.

8. N. Pelz, (Daimler-Benz A.G. Stuttgart Unterturkheim, West Germany), Private Communication.

9. J.P. Redfern (Stanton Redcroft), "25 Years of Laboratory Practice." Vol. 26, No. 5, pp. 395-397, 1977.

10. A.D. Little Inc., Cambridge, Massachusetts, "Evaluation of Filter Media for Quantitative Collection of Particulate Matter from Engine Exhaust." EPA Contract No. 68-02-1211.

11. W.F. Ball, Monthly Technical Progress Report No. 7, January 1979. DOT-TSC Contract 1545.

12. W.C. Gergel and J.E. Riester, "Diesel Engine Oil Consumption Studies." Society of Automotive Engineers Paper 740525.

13. D. Collins, R.D. Cuthbertson, R.D. Gawen and R.W. Wheeler, "The Use of Constant Volume Sampler and Dilution Tunnel to Compare the Total Particulates from a Range of Automotive Engines." Society of Automotive Engineers Paper 750904.

Unregulated Emissions Measurement Methodology*

Harry E. Dietzmann
Southwest Research Inst.

Frank M. Black
U.S. Environmental Protection Agency

ABSTRACT

Analytical procedures for the measurement of unregulated emissions in automotive exhaust have been developed. Compounds in the emissions characterization inventory include organic sulfides, nickel carbonyl, aldehydes, ammonia, sulfur dioxide, hydrogen sulfide, organic amines, nitrous oxide, individual hydrocarbons, total cyanide, sulfate, phenol, particulate and others.

Validation experiments were conducted with selected methods to develop a well-defined analytical procedure. Qualification of the procedure involved injection of known amounts of the individual species into a dilution tunnel - CVS system. These analytical procedures are currently being employed to characterize unregulated emissions from a variety of low NO_x vehicles, advanced powerplants and vehicles operating under malfunction conditions.

THE PURPOSE of this paper is to describe the methodology employed in the development and implementation of analytical procedures used in the chemical characterization of automotive exhaust. Compounds included in the chemical characterization inventory are either irritants, toxic, suspected carcinogens or adversely affect the quality of the atmosphere.

During the early years of chemical characterization of exhaust, very little work was conducted to qualify analytical procedures with real exhaust. Most procedures were ambient air procedures modified to eliminate known interferences in exhaust. This paper describes the selection criteria, validation and qualification experiments, and the procedures currently used in the chemical characterization of unregulated emissions.

LITERATURE SEARCH

The first step in the unregulated emissions measurement methodology was to conduct a thorough literature search for candidate analytical methods. Although typically called a literature search, efforts were extended to contacting key experienced individuals in government and industry in a quest to make the appropriate procedure selection. Evaluation criteria for the selection of the candidate analytical methods was based on the following items:

- sensitivity (MDL)
- free of interferences
- repeatability and reproducibility
- continuous or grab sample (i.e. collection technique)
- possibility of success
- previous application to exhaust
- cost effectiveness of procedure
- general analytical considerations

In several instances, more than one candidate method was selected to undergo a variety of validation experiments. The term "validation" experiments relative to unregulated emissions refers to those experiments relating to the development of a well-defined analytical procedure.

In some cases, procedures were available that had been successfully applied to Diesel and gasoline exhaust. An example of this was aldehyde measured with the 2,4-dinitrophenylhydrazine (DNPH) procedure.

Each procedure was reviewed and evaluated for its potential success in being applied to automotive exhaust. A summary of the unregulated emissions included in the chemical char-

*Paper 790816 presented at the Off-Highway Vehicle Meeting and Exposition, Milwaukee, Wisconsin, September, 1979.

acterization inventory is presented in Table 1, along with the chemical formula, molecular weight, and reasons for interest.

VALIDATION EXPERIMENTS

After selection of the candidate analytical methods, a variety of experiments were conducted with each procedure to define the method. Each type of collection and detection system required specific experiments to validate each step from the collection to the finish of the procedure. There are four basic types of sample collections used in the chemical characterization of unregulated emissions. These are direct bag analysis, wet chemical collection, adsorbent traps and filters.

Minimum considerations during the validation experiments were:

- establishing the minimum detection limits (MDL)
- generating calibration curves
- defining interferences
- bag storage stability
- sample storage stability
- reagent stability
- establishing flow rates, collection media and temperature for absorbing reagent
- establishing collection efficiency
- establishing the repeatability and reproducibility
- calibration techniques required
- applicability to FTP testing
- applicability to gasoline and Diesel exhaust
- establishing quality control techniques for procedures

The aforementioned considerations may apply to specific collection techniques and not all will apply to each collection or measurement. An example of bag storage stability is shown for N_2O in Figure 1. In those cases where the samples were collected directly from the CVS, this was an important criteria in the acceptability of this collection technique.

Wet collection techniques required evaluation of sampling parameters, such as sample flow rate, temperature of absorbing reagent, strength of absorbing reagent, collection efficiency and number of impingers required. In addition, reagent stability, preparation and stability of standards, and interferences were evaluated. Similar experiments were conducted using adsorbent traps, however, only limited work was performed with filter collection due to the available existing data. Table 2 summarizes the unregulated emission measurement procedures, including collection techniques, detection methods, and minimum detection values. These procedures were then qualified as described in the next section.

Table 1 - Chemical Characterization Inventory

Unregulated Emission	Chemical Formula	Molecular Weight	General Concern
Hydrogen sulfide	H_2S	34.08	toxic gas
Organic sulfides			
carbonyl sulfide	COS	60.08	irritants
methyl sulfide	CH_3SCH_3	62.13	
dimethyl disulfide	CH_3SSCH_3	94.20	
ethyl sulfide	$C_2H_5SC_2H_5$	90.19	
Sulfur dioxide	SO_2	64.06	irritant
Sulfate	$SO_4^=$	96	irritant
Hydrogen cyanide	HCN	27.03	toxic gas
Ammonia	NH_3	17.03	irritant
Nitrous oxide	N_2O	44.01	reacts with ozone
N-nitrosodimethylamine	$(CH_3)_2NNO$	74.08	chemical carcinogen
Organic amines			
methyl amine	CH_3NH_2	31.06	amines are possible carcinogen precursors
dimethyl amine	$(CH_3)_2NH$	45.06	
trimethyl amine	$(CH_3)_3N$	59.11	
ethyl amine	$C_2H_5NH_2$	45.09	
diethyl amine	$(C_2H_5)_2NH$	73.14	
triethyl amine	$(C_2H_5)_3N$	101.19	
Aldehydes (& Ketones)			
formaldehyde	CH_2O	30.03	irritants
acetaldehyde	CH_3CHO	44.05	
acrolein	CH_2CHCHO	56.07	
propanal	CH_3CH_2CHO	58.08	
acetone	CH_3COCH_3	58.08	
i-butanal	$CH_3CH(CH_3)CHO$	72.11	
methylethyl ketone	$CH_3COCH_2CH_3$	72.11	
crotonal	$CH_3CHCHCHO$	70.09	
benzaldehyde	C_6H_5CHO	106.13	
hexanal	$CH_3(CH_2)_4CHO$	100.13	
Individual hydrocarbons			
methane	CH_4	16.04	some are photochemically reactive
ethylene	C_2H_4	28.05	
ethane	C_2H_6	30.07	
acetylene	C_2H_2	26.04	
propane	C_3H_8	44.10	
propylene	C_3H_6	42.08	
benzene	C_6H_6	78.12	
toluene	C_7H_8	92.15	
Nickel carbonyl	$Ni(CO)_4$	170.75	highly toxic

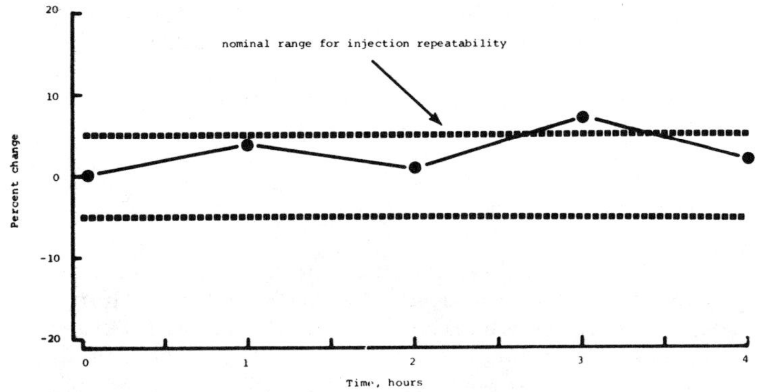

Fig. 1 - N_2O bag sample storage stability (short term)

QUALIFICATION EXPERIMENTS

Upon completion of the validation experiments, these procedures were then qualified with gasoline and/or Diesel exhaust. The qualification experiments have been completed for gasoline exhaust and are nearing completion for Diesel. The term "qualification" experiments refers to a sequence of tests using the dilution tunnel, test vehicle and the specific analytical procedure being qualified. The di-

Table 2 - Summary of Unregulated Emission Measurement Procedures

Unregulated Emission	Collection Technique	Detection Method	Proecdural Minimum Detection Values, ppm Instrument	Overall[a]
Hydrogen sulfide	impinger	colorimetric	0.01	0.01
Organic sulfides	traps	GC-FPD	0.05	0.0002
Sulfur dioxide	impingers	IC	0.01	0.005
Sulfate	filters	LC-BCA		
Individual HC	bag	GC-FID	0.05	0.05
Aldehydes	impingers	GC-FID	2.0	0.005
Nitrous oxide	bag	GC-ECD	0.01	0.01
Hydrogen cyanide	impingers	GC-ECD	0.01	0.005
Ammonia	impingers	IC	0.01	0.005
Organic amines	impingers	GC-NPD	0.01	0.002
DMNA	traps	GC-MS		
Nickel carbonyl	bag	CL	<0.001	<0.001
Metals	filter	X-ray	-	-
Phenols	impingers	GC-FID	-	-
DOAS	traps	LC-UV	-	-
Organic solubles	filters	HPLC	-	-

[a] 23 minute sampling period (i.e. cold/hot start, 505 sec + stabilized, 867 sec)

Table 3 - Summary of Injection Recovery Experiments

Unregulated Emissions	% Recovery Tunnel Only	% Recovery with Tunnel and Exhaust
Organic sulfides		
COS	82	114
CH_3SCH_3	74	12
$C_2H_5SC_2H_5$	68	21
CH_3SSCH_3	48	4
Organic Amines		
CH_3NH_2	6.2, 10.1, 13.1	-
CH_3NHCH_3	18, 29, 36	-
$(CH_3)_3N$	76, 89, 93	-
Sulfur dioxide	101	98
Nitrous oxide	102	98, 107
Hydrogen cyanide	103	93
Individual HC	100	100
Aldehydes	103	100
Hydrogen sulfide	108	90
N-nitrosodimethylamine	not required	not required
$Ni(CO)_4$	not required	not required
Sulfate	not required	not required

lution tunnel used in the qualification experiments is shown schematically in Figure 2. The first step was to establish any tunnel losses without any vehicle exhaust. This was performed much in the same manner as a routine propane check. A known amount of a specific compound was injected as a gas or aerosol into the dilution tunnel at the point where the exhaust would normally enter the tunnel. Recoveries for these tests are summarized in Table 3. In general, recoveries without exhaust were essentially 90-100 percent with two exceptions. The recoveries of organic sulfides ranged from 48-82 percent depending on the sulfide whereas organic amines exhibited recoveries from 10-93 percent. The organic amines have a great affinity for the stainless steel tunnel and the recoveries were dependent to a great extent on tunnel history (i.e. previous amine injections) and the specific amine.

Three unregulated emissions were not required to undergo qualification testing. These include nickel carbonyl, N-nitrosodimethylamine, and sulfate. Due to the extreme health hazards in working with nickel carbonyl, it was decided not to perfrom any tunnel injections with nickel carbonyl but rather limit studies to validation experiments. The N-nitrosodimethylamine secondary dilution system and analytical method had previously been qualified by EPA and requalification was not required.

ANALYTICAL PROCEDURES

This section briefly describes the analytical procedures developed as a result of the literature search, validation and qualification experiments. Each procedure is described from

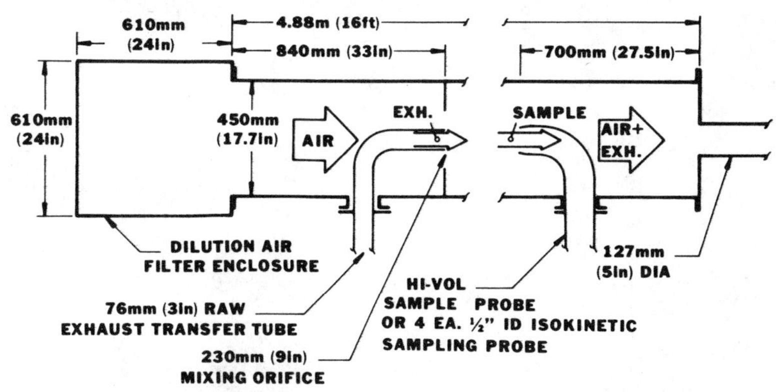

Fig. 2 - Schematic section of dilution tunnel for Diesel particulate sampling

the collection to the analysis. Additional details for these methods are available(1)*.

DIRECT BAG ANALYSIS - Three unregulated emissions are measured directly from Tedlar bags collected during the FTP, HFET, SET-7 and NYCC driving cycles. Bag samples and backgrounds are collected using the multi-bag sampling system illustrated in Figure 3. By collecting separate bags for the measurement of N_2O, $Ni(CO)_4$ and individual HC, simultaneous analysis can be effectively accomplished upon completion of the emissions test. The sensitivity of these three analytical procedures is sufficiently low to allow direct bag analysis.

Nitrous Oxide - The measurement of N_2O is accomplished using a Perkin-Elmer 3920B gas chromatograph equipped with an electron capture detector(2). The system employs two pneumatically

* Numbers in parentheses designate References at end of paper.

operated electrically controlled Seiscor valves, an analytical and a stripper column. The gas chromatograph separation is accomplished at room temperature. A special control console was developed to house the entire system except for the electron capture detector. The nitrous oxide analytical system is shown in Figure 4.

The gas chromatograph flow schematic is illustrated in Figure 5 and a typical gas chromatograph calibration trace is shown in Figure 6. A gasoline-fueled engine exhaust trace is presented in Figure 7. This analysis requires 4-5 minutes and the N_2O response with an electron capture detector is linear in the typical operating range for exhaust (0-1 ppm).

Individual Hydrocarbons - A multi-column gas chromatograph analytical technique was selected for the measurement of selected individual hydrocarbons in exhaust(3). Those specific hydrocarbons included in this analysis are methane (CH_4), ethylene (C_2H_4), ethane (C_2H_6), acetylene (C_2H_2), propane (C_3H_8), propylene (C_3H_6), benzene (C_6H_6) and toluene (C_7H_8). Bag samples are collected during the CVS tests and analyzed in the gas chromatograph system containing four separate columns and a flame ionization detector (FID). The peak areas are compared to an external calibration blend and individual hydrocarbon concentrations are analyzed with a Hewlett-Packard 3354 computer system.

The analysis for individual hydrocarbons is conducted with a Varian 1440 gas chromatograph using a flame ionization detector (FID). Four separate packed columns are used to resolve these individual hydrocarbons. An elaborate system of timers, solenoid valves, and gas sampling valves are used to direct the flow of

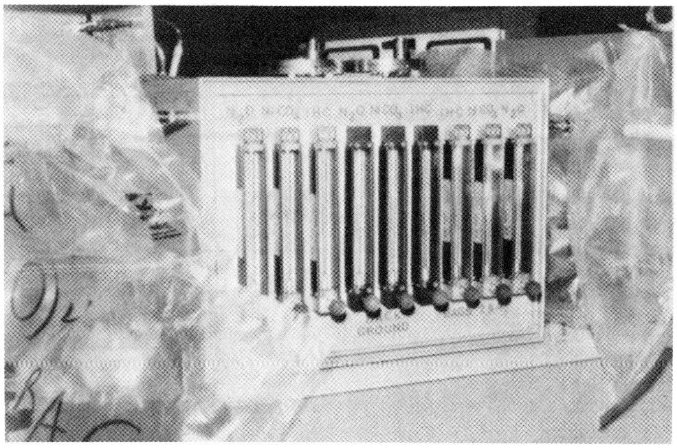

Fig. 3 - Bag sampling system for N_2O, $Ni(CO)_4$, IHC

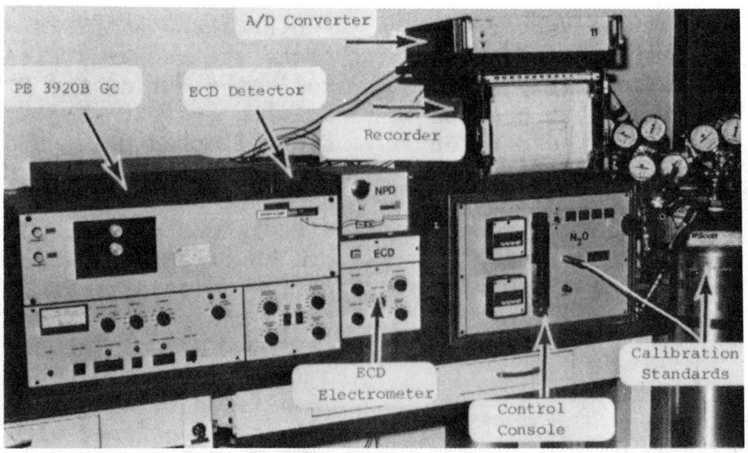

Fig. 4 - Nitrous oxide analytical system

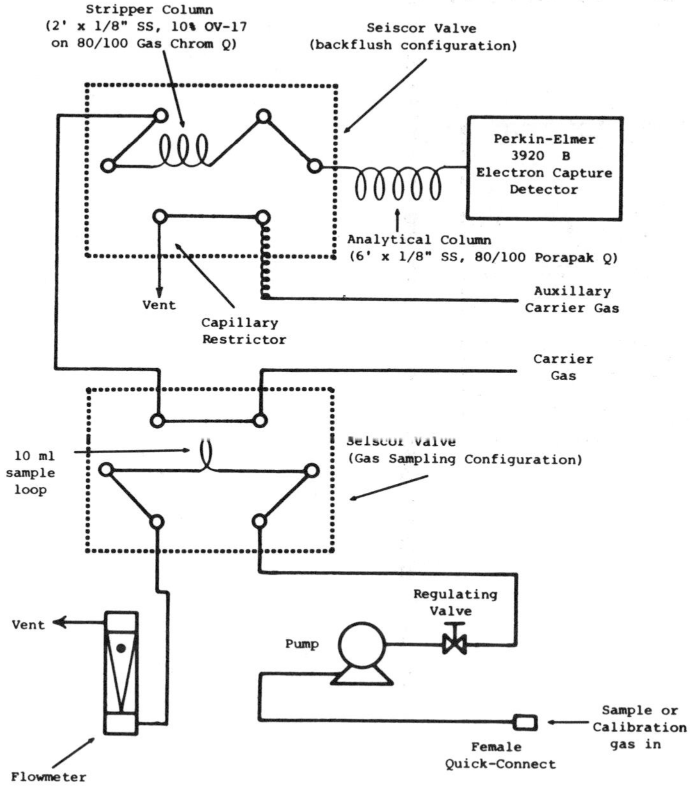

Fig. 5 - Flow schematic of nitrous oxide analytical system (Step 1 - Purge of sample loop of GSV)

the sample through the system. The actual analytical system is shown in Figure 8. The gas chromatographic flow schematic is illustrated in Figure 9 and a typical calibration trace is presented in Figure 10. Additional operational parameters such as columns, flow rates, temperature programming rates, etc., are described in more detail(3).

Nickel Carbonyl - The measurement is accomplished using a NO_x chemiluminescent analyzer modified with a Kodak No. 54 Wratten filter(4). In addition to the internal instrument modification, a special interface was developed to permit calibration using nickel carbonyl calibration blends. A flow schematic of the nickel carbonyl analytical system is shown in Figure 11 with a composite view of the chemiluminescent analyzer, sampling and dilution system presented in Figures 12 and 13.

The exhaust sample is pumped out of the bag through a fiber glass filter to remove any nickel particulate which might be present in the exhaust sample. It is then mixed with an equal volume of 100 percent carbon monoxide which has been previously passed through a quartz furnace. The mixture of exhaust and carbon monoxide enters the chemiluminescent analyzer where nickel carbonyl reacts with ozone and carbon monoxide to produce a light emitting species.

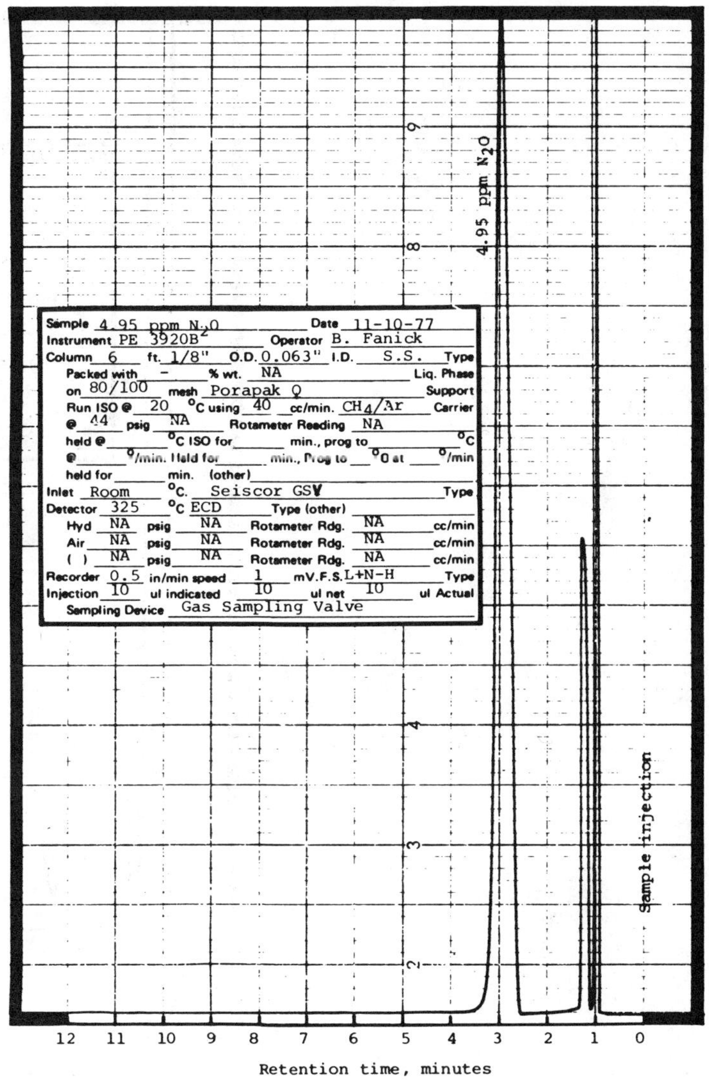

Fig. 6 - Typical N_2O calibration blend gas chromatograph trace

The light emitted by nitric oxide (NO) and olefins is filtered out by the use of a Kodak No. 54 Wratten filter. The response of the detector to the emitted light is then compared to a diluted standard nickel carbonyl blend. This comparison allows the concentration of nickel carbonyl to be determined.

WET COLLECTION - Six unregulated emissions are collected in glass impingers followed by analysis using colorimetric, gas chromatographic or ion chromatographic analysis. Those six unregulated emissions are total cyanide, sulfur dioxide (SO_2), ammonia (NH_3), organic amines, hydrogen sulfide (H_2S) and aldehydes. A brief description of the analytical hardware and the measurement procedure are described as follows.

Total Cyanide - The measurement of total cyanide (hydrogen cyanide and cyanogen) in exhaust is accomplished by collecting exhaust in glass impingers. A 1.0 N potassium hydroxide absorbing solution is maintained at ice bath

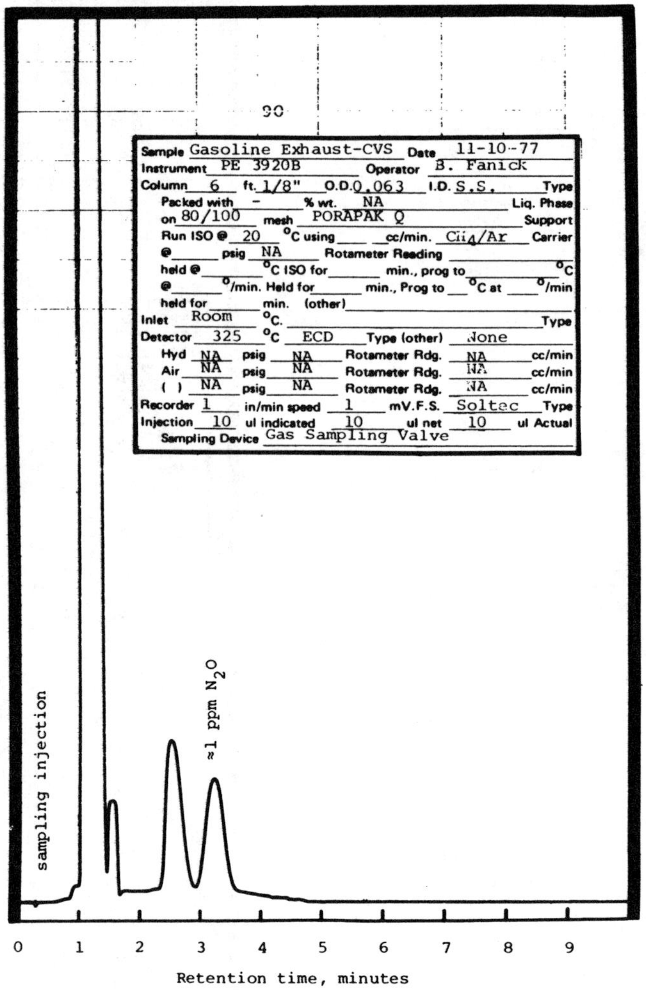

Fig. 7 - Typical gasoline-CVS exhaust sample for N_2O analysis

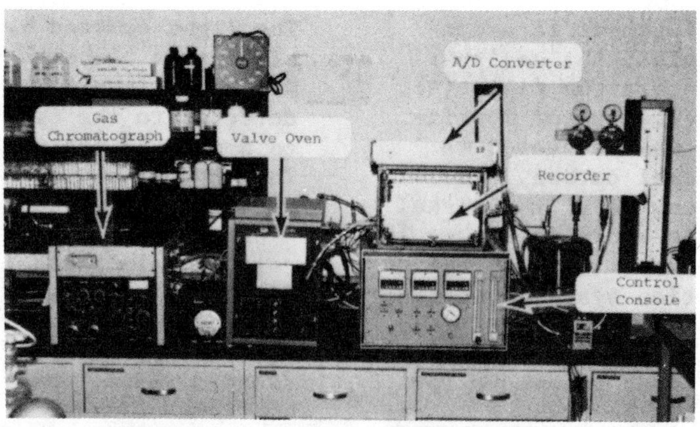

Fig. 8 - Analytical system for individual hydrocarbons

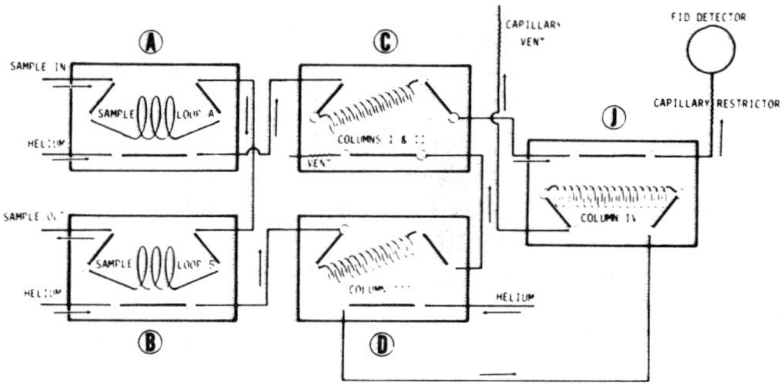

Fig. 9 - Sample purge position prior to sample injection

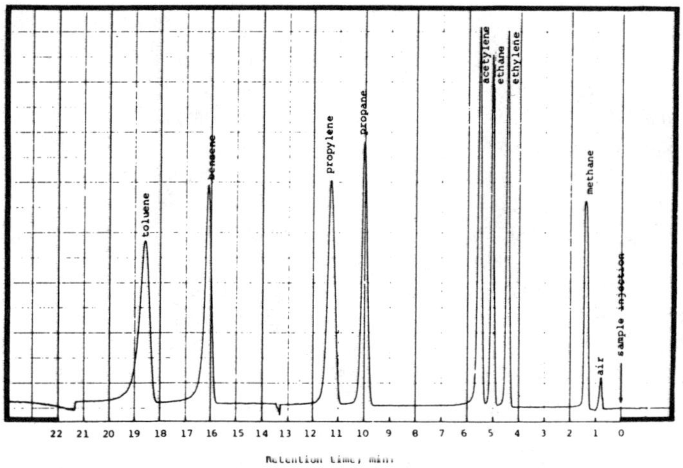

Fig. 10 - Calibration blend for specific hydrocarbons

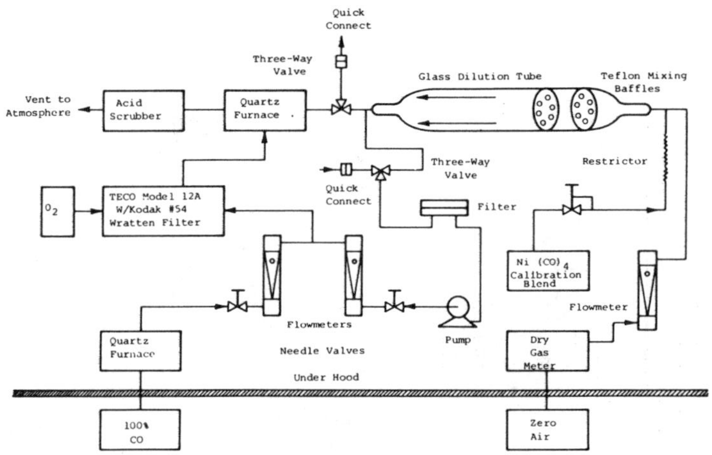

Fig. 11 - Nickel carbonyl analysis and dilution system flow schematic

Fig. 12 - Glass dilution chamber

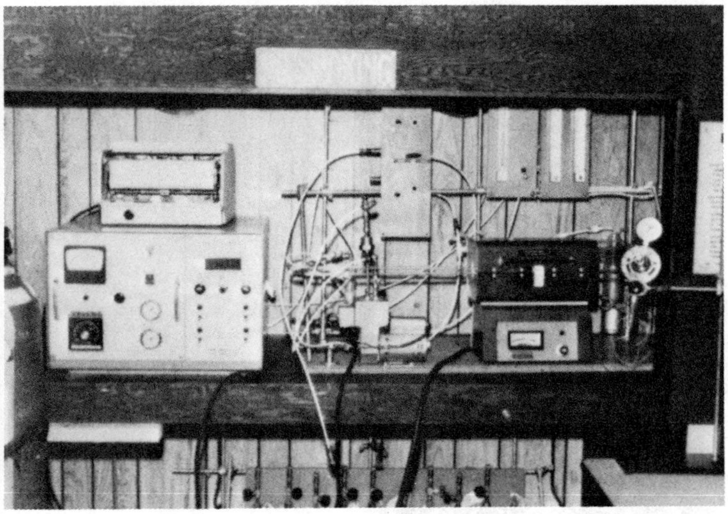

Fig. 13 - Chemiluminescent detector, sampling and dilution system

temperature. An exhaust sample is collected continuously at 4 liters per minute during the test cycle. Upon completion of the test, an aliquot of the absorbing reagent is treated with KH_2PO_4 and chloramine-T. A portion of the resulting cyanogen chloride is injected into a Perkin-Elmer 3920B gas chromatograph equipped with an electron capture detector (ECD)(5). External CN^- standards are used to quantify results. The sample collection flow schematic is presented in Figure 14 and analytical system is illustrated in Figure 15. A typical gas chromatograph trace for a standard is shown in Figure 16.

Sulfur Dioxide - Sulfur dioxide is collected in wet chemical impingers and the resulting solution is analyzed using an ion chromatograph (1). A three percent hydrogen peroxide solution is used as the absorbing reagent and is maintained at ice water temperatures during the sample acquisition. The flow schematic for the sample collection for sulfur dioxide is illustrated in Figure 17. The ion chromatograph employed in this analysis is shown in Figure 18 and a typical sample ion chromatograph trace is presented in Figure 19.

The hydrogen peroxide oxidizes the SO_2 to $SO_4^=$ and the ion chromatograph actually measures the sulfate in the absorbing solution. The sample passes through a Fluoropore filter to remove the sulfate in the gas stream prior to entering the absorbing reagent. Sample concentrations are determined by direct comparison against external standards.

Orgánic Amines - The measurement of organic amines in exhaust is accomplished by bubbling the exhaust through glass impingers containing dilute sulfuric acid. The amines are complexed by the acid to form stable sulfate salts which remain in solution. An aliquot of this solution is injected into a gas chromatograph equipped with an Ascarite pre-column and a nitrogen phosphorus detector (NPD)(6). The sample collection system is quite similar to the total cyanide flow schematic shown in Figure 14 except that the sample line is heated. The gas chromatograph analytical system is shown in Figure 20 and an amines gas chromatograph trace is illustrated in Figure 21. Organic amines that are included in this analysis are monomethylamine, dimethylamine, trimethylamine, monoethylamine, diethylamine and triethylamine. External amine standards in dilute sulfuric acid are used to quantify the results.

Aldehydes and Ketones - The measurement of aldehydes and ketones is accomplished by bubbling the exhaust through glass impingers co--taining 2,4 dinitrophenylhydrazine (DNPH) in dilute hydrochloric acid(7). The aldehydes and ketones react with DNPH to form their respective phenylhydrazone derivatives. These derivatives are insoluble or only slightly soluble in the DNPH/HCl solution and are removed by filtration followed by pentane extraction. The precipitate and extract is evaporated to dryness and dissolved in a quantitative volume of toluene containing a known amount of anthracene as an internal standard. A portion of this dissolved extract is injected into a gas chromatograph equipped with dual flame ionization detectors.

The glass impingers are shown in Figure 22 and the filtration of the absorbing solution is illustrated in Figure 23. The wrist action

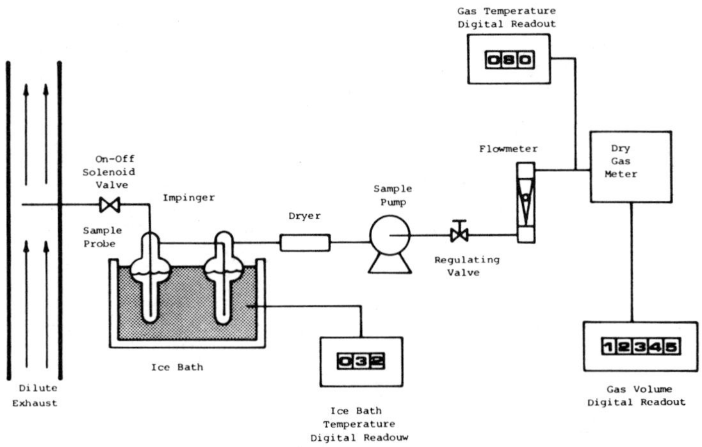

Fig. 14 - Total cyanide sample collection flow schematic

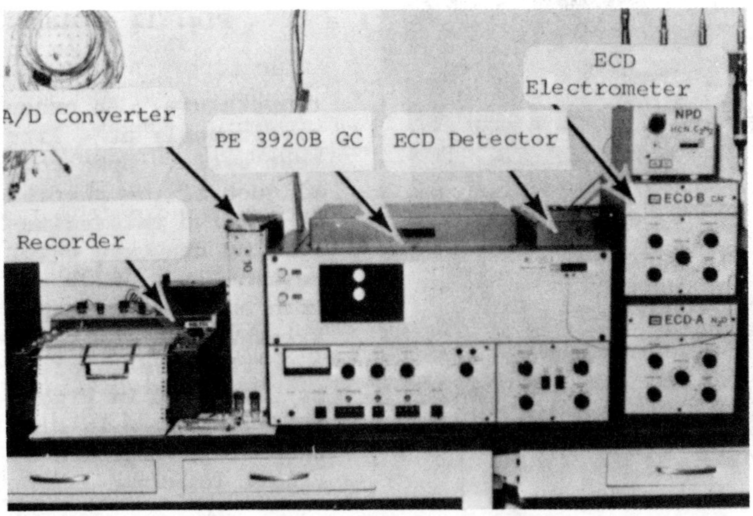

Fig. 15 - Total cyanide analytical system

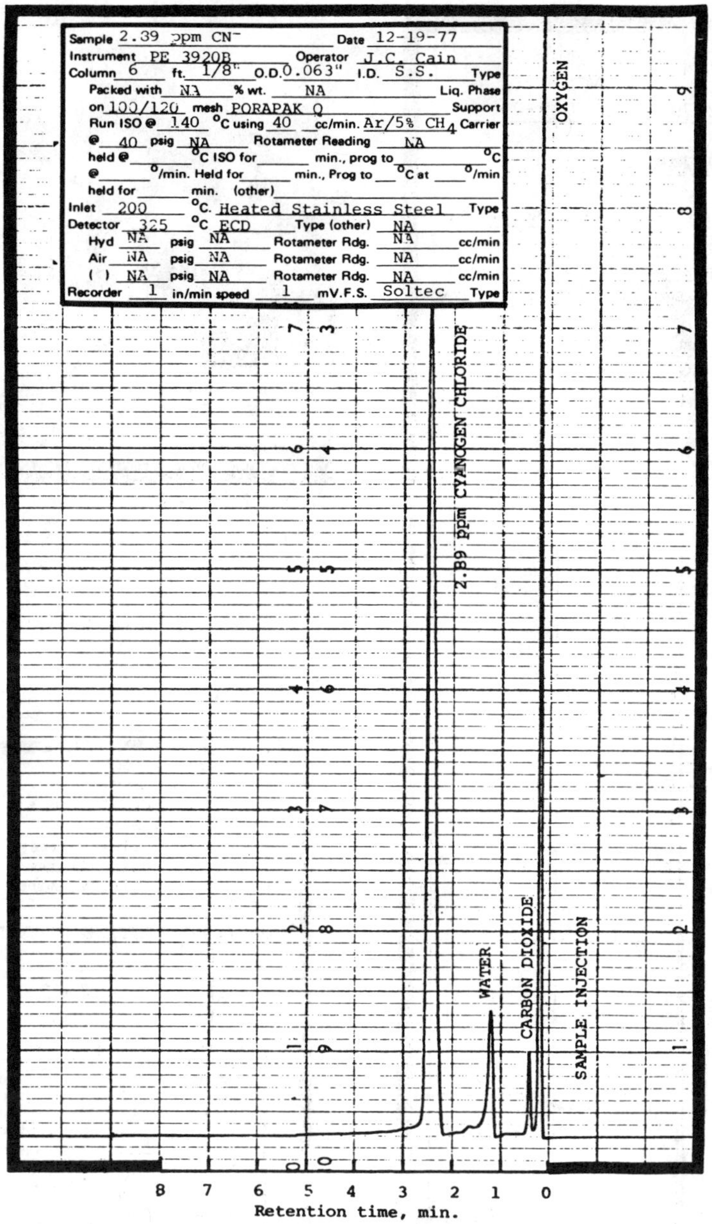

Fig. 16 - Typical trace for a standard

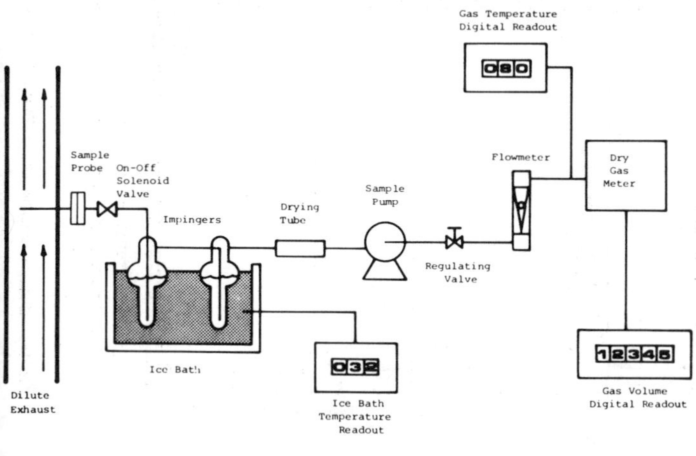

Fig. 17 - SO_2 sample collection flow schematic

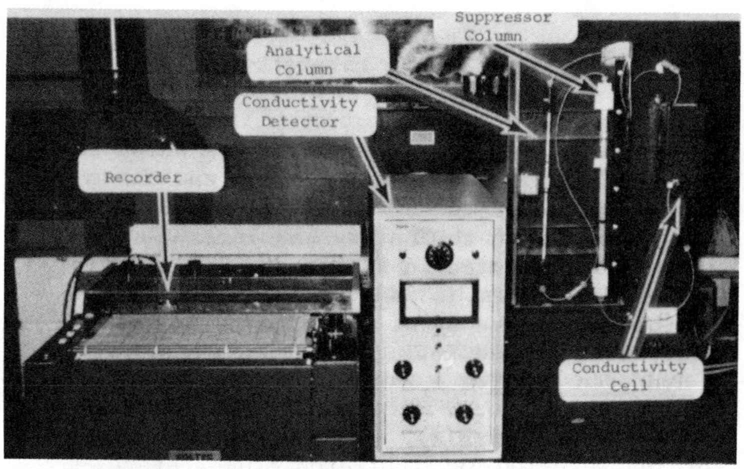

Fig. 18 - SO_2 ion chromatograph

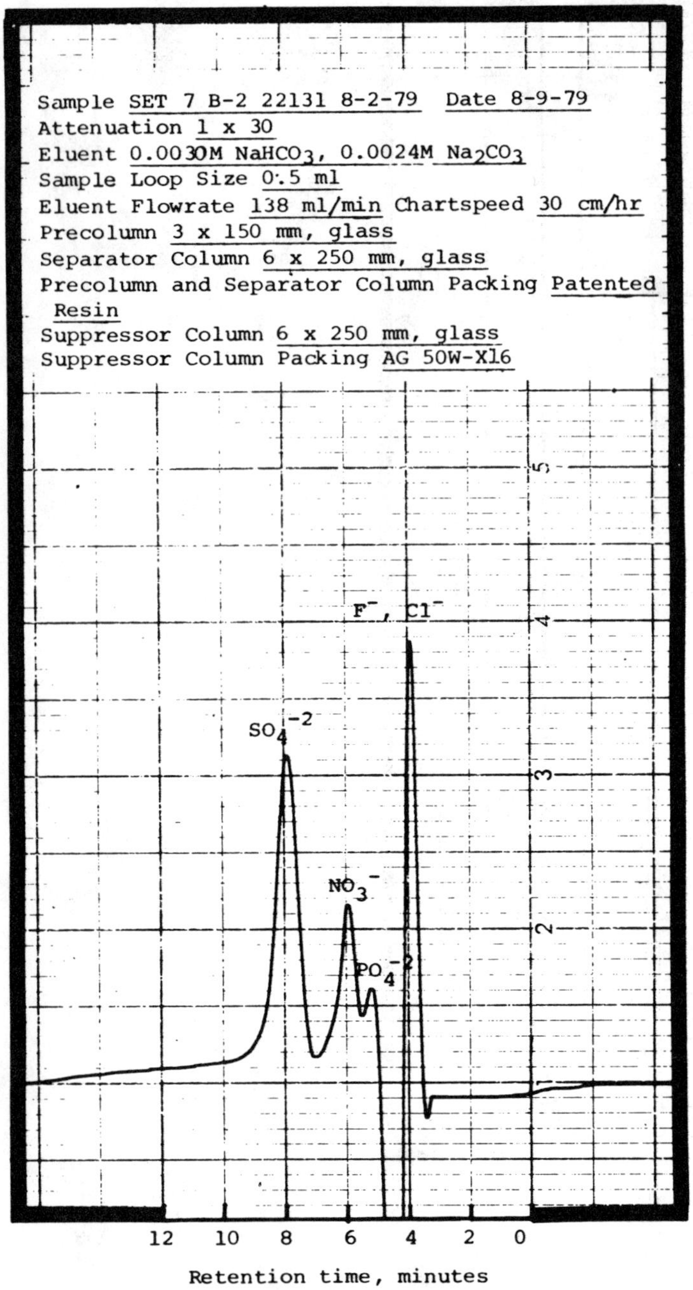

Fig. 19 - Sample chromatogram

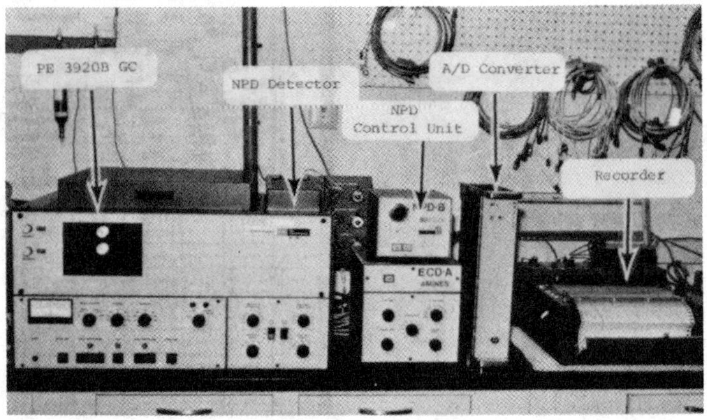

Fig. 20 - Organic amine analytical system

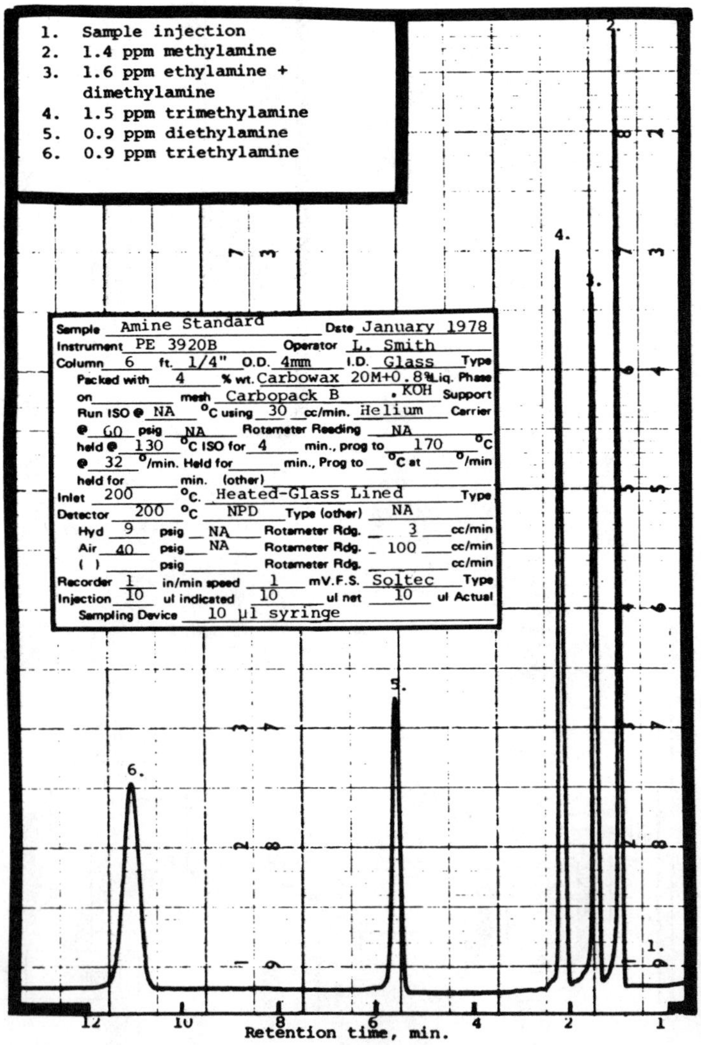

Fig. 21 - Chromatograph of amine standard

automatic shaker and vacuum oven are shown in Figures 24 and 25 respectively. A gas chromatograph trace of the standard blend of aldehydes and ketones with the anthracene internal standard is illustrated in Figure 26. The sample collection system is identical to that described in Figure 14 of the organic amines analytical procedure.

Hydrogen Sulfide - The measurement of H_2S is accomplished by the methylene blue colorimetric method of analysis(8). Figure 14 illustrates the sample collection flow schematic. The absorbing solution is a buffered zinc acetate solution which traps the sulfide ion as zinc sulfide. The exhaust is collected continuously

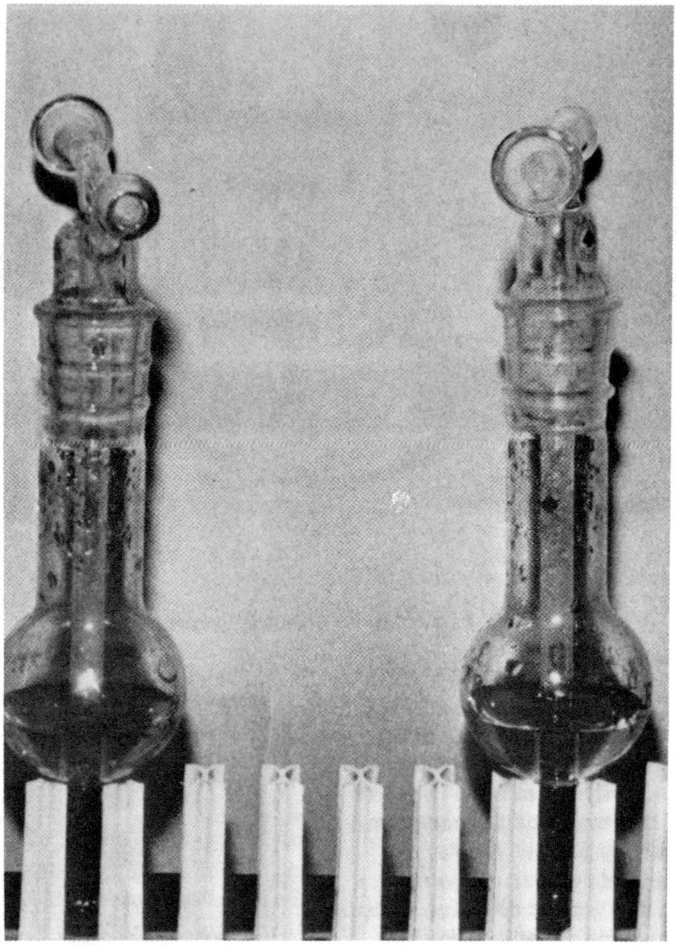

Fig. 22 - Impingers containing HCl/DNPH absorbing solution

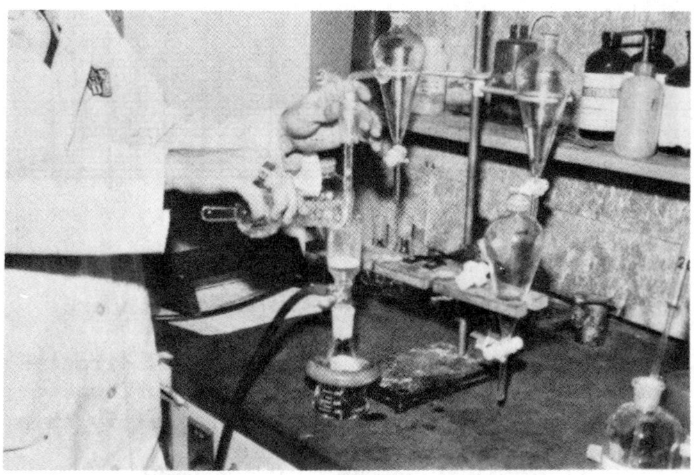

Fig. 23 - Filtration of absorbing solution

Fig. 24 - Automatic shaker

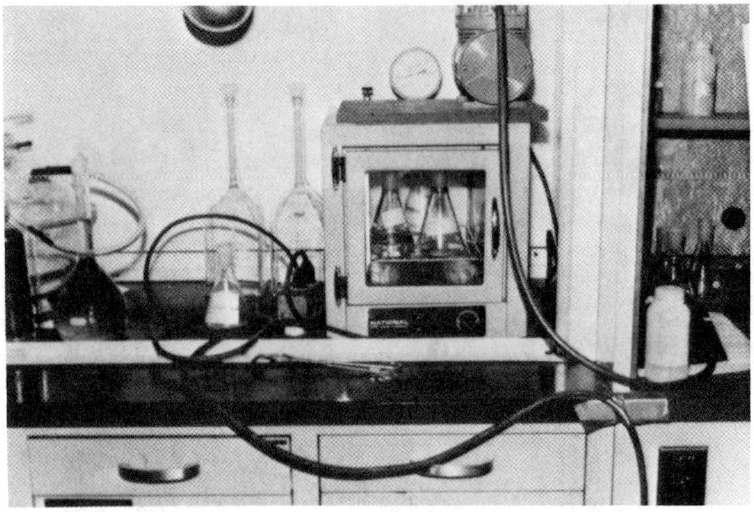

Fig. 25 - Vacuum oven

during the entire test cycle. Upon completion of the test, the absorbing solution is treated with N,N dimethyl-paraphenylene diamine sulfate and ferric ammonium sulfate. Cyclization occurs forming the highly colored heterocyclic compound methylene blue (3,9-bis-dimethylaminophenozothionium sulfate). The resulting solution is analyzed in a Beckman Model 25 spectrophotometer at 667 mm with a 1 cm or 4 cm pathlength cell, depending on the concentration. The minimum detection limit is 0.01 ppm H_2S.

Ammonia - The ammonia in exhaust is collected using a dilute sulfuric acid absorbing reagent in glass impingers(1). The ammonia is ionized to NH_4^+ during the collection process. An aliquot of the absorbing reagent is injected into a Dionex Model 10 Ion Chromatograph. By using external standards of $(NH_4)_2SO_4$ the concentration of ammonia in exhaust can be calculated. An overall view of the ion chromatograph system is presented in Figure 27 and a typical exhaust ion chromatograph trace is shown in Figure 28.

Phenols - A procedure for the measurement of phenols in Diesel exhaust is currently undergoing qualification testing. This procedure involves collection of the phenols in a dilute potassium hydroxide absorbing solution(9). After collection, the phneols are extracted from the absorbing reagent (after acidification) with diethyl ether. The $C_2H_5OC_2H_5$ is evaporated to about 1 ml and the sample volume is brought up to exactly 2 ml. An aliquot of this solution is injected into a gas chromatograph equipped with a flame ionization detector. Using external standards it is possible to determine the exhaust concentrations of the lower molecular weight phenols included in this analysis. Specific details of the sample collection and analysis will be published once they are final-

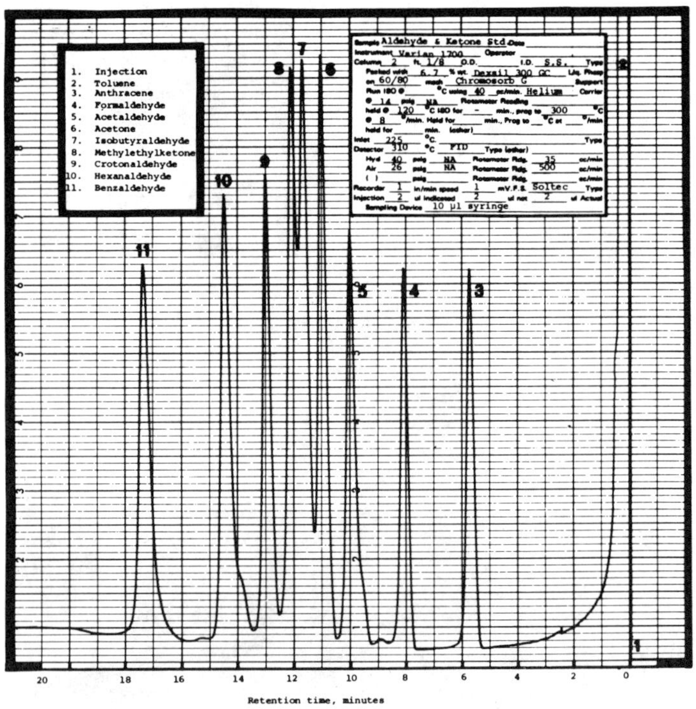

Fig. 26 - Chromatograph of standard

ized. The measurement of phenols is directed toward Diesel exhaust only and no work was or will be conducted to validate or qualify this procedure for gasoline exhaust.

General Comments - The flow rate for collection of H_2S, SO_2, HCN, organic amines, aldehydes and NH_3 is 4 liters per minute. Due to the extremely low concentration of phenols in Diesel exhaust, it is estimated the sample flow rate will approach 1 CFM. All impingers are immersed in an ice water bath at 0-5°C. A composite

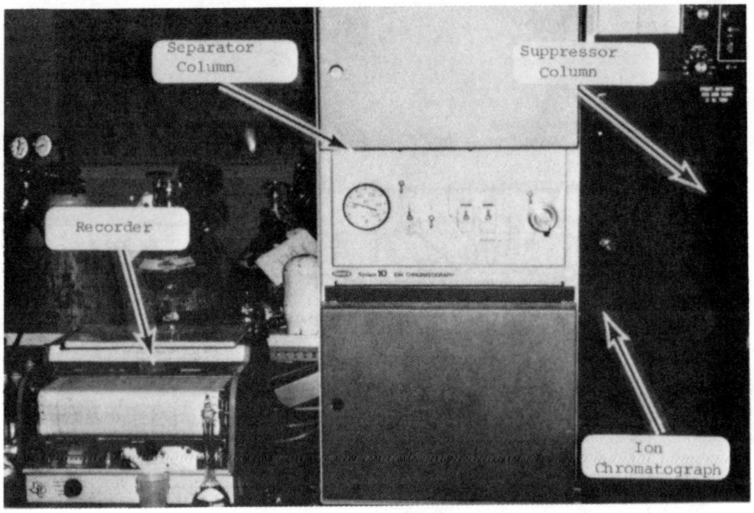

Fig. 27 - NH_3 ion chromatograph

background is obtained during the various test sequences to allow for any background concentration, solvent impurities, etc.

TRAPS - Several unregulated emissions are collected on adsorbent traps during the emission tests. Those emissions included in this category include organic sulfides, N-nitrosodimethylamine, Diesel odor analysis system (DOAS) and boiling point distribution. Of these four groups of unregulated emissions, only the organic sulfides analytical procedure was qualified at Southwest Research Institute. The N-nitrosodimethylamine sample collection and analytical system was qualified by the Environmental Protection Agency at Research Triangle Park(10). The remainder of this section briefly describes those analytical procedures employing traps as the collection media.

Organic Sulfides - The measurement of organic sulfides in exhaust is accomplished by passing the exhaust through Tenax-GC traps at -76°C(1). The organic sulfides are removed from the exhaust by the traps at -76°C. Sulfides that are included in this analysis are carbonyl sulfide (COS), methyl sulfide $(CH_3)_2S$, ethyl sulfide $(C_2H_5)_2S$, and dimethyl disulfide $(CH_3)_2S_2$. The organic sulfides are thermally desorbed from the traps into a gas chromatograph equipped with a flame photometric detector for analysis. External organic sulfide standards generated from permeation tubes are used to quantify results. Detection limits for this procedure are on the order of 0.1 ppb.

Figure 29 illustrates the sample flow schematic for the collection of organic sulfides. The GC flow schematic during the calibration mode is illustrated in Figure 30 and analytical mode in Figure 31. An overall view of the organic sulfide analytical system is shown in Figure 32. A typical gas chromatograph trace of the permeation calibration blend is illustrated in Figure 33.

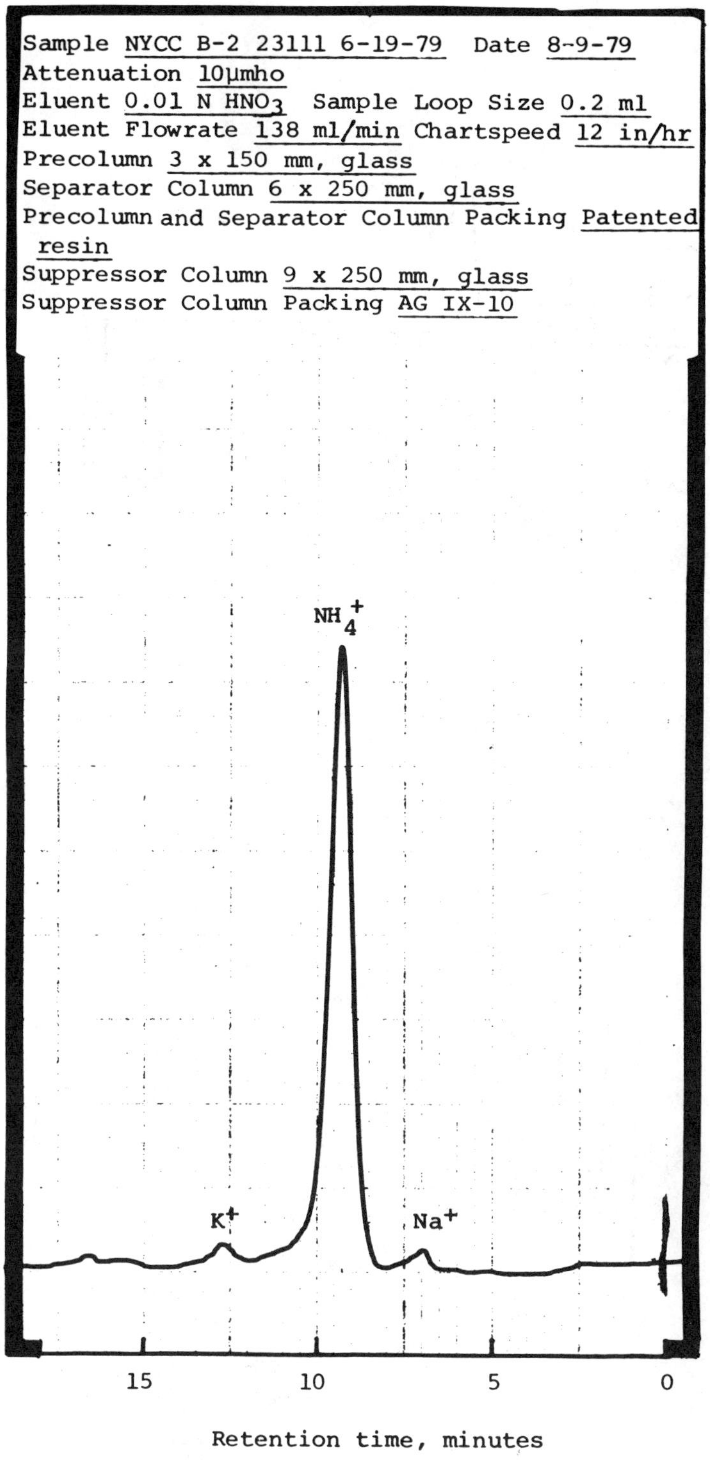

Fig. 28 - Sample chromatogram

N-nitrosodimethylamine - The measurement of N-nitrosodimethylamine is accomplished by collection of diluted exhaust on cartridges filled with Tenax-GC absorbent. The N-nitrosodimethylamine is thermally desorbed into a gas chromatograph-mass spectrometer (GC-MS) analytical system. External standards are used to quantitate results(10). In order to eliminate possible artifact formation, a secondary dilution system was developed by EPA-RTP to keep

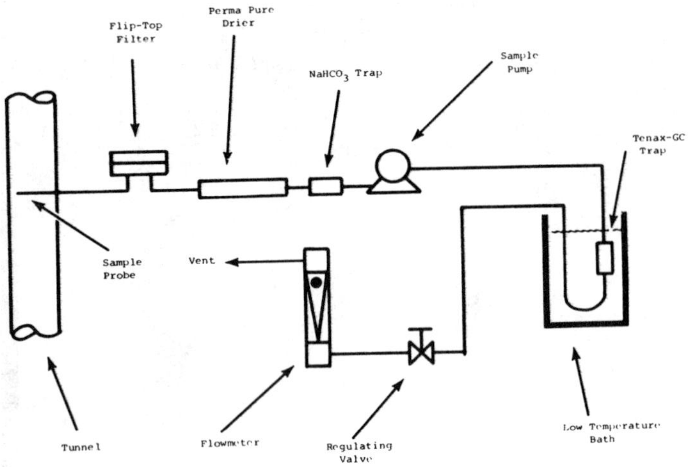

Fig. 29 - Organic sulfide sample flow schematic

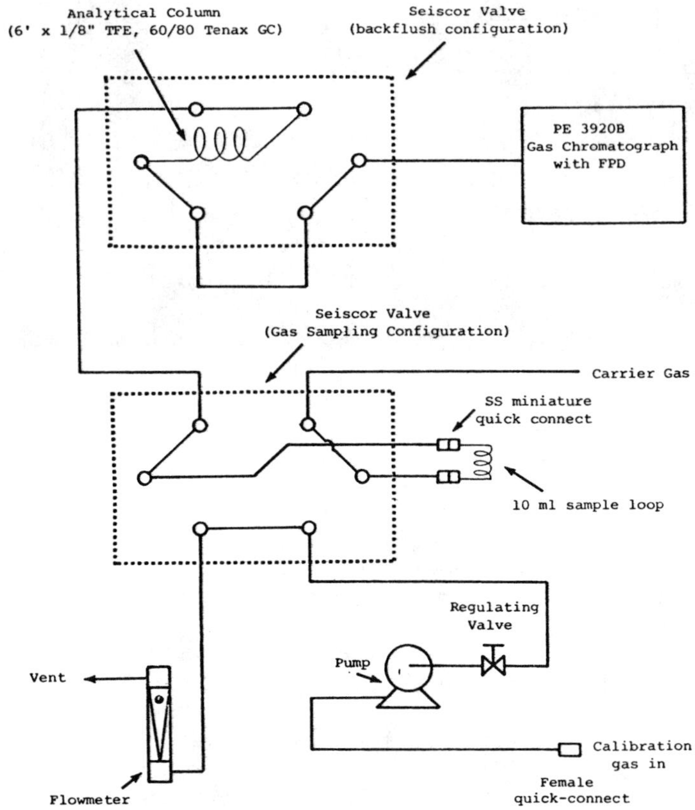

Fig. 30 - Flow schematic of organic sulfide calibration system (Step 2 - inject calibration gas into GC system)

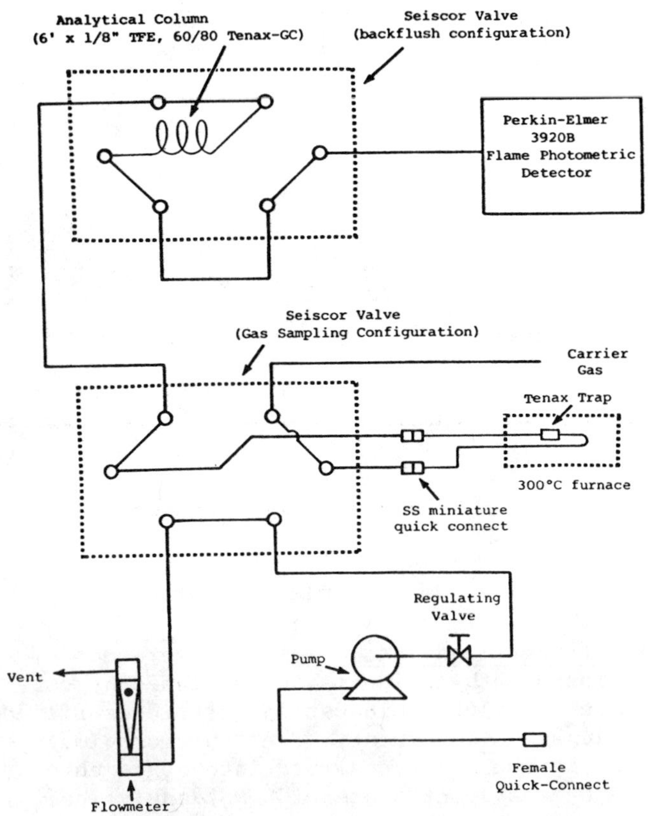

Fig. 31 - Flow schematic of organic sulfide analysis system (Step 2 - inject Tenax trap contents into GC system)

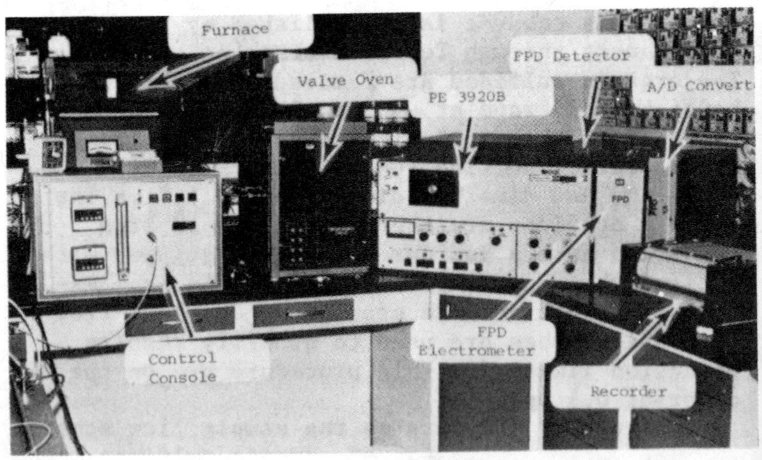

Fig. 32 - Organic sulfide analytical system

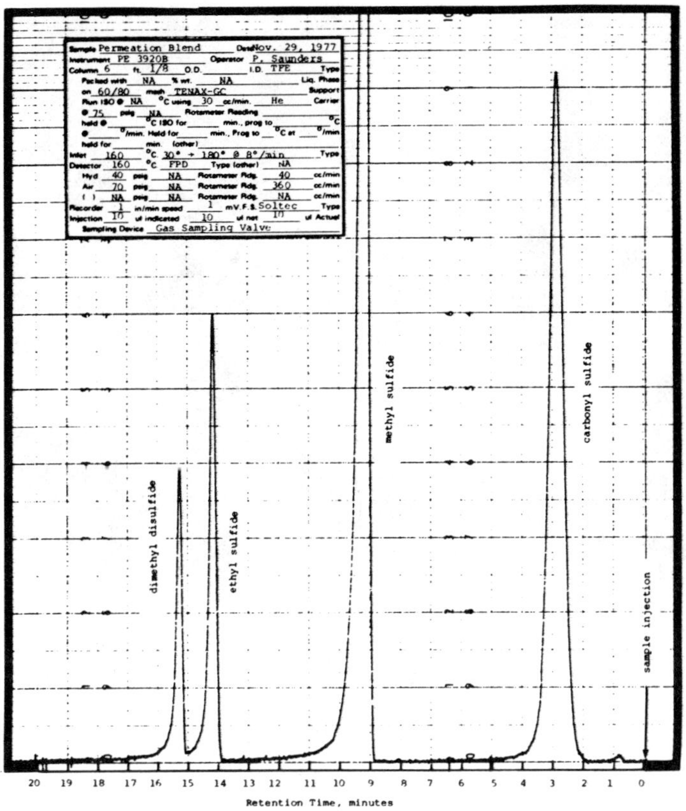

Fig. 33 - Chromatogram of organic sulfide standard

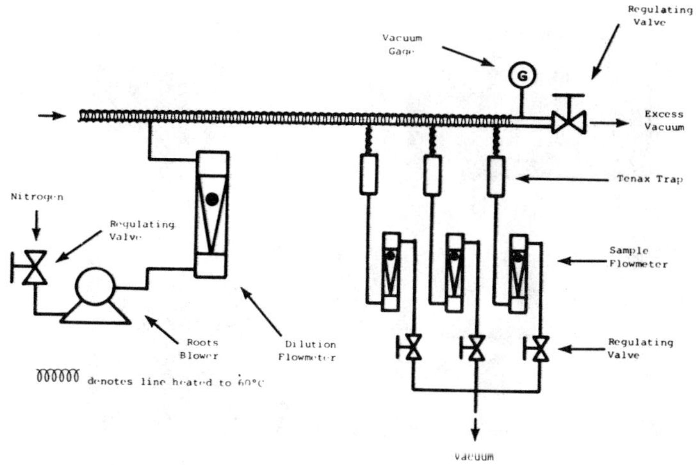

Fig. 34 - Secondary diluter

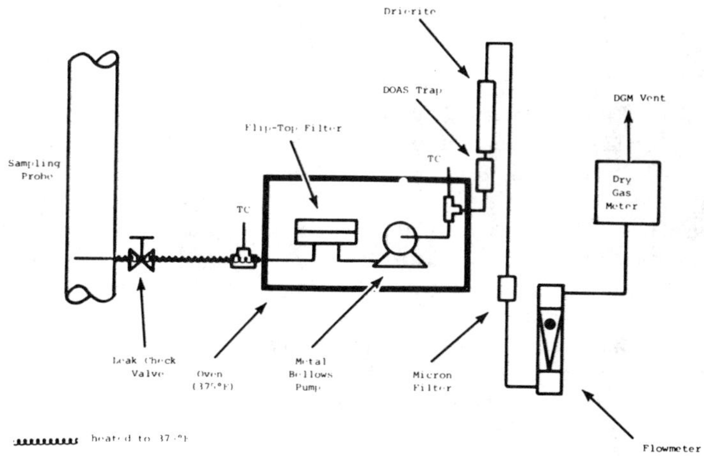

Fig. 35 - DOAS sampling system

the NO levels less than 5 ppm. At levels higher than 5 ppm NO, it was found that it was possible to form N-nitrosodimethylamine on the Tenax material under certain conditions. A schematic of the secondary dilution system is presented in Figure 34.

The analysis of the Tenax-GC cartridges for N-nitrosodimethylamine is routinely performed by Research Triangle Institute using the GC-MS techniques described in Reference 10. This procedure is quite sensitive and allows positive identification of other nitrosoamines as well as N-nitrosodimethylamine and can be applied to both Diesel and gasoline exhaust.

Diesel Odor Analysis System (DOAS) - In the early 1970's, Arthur D. Little, Inc. developed an instrumental method for the measurement of Diesel exhaust odor. This work was sponsored by the Coordinating Research Council (CRC) and is routinely used at SwRI along with the human odor panel evaluations of Diesel exhaust(11). An exhaust sample is collected on a Chromasorb 102 trap using the flow schematic illustrated in Figure 35. The exhaust collected on the trap is eluted from the trap using cyclohexane. The compounds believed responsible for Diesel exhaust odor are eluted in the first two ml. An aliquot of the exhaust concentrate is injected into the DOAS instrument for subsequent analysis for aromatics and oxygenates. By employing external standards, the aromatics and oxygenates in exhaust can be quantitatively determined and by using an equation developed by ADL, an exhaust odor intensity can be calculated. The DOAS shown in Figure 36 is essentially an automatic liquid chromatograph with a uv detector, that was designed to perform a specific function. This technique was developed for Diesel exhaust, but it has been used for gasoline exhaust on occasion.

Boiling Point Distribution - The boiling point distribution of gaseous hydrocarbons from Diesel exhaust is accomplished using the sample collection technique described in the DOAS collection system. Instead of eluting with cyclohexane, the organics are eluted using carbon disulfide. An aliquot is injected into a gas chromatograph equipped with a FID. There are several GC procedures that have been satisfactory, i.e., packed and capillary columns.

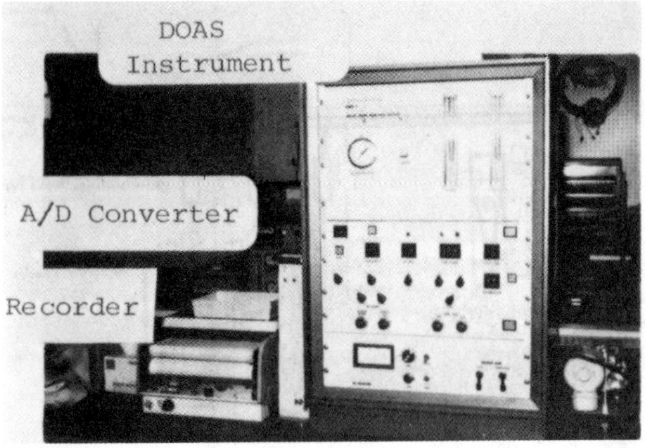

Fig. 36 - DOAS instrument

Fig. 37 - Dilution tunnel for particulate/ sulfate sample collection

FILTERS - Unregulated emissions that are typically collected on filters include sulfate, metal and total particulate mass. These filters are collected during the driving cycle using an 18 inch stainless steel dilution tunnel. A schematic of the dilution tunnel is illustrated in Figure 2 and an overall view of the tunnel is shown in Figure 37. The filters are collected simultaneously under isokinetic sampling conditions. A brief description of the analytical procedures is presented below.

Sulfate - During 1974-1975, EPA conducted an intensive program in the area of sulfate emission measurements. This work was conducted in-house as well as by independent contractors. Two immediate results of this work were the development of a driving cycle specifically for the measurement of sulfate and the barium chloranilate (BCA) procedure for measuring sulfate (12). The driving cycle was called the sulfate emissions test cycle seven, commonly called SET-7. The BCA procedure involves isokinetically collecting the sulfate on 47 mm Fluoropore filters followed by ammoniation. The soluble sulfates are leached from the filter with a measured volume of 60 percent isopropyl alcohol - 40 percent water solution (60 percent IPA). A fixed volume of the sample extract is injected into a high pressure liquid chromatograph system (HPLC). The sample and solvent are pumped through a column of strong cation-exchange resin in Ag^+ form to scrub out the halides (Cl- and Br^-) then through a strong cation-exchange (H^+ form) column to remove the cations and convert the sulfate to sulfuric acid. The sample $(H_2SO)_4$ then enters a reactor column of barium chloranilate crystals to precipitate out barium sulfate and release the highly colored chloranilate ions. The amount of chloranilate ions released is proportional to the sulfate in the

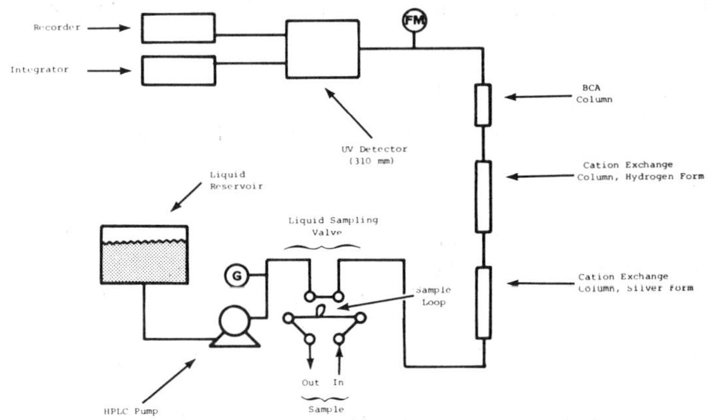

Fig. 38 - Schematic of an automated BCA sulfate set-up

sample and is measured by a uv detector at 310 nanometers. A flow schematic of the BCA analytical system is shown in Figure 38. The system was originally qualified for gasoline exhaust and has been applied to Diesel particulate. Some procedural modifications may be required for qualification of Diesel exhaust.

Particulate - During the past 12 months there has been an accelerated effort by EPA and industry in Diesel particulate measurements. This is evidenced by the routine acquisition of 90 mm and 20 x 20 inch Pallflex T60A20 filters in addition to the standard 47 mm and 8 x 10 inch filters. Each filter media and size has a specific role to fulfil. A schematic of the possible filter combinations on a given test is illustrated in Figure 39. Most filters require weighing before and after the test,

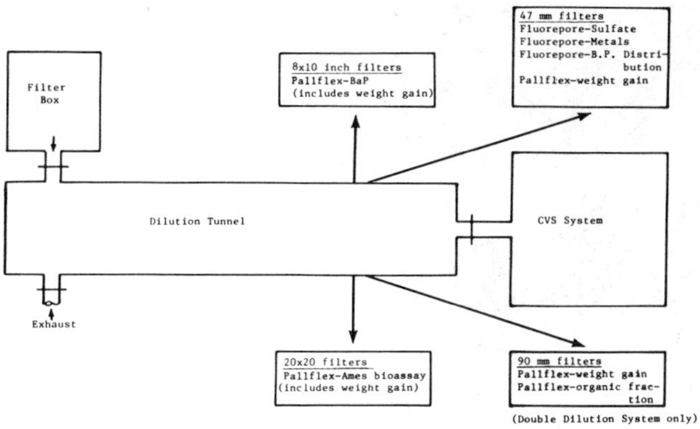

Fig. 39 - Distribution of various particulate filters

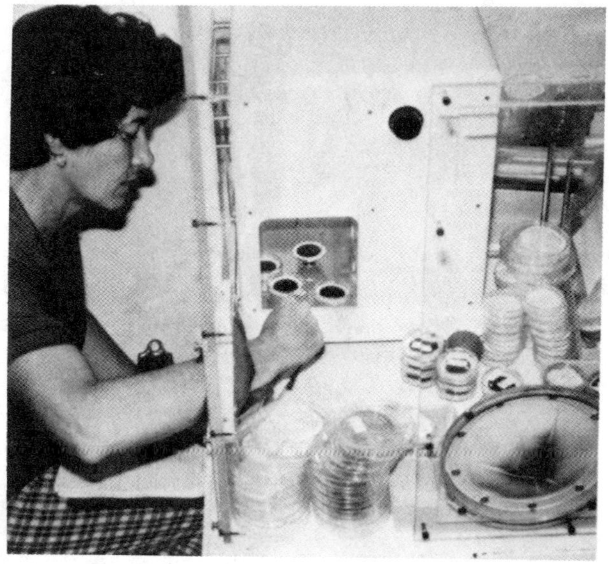

Fig. 40 - Environmentally controlled particulate filter weighing system

performed with balance (Mettler M-5) and weighing chamber shown in Figure 40. The balance, Soxhlet extractors and HPLC system are used under yellow lights to prevent sample deterioration.

The Soxhlet extractors are used to the organic solubles from the total particulate mass by gentle continuous washing with an organic solvent. A series of large Soxhlet extractors used for 20 x 20 filters is presented in Figure 41. This extraction allows determination of the organic fraction of the total particulate mass. In addition, the organic solubles may be subjected to a battery of independent tests. These tests include the Ames bioassay analysis, boiling point distribution, analysis for Benzo (α)pyrene, high pressure liquid chromatography (HPLC) fractionation and others. With the advent of the 20 x 20 inch filter collection system, it is now possible to collect large quantitites of particulate in a relatively short period of time, particularily if the entire exhaust is filtered as shown in Figure 42.

Metals - The determination of metals in exhaust is accomplished by collecting the particulate on 47 mm Fluoropore filters. These filters are then analyzed directly using a Siemens Model MRS-3 simultaneous spectrometer(13). This analysis was performed by the Emissions Measurement and Characterization Division in the Environmental Sciences Research Laboratory of the Environmental Protection Agency at Research Triangle Park, North Carolina.

DISCUSSION

A variety of analytical procedures currently used in the chemical characterization of unregulated emissions have been presented. Most of these procedures have completed validation and qualification experiments and apply to both gasoline and Diesel exhaust. Some procedures (i.e., organic solubles, DOAS, etc.) have been developed specifically for Diesel exhaust. The procedures are currently being applied to mea-

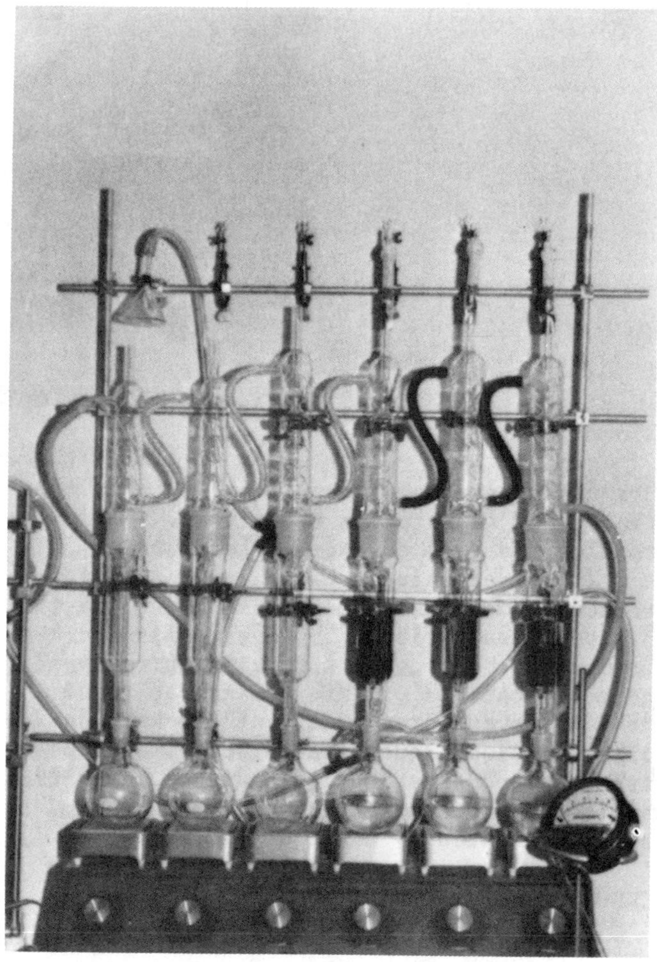

Fig. 41 - Soxhlet extractors for 20 x 20 filters

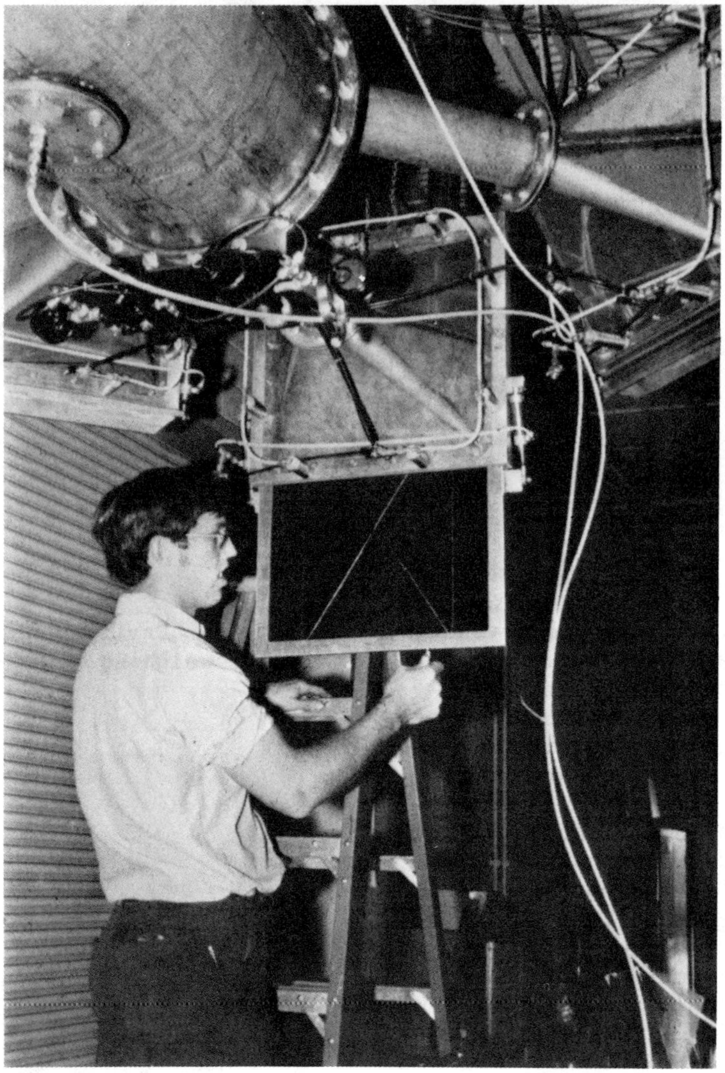

Fig. 42 - 20 x 20 inch filter collection system

sure unregulated emissions from vehicles operating under malfunction conditions, low NO_x vehicles and from advanced powerplants.

It is recognized that these analytical procedures were "state-of-the-art" at the time this paper was written, however, emission measurement technology is advancing quite rapidly and modifications or imporvements to these methods are expected. Many of the wet chemical collection techniques will probably be replaced with continuous methods of analysis. New Diesel particulate collection techniques are being currently evaluated by EPA and SwRI.

ACKNOWLEDGEMENT

This paper is based on work performed by Southwest Reserach Institute under EPA Contracts 68-02-2497 and 68-02-2703. The authors wish to acknowledge the grateful technical assistance of Doctors Ronald Bradow, Roy Zweidinger and Silvestre Tejada of EPA-RTP and Robert Garbe of EPA-AA. Primary researchers at SwRI were Dr. Lawrence Smith, Mary Ann Parness and E. Robert Fanick.

REFERENCES

1. H.E. Dietzmann, L.R. Smith, M.A. Parness, and E.R. Fanick, "Analytical Procedures for Characterizating Unregulated Pollutant Emissions from Motor Vehicles." Final Report to Environmental Protection Agency under Contract No. 68-02-2497, February, 1979.

2. Private communication between Harry Dietzmann of Southwest Research Institute and Dr. Roy B. Zweidinger and Frank M. Black of the Environmental Protection Agency, March, 1977.

3. F.M. Black, L.E. High, and J.E. Sigsby, "Methodology for Assignment of the Hydrocarbon Photochemical Reactivity Index for Emissions

from Mobile Sources." Final Report to the Environmental Protection Agency, March, 1975.

4. D.H. Stedman, and D.A. Tamaro, Analytical Letters, Vol. 9 (1), 1976; page 81.

5. J.C. Valentour, V. Aggarwal, and I. Sunshine, Analytical Chemistry, Vol. 46, page 924, 1974.

6. C.E. Andre and A.R. Mosier, Analytical Chemistry, Vol. 215, page 1971, 1973.

7. F. Stump, Oxygenated Compounds in Automobile Exhaust. Gas Chromatographic Procedure, MSERB - ESRL-EPA, Research Triangle Park, N.C.

8. D.L. Flamm and R.E. James, Environmental Science and Technology, page 159, 1975.

9. R.A. Brown, T.D. Searl, W.H. King, Jr., W.A. Dietz, and J.M. Kelliher, Rapid Methods of Analysis for Trace Quantities of Polynuclear Aromatic Hydrocarbons and Phenols in Automobile Exhaust, Gasoline and Crankcase Oil. Final Report CRC-APRAC Project CAPE-12-68, December, 1971.

10. E.D. Pellizzari, "Analysis of N-nitrosodimethylamine in Exhaust Gases Using a Tenax GC Cartridge and Gas Chromatography/Mass Spectrometry/Computer." EPA Contract No. 69-02-2767, November 27, 1978.

11. P.L. Levins and D.A. Kendall, "Application of Odor Technology to Mobile Source Emission Instrumentation." CRC Project CAPE 7-68 Contract No. 68-02-0561, September 1973.

12. S.B. Tejada, J.E. Sigsby, Jr., and R.L. Bradow, "Determination of Soluble Sulfates in Automobile Exhaust by Automated HPLC Modification of the Barium Chloranilate Method." EPA Internal Procedure, 1975.

13. J. Wagman, R.L. Bennett and K.T. Knapp, "Simultaneous Multiwavelength Spectrometer for Rapid Elemental Analysis." Reprint from X-Ray Fluorescence Analysis of Environmental Samples, Ann Arbor Science Publishers, Inc., Ann Arbor, Michigan, 1977.

Emissions from Trucks by Chassis Version of 1983 Transient Procedure**

Harry E. Dietzmann and Mary Ann Parness
Southwest Research Institute

Ronald L. Bradow
U.S. Environmental Protection Agency

IN PREVIOUS YEARS, heavy-duty diesel engine certification tests were conducted with procedures described in the Federal Register(1-4)*. These procedures were steady-state and were limited to stationary dynamometer testing. Recently the U.S. Environmental Protection Agency (EPA) proposed a transient test procedure for the measurement of gaseous emissions from heavy-duty engines(5). A finalized test procedure for the measurement of gaseous emissions from 1984 and later diesel engines was published in the Federal Register(6-8). A chassis version of the heavy-duty transient cycle was developed by EPA to allow testing of in-use vehicles(9). Characterization of regulated gaseous, particulate and unregulated emissions has been conducted on heavy-duty diesel engines using steady-state operation(10-12) and on light-duty diesel(13-14) with the FTP. This project employed the chassis version of the 1983 transient cycle to characterize gaseous and particulate emissions from four heavy-duty diesel engines operating with several fuels.

ENGINE AND FUEL DESCRIPTION

Table 1 lists the particulars that describe the four heavy-duty diesel engines (and corresponding vehicles) included in this study. The Caterpillar 3208, Mack ENDT 676 and Cummins Formula 290 were obtained in vehicles from local leasing agencies. The Detroit Diesel 8V-71 was in a city bus obtained from the VIA Metropolitan Transit of San Antonio, Texas. All engines were deemed "road worthy" upon receipt and no maintenance or repair was required prior to testing.

Table 2 lists the properties of the five test fuels employed in this study. Three fuels (EM-329-F, EM-241-F and EM-242-F) were used with the first three engines. Two additional fuels (EM-400-F and EM-401-F) were included with the city bus. EM-329-F was a national average DF-2, EM-241-F was a minimum quality

*Numbers in parentheses designated references at end of paper.

**Paper 801371 presented at the Fuels and Lubricants Meeting, Baltimore, Maryland, October, 1980.

ABSTRACT

Regulated gaseous, particulate and several unregulated emissions are reported from four heavy-duty diesel engines operated on the chassis version of the 1983 transient procedure. Emissions were obtained from Caterpillar 3208, Mack ENDT 676, Cummins Formula 290 and Detroit Diesel 8V-71 engines with several diesel fuels. A large dilution tunnel (57' x 46" ID) was fabricated to allow total exhaust dilution, rather than the double dilution employed in the stationary engine version of the transient procedure. A modal particulate sampler was developed to obtain particulate data from the individual segments of the 1983 transient procedure. The exhaust gas was analyzed for benzo(a)pyrene, metals, N_2O, NO_2, individual hydrocarbons and HCN. Sequential extractions were performed and measured versus calculated fuel consumptions were obtained.

Table 1. Description of Test Engines/Vehicles

Engine Number	1	2	3	4
Engine Make	Caterpillar	Mack	Cummins	Detroit Diesel
Engine Model	3208	ENDT 676	Formula 290	8V-71
Strokes/cycle	4	4	4	2
Cylinder Arrangement	V-8	I-6	I-6	V-8
Displacement, liters	10.42	11.08	14.01	9.31
cubic in.	636	676	855	568
Year	1979	1979	1979	1977
Rated Speed, rpm	2,800	2,100	1,900	2,100
Power at Rated Speed, kw	149	211	216	209
hp	200	283	290	280
Vehicle Make	Ford	Mack	Ford	GMC
Vehicle Model	7,000	Conv. Cab	9,000	RT
Vehicle Description	18' Bobtail Van	Tractor	Tractor	City
Drive Train	single axle	dual drive tandem axle	dual drive tandem axle	single axle
Odometer, miles	7,000	69,000	26,000	60,000
GVW, lbs	27,500	80,000	80,000	36,900
Dynamometer inertia				
Front rolls, lbs	16,000	24,500	24,500	32,000
Back rolls, lbs	none	24,500	24,500	none

Table 2. Properties of the Five Diesel Test Fuels

	Fuel Code				
	EM-329-F	EM-241-F	EM-242-F	EM-400-F	EM-401-F
Fuel Description	National Average DF-2	Minimum Quality DF-2	Premium Quality DF-2	DF-1 Emissions Test Fuel	JP-7
Properties					
Density, g/mℓ	0.8369	0.8613	0.8324	0.8118	0.800
Gravity, °API	37.5	32.7	38.4	42.9	45.3
Cetane, (D-976)	48.0	42.0	52.1	49.0	56
Viscosity, CS (D-445)	2.36	2.46	2.58	1.69	1.69
Flash point, °C	65	71	76	70	74
Sulfur, wt.% (D-1266)	0.24	0.25	0.24	0.18	nil
Gum, mg/100 mℓ (D-481)	21.6	70.2	42.5	4.6	0.5
Carbon, wt. %	85.80	87.21	86.14	86.37	84.71
Hydrogen, wt. %	13.01	12.41	13.77	13.54	14.20
Nitrogen, wt. %	0.040	0.610	0.046	0.006	0.025
FIA:					
Aromatics, %	21.3	35.8	16.9	10.5	2.7
Olefins, %	1.7	1.6	1.9	1.5	1.9
Saturates, %	77.0	67.6	81.2	88.0	95.4
Distillation (D-86)					
IBP, °C	171	179	182	190	193
10% point, °C	209	213	206	203	206
20% point, °C	224	228	222	207	210
30% point, °C	236	240	232	209	213
40% point, °C	246	249	242	212	217
50% point, °C	254	258	253	214	221
60% point, °C	264	267	264	217	224
70% point, °C	274	276	276	221	228
80% point, °C	286	286	287	227	233
90% point, °C	302	299	303	238	239
95% point, °C	317	316	317	258	244
EBP, °C				293	261
recovery, %	98	97	97	99	99
residue, %	2	3	3	1	1
loss, %	0	0	0	0	0

DF-2, EM-242-F was premium quality DF-2, EM-400-F was a DF-1 emissions test fuel and EM-401-F was JP-7.

DESCRIPTION OF FACILITIES

Gaseous and particulate emissions were obtained with a large single dilution CVS utilizing a 45.75 in (1.162 m) inside diameter stainless steel tunnel with a 57 foot (17.37 m) length. This facility was designed to service both a stationary 500 hp (373 kW) dynamometer and the tandem axle chassis dynamometer. Constant volume tunnel flows from 1000 SCFM to 12,000 SCFM, in increments of 1000 SCFM, can be selected from multiple calibration of the four 3000 SCFM Roots type blowers.

The capacity and flexibility provided adequate dilution air to maintain the 125°F (51.7°C) sample zone temperature limit for very large truck engines tested with the transient cycle. The sampling zone contained probes for collection of gaseous samples by bag, continuous HC by HFID and NO_2 by CL, four individual 47 mm particulate samples, three independent 20 x 20 particulate samples and a single probe with sequential sampling to four 47 mm filter positions. Several views of the dilution tunnel and related sampling hardware are shown in Figure 1.

Prior to emission testing, a variety of experiments were conducted to confirm that satisfactory mixing was achieved. Propane checks were performed prior to testing for each engine.

The tandem axle chassis dynamometer has been described in several publications(15-17). Inertia weights were selected using EPA Technical Publications(9) as a guideline. Several views of the chassis dynamometer are illustrated in Figure 2.

TEST PLAN AND PROCEDURES

The chassis and stationary heavy-duty engine transient cycles were developed from the CAPE-21 survey data. Engines produced for sale beginning in the year 1985, will be required to meet emission standards using the stationary dynamometer version of the 1983 transient cycle. There are more requirements regarding engine load/speed traces, statistics, etc. for the engine dynamometer tests. The emission rates from the chassis tests are expressed in g/km whereas the stationary dynamometer engine test results are expressed in g/BHP-hr. The scope of this project did not provide for testing engines on a stationary dynamometer and as such no comparisons between engine and vehicle testing can be made.

The four heavy-duty engines were operated over the chassis version of the 1983 transient cycle. The 1983 heavy-duty transient cycle consists of a cold start followed by a 20-minute soak then a hot start. Other than the starts, the cold and hot start portions of the test are identical. The chassis version of the 1983 heavy-duty transient cycle consists of four individual segments described in Table 3. Although no statistical analysis of driver performance was required, cursory evaluation of the driver aids indicated satisfactory driver performance according to the chassis procedure. Tedlar bags were used to collect CVS samples for CO, CO_2 and NO_x. Although NO_x was measured in accordance with recommended practice(9), there is some concern over NO_x losses in CVS bags containing dilute diesel exhaust. These losses are typically from 5-10 percent depending on the absolute level of NO_x in the bag.

In order to meet the needs of this research effort, additional hot start tests were required. Each test consisted of a cold start followed by

Fig. 1A - Front view of vehicle on chassis dynamometer

Fig. 1B - Rear view of vehicle on chassis dynamometer

Fig. 1C - Particulate collection from large dilution tunnel

Fig. 1D - Close-up view of individual and modal 47 mm sampling systems

Fig. 1 - Several views of giant dilution tunnel and sampling hardware

Fig. 2 - Several views of chassis dynamometer

Table 3. 1983 Transient Cycle Individual Segment Description
Chassis Emission Testing of Heavy-Duty Diesel and Gasoline Engines

Segment	Time, sec	Distance, km (miles)	Average Speed km/hr (mph)
NYNF	254	0.83 (0.52)	11.76 (7.31)
LANF	285	1.91 (1.19)	27.07 (16.82)
LAF	267	5.33 (3.31)	75.54 (46.91)
NYNF	254	0.83 (0.52)	11.76 (7.31)
Composite	1060	8.90 (5.54)	30.23 (18.79)

three replicate hot starts with a 20-minute soak between tests. A minimum of two tests were conducted with each engine-fuel combination. A summary of the test plan for gaseous and particulate sample acquisition is shown in Table 4. Replicate hot starts were required to collect sufficient particulate on the 4 x 47 mm modal filters. Additional 20 x 20 filters were also collected during the replicate hot starts for sequential organic extractions. Sequential extractions of the diesel particulate were conducted using cyclohexane, methylene chloride and acetonitrile to determine the relative polarity distribution of the organic solubles. Cyclohexane will extract the non-polars, methylene chloride extracts the compounds of intermediate polarity while acetonitrile removes the polar organic materials in the diesel particulate.

Several gaseous unregulated emissions were measured in the CVS bags, including N_2O, HCN and individual hydrocarbons (IHC)(18-20). Bag storage stability experiments for the N_2O and IHC analysis were conducted during the qualification and validation of the procedures(18,19). No losses of N_2O or IHC were observed with bag sampling. Bag experiments conducted for HCN stability in highly dilute exhaust indicate losses less than 10 percent during typical analysis. Those compounds included in the IHC analysis were methane, ethylene, ethane, acetylene, propane, propylene, benzene and toluene. Filters were collected for BaP, sulfate and metals analysis. Although good engineering practices were employed in the collection of filters for BaP, sulfate and metals, no validation or qualification of the sampling techniques were attempted in this project. Minimum detection limits for HCN are 0.01 ppm and 0.1 ng for BaP.

RESULTS

This section describes gaseous, particulate and chemical characterization results obtained from four heavy-duty diesel engines operating on several fuels. Engine-to-engine, and fuel-to-fuel comparisons are made for those results of interest. It should be emphasized that the Caterpillar 3208, Cummins Formula 290 and Mack ENDT 676 were all 1979 models and were required to meet more stringent gaseous emission standards than the 1977 Detroit Diesel 8V-71 engine.

GASEOUS AND PARTICULATE - Regulated gaseous and particulate emission rates for all tests with each fuel are presented in Appendix A. These emission rates are summarized on Table 5. Figure 3 illustrates the NO_x emission rates for all engines and fuels in the test plan. The NO_x rates for the Caterpillar 3208 were nominally 11-12 g/km with the Mack ENDT 676 and Cummins Formula 290 ranging from 15-18 g/km. The city bus with the DD 8V-71 was slightly higher at 22-27 g/km. There were no obvious NO_x trends regarding fuel composition on any of the four engines tested.

The bar chart presented in Figure 4 illustrates a comparison of HC emission rates for the four engines tested. The HC emission rates were slightly higher with EM-241-F (minimum quality DF-2) for the three 4-stroke engines (Caterpillar 3208, Mack ENDT 676 and Cummins Formula 290). No fuel effects were observed with the 2-stroke DD 8V-71 engine for HC emissions.

Figure 5 is a bar chart comparison of the CO emission rates for the four engines evaluated. CO emissions from the Caterpillar 3208 ranged from 3-4 g/km while the Mack ENDT 676 was nominally 5-6 g/km and the Cummins Formula 290 ranged from 4-5 g/km. The 1977 DD 8V-71 engine produced consistently higher CO emission rates (43-45 g/km) than the 1979 4-stroke engines. No effect on CO emissions could be attributed to fuel composition for the three primary DF-2 test fuels (EM-329-F, EM-241-F and EM-242-F). However, the CO emissions were reduced 25-30 percent with EM-400-F (DF-1 emissions) and EM-401-F (JP-7).

Particulate emission rate comparisons are presented in Figure 6. The Caterpillar 3208 emitted the lowest particulate at about 0.6 g/km followed by the Cummins Formula 290 with 1 g/km and the Mack ENDT 676 with a nominal particulate rate of 1.2 g/km. The 2-stroke 1977 8V-71 had a particulate rate almost twice ($\approx$2 g/km) that of the other three engines for the three primary DF-2 test fuels. A substantial reduction in particulate was observed with tests conducted with the two narrow boiling range fuels (EM-400-F and EM-401-F). A 30-40 percent reduction in particulate was observed with the DF-1 emissions test fuel and the JP-7.

MODAL PARTICULATE - Individual filters were collected for each segment of the transient cycle. Modal filters were collected to determine the particulate contribution of each segment of the transient cycle. These results are summarized in Table 6. As illustrated in Figure 7, all engines followed the same particulate distribution pattern, i.e., LAF>LANF>NYNF(1)>NYNF(2). In general, the LAF segment of the transient cycle accounted for 40-55 percent of the particulate, the LANF accounted

Table 4. Test Plan for Gaseous and Particulate Sample Acquisition During Heavy-Duty Diesel Transient Cycle Operation

Exhaust Measurement	Cold Start				Hot Start				Hot Start (Replicate)			
	NYNF	LANF	LAF	NYNF	NYNF	LANF	LAF	NYNF	NYNF	LANF	LAF	NYNF
CO	bag	bag	bag	bag	bag	bag	bag	bag	NR	NR	NR	NR
CO_2	bag	bag	bag	bag	bag	bag	bag	bag	NR	NR	NR	NR
NO_x	bag	bag	bag	bag	bag	bag	bag	bag	NR	NR	NR	NR
NO_2	cont	cont	cont	cont	cont	cont	cont	cont	NR	NR	NR	NR
HC	cont	cont	cont	cont	cont	cont	cont	cont	NR	NR	NR	NR
N_2O	bag	bag	bag	bag	bag	bag	bag	bag	NR	NR	NR	NR
IHC	bag	bag	bag	bag	bag	bag	bag	bag	NR	NR	NR	NR
HCN	bag	bag	bag	bag	bag	bag	bag	bag	NR	NR	NR	NR
47 mm Pallflex gravimetric	one filter for cold start				one filter for hot start				NR	NR	NR	NR
47 mm Fluoropore metals	one filter for cold start				one filter for hot start				NR	NR	NR	NR
47 mm Fluoropore sulfate	one filter for cold start				one filter for hot start				NR	NR	NR	NR
47 mm Pallflex Position 1	Yes	NR	NR	NR	Yes	NR	NR	NR	Yes	NR	NR	NR
(modal) Position 2	NR	Yes	NR	NR	NR	Yes	NR	NR	NR	Yes	NR	NR
Position 3	NR	NR	Yes	NR	NR	NR	Yes	NR	NR	NR	Yes	NR
Position 4	NR	NR	NR	Yes	NR	NR	NR	Yes	NR	NR	NR	Yes
20x20 Pallflex Position 1	one filter for cold start				one filter for hot start				one filter for hot start			
20x20 Pallflex Position 2	one filter for cold start				one filter for hot start				one filter for hot start			
20x20 Pallflex Position 3	one filter for cold start				one filter for hot start				one filter for hot start			

NOTE 1: 20x20 Pallflex filters will be used for BaP (@EPA-RTP), and sequential organic extractions
NOTE 2: Additional hot start tests may be required for sufficient loading of 47 mm Pallflex for model analysis and traps
NOTE 3: NR means not required

Table 5. Summary of Gaseous and Particulate Emission Rates of Four Heavy-Duty Diesel Engines Operating on 1983 Chassis Transient Cycle with Several Test Fuels

Vehicle	Fuel	Emission Rates, g/km			
		HC	CO	NO_x	Particulate
Caterpillar 3208	EM-329-F	1.18	3.16	11.98	0.56
(1979)	EM-241-F	1.50	4.31	11.82	0.59
	EM-242-F	1.02	2.97	10.86	0.55
Mack ENDT 676	EM-329-F	0.89	5.88	18.48	1.21
(1979)	EM-241-F	1.16	5.37	18.12	1.23
	EM-242-F	0.94	5.62	17.57	1.18
Cummins 290	EM-329-F	1.44	4.99	16.99	1.00
(1979)	EM-241-F	1.64	4.90	15.18	1.10
	EM-242-F	1.34	4.24	15.17	0.96
DD 8V-71	EM-329-F	1.32	46.66	22.07	2.07
(1977)	EM-241-F	1.37	46.21	24.29	1.94
	EM-242-F	1.41	43.57	25.00	1.96
	EM-400-F	1.45	33.17	26.35	1.39
	EM-401-F	1.46	30.55	23.74	1.16

for 20-26 percent of the particulate and the NYNF produced 10-20 percent.

A comparison of the modal and particulate emission rates are presented in Table 7. The modal filters were collected over 1 cold and 3 hot start transient cycles. The sum of the individual modal particulate was within 5 percent of the composite particulate for all engine-fuel combustions. This implies that there were little or no tunnel or sampling losses during the collection of modal filters.

SEQUENTIAL EXTRACTIONS - A three solvent sequential extraction was conducted using cyclohexane, methylene chloride, and acetonitrile. The results of these extractions are summarized in Table 8. In all cases, the majority of the organic extractables were removed by cyclohexane. The cyclohexane extract, containing non-polar compounds, accounted for 60-80 percent of the organics removed from the particulate using the sequential extraction solvent system. With the three priamry DF-2 fuels, the Mack ENDT 676 had the lowest extractables (13-16 percent), followed by DD 8V-71 at 20 percent and Cummins Formula 290 at 25-26 percent. The Caterpillar 3208 had the highest total organic solubles at 56-63 percent. A noticeable increase in the total organic solubles of the DD 8V-71 was observed with EM-400-F and EM-401-F.

Additional chemical characterization of the extracts from the various sequential solvent extractions are in progress and will be available at a later date(21).

UNREGULATED GASEOUS EMISSIONS - Several unregulated gaseous emissions were measured with each engine/fuel combination, including HCN, N_2O and individual hydrocarbons. Of the individual hydrocarbons measured (CH_4, C_2H_4, C_2H_6, C_2H_2, C_3H_6, C_3H_2, C_6H_6, and C_7H_8) only C_2H_4 was consistently produced by each engine-

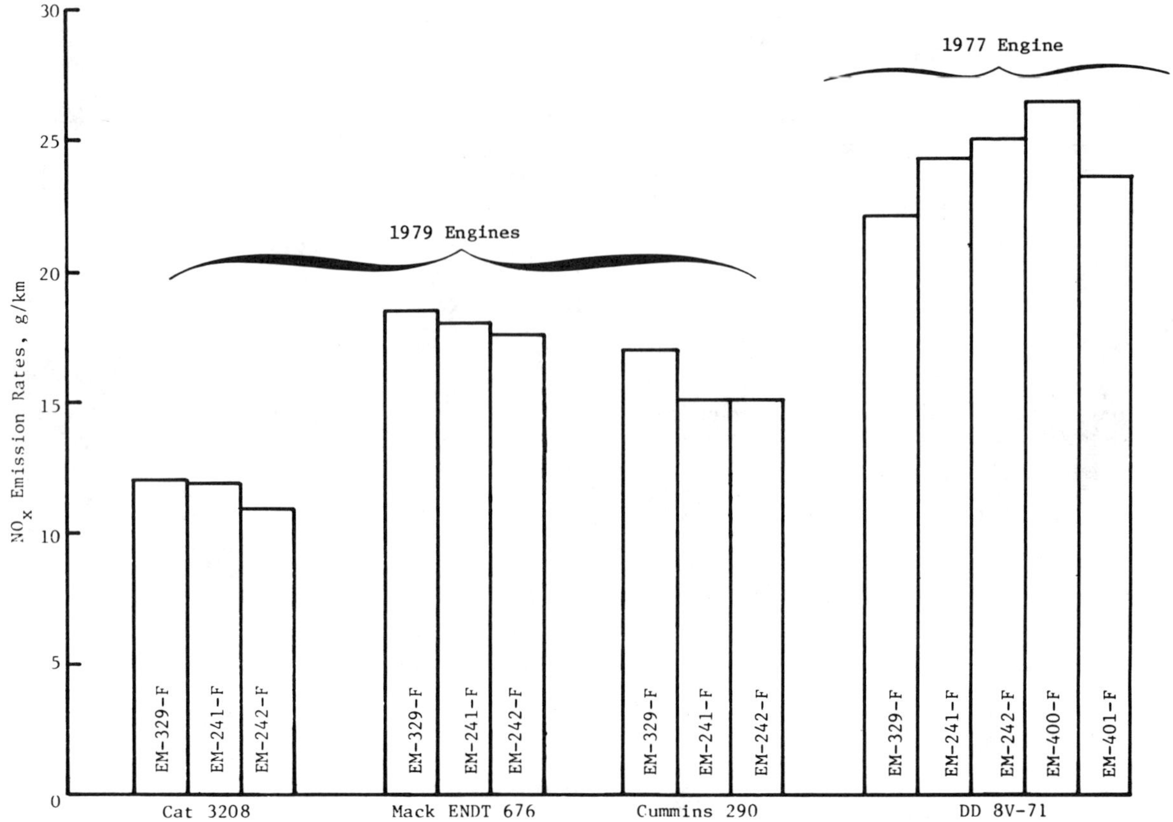

Fig. 3 - NO_x emission rates from four HD diesel engines operated on the 1983 transient cycle (chassis) with several fuels

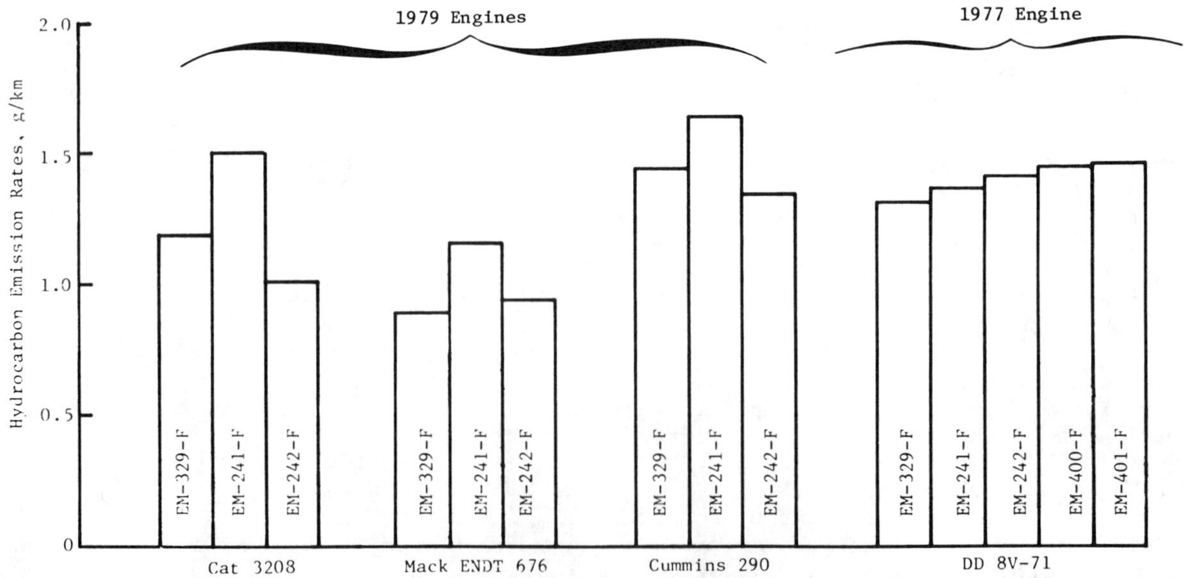

Fig. 4 - Hydrocarbon emission rates from four HD diesel engines operated on the 1983 transient cycle (chassis) with several fuels

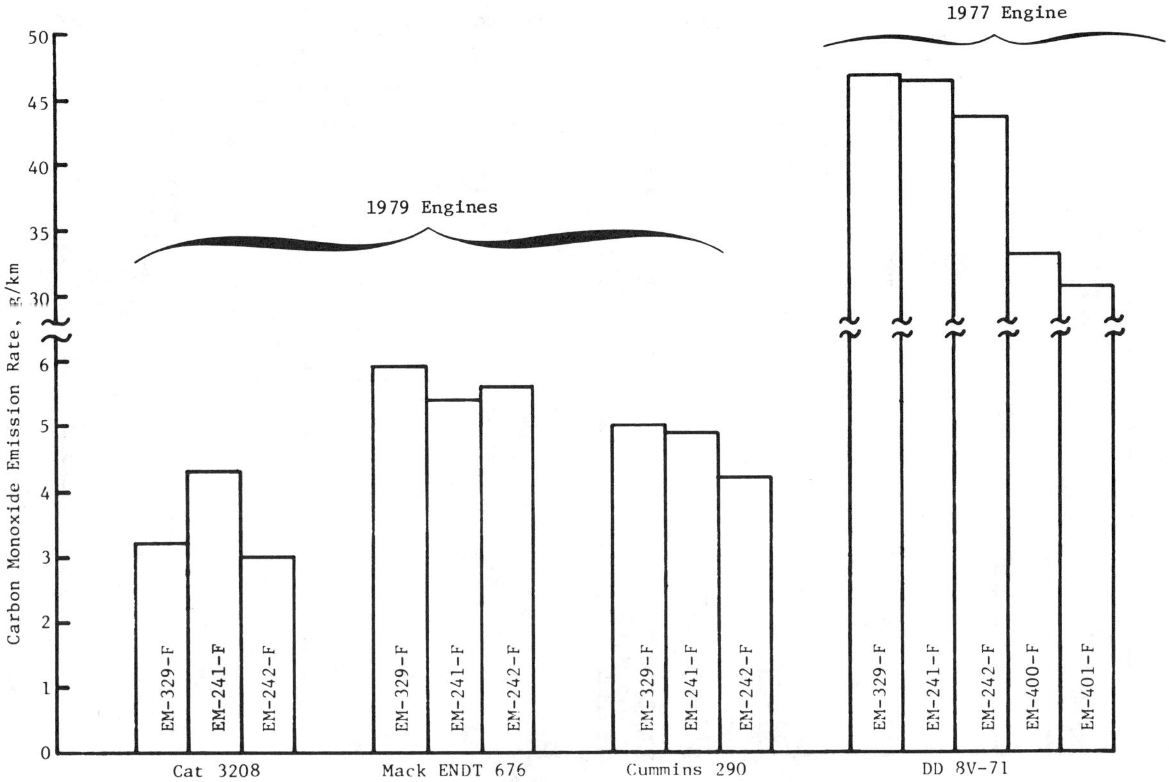

Fig. 5 - Carbon monoxide rates from four HD diesel engines operated on the 1983 transient cycle (chassis) with several fuels

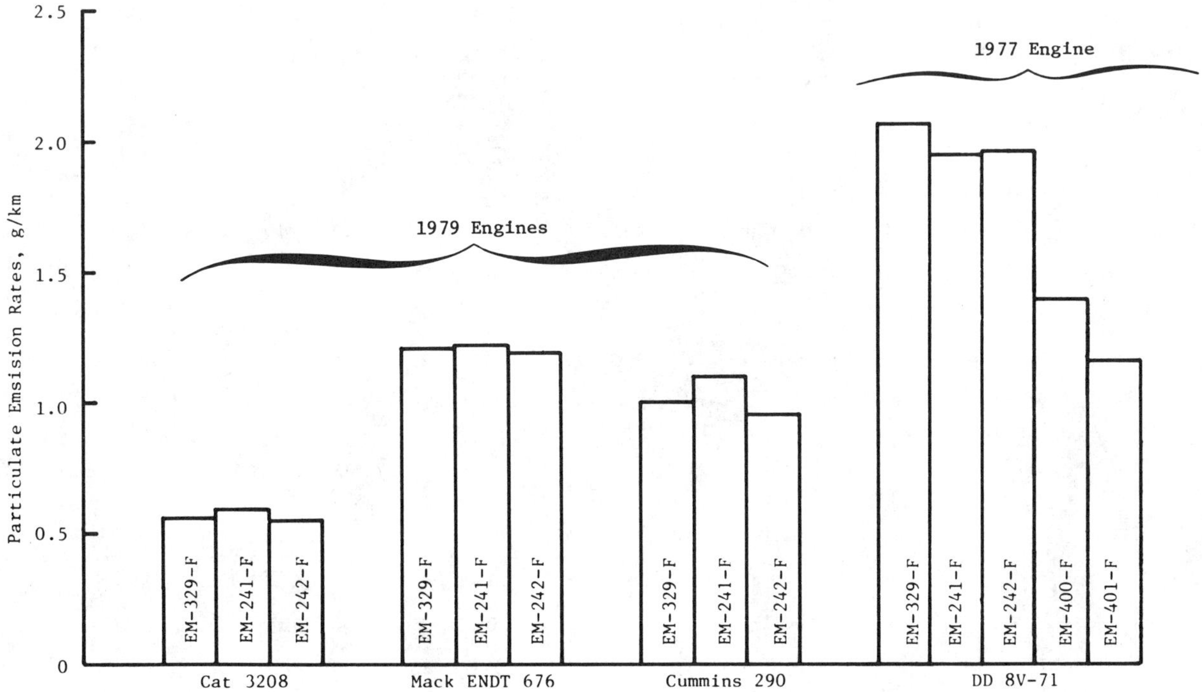

Fig. 6 - Particulate emission rates from four HD diesel engines operated on the 1983 transient cycle (chassis) with several fuels

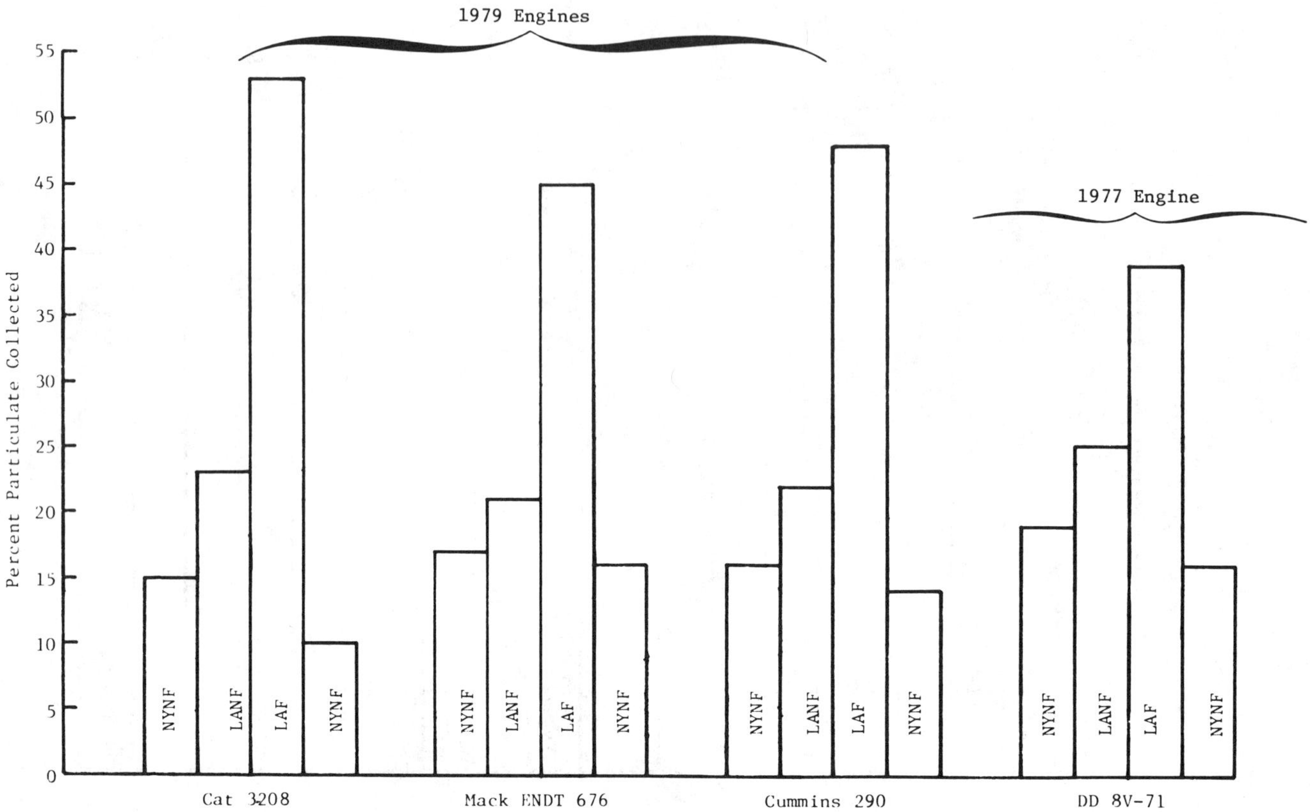

Fig. 7 - Percent particulate collected during each segment of transient cycle

Table 6. Percent Particulate Collected During Each Segment of Transient Cycle

Vehicle	Fuel	Percent Particulate Collected During Cycle NYNF	LANF	LAF	NYNF
Caterpillar 3208	EM-329-F	14	21	55	10
Caterpillar 3208	EM-241-F	16	24	52	11
Caterpillar 3208	EM-242-F	14	23	53	10
Mack ENDT 676	EM-329-F	18	21	45	16
Mack ENDT 676	EM-241-F	17	20	47	15
Mack ENDT 676	EM-242-F	17	21	44	18
Cummins Formula 290	EM-329-F	16	23	47	15
Cummins Formula 290	EM-241-F	15	23	48	13
Cummins Formula 290	EM-242-F	16	21	50	13
Detroit Diesel 8V-71	EM-329-F	20	26	38	17
Detroit Diesel 8V-71	EM-241-F	20	25	39	17
Detroit Diesel 8V-71	EM-242-F	17	26	39	18
Detroit Diesel 8V-71	EM-400-F	19	24	41	16
Detroit Diesel 8V-71	EM-401-F	19	26	40	14

Table 7. Comparison of Modal and Composite Particulate Emission Rates (g/test segment or g/test)

Vehicle	Fuel	Particulate Rate, g						ΣModal÷ Composite
		NYNF	LANF	LAF	NYNF	ΣModal	Composite	
Caterpillar 3208 (1979)	EM-329-F	0.65	0.95	2.46	0.45	4.50	4.44	1.01
	EM-241-F	0.74	1.13	2.43	0.52	4.82	4.57	1.05
	EM-242-F	0.64	1.03	2.38	0.44	4.48	4.32	1.04
Mack ENDT 676 (1979)	EM-329-F	1.85	2.22	4.73	1.69	10.48	11.03	0.95
	EM-241-F	1.81	2.14	4.76	1.59	10.51	11.10	0.95
	EM-242-F	1.79	2.23	4.75	1.19	10.68	10.73	1.00
Cummins 290 (1979)	EM-329-F	1.35	1.99	4.01	1.26	8.62	9.11	0.95
	EM-241-F	1.50	2.24	4.76	1.32	9.82	10.17	0.97
	EM-242-F	1.49	1.93	4.51	1.19	9.11	9.38	0.97
DD 8V-71 (1977)	EM-329-F	3.65	4.75	6.93	3.11	18.43	18.60	0.99
	EM-241-F	3.56	4.60	7.03	3.03	18.21	17.85	1.02
	EM-242-F	2.80	4.36	6.42	2.94	16.59	17.13	0.97
	EM-400-F	2.50	3.13	5.28	2.02	12.93	12.85	1.01
	EM-401-F	2.01	2.79	4.24	1.50	10.54	10.64	0.99

Table 8. Sequential Extraction with Cyclohexane, Methylene Chloride, and Acetonitrile for Four Diesel Engines with Several Fuels

Engine	Fuel	Sequential Extraction Solvents[b]			
		Cyclohexane[a]	Methylene Chloride[a]	Aceto-nitrile[a]	Total
Caterpillar 3208 (1979)	EM-329-F	50.39%	3.72%	9.27%	63.38%
	EM-241-F	49.28	4.64	8.46	62.38
	EM-242-F	50.58	3.75	2.02	56.35
Mack ENDT 676 (1979)	EM-329-F	7.31	2.14	3.61	13.06
	EM-241-F	9.53	1.78	4.56	15.87
	EM-242-F	7.93	1.96	3.08	12.97
Cummins Formula 290 (1979)	EM-329-F	17.63	3.22	5.42	26.27
	EM-241-F	19.55	4.40	2.92	26.87
	EM-242-F	19.42	2.72	3.51	25.65
Detroit Diesel 8V-71 (1977)	EM-329-F	13.85	4.04	2.48	20.37
	EM-241-F	15.68	2.77	2.46	20.91
	EM-242-F	12.65	4.42	2.74	19.81
	EM-400-F	20.89	3.74	3.29	27.92
	EM-401-F	25.79	3.09	2.87	31.75

[a] Order of sequential extraction: cyclohexane, methylene chloride, acetonitrile

[b] Cyclohexane extracts non-polar compounds, methylene chloride extracts intermediate polarity compounds and acetonitrile extracts the polar compounds in diesel particulate

fuel combination. Methane was generally present, however, after correcting for background, the CH_4 emission rates were generally zero or negligible. The emission rates of HCN, N_2O and C_2H_4 are illustrated in Table 9. HCN emission rates were higher with the 2-stroke DD 8V-71 than the three 4-stroke engines, almost by a 2:1 ratio. The N_2O emission rates of the DD 8V-71 were slightly higher than the Caterpillar 3208, Mack ENDT 676 and Cummins Formula 290. The C_2H_4 emission rates were lowest with the Mack ENDT 676 with the three other engines slightly higher. No trends were apparent that could be attributed to fuel composition for HCN, N_2O, or C_2H_4.

METAL EMISSION RATES

Fluoropore filters were collected during the cold and hot portions of the transient cycle for metal analysis. Filters were shipped to EPA-RTP ESRL for metals analysis using a Siemins x-ray fluorescence analyzer. The results for the four heavy-duty diesel engines with several test fuels are summarized in Table 10. The primary element detected with the three 4-stroke engines was sulfur which accounted for 75-95 percent of all those elements measured (excluding iron). The majority of the sulfur was determined to be in the form of sulfate.

Metal emissions from the 2-stroke 8V-71 were nominally 2-4 times greater than the three 4-stroke engines. This was primarily due to the increased oil consumption as evidence by the appearance of significant quantities of calcium and phosphorus. In this case, the sulfur accounted for 14-16 percent of the total metals measured, except for the low sulfur fuel EM-401-F. Calcium and phosphorus emission rates were consistent, independent of the fuel tested, and accounted for nominally 49 and 23 percent, respectively of the metal emission rates.

Figure 8 illustrates the relative contribution of the total metal emission rates to the particulate rate of the engine. The metals in the particulate of the DD 8V-71 accounted for about 8 percent of the weight, with metal emissions from 4-stroke engines ranging from 3.1 to 4.6 percent of total particulate.

NO_2 MODAL CONCENTRATIONS

The primary concern of NO_2 is the possibility of artifact formation on the particulate during collection. Preliminary experiments conducted by EPA-RTP indicate that NO_2 concentrations below 10 ppm did not produce artification formation(22). With the increasing emphasis on Ames bioassay analysis, NO_2 concentrations were monitored continuously during the complete transient cycle. The NO_2 concentrations for each segment of the cold and hot starts of the transient cycle were then computed and are summarized in Table 11. The LAF segment of the 1983 transient cycle produces the greatest NO_2 concentrations with the three 4-stroke engines. These NO_2 values were generally more than twice those of other segments of the transient cycle. The 2-stroke DD 8V-71 bus engine produced NO_2 which did not appear to be cycle dependent. In all cases, the NO_2 concentrations in the dilute exhaust were 5 ppm or less with no apparent fuel effects.

Other emissions data is presented on a mass basis (i.e., g/km) whereas the NO_2 is expressed in terms of ppm concentration. Engine-to-engine comparisons are not valid since different dilutions were required to keep gas temperatures at the filter face below 125°F.

BENZO(a)PYRENE - BaP emission rates for all four heavy-duty diesel engines are listed in Table 12. For three of the four engines (Mack ENDT 676, Cummins Formula 290, and DD 8V-71), the BaP emission rates were substantially greater with the minimum quality DF-2 fuel than the two other primary test fuels. The Cummins Formula 290 produced much greater quantities of BaP than any of the other engines with all three fuels. The premium quality DF-2 (EM-242-F), DF-1 emission test fuel (EM-400-F)

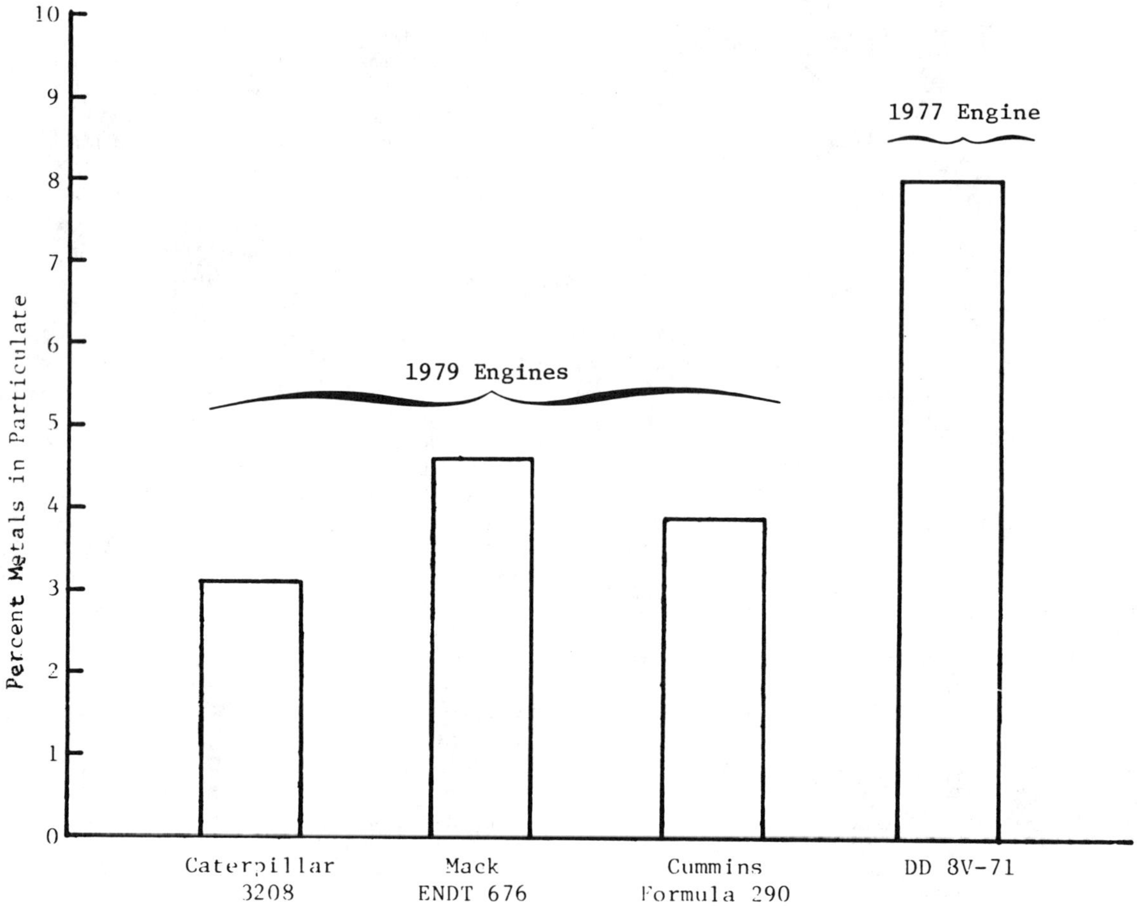

Fig. 8 - Percent metals in particulate of four heavy-duty diesel engines

Table 9. Summary of HCN, N_2O, and C_2H_4 Emission Rates

Engine	Fuel	Emission Rates, mg/km HCN	N_2O	C_2H_4
Caterpillar 3208	EM-329-F	14	22	66
(1979)	EM-241-F	12	21	91
	EM-242-F	13	14	70
Mack ENDT 676	EM-329-F	23	35	56
(1979)	EM-241-F	19	36	48
	EM-242-F	19	31	52
Cummins 290	EM-329-F	17	22	83
(1979)	EM-241-F	21	23	76
	EM-242-F	17	29	66
DD 8V-71	EM-329-F	40	43	96
(1977)	EM-241-F	46	48	89
	EM-242-F	44	47	84
	EM-400-F	44	53	93
	EM-401-F	32	42	95

Table 10. Metal Emission Rates for Four Heavy Duty Diesel Engines Operating on Several Fuels

Engine	Fuel	Metal Emission Rates, mg/km									
		Ca	Cd	Cl	Fe	K	P	S	Si	Zn	Total[b]
Caterpillar 3208 (1979)	EM-329-F	1.10	ND[a]	ND	ND	0.21	0.45	14.16	ND	ND	15.91
	EM-241-F	1.66	ND	0.05	ND	0.24	0.75	15.52	1.66	ND	20.54
	EM-242-F	0.79	0.66	0.04	ND	0.27	0.57	13.31	ND	ND	16.03
Mack ENDT 676 (1979)	EM-329-F	0.28	ND	0.14	8.38	0.46	ND	24.83	0.12	0.60	62.93
	EM-241-F	0.18	ND	0.03	20.41	0.41	0.56	30.29	0.12	3.25	60.48
	EM-242-F	0.13	ND	ND	11.71	0.37	ND	28.25	0.91	0.40	42.079
Cummins Formula 290 (1979)	EM-329-F	3.72	ND	0.90	12.08	0.61	3.19	24.39	1.58	ND	48.25
	EM-241-F	1.80	ND	0.47	3.84	0.55	4.22	25.51	ND	0.43	36.83
	EM-242-F	2.24	ND	0.35	ND	0.59	4.40	23.75	ND	ND	34.79
DD 8V-71 (1977)	EM-329-F	64.40	0.10	0.54	1.04	0.54	29.33	20.14	1.30	5.73	140.32
	EM-241-F	68.10	1.51	0.41	ND	0.33	31.60	23.35	0.98	5.07	141.08
	EM-242-F	63.25	5.58	0.70	ND	0.41	29.15	17.69	1.40	5.46	142.67
	EM-400-F	63.24	1.50	0.70	ND	0.51	28.35	17.05	1.36	ND	124.75
	EM-401-F	66.60	0.08	0.97	ND	1.19	30.91	4.71	ND	0.67	114.04

[a] ND donates not detected

[b] Total metal particulate emission rates may not agree with sum of individual metal emission rates listed on table since only primary metals were included. Occasional appearances of Br, Ba, Mg, Na, Pb, Sb, and Ti were not included in this summary.

Table 11. NO_2 Modal Concentrations for Four Heavy Duty Diesel with Various Fuels

Vehicle	Fuel	NO_2 Modal Concentrations, ppm									
		Cold Start					Hot Start				
		NYNF	LANF	LAF	NYNF	CFTP	NYNF	LANF	LAF	NYNF	HFTP
Caterpillar 3208 (1979)	EM-329-F	2.47	2.48	5.34	1.35	2.92	1.22	1.55	3.19	0.45	1.62
	EM-241-F	2.50	2.78	5.02	0.62	2.75	1.13	1.82	3.88	0.34	1.81
	EM-242-F	1.56	1.85	4.86	0.60	2.23	0.60	1.06	3.32	0.28	1.30
Mack ENDT 676 (1979)	EM-329-F	0.40	0.74	2.83	0.58	1.15	0.20	0.56	2.64	0.51	0.98
	EM-241-F	0.51	0.81	4.37	0.34	1.52	0.23	0.61	--	--	--
	EM-242-F	--	--	--	--	--	0.28	0.78	3.40	0.34	1.21
Cummins Formula 290 (1979)	EM-329-F	0.50	0.62	2.59	0.17	0.98	0.34	0.40	1.62	0.17	0.64
	EM-241-F	0.48	0.74	1.86	0.28	0.85	0.37	0.74	2.03	0.31	0.87
	EM-242-F	0.51	0.81	2.75	0.37	1.12	0.43	0.79	2.19	0.46	0.97
DD 8V-71 (1977)	EM-329-F	2.05	1.99	2.37	2.28	2.17	2.21	2.25	1.69	2.59	2.18
	EM-241-F	2.13	2.40	2.51	2.55	2.40	2.13	2.34	2.70	2.52	2.42
	EM-242-F	2.46	2.54	2.39	2.68	2.52	2.89	2.88	3.11	3.00	2.97
	EM-400-F	2.88	2.91	3.13	3.06	2.99	2.92	3.01	3.44	3.07	3.11
	EM-401-F	2.89	2.86	3.29	3.06	3.02	2.75	2.73	3.29	2.84	2.90

Table 12. Benzo(a)Pyrene Emission Rates for Four Heavy Duty Diesel Engines Operating on Several Fuels

Fuel	BaP Emission Rate, μg/km			
	Cat 3208	Mack ENDT 676	Cummins Formula 290	Detroit Diesel 8V-71
EM-329-F	1.004	0.573	3.292	0.826
EM-241-F	0.889	1.286	6.203	1.139
EM-242-F	0.510	0.280	2.864	0.443
EM-400-F	--	--	--	0.563
EM-401-F	--	--	--	0.433

and JP-7 (EM-401-F) produced the least amount of BaP for all engines.

MEASURED VERSUS CALCULATED FUEL CONSUMPTION - A summary of the calculated and measured fuel consumed is presented in Table 13. A bar chart comparison of calculated divided by measured fuel consumed for four test engines is shown in Figure 9. As observed, the calculated fuel consumption for the Caterpillar 3208 was ≈8 percent high, while the Mack ENDT had calculated and measured fuel consumption essentially equivalent. The Cummins Formula 290 and DD 8V-71 calculated fuel consumption was 2-3 percent lower than the measured fuel consumed.

SUMMARY

A brief summary of regulated gaseous, particulate and pertinent chemical characterization data is presented below:

- HC emission rates were higher with the minimum quality DF-2 with the 4-stroke engines.
- Particulate and CO emissions were 30-40 percent less with DF-1 and JP-7 than the three DF-2 fuels with the DD 8V-71.
- NO_x emissions were unaffected by fuel consumption.
- The majority (40-55 percent) of the composite particulate was produced during the LAF segment of the transient cycle.
- Excellent agreement was obtained between the sum of the individual modal and composite particulate filters.
- HCN and N_2O emissions were highest with the DD 8V-71.
- Metal emissions accounted for 8 percent of the particulate emissions for the DD 8V-71, 3-5 percent for the other engines.
- NO_2 concentrations were 5 ppm or less during the transient cycle, with the LAF having more than twice the NO_2 concentrations of the other segments of the transient cycle (4-stroke only).
- BaP emissions were higher with the minimum quality DF-2 for three of the four engines tested. The Cummins Formula 290 had BaP emissions noticeably higher than the other three engines.
- Total organic extractables using C_6H_{12}, CH_2Cl_2, and CH_3CN were highest (~60 percent) with the Caterpillar 3208 and lowest with the Mack ENDT 676 (~15 percent).
- Calculated and measured fuel consumption agreed.

In conclusion, this paper has presented gaseous and particulate emission rates from four heavy-duty diesel engines operating over the 1983 transient cycle with several fuels. A giant tunnel was employed in conjunction with a variable dilution CVS system to allow total exhaust dilution.

ACKNOWLEDGEMENT

This paper is based on work performed by Southwest Research Institute under EPA Contract 68-02-2993. The authors wish to thank Dr. Robert Jungers and R.E. Bumgarner of EPA EMSL for conducting the benzo(a)pyrene analysis and Robert Kellogg of EPA EMSL for conducting the metals analysis. In addition, the technical assistance of Charles M. Urban and Terry Ullman of SwRI is gratefully acknowledged.

Table 13. Summary of Calculated and Measured Fuel Consumption for All Test Engines

1979 Caterpillar 3208	Fuel Consumed, ℓ/100 km	
Test	Calc.	Measured
111	30.96	--
112	30.78	--
---	--	--
Avg	30.87	--
121	30.44	--
122	29.99	--
123	29.71	27.84
Avg	30.05	27.84
131	31.17	28.99
132	31.26	29.02
133	30.10	28.58
Avg	30.84	28.86

1979 Mack ENDT 676	Fuel Consumed, ℓ/100 km	
Test	Calc.	Measured
211	46.82	47.33
212	48.19	47.93
213	48.01	47.78
Avg	47.67	47.68
221	48.05	48.01
222	45.56	47.45
223	45.92	46.30
Avg	46.51	47.25
231	49.41	47.91
232	46.13	45.94
233	48.37	47.54
Avg	47.97	47.13

1979 Cummins Formula 290	Fuel Consumed, ℓ/100 km	
Test	Calc.	Measured
311	51.31	51.69
312	51.20	51.60
313	48.96	51.44
Avg	50.49	51.58
321	48.18	50.26
322	48.49	50.24
323	49.53	50.15
324	49.84	50.52
Avg	49.01	50.29
331	51.05	50.67
332	50.54	51.08
333	51.52	51.26
Avg	51.04	51.00

1977 Detroit Diesel 8V-71	Fuel Consumed, ℓ/100 km	
Test	Calc.	Measured
414	60.96	63.85
415	60.43	63.67
---	--	--
Avg	60.69	63.76
421	60.26	63.21
422	62.10	63.24
423	60.20	63.15
Avg	60.85	63.20
431	62.90	64.12
432	63.00	64.94
433	62.12	63.29
Avg	62.67	64.12
441	59.04	63.32
442	62.66	63.50
443	60.11	62.85
444	62.35	64.35
Avg	61.04	63.51
451	60.91	64.33
452	61.54	62.84
453	60.75	62.11
454	60.97	62.05
Avg	61.04	62.83

Appendix A. Regulated and Particulate Emission Rate Summary for Caterpillar 3208 and Mack ENDT 676 with Three Test Fuels

			Emission Rate, g/km											
			HC			CO			NO_x			Particulate		
Test[a]	Fuel	Date	Cold	Hot	Total	Cold	Hot	Total	Cold	Hot	Total	Cold	Hot	Total
111	EM-329-F	2-21-80	1.32	1.16	1.18	4.25	3.07	3.24	13.85	13.16	13.26	0.62	0.54	0.55
112	EM-329-F	2-22-80	1.45	1.14	1.18	4.08	2.91	3.08	11.44	10.79	10.88	0.59	0.57	0.57
		Avg.	1.39	1.15	1.18	4.17	2.99	3.16	12.65	11.98	11.98	0.61	0.56	0.56
121	EM-241-F	2-26-80	1.95	1.48	1.55	5.93	3.75	4.06	12.98	12.13	12.25	0.67	0.54	0.56
122	EM-241-F	2-27-80	1.80	1.42	1.47	6.77	4.17	4.54	12.00	11.72	11.76	0.65	0.60	0.61
123	EM-241-F	2-28-80	1.95	1.40	1.48	6.34	4.00	4.33	11.91	11.37	11.45	0.68	0.59	0.60
		Avg.	1.90	1.43	1.50	6.35	3.97	4.31	12.30	11.74	11.82	0.67	0.58	0.59
131	EM-242-F	3-3-80	1.15	0.98	1.00	3.98	2.76	2.93	11.83	11.14	11.24	0.57	0.56	0.56
132	EM-242-F	3-4-80	1.17	1.06	1.08	4.14	2.96	3.13	10.87	10.37	10.44	0.53	0.55	0.55
133	EM-242-F	3-5-80	1.08	0.97	0.99	3.83	2.68	2.84	11.47	10.81	10.91	0.54	0.54	0.54
		Avg.	1.13	1.00	1.02	3.98	2.80	2.97	11.39	10.77	10.86	0.55	0.55	0.55
211	EM-329-F	3-11-80	--	--	--	--	--	--	--	--	--	--	--	--
212	EM-329-F	3-12-80	0.86	0.81	0.82	6.41	6.03	6.08	19.59	18.28	18.47	1.22	1.27	1.26
213	EM-329-F	3-13-80	0.98	0.96	0.96	6.83	5.49	5.68	19.78	18.26	18.48	1.21	1.15	1.16
		Avg.	0.92	0.89	0.89	6.62	5.76	5.88	19.69	18.27	18.48	1.22	1.21	1.21
221	EM-241-F	3-14-80	0.88	1.00	0.98	6.97	5.54	5.75	20.93	18.86	19.16	1.23	1.34	1.32
222	EM-241-F	3-17-80	0.98	1.48	1.41	7.15	4.73	5.08	21.52	16.61	17.31	1.31	1.22	1.23
223	EM-241-F	3-18-80	1.03	1.09	1.08	6.65	5.04	5.27	20.20	17.15	17.90	1.10	1.15	1.14
		Avg.	0.96	1.19	1.16	6.92	5.10	5.37	20.88	17.66	18.12	1.21	1.24	1.23
231	EM-242-F	3-19-80	1.08	0.95	0.97	6.46	5.83	5.92	18.42	16.64	16.90	1.30	1.17	1.19
232	EM-242-F	3-20-80	1.21	0.95	0.99	6.19	5.37	5.49	19.33	17.89	18.10	1.23	1.12	1.14
233	EM-242-F	3-21-80	0.75	0.89	0.87	5.81	5.40	5.45	17.35	17.76	17.70	1.15	1.21	1.20
		Avg.	1.01	0.93	0.94	6.15	5.53	5.62	18.37	17.43	17.57	1.23	1.17	1.18

[a] 100 series designates Caterpillar 3208, 200 series designates Mack ENDT 676

Appendix A (Cont'd). Regulated and Particulate Emission Rate Summary for Cummins Formula 290 and Detroit Diesel 8V-71 with Three Test Fuels

			Emission Rate, g/km											
			HC			CO			NO_x			Particulate		
Test[a]	Fuel	Date	Cold	Hot	Total	Cold	Hot	Total	Cold	Hot	Total	Cold	Hot	Total
311	EM-329-F	4-2-80	1.53	1.52	1.52	5.48	5.35	5.37	15.17	17.01	16.75	1.12	1.02	1.03
312	EM-329-F	4-3-80	1.71	1.40	1.44	4.58	5.23	5.14	15.18	15.95	15.84	1.16	0.97	1.00
313	EM-329-F	4-4-80	1.51	1.35	1.37	4.52	4.46	4.47	17.87	18.46	18.38	1.07	0.95	0.97
		Avg.	1.58	1.42	1.44	4.86	5.01	4.99	16.07	17.14	16.99	1.12	0.98	1.00
321	EM-241-F	4-14-80	1.86	1.57	1.61	5.17	5.02	5.04	14.48	14.88	14.82	1.37	1.01	1.06
322	EM-241-F	4-15-80	2.07	1.69	1.74	5.09	4.77	4.81	15.44	15.53	15.52	1.40	1.09	1.13
323	EM-241-F	4-16-80	1.98	1.58	1.64	5.30	4.90	4.96	13.28	13.95	13.86	1.35	1.08	1.12
324	EM-241-F	4-17-80	1.77	1.55	1.58	4.98	4.89	4.90	15.89	16.62	16.52	1.25	1.04	1.07
		Avg.	1.92	1.60	1.64	5.14	4.90	4.90	14.77	15.25	15.18	1.34	1.06	1.10
331	EM-242-F	4-21-80	1.41	1.26	1.28	4.52	4.12	4.18	15.46	15.39	15.40	1.07	0.90	0.92
332	EM-242-F	4-22-80	1.44	1.39	1.40	3.96	4.35	4.29	15.09	14.91	14.94	1.15	0.95	0.98
333	EM-242-F	4-23-80	--	--	--	--	--	--	--	--	--	--	--	--
		Avg.	1.43	1.33	1.34	4.24	4.24	4.24	15.28	15.15	15.17	1.11	0.93	0.96
414	EM-329-F	5-15-80	1.43	1.29	1.31	56.71	46.73	48.16	22.71	20.37	20.71	2.44	2.06	2.11
415	EM-329-F	5-16-80	1.12	1.37	1.33	50.57	44.25	45.16	25.29	23.11	23.43	2.54	1.95	2.03
		Avg.	1.28	1.33	1.32	53.64	45.49	46.66	24.00	21.74	22.07	2.49	2.01	2.07
421	EM-241-F	5-19-80	1.30	1.38	1.37	52.29	44.40	45.53	25.39	22.88	23.24	2.32	1.88	1.94
422	EM-241-F	5-20-80	1.40	1.28	1.30	52.46	44.76	45.87	26.92	25.01	25.29	2.29	1.86	1.92
423	EM-241-F	5-22-80	1.44	1.43	1.43	56.33	45.72	47.24	25.73	24.10	24.34	2.25	1.92	1.97
		Avg.	1.38	1.36	1.37	53.69	44.96	46.21	26.01	24.00	24.29	2.29	1.89	1.94
431	EM-242-F	5-23-80	1.27	1.32	1.44	51.78	43.71	44.87	25.04	23.43	23.66	2.23	1.89	1.94
432	EM-242-F	5-26-80	1.37	1.24	1.26	52.89	43.76	45.07	30.58	27.30	22.77	2.27	1.81	1.88
433	EM-242-F	5-27-80	1.33	1.58	1.54	45.17	40.04	40.78	25.30	23.26	23.56	2.06	2.07	2.07
		Avg.	1.32	1.38	1.41	49.95	42.50	43.57	26.97	24.66	25.00	2.19	1.92	1.96
441	EM-400-F	5-28-80	1.47	1.52	1.51	39.87	33.22	34.17	28.28	26.75	26.97	1.72	1.33	1.39
442	EM-400-F	5-29-80	1.49	1.26	1.29	39.06	33.83	34.58	28.75	26.60	26.91	1.76	1.38	1.43
443	EM-400-F	5-30-80	1.57	1.54	1.54	40.25	29.18	30.76	26.56	24.94	25.17	1.71	1.30	1.36
		Avg.	1.51	1.44	1.45	39.73	32.08	33.17	27.86	26.10	26.35	1.73	1.34	1.39
451	EM-401-F	6-10-80	1.41	1.52	1.50	38.37	31.64	32.60	24.89	23.66	23.84	1.24	1.05	1.08
452	EM-401-F	6-11-80	1.54	1.35	1.38	35.92	28.56	29.61	25.11	23.53	23.75	1.52	1.16	1.21
453	EM-401-F	6-12-80	1.62	1.49	1.51	35.87	28.37	29.44	25.43	23.25	23.56	1.47	1.14	1.19
		Avg.	1.52	1.45	1.46	36.72	29.52	30.55	25.14	23.48	23.72	1.41	1.12	1.16

[a] 300 series designates Cummins Formula 290, 400 series designates Detroit Diesel 8V-71

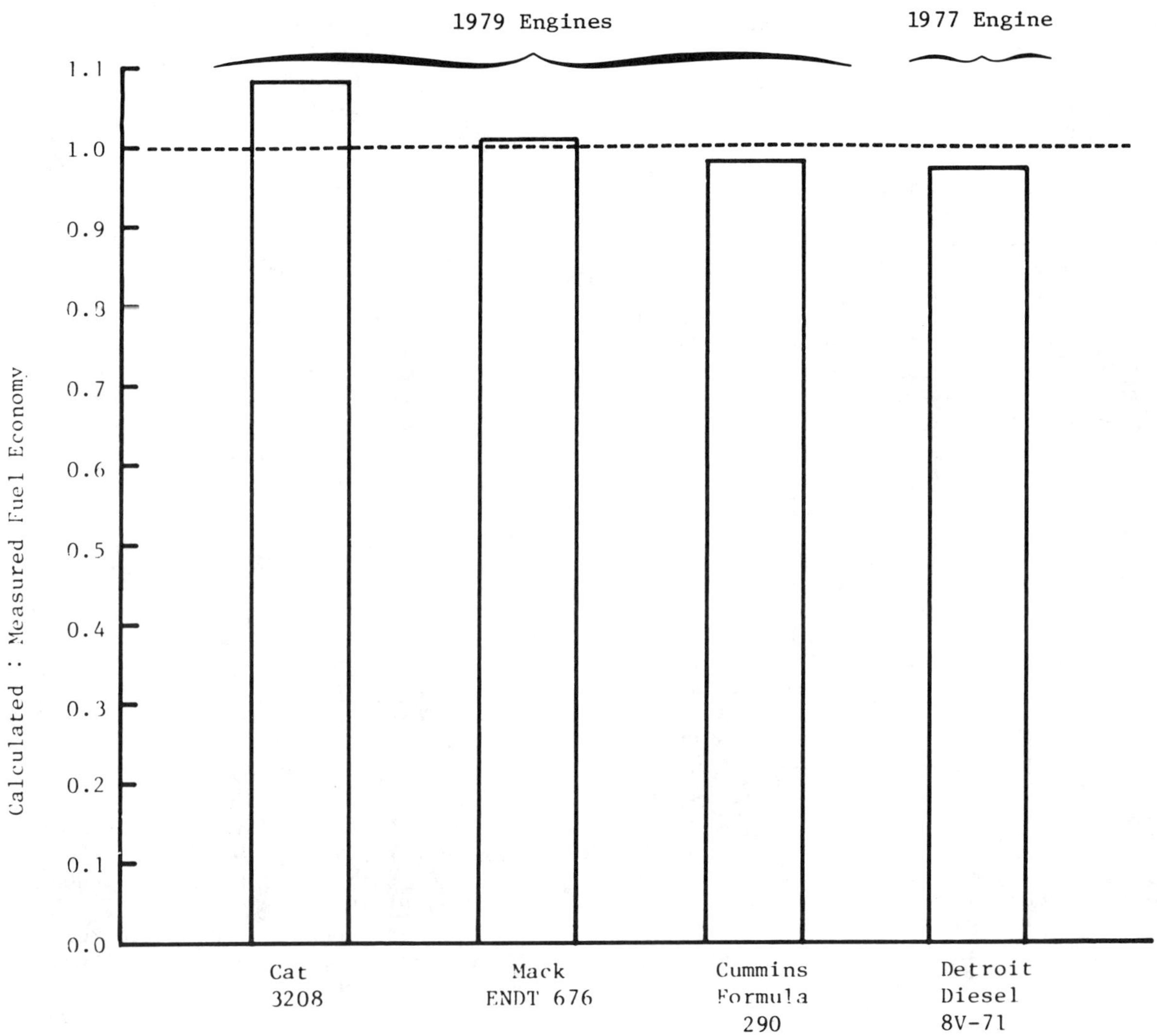

Fig. 9 - Comparison of calculated and measured fuel economy

REFERENCES

1. Federal Register, "Control of Air Pollution from New Motor Vehicles and Engines, Heavy Duty Engines," Part II, 40 CFR, Part 85, Volume 37, Number 175, September 8, 1972.

2. Society of Automotive Engineers Recommended Practice J215, "Continuous Hydrocarbon Analysis of Diesel Emissions," 1972 SAE Handbook, pages 867-871.

3. Society of Automotive Engineers Recommended Practice J117, "Measurement of Carbon Dioxide, Carbon Monoxide, and Oxides of Nitrogen in Diesel Exhaust," 1972 SAE Handbook, pages 852-857.

4. Federal Register, "Emission Regulations for Gasoline-Fueled and Diesel Heavy-Duty Engines; Gaseous Exhaust Test Procedures," Part II, 40 CFR, Part 85, Volume 42, Number 174, September 8, 1977.

5. Federal Register, "1983 and Later Model Year Heavy-Duty Engines, Proposed Gaseous Emission Regulations, Part II," 40 CFR, Part 86, Volume 44, Number 31, February 1979.

6. Federal Register, "Gaseous Emission Regulations for 1984 and Later Model Year Heavy-Duty Engines, Part II," 40 CFR, Part 86, Volume 45, Number 14, January 1980.

7. Danielson, Eugene, "Recommended Practice for Measurement of Gaseous and Particulate Emissions from Heavy-Duty Diesel Engines under Transient Conditions - DRAFT, EPA Technical Report, SDSB-79-18, April 1979.

8. France, Chester J., William Clemens, and Tad Wysor, "Recommended Practice for Determining Exhaust Emissions from Heavy-Duty Vehicles Under Transient Conditions," EPA Technical Report SDSB-79-08, February 1979.

9. France, Chester J., "Transient Cycle Arrangement for Heavy-Duty Engine and Chassis

Emission Testing," EPA Technical Report HDV 78-04, August 1978.

10. Springer, Karl J. and Ralph C. Stahman, "Unregulated Emissions from Diesels used in Trucks and Buses," SAE Paper No. 770258 presented at the 1977 Automotive Engineering Congress and Exposition, Detroit, February 28-March 4, 1977.

11. Hare, Charles T., and Bradow, Ronald L., "Characterization of Heavy-Duty Diesel Gaseous and Particulate Emissions, and Effects of Fuel Composition," SAE Paper No. 790490 presented at the Congress and Exposition, Detroit, February 26-March 2, 1979.

12. Ullman, T., Springer, K., and Baines, T., "Effects of Six Variables on Diesel Exhaust Particulate." ASME Paper 80-DGP-42 presented at the Energy Technology Conference and Exhibition, New Orleans, LA, February 1980.

13. Hare, Charles T., "Characterization of Gaseous and Particulate Emissions from Light-Duty Diesels Operated on Various Fuels." Final Report to the Environmental Protection Agency under Contract 68-03-2440, July 1979.

14. Hare, Charles T., and Baines, Thomas M., "Characterization of Particulate and Gaseous Emissions from Two Diesel Automobiles as Functions of Fuel and Driving Cycle." SAE Paper No. 790424 presented at the Congress and Exposition, Detroit, February 26-March 2, 1979.

15. Springer, Karl J., "An Investigation of Diesel Powered Vehicle Odor and Smoke." Final Report Part I to the Department of Health, Education, and Welfare under Contract No. PH 86-66-93, March 1967.

16. Springer, Karl J., "An Investigation of Diesel Powered Vehicle Odor and Smoke." Final Report Part II to the Department of Health, Education, and Welfare under Contract No. PH 86-67-72, February 1968.

17. Stahman, Ralph C., Kittredge, George D., and Springer, K.J., "Smoke and Odor Control for Diesel-Powered Trucks and Buses. SAE Paper No. 680443 presented at SAE Mid-Year Meeting, Detroit, May 20-24, 1968.

18. Smith, Lawrence R., Parness, Mary Ann, Fanick, E. Robert, and Dietzmann, Harry E., "Analytical Procedures for Characterizing Unregulated Emissions from Vehicles Using Middle-Distillate Fuels." Interim Report to the Environmental Protection Agency under Contract No. 68-02-2497, April 1980.

19. Dietzmann, Harry E., Smith, Lawrence R., Parness, Mary Ann, and Fanick, E. Robert, "Analytical Procedures for Characterizing Unregulated Pollutant Emissions from Motor Vehicles." Interim Report under Contract No. 68-02-2497 to the Environmental Protection Agency, Publication No. EPA-600/2-79-017, February 1979.

20. Dietzmannm Harry E., and Black, Frank M., "Unregulated Emissions Measurement Methodology." SAE Paper No. 790815 presented at the Off-Highway Vehicle Meeting and Exposition, MECCA, Milwaukee, September 10-13, 1979.

21. Dietzmann, Harry E., and Parness, Mary Ann, "Study of Emissions from Trucks Over Transient Driving Cycles," Final Report to the Environmental Protection Agency under Contract 68-02-2993, 1980.

22. Bradow, Ronald L. and Zweidinger, Roy, "Effect of NO_2 on Ames Results," April, 1980, (Personal Communication).

CHARACTERIZATION, ENGINE CONTROLS & FUEL EFFECTS

Light-Duty Diesel NO_X-HC-Particulate Trade-Off Studies**

W. R. Wade
Ford Motor Co.
Dearborn, MI

In passenger car and light truck applications, currently available divided chamber Diesel engines, which were developed for optimum fuel economy with relatively high speed, high specific power output, have the potential to provide improvements in volumetric fuel economy (miles per gallon of Diesel fuel) over the EPA combined driving cycle, relative to the homogeneous charge gasoline engine. These improvements in volumetric fuel economy of the Diesel engine have usually been determined at emission levels which exceed future regulated or proposed requirements. Forecasted emission regulations will require a significant reduction in NO_X, HC and particulate emission levels of current Diesel engines which are designed for optimum fuel economy and relatively high specific power output. The minimum emission capability of the Diesel combustion process has not yet been determined although current worldwide research efforts are underway to define this capability.(1-19)* This paper will discuss the potential emission control capability of typical divided chamber, light duty Diesel engines.

The following aspects of Diesel emission control capability will be examined:

1. The emission capability of currently available Diesel engines without internal or external emission control measures will be defined as a function of vehicle and powertrain design parameters.
2. The emission control capability of a typical divided chamber Diesel engine with external modifications applied to its operating conditions will be defined.
3. The capability of limited modifications to the combustion system of a typical divided chamber Diesel engine to improve its emission characteristics will be described.
4. Techniques for collecting and periodically oxidizing particulate emissions will be discussed.

THE DIESEL ENGINE

Relatively high speed, high specific output, light-duty Diesel engines currently use the divided combustion chamber configuration. In this configuration, fuel is injected in the form of finely atomized droplets into a pre-combustion chamber. The precombustion chamber configurations commonly used in light-duty Diesels are shown in Figure 1. Although different geometric configurations are used, the pre-combustion chamber and inlet throat(s) are designed with the goal of producing vigorous mixing of the air and fuel. After combustion has been initiated in the precombustion chamber, the combustion products exit at high velocity through the throat(s) of the pre-combustion chamber into the main combustion chamber where they mix with the remaining air charge to complete the combustion process.

The emission levels of a Diesel engine are determined by its combustion process, which was originally developed for the combustion of lower grade fuels with improved fuel economy. The following reasons for the fuel economy improvement of the Diesel engine relative to the homogeneous charge gasoline engine provide some insight into the emission characteristics of the Diesel combustion process:

1. The compression ratio of the light duty, divided chamber Diesel, which is over 20:1 in contrast to the 8:1 compression ratio of a homogeneous charge gasoline

* Numbers in parentheses refer to references listed in the Reference Section at the end of this paper.

**Paper 800335 presented at the Congress and Exposition, Detroit, Michigan, February, 1980.

ABSTRACT

The emission control potential of typical divided chamber, light-duty Diesel engines was investigated by using engine dynamometer mapping tests, vehicle tests with engines modified to implement selected control strategies for reduced emission levels, experiments with combustion system modifications, and evaluations of techniques for the exhaust treatment of particulate emissions.

A dynamometer mapping program was conducted on a Diesel engine with a swirl chamber combustion system to determine the emission control capability with modulated EGR and fuel injection timing. Emission projections from mapping tests, confirmed by selected vehicle test results, indicated that the low mileage engineering objectives assumed for the .41/3.4/1.0/.6 gm/mi HC/CO/NO_X/particulate emission level may be approached in experimental laboratory vehicles up to 3000 lb. inertia weight with optimized control systems. Particulate emissions limit the capability to meet the emission objective levels at higher inertia weights. Experimental Diesel combustion system modifications did not result in significant improvements in NO_X, HC, and particulate emissions but caused a loss in fuel economy and/or power.

Techniques for collecting and periodically oxidizing particulate emissions were investigated. The regeneration of a particulate trap was achieved at a high load engine dynamometer operating condition during a steady state simulation of the AMA cycle.

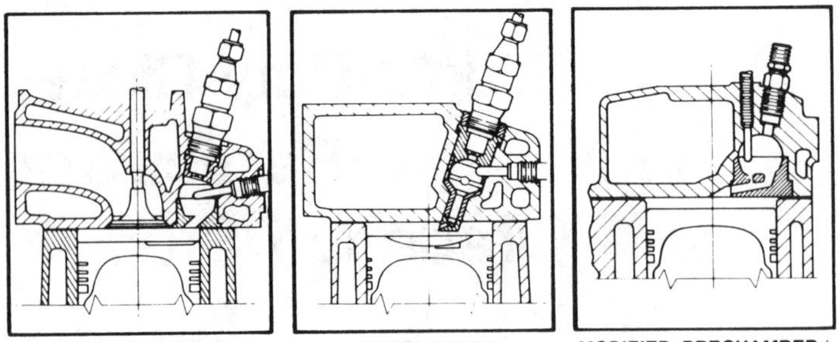

Figure 1 — Light Duty Diesel Divided Combustion Chamber Configurations

engine, provides a portion of the fuel economy advantage.

2. The lean air fuel ratios of the Diesel, which range from 16:1 to 90:1, in contrast with the near stoichiometric air fuel ratios of the homogeneous charge gasoline engine, provide a portion of the fuel economy advantage.

   Reduced dissociation losses, reduced heat loss, and the absence of throttling losses are responsible for the fuel economy advantage due to lean air fuel ratio operation.

3. The higher specific gravity of diesel fuel, which provides 11-13% more energy per gallon, accounts for a portion of the apparent volumetric fuel economy advantage. If an energy-based fuel economy comparison is made, then this advantage for the Diesel engine would be eliminated.

Although the improved fuel economy of the Diesel provides an incentive for its use, exhaust emission characteristics may limit the future application of the Diesel for light duty vehicle use. The CVS-CH* emission levels of several production Diesel vehicles, as reported by EPA(20,21,22,23) during experimental or certification tests or as measured by Ford, are shown in Figure 2 and are compared to the assumed low mileage engineering objective levels for the 1981 .41/3.4/1.0 gm/mi HC/CO/NO_X emission standards and the proposed 1981 and 1983 .6 and .2 gm/mi particulate requirements.(24) The low mileage engineering objective levels are assumed to be equal to 80% of the emission standards except for HC emissions where 73% is applied to account for the greater variability in measuring Diesel HC emissions. Although the HC emissions of some of the vehicles shown in Figure 2 are below the .30 gm/mi HC objective level, three Diesel vehicles show HC emissions above this objective level. CO emissions of all Diesels are well below the 2.7 gm/mi CO objective level requirement. In contrast, NO_X emissions of all vehicles shown in Figure 2 exceed the .80 gm/mi NO_X objective level by up to 150% for the heavier inertia weight vehicles. Likewise, each vehicle shown exceeds the objective level for the 1983 .2 gm/mi particulate requirement by up to 400% for one of the heavier inertia weight

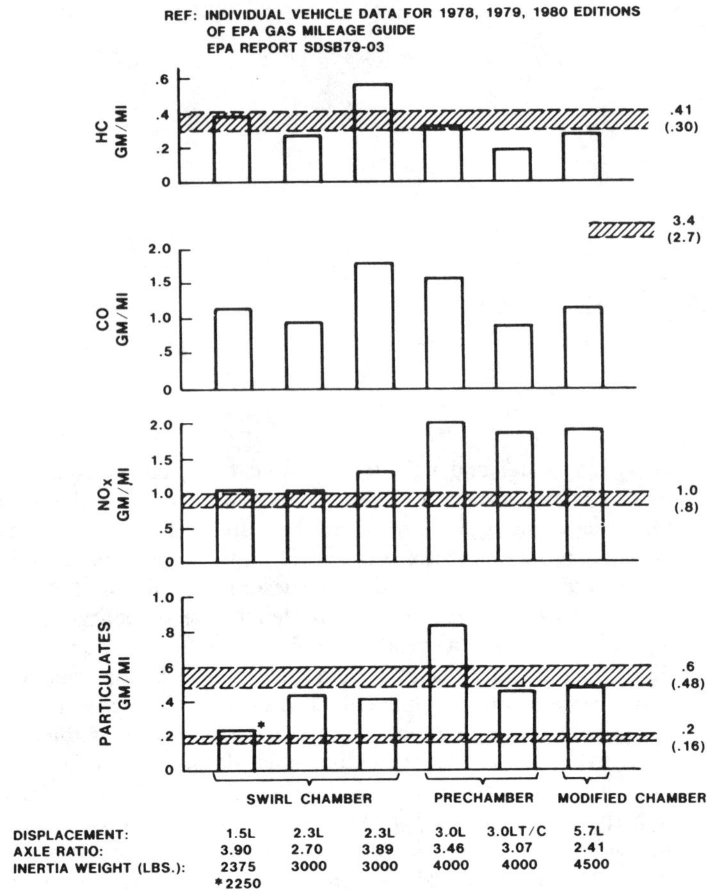

Figure 2 — CVS-CH Emissions of Typical Passenger Cars with Different Diesel Combustion Systems

vehicles. The particulate emissions of only one vehicle exceed the objective level for the 1981 .6 gm/mi particulate requirement. However, as discussed in a later section of this paper, particulate emissions may increase significantly when NO_X emissions are reduced to approach the .8 gm/mi NO_X objective level.

The relatively low levels of gaseous exhaust emissions of current Diesel engines shown in Figure 2 are obtained without the use of exhaust after-treatment devices which are usually applied to homogeneous charge gasoline engines to meet similar emission levels. HC emissions of the Diesel combustion process are relatively low because droplet burning avoids wall quench and crevice area HC emissions typical of a homogeneous charge gasoline engine. Additionally, the high compression ratios of the Diesel provide favorable oxidation conditions to eliminate a portion of the potential mid-air quench HC emissions which may result from very lean fuel air mixtures in the main chamber. CO emissions are low due to the lean air fuel ratios of the Diesel during operation over the CVS cycle. NO_X emissions are lower than the NO_X emissions of homogeneous charge gasoline engines without external emission controls for the following reasons: locally rich droplet combustion which suppresses NO_X formation due to the lack of oxygen availability; two-stage combustion consisting of the rich pre-chamber and lean main chamber which suppresses NO_X formation due to limited oxygen availability and moderate temperature levels; and the large exhaust residual in the prechamber which reduces oxygen availability in the prechamber. In contrast,

---

* CVS is used in this paper to refer to the urban driving cycle specified by the Federal emission test procedure. CVS is an abbreviation for the constant volume sampling technique specified by this procedure. CH refers to the weighted cold/hot driving cycle emissions.

particulate emissions of the Diesel combustion process are significantly higher than the particulate emissions of a homogeneous charge gasoline engine because insufficient oxygen has been mixed with a portion of the fuel at the time of combustion. Therefore, the features which provide relatively low gaseous emission levels in the Diesel are directly responsible for the high particulate emission levels of the Diesel.

BASELINE DIESEL EMISSION CAPABILITY

Although vehicle emission data is shown in Figure 2 for each of the three light-duty Diesel pre-combustion chamber configurations shown in Figure 1, significant differences in engine displacements and vehicle inertia weights preclude a direct comparison of the emission characteristics of the various combustion systems. To accurately determine the Diesel combustion system configuration with the best emission characteristics at similar engine speed and load operating conditions requires extensive engine dynamometer mapping tests on each candidate system. However, estimates of combustion system performance can be derived from the vehicle emission data shown in Figure 2 by using the emission index concept to eliminate engine and vehicle size effects. The emission index concept, which is defined as the grams of emission species per kg of fuel consumed by the combustion system, provides a convenient method to assess Diesel combustion system characteristics from CVS vehicle emission data. An overall CVS-CH emission index (EI) is defined as follows, using HC emissions as an example:

$$\text{HCEI}\ \frac{\text{gm}}{\text{kg fuel}} = \text{CVS-CH HC}\ \frac{\text{gm}}{\text{mi}}\ \text{X}\ \text{MPG}\ \frac{\text{mi}}{\text{gal}}\ \text{X}\ \frac{1}{\text{D}}\ \frac{\text{gal}}{\text{kg fuel}}$$

where: D = fuel density in kg/gal

The overall CVS-CH emission indices derived from the emission data of the six vehicles shown in Figure 2 are plotted in Figure 3 for HC, CO, NO_X and particulate emissions. A cursory comparison of Diesel combustion systems based on the emissions produced from a unit quantity of fuel consumed can be obtained from the bar charts shown in this figure. However, the data in Figure 3 indicate that none of the combustion systems are clearly superior for each of the emission constituents considered here. The ranges of emission indices, shown in Table 1, indicate that the largest variability occurs in the HC emission index, which is probably more strongly influenced by the fuel injection system characteristics than by the combustion chamber configuration.

A shortcoming of the emission index method of comparison of combustion systems arises when vehicles have significantly different engine displacement/vehicle inertia weight ratios or N/V (engine speed/vehicle speed) ratios which may cause the engine to operate at either more or less favorable conditions for emissions for a different percentage of the CVS cycle operating time. The effects of axle ratio are shown in Figure 3 for a 2.3L Diesel powered vehicle which was operated at two different axle ratios. Note that the lower axle ratio causes reductions in the gaseous emission levels but a slight increase in particulate emissions. Effects of engine displacement/vehicle inertia weight ratio and N/V ratio will be discussed more extensively in a later section of this paper.

In addition to providing a comparison of combustion systems, emission indices can be used to estimate the required CVS cycle fuel economy necessary to approach selected vehicle gm/mi emission requirements. Representative CVS-CH emission indices

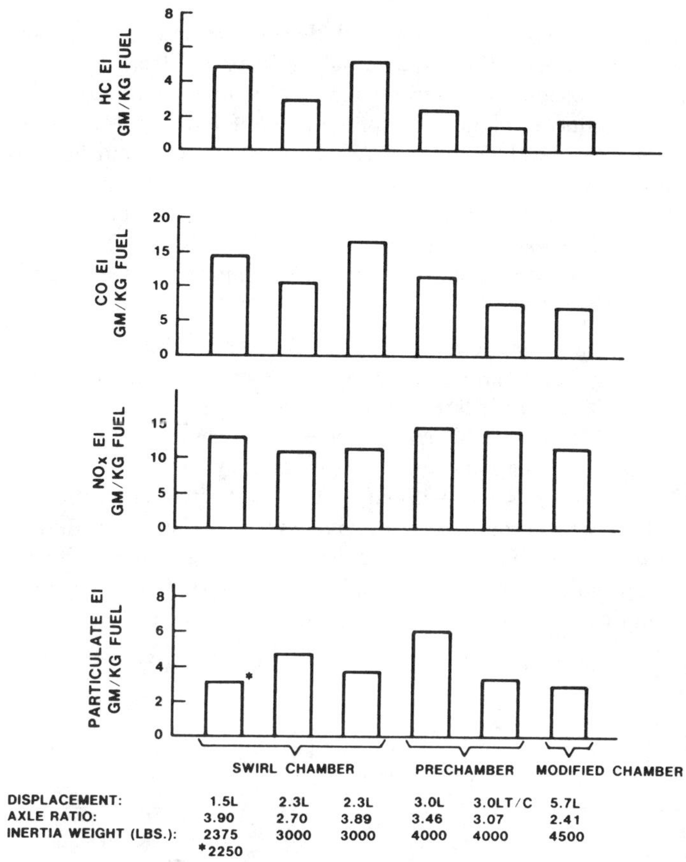

Figure 3 — CVS-CH Emission Indices of Typical Passenger Car Diesel Combustion Systems

Table 1

LIGHT-DUTY DIVIDED CHAMBER DIESEL ENGINE CVS-CH EMISSION INDICES

Emission	Emission Index gm/kg fuel
HC	1.3 - 5.0
CO	6.8 - 16.6
NO_X	11.0 - 14.5
Particulates	2.8 - 6.0

obtained from vehicle projections using mapping data from a swirl chamber Diesel were used to compute gm/mi emission levels for a range of CVS fuel economy values. The results of these calculations are shown in Figure 4 where CVS-CH emissions are plotted against CVS-CH fuel economy. The emission indices used for these calculations are shown on the figure. This figure clearly shows the projected increase in vehicle emission levels in gm/mi as the vehicle CVS-CH fuel economy is reduced. Lighter vehicles with higher fuel economy will have lower emissions in gm/mi when similar emission indices for the combustion system are retained.

The assumed engineering emission objectives for a .41/3.4/1.0/.6 gm/mi HC/CO/NO_X/particulate emission level are shown in Figure 4. To approach these emission objective levels without the application of external emission controls, a vehicle CVS fuel economy of over 40 mpg is required. This CVS fuel economy value is required to approach the .80 gm/mi NO_X emission objective level. A lower CVS fuel economy of about 25 mpg is sufficient to approach either the .30 gm/mi HC or the .48 gm/mi particulate emission objective levels. Therefore, to approach these emission objective levels, NO_X emissions dictate the fuel economy requirement for the Diesel combustion system without external emission controls.

Exhaust gas recirculation (EGR) and fuel injection timing modulation can significantly affect the emission levels of a swirl chamber Diesel. In a later section of this paper, the emission reductions obtained by the application of these control measures are projected from mapping data. Approximately 50% reductions in NO_X and HC emissions without a significant effect on particulate emissions were projected from mapping data for an optimized control system. These projections were used to calculate optimized emission indices for a swirl chamber Diesel with EGR and injection timing modulation. These optimized emission indices were used to calculate the optimized CVS-CH emissions for a swirl chamber Diesel as a function of CVS-CH fuel economy, and the results are shown by the dotted lines in Figure 4. To approach the engineering objectives for a .41/3.4/1.0/.6 gm/mi HC/CO/NO_X/particulate emission level, a vehicle CVS fuel economy of about 30 mpg is required. In contrast to the NO_X limitation of the baseline case, this CVS fuel economy value is required to approach the .48 gm/mi particulate emission objective level. Lower fuel economy values are sufficient to approach HC and NO_X emission objective levels. To approach the .2 gm/mi particulate emission level, the calculations in Figure 4 indicate that a vehicle fuel economy over 50 mpg would be required for the current swirl chamber Diesel.

A plot of CVS-CH fuel economy for light-duty Diesel vehicles[20-22] as a function of vehicle inertia weight is shown in Figure 5. By combining the vehicle emissions vs. fuel economy calculations shown in Figure 4 with the fuel economy vs. vehicle inertia weight data shown in Figure 5, an estimate of baseline and optimized emission levels for a swirl chamber combustion system as a function of vehicle inertia weight can be derived and the results are shown in Figure 6. The .48 gm/mi particulate emission objective level limits the vehicle inertia weight to approximately 3000 lbs. with an optimized emission control system. In contrast, the .80 gm/mi NO_X emission objective level without a particulate requirement is projected to be approached by a 4000 lb. inertia weight vehicle when an optimized emission control system is applied. The capabilities of the optimized emission control system are based on mapping projections discussed later in this paper. Verification of these capabilities in actual vehicles has not been demonstrated.

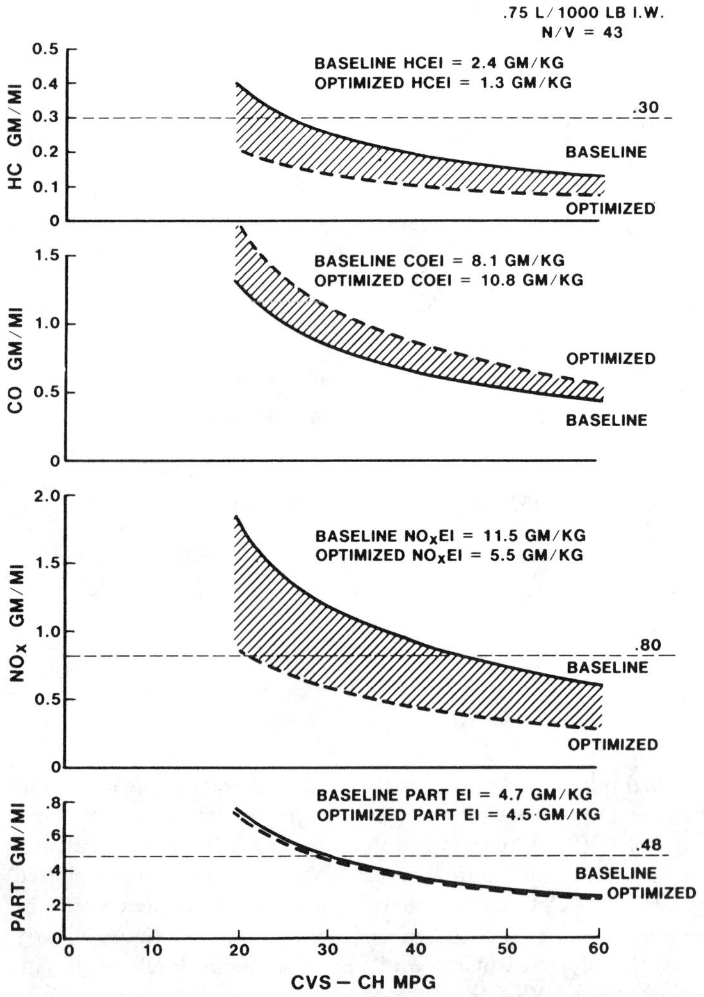

Figure 4 — CVS-CH Emissions in GM/MI for Diesel Passenger Cars As Estimated from Emission Indices vs. CVS-CH Fuel Economy

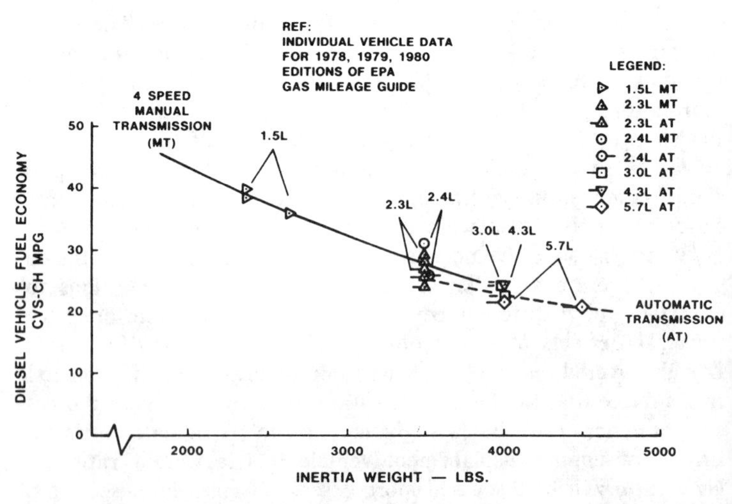

Figure 5 — CVS-CH Fuel Economy vs. Inertia Weight for Typical Diesel Powered Passenger Cars

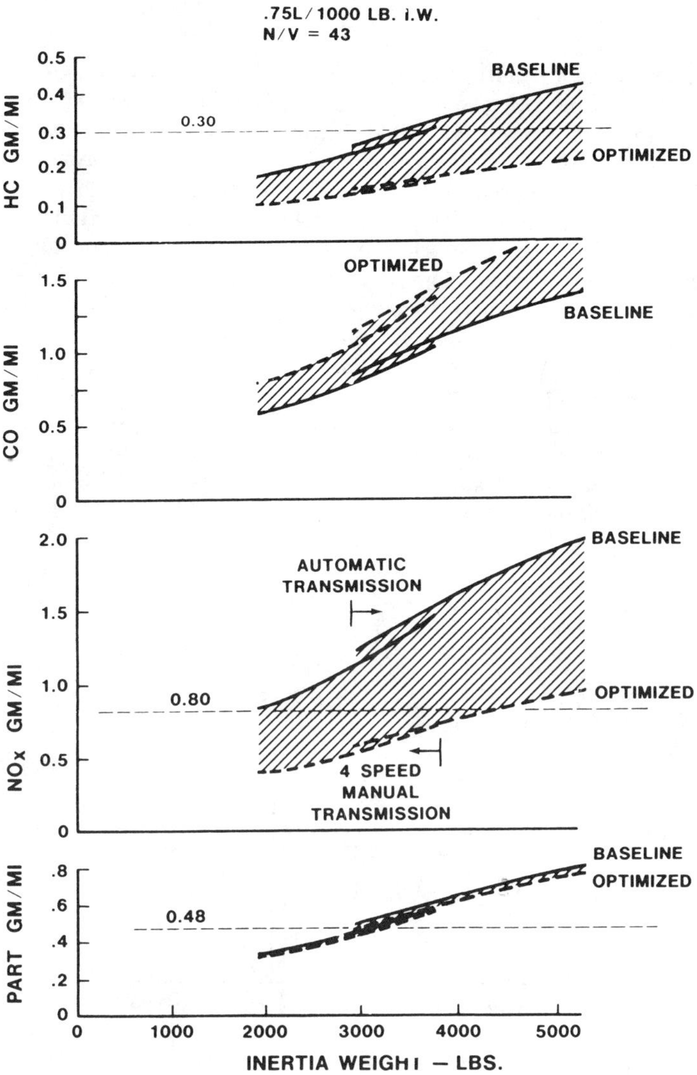

Figure 6 — CVS-CH Emissions in GM/MI for Diesel Passenger Cars As Estimated from Emission Indices vs. Vehicle Inertia Weight

EFFECT OF VEHICLE PARAMETERS ON DIESEL EMISSIONS

To determine the effects of vehicle and drivetrain parameters on light-duty Diesel vehicle emission capability, emission island maps for a typical baseline Diesel engine with a swirl chamber combustion system were generated and applied to a vehicle simulation computer program. The vehicle simulation computer program was used to calculate the second-by-second power requirements of a vehicle operated according to the driving cycle requirements (speed vs. time) on the chassis dynamometer.[25] These vehicle power requirements are translated into engine speed and torque requirements by accounting for the operating characteristics and efficiencies of the tires, rear axle, transmission, torque convertor and accessories. Emissions at each second-by-

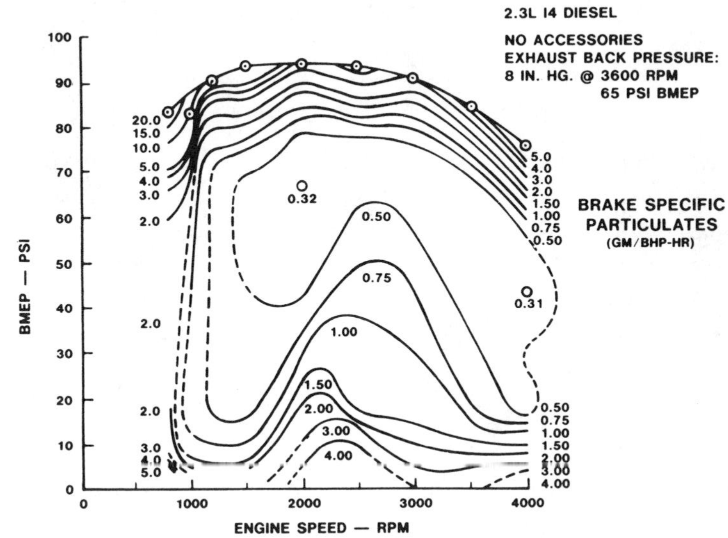

Figure 7 — Brake Specific Particulate Island Map for a 2.3L I4 Swirl Chamber Diesel

second engine speed and torque requirement were determined from emission island maps and these emissions were summed over the 1372 seconds of the CVS cycle to give projected gm/mi emission values. An example of the brake specific particulate emission island map is shown in Figure 7 for a production 4 cylinder Diesel engine with swirl chamber combustion system, a displacement of .58 L /cylinder and a bore/stroke ratio of 1.08. Small correction factors to account for transient operation and for cold start emissions were obtained by comparing vehicle emission projections based on steady state engine dynamometer mapping data with actual vehicle emission data for several selected vehicles.

The effects of load on emissions, smoke, air fuel ratio, and exhaust temperature are shown in Figure 8 which was obtained from the emission island maps at a constant 1500 RPM engine speed. Brake specific HC emissions show significant increases at light loads. In contrast, brake specific particulate emissions increase significantly at heavy loads where the air fuel ratio is richer than 20:1. Brake specific NO_X emissions show the unusual characteristic of decreasing as the load increases. This trend of lower brake specific NO_X emissions as the load is increased is opposite to most open chamber or homogeneous charge engines because of the two-stage combustion process of the divided chamber Diesel engine. When the load is over approximately 50% of full load, the prechamber will operate at an average air fuel ratio which is richer than stoichiometric, thereby tending to suppress the formation of NO_X due to the lack of excess oxygen and the lower average gas temperatures relative to stoichiometric combustion.

The vehicle parameters that were investigated for their effect on vehicle emission levels using the emission island maps and the vehicle simulation computer program included: inertia weight, engine displacement/vehicle inertia weight ratio and axle ratio. The effects of vehicle inertia weight and engine displacement/inertia weight ratio on swirl chamber Diesel HC, CO, NO_X and particulate emissions, fuel economy and performance as indicated by the 0-60 mph acceleration time criteria are shown in Fig-

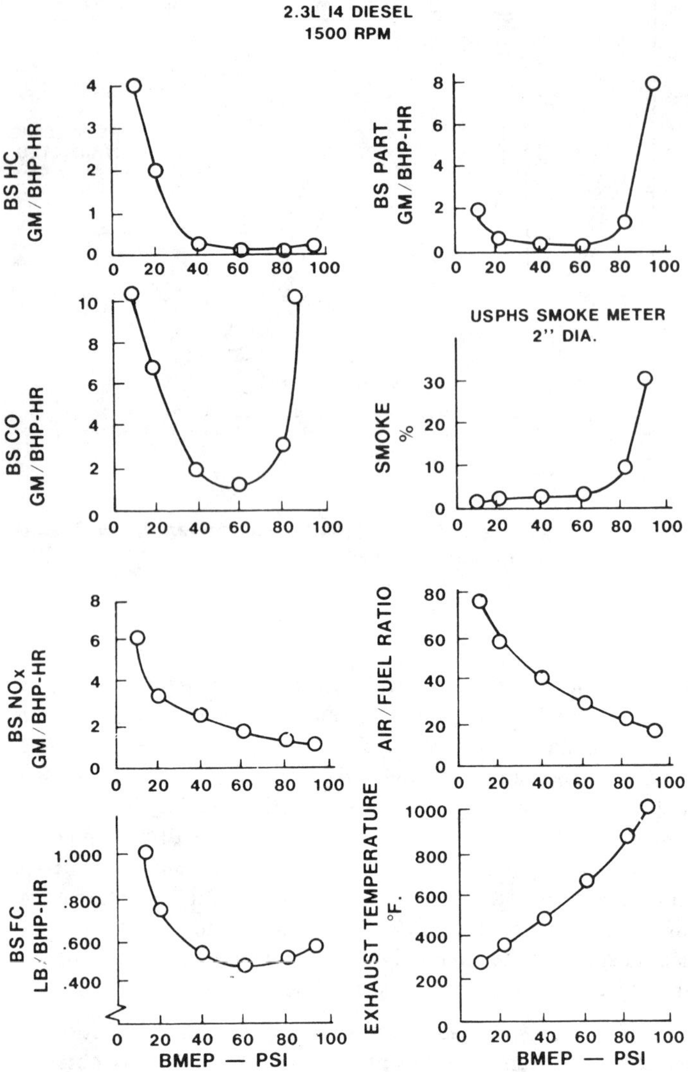

Figure 8 — Effect of Load on Emissions and Exhaust Temperature of a 2.3L I4 Swirl Chamber Diesel at 1500 RPM

ure 9 for a 3.0 axle ratio. HC, CO, NO_X and particulate emissions are shown to increase as the vehicle inertia weight is increased at a constant engine displacement/inertia weight ratio. An increase in vehicle inertia weight at a constant engine displacement/inertia weight ratio does not change the engine RPM-BMEP operating conditions, but the total grams of emissions increase due to the increased engine work (hp-hrs)* required to drive the CVS cycle. Decreasing the engine displacement/inertia weight ratio is shown to decrease HC, CO and NO_X emissions. A decrease in engine displacement/inertia weight ratio causes the engine to operate at higher BMEP levels which result in reductions in HC, CO and NO_X, as shown by the emission data plotted as a function of BMEP in Figure 8. Brake specific particulate emissions are shown in Figure 8 to be nearly constant at moderate loads. As a result, particulate emissions are not significantly affected by a reduction in engine displacement/inertia weight ratio until values below 1.0 L/1000 lbs. inertia weight are encountered. A reduction in displacement/inertia weight ratio is usually limited by the requirements to maintain acceptable vehicle performance. As shown in Figure 9, performance can be maintained over a broad range of vehicle inertia weights if the engine displacement/inertia weight ratio is maintained constant. Fuel economy is strongly influenced by vehicle inertia weight, as shown in Figure 9, whereas reductions in the engine displacement/inertia weight ratio exhibit less influence on fuel economy. These projections have indicated that for a specified vehicle inertia weight, the engine displacement has an important effect on emission levels. Reducing engine displacement tends to reduce the vehicle gm/mi emission levels more than would be obtained from a reduction in fuel consumption at constant emission indices.

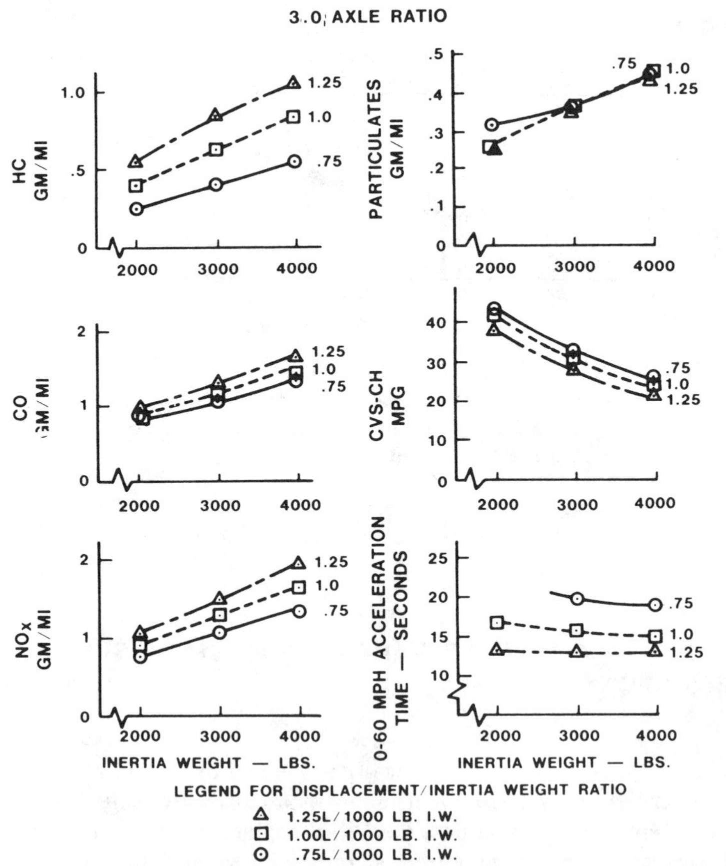

Figure 9 — Effect of Vehicle Inertia Weight and Engine Displacement/Vehicle Inertia Weight Ratio on Emissions, Fuel Economy and Performance of Swirl Chamber Diesel Vehicles (Calculated from Island Maps)

* Engine work (hp-hrs) required to drive the CVS cycle is calculated by summing the second-by-second products of engine hp and the one second time increment for the total 1372 seconds in the CVS cycle.

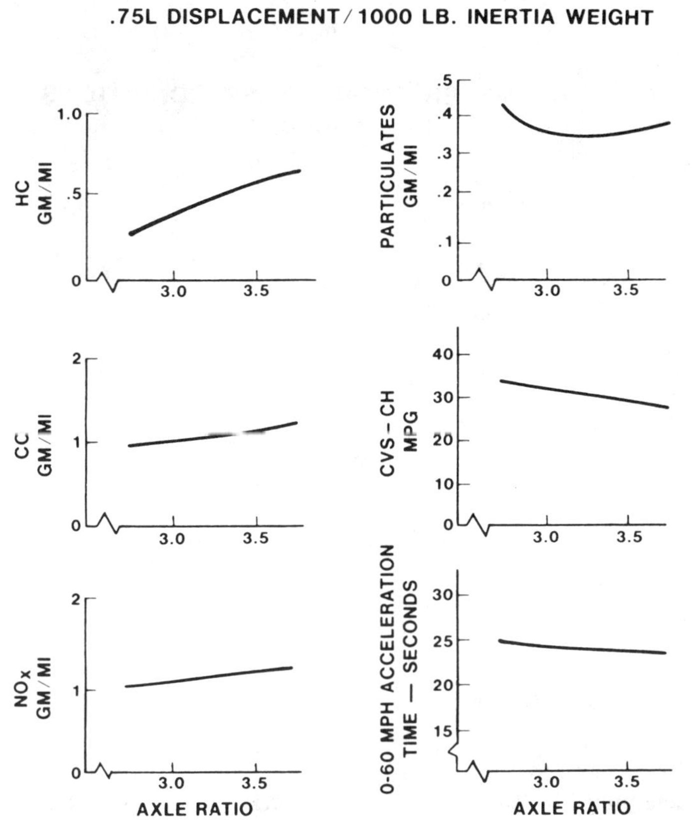

Figure 10 — Effect of Axle Ratio on Emissions, Fuel Economy and Performance of Swirl Chamber Diesel Vehicles (Calculated from Island Maps)

The effects of axle ratio on emissions, fuel economy and performance are shown in Figure 10. Reductions in axle ratio have similar effects to reductions in engine displacement/vehicle inertia weight ratio. A decrease in axle ratio causes the engine to operate at the same power level but at a lower RPM and higher BMEP level which also results in different emission characteristics depending on the individual emission island map. Lower axle ratio provides improvements in HC, CO and NO_x emissions. Particulate emissions increase when the axle ratio is reduced to very low values. Lower axle ratio is shown to improve fuel economy, but at the expense of reduced performance.

LOW EMISSION POTENTIAL OF THE DIESEL

A Diesel engine dynamometer mapping program was conducted to determine the potential of exhaust gas recirculation and fuel injection timing (which controls the start of combustion timing) to reduce the emission levels of the Diesel combustion process below the baseline levels described in the previous sections for a range of engine sizes and vehicle weights. Mapping is a technique which has previously been applied to homogeneous charge and PROCO engines for projecting vehicle emissions and fuel economy from engine dynamometer test data obtained from a number of carefully selected steady-state engine speed-load operating conditions.[25,26,27] After the test data have been obtained for a wide range of operating conditions at these test points, an optimization computer program is used to select the control parameters (EGR and injection timing values) at each test condition to provide the maximum fuel economy at specified emission levels.

A production 2.3L 4-cylinder passenger car Diesel engine, which was previously used to generate the emission island maps, was selected for the mapping tests. This engine, which incorporates a swirl chamber combustion system, was considered to be a typical high speed, high specific output, indirect injection Diesel engine. Since the emission levels of this engine were considered to be representative of the swirl chamber combustion system, the emission test results at the same RPM-BMEP conditions were used to represent the emission levels of a wide range of engine sizes.

A limited number of engine dynamometer speed and load mapping conditions were selected to be representative of the engine operating conditions encountered in a wide range of vehicles operated over the CVS emission test driving cycle. The number of test points was carefully limited so that extensive EGR and injection timing conditions could be tested at each mapping point within a reasonable time. To select these mapping points, the vehicle simulation computer program, described earlier, was used to produce an engine BMEP-engine speed-time matrix for the CVS driving cycle. By superimposing the engine BMEP-engine speed envelope of the CVS driving cycle of all of the vehicles being simulated on a single BMEP-engine speed map, adequate engine dynamometer test points were selected to represent the entire region of BMEP and engine speed for all vehicles. By using BMEP to define engine load, engine displacement variations could easily be accommodated at reasonable engine displacement/inertia weight ratios and N/V ratios without changing the definition of the engine speed-load operating test conditions.

The vehicles that were simulated prior to selecting the engine mapping conditions are shown in Table 2. These engine displacements, inertia weights and axle ratios were selected in order

Table 2

SIMULATED DIESEL VEHICLES USED FOR MAPPING PROJECTIONS

Engine Displacement (L)	Inertia Weight (lb)	PAU (hp)	Transmission	Axle Ratio
1.6	2000	7.3	4 Speed Manual	2.70
1.6	2000	7.3	4 Speed Manual	3.89
2.3	3000	10.3	4 Speed Manual	2.70
2.3	3000	10.3	4 Speed Manual	3.89
3.2	4250	11.3	Automatic	2.50
3.2	4250	11.3	Automatic	3.08
3.2	4250	11.3	Automatic	3.75

to project the emission control potential of the Diesel engine for a wide range of light duty vehicles having typical engine displacement/inertia weight ratios. Examples of the envelopes of the CVS BMEP-speed regions for a 2.3L displacement engine in 3000 lb. inertia weight vehicles with 2.70 and 3.89 axle ratios are shown in Figure 11. Although not shown for the sake of clarity, similar overall envelopes are obtained when the CVS BMEP-engine speed regions of all seven simulated vehicles are plotted on the BMEP-engine speed coordinates shown in Figure 11. The nine BMEP-engine speed test points selected for mapping on the dynamometer engine are listed in Table 3. These dynamometer engine test points, which were selected to represent the overall BMEP-engine speed region for all seven vehicles, are also shown on Figure 11. These engine test conditions are specified for the dynamometer engine operating without the cooling fan or other accessories but with an arbitrarily selected simulated vehicle exhaust back pressure of 8 in. Hg at 3600 RPM — 65 psi BMEP. The engine BMEP-speed-time matrix calculated from the vehicle simulation computer program shows gross engine BMEP (except for exhaust back pressure losses) since allowances are made in the computer program for accessory losses appropriate for the different vehicles being simulated. The selection of mapping test conditions with several common engine speeds and loads provides for a comparison of engine emissions as a function of speed and load and for the construction of speed and load dependent control system calibration functions.

The nine mapping test points were applied to the BMEP-speed-time matrix of each of the seven vehicle simulations listed in Table 2 to obtain CVS cycle weighting factors. BMEP-speed regions, which can be represented by the engine dynamometer test points, were selected by using judgment to satisfy as best as possible the requirements listed in Reference 25. The two most important requirements state that the total time within the nine regions must equal the total cycle time and that the total engine work (hp-hrs) of the nine regions represented by the mapping points must equal the total second-by-second integrated engine work obtained from the vehicle simulation computer program. Weighting factors defined as the bhp-hrs contained within each region were determined for each of the seven vehicles. Weighting factors were obtained by adding all of the times within a region and multiplying this total time by the hp of the BMEP-engine speed mapping point. An example of a torque-speed-time matrix with the regions to be represented by the mapping points superimposed is shown in Figure 12 for a 2.3L displacement engine in a 3000 lb. inertia weight vehicle with a 4-speed manual transmission and a 3.89 axle ratio.

The mapping test program on the dynamometer engine consisted of measuring emissions and other pertinent engine data over a wide range of values for EGR and injection timing at each

Table 3

DYNAMOMETER ENGINE TEST CONDITIONS FOR MAPPING

Engine Speed RPM	Load-BMEP PSI
600	22
1000	15
1250	22
1250	37
1250	75
1700	37
1700	75
2100	37
2400	75

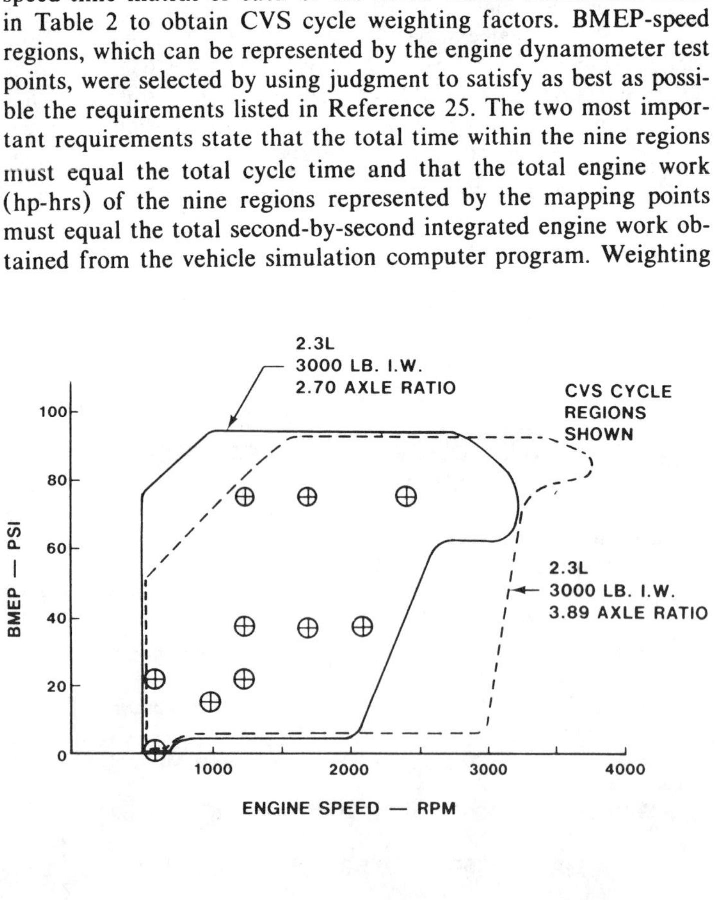

Figure 11 — Envelopes of CVS BMEP-Engine Speed Regions for 2.3L, 3000 lb. Inertia Weight Vehicles with 2.70 and 3.89 Axle Ratios

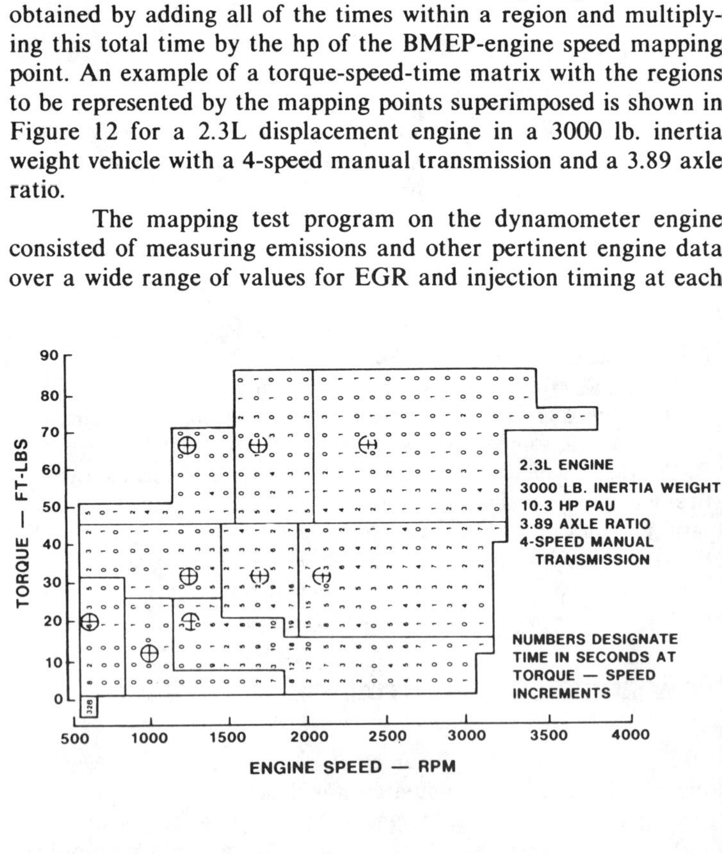

Figure 12 — A Torque-Speed-Time Matrix with Mapping Point Regions Superimposed

speed-load condition. The percent EGR in these tests is defined as follows:

$$\% \text{ EGR} = \frac{W_{EGR}}{W_{air} + W_{fuel}} \times 100$$

Changes in injection timing refer to changes from the base timing schedule of the fuel injection pump used on the mapping engine. Incremental changes in injection timing from the base schedule were made at each mapping condition.

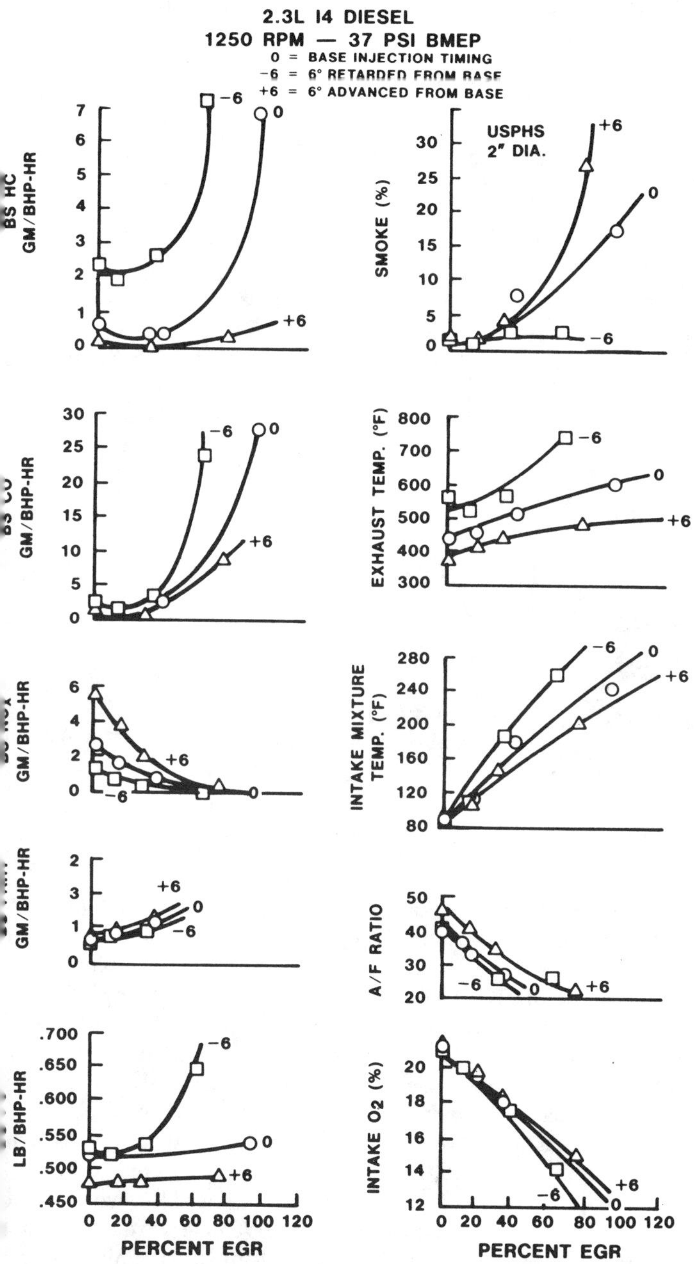

Figure 13 — The Effects of EGR and Injection Timing on Emissions, Fuel Consumption and Operating Parameters for a Swirl Chamber Diesel at Moderate Load (1250 RPM — 37 PSI BMEP)

DYNAMOMETER TEST RESULTS

An example of the dynamometer mapping data showing the effects of EGR at several values of injection timing for a moderate load condition is shown in Figure 13. Emissions, fuel consumption, smoke, intake and exhaust temperatures, air fuel ratios and intake oxygen concentration are shown in the figure. At the base injection timing, EGR is shown to provide significant reductions in NO_X emissions. The amounts of EGR required to provide a constant percentage reduction in NO_X emissions differs significantly at the different engine conditions which have widely different baseline air fuel ratios. These different air fuel ratios result in different amounts of oxygen contained in the recycled exhaust gas. However, similar reductions in NO_X emissions are obtained at similar intake charge oxygen concentrations, which, in turn, reflect the amount of inert diluent recycled into the combustion system. The intake charge oxygen concentrations shown in the figure are calculated from the following equation:

$$\text{Intake Charge } O_2 \text{ Concentration} = \frac{1(\text{int } O_2) + \frac{\% \text{ EGR}}{100}(\text{exh } O_2)}{1 + \frac{\% \text{ EGR}}{100}}$$

where: $(\text{int } O_2) = .21$ = oxygen concentration mole fraction in the intake air.

Moderate amounts of EGR provide a reduction in brake specific HC emissions at some conditions. This reduction in mass HC emissions occurs primarily due the reduction in exhaust flow rate when EGR is used to displace a portion of the intake excess air. A slight deterioration in fuel consumption occurs with the use of EGR.

Particulate, CO and smoke emissions increase significantly as the EGR rate is increased. The use of EGR decreases the intake charge oxygen concentration which reduces the amount of oxygen available to premix with the injected fuel prior to combustion. This reduced oxygen availability probably contributes to the increase in particulate emissions. Reduced intake charge oxygen concentration also reduces the amount of oxygen available for the oxidation of the particulate emissions during the expansion stroke. Rapid increases in particulate emissions and/or smoke tend to limit the amount of EGR that can be used in the Diesel combustion system at heavy loads, whereas rapid increases in HC emissions tend to limit the amount of EGR that can be used at light to moderate loads.

The effects of injection timing on emissions, fuel consumption and operating parameters are also shown in Figure 13. Advancing injection timing results in increases in NO_X emissions since a greater portion of burned gases experience higher peak temperatures near TDC. HC emissions are slightly reduced as the injection timing is advanced from the base timing since a longer residence time at high temperature is provided for HC oxidation. Retarded injection timing causes rapid increases in HC emissions due to the late burning and subsequent early quenching of the combustion products during the expansion stroke. At moderate loads, retarded injection timing provides a very slight reduction in particulate emissions. This may be due to the increased amount of air that is present in the prechamber at the initiation of combustion when injection timing is retarded.

As shown by the data in Figure 13, several strategies are available to reduce the NO_X emissions to a specified objective level. EGR can be used with the base injection timing to reduce the NO_X emissions but HC emissions may increase beyond baseline levels. At light to moderate loads, advanced injection timing can be used to reduce HC emissions although a larger amount of EGR is required to meet a NO_X objective level. At heavy load, slightly retarded injection timings can be used to reduce the NO_X emissions without incurring a significant increase in HC or particulate emissions.

DIESEL VEHICLE PROJECTIONS

Vehicle inertia weight strongly influences the emission levels of light-duty Diesel powered vehicles, as was previously shown in Figure 6. Likewise, the inertia weight strongly influences the amount of control required to meet future emission objective levels. The capability to meet the engineering objective levels of .30/2.7/.80/.48 gm/mi HC/CO/NO_X/particulates assumed to be required for the .41/3.4/1.0/.6 gm/mi HC/CO/NO_X/particulate emission requirements was examined over a wide range of inertia weights. As previously shown in Table 2, vehicles with inertia weights of 2000 lb., 3000 lb. and 4250 lb. were selected for a detailed analysis of the capability to meet future emission objective levels. Engine sizes of 1.6L, 2.3L and 3.2L were selected for the 2000, 3000 and 4250 lb. inertia weight vehicles, respectively, as being representative of the engine displacement/vehicle inertia weight ratios (about .75 to .80 L/1000 lb. I.W.) of typical light duty Diesel vehicles. For each vehicle, several axle ratios were considered because of the significant effect of axle ratio on emissions, fuel economy and performance.

Mapping emission data was scaled to the appropriate engine displacement by assuming that the brake specific emission levels for the different size engines were identical at the same engine speed and BMEP operating conditions. This assumption is most nearly correct when the engine friction is proportional to engine displacement and the displacement per cylinder and bore/stroke ratio remain constant for different size engines. Different total engine displacements can be obtained by changing the number of cylinders. However, for the purpose of this analysis, the effect of displacement per cylinder on emissions was assumed to be a secondary effect.

For each vehicle and axle ratio combination, the following emission projections were made from the mapping data to assess the capability to reach the engineering objective levels of .30/2.7/.80/.48 gm/mi HC/CO/NO_X/particulates:

1. Baseline emission levels at the base injection timing for best fuel economy and no EGR
2. Emission levels with modulated injection timing and no EGR
3. Emission levels with modulated EGR and base injection timing
4. Emission levels with modulated EGR and modulated injection timing

Each projection was made using an optimization computer program which selects the mapping point conditions to provide the optimum fuel economy at the emission objective levels. The results of these emission projections are shown in Figures 14, 15 and 16 for 2000, 3000 and 4250 lb. inertia weight vehicles, respectively. Although all of the vehicles shown in these figures are projected to be capable of approaching the .30/2.7/.80 gm/mi HC/CO/NO_X emission objective levels, the extent of the control measures required to achieve these levels increases at the higher inertia weights.

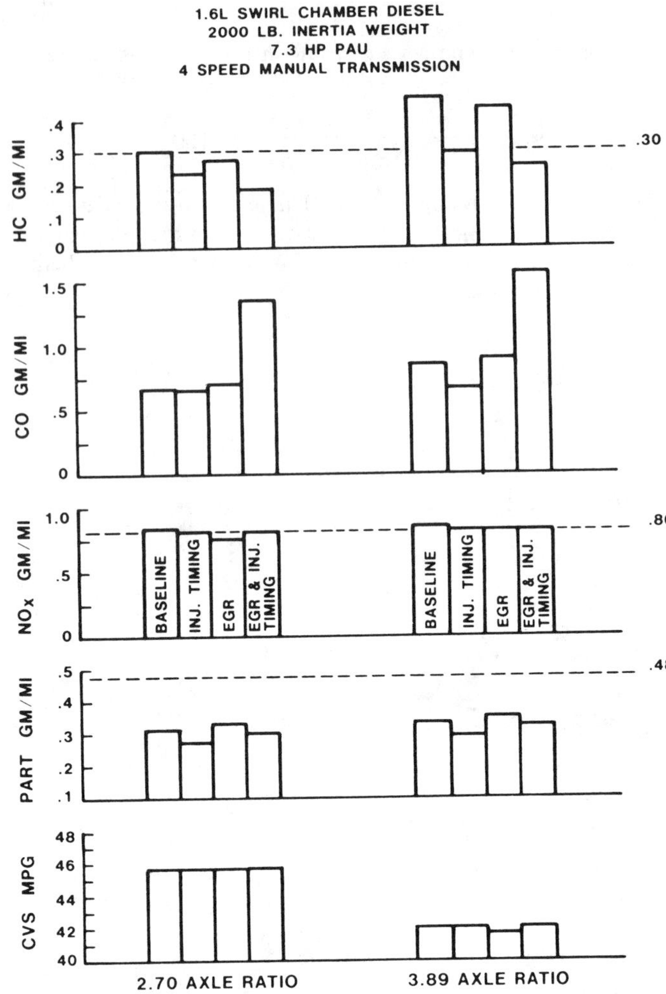

Figure 14 — Emission and Fuel Economy Projections for a 1.6L Swirl Chamber Diesel in a 2000 Lb. Inertia Weigh Vehicle

For 2000 lb. inertia weight vehicles shown in Figure 14 either injection timing modifications or small amounts of EGR are projected to be required to approach the emission objective levels. The use of EGR with modulated injection timing provides the additional reduction in HC emissions required for the 3.89 axle ratio. At this inertia weight, particulate emissions are below the .48 gm/mi objective level and are not significantly affected by the control measures applied for the gaseous emissions.

For a 3000 lb. inertia weight vehicle with the 2.70 axle ratio, shown in Figure 15, modulated EGR is projected to be adequate to approach the emission objective levels. However, due to the higher baseline HC emissions of the vehicle with the higher 3.89 axle ratio, both EGR and modulated injection timing are required to approach the emission objective levels with this axle ratio. Particulate emissions for the vehicle with a 2.70 axle ratio are projected to be below the .48 gm/mi objective level. However, due to the higher baseline NO_X level of the vehicle with the 3.89 axle more extensive control measures are required which cause the particulate emissions to slightly exceed the .48 gm/mi objective level when the gaseous emission objective levels are approached.

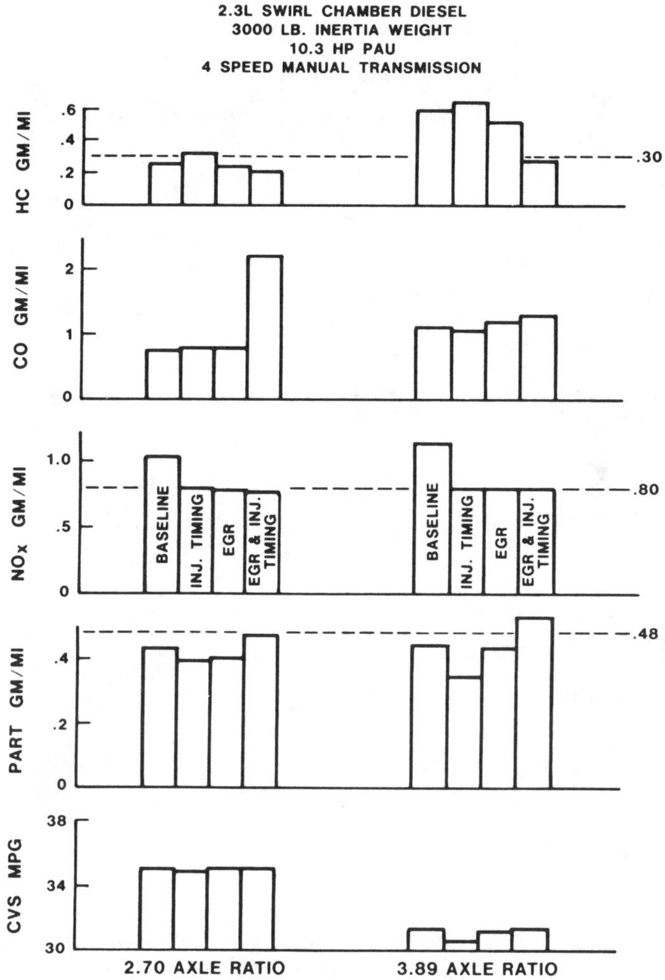

Figure 15 — Emission and Fuel Economy Projections for a 2.3L Swirl Chamber Diesel in a 3000 Lb. Inertia Weight Vehicle

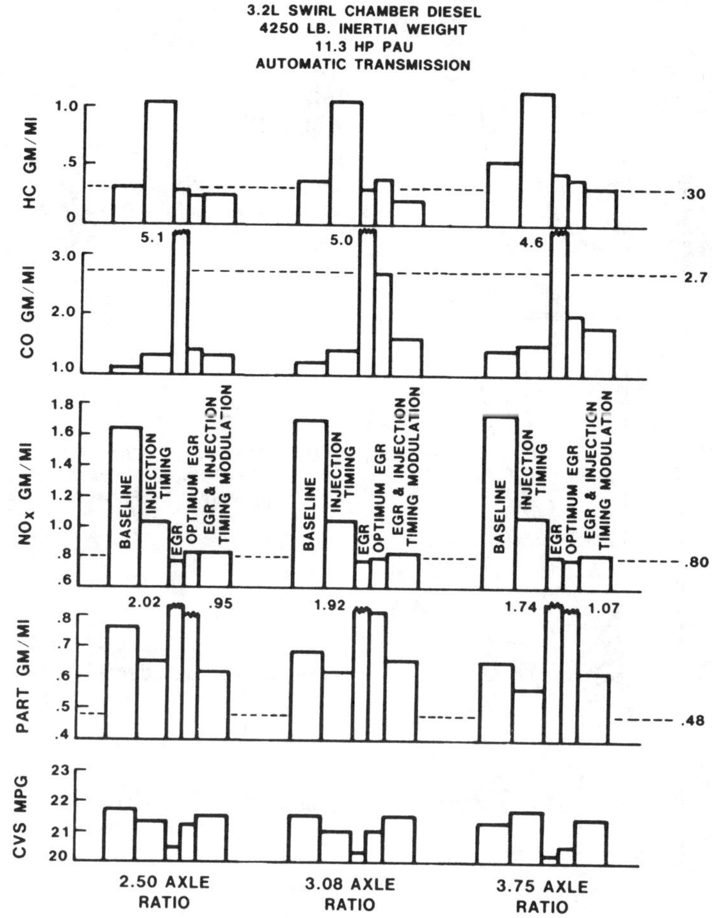

Figure 16 — Emission and Fuel Economy Projections for a 3.2L Swirl Chamber Diesel in a 4250 Lb. Inertia Weight Vehicle

For 4250 lb. inertia weight vehicles shown in Figure 16, accurately modulated EGR and injection timing are projected to be required to approach the HC emission objective level at the two higher axle ratios. In the case of this heavy vehicle with different weighting factors for the mapping points and with more stringent control of gaseous emissions, modulated EGR and injection timing are projected to provide improvements in particulate emissions relative to other cases shown in Figure 16 with less complex control systems. However, even with the optimized control system, the particulate emissions of the 4250 lb. inertia weight vehicles exceed the .48 gm/mi engineering objective level by 30-40%. The effect of optimizing the EGR strategy on particulate emissions is also shown in Figure 16. Without an optimum EGR strategy (at the base injection timing), particulate emissions are projected to be 70-100% higher than the levels obtained with the optimized EGR strategy.

Small fuel economy losses of less than 2% were projected for the best calibrations to meet the gaseous emission objective levels with minimized particulate emissions. Greater fuel economy losses were incurred with the non-optimized calibration strategies. These projected calibration strategies did not take into account engine noise levels or the effects of EGR on engine durability. Therefore, extensive vehicle tests would have to be conducted to determine if the projected emission levels could be achieved and to examine all other aspects of engine operation.

The selection of an optimum calibration depends significantly on the emission levels required at each of the mapping points. Because of the increased engine work (bhp-hrs) required to operate the higher inertia weight vehicles over the CVS cycle, significantly lower average brake specific emission levels are required for these vehicles. The effects of the lower average brake specific emission objectives required for higher inertia weight vehicles on the resulting calibration strategy are shown in Figure 17. At the $BSNO_X$ objective level required for .8 gm/mi NO_X in a 3000 lb. inertia weight vehicle, only moderate amounts of EGR are required at an advanced injection timing setting to meet the NO_X objective. The advanced injection timing is selected to provide optimum fuel economy with minimum HC emissions. However, at the lower $BSNO_X$ objective level required for the 4250 lb. inertia weight vehicle, the injection timing is retarded to the base level to minimize the increase in particulate emissions. As a result, the increase in the EGR requirement to meet the $BSNO_X$ objective level is minimized.

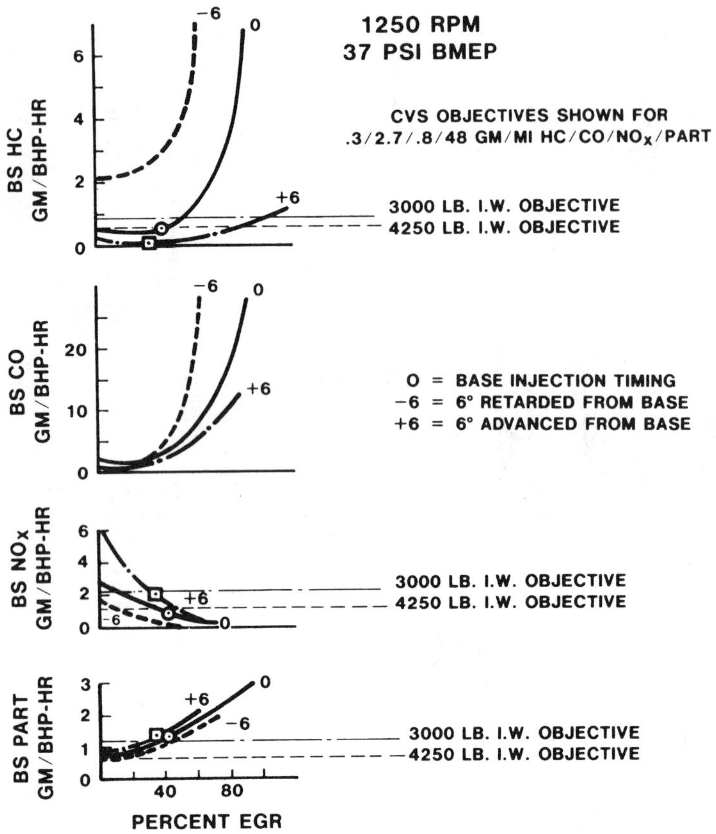

Figure 17 — **Effect of Vehicle Inertia Weight on Brake Specific Emission Objectives and EGR and Injection Timing Calibration Trends**

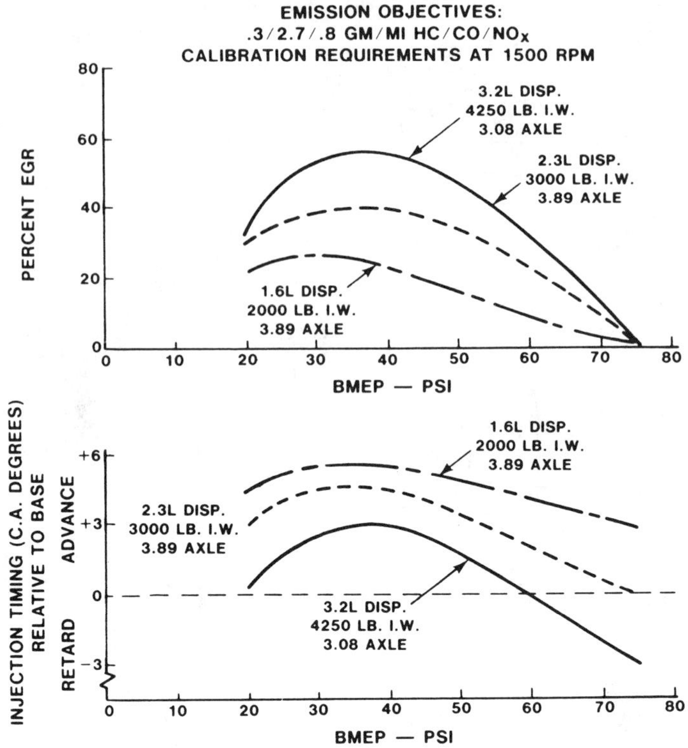

Figure 18 — **Typical Calibration Requirements for the .41/3.4/1.0 gm/mi HC/CO/NO_X Emission Requirements in 2000, 3000 and 4250 Lb. Inertia Weight Diesel Powered Vehicles**

Typical calibration requirements at 1500 RPM engine speed for the 2000, 3000 and 4250 lb. inertia weight vehicles are shown in Figure 18. Generally, the calibration strategy trends shown in Figure 17 are evident in the overall calibration strategies shown in Figure 18. The EGR requirement increases as the inertia weight increases. As inertia weight increases, the amount of injection advance decreases.

The general requirements for a low emission calibration for a swirl chamber Diesel are shown in Figure 19 and described below:

1. EGR is used to displace a portion of the intake air charge. The amount of EGR is usually limited by the smoke or particulate emissions at heavy loads or by HC emissions at light loads so that an air fuel ratio within the range of 20 to 40 is maintained. Richer air fuel ratios usually result in rapid increases in particulate emissions.
2. Advanced injection timing is used for HC control at light loads whereas retarded injection timing is used for NO_X control at heavy loads where the amount of EGR is limited by the small amount of excess air that can be replaced with EGR.

A summary of the vehicle emission projections is shown in Figure 20 on a plot of particulate emissions vs. NO_X emissions for the 2000, 3000 and 4250 lb. inertia weight vehicles. Particulate emissions are shown to increase significantly for the higher inertia weight vehicles as the NO_X is reduced to approach the .8 gm/mi objective level. However, by using an optimum schedule for modulated EGR and injection timing, these projections indicate that the particulate emissions may be controlled to nearly the baseline levels.

Also shown on Figure 20 are the engineering objectives assumed to be required to meet the .6 gm/mi particulate level and the 1.0 gm/mi NO_X level. A 3000 lb. vehicle with a 2.3L engine and an optimized calibration is projected to approach the 1.0 gm/mi NO_X and .6 gm/mi particulate objective levels. Vehicles of less than 3000 lb. inertia weight are also projected to meet the 1.0 gm/mi NO_X and .6 gm/mi particulate objective levels. None of the vehicles analyzed with current swirl chamber Diesel engines is projected to approach the engineering objectives for the . gm/mi particulate level.

In computing the optimized vehicle emission projections, trend was observed to indicate that particulate emissions tend t

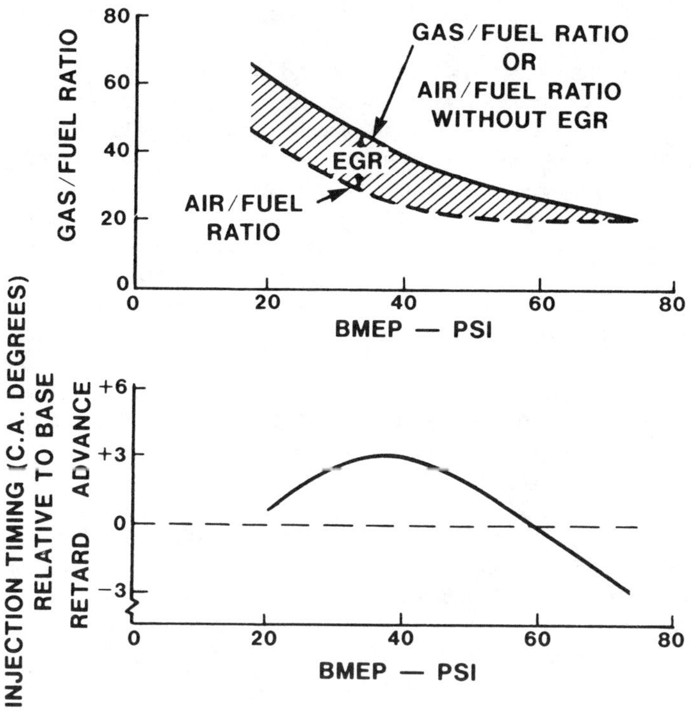

Figure 19 — General Requirements of a Low Emission Calibration for a Swirl Chamber Diesel

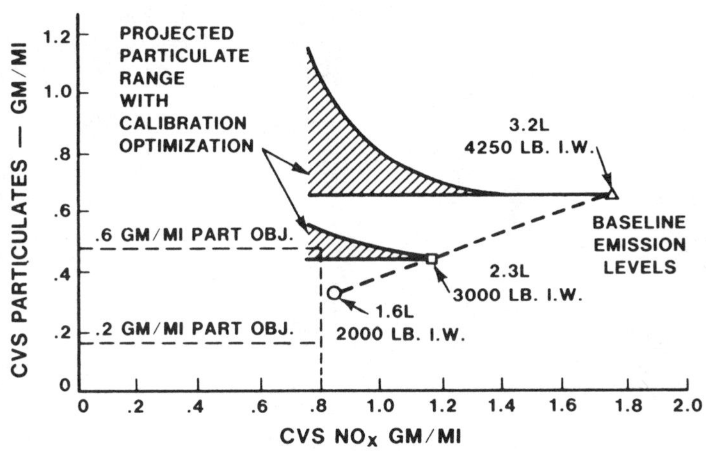

Figure 20 — Summary of Projections of Particulate Emissions vs. NO_X Emissions for a Range of Vehicle Inertia Weights

increase as the HC emissions are reduced at a controlled NO_X level. These trends are shown in Figure 21 for two NO_X levels. This trade-off between particulate and HC emissions at a controlled NO_X level occurs when injection timing advance is used to reduce HC emissions. With advanced injection timing, particulate emissions increase due to the increased EGR requirement and the reduced air charge in the prechamber at the time of fuel injection and combustion.

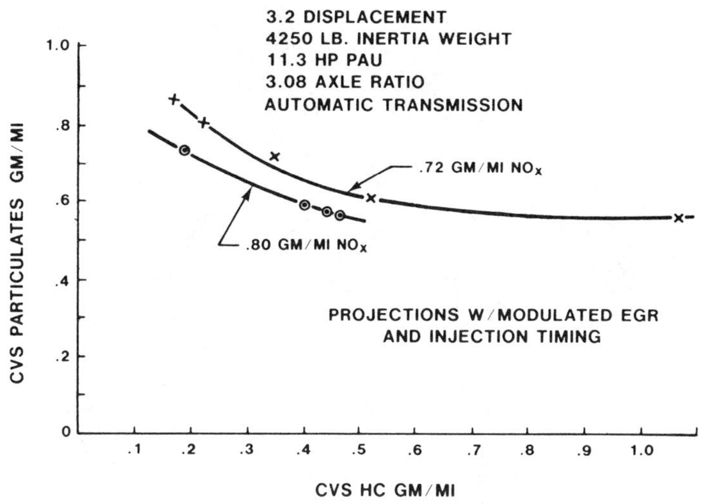

Figure 21 — Projected Particulate Emissions vs. HC Emissions for Two NO_X Levels

VEHICLE RESULTS

To verify selected vehicle emission projections, chassis dynamometer vehicle emission tests were conducted on modified Diesel vehicles. Experimental emission control systems were developed for a 2.1L and a 2.3L Diesel engine, which was identical to the 2.3L mapping engine, and these engines were installed and tested in vehicles. In addition to chassis dynamometer emission tests on the actual vehicles, a scaling technique was developed which allowed these vehicles to be operated as a portion or a multiple of a larger or smaller vehicle, respectively. The vehicle test results at these conditions were scaled up or down to the appropriate scaled or simulated vehicle.

Tests were conducted on the following vehicles or scaled vehicles:

1. Actual test on a 2.3L engine in a 3000 lb. inertia weight vehicle.
2. A 3.2L engine in a 4250 lb. inertia weight vehicle in which the test results were obtained by scaling the data obtained from tests on the vehicle with the 2.1L engine.
3. A 1.6L engine in a 2000 lb. inertia weight vehicle in which the test results were obtained by scaling the data obtained from tests on the vehicle with a 2.3L engine.

The technique for selecting the scaled vehicle chassis dynamometer test parameters is shown in Figure 22 for the case of a 3.2L engine in a 4250 lb. inertia weight vehicle scaled from test results on a vehicle with a 2.1L engine with similar combustion characteristics. The following procedures were used in performing these scaled vehicle tests:

1. The ratio of actual engine displacement to scaled engine displacement determined the scaling factor which was subsequently applied in the determination of inertia weight and road load for the test vehicle.
2. The engine displacement/vehicle inertia ratio for the test vehicle was selected to equal the engine displacement/vehicle inertia weight ratio of the scaled vehicle.

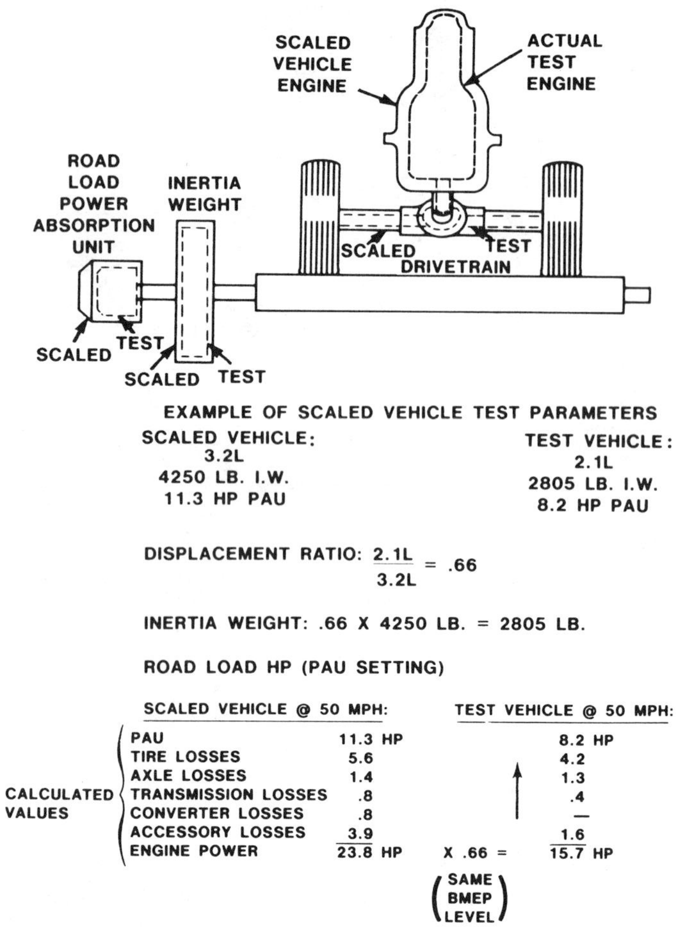

Figure 22 — Technique for Selecting Scaled Vehicle Chassis Dynamometer Test Parameters

Therefore, the inertia test weight was selected as follows to ensure correct scaling of this ratio:

$$IW_{test} = IW_{scaled} \times \frac{\text{Engine Disp}_{test}}{\text{Engine Disp}_{scaled}}$$

3. The chassis dynamometer PAU* HP was selected as follows to ensure that the test engine operated at the same BMEP level as the scaled engine:

$$\text{Engine HP}_{scaled} = \text{PAU HP}_{scaled} + \text{Drivetrain losses}_{\text{scaled vehicle}}$$

$$\text{Engine HP}_{test} = \text{Engine HP}_{scaled} \times \frac{\text{Engine Disp}_{test}}{\text{Engine Disp}_{scaled}}$$

$$\text{PAU HP}_{test} = \text{Engine HP}_{test} - \text{Drivetrain losses}_{\text{test vehicle}}$$

* PAU is the horsepower absorbed by the power absorption unit of the chassis dynamometer at 50 mph.

Drivetrain losses include tire, axle, transmission and accessory losses.

4. The measured emissions and fuel consumption of the test vehicles were multiplied by the ratio of scaled to test vehicle engine displacements to compute the emissions and fuel consumption for the scaled vehicles. Since small differences existed in some cases between the measured "scaled" fuel economy and the projected fuel economy for the scaled vehicle, a correction factor was developed which was equal to the ratio of the projected fuel consumed to the actual fuel consumed for the scaled test vehicle. This correction was used since the gm/mi emission levels are directly related to the quantity of fuel consumed during the emission test. Once this factor was developed from baseline vehicle tests, it was subsequently applied to all emission and fuel economy test results obtained from the test vehicle operated at the scaled test parameters.

The Diesel engines in the test vehicles were equipped with either a vacuum-mechanical or an electronic EGR control system to test strategies for selected vehicle projections using only EGR at the base injection timing schedule. The vacuum-mechanical EGR control system was designed to provide a scheduled EGR flow rate as a function of engine load. The fuel injection pump was modified to accommodate this control system. The electronic control system was developed to provide greater flexibility in the functions provided by the vacuum mechanical system and this system was used on one of the experimental test vehicles.

EGR calibration strategies were developed from the mapping projections for the 2000, 3000 and 4250 lb. vehicles. The scaled vehicle and actual vehicle test results for selected axle ratios are compared to vehicle projections in Figure 23. The agree-

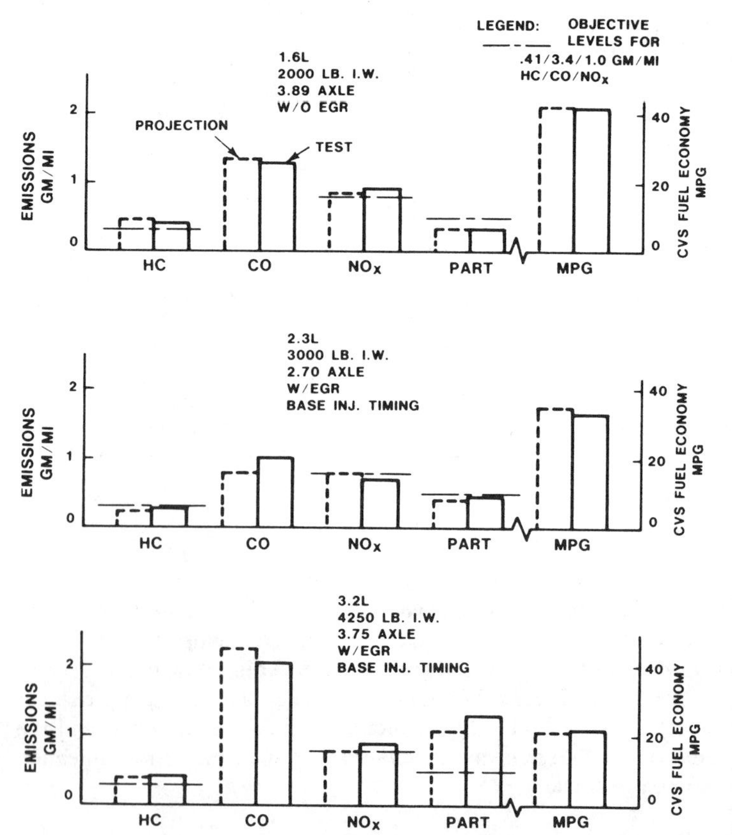

Figure 23 — Comparison of Selected Mapping Projections with Scaled or Actual Vehicle Test Data

ment between these vehicle test results, shown by the solid bars, and mapping projections, shown by the dashed bars, is generally good. The differences shown between test results and projections are generally within the range of variations experienced with Diesel vehicle emissions measured at different times or on different chassis dynamometer test facilities. The relatively good agreement between projections and vehicle test results for selected simple calibration strategies gives an indication of the accuracy of the projections and provides encouragement that the projections for more complex control strategies may also be achieved if suitable control system hardware were developed and no additional adverse transient effects were incurred.

NO_X — PARTICULATE RELATIONSHIP

Published Diesel vehicle emission data,[23, 28, 29] test results from experimental Diesel vehicles equipped with EGR and mapping projections were used to construct a plot of CVS-CH particulate emissions vs. NO_X emissions for a range of vehicle inertia weight classes ranging from 2250 lb. to 4500 lb. This graph, shown in Figure 24, indicates that NO_X and particulate emissions of production vehicles, represented approximately by the right hand termination points of the curves, increase significantly as the inertia weight is increased. Reductions in NO_X emission levels from the baseline levels shown for production vehicles are obtained by the use of EGR. As EGR is applied to reduce NO_X emissions toward the .8 gm/mi NO_X engineering objective, particulate emissions increase. Near the .8 gm/mi NO_X objective level, the particulate emission increase is greatest for the 4500 lb. inertia weight vehicle since the largest amount of EGR is required to provide the largest reduction from the high baseline NO_X emission level. An experimental combustion system in a 4500 lb. inertia weight vehicle, reported to the Environmental Protection Agency[29], is shown by the dotted line on this graph. These results indicate a potential for improving the current particulate vs. NO_X emission relationship under carefully controlled laboratory conditions. Mapping projections, shown earlier in Figure 20, indicate potential for a slight reduction from the particulate emission levels at a .8 gm/mi NO_X level, shown in Figure 24, with the development of complex control systems to implement optimized control strategies for EGR and injection timing.

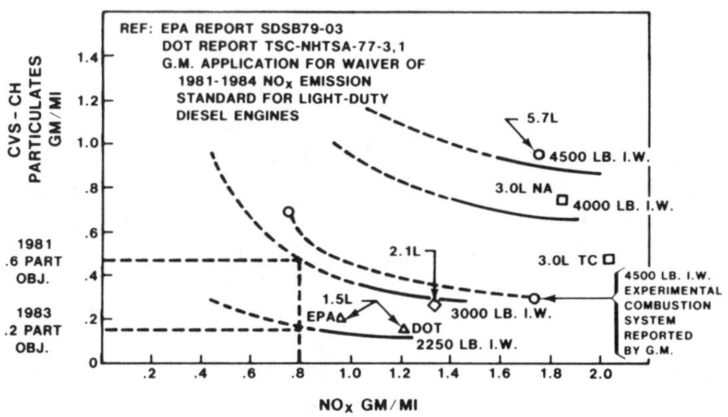

Figure 24 — CVS-CH Particulate Emissions vs. NO_X Emissions for a Range of Vehicle Inertia Weights from 2250 Lb. to 4500 Lb.

Shown on the particulate vs. NO_X emission plot in Figure 24 are the estimated engineering objectives required to meet the 1.0 gm/mi NO_X emission level and the .6 gm/mi and .2 gm/mi particulate emission levels proposed for 1981 and 1983, respectively[24]. As indicated previously by the projections shown in Figure 20, a 2.1L Diesel equipped with EGR in a 3000 lb. inertia weight vehicle marginally approaches the low mileage objectives assumed to be required for a .6 gm/mi particulate level at a 1.0 gm/mi NO_X level. However, limited evidence has been reported to suggest that EGR may have an adverse effect on engine durability[29]. Below 2000 lb. inertia weights, the low mileage emission objectives for the .2 gm/mi particulate level at a 1.0 gm/mi NO_X level may be approached, although no vehicle data are available to confirm this extrapolation from the data shown in Figure 24.

COMBUSTION SYSTEM MODIFICATIONS

The particulate emission vs. NO_X emission trends shown in Figure 24 indicate that only a very limited range of Diesel engine displacements and inertia weights are projected to comply with forecasted NO_X and particulate regulations. In order to broaden the range of Diesel vehicles which could comply with future requirements, combustion system modifications were investigated for their potential to provide simultaneous reduction of HC, NO_X and particulate emissions. These combustion system modifications were designed with the goal of improving the mixing of the injected fuel and air prior to and during combustion in order to avoid particulate formation conditions.

In the Diesel engine, mixing time is limited by the ignition delay time which, in turn, is influenced by the rate at which an ignitable fuel air mixture can be formed after injection of the fuel into the combustion chamber. Therefore, the possibility of achieving improved mixing without directly reducing the ignition delay time appears to be doubtful since ignition is likely to occur whenever a similar degree of mixing, at similar temperatures, is obtained. Nevertheless, the following engine modifications were estimated to have some limited potential to provide low particulate emissions:

1. Turbocharging to provide an increase in the oxygen availability and intake air temperature.
2. Improved fuel atomization to provide smaller droplets and shorter vaporization times.
3. Modified cylinder air charge fluid motions to enhance mixing.
4. Heat conservation and improved mixing during the expansion stroke to improve the oxidation of particulates that were formed in the combustion process.

TURBOCHARGING

Turbocharging has been proposed as a technique for reducing particulate emissions for the following reasons:

1. Turbocharging provides more oxygen which has the potential to improve fuel-air mixing and combustion and to enhance particulate oxidation during the expansion stroke.
2. Turbocharging provides higher intake charge temperatures which have the potential to improve fuel vaporization.

Tests on several turbocharged Diesel passenger cars have shown that only low levels of boost pressure are attained by the turbocharger during limited portions of a CVS emission test.

These results are illustrated in Figure 25 where the CVS cycle operating region is shown on a BMEP-engine speed map with the turbocharger boost pressure characteristics superimposed for a swirl chamber Diesel. As shown in the figure, the higher boost pressure levels are achieved outside the CVS cycle operating region. Since the boost pressure levels achieved in the CVS cycle are low, no significant changes in CVS emission levels due to turbocharging have been measured on several experimental Diesel test vehicles.

To determine the potential of elevated intake charge pressures at typical CVS operating conditions for reducing particulate emissions, dynamometer engine mapping tests were conducted at elevated intake charge pressures supplied by an electric motor driven positive displacement blower and at a back pressure schedule to simulate the back pressure imposed by the exhaust turbine of the turbocharger. Results from these tests are shown in Figure 26 for the 1700 RPM-75 psi BMEP engine operation condition. Emission, smoke, fuel consumption, air fuel ratio and exhaust temperature data are shown as functions of EGR for the naturally aspirated condition and for intake charge pressures of 2 and 4 in. Hg. These results indicate that without EGR, elevated intake charge pressure reduces HC emissions and increases NO_X emissions. The NO_X emissions increase due to the higher pressure and temperature conditions within the cylinder during combustion. The decrease in HC emissions is probably due to the beneficial effects of the increased intake charge temperature on the HC oxidation processes. Particulate emissions are not significantly affected by the elevated intake charge pressure without EGR.

Control of NO_X emissions to a desired objective level under turbocharged conditions requires an increase in the EGR rate as shown in Figure 26 for an arbitrarily assumed $BSNO_X$ objective level. At this controlled $BSNO_X$ level, HC emissions are lower at the simulated turbocharged conditions. The particulate emission levels at the controlled NO_X level show no significant effect due to turbocharging. However, at significantly lower $BSNO_X$ levels, turbocharging shows some potential for providing lower particulate emissions relative to the levels obtained at naturally aspirated conditions. At light loads, limited data at elevated intake charge pressures indicated that HC emissions were higher relative to naturally aspirated conditions. This HC emission increase may be due to an increase in the mid-air quench HC emission sources caused by the leaner air-fuel ratios provided by turbocharging.

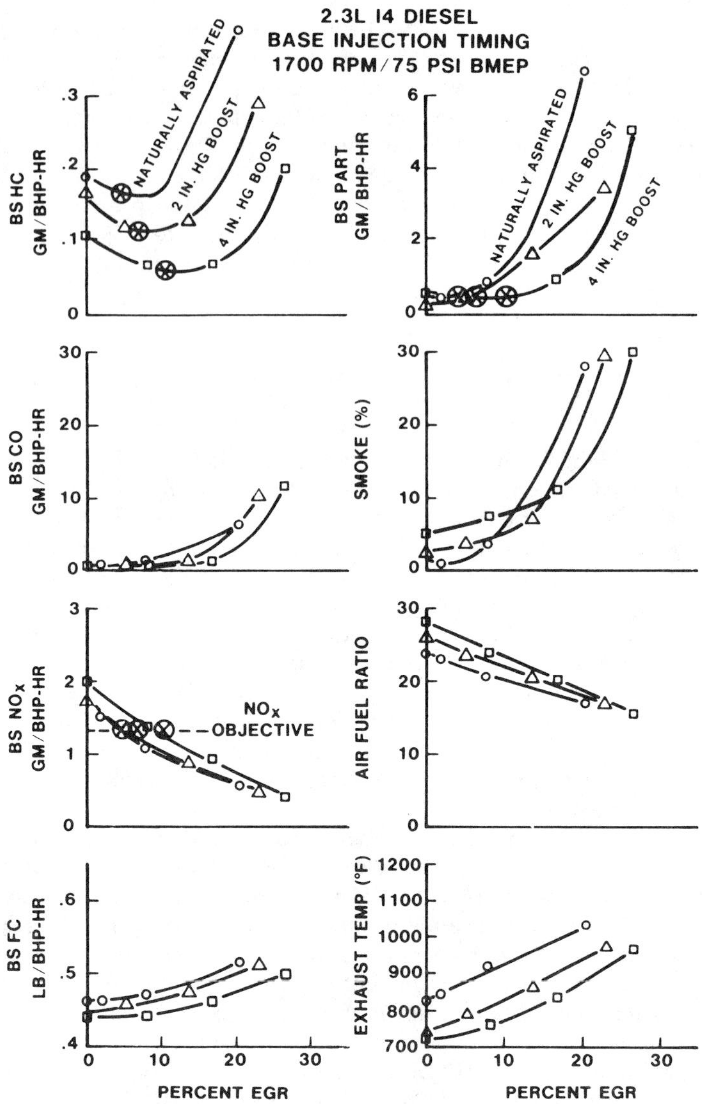

Figure 26 — Effect of Turbocharging and EGR on Emissions, Fuel Consumption and Operating Conditions of a Swirl Chamber Diesel Engine

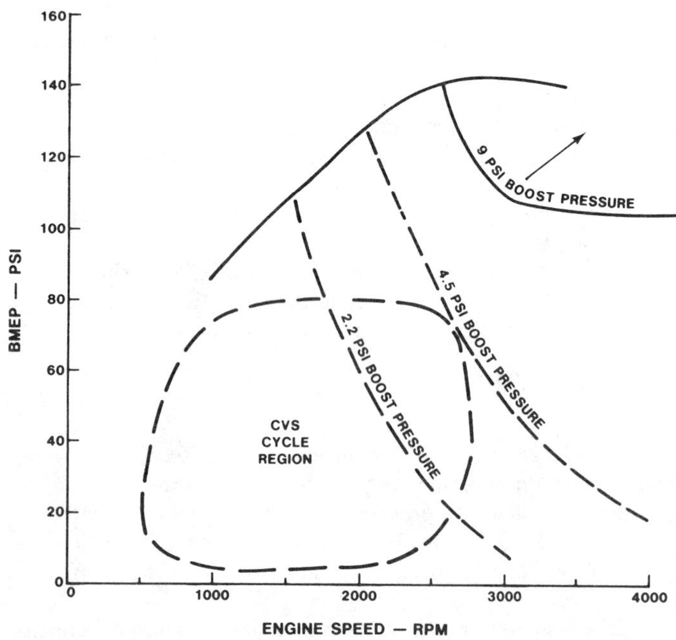

Figure 25 — CVS Cycle Operating Region Shown on an Engine BMEP-Speed Map with Typical Turbocharger Boost Pressure Characteristics Superimposed for a Swirl Chamber Diesel Engine

COMBUSTION SYSTEM TESTS

Numerous engine modifications with the potential for lowering the in-cylinder particulate emissions have been tested in an experimental dynamometer test engine similar to the engine used for the mapping tests described earlier. The modifications that were tested included the following:

1. Wider spray angle pintle injector
2. Two and three hole injectors with wider spray angles
3. Long throat prechamber
4. Low swirl prechamber
5. Heated fuel
6. Emulsified fuel
7. Heated air
8. Faster and slower injection rates
9. Thermally insulated prechamber
10. Higher coolant temperatures

The results of these tests are shown in Figure 27. To assess the extent of improvement obtained with a modification to the combustion system, particulate emissions are plotted as a function of HC emissions at a controlled $BSNO_X$ level which is usually achieved by use of EGR. At this controlled $BSNO_X$ level of 2.0 gm/bhp-hr, the particulate vs. HC emission relationship for the production 2.1L swirl chamber Diesel engine is shown by the solid line in Figure 27. As indicated by the data points plotted on the figure, none of the combustion system modifications achieved a significant improvement in the particulate vs. HC emission relationship. Most of the modifications resulted in a deterioration in the particulate vs. HC emission relationship. In addition, many of the modifications resulted in a loss of fuel economy and/or power.

Some of the apparent reasons for not realizing improvements with these modifications are discussed below. Attempts to achieve improved atomization with injector modifications and heated fuel did not yield lower particulate emissions because it is believed that the high air swirl motion in the prechamber was not matched to the modified fuel spray characteristics. As a result, the fuel spray could not penetrate the air swirl to achieve good mixing, but instead the fuel was swept against the prechamber wall forming rich fuel pockets which caused an increase in particulate and HC emissions. It is believed that the attempts to increase the temperature of the combustion conditions did not yield improvements because the modifications applied did not result in changes to the combustion gas temperatures which were adequate to modify the combustion conditions. In addition, increases in temperature could also decrease the ignition delay time which could increase the amount of droplet fuel burning which tends to increase particulate emissions.

Reductions in air swirl showed a deterioration in particulate emissions. Therefore, an increase in swirl or perhaps turbulence level of the air charge could possibly be developed to yield limited improvements in the HC vs. particulate emission relationship. More extensive modifications to the combustion chamber, fuel injection system and operating conditions may have the potential for further improvements in the HC vs. particulate relationship.

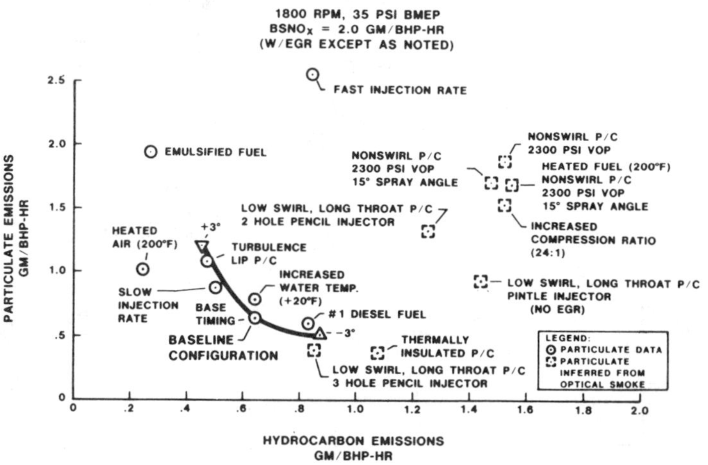

Figure 27 — Effect of Combustion System Modifications on Particulate and HC Emissions at a Controlled NO_X Level for a Swirl Chamber Diesel Engine

EXHAUST TREATMENT

Diesel combustion system modifications have not yet shown potential for significant reductions in particulate emissions at low levels of gaseous emissions. Therefore, to approach forecasted HC, NO_X and particulate emission requirements in moderate to larger size vehicles, treatment of the combustion gases after being exhausted from the cylinder may be required.

The potential for reducing particulates in the exhaust gas by the use of a manifold reactor for particulate oxidation was investigated. The exhaust gas from a single cylinder Diesel engine was electrically heated in a 50 ft. sample line in which particulate emissions were measured at 5 ft. intervals. The results of these tests, shown in Figure 28, are compared to a theoretical calculation for the oxidation of carbon(30,31). These results indicate that a temperature range well above 1500°F would be required to achieve significant oxidation of the particulate emissions within the residence time provided by an exhaust reactor with a volume equal to the engine displacement.

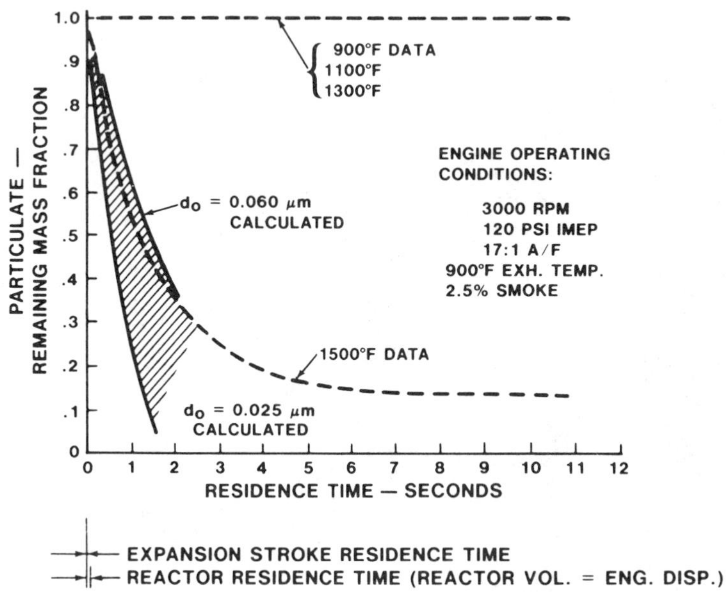

Figure 28 — Measured and Calculated Particulate Emission Oxidation Rates in a 50 Ft. Heated Exhaust Sample Line

A steady state exhaust temperature map for the 2.3L Diesel engine used for the mapping tests described earlier is shown in Figure 29. The CVS operating region for a typical Diesel vehicle is superimposed on this exhaust temperature map. Exhaust temperatures approaching 1100°F may be reached for short periods on the CVS cycle but this temperature is insufficient to provide particulate oxidation within a reasonable volume exhaust reactor. As a result of these exhaust temperature limitations, without external heat addition to a manifold reactor, a continuous oxidation concept for Diesel particulate control is not feasible due to this chemical kinetic limitation.

Since continuous oxidation of the particulate emissions is not feasible, techniques to trap or collect particulates and to oxidize them periodically at a later time have been investigated. The generalized plot of particulate emissions vs. NO_X emissions for a range of vehicle inertia weights, shown in Figure 24, indicates that collection efficiencies as high as 85% may be required of a particulate trap depending on the vehicle inertia weight and the combustion system. This efficiency level must be retained throughout the 50,000 mile AMA durability cycle.

Blockable or full filtration type and non-blockable or diffusion type Diesel particulate traps have been evaluated. Blockable traps usually result in higher back pressure which adversely affects fuel economy and performance. On the other hand, blowoff of collected particulates can be a problem with non-blockable traps. None of the traps evaluated to date have met all of the requirements for collection efficiency, pressure drop, volume and durability.

During experimental laboratory evaluations, traps loaded with particulates have been regenerated by placing them in an oven for 12 hours at 1000°F to oxidize the collected particulates. Experimental traps have also been regenerated on a dynamometer engine operated at a high load, high exhaust temperature condition. To evaluate regeneration of a particulate trap on an engine under controlled conditions, a simulated steady state representation of the AMA durability cycle for engine dynamometer operation was developed. Steady state engine speed-load and time conditions, selected to represent the AMA cycle on a dynamometer engine, are shown in Figure 30. The entire cycle is approximately two hours in duration with wide open throttle operation occurring only in the last few minutes of the cycle. A blockable type particulate trap was installed as close as possible to the exhaust ports of a 2.3L Diesel engine on the dynamometer, and the engine and trap system were tested over the simulated AMA cycle. The pressure drop across the particulate trap vs. cycle time obtained over the simulated AMA cycle are shown in Figure 30. Particulates are collected at varying rates for the different operating conditions as indicated by the different rates of increase in pressure drop vs. time. These collection rates depend on the particulate concentrations and exhaust flow rates at the light to moderate load conditions. On the first simulated wide open throttle acceleration condition, the collected particulates were rapidly oxidized at the 1260°F exhaust temperature obtained at the inlet to the trap. Trap regeneration is shown by the sharp drop in the trap pressure drop after the beginning of the first wide open throttle acceleration. Complete regeneration of the trap was demonstrated by the restoration of the initial value of the light load pressure drop at the beginning of the next simulated AMA cycle. The feasibility of achieving on-board regeneration during transient vehicle operation on the AMA cycle or in normal driving remains an open issue.

CONCLUSIONS

1. Projections of Diesel vehicle emissions from steady state engine dynamometer mapping data have shown generally good agreement with actual vehicle emission test results.
2. HC, CO, NO_X and particulate average emission indices of the Diesel swirl chamber combustion system over the CVS-CH emission test cycle were used to estimate vehicle inertia weights that could approach forecasted emission requirements. Without emission controls, vehicle inertia weights of 2000 lb. and less are projected to approach the low mileage engineering objectives for the .41/3.4/1.0/.6 gm/mi HC/CO/NO_X/particulate emission level. Without emission controls, NO_X is the emission which limits the highest vehicle inertia weight which can be used to meet these objectives.

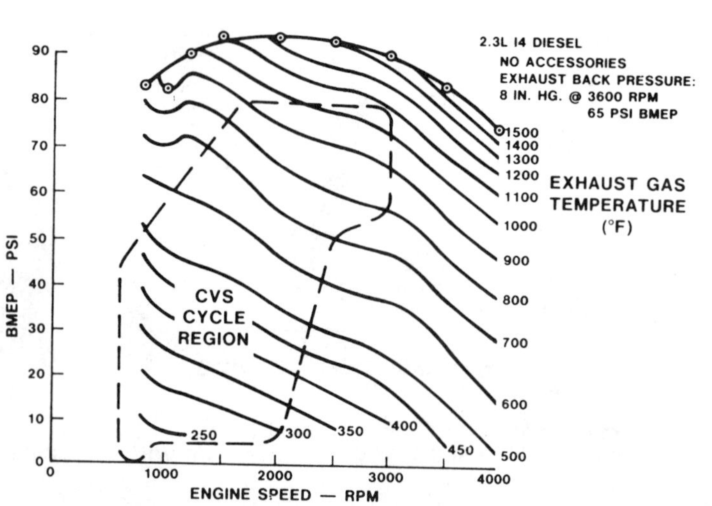

Figure 29 — Exhaust Temperature Map for a 2.3L I4 Swirl Chamber Diesel with a Typical CVS Operating Region Superimposed

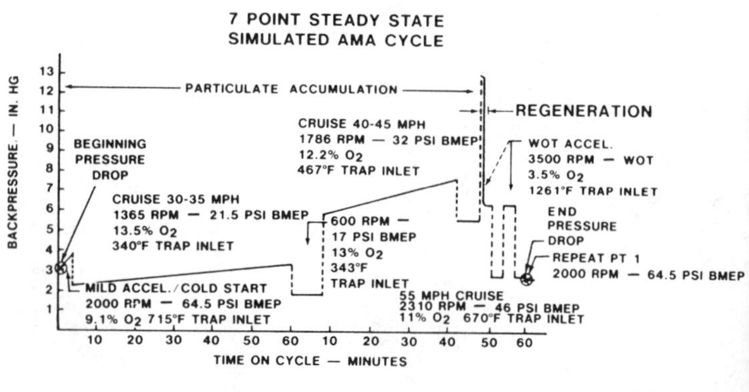

Figure 30 — Pressure Drop Across a Diesel Particulate Trap Operated Over a Steady State Simulated AMA Cycle with Regeneration Occurring During the First WOT Acceleration

3. The emission levels of the swirl chamber Diesel are significantly influenced by the engine displacement/vehicle inertia weight ratio and the vehicle axle ratio. Reductions in both parameters provide reductions in HC, CO and NO_X emissions without significant adverse effects on particulate emissions.
4. Mapping projections for swirl chamber Diesel engines indicate that low mileage engineering objectives for the .41/3.4/1.0 gm/mi HC/CO/NO_X emission level may be approached with complex, optimized control systems for modulated EGR and injection timing in vehicles up to 4250 lb. inertia weight. However, at the 4250 lb. inertia weight, particulate emissions substantially exceed the engineering objective required for the .6 gm/mi emission level. At 3000 lb. inertia weight, the low mileage engineering objectives for the .41/3.4/1.0/.6 gm/mi HC/CO/NO_X/particulate emission level are projected to be approached if a suitable control system for EGR and injection timing were developed. The achievement of these projected emission levels in vehicles will require the development of innovative engine control systems and the resolution of the possible adverse effects of EGR on engine durability that has been reported in Reference 29.
5. Current Diesel vehicle test results indicate that the particulate emissions increase significantly as EGR is used to reduce NO_X emissions toward the low mileage engineering objective assumed for a 1.0 gm/mi NO_X emission level. Mapping projections indicate that improvements in particulate emissions may be achievable with an optimized control system for EGR and injection timing.
6. Turbocharging does not appear to have the potential for reducing particulate emissions of a vehicle with a swirl chamber Diesel. Current turbochargers are largely ineffective in providing significant intake charge pressures during the CVS emission test cycle. At steady state engine dynamometer conditions, particulate emissions at moderate levels of controlled NO_X emissions were not affected by turbocharging.
7. Modifications to the swirl chamber Diesel combustion system have not resulted in a significant improvement in the HC, NO_X or particulate emissions and many of the modifications resulted in a loss of fuel economy and/or power.
8. The gas temperature required for thermal oxidation of particulates during the residence time of a manifold reactor exceeds 1500°F, which is significantly above the 1100°F maximum exhaust temperature of a swirl chamber Diesel operated over the CVS emission test cycle. Without external heat addition, a continuous oxidation concept for Diesel particulate control is not feasible.
9. A particulate trap device, which has not yet been defined, may be required to approach forecasted HC, NO_X and particulate emission requirements. An experimental trap has been regenerated by oxidizing the collected particulates at a high load engine dynamometer operating condition during a steady state simulation of the AMA cycle where the exhaust temperature exceeded 1260°F. The feasibility of achieving on-board regeneration during transient vehicle operation while retaining a low exhaust back pressure and adequate trap mechanical and thermal durability remains an open issue.
10. The future use of the Diesel engine in a full range of inertia weight class vehicles depends on the:
    a. Further development of a Diesel combustion system with low HC, NO_X and particulate emissions.
    b. Optimization of control systems for modulated EGR and modulated injection timing.
    c. A durable and efficient regenerative trap with low back pressure.

ACKNOWLEDGEMENT

The author acknowledges the many people who have contributed to this project. Many of the engine dynamometer mapping tests were conducted at the Southwest Research Institute, under contract from Ford Motor Company. The contributions of Messrs. C. Jones, I. Garwin, E. King, D. Kosacheff, J. Florek and J. Monty made the timely completion of this paper possible. In addition, many other people made valuable contributions during the preparation of this paper.

REFERENCES

1) H. W. Barnes-Moss and W. M. Scott, "The Light Duty Diesel Engine for Private Transportation," SAE Paper No. 750331, 1975.
2) M. L. Monaghan and J. J. McFadden, "A Light Duty Diesel for America?", SAE Paper No. 750330, 1975.
3) T. Suzuki and K. Usami, "A Modification of Combustion Systems for Low Exhaust Emission and Its Effects on Durability of Prechamber Diesel Engine," SAE Paper No. 760213, 1976.
4) R. S. Radovanovic and M. S. Djordjevic, "Diesel Precombustion Chamber Design and Its Influence on the Engine Performance and Exhaust Pollutants," SAE Paper No. 790496, 1979.
5) C. J. Walder, "Reduction of Emissions from Diesel Engines," SAE Paper No. 730214, 1973.
6) M. Amano, H. Sami, S. Nakagawa, and H. Yoshizaki, "Approaches to Low Emission Levels for Light-Duty Diesel Vehicles," SAE Paper No. 760211, 1976.
7) C. J. Walder, "Some Problems Encountered in the Design and Development of High Speed Diesel Engines," SAE Paper No. 978A, 1965.
8) D. R. Nightingale, "A Fundamental Investigation into the Problem of NO Formation in Diesel Engines," SAE Paper No. 750848, 1975.
9) R. Pischinger and W. Cartellieri, "Combustion System Parameters and Their Effect Upon Diesel Engine Exhaust Emissions," SAE Paper No. 720756, 1972.
10) K. Oblander and H. G. Schmidt, "Influence of Engine Design and Fuel Quality on Diesel Emissions, I Engine Design, II Fuel Quantity," Proceedings of the 9th World Petroleum Congress, London, Volume 6, 1975.
11) M. F. Russell, "Recent CAV Research into Noise, Emissions and Fuel Economy of Diesel Engines," SAE Paper No. 770257, 1977.
12) G. Greeves, I. M. Khan, C. H. T. Wang, and I. Fenne, "Origins of Hydrocarbon Emissions from Diesel Engines," SAE Paper No. 770259, 1977.

13) V. K. Duggal, T. Priede, and I. M. Khan, "A Study of Pollutant Formation within the Combustion Space of a Diesel Engine," SAE Paper No. 780227, 1978.

14) K. T. Rhee, P. S. Myers, and O. A. Uyehara, "Time and Space Resolved Species Determination in Diesel Combustion Using Continuous Flow Gas Sampling," SAE Paper No. 780226, 1978.

15) J. E. Bennethum, J. M. Mattavi and R. R. Toepel, "Diesel Combustion Chamber Sampling — Hardware, Procedures and Data Interpretation," SAE Paper No. 750849, 1975.

16) I. A. Voiculescu and G. L. Borman, "An Experimental Study of Diesel Engine Cylinder — Averaged NO_X Histories," SAE Paper No. 780228, 1978.

17) W. M. Scott, "Looking in on Diesel Combustion," SAE Paper No. 690002, 1969.

18) A. B. Allan and A. Khin, "Investigation of Combustion Phenomena in a Swirl Chamber Compression Ignition Engine Using Schlieren Techniques," SAE Paper No. 700500, 1970.

19) C. T. Vuk, M. A. Jones and J. H. Johnson, "The Measurement and Analysis of the Physical Character of Diesel Particulate Emissions," SAE Paper No. 760131, 1976.

20) Individual Vehicle Data for 1980 Edition of EPA Gas Mileage Guide.

21) Individual Vehicle Data for 1979 Edition of EPA Gas Mileage Guide.

22) Individual Vehicle Data for 1978 Edition of EPA Gas Mileage Guide.

23) E. Danielson, "Particulate Measurement — Light Duty Diesel Particulate Baseline Test Results", EPA Technical Support Report SDSB 79-03, January, 1979.

24) "Notice of Proposed Rulemaking, Particulate Regulation for Light Duty Diesel Vehicles," Federal Register, Vol. 44, No. 23, February 1, 1979.

25) A. J. Scussel, A. O. Simko and W. R. Wade, "The Ford PROCO Engine Update," SAE Paper No. 780699, August, 1978.

26) R. E. Baker and E. E. Daby, "Engine Mapping Methodology," SAE Paper No. 770077, February, 1977.

27) J. E. Auiler, et al, "Optimization of Automotive Engine Calibration for Better Fuel Economy — Methods and Applications," SAE Paper No. 770076, February, 1977.

28) "Data Base for Light-Weight Automotive Diesel Power Plants", Prepared by Volkswagenwerk for V. S. Department of Transportation, Report No. DOT TSC-NHTSA-77-31, September, 1977.

29) "General Motors Application for Waiver of the 1981-1984 NO_X Emission Standard for Light-Duty Diesel Engines," submitted to EPA, May 1, 1979.

30) J. Nagle and R. F. Strickland-Constable, "Oxidation of Carbon Between 1000°C-2000°C," Proceedings of the Fifth Carbon Conference, Vol. 1 pp. 154-164 (1962).

31) D. L. Stivender and C. A. Amann, "The Formation and Oxidation of Diesel Soot — A Rudimentary Study," Paper No. CSS/C1-79-15, Spring Meeting, Central States Section, The Combustion Institute, April, 1979.

Diesel Engined, Light Duty Vehicles for an Emission Controlled Environment

C. C. J. French and D. A. Pike
Ricardo Consulting Engineers, Ltd.
Bridge Works, Shoreham-by-Sea

THE PURPOSE OF THE FIRST PART of the paper is to show the interaction between fuel economy, gaseous emissions, and exhaust particulate emissions of current diesel engined light duty vehicles and to estimate the levels which it might be possible to achieve in the future.

The second part describes and gives the results of an extensive programme of tests on a Mercedes 300D car fitted with a turbocharger and with modulated exhaust gas recirculation (EGR) to establish the feasibility of achieving very low levels of NOx emissions and to quantify the penalties which would result with respect to fuel economy and particulate emissions.

PART 1 - INTERACTION BETWEEN EXHAUST EMISSION LEVELS AND FUEL ECONOMY

Diesel engined light duty vehicles have a much longer history than is generally known in the United States. Diesel cars were introduced by Citroen and Mercedes in the 1930s, and from the 1950s they came into wider and wider use in Europe, especially in the field of taxi operation where, with much start/stop operation, the fuel saving of up to 50% became vital to the taxi service.

In the United States unit fuel costs have always been very low since fuel taxes have never been used as a major source of revenue. As a result there has never been the incentive to switch to diesel engines in high mileage light duty vehicles and it is only relatively recently, with the need to conserve petroleum fuel and with the consequent advent of fuel economy legislation, that American vehicle manufacturers have turned their attention to the development of high speed diesel engines.

The indirect injection diesel engine, be it either prechamber or swirl chamber, has naturally low levels of exhaust emissions. It has for example been widely used for mines service due to low NOx, and the relative absence of exhaust odour due to low hydrocarbons was responsible for the choice of such engines for London buses in the 1930s. In light duty service the NOx levels have been further reduced by the desirability of running with retarded injection timings to minimise engine noise.

*Paper 790761 presented at the International West Coast Meeting, Portland, Oregon, August, 1979.

ABSTRACT

The diesel engined light duty vehicle shows a substantial improvement in fuel economy over its gasoline engined counterpart. It will however have very great difficulty in meeting proposed gaseous and particulate exhaust emission levels.

The interactions between fuel economy, NOx emissions, and exhaust particulate emissions as demonstrated by an extensive series of tests on a Mercedes Benz 300D car with modulated exhaust gas recirculation are correlated with published data from other diesel engined vehicles to predict levels which could be achieved with current technology and to estimate the fuel economy penalties.

Due to a high friction level at high engine speeds, the diesel engine shows little fuel economy advantage over the gasoline engine under full power conditions. With the high power/weight ratio of light duty vehicles however, and especially with low maximum permitted speeds on our highways, today's engines do not run at high speeds in service, and there is a resulting fuel economy advantage for the diesel engined vehicle of 25-30% over the Federal Test Procedure cycle, of which some 11% or so is due to the higher volumetric calorific value of the fuel.

The fuel economy of modern diesel cars, as exhibited by 1979 EPA certification data (1)*, is shown in Fig.1. The fuel consumption is of course strongly affected by vehicle weight, as shown by the 45° line drawn on the figure. The majority of the points will be seen to lie close to the mean line, but three vehicles give about 18% better economy. One of these vehicles, with 2250 lb inertia weight, is a special turbocharged Volkswagen Rabbit to a US Department of Transportation specification, and the other two are pick-up type vehicles. While insufficient information is given to be certain, it is quite possible that these three vehicles have a lower gearing between the engine and the rear wheels, which reduces vehicle performance but is known to improve the fuel economy.

The NOx and particulates emissions for these vehicles are shown in Fig.2. It can be seen that there is a greater spread of this data than in the case with fuel economy but that, once again, vehicle weight is an important parameter. This is shown clearly for NOx where once again a 45° mean line would seem to be appropriate. The very large scatter for particulates makes the slope of the line less certain, although again 45° seems not to be unreasonable. The results for the turbocharged VW Rabbit and Mercedes 300 SD cars indicate that turbocharging appears to be very effective in reducing particulates if the baseline level is high.

The hydrocarbon levels have not been plotted as they exhibit such a large scatter that it is difficult to detect a definite trend. Part at any rate of the large scatter is probably due to the presence of secondary injections as these are known, from unpublished Ricardo work, to have a serious effect on hydrocarbon emissions. The presence of turbocharging reduces HC levels, and it is noteworthy that the two vehicles with the lowest levels have turbocharged engines.

FUTURE EMISSIONS ENVIRONMENT - While the vehicles whose performance has been described in the previous section can meet 1979 legislation, this legislation is of course to be tightened in subsequent years. For 1980, with a move to 0.4 g/mile hydrocarbons, it is clear that work will be necessary on many of the vehicles. Levels well inside this are however being achieved currently on 4000 lb and 4500 lb vehicles, and by close attention to the combustion system and especially to the injection system, it should be possible to meet legislative limits on all vehicles. The problem is however more severe in California, where 1.0 g/mile NOx is legislated in 1980, since the majority of diesel engined vehicles do not currently meet 1.5 g/mile NOx. In the future, from current US proposals, we shall have to meet tighter NOx levels together with particulate and fuel economy legislation. It is possible that very low NOx or particulate levels may be set, so that diesel engined vehicles may just not be able to meet them technically, but before this point is reached it seems likely that the heavier vehicles will be ruled out.

Much of the problem is one of interaction between various limits, as demonstrated by the results achieved in the programme on the Mercedes 300D car described in the second part of this paper. The results for this 4000 lb inertia weight vehicle with a low level of turbocharging , and as NOx is reduced by introducing modulated EGR, and by injection retard, are summarised in Figs. 3 and 4.

Fig. 3 shows the interaction between NOx and HC. It can be seen that while there is a unique relationship, it should be possible to achieve the 0.7 g/mile NOx and 0.3 g/mile HC which would be required to meet production levels of 1 g/mile NOx and 0.41 g/mile HC. The two difficulties which are likely to arise however are firstly that the durability of EGR systems is not yet established, and indeed there is quite a lot of evidence that fouling of the engine inlet tracts by exhaust particulate deposits will be a major problem, and secondly, the broad (42-50) cetane spread of US No.2D fuel, and in particular the low cetane rating of the EPA test fuel at 42 cetane. The programme was carried out with 50 cetane fuel, which indeed is also typical of two samples of US fuel which have been used by Ricardo. With 42 cetane fuel, an increase in HC levels is probable, and it will not be possible to go so low in NOx for a given HC limit.

Fig. 4 shows the relationship between NOx and fuel consumption, and particulates and fuel consumption. It can be clearly seen that reducing NOx levels worsens fuel consumption and that, although the data is somewhat scattered, it also increases particulates. From these test results it would not be possible to run at

* Numbers in parenthesis designate References at end of paper.

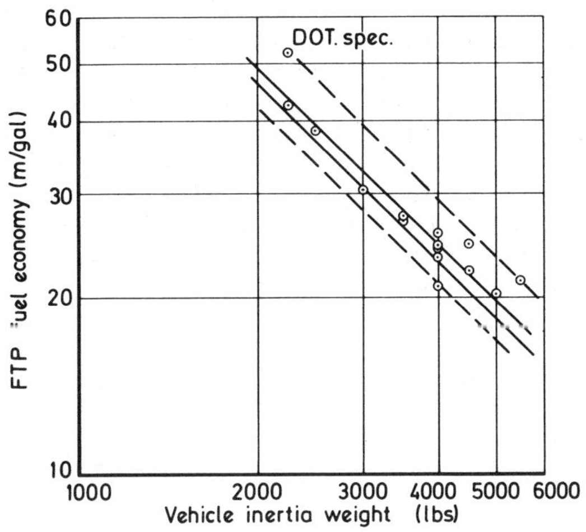

Fig. 1 - Fuel economy of diesel cars

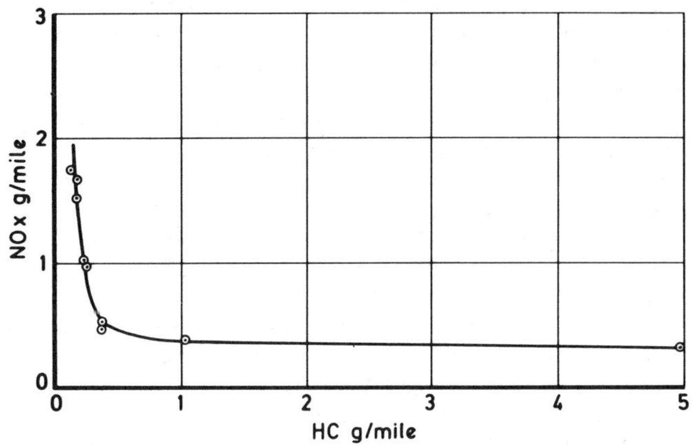

Fig. 3 - Interaction between NOx and HC emissions

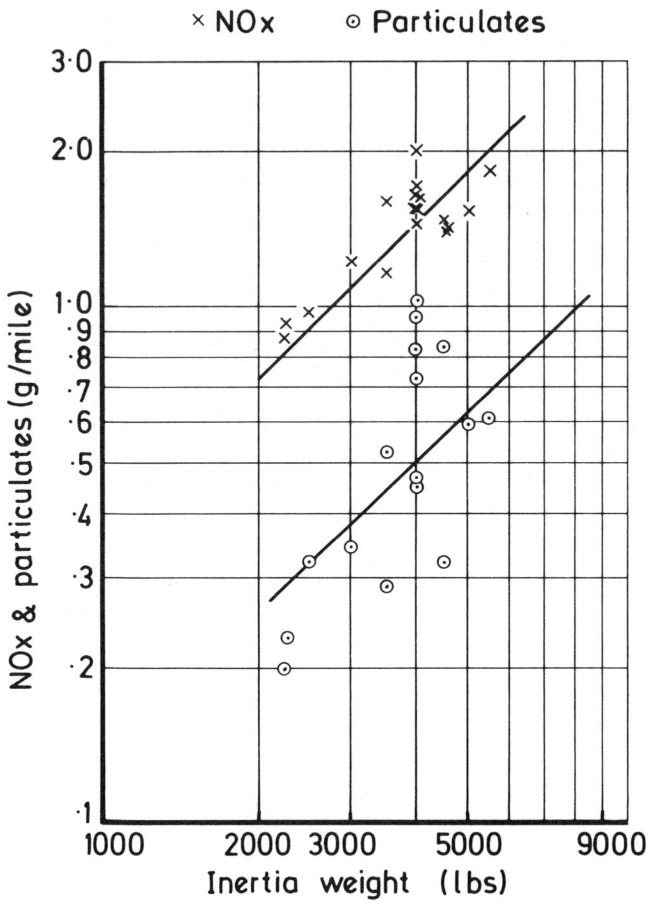

Fig. 2 - NOx and particulates emissions for diesel cars

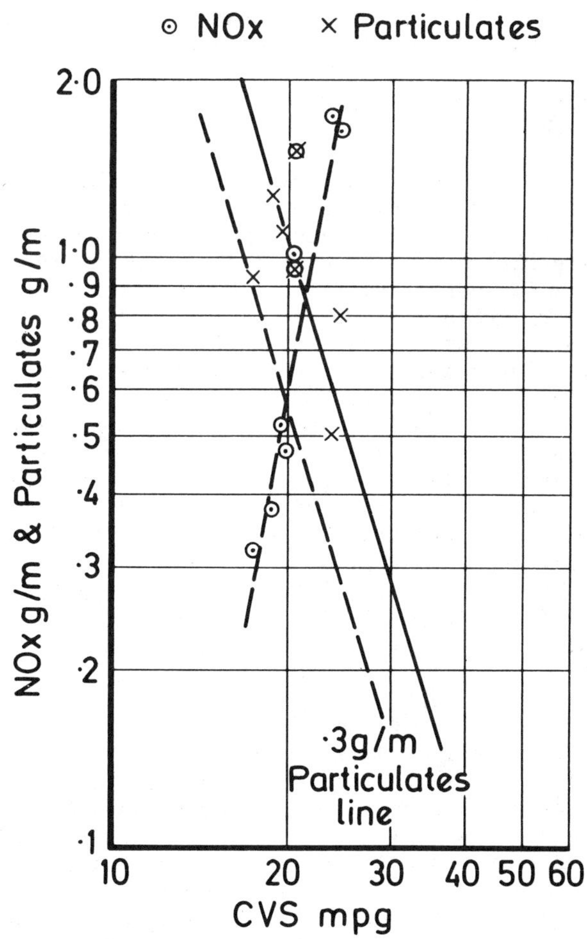

Fig. 4 - Relationship between NOx and fuel consumption and between particulates and fuel consumption lower than 1.5 g/mile NOx with a corresponding particulates level of 0.6 g/mile. In neither case does this leave any scope for production tolerances.

Fig.2 indicates however that it should be possible, with the best of today's technology, to lower the particulates level of a 4000 lb car with today's injection timings to just over 0.3 g/mile. Taking this as a starting point, it would appear that with EGR and injection retard it should be possible to achieve 0.5 g/mile particulates, 0.75 g/mile NOx, and 0.28 g/mile HC, i.e. to meet the production limits of 0.6 g/mile particulates, 1 g/mile NOx, and 0.4 g/mile HC. There would however be a fuel economy penalty of 12% or so.

With current technology it would appear to be very difficult to meet a particulate level of 0.2 g/mile. Hang-on devices have been tested with particulate reduction efficiencies of 30% to 40% (2) but without proven durability. With the aid of one of these devices, a 2250 lb production car should meet 0.2 g/mile particulates with 1 g/mile NOx, and if so, a 3500 lb test car could possibly just get inside 0.2 g/mile particulates, but with about 1.4 g/mile NOx and no allowance for production scatter.

Part II - MODULATED EXHAUST GAS RECIRCULATION PROGRAMME

The three well known ways of reducing NOx are retard of injection timing, exhaust gas recirculation, and water injection either with the fuel as an emulsion or separately.

The work reported in this section of the paper, which was carried out on behalf of the U.S. EPA, was aimed at quantifying the effects of using modulated exhaust gas recirculation (EGR) as a means of NOx reduction.

EGR has been demonstrated by many workers as an effective means of reducing NOx levels, acting partly by lowering the cycle temperatures and partly by reducing the oxygen available for combination with the nitrogen. It had also been demonstrated by Daimler-Benz, and reported to EPA in 1973, that with high levels of EGR a 147 CID diesel engine can meet 0.4 g/mile NOx levels, although with a serious loss of performance and unacceptably high levels of exhaust smoke.

The aims of the Ricardo/EPA programme were to design, develop, and demonstrate a modulation system for EGR to obtain the best practical reduction in NOx formation. A turbocharger was available to minimise any loss of performance due to the use of EGR, and the ratio of EGR employed was to be applied preferentially to the operating modes shown to contribute the largest amounts of NOx when measured during the driving cycle of the Federal Test Procedure (FTP).

For the purposes of the test, a Mercedes 300D passenger car was loaned by Daimler-Benz, and the baseline emission characteristics were established with this vehicle. The engine was subsequently installed on a test bed and its EGR response determined before it was re-installed in the car fitted with a modulated EGR system, which was developed for this use and which was sufficiently flexible to be tested under a range of EGR conditions. The levels of exhaust emissions obtained from the car were compared with levels which had been predicted by computer synthesis from the test bed results.

The vehicle engine details together with the instrumentation employed are listed in the Appendix.

VEHICLE BASELINE TESTS - The baseline emission and fuel economy data, obtained from chassis dynamometer tests on the vehicle in its as-delivered form, are given in Table 1 together with the published data for this vehicle type. It can be seen that there was quite good correlation between the two sets of data.

Exhaust smoke measurements over the driving cycle gave a peak instantaneous level of 20% opacity during cold acceleration, reducing to 10% when hot.

EFFECT OF TURBOCHARGER - The turbocharger was intended to offset the adverse effects on performance of employing EGR and as such it was not intended to add additional fuel to increase the engine power.

The Garrett AiResearch T3 turbocharger was fitted to the engine in the vehicle, using manifolds supplied by Daimler Benz. Tests were carried out using two alternative waste gate springs. For the major part of the programme the lighter spring was chosen, giving a boost pressure of 0.25 bar.

With no additional fuelling, the turbocharger was found to act as a parasitic loss, and as a result there was a depreciation in engine performance as shown in the data given in Table 2.

Table 1

	Ricardo Tests			Published D/B Results
Emissions 1975 FTP	Test 1	Test 2	Mean	(Reference 3)
HC g/mile	0.189	0.176	0.183	0.229
NOx g/mile	1.671	1.654	1.660	1.550
CO g/mile	0.943	0.929	0.940	1.430
Particulates g/mile	0.775	0.832	0.804	
Fuel Economy				
CVS Test m/gal (US)	24.77	23.88	24.30	23.9
Highway test m/gal(US)	28.96	30.94	29.95	32.4
Vehicle Performance				
Acceleration time sec.		Mean		
0-40 mph		9.6		
0-60 mph		21.2		19.0 (0-62 mph)
20-60 mph		18.7		
Drive-by Noise (SAE J986a)				
30 mph low gear		74.2 dBA		

Table 2

	Turbocharged			
Emissions 1975 FTP	Test 1	Test 2	Mean	As-Received
HC g/mile	0.141	0.14	0.14	0.183
NOx g/mile	1.773	1.72	1.75	1.66
CO g/mile	1.032	0.849	0.94	0.94
Particulates g/mile	0.498	0.501	0.5	0.8
Fuel Economy				
CVS test m/gal (US)	24.15	23.48	23.8	24.3
Highway test m/gal (US)	25.3	25.83	25.6	29.95
Vehicle Performance				
Acceleration time sec.		Mean		
0-40 mph		9.9		9.6
0-60 mph		22.4		21.2
20-60 mph		19.2		18.7
Drive-by Noise Test (SAE J986a)				
30 mph(low gear)		74.5 dBA		74.2 dBA

IDENTIFICATION OF CRITICAL OPERATING MODES - In order that the critical modes of engine operation could be subsequently reproduced on the test bed, it was necessary to indicate the engine speed and the fuel delivery to the engine while the vehicle was being driven over the operating cycle.

Due to difficulties in fitting a suitable transducer to the fuel injection pump rack, a linear characteristic rotary potentiometer was fitted to the fuel pump linkage. Since the fuel pump governor was of the idle/over-speed type, it was believed that this would give an accurate indication of rack position, a belief which was subsequently confirmed by checks when the engine was on the test bed.

In order to relate the emission traces to engine modes, it was also necessary to establish the delay times in the various sample lines to the gas analysers. These times were obtained by injecting suitable gases into the sample lines and measuring the time taken for a particular analyser to respond.

Tests were then carried out over the FTP cycle on the turbocharged vehicle, obtaining the following chart readings:-

Engine speed, injection pump linkage position, exhaust temperature, dilute HC, raw HC, dilute CO, raw CO, dilute CO_2, raw CO_2, dilute NOx and raw NOx. From these results selected traces were redrawn by hand on a common base, taking account of the various delay times of the sample lines. Typical traces for dilute NOx, dilute CO_2, pump linkage position, and engine speed are given in Fig.5, which shows part of the trace for the stabilised portion of the driving cycle.

An examination of the factors controlling NOx concentration indicated that the best correlation was obtained with linkage position x engine speed, i.e. torque x speed, or engine power. A typical relationship, also for the stabilised phase, is shown in Fig.6.

Since one of the aims of the programme was to reduce the NOx levels to as low a value as possible, it was necessary to calculate the weighted level of the NOx trace in g/mile and make sure that this agreed with the bag values. This correlation was carried out by integrating the areas under the traces which gave a value of 1.8 g/mile, which agreed very well with 1.75 given by the CVS test.

Reducing, by calculation, the mean NOx ppm values obtained from the three phases of the FTP, mean absolute levls (in ppm) required to achieve 1.0 g/mile, 0.5 g/mile, and 0.4 g/mile NOx respectively, were established. The results of these calculations are given in Table 3.

These figures show that to achieve the lowest level of NOx proposed, it would be necessary to reduce the mean NOx level of the test phases to values which are below those given by the unmodified engine when idling; currently shown to be 15, 15, and 14 ppm respectively for each of the three phases of the FTP.

It was clear from these results that while the peaks of NOx are formed during acceleration of the vehicle, with levels which are largely proportional to the power output, it is necessary to reduce the NOx levels for the whole cycle, including idling modes, if there was to be any hope of achieving very low levels of overall NOx. This would be more advantageously achieved by applying a maximum of EGR at idle, with reducing percentages as speed and load are increased in order to limit loss of performance and to prevent an increase in exhaust smoke.

BASELINE ENGINE TESTS - As a preliminary to the engine mapping tests, the engine was mounted on a test bed with an electric dynamometer and baseline engine performance maps were obtained covering the whole range of loads and speeds, both with and without the turbocharger fitted.

Typical full throttle engine performance data, both naturally aspirated and turbocharged are shown in Fig. 7, which confirms the loss in power under the turbocharged condition with constant fuelling.

Full emission maps were taken, and these showed that the NOx levels were increased by turbocharging (Fig.8). It was found however that, as might be expected, the HC levels were reduced. One factor noted from the motoring tests to measure engine friction was that the fuel did not shut off on the over-run, as both NO and HC could be measured in the exhaust, as shown in Fig. 9.

Since it was intended that retard of injection timing would be used in conjunction with EGR to provide more extensive NOx reduction, performance and emission measurements were also carried out over a range of injection timings. Typical results from these tests at 35 rev/s are shown in Figs. 10 and 11. These demonstrate an increase in HC as the timing is retarded but show that it is necessary to retard the timing by nine degrees or so to give a reduction of NOx under all

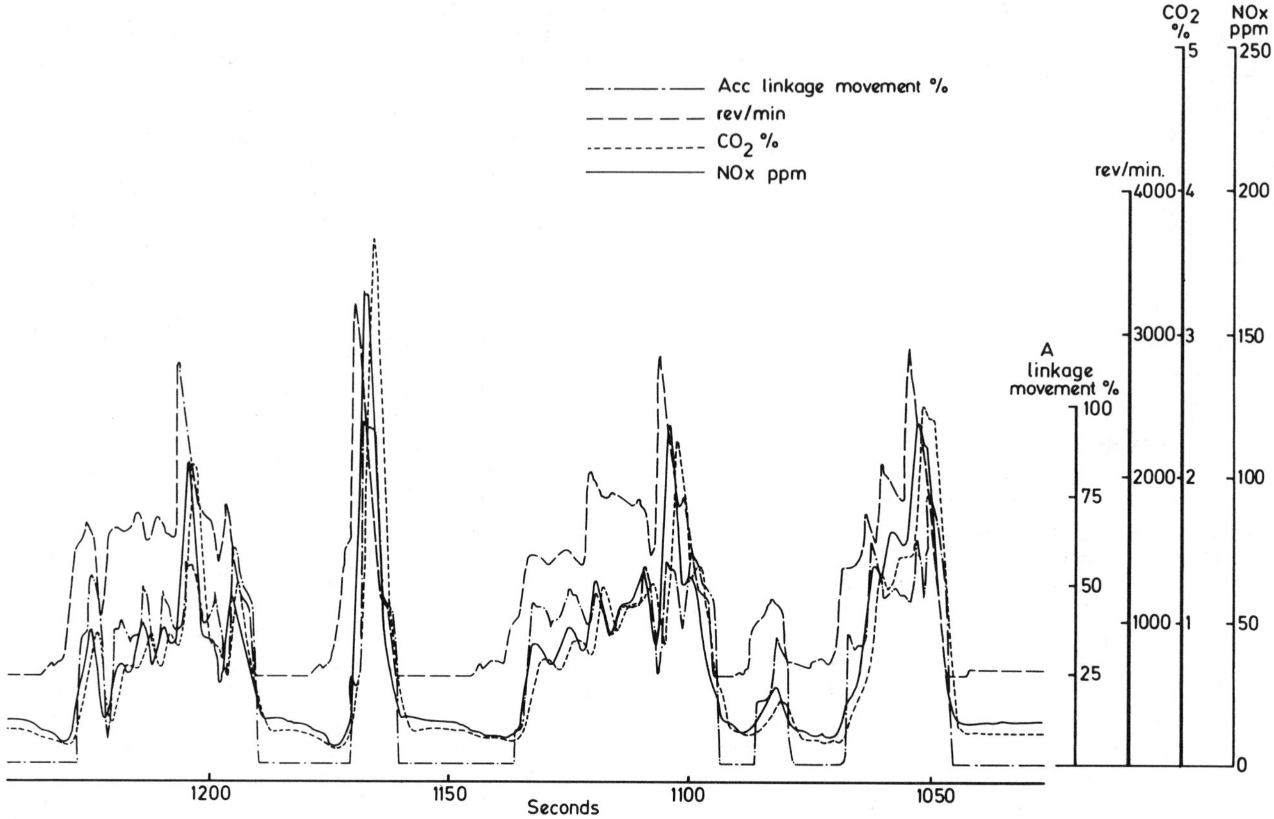

Fig. 5 - Typical stabilised traces of NOx, CO, engine speed and accelerator movement

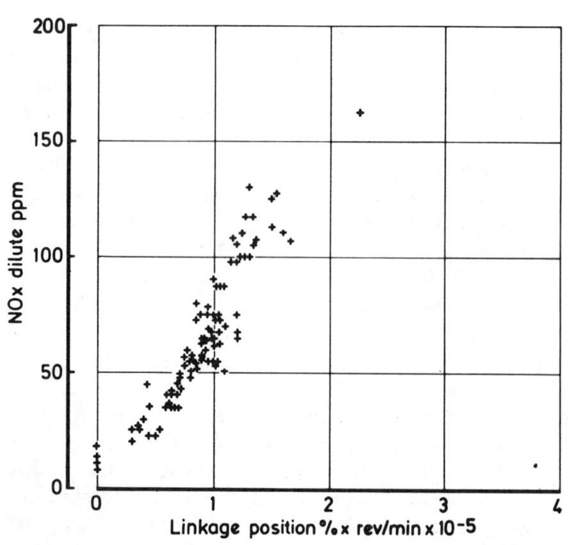

Fig. 6 - Relationship between NOx emission and combined linkage position and engine revolutions for stabilised phase of FTP

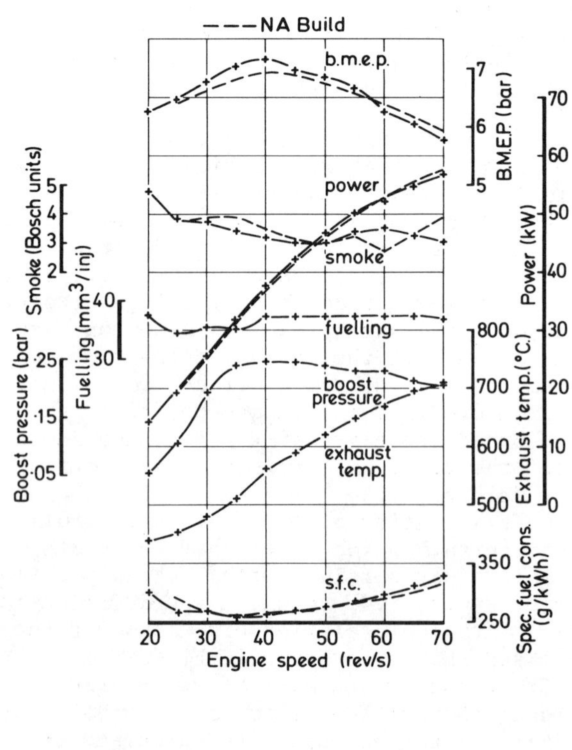

Fig. 7 - Power curve with turbocharger fitted

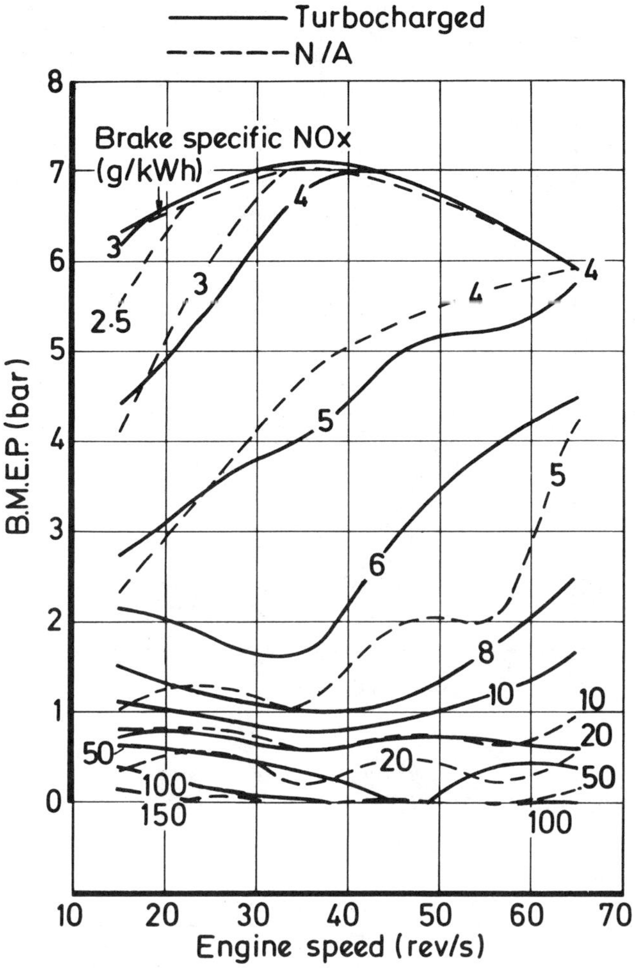

Fig. 8 - Composite map of brake specific NOx emission

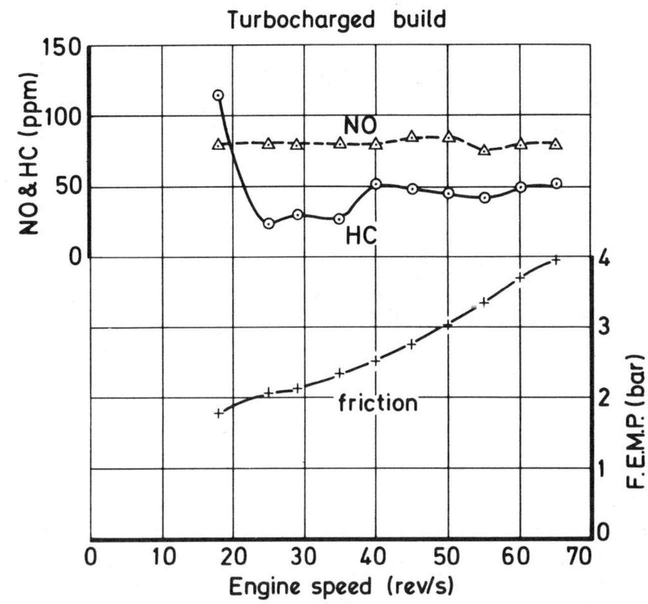

Fig. 9 - Motoring friction and emissions under over-run conditions - turbocharged build

conditions. Smaller amounts of retard do reduce smoke levels however, which might be useful as a means of reducing smoke with EGR.

BASELINE EGR TESTS - The three possible ways of introducing EGR are shown in Fig.12. Since (a) has the advantage of lower EGR temperature and good mixing in the turbocharger compressor, it was decided to employ this system. The EGR gas was tapped off from the main exhaust system at a position approximately six feet from the engine to provide some measure of cooling for the gas.

It was thought initially that it would be desirable to fit two butterfly valves, as shown in Fig.12, one to control the EGR flow and one to produce an artificial intake depression, for use when the EGR flow was insufficient without its use. These two valves were fitted when the baseline EGR tests were carried out. EGR quantities were defined as the reduction in air flow to the inlet of the engine, measured by means of a viscous flow air meter, when EGR was employed.

Typical performance and emissions data when using varying degrees of EGR are shown in Figs. 13 and 14.

EGR levels of up to 80% were achieved at some load and speed combinations, but in practice these high levels were not acceptable at other than light load at most speeds due to the onset of misfire. Also, as the level of EGR was increased, the maximum acceptable load of the engine was limited by one or more of the following factors:-

a. Excessively black exhaust smoke.
b. An exhaust CO level of over 0.2%.
c. Inlet manifold temperature limited to 220°C approximately.
c. Turbine entry temperature limited to 750°C approximately.

COMPUTER SIMULATION PROGRAM - To relate these results to vehicle emissions, a CVS simulation program was used which could accommodate a modulation of EGR based on the engine load and speed requirements of the 1975 FTP drive cycle. It was appreciated that the use of steady state engine data would not lead to very accurate predictions of exhaust emissions, but it was believed that the use of the program would enable trends to be predicted.

To assess the validity of the method, data from the zero EGR steady state tests were used to simulate the vehicle performance to give data which could be compared with the actual vehicle tests. The results of these computations gave emission levels higher than

Table 3

	NOx g/mile	Calculated NOx Mean Level ppm Cold Transient	Stabilised	Hot Transient
Current calculated level	1.8	63.1	37.4	59.8
Reduced level	1.0	35.2	20.9	33.6
Reduced level	0.5	17.6	10.5	16.8
Target level	0.4	14.1	8.4	13.4

Table 4

	HC g/m	NOx g/m	CO g/m	Fuel Cons. m/gal(US)
Original vehicle T/C zero EGR	0.14	1.75	0.94	23.8
Computed zero EGR	0.16	2.58	1.29	21.2
% Increase	14.3	47.4	37.2	11.3

those of the original vehicle as can be seen from Table 4.

While the agreement was not perfect, it was believed that it was sufficiently accurate to predict trends, and indeed this proved to be the case. It is assumed that the differences between the actual and the predicted vehicle results are due partly to the substitution of steady state results for transient operation and partly to the absence of cold start data for the computed results.

DEVELOPMENT OF MODULATED EGR SYSTEM - At the start of the programme it had been agreed that an EGR system would be developed for the vehicle which would enable tests to be carried out with modulated EGR and with injection timings which would cover the following Exhaust Emission Constraints:-

1. Lowest NOx while maintaining other performance parameters such as HC,CO,CO_2, and exhaust smoke with minimum trade offs.
2. Minimised NOx emissions with no more than an estimated level of 0.41 g/m HC.
3. Minimised NOx emissions with the only constraint being the retention of 85% of the peak bhp of N/A engine.
4. NOx emissions of 0.3-0.4 g/m, if 2. and 3. are estimated to result in lower than 0.4 g/m NOx.
5. The EGR level estimated to achieve 1.0 g/m NOx.

In order to develop the EGR control system it was obviously necessary first to establish the required EGR maps corresponding to these constraints.

To establish the feasibility of the project it was decided to attempt to carry out the tests corresponding to Constraint 2 above. As the HC emission of the baseline vehicle was low (about one third of the maximum limit allowed) EGR levels were chosen which produced a three times increase in HC level at each point of the map.

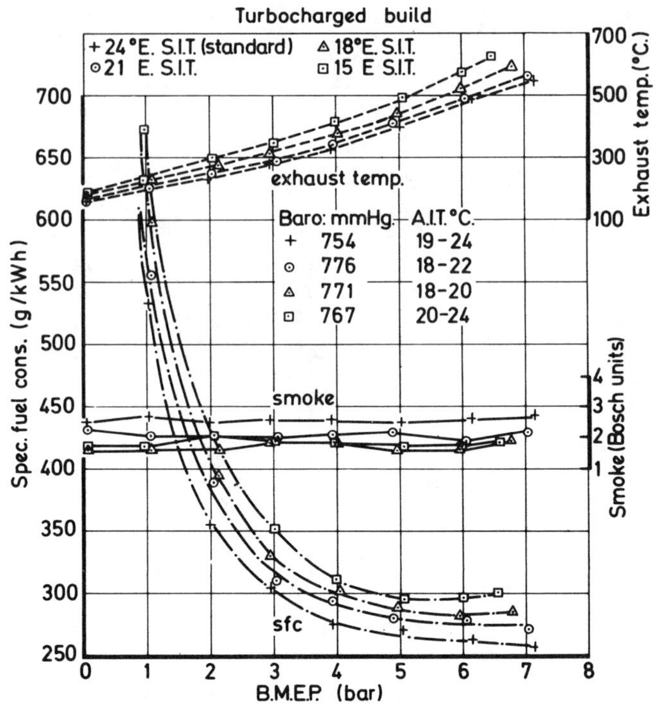

Fig. 10 - Load range performance at various injection timings:35 rev/s - turbocharged build

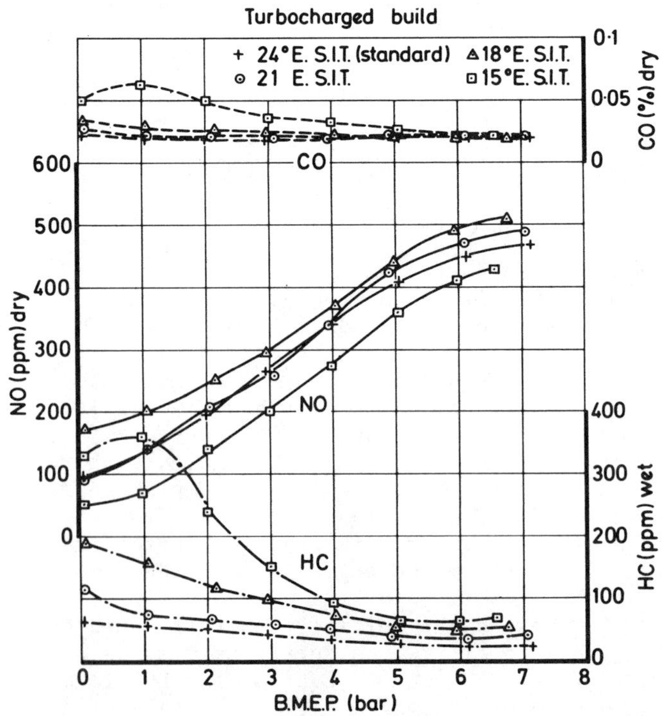

Fig. 11 - HC, NO and CO emissions at various injection timings:35 rev/s - turbocharged build

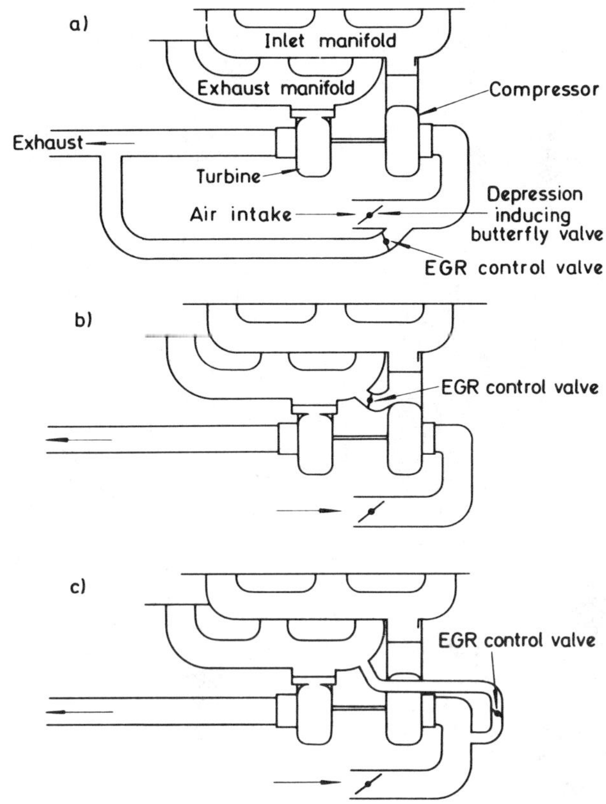

Fig. 12 - Alternative EGR circuits

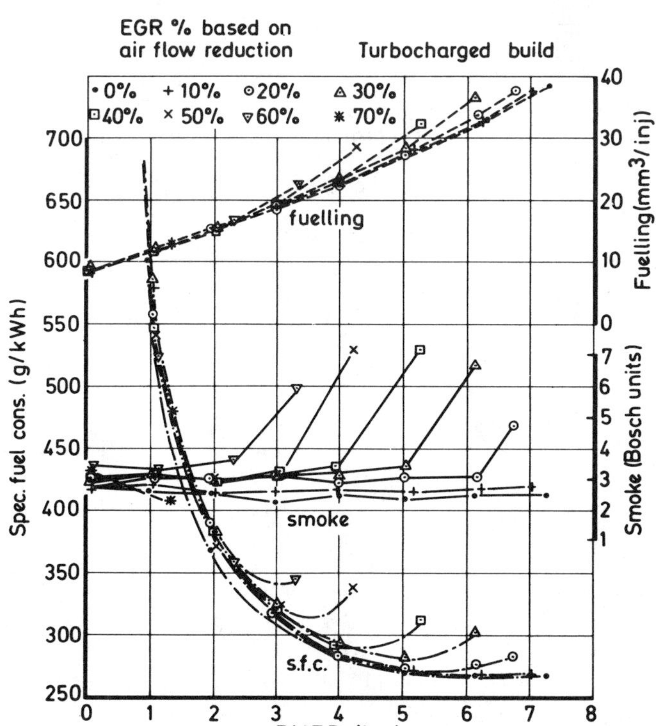

Fig. 13 - Effect of EGR on engine performance at 35 rev/s - turbocharged build

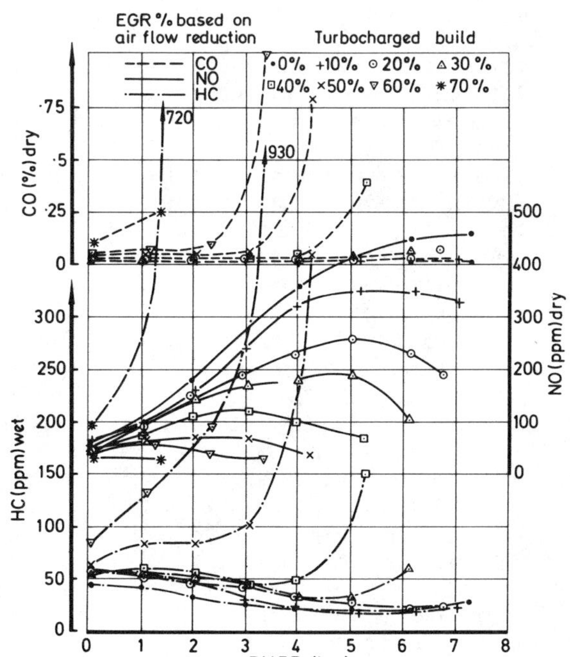

Fig. 14 - Effect of EGR on NO, HC, and CO emissions at 35 rev/s - turbocharged build

Using this procedure, a complete map of EGR levels was drawn up for each load and speed but modified to give zero EGR at full load in order to prevent any loss of full load performance. The EGR map produced by this method is shown on Fig. 15, and using these data the modified levels of the other parameters were identified and prepared for computation. The following parameters were recorded at each of the six speeds at 1 bar load steps:- EGR %, NO ppm, HC ppm, CO %, air meter reading, fuelling mm^3/injection, and pump rack position.

The latter three parameters were obtained from separate sets of curves, and in addition to the above a bmep/rack position matrix had to be produced to provide modified load data for the simulated FTP cycle to allow for the fall off in mid load power for a given fuelling level caused by the application of EGR. Similarly, an allowance was made for the changes in idle emission levels with EGR. For over-run conditions the emission levels obtained during the motoring tests without EGR were used.

The result of the computation from the EGR modulation of Fig.15 (designated Mod.1) is shown in Table 5 with the simulated zero EGR results for comparison.

The figures show that for the particular EGR modulation shown in Fig.15 the predicted NOx emission would be reduced to one fifth of the zero EGR condition, with the HC and CO emissions almost doubled. These results were only indications of likely trends produced by the EGR modulation and were not indicative of actual vehicle results, which would only be apparent if that particular EGR map could be applied to the vehicle in future tests.

Having established a technique for the simulation of vehicle emissions, which could be applied to any modulation of EGR, it was necessary to attempt to apply the above modulation to the engine. To this end the EGR map, Mod. 1, was transferred point by point to a series of curves relating fuel pump linkage position (i.e. engine load) to EGR valve position at each of the six engine speeds. These curves are shown on Fig.16, and it can be seen that the relationship between valve position and fuel pump linkage position was reasonably close at 25, 35, 45, and 55 rev/s but was different at the extreme speeds of 15 and 65 rev/s. This factor is indicative that in order to reproduce exactly this particular EGR modulation on the engine, some speed sensitive device would be necessary to modify any direct EGR valve to fuel pump linkage at extremes of the speed range.

In order to investigate the effects of employing only direct connection between the EGR valve and injection pump linkage positions, two additional curves based on those in Fig. 16 were produced, each showing all six speeds as a common relationship. These curves are shown on Fig. 17 and were subsequently transferred to full EGR modulation maps on Figs. 18 and 19 to be designated Mod.2 and Mod.3 respectively.

At the same time it was clear that for Mod.1 the level of EGR indicated at zero load (and at idle) of 60% to 70% would be impractical due to excessive white smoke and difficulty in controlling the engine. Therefore it was decided to recompute the emission results using a maximum EGR level of 50%. The result of these changes raised the NOx figures slightly and, surprisingly, also increased the HC and CO predicted levels.

The results are shown in Table 6 with those of the original Mod. 1, and for Mods.2, 3, and 4. The zero EGR computed figures are also shown.

Mod. 4 in the above table was for a calculated fixed orifice equivalent to producing a 15% reduction in full load power to comply with one of the objectives in the original programme. This fixed orifice was equal to an EGR rate of 10% at full speed, increasing to approximately 20% at low speed with no variation with load. The NOx emission at 1.2 g/m represents 46% of the zero EGR level.

The NOx results for Mod.2 and Mod.3 were reduced to 50% and 30% of the zero EGR levels respectively. Although both modulations were derived from a theoretical direct linkage between valve and pump linkage, the deviations shown on Fig. 17 at mid load caused considerable changes in the rate of application of EGR, which can be seen by comparison between the Mod.2 and Mod.3 maps on Figs. 18 and 19. The effects of these differences are shown in Table 6 above, with the Mod.3 figures being comparable with those of the reduced EGR Mod.1 figures.

Table 5

	HC g/m	NOx g/m	CO g/m	Fuel m/gal(US)
Zero EGR	0.16	2.58	1.29	21.1
Mod.1	0.2	0.57	2.33	20.9

Table 6

	HC g/m	NOx g/m	CO g/m	Fuel m/gal(US)
Zero EGR	0.16	2.58	1.29	21.1
Mod.1	0.2	0.57	2.33	20.9
Mod.1 (50% max.EGR)	0.22	0.61	2.46	20.94
Mod.2	0.17	1.31	1.6	20.9
Mod.3	0.21	0.76	2.22	20.37
Mod.4 (Fixed Orifice)	0.14	1.2	1.6	21.0

EGR VALVE CONFIGURATIONS - From an analysis of all the above computer results it was evident that the existing arrangement of EGR control on the engine employing two butterfly

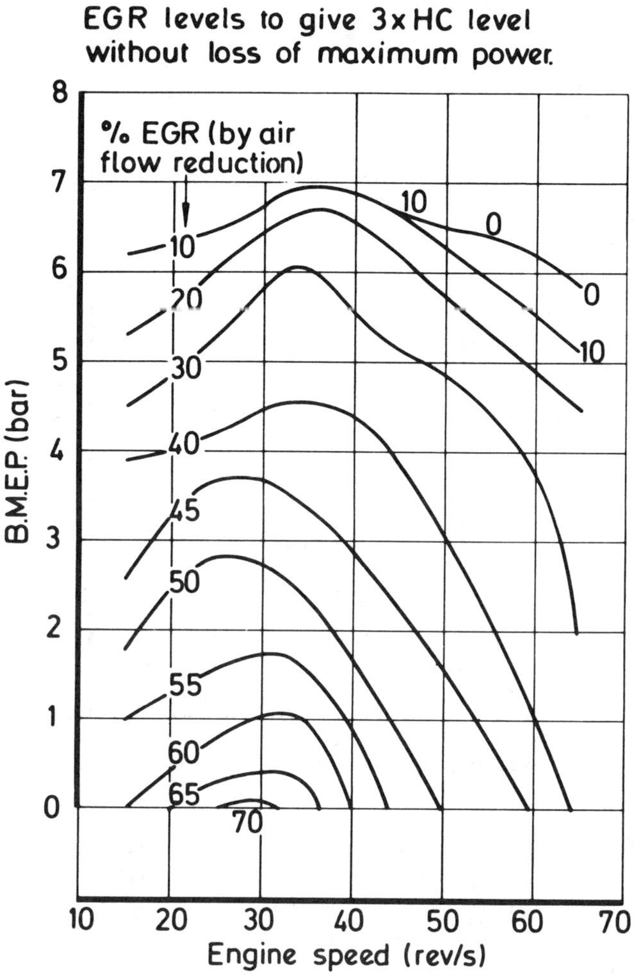

Fig. 15 - Composite map of EGR modulation 1. EGR levels to give 3 x HC level without loss of maximum power

valves was not ideal. The butterfly valve controlling the EGR flow also operated at high temperature, and in consequence suffered considerably from distortion problems which caused difficulty in adjustment and in fully closing the valve for zero EGR conditions. In addition, the test bed results indicated that more than sufficient EGR for practical results could be obtained without the use of the depression inducing butterfly valve in the engine intake system. Also from the above computer simulations it was evident that an EGR level directly controlled by accelerator linkage could be satisfactory although with a non-linear relationship. In view of these findings, a new valve having a poppet or obturator type valve head giving a continuously variable EGR function was developed.

Experiments to operate such a valve by suction, using the vehicle's vacuum system, were carried out. Such a system would have the advantage of simplicity, being similar to that used on many gasoline engines, but would of necessity have to be large because of the much greater quantity of EGR involved. These tests showed however that the response time of a vacuum cell was rather slow in comparison with a direct mechanical control which would give an instant response, and this system was abandoned in favour of the mechanical linkage.

The new valve was therefore designed and constructed together with an eccentric cam operating mechanism connected by a flexible cable to the injection pump control linkage. The design of this valve was such that the valve head could be assembled to open or close against exhaust gas pressure, as shown in Fig. 20.

By adjustment of the position of the cam relative to accelerator position and by change of profile of the valve head, a very extended range of EGR modulation with load could be achieved. In addition, with the valve clos-

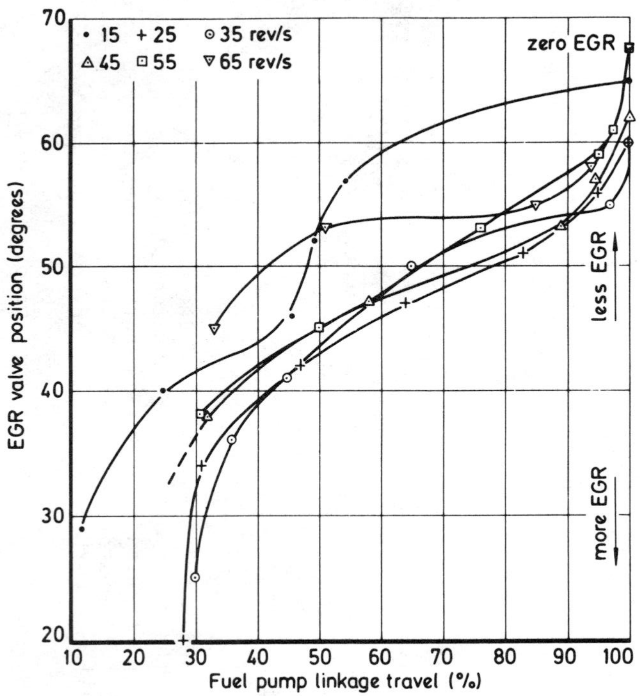

Fig. 16 - Relationship between EGR valve position and fuel pump linkage position for EGR modulation 1 (based on 3 x HC levels)

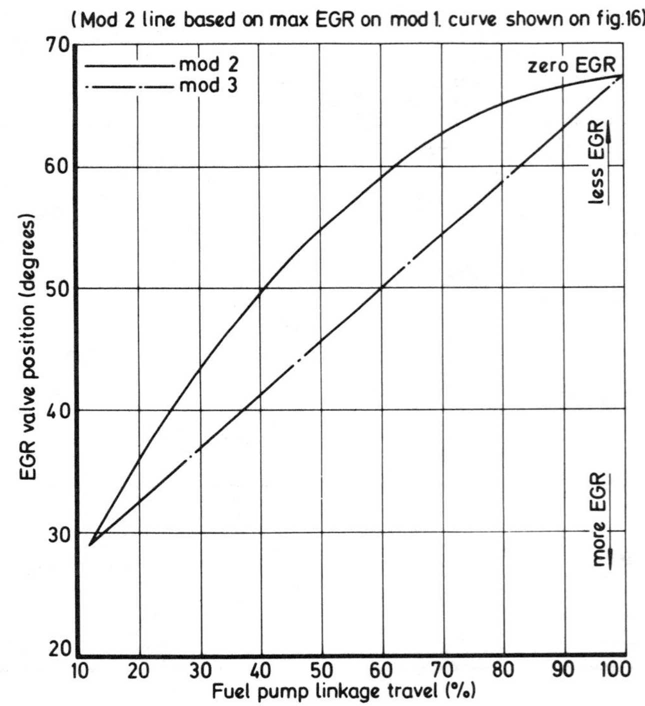

Fig. 17 - Relationship between EGR valve position and fuel pump linkage position for EGR modulations 2 and 3 at all speeds

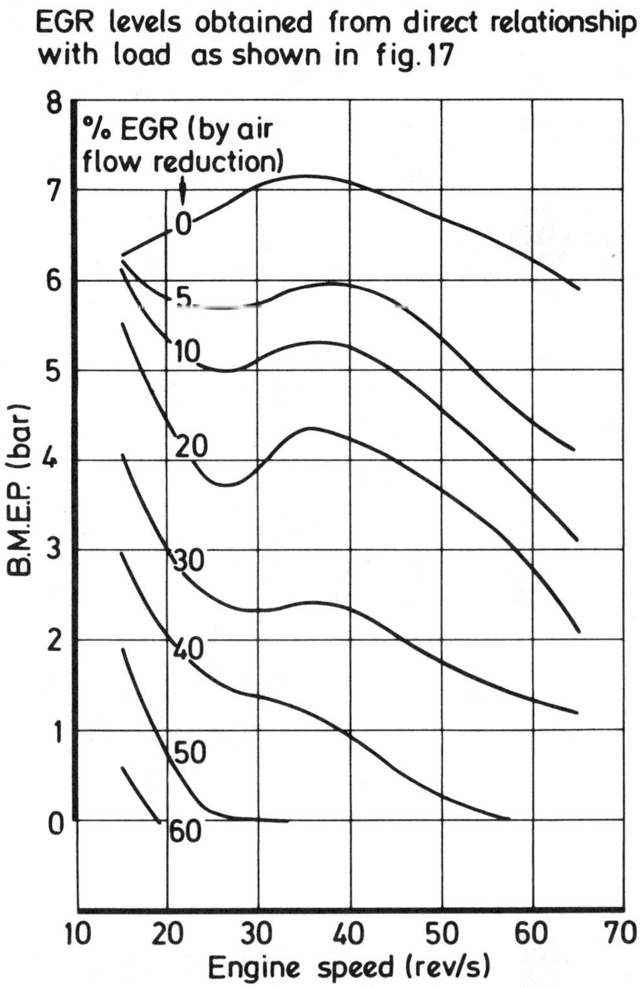

Fig. 18 - Composite map of EGR modulation 2
EGR levels obtained from direct relationship with load as shown in Fig. 17

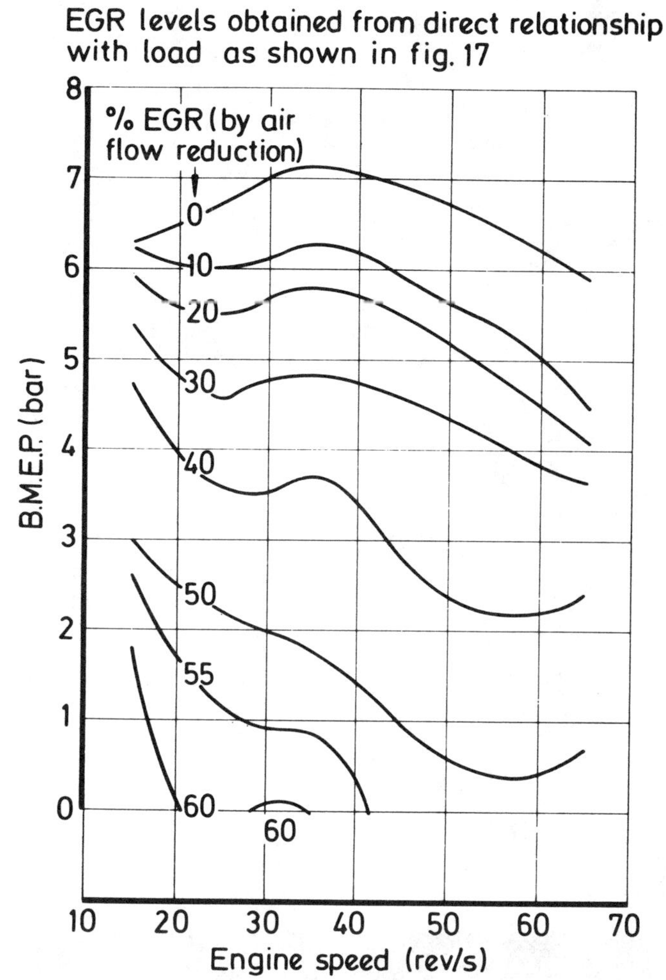

Fig. 19 - Composite map of EGR modulation 3
EGR levels obtained from direct relationship with load as shown in Fig. 17

ing by gas pressure, the rate of closure could be controlled by setting a backlash adjustment on the cam which, because the exhaust pressure changed with speed, would give a measure of speed modulation of the EGR. Figures 21 (a) and (b) show the installation of the new valve on the engine and the prototype operating mechanism.

Fig. 22 shows the first EGR modulation map obtained from the engine with the arrangement shown in Figures 20 and 21 with the valve set to lift 20mm from its seat at 0% accelerator link travel (idle condition) and zero lift at 100% accelerator link travel. This EGR modulation (designated Mod.A) is comparable to Mod. 2 or 3, shown in Figs. 18 and 19, and the simulated vehicle emission results from these Mods. are shown in Table 7.

It can be seen that the Mod.A result falls between Mods.2 and 3 in respect of NOx emission but is improved in respect of HC, CO, and fuel consumption.

Having established the method of developing the geometry of the EGR system, the procedure was repeated for the various other exhaust emission constraints, the appropriate engine maps were obtained, and the vehicle simulations were carried out.

The variations in EGR modulation were produced by combinations of EGR valve seat diameter, valve head profile, valve lift, and changes from upward to downward closing valves. Full details of the various builds are given in (3).

The simulated vehicle results for each of these EGR modulations are shown in Table 8 with the zero EGR prediction and the actual vehicle baseline results shown for comparison.

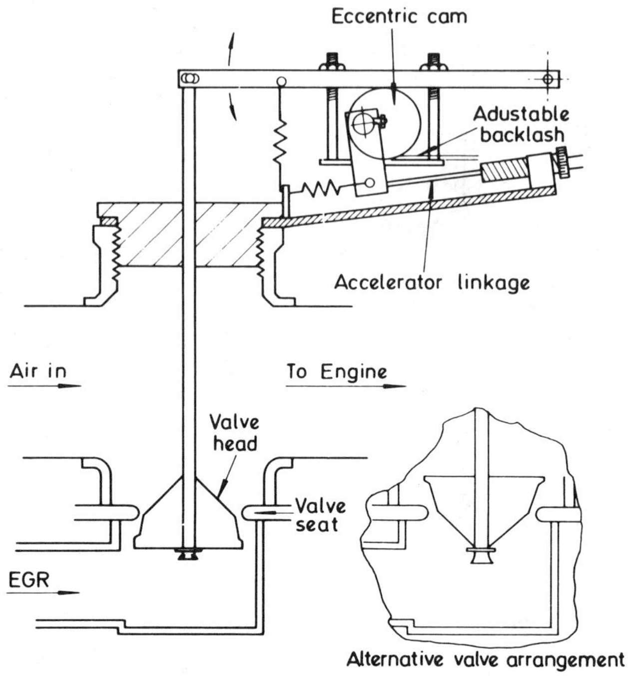

Fig. 20 - Diagrammatic sketch of EGR valve

Fig. 21 - a. EGR valve installation -
b. Installation of valve actuator

Table 7

	HC g/m	NOx g/m	CO g/m	Fuel m/gal(US)
Zero EGR	0.16	2.58	1.29	21.1
Mod.2	0.17	1.31	1.6	20.9
Mod.3	0.21	0.76	2.22	20.37
Mod.A	0.145	1.13	1.57	21.26

Table 8

	HC g/m	NOx g/m	CO g/m	Fuel m/gal(US)
Mod.A	0.145	1.13	1.57	21.26
Mod.B	0.42	0.689	8.535	20.97
Mod.C	0.269	0.589	4.168	20.84
Mod.D	0.147	1.131	1.607	21.11
Mod.E	0.179	0.714	2.2	21.05
Mod.F	0.163	0.879	1.724	20.9
Mod.G	0.23	0.647	2.896	21.12
Zero EGR	0.16	2.58	1.29	21.1
Original Vehicle Tests	0.14	1.75	0.94	23.8

Table 9

Vehicle Build	Gaseous Emissions g/m HC	NOx	CO	Part. g/m	Fuel mpg (US) CVS	HEFT	Noise dBA	Acceleration Seconds (mph) 0-40	0-60	20-60
As-received	0.183	1.66	0.94	0.8	24.3	29.95	74.2	9.6	21.2	18.7
Turbocharged	0.14	1.75	0.94	0.5	23.8	25.6	74.5	9.9	22.4	19.2
EGR Modulation G	0.395	0.528	2.408	1.11	19.47	24.94	78.0*	10.6	23.5	20.1
EGR Modulation C+	0.382	0.471	2.379	-	19.95	-	-	-	-	-
EGR Modulation C+ 7° Retard	4.974	0.32	6.129	0.93	17.15	-	-	11.6	28.6	22.8
EGR Modulation C+ 4° Retard	1.053	0.379	2.956	1.228	18.86	24.43	76.2	11.0	25.3	22.4
EGR Modulation D	0.226	0.961	1.563	1.59	20.08	25.78	-	10.2	22.6	19.1
EGR Modulation D- (7mm lift)	0.245	1.004	1.392	-	20.17	-	-	-	-	-
EGR Modulation D- (2mm lift)	0.171	1.52	1.329	0.964	20.28	25.87	76.4	9.4	20.5	19.1

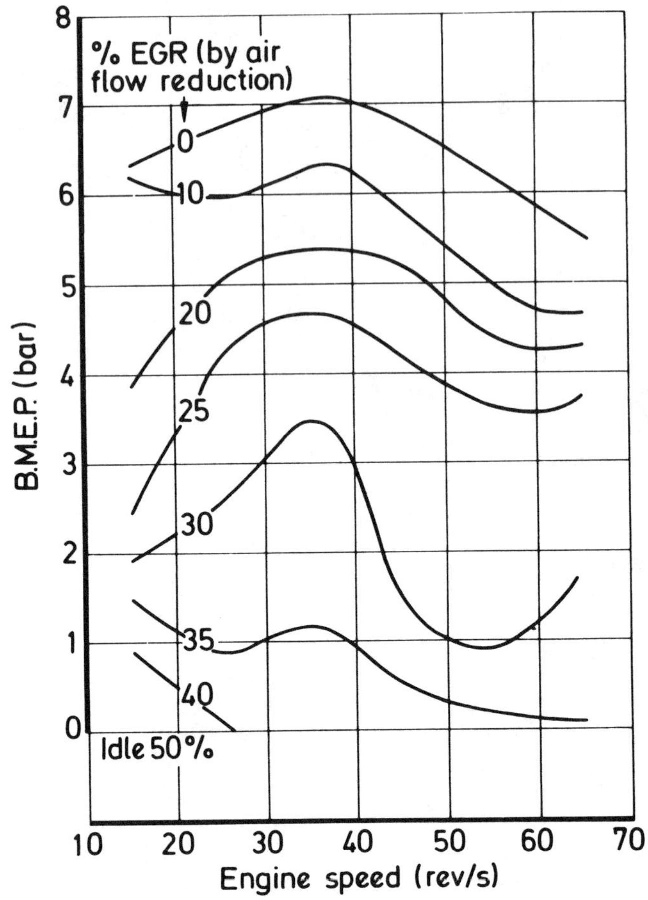

Fig. 22 - Composite map showing EGR modulation A

VEHICLE TESTS WITH MODULATED EGR - The engine, complete with EGR system and automatic transmission, was then installed in the vehicle.

The installation of the EGR system in the vehicle was accomplished without difficulty, but certain modifications had to be carried out to the system as used on the test bed. These changes were mainly of a minor nature, the most important of which was the re-construction of the EGR pipe to avoid contact with other parts of the vehicle installation. This pipe also had to be shortened on the vehicle, being tapped into the main exhaust pipe at approximately 4 feet from the cylinder head, compared with 6 feet on the test bed. This may raise the temperature of the EGR at the valve, but the effect would be offset by the presence of cooling air resulting from the motion of the vehicle. In addition, the more transient nature of vehicle operation would prevent such high temperatures as were achieved under test bed steady state conditions.

Assessment of the Mercedes 300D vehicle with EGR was carried out in the same way as in the baseline tests, viz:-

1. 1975 FTP tests for gaseous emissions
2. EPA HFET
3. Exhaust smoke obscuration tests
4. Particulate emissions measurement
5. Drive-by noise measurement
6. Vehicle performance assessments

Replicate tests were made, where possible, the test series being repeated with EGR levels and injection timings set for each of the constraints set out earlier. To establish the exact EGR valve build to reach these constraints, several additional tests were carried out with various modifications to the modulation system.

While the detailed results are given in (3) they are summarised in Table 9.

The results show that although very low NOx figures were achieved, these were at the expense of large increases in the emissions of HC and CO. Also, not shown in Table 9 was the greatly impaired driveability of the vehicle with the larger amount of EGR which had to be applied, particularly noticeable at low speed steady state driving. At these conditions persistent misfire was apparent, and with the retarded injection timing necessary for the lowest NOx builds, white smoke emissions were also present, although black smoke under these retarded conditions was less of a problem.

Fuel economy of the vehicle was also considerably worsened by both EGR and retarded timings when compared with the zero EGR build of the vehicle, either naturally aspirated or turbocharged. However, with the lower EGR builds for constraints numbers 1 and 5, the fuel consumption figures were comparable for the HFET but were approximately 16% worse for the 1975 FTP when compared with the original turbocharged version of the vehicle.

The exhaust smoke opacity, when observed under road driving conditions, was worsened by the application of EGR but not excessively so. This was also evident on cold start up when very high instantaneous peaks were shown on the recording traces made during the 1975 drive cycles.

The acceleration of the vehicle was impaired for all configurations of the EGR valve, apart from those appropriate for the minimum EGR levels corresponding to constraint number 1 (see earlier list of constraints). With the levels of EGR and injection retard required for constraint 4, the time for the acceleration from 0-60 mph was increased by 7 and 4 seconds respectively for 7 and 4 crank degrees injection timing retard.

Particulate emissions were high for each build of the EGR valve when compared with the zero EGR level for the turbocharged version of the vehicle. The lowest particulate level of 0.93 g/m for EGR Mod.C + 7° injection retard was almost twice the level of 0.5 g/m for the normal turbocharged version of the vehicle. A similar level was shown for the minimum EGR build for constraint No.1 (Mod.D-). For the EGR modulation derived for 1 g/m NOx, a particulate emission of 1.59 g/m was shown, more than three times that for the turbocharged version and twice that recorded for the naturally aspirated version of the vehicle without EGR.

Drive-by noise tests were not obtained for all of the EGR modulations, but the results that were obtained showed noise levels to be higher than for the zero EGR versions. The lowest level of EGR produced a noise increase of 2dBA as did the EGR +4° retard condition. It was not possible to carry out noise tests for the 1 g/m or lowest possible NOx levels of EGR due to a prolonged period when road conditions were unsuitable.

Subjectively the noise of the vehicle was little changed from the original build on full acceleration drive-by conditions when the EGR valve was closed. However, it was noticeable that induction noise increased at light load conditions when the EGR valve effectively opened the unsilenced exhaust system into the engine inlet. Also cold combustion could be noisy and harsh under conditions of high EGR, although the normal effect of EGR, when the engine was hot, was to reduce combustion noise.

At idle conditions, when the EGR valve was opened to the maximum, combustion was particularly quiet with a hot engine. Some mechanical noise was evident at idle however, due to the self-cleaning mechanism of the EGR valve causing a metallic rattle. This effect could be eliminated by modification to the valve.

For the highest levels of EGR for constraints 2, 3, and 4, the driveability of the vehicle was impaired at steady state low speed conditions because of persistent misfire. This was not apparent at maximum acceleration or at the conditions of lower rates of EGR. Driving of the 1975 FTP cycle was not affected, although cold acceleration was impaired.

Although durability was not an integral part of the test programme, the effects on engine performance etc. of inducing large quantities of exhaust gas were noted.

During the test bed tests, a gradual fall in boost pressure was observed, and to assess this the EGR valve was removed and the supply pipe blanked off, to eliminate the chance of exhaust gas leakage into the intake system, and a full load power curve over the speed range of the engine was obtained. This was shown to give a 40% lower boost pressure at 65 rev/s with 6.5% lower power. At lower speeds the boost pressure, power, and bmep were all correspondingly lower than for the original power curve. It was suspected that the prolonged running with relatively high levels of EGR may have affected the turbocharger, and to investigate this it was removed for inspection, On dismantling, the turbocharger compressor was found to be lightly coated with an oily, sooty deposit on the vanes and surface of the compressor housing. This was easily removed. The wastegate valve, spring, and diaphragm were lightly coated with soot but apparently in good working order, although some components had obviously been overheated in some degree. In order to increase the boost pressure to its original level, shims were placed under the wastegate spring to an amount calculated from the rate of the spring and the increase in pressure required. The turbocharger was reassembled without further change. The engine had run for approximately 45 hours at varying levels of EGR up to this time.

Before re-installing the turbocharger on the engine, the inlet manifold was removed and the condition of the inlet ports and valves examined in situ. The ports and rear of the valve heads were slightly covered with the same oily soot deposit as on the turbocharger compressor. This was also present in the inlet manifold but was not considered to be detrimental to the engine performance. The turbocharger was therefore re-fitted to the engine without further change. An additional full load power curve was obtained from the engine which resulted in a complete restoration of performance to that of the original tests. The boost pressure was shown to have risen to 10% above that of the original test due to the combined effects of cleaning the turbocharger compressor and increasing the wastegate spring pressure, and it was decided that this would be acceptable for future tests as the level of boost was in any case relatively low.

Further evidence of EGR causing deposits to build up were shown on the EGR butterfly valves used for the majority of the tests and of near blockage of the sample pipe from the inlet manifold to the CO_2 analyser.

The effects of prolonged running with EGR were not established from these tests but would be likely to cause detriment to the performance of the engine.

During the EGR tests on the vehicle approximately 500 miles were driven; the effect of such low mileage on carbon deposits was not expected to be great, and no inspection of the engine was carried out during this period.

CONCLUSION OF MODULATED EGR PROGRAMME - It would appear that the original concept of using the turbocharger to supplement any loss of performance caused by the application of EGR may have been inappropriate, the fitting of the turbocharger in itself causing a loss in performance. Tests on the engine with EGR and no turbocharger were not carried out, and therefore no conclusions in this respect can be drawn. However, there is little doubt that the use of the turbocharger was of considerable assistance in promoting even distribution of the EGR by the chosen method of its introduction into the compressor intake.

For the most efficient application of EGR, tests to determine areas of the 1975 FTP cycle which contributed disproportionately high levels of NOx emissions to the weighted results were carried out. The conclusions derived from these tests indicated that all areas of the FTP cycle would require treatment with EGR if the lowest NOx levels of the target constraints were to be achieved.

The lowest level of NOx achieved was 0.32 g/m. This was achieved by heavily retarding the injection and by applying EGR at full load, thus causing a reduction in vehicle performance. Under these conditions the HC and CO emissions were above the legislated limit, and the driving of the vehicle was severely affected due to misfire at steady speeds below 40 mph.

With a level of EGR modulated to achieve 1.0 g/m NOx, HC and CO levels were within reasonable limits, and driveability was not affected. Noise and performance were however inferior to those of the original vehicle.

For any condition of EGR the particulate emissions of the vehicle were severely affected, being increased up to twice the value recorded for the original naturally aspirated version of the vehicle.

The project however showed that:-

1. 0.4 g/m of NOx can be achieved with this engine/vehicle combination if no other considerations are taken into account.
2. 1.0 g/m NOx can be achieved with less detriment to other parameters.
3. Particulate emissions are severely increased by EGR.
4. Fuel economy is adversely affected by EGR.
5. Durability of the intake system of the engine is adversely affected by EGR.

ACKNOWLEDGEMENTS

The Modulated EGR programme on the Mercedes-Benz 300D car was carried out under sponsorship of the U.S. Environmental Protection Agency, Contract No. 69-03-2465. We should like to ackowledge our indebtedness to them, to the Project Officer, Mr. F.P. Hutchins, and to Daimler-Benz, for help given during the course of the programme.

REFERENCES

1. Eugene Danielson, "Particulate Measurement - Light Duty Diesel Particulate Baseline Test". EPA Technical Support Report SDSB 79-03.

2. D. Collins, R.D. Cuthbertson, R.W. Gawen, R.W. Wheeler, "The Use of Constant Volume Sampler and Dilution Tunnel to Compare the Total Particulates from a Range of Automotive Engines". SAE 750904.

3. D.A. Pike "Modulated Exhaust Gas Recirculation System for Light Duty Vehicles", EPA Report No. 460/3-78-007.

APPENDIX

ENGINE AND EQUIPMENT - Daimler-Benz Mercedes 300D passenger car (USA version) fitted with automatic transmission and having an inertia weight of 4000 lb.

Engine type OM617, five cylinders of 3.005 litre capacity, 91mm bore by 92.4mm stroke. Indirect injection using pre-chamber combustion system. Standard fuel injection system by Robert Bosch with DNOSD220 nozzles. The turbocharger employed was by Garrett AiResearch - Model T3.

Exhaust Emissions - were measured using the normal standard equipment, viz.:- HC by heated FID (by IPM) with 180°C sample line;

CO by IRGA; NO and NOx by chemiluminescence (for the test bed work NO was measured by IRGA).

Particulate Emissions - were measured using a Ricardo tunnel as described in (2).

Exhaust Smoke - Vehicle tests employed a full flow U.S. PHS obscuration smoke meter with pen recorder. For test bed work, the Bosch sample system was employed.

Noise - Vehicle noise was assessed using a Bruel and Kjaer portable sound level meter type 2203 under SAE drive-by conditions to J986a. Test bed noise and vehicle idle noise were assessed subjectively.

Exhaust Emission Patterns from Two Light-Duty Diesel Automobiles**

Peter A. Gabele, Frank M. Black, Foy G. King, Jr., Roy B. Zweidinger, and Rex A. Brittain
Environmental Protection Agency
Research Triangle Park, NC

ABSTRACT

Particulate and gaseous emissions from two light-duty diesel automobiles were examined over six operating cycles. Particulate characterizations included mass emission rate, soluble organic content, and trace element content determinations. The particulate matter was sampled using both a total filtration system having 20 X 20 inch filters and a conventional probe arrangement employing 47 mm filters. Bioassay of selected diesel particulate extract was carried out using the Ames test procedure. Gaseous characterization included real-time analysis of regulated exhaust emissions. A special operating schedule was developed for each test car using the real time system to modally determine the effect of acceleration rate on regulated gaseous emissions.

INTRODUCTION

Exhaust emissions from light-duty diesel automobiles have been under investigation over the past several years. When the diesel engine became a candidate for high volume production in the United States, much of the work focused on the emission of regulated gaseous pollutants (1-5)* In addition to demonstrating itself as a fuel efficient engine possessing performance characteristics compatible for automotive use, the light-duty diesel became recognized for its low emission of regulated gaseous pollutants.

Later, as the diesel automobile increased in popularity, investigators began to examine diesel exhaust in greater detail. Efforts were undertaken by some to characterize both the unregulated gaseous and particulate fractions of light-duty diesel exhaust. (6,7,8) Because findings had shown particulate emission rates from diesels to be relatively high, most of the attention was directed at the character of these emissions.

* Numbers in parentheses designate references at end of paper.

**Paper 810081 presented at the International Congress and Exposition, Detroit, Michigan, February, 1981.

Diesel exhaust particles predominate in size ranges less than 10 microns. They are generally composed of carbonaceous material with substantial quantities of extractable organics. These small particles are able to survive transport to lung tissue where deposition presumably occurs. Two serious concerns are the health consequences associated with deposition in the lung and the extent to which the organics participate in photochemical reactions within the atmosphere. Reflecting these concerns, a great deal of emphasis is currently being placed upon the further chemical characterization of diesel exhaust particles and the assessment of any significant health risks associated with the organic fraction.

The study reported in this paper investigates gaseous and particulate matter emissions trends from two diesel automobiles. Regulated gaseous emissions are examined utilizing a real-time analysis system recently developed at the U.S. Environmental Protection Agency, Research Triangle Park, North Carolina. Using this system, an effort is made to modally determine the effect of different acceleration rates upon exhaust emissions. Particulate matter emission rates are examined over six separate test cycles using two different collection techniques: the total diluted gas filtration method and the more conventional probe method. The characterizations of particulate matter show the quantitative presence of the soluble organic fraction (SOF), trace metals, and SOF molecular weight distributions. Bioassay test results obtained for a select portion of the particulate organic extractions are also presented.

VEHICLE AND FUEL DESCRIPTION

The two test vehicles used in this study are decribed in Table 1. The Rabbit diesel was purchased new by EPA and underwent a break-in period with mileage accumulation prior to testing. The Oldsmobile diesel was on loan to EPA from General Motors and had been operated on the test fuel for over 1000 miles before being tested.

Table 1 - Description of Test Vehicles

	OLDSMOBILE DIESEL	VW RABBIT DIESEL
MODEL YEAR	1980	1980
VEHICLE SIZE	FULL-SIZE	SUBCOMPACT
NUMBER OF DOORS	4	2
NUMBER OF PASSENGERS	6	4
ODOMETER, mi.	11,500	1,500
NUMBER OF CYLINDERS	V-8	I-4
DISPLACEMENT, liters	5.7	1.5
COMPRESSION RATIO	23:1	23:1
OUTPUT POWER, HP	–	48
TRANSMISSION TYPE	AUTO	MANUAL
SPEEDS	3	4
INERTIA TEST WT., lb.	4500	2250
ROAD LOAD, HP AT 50 mph	11.4	6.8
EGR EQUIPPED	YES	NO
COMBUSTION CHAMBER	PRECHAMBER	PRECHAMBER

The diesel fuel which was used in this study was obtained through a local commercial vendor. Inspection results for the fuel are shown in Table 2. Aside from this fuel's rather high aromatic content, it resembles the National Average Number 2 diesel fuel. National Average fuel has a Cetane Number of 48.7, a sulfur content of 0.23 percent by weight, and an End Point of 638°F. The fuel used in this study has corresponding values of 46, 0.18 percent, and 638° F. Its aromatic content of 39.9 percent is somewhat higher than the 29.8 percent for the National Average.

FACILITY DESCRIPTION

This test program was conducted within a test cell which contains a chassis dynamometer for vehicle road load simulation. The dynamometer, manufactured by Burke Porter, simulated road load by means of a DC electric motor-generator directly coupled to the front rolls of the dynamometer. A Reliance analog control system was used to vary armature current to achieve the desired motor torque. The electric drive can add or subtract torque to produce simulated inertias from 1,500 to 5,500 pounds at 250 pound increments. The dynamometer rolls are 9.5 inches in diameter and the front rolls were chain coupled to the dynamometer's rear rolls. The roll coupling feature was added to reduce the vehicles' tendency to rock within the rolls during heavy acceleration and deceleration.

Exhaust gases from the test vehicle were directed via a 3 in diameter flexible stainless steel line to an 8 in diameter dilution tunnel (Figure 1). Within the dilution tunnel raw exhaust gas was thoroughly mixed with filtered diluent and the diluted mixture drawn through the system by two positive displacement pumps located downstream. These pumps operated either separately or in parallel when maximum flow was desired. Flowrates through this constant volume system (CVS) were capable of being varied from about 170 standard cubic feet per minute (SCFM) to about 1200 SCFM, thus making it possible to vary dilution ratios within the tunnel. This capability was desirable as a means of controlling temperature within the tunnel to levels required for sampling particulate matter. It also permitted some control over species concentrations within the sampling system. For example, control over the level of nitrogen dioxide (NO_2) in the diluted gas stream was essential in order to reduce the effect of nitrated artifacts on bioassay test results.

Table 2 - Description of Test Fuel

SOURCE	AMOCO
TYPE	2-D
GRAVITY, API	33.9
DENSITY, g/ml.	0.8551
SULFUR CONTENT, wt. %	0.18
CETANE NUMBER	46.0
DISTILLATION TEMP., °F	
IBP	357
10%	428
50%	509
90%	597
EP	638
FIA, %	
AROMATICS	39.9
OLEFINS	2.0
SATURATES	58.1

TEST SEQUENCE AND CYCLE DESCRIPTION

RUN SCHEDULE-Both the Rabbit and Oldsmobile diesels were tested in similar test plans which were designed to obtain particulate samples over six different cycles and gaseous modal data from three cycles. The six cycles were the 1975 Federal Test Procedure (75 FTP), the Highway Fuel Economy Test (HFET), the Congested Urban Expressway Cycle (CUE), the New York City Cycle (NYCC), the Special Cycle (SC), and a 10 minute 45 mph steady state test (45 SS). The test sequence and sampling effort is summarized in Table 3. Except for the FTP's, each test cycle followed the previous test cycle by a fifteen minute hot soak period.

Real-time gas analysis was performed over the 1975 Federal Test Procedure (FTP), the "unclipped" Federal Test Procedure (FTP-U), and the Special Cycle (SC). The Special Cycle is a 24-mode cycle developed within the program to evaluate the effects of different acceleration rates on regulated gaseous emissions. The "unclipped" FTP was conducted like the 1975 FTP except that the original speed-time trace was used (9). This original trace was modified by reducing a few of the heavy accelerations to obtain the FTP version which is currently the standardized speed-time trace (10). For example, the FTP-U contained a maximum

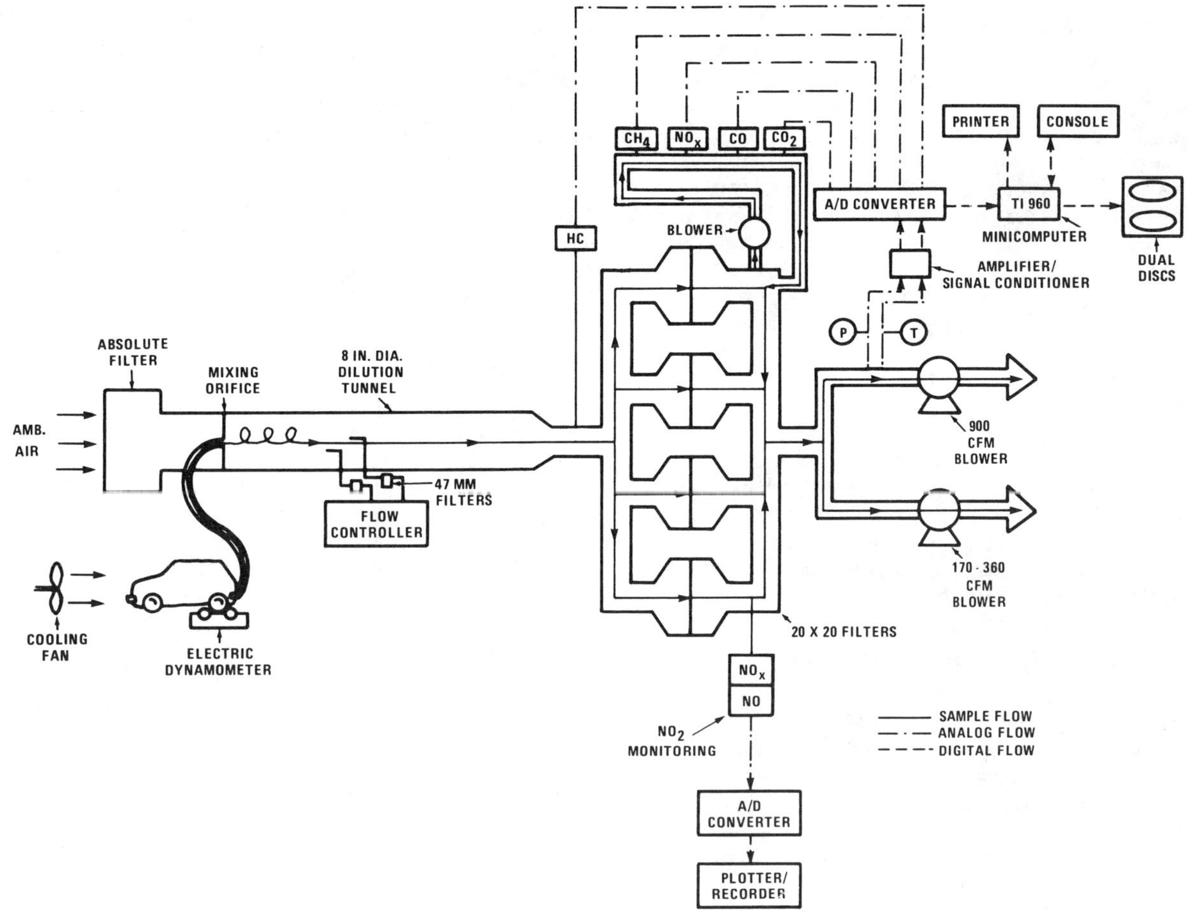

Fig. 1 - The sampling scheme

Table 3 - Test Sequence and Sampling Effort

	DATA ACQUISITION EFFORT*	ACTIVITY
DAY 1		PREP VEHICLE
DAY 2	A, B, C, D, E	FTP
	B, C, D, E	HFET
	B, C, D, E	CUE
	B, C, D, E	NYCC
	B, C, D, E	SC
	B, C, D, E	45 mph S.S.
DAY 3	A, C, E	FTP
	A, C, E	SC
	A, C, E	SC
DAY 4	A, C, E	FTP (u)
	A, C, E	SC
	A, C, E	SC
DAY 5	A, C, E	FTP (u)
	A, C, E	SC
	A, C, E	SC

*KEY: A - REAL TIME GAS ANALYSIS
B - 47 mm FILTERS
C - 20 in. x 20 in. FILTERS
D - NO_2 MONITOR
E - BAG GAS ANALYSIS AND FUEL ECONOMY

acceleration of 5.6 mph per second while the FTP contained no acceleration in excess of 3.3 mph per second. This "clipping" of acceleration rates was effected to avoid operating in excess of specification limits for early dynamometers.

SPECIAL CYCLE DESCRIPTION AND DEVELOPMENT-In order to better evaluate the effect of varying acceleration rates upon exhaust gas emissions, a Special Cycle was devised. The cycle contains 24 distinct modes including two accelerations at wide open throttle (WOT), two at 85 percent of the WOT rate, and two at 70 percent of the WOT rate. The decelerations were made equal but opposite to the accelerations which immediately preceded them. A speed-time trace for the Special Cycle is shown in Figure 2. The idle and 50 mph cruise modes extended for a long enough period to insure time for the gas analyzers to settle out. The total distances covered were 6.9 miles for the VW cycle and 7.3 miles for the Oldsmobile cycle. Different cycles were required for each car since their WOT acceleration rates were different.

Because all acceleration and deceleration modes within the cycle were derived from the WOT acceleration rate, WOT acceleration curves were needed for each car. They were obtained by accelerating at WOT in both directions over a level section of road on a day when wind speeds were negligible. The cars were equipped with a fifth wheel and strip chart recorder for data collection. From these data an average WOT acceleration curve was plotted and the associated 85 percent and 70 percent curves were drawn using a Hewlett Packard 9100 series plotter and calculator unit. The 85 and 70 percent WOT curves represent, respectively, curves having a slope at each given speed value equal to 85 and 70 percent of the slope obtained at WOT.

The WOT data were generated in an actual road test because the dynamometer was not available at the time. As it happened, the speed-time trace recorded at WOT on the road agreed well with that recorded on the dynamometer, thus demonstrating that the road and inertia loads were being properly simulated by the dynamometer.

GASEOUS EMISSION MEASUREMENT EQUIPMENT AND PROCEDURES -Regulated gaseous emissions, CO_2, and methane (CH_4) were determined using standard bag sampling and analysis procedures in accordance with the Federal Register (11). In numerous test runs, these same emissions were measured to obtain modal emissions data using a real-time system centered around a TI-960 B minicomputer.

When the real-time system was in operation, diluted exhaust gases were branched off from the CVS system through a 1 in diameter line. A small blower located within this line transported the gases to the vicinity of the exhaust gas analyzers. This arrangement reduced delay times associated in moving the sample from the main CVS line to the analyzers.

Analog outputs from the gas analyzers were directed to the computer through analog to digital A/D converters (see Figure 1). In addition to gas data, temperature and pressure levels at the inlet to the CVS and air flowrates into the diesel engine were also monitored continuously. These inputs to the computer made possible the modal calculation of CVS flowrates corrected to standard atmospheric conditions and diluent to raw exhaust (dilution ratio) ratios within the CVS - tunnel system.

The A/D converter system used a programmable gain amplifier to provide 13 full scale input signal ranges from 2.5 mv to 10.24 v. It converted the analog signals to 13-bit binary numbers. Although the converter was configured with forty input channels, only sixteen were programmed for use. Each channel was sampled at a rate of ten samples per s.

The digital data outputted from the A/D converter was stored in one of two memory buffers of the computer. The two buffers alternately filled and dumped their contents onto a disc unit. Thus, the data was gathered continuously and stored serially within a large disc file during test runs.

At the completion of a run, a program was loaded and information was retrieved from the disc for processing. The data retrieval and processing program basically integrated all data points within the defined modes and calculated average diluted exhaust concentrations, dilution ratios, and CVS flowrates. It also calculated the modal emission rate in grams per mile for hydrocarbons (HC), carbon monoxide (CO), nitrogen oxides (NO_x), CO_2, and CH_4. The mode definitions were made at the beginning of the program by inputting mode type, time, and distance information which were stored on cassette tape. Because all gas data generated during the test remained stored on the disc, the same run could be evaluated using different modal analysis schemes. It was also possible to concentrate on certain portions of the cycle not presently defined within the modal scheme.

When utilizing the real-time system during the 1975 FTP and the FTP-U cycles, a modal scheme developed at General Motors was used. A breakdown showing the modes and their duration is shown graphically with the FTP speed-time trace in Figure 3. Data on regulated emissions, CO_2, and CH_4 was integrated by mode such that average emission rates for each mode were calculated. The

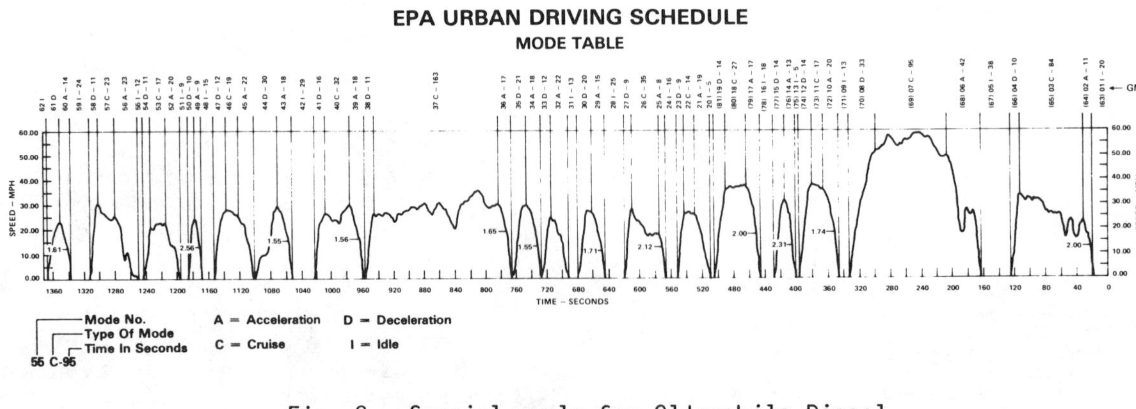

Fig. 2 - Special cycle for Oldsmobile Diesel

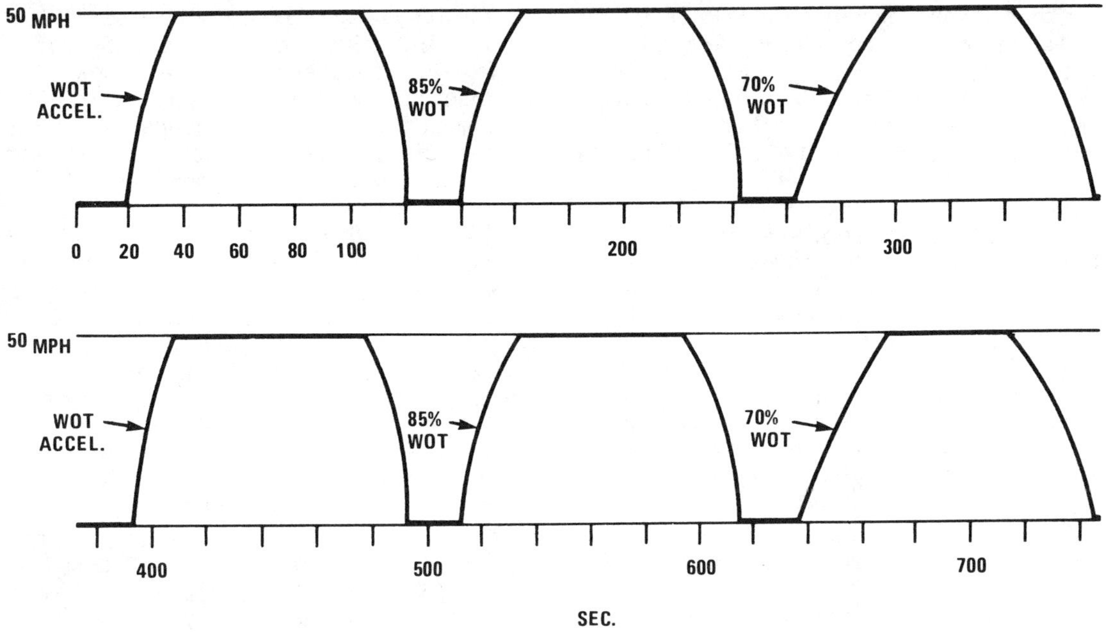

Fig. 3 - FTP mode table

processed data also calculated the emission rates for each of the three phases of the FTP.

PARTICULATE ANALYSIS AND MEASUREMENT EQUIPMENT AND PROCEDURES

PARTICULATE SAMPLING STATIONS-Particulate matter in diluted exhaust was sampled using both a conventional method employing twin probes with 47mm filters and a total diluted gas filtration system employing four rather massive 20 inch by 20 inch filters in parallel. Both sampling methods were operated concurrently during test runs.

Sample flowrates through the probe filters were kept constant with a flow controller, a method that involved varying a flow regulator downstream to maintain a constant pressure drop acrossed a flow orifice. Each probe contained a primary and back-up filter and sampling of particulate matter was done in accordance with the EPA Standard for Emission of Particulate Regulation of Diesel-Fueled Light-Duty Vehicles and Light Duty Trucks (12). Because of the twin probe arrangement, duplicate filters were collected whenever possible during tests.

The 20-in by 20-in filters were used to collect large quantities of diesel particulate matter for extract to be used in Ames testing. The filters were configured and sized to permit sampling of the total diluted exhaust gas stream. With these methods, all the exhaust particles generated by the vehicle during testing were captured.

Particles sampled for particulate emission rate determination and for work involving the extraction of organic material were collected on Pallflex T60A20 teflon-coated glass fiber filters. The Pallflex filter was selected because the fibers do not appear to interact with gaseous organics and the pressure drop across the filter does not become excessively high during collection (13). Particles destined for x-ray analysis in trace metal determination were collected on Fluoropore filters, which have been found to be well suited for this procedure.

47MM FILTER HANDLING - Prior to testing, the filters were weighed and placed within filter holders which were situated in-line on the sample probes (see Figure 1). Sample flowrates were maintained at about 1 SCFM with teflon-coated glass fiber filters and at about 0.6 SCFM with the Fluoropore filters. Following the test, the filters were removed, inserted in petri dishes, and transported to a temperature and humidity controlled weighing room for equilibration. Sample weight was later recorded and the gram per mile emission rate was calculated. For the FTP, a "weighted" emission rate was calculated by breaking out the emission rates for each of the three test phases and applying weighting factors. This procedure was made possible by the fact that a separate filter had been collected for each test phase.

After the sample weights had been determined, the teflon-coated glass fiber filters were placed within separate soxhlets for extraction of organic materials. The extraction process was carried out for 8 h using methylene chloride as the solvent. The amount of organic extract present was determined by two methods: the direct method and the Delta Filter Method.

The direct method involved three stages. Initially the filters were placed within the soxhlets for extraction of the organics, as explained earlier. The extracted solution was "blown down" under vacuum at a temperature of about 30°C removing much of the highly volitile solvent. The highly concentrated solution was then transferred to vials (that had been weighed prior to the extraction process) and dried rapidly by bathing in gaseous nitrogen. After equilibrating in the weighing room, the vials were reweighed and the precent extractable derived by taking the difference of the two weights.

The second method, the Delta Filter Method, was considerably shorter requiring only the initial extraction stage. After the filters had been extracted for 8 h they were removed from the soxhlets and placed into petri dishes. Then the filters were vacuum dried 1 h in an oven at a temperature of about 85°C. After drying, the filters were removed to the weighing room and allowed to equilibrate overnight. The next day the filters were reweighed and a new sample weight was recorded. The percent extractable was derived by subtracting the sample weight after extraction from the sample weight before extraction.

X-ray analysis was carried out on the Fluoropore filters to determine the quantity of various trace elements in the particulate matter. The instrumentation used for these analyses was a Siemens MRS-3 x-ray fluorescence spectrometer. It is automated and computer-controlled, with 16 fixed monochromators and one scanning monochromator. Counting intervals are normally 100 s for the fixed monochromators and 20 s for the scanning monochromator.

20-IN x 20-IN FILTER HANDLING - The large 20-in square teflon-coated glass fiber filters were used in groups of four (one per filter holder) per test. Except for those gas streams sampled by the 47 mm filters, the 20 x 20's sampled the entire diluted exhaust gas stream. The filters were weighed in groups of four both before and after the test. The particulate emission rate was easily calculated as the sample weight on the filters divided by the test cycle distance. Emission rates so calculated were later compared with those calculated from 47 mm filter data.

Large quantities of extracted material necessary for Ames testing were obtained from the large 20-in x 20-in filter samples collected during each test. These were immediately weighed and stored at -20°C, prior to preparation of organic extracts of exhaust particulates for Ames testing and chemical characterization. Typically, two 20-in x 20-in filters were soxhlet extracted with methylene chloride overnight and the extract filtered through a 0.2 um Millipore filter. The solvent was then reduced to 100 ML on a rotary evaporator, and a 2 ML aliquot evaporated under nitrogen to determine the extracted mass. A larger aliquot for the Ames bioassay was similarly prepared.

BIOLOGICAL TESTING-Ames testing of extracts was conducted by Northrop Services, Inc., RTP, N.C. Samples were tested in strain TA 98 of the Ames Salmonella typhimurium mutagenisis bioassay. (14) The extract samples, which were supplied dried and under nitrogen, were diluted to 2 mg per ml in dimethyl-sulfoxide and assayed at doses of 50, 100, 300, and 500 ul. Specific activities were calculated from the linear portion of the dose response curve, which in all samples was limited to the spontaneous revertant level plus the 50 and 100 ul doses. All samples were tested in a single experiment.

MONITORING NO_2-An NO_2 monitor was operated on days when the complete six cycle sequence was run (i.e., day 2 of Table 3). It was operated only as necessary to define the average and peak levels of NO_2 in the diluted exhaust sample. This action was taken to document NO_2 levels because potential for interaction between NO_2 and poly-nuclear aromatic compounds to yield nitro species which are direct mutagens in the Ames bioassay procedure has been reported (15). Furthermore, NO_2 emissions from diesel engines have been reported as representing at times a significant fraction of the overall NO_x emissions (16). Forthcoming purblications from this laboratory will address more directly the nitrate artifact problem relative to diesel engine exhaust sampling.

The levels in diluted exhaust gas were monitored continuously by recording the difference between outputs from a NO_x analyzer and an NO analyzer. Both analyzers were identical TECO chemiluminescent types, except that the NO analyzer was operated without a converter. The response time of each analyzer to a pulse of NO injected into the dilution tunnel was equated by adjusting sample line lengths to the analyzers. This precaution insured that the analyzers detected the same sample when the difference measurement was taken.

Analog outputs from each analyzer were inputted into a Bascom Turner Model 8110 analog to digital recorder/plotter. Data was digitized, loaded into a buffer, dumped onto a floppy disc, and later retrieved for plotting and integrating. The recorder automatically calculated the difference measurement, plotted the NO_2 level versus time, and integrated to obtain the average level over the cycle.

GASEOUS EMISSION RESULTS

CONSTANT VOLUME SAMPLING SYSTEM-The regulated gaseous emissions and fuel economy data were obtained routinely by the conventional CVS procedure described earlier. The results (Table 4) represent the mean levels determined for the two diesel cars. The number of tests conducted to arrive at the mean and the standard deviations are also reported. Emissions data obtained for the Rabbit diesel during a period when fuel injector problems existed have not been averaged into the data reported in Table 4.

Regulated emissions from the Oldsmobile diesel are below the 1980 emission standards level, while hydrocarbon emissions from the Rabbit diesel are somewhat higher than the 0.4 g/mi level. As mentioned earlier, the Rabbit diesel had been purchased only a month before

Table 4 - Comparison of 75 FTP, FTP-U, HFET, CUE, NYCC, SC, and 45 mph SS Gaseous Emissions Data, g/mi

VEHICLE/TEST		HC	CO	NO_x	CH_4	FUEL ECONOMY, mpg	NO. OF TESTS
OLDSMOBILE DIESEL							
75 FTP	$\bar{x}$	0.255	1.244	1.541	0.001	19.671	7
	s	0.035	0.172	0.195	0.002	1.281	
FTP-u	$\bar{x}$	0.281	1.390	1.601	0.001	16.956	3
	s	0.033	0.099	0.111	0.000	0.040	
HFET	$\bar{x}$	0.139	0.846	1.014	–	27.101	3
	s	0.020	0.030	0.085	–	1.276	
CUE	$\bar{x}$	0.171	1.048	1.096	0.002	26.466	3
	s	0.034	0.084	0.171	0.001	3.398	
NYCC	$\bar{x}$	0.734	3.195	3.583	0.001	9.430	3
	s	0.135	0.386	0.312	0.023	0.388	
SC	$\bar{x}$	0.193	1.501	1.178	0.005	22.280	10
	s	0.031	0.165	0.081	0.003	0.700	
45 mph. ss	$\bar{x}$	0.091	0.444	1.114	–	34.961	3
	s	0.010	0.021	0.147	–	1.241	
VW RABBIT DIESEL							
75 FTP	$\bar{x}$	0.505	1.311	0.989	0.027	39.363	4
	s	0.042	0.019	0.076	0.014	1.568	
FTP-u	$\bar{x}$	0.447	1.603	1.130	0.033	37.307	3
	s	0.027	0.359	0.291	0.010	4.569	
HFET	$\bar{x}$	0.532	1.210	1.020	0.048	40.805	3
	s	0.068	0.092	0.044	0.024	0.436	
CUE	$\bar{x}$	0.613	1.260	1.011	0.008	42.841	3
	s	0.054	0.154	0.118	0.020	2.396	
NYCC	$\bar{x}$	1.026	2.507	1.703	0.013	26.048	3
	s	0.076	0.066	0.226	0.004	1.262	
SC	$\bar{x}$	0.506	1.360	0.883	0.036	42.779	6
	s	0.027	0.105	0.120	0.009	0.861	
45 mph. ss	$\bar{x}$	0.266	0.839	0.932	0.013	50.829	3
	s	0.019	0.062	0.122	0.000	1.393	

the test program began and initial emissions results indicated an engine malfunction condition (see Table 5). The car was returned to the dealer for examination of the injector spray patterns and engine timing. After determining that the injector timing had been somewhat retarded, it was advanced three degrees and the test program was resumed. As the results in Vehicle Condition "B" of Table 5 demonstrate, emissions generally deteriorated after the timing was advanced. For this reason, the vehicle was returned to the dealer for recheck of the injector spray patterns. This time one

Table 5 - Emission Trends from Rabbit Diesel

VEHICLE* CONDITION	TYPE TEST	HC, gm/mi	CO, gm/mi	NO_x, gm/mi	PARTICULATE, gm/mi	FUEL ECONOMY, mpg
A	FTP	0.68	1.69	1.23	0.54	39.8
B	FTP	0.74	1.70	1.28	0.76	36.5
C	FTP	0.51	1.31	0.99	0.43	39.4
CERT.	FTP	0.38	1.16	1.05	e	40.2
A	SC	0.73	2.01	0.96	0.60	43.1
B	SC	0.96	1.86	0.88	0.55	39.0
C	SC	0.51	1.36	0.88	0.48	42.8

*A - TIMING RETARDED/BAD INJECTOR
B - BAD INJECTOR
C - INJECTOR TIP REPLACED
CERT - CERTIFICATION RESULTS FOR 1980 DIESEL RABBIT
e - NOT AVAILABLE

of the injector tips was defective and it was replaced. A subsequent reduction (condition "C," Table 5) in both gaseous and particulate emissions was noted, although the HC and CO values were still significantly higher than values obtained by EPA in certification.

Concern that the new engine was not broken in properly and that blow-by emissions might exist prompted an examination of the SOF molecular weight distributions. This examination identified the principle sources of the organic mass by use of a gas chromatograph, which Black and High have reported to be a useful tool in detecting lubricant consumption (13). The distributions were examined to insure that the engine had been broken-in from an emission standpoint. Figure 4 shows the distribution for the Rabbit. This molecular weight distribution curve and the one shown in Figure 5 for the Oldsmobile represent the average distributions obtained over the six test cycles examined. The lack of abnormally high quantities of the lube oil components (C-25 to C-40) implies that lube oil consumption was not a problem for either vehicle. For the Rabbit, however, high levels of the fuel component were evident out to about C-25, suggesting that a fuel delivery problem continued to exist. The somewhat elevated HC and CO emissions from the Rabbit diesel were attributed to this problem.

REAL-TIME SAMPLING AND ANALYSIS - Regulated gaseous emissions were determined modally over the 75 FTP, FTP-U, and the SC tests using the real-time sampling system described earlier. Table 6 presents the percentage difference in emissions determined by using the CVS procedure versus the real-time system. In general there appears to be good agreement between the two methods. Table 7 shows the comparison between standard deviations (standard deviations are expressed as a percent of the mean) obtained when using the CVS procedure and the real-time system. The repeatability of the real-time system was even better in most cases than the CVS procedure, which was in itself satisfactory. Because air flowrate into the diesel engine was monitored on a continuous basis with the real-time system, modal determination of the actual dilution ratios within the CVS was possible. The average dilution ratio over each test cycle or test phase could also be determined; hence, correction for background emissions which includes this dilution ratio value could be calculated more directly. Because the calculation of dilution ratio in accordance with the Federal Register procedure (11) assumes stoichiometric combustion, which for diesels is an unrealistic assumption, it was of interest to compare results obtained using the Federal Register procedure with those obtained using the direct measurement method. The results in Table 8 are the same for all practical purposes. The percent difference only becomes significant when the emissions level becomes very low. For example, the HC emission rate for Phase 2

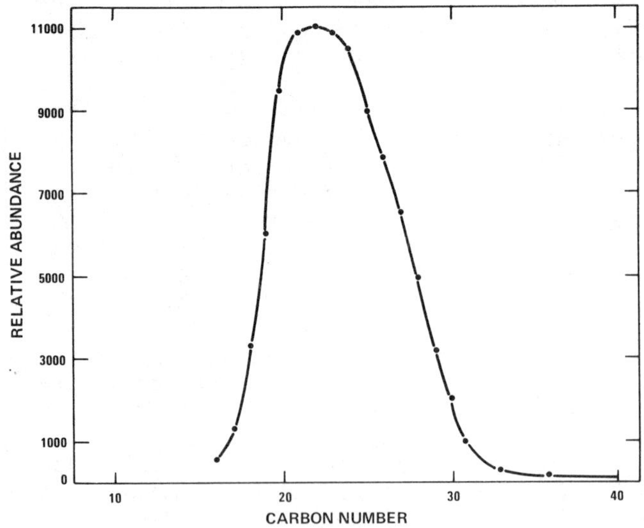

Fig. 4 - Molecular weight distribution for particulate SOF - VW Rabbit

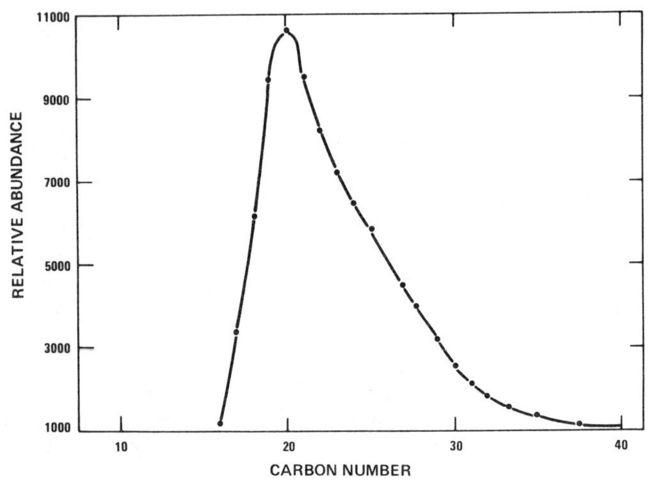

Fig. 5 - Molecular weight distribution for particulate SOF - Oldsmobile

was 0.25 g per mi with the calculated dilution ratio versus 0.27 g per mi with the measured ratio, a difference of about 8 percent. The calculated and measured dilution ratio values are also shown in Table 8 for each of the three phases of the FTP.

The modal analysis system made it possible to determine the effect of differing acceleration rates upon gaseous emission patterns and to examine the difference in emissions between the FTP and the FTP-U. Because the FTP-U contained a few higher accelerations than the FTP, some knowledge of the emission rate differences between the two cycles was desired.

While the gaseous data presented in Table 4 for the Oldsmobile do support expectations for an increase in emissions and fuel consumption over the FTP-U test cycle, the data for the Rabbit diesel do not. To better understand what was happening, it was necessary to examine the modal data.

Mode 6 of the FTP-U contained the highest acceleration rate encountered throughout the test. The rate from 0 to 15 mph had a value of 4.7 mph per s as compared with a value of 3.3 mph per s for the FTP. Ten s following the start of the acceleration, the speeds leveled off to an approximate value of 26 mph in both cycles.

Table 6 - Percentage Difference in Emissions Determinations - CVS Procedure vs. Real-Time.

VEHICLE/TEST	Δ HC, percent	Δ CO, percent	Δ NO_x, percent
OLDSMOBILE DIESEL			
75 FTP	+ 2.4	- 5.5	- 3.6
FTP-u	+ 2.5	- 7.9	- 6.7
SC	+ 3.7	- 8.6	- 1.3
VW RABBIT DIESEL			
75 FTP	+ 6.4	- 5.7	+ 0.5
FTP-u	+ 3.8	+ 10.4	+ 6.1
SC	+ 5.2	- 1.4	+ 3.7

Table 7 - Comparison of Test-to-Test Repeatability* for Oldsmobile - CVS Procedure vs. Real-Time

TEST/METHOD	HC	CO	NO_x	CO_2
75 FTP				
CVS	14	14	12	6
REAL TIME	8	8	14	5
FTP-u				
CVS	11	7	7	2
REAL TIME	13	1	2	3
SC				
CVS	16	11	7	3
REAL TIME	10	12	4	1

*REPEATABILITY EXPRESSED AS THE PERCENT OF THE STANDARD DEVIATION OF THE MEAN: [(STANDARD DEVIATION)/MEAN] X 100

Table 8 - Comparison of Oldsmobile FTP Data Obtained with Calculated Versus Measured Dilution Ratio

TEST PHASE/METHOD	CO_2, g/mi	HC, g/mi	CO, g/mi	NO_x, g/mi	DILUTION RATIO
NO. 1					
CALCULATED	615.31	0.44	1.85	1.35	8.59
MEASURED	616.72	0.45	1.86	1.35	4.23
NO. 2					
CALCULATED	628.03	0.25	1.34	1.97	12.64
MEASURED	630.08	0.27	1.35	1.95	5.08
NO. 3					
CALCULATED	555.60	0.24	1.47	1.49	8.59
MEASURED	556.97	0.25	1.48	1.49	4.05

A look at the Rabbit modal data revealed no substantial differences in emissions between the FTP and FTP-U versions of Mode 6. A closer examination of the WOT acceleration rate for the Rabbit indicated that the rate from 0 to 15 mph was 3.7 mph per s. This meant that the vehicle was being operated at a near WOT acceleration rate in the FTP and that at WOT in the FTP-U, the vehicle was unable to keep up with the trace. Thus, emission trends resulting from differing acceleration rates could not be successfully understood by comparing FTP data with FTP-U data when the vehicle was unable to accelerate at rates required by the cycle. (Although more gradual accelerations existed in both cycles, those were of less impact.)

When emissions from the Oldsmobile were examined s by s over the first 10 s of Mode 6, the difference between FTP and FTP-U emissions of CO, HC, NOx, and CO_2 were, respectively, 6, 4, 2, and 1 percent. While the emissions differences cited in Table 4 do not appear significant, the difference in fuel economy does. From the data collected within this program, the reason for this marked difference is not apparent.

The effect of varying acceleration rates upon exhaust emissions was examined with greater focus over the Special Cycle (SC) test runs. The SC, which was described earlier, contained accelerations at WOT, 85 percent of WOT, and 70 percent of WOT. Figures 6 and 7 show emissions trends observed from the Rabbit and Oldsmobile diesels.

CO emissions from the Rabbit show the greatest sensitivity to increases in the acceleration rate, while NOx emissions appear almost insensitive to changes. CO emission increased about 60 percent when going from the 70 percent rate to WOT. HC emissions showed virtually no increase when the acceleration rate rose from 70 to 85 percent of WOT; however, the increase from the 85 percent rate to WOT affected a 54 percent increase in emissions. CO_2 emissions, which reflect the vehicle's fuel consumption rate, increased by about 5.5 percent when testing at the 85 percent rate,as opposed to the 70 percent rate. Another increase of 22 percent was noted when going from the 85 percent rate to the WOT rate.

The emission trends observed from the Oldsmobile were more sensitive to increases in the acceleration rate. Emissions of HC, CO, and NO_x increased 128, 52, and 39 percent, respectively, as the acceleration rate was increased from the 70 percent rate to the WOT rate. CO_2 emissions exhibited a rather steady increase of 43 percent as the acceleration rate was increased over the range.

It should be remembered that the emissions in Figures 6 and 7 are expressed on a g per mi basis. Because the vehicle was accelerated at different rates to a cruise speed of 50 mph, the distances and times over the differing accelerations varied. This resulted in portrayals which made the emissions at higher accelerations appear somewhat exaggerated. For

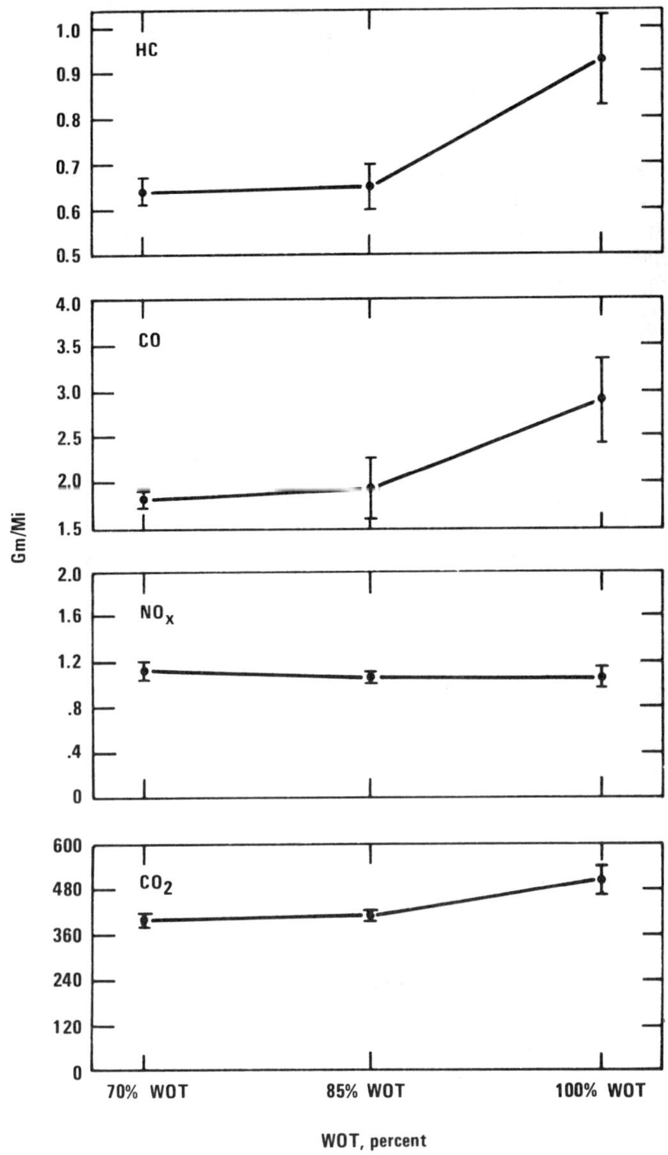

Fig. 6 - VW Rabbit emission trends with change of acceleration rate

example, the mass emission of HC from the Rabbit over a 1 mi course was 0.582 g when the vehicle was accelerated at WOT then operated at a 50 mph cruising speed. The mass emission was reduced to 0.495 g when the acceleration was lowered to 70 percent WOT and the remainder of the 1 mi course was driven at 50 mph. But this 15 percent reduction in mass emission appears small compared to the 54 percent reduction cited as the change in emission rate.

PARTICULATE MATTER EMISSIONS RESULTS

PARTICULATE MATTER EMISSION RATES - The emission rate of particulate matter from the test vehicles was principally determined by the sample probe method of collecting material on 47 mm filters. Because large 20-in by 20-in square filters were collected concurrently, emission rate determinations were worked up from these as well. Table 9 shows the results with the means and standard deviations listed for each car.

The average of standard deviations within the sampling method using 47 mm filters was 11 percent and that within the 20 by 20 method was 17 percent. Considering this, agreement between the two sampling methods was good with the average difference equal to 14 percent. Emission rates were highest for each vehicle when tested on the New York City Cycle (NYCC). The NYCC has a duration of about 10 min, a distance of 1.13 mi, and an average speed of 7.1 mph. Hence, because of the low average speed of the cycle coupled with its "stop and go" nature, high emission rate values would be expected when expressed in units of g per mi. Emission rates from both vehicles appeared relatively insensitive to test cycle. The average emission rate from the Oldsmobile over all six test cycles was 0.54 g per mi with a standard deviation of 0.09 g per mi. For the Rabbit the average was 0.51 g per mi with a standard deviation of 0.11 g per mi. In both cases the standard deviations obtained from cycle to cycle are only slightly higher than the standard deviations obtained when testing within the same cycle.

Table 9 - Particulate Emission Rates

VEHICLE/TEST		SAMPLING METHOD: PROBE SAMPLING, gm/mi	SAMPLING METHOD: 20" X 20" SAMPLING, gm/mi
OLDSMOBILE DIESEL			
75 FTP	$\bar{X}$	0.54	0.54
	s	0.15	0.11
HFET	$\bar{X}$	0.58	0.46
	s	0.09	0.15
CUE	$\bar{X}$	0.53	0.41
	s	0.09	0.08
NYCC	$\bar{X}$	0.68	0.65
	s	0.05	0.07
SC	$\bar{X}$	0.53	0.44
	s	0.05	0.07
45 ss	$\bar{X}$	0.39	0.41
	s	0.04	0.07
VW RABBIT DIESEL			
75 FTP	$\bar{X}$	0.43	0.45
	s	0.03	0.05
HFET	$\bar{X}$	0.54	0.37
	s	0.11	0.21
CUE	$\bar{X}$	0.61	0.36
	s	0.03	0.03
NYCC	$\bar{X}$	0.64	0.66
	s	0.04	0.00
SC	$\bar{X}$	0.48	0.43
	s	0.01	0.01
45 ss	$\bar{X}$	0.35	0.36
	s	0.03	0.03

Table 10 - Soluble Organic Fraction of Particulate Matter

VEHICLE/TEST		% EXTRACTED BY DIRECT WEIGHING	% EXTRACTED "DELTA FILTER" METHOD	% DIFFERENCE BETWEEN METHODS
OLDSMOBILE DIESEL				
75 FTP	$\bar{X}$	15.20	15.22	3.30
	s	4.04	3.85	
HFET	$\bar{X}$	15.96	16.12	4.51
	s	2.46	2.39	
CUE	$\bar{X}$	14.35	14.60	6.13
	s	1.54	1.71	
NYCC	$\bar{X}$	15.05	14.75	6.05
	s	0.81	1.56	
SC	$\bar{X}$	11.07	11.39	6.59
	s	1.72	2.42	
45 ss	$\bar{X}$	28.94	28.87	3.25
	s	2.24	2.04	
VOLKSWAGEN DIESEL				
75 FTP	$\bar{X}$	27.13	26.22	5.46
	s	6.09	6.20	
HFET	$\bar{X}$	27.36	26.75	2.89
	s	5.24	5.53	
CUE	$\bar{X}$	29.63	29.03	3.01
	s	4.70	4.42	
NYCC	$\bar{X}$	26.79	24.16	13.79
	s	2.32	2.61	
SC	$\bar{X}$	27.69	27.31	4.89
	s	3.71	2.90	
45 ss	$\bar{X}$	31.93	31.92	4.30
	s	3.26	3.14	

SOF CONTENT - Table 10 shows the extractable organic as a percent of the total particulate matter from which it was extracted. Each vehicle registered its highest SOF's in the 45 mph steady state tests; however, levels from the Oldsmobile were significantly lower than the Rabbit's. Again the Rabbit's emission levels were relatively insensitive to the cycle while those from the Oldsmobile were significantly higher for the 45 mph steady state and lower for the Special Cycle.

As mentioned previously, the SOF content was determined by two different methods: the direct method and the Delta filter method. The direct method essentially weighs the extract directly while the Delta filter method takes the difference of the filter weight before and after extraction. Agreement between the two methods has been previously cited as being quite good (17). Although the samples obtained on 47 mm filters were considerably lighter in this study (the average sample weight equalling 8.3 mg), agreement between the two methods was within 8 percent for all filters examined. The difference expressed as a percent of the sample weight for all test cycle and vehicle combinations examined is given in Table 10. High difference values cited for the Rabbit when tested on the NYCC are mostly due to the extremely small sample quantities obtained over that cycle. The average sample weight for these tests was about 1.5 mg compared with an average of 13 mg for the HFET. Considering the small sample quantities involved, agreement is very good overall. Because great care must be taken to insure that the extract is properly dried before weighing in the direct method, the Delta filter method is somewhat easier to perform.

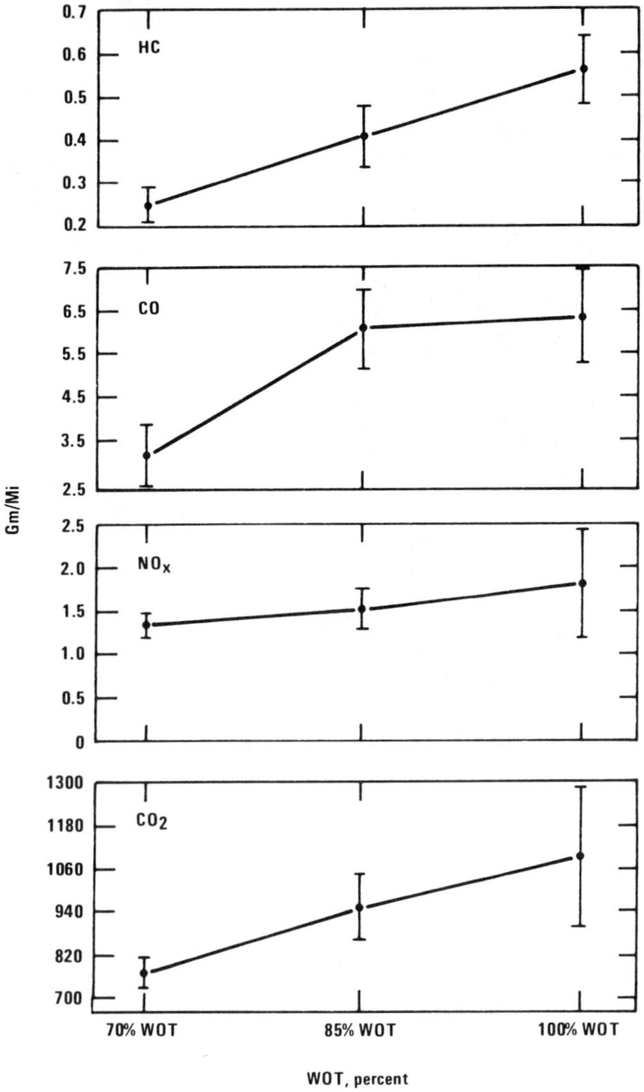

Fig. 7 - Oldsmobile emission trends with change of acceleration rate

TRACE METALS EMISSIONS - Tables 11 and 12 show the emission of trace metals or elements contained within the particulate matter. The predominant elements emitted from both cars were Na, Fe, Ca, S, Zn, and Sb. Of these, Na, P, Ca, and Zn are probably associated with the lubricating oil and its additives. Although the SOF molecular weight distributions (Figures 4 and 5) suggest that excessive oil consumption was not a problem, the x-ray data implies that lube oil consumption did exist to some extent. Fe emissions likely originated from exhaust system deterioration and S emissions from the fuel sulfur. Assuming that the particulate sulfur was contained as sulfate material ($=SO_4$), the sulfate represents 1.8 percent and 1.4 percent by mass of the total particulate emitted from the Oldsmobile and VW Rabbit, respectively. These percentages were fairly consistent regardless of test cycle since the relative standard deviations of the results were 21 percent for the Oldsmobile and 17 percent for the Rabbit over all six cycles.

The weight percent sulfur content of the fuel was 0.18 which is close to the 0.23 National Average value.

For the FTP, the total elemental weight percentages of the total particulate matter emission from the Rabbit and Oldsmobile were 1.8 percent and 2.0 percent, respectively. These values agree closely with those cited by Hare and Baines (18) in their examination of emissions from two diesel automobiles. In that study, the total elemental weight percentage obtained from a Rabbit diesel tested within that study was 1.9 percent over the FTP using the National Average diesel fuel.

AMES TEST RESULTS - A matrix of 18 particulate extract samples was tested and consisted of one sample for each test cycle from both the Oldsmobile and Rabbit. An addition-

Table 11 - Trace Metal Emissions, mg/mi, Oldsmobile Diesel

EMISSIONS/TEST	FTP	FET	CUE	NYCC	SC	45-SS
NA	0.2113	0.2291	0.0523	0.3219	0.1155	0.2593
MG	0.1238	0.5470	0.0520	0.2522	0.0040	0.1045
AL	0.0423	0.0179	0.0159	0.0225	0.0076	0.0146
SI	0.0338	0.0316	0.0246	0.0491	0.0318	0.0282
P	0.9411	0.9064	0.7238	1.8202	0.7094	1.0938
S	2.9565	3.5895	2.3212	4.1592	3.1590	3.2813
CL	0.0245	0.0243	0.02749	0.0485	0.0351	0.0157
CA	1.6567	1.5099	1.1842	3.1539	1.2001	1.9042
CR	0.2725	0.1912	0.1796	0.2248	0.0686	0.3644
MN	0.2571	0.1437	0.1274	0.3519	0.0809	0.3586
FE	0.2207	0.2528	0.1007	0.2960	0.2019	0.0754
CO	0.0621			0.0062	0.0219	0.0311
ZN	3.1121	3.2319	2.1517	4.3625	2.8716	3.8757
CD	0.0143	0.0195	0.0121	0.0174	0.0193	0.0224
SN	0.6254		0.0460	0.0638	0.0415	0.0925
SB	0.2319	0.2135	0.1645	0.4339	0.1617	0.2755
K	0.0091	0.0138	0.0052	0.0257	0.0144	0.0137
CU					0.0830	

Table 12 - Trace Metal Emissions, mg/mi, VW Rabbit Diesel

EMISSIONS/TEST	FTP	FET	CUE	NYCC	SC	45-SS
NA	.3340	.1216	.1655	1.546l	.2723	.0590
MG	.1654	.6901	.0904	.4024	.1793	.0188
AL	.5306	.0039	-	.2087	.0118	.0028
S	4.1595	2.1492	2.3982	7.2862	3.9790	.8701
CA	3.1206	1.5705	1.1849	16.8164	3.059	.6830
P	1.2303	.9128	.4186	8.7967	1.7660	.4068
FE	.8747	.0423	.3555	-	.3178	
ZN	.4156	2.2444	2.5460	8.3516	4.4446	.9741
CD	.0067	.0050	.0161		.0330	.0047
SB	.4040	.1960	.1235	2.3254	.3865	.0952
BA	.0392	-	.0053	.1827	.0442	
PB	1.5789	1.1749	.4860		1.2587	.3285
CL		.0207	.0536	-		
CU		-	.0251	-	.0497	.0061
BR		.2370	.1465	2.1528	.1331	.0140
TI			.0063		.3757	.0667

al six FTP samples from the Rabbit were tested; of these, four reflected an injector malfunction (a faulty spray pattern existed on one injector.) Initial results were obtained only in strain TA 98 without microsomal activation. Experience to date with diesel samples has shown that tests using strain TA 98 without activation have produced higher revertant levels than tests with activation. By testing the matrix in a single experiment, data in the following areas was sought relative to mutagenicity:

1. Day to day variability of FTP test cycle.
2. Variations among cycles and comparison of cycles between the Oldsmobile and Rabbit.
3. Effects of injector malfunction.

Figure 8 shows the results obtained for a series of FTP's run on the Rabbit. During the first four test dates, the car had an injector malfunction. The last three test were run following injector tip replacement. Both sets of runs showed fair repeatability with an average relative standard deviation of 10.5 percent. However, injector malfunction resulted in a 73 percent increase in mutagenicity. Recent work by the Nissan Motor Company has also demonstrated that significant changes in the mutagenicity of particulate extracts can be attributed to injector malfunction (19).

Figures 9 and 10 are comparisons of the Oldsmobile and Rabbit for the six different test cycles studied. In Figure 9, the mutagenic activity is expressed in terms of revertants per microgram of extract and shows the Oldsmobile to be higher for the FTP, NYCC and SC, while the Rabbit is higher in the HWFET, CUE and 45 SS. However, when the mutagenic activity is normalized with respect to the extractable organics and particulate emissions, i.e., activity is expressed in revertants per mile, the Rabbit is higher for all test cycles except the NYCC as shown in Figure 10. However, the high activity calculated for the Oldsmobile in the NYCC is likely an atypical value since the particulate emissions for that individual test were more than double that normally observed.

Chemical characterization of the organic extracts is currently in progress. Initial efforts are being directed toward quantifying the nitro-aromatics, especially the nitropyrenes which have recently been reported to be mutagenic (20) and have also been identified in diesel extracts (21). The concentration of these compounds and their related mutagenic activity will then be correlated with the observed Ames test results.

CONCLUSIONS

1. Emissions of particulate matter and regulated gases measured from the Rabbit diesel when tested with

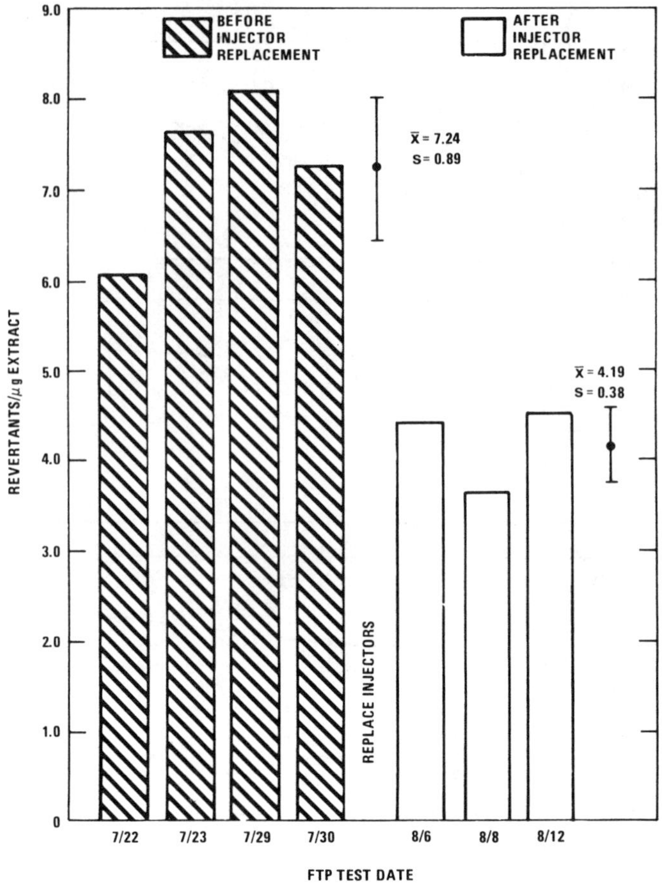

Fig. 8 - Comparison of extract mutagenicity for VW Rabbit diesel before and after injector replacement (Ames Test, TA 98-S9)

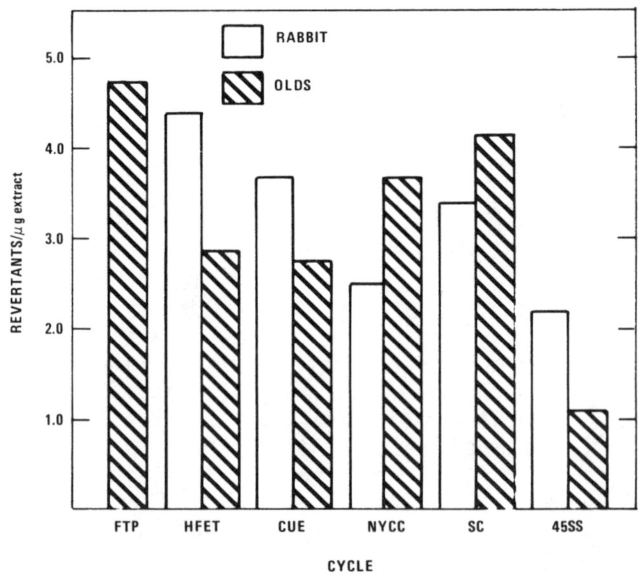

Fig. 9 - Comparison of extract mutagenicity for Oldsmobile diesel vs. VW Rabbit diesel (Ames Test, TA 98-S9)

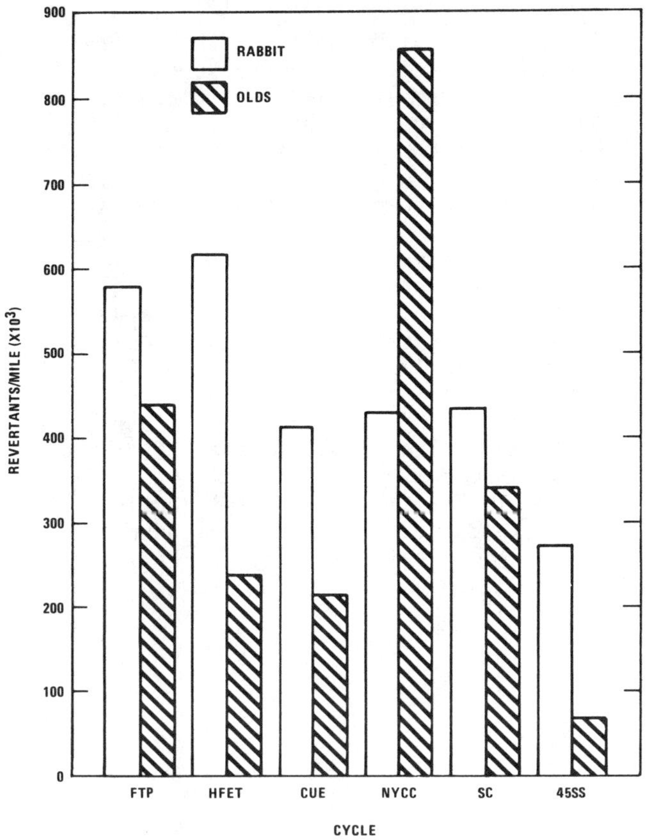

Fig. 10 - Comparison of mutagenicity expressed in revertants/mi, Oldsmobile diesel vs. VW Rabbit diesel (Ames Test, TA 98-S9)

one faulty injector tip were significantly higher than those measured after the tip was replaced.

2. The injector malfunction had a larger impact on mutagenic emissions than the cycle over which the vehicle was operated.
3. Differences in emissions and fuel economy for the FTP and FTP-U cycles were not detectable for the two diesel cars examined modally.
4. For the two vehicles examined, CO and HC g per mi emission rates were most sensitive and NOx emissions the least sensitive to changes in acceleration rate.
5. SOF extract quantities obtained using the Delta Filter Method agreed well with those determined using the direct weighing procedure.
6. Regulated gaseous emission rates obtained using the CVS system method agreed well with those obtained using the real-time method.
7. Trace metal emissions represented an insignificant portion of the total particulate mass.
8. Particulate sulfate emissions expressed as a percentage of the total particulate matter emitted were not significantly affected by the driving cycle.

ACKNOWLEDGMENT

The authors wish to express gratitude to Robert Bury for his timely assistance in processing the large volume of data associated with this study and to Dr. Larry Claxton for his assistance in interpreting Ames test data.

REFERENCES

1. E. Shama and T.O. Wagner, "Fuel Quality or Engine Design: Which Controls Diesel Emissions?" SAE Paper 730168, January, 1973.
2. H.W. Barnes-Moss and W.M. Scott, "The Light Duty Diesel Engine for Private Transportation." SAE Paper 75033, February, 1975.
3. M.F. Russell, "Recent CAV Research into Noise, Emissions, and Fuel Economy of Diesel Engines." SAE Paper 770257, February, 1977.
4. G. Greeves et al, "Origins of Hydrocarbon Emissions from Diesel Engines." SAE Paper 770259, February, 1977.
5. K.J. Springer and H.A. Ashby, "The Low Emission Car for 1975 - Enter the Diesel." SAE Paper 739133, August, 1973.
6. J.N. Braddock and R.L. Bradow, "Emissions Patterns of Diesel Powered Passenger Cars." SAE Paper 750682, June, 1975.
7. C.T. Vuk, M.A. Jones and J.H. Johnson, "The Measurement and Analysis of the Physical Character of Diesel Particulate Emission." SAE Paper 760131, February, 1976.
8. J.N. Braddock and P.A. Gabele, "Emission Patterns of Diesel-Powered Passenger Cars - Part II." February, 1977.
9. G.C. Haas, M.P. Sweeney, and J.N Pattison. "Laboratory Simulation of Driving Conditions in the Los Angeles Area", Published in "Vehicle Emissions, Part II." SAE Progress in Technology Series, Vol. 12. New York. Society of Automotive Engineers, Inc., 1968, pp. 317-324.
10. R.E. Kruse and T.A. Huls, "Development of a Federal Urban Driving Schedule." SAE Paper 730553, Detroit, Michigan, May 1973.
11. Federal Register, Vol. 37, No. 221, November, 1972.
12. Federal Register, Vol. 45, No. 45, March, 1980.
13. F. Black and L. High, "Methodology for Determining Particulate and Gaseous Diesel Hydrocarbon Emissions." SAE Paper 790422, February, 1979.

14. B.N. Ames, J. McCann, and E. Yamasaki, "Methods for Detecting Carcinogens and Mutagens with the Salmonella/Mammalian Microscope Mutagenicity Tests." Mutation Research 31:347-364 (1975).

15. J.N. Pitts, Jr., K.A.Van Cawvenberghe, D. Grosjean, J.P. Schmid, D.R. Fitz, W.L. Belser, Jr., G.B. Knudson, and P.M. Hynds, "Atmospheric Reactions of Polycyclic Aromatic Hydrocarbon: Facile Formation of Mutagenic Nitro Derivatives" Science 202, 515 (1978).

16. J.C. Hilliad and R.W. Wheeler, "Nitrogen Dioxide in Engine Exhaust." SAE Paper 790691, February, 1979.

17. E.F. Funkerbusch, D.G. Leddy, and J.H. Johnson, "The Characterization of the Soluble Organic Fraction of Diesel Particulate Matter." SAE Paper 790418, February, 1979.

18. C.T. Hare and T.M. Baines, "Characterization of Particulate and Gaseous Emissions from Two Diesel Automobiles as Functions of Fuel and Driving Cycle," SAE Paper 790424, February, 1979.

19. Nissan Motor Company, Private Communication.

20. H.S. Rosenkranz, E.C. McKoy, D.R. Sanders, M. Butler, D.K. Kiriazides, R. Mermelstein, "Nitropyrenes: Isolation, Identification, and Reduction of Mutagenic Impurities in Carbon Black and Toners." Science 209:1039 (1980).

21. D. Schuetzle, F.S.C. Lee, J.T. Prater, S. Tejada, "The Identification of Polyaromatic Hydrocarbon Derivations in Mutagenic Fractions of Diesel Particulate Extracts." International Journal of Environmental Chemistry, In Press.

Characterization of Particulate Emissions from In–Use Diesel Vehicles**

Richard E. Gibbs, James D. Hyde and Stanley M. Byer
New York State Department of Environmental Conservation

ABSTRACT

Particulate emissions data are presented from a group of 19 in-use diesel automobiles. Five driving cycles and three fuel/lubricating oil combinations have been used to obtain particulate emissions data and also to collect particulate samples for chemical and bioassay characterization studies. Particulate samples were solvent extracted with dichloromethane to obtain the soluble organic fraction (SOF) of the emitted particulate. Particulate and SOF data are presented for emission/distance, emission/time and emission/fuel consumed basis by vehicle manufacture, driving cycle, and fuel/oil subcategory. Some SOF samples have been used in the bioassay test of Ames, and these results are presented for different vehicles and driving cycles. Particulate emissions, SOF emissions and bioassay activity results are presented in terms of mileage accumulation effects.

INTRODUCTION

Background of the Problem - The importance of resource conservation is becoming generally felt as a relentless fact, and this presents new opportunities for fuel efficient combustion technology, such as the diesel engine. The number of diesel powered vehicles in all weight classes is thus expected to rise based on the inherent fuel consumption advantages of the diesel relative to comparable spark-ignition engines. However, the time scale for implementing major changes in combustion technology is large, even when measured in years, and the characteristic particulate emission of the diesel dictate the present need to carefully evaluate the impact of a trend toward diesel combustion. The chemical complexity of diesel particulate has impeded progress in even identifying important considerations, while biological screening assays, such as the Ames test, have increased awareness that some of the myriad of compounds carried by the particulate should be prudently considered as possible carcinogens. The lack of a definitive chemical basis for understanding the observed bioassay indication hinders constructive interaction among various areas impacting the overall issue, such as health effects, atmospheric chemistry, control technology options and effectiveness, regulatory action, and especially the underlying need for sampling and measurement methodologies for the bio-active compounds themselves.

Previous Experimental Work - The physical and chemical characterization and sampling and measurement methodologies for diesel emissions have been extensively reported for various engines, fuels and test conditions (1-15)*. The use of air dilution of vehicle exhaust followed by collection of particulate samples by filtration has been examined for application to diesel particulate by various investigations (16-20) and is the specified method for particulate measurement certification tests (21). Little chemical characterization data is presently available from in-use light-duty diesel engines, although gaseous, particulate and fuel economy results from two groups of General Motors 5.7 liter diesels have been recently reported (22,23).

The soluble organic fraction (SOF) of diesel particulate has been examined by the Ames bioassay

* Numbers in parentheses designate References at end of paper.

**Paper 801372 presented at the Fuels and Lubricants Meeting, Baltimore, Maryland, October, 1980.

screening test and found to contain mutagenic substances (24-26). At a recent EPA conference on diesel emissions (27) several authors contributed in this area. Presently little is known of the chemical identity of the mutagens; and evidence has been reported (28) demonstrating an increase in mutagenic response with increased NO_2 levels in the exhaust dilution apparatus and the formation mutagen artifacts in particulate samples from systems to which NO_2 has been added. It is therefore difficult to assess the effects of sampling conditions on mutagenic response. Sampling procedures valid for emissions measurement cannot be extrapolated to be quantitative sampling procedures for bioactive compounds, even though nothing better may be available until further work in this area is performed.

General Plan of the New York State Study - The study underway at New York State recognizes the need for emission factors from in-use diesel vehicles. Particulate emissions samples are quantitatively collected, solvent extracted and bio-assayed to determine the effects of vehicle design, fuels, oils and test conditions. Emissions and fuel consumption data are being collected from a sample of in-use diesel cars, which currently number 19 vehicles. A test protocol has been adopted to establish the sample group that involves about one week of dynamometer testing to obtain gaseous and particulate emissions data, and includes collection of particulate samples of sufficient size from which to perform chemical characterization and bio-assay tests. Sample group vehicles are retested to accumulate information on how vehicle ageing changes measured parameters. The vehicle test protocol adopted includes five vehicle operating modes (driving cycles) repeated with three combinations of engine fuel and lubricating oil. The fuel/oil combinations include testing the vehicle with 1) "as received" fuel and oil, 2) project reference fuel and "as received" oil, 3) project reference fuel and fresh lubricating oil. Particulate samples are collected by a standard dilution tunnel technique, followed by solvent extraction with dichloromethane to obtain the soluble organic fraction (SOF), or particulate extract.

A parallel effort to the in-use vehicle testing is directed toward developing SOF chemical characterization data to identify the mutagenic compounds present. The chemical complexity of the SOF, and the fact that the mutagenic activity is not attributable to the major constituents, makes identification of the mutagens difficult, and thus detailed chemical studies in this project are presently limited to a few select samples. To initiate chemical studies, three vehicles (Oldsmobile Cutlass wagon 5.7 liter V8, VW Rabbit 4-speed, and Mercedes-Benz 300D automatic) were operated on project reference fuel for approximately 50 days to collect large samples of particulate (50-100 g) for extraction, continuing chemical characterization and bioassay. Progress on these chemical studies has been reported elsewhere, (29-34) and is not included in this paper.

The in-use vehicle emissions fuel economy test results, chemical characterization of large particulate samples, and bioassay results from both in-use vehicle samples and the large samples are major complementary elements of the overall project.

VEHICLE EXPERIMENTAL PROCEDURES

Vehicle Testing Equipment - The major thrust of particulate characterization has been the collection of particulate samples according to 1981 EPA procedures (21) for new vehicle certification using a standard Constant Volume Sampler (CVS) air dilution of the vehicle exhaust followed by particulate sampling in a dilution tunnel. To obtain samples of sufficient size for subsequent chemical and bioassay analyses from discrete vehicle tests, a scaled-up particulate sampling system was added to the dilution tunnel. Fig. 1 is a schematic of

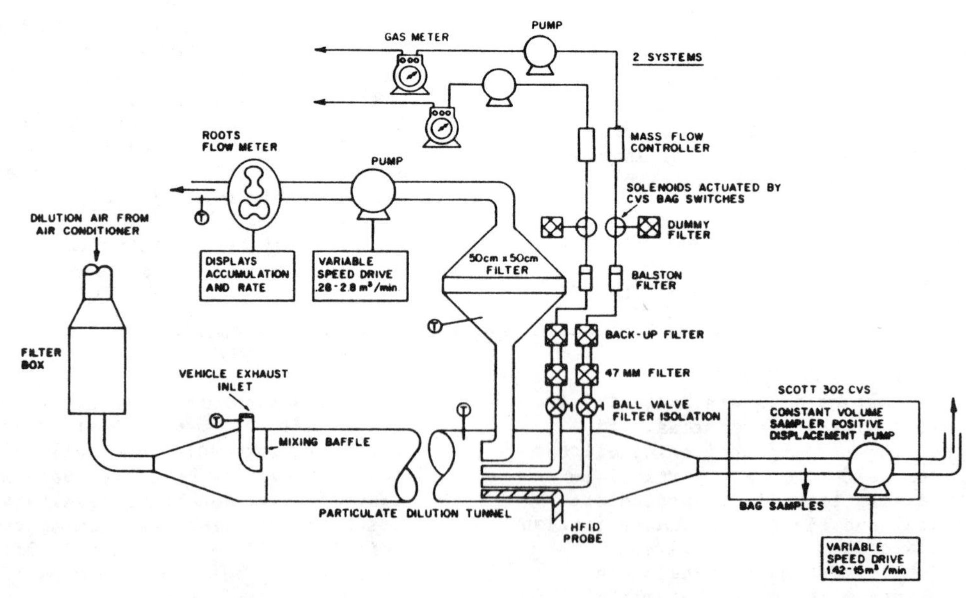

Fig. 1-Schematic of Particulate Collection Apparatus.

the exhaust dilution and particulate sampling equipment. Particulate mass emissions for vehicle tests are determined by proportional sampling of dilute exhaust by either of the two 47 mm filter systems. Each filter can be isolated from the probe and dilution tunnel by smooth-bore ball valves to permit changing of clean and loaded filters during vehicle testing. Each 47 mm filter is equipped with an identical 47 mm backup filter immediately downstream of the primary filter. Constant sampling flow is maintained as particulate is progressively deposited on the filter by means of a mass flow controller (Tylan Model FC-202) capable of maintaining constant mass flow over a wide range of gas temperatures and pressures. These flow controllers operate from a common power supply and display panel, and have individually adjustable mass flow rate set-points. These probe systems are flow calibrated each week with standard volumetric gas meters.

The particulate sampling system for obtaining bulk soot samples for extraction uses a filter assembly developed by EPA employing a 50 cm x 50 cm (20 in x 20 in) square filter mat. A nominal flow of 2.8 m^3/min (100 cfm) can be drawn through this filter and this provides 0.5 to 2g of particulate from typical vehicle tests. Although not used for quantitative measurement of total particulate mass emissions, this system must include accurate flow measurement since the volumes filtered are a significant contribution to the total dilute exhaust volume, and this sampler is used during vehicle tests including other gaseous and particulate emissions determinations. The dilute exhaust volume passing through the 50 cm x 50 cm filter is measured by a Roots-type totalizing flow meter (Dresser Industries model 5M125) mounted on the vent side of the blower used to draw sample through the filter (Fig. 1). Continuously recorded temperature measurement at the Roots meter exit is used to correct the volumetric data to standard conditions. Propane tests with both CVS and 50 cm x 50 cm operating verify the flow measurement calibration of the complete system.

The filter media used in both the 47 mm and 50 cm x 50 cm filter systems is Pallflex Products Corporation type T60A20 teflon coated glass fiber. The nominal flows of the 47 mm and 50 cm x 50 cm system are 9 l/min and 2.26 x 10^3 l/min respectively, referenced to 0°C. Dilute exhaust flows per area of filter exposure are 0.014 l/sec/cm^2 and 0.018 l/sec/cm^2 for the 47 mm and 50 cm x 50 cm filters respectively, both within the EPA specified range for particulate measurement (.00951 $<$ flow $<$ .0272 l/sec/cm^2 at 0°C). Operation of the 50 cm x 50 cm filter in conjunction with the constant volume sampler (CVS) requires the 50 cm x 50 cm system to be started before the CVS to avoid filter tearing.

The volume filtered through the 50 cm x 50 cm system during emissions testing is obtained from a digital accumulator on the Roots meter actuated in syncronization with other sampling devices by the CVS bag fill switches.

The maximum allowable temperature of dilute exhaust for particulate sampling by EPA procedure is 52°C, and this requirement directly affects the dilution air volume requirements necessary to cool the vehicle exhaust prior to sampling. Sampling requirements for the variety of vehicles, engine sizes, inertia weights, and vehicle operating modes encountered in this study are best met with a range of dilution air flows. The need to obtain measurable levels of gaseous species simultaneous with particulate sampling prohibits the dilution air flow from being set at one high value to satisfy only the particulate sampling requirements. To accommodate these needs, an existing positive-displacement pump (PDP) CVS was modified by addition of a 7.50 kw variable frequency 3Ø motor controller to give a variable speed motor drive providing continuous adjustment of CVS flow between 1.5-15 m^3/min (50-525 cfm). In combination with the 2.8 m^3/min (100 cfm) bulk particulate sampler, and the conditioned-air inlet to the dilution air filter box, the range of CVS flows available has satisfied the 52°C maximum temperature requirement for all vehicle test conditions. CVS + 50 cm x 50 cm filter flows have been consistently set to give a total dilute exhaust flow of 11.4 m^3/min (400 cfm) and 17 m^3/min (600 cfm) for Volkswagen and General Motors vehicles respectively. Strip chart recordings of dilute exhaust and Roots meter temperature are obtained for all tests. The PDP CVS blower is mass flow calibrated as a function of blower speed, and 132 daily propane tests since the start of in-use vehicle testing have averaged 0.75% $\pm$ 1.8% error.

Connection between the vehicle tailpipe and the dilution tunnel is either by a short length (approximately 1 m) of 7.6 cm (3 in) uninsulated flexible exhaust tube for vehicles with tailpipe exit near the tunnel (passenger side) or by a longer connector to vehicles with a tailpipe connector on the side of the vehicle opposite the tunnel location (driver side). Two types of connector have been utilized in this latter case. Through the first 17 vehicle tests, this connector was of the same type as above, but of sufficient length (4 m) to reach the tailpipe. For subsequent vehicle tests a smooth stainless steel tube (7.6 cm x 3 m) lagged with 5 cm of fiberglass insulation was used to link the tunnel to the tailpipe with short flexible uninsulated sections to make the connections. Particulate emission tests with the insulated and uninsulated systems showed approximately a 5% increase for the insulated case. EPA (35) has reported on the effect of tailpipe hookup on observed particulate emissions. After nine months of daily testing the dilution tunnel was carefully cleaned. From the weight of tunnel scrapings and an estimate of the total mass of particulate entering the tunnel in the course of testing, the particulate losses in the tunnel are estimated to be 3%.

Gaseous emissions and fuel consumption are measured by the 1979 FTP simultaneously with particulate sampling by the 1981 FTP. The emissions chassis dynamometer used is a Clayton CTE-50, equipped with automatic road-load and direct-drive variable inertia (250 lb inertia increments to 3,000 lb, 500 lb inertia increments above 3,000 lb). The vehicle test bay is maintained at constant temperature by two air conditioning/heating/humidification/dehumidification units with a combined capacity of 106 kw and circulating 425 m^3/min conditioned air onto the vehicle positioned on the dynamometer. In order not to abuse privately owned vehicle tires on the dynamometer, and to aid in uniformity of test conditions, all testing was performed on duplicate tires, identical to vehicle manufacturer specifications, inflated to 310 KPa (45 psig).

Test Cycle Sequence - Vehicles included in the sample group are brought to the Department of Environmental Conservation's Automotive Emissions Laboratory (AEL) in Latham, New York for a week-long comprehensive battery of tests to complete one "vehicle-test." Thirty-four such vehicle tests on nineteen vehicles constitute the data base of this paper. Each vehicle test is broken down into three repetitive "test-phases" according to the fuel and lubricating oil combination used. The three combinations of fuel and lubricating oil corresponding to each test phase are: Phase I - vehicle tested in as-received condition, Phase II - vehicle tested with fuel stored at AEL as a project reference fuel, and Phase III - vehicle tested with reference fuel after crankcase oil and oil filter are replaced with new oil and filter of manufacturer specification.

Five different vehicle driving schedules were used to obtain data for different vehicle operating modes. (Reference (5) gives speed-time traces and general cycle characteristics) The test cycles used were: 1979 Federal Test Procedure (FTP), Congested Freeway Driving Schedule (CFDS), Highway Fuel Economy Test (HFET), 50 mph cruise (50C) and Idle.

During the three segments of the FTP individual 47 mm diameter particulate samples were obtained for each bag; however, a single composite bulk particulate sample was collected on the 50 cm x 50 cm filter during all three bags. CFDS test segments included particulate collection on both 47mm probes to verify system balance and operation. The HFET cycles were always run in blocks of three back-to-back cycles separated by 30 s idle periods. A single bulk sample was always collected from all three HFET cycles in any one block, but only during the middle HFET were the 47 mm filter samples and gaseous measurements performed.

Table 1-Sequence For One Complete Vehicle Test

Test Segment	Particulate Filter	Gaseous Analyses	Test-Phase	Day
Procure Vehicle				ONE
Prepare for Testing				
30 min @ 50 mph pre-test conditioning			I = as delivered fuel and oil	
30 min @ 50 mph	Yes			
CFDS	Yes			
HFET x 3	Yes			
Overnight soak				
FTP	Yes	Yes		TWO
CFDS	Yes	Yes		
HFET x 3	Yes	Yes		
30 min @ idle	Yes			
Change Fuel System to AEL Control Fuel			II = control fuel, as delivered oil	
30 min @ 50 mph pre-test conditioning				
30 min @ 50 mph	Yes			
CFDS	Yes			
HFET x 3	Yes			
Overnight soak				
FTP	Yes	Yes		THREE
CFDS	Yes	Yes		
HFET x 3	Yes	Yes		
30 min @ idle	Yes			
Change Oil to Fresh Oil of Manufacturers Recommendation			III = control fuel, fresh oil	
30 min @ 50 mph cruise pre-test conditioning				
30 min @ 50 mph cruise	Yes			
CFDS	Yes			
HFET x 3	Yes			
Overnight soak				
FTP	Yes	Yes		FOUR
CFDS	Yes	Yes		
HFET x 3	Yes	Yes		
30 min @ idle	Yes			
Prepare Vehicle for Owner				
Return Vehicle to Owner				

The organization of these vehicle testing modes into a single complete vehicle-test is shown in Table 1. Each of the three test-phases begins with an afternoon portion of testing in which only particulate emissions (no gaseous or fuel consumption) and particulate bulk samples are obtained. The afternoon portion begins with 30 minutes of 50 mph cruise with no sampling to stabilize the vehicle after fuel/oil change. As indicated in Table 1, the remainder of the afternoon follows a sequence of (1) 50 mph cruise, (2) CFDS, (3) 3x HFET; at this time the vehicle is placed in an overnight "cold soak" in preparation of the next mornings' FTP. The morning of testing begins with an 1979 FTP with 1981 diesel particulate sampling, followed by (1) CFDS, (2) 3x HFET, and (3) Idle. Gaseous measurements are made for all the morning cycles with exception for the final Idle period.

This test protocol generates 21 bulk particulate samples (50 cm x 50 cm) and 33 47 mm particulate emissions filters (not including back-up filters) from each complete vehicle-test.

It was not possible to use a single reference oil for Phase III of each privately owned vehicle test considering the manufacturers' recomendations for the various vehicle types and the seasonal changes in Albany, New York. The third phase was thus run with oil and filter of vehicle manufacturer specification, selecting from either Castrol 30 W, 10W30, or 20W50.

Vehicle Sample Group - The 19 vehicles presently in the sample group are listed in Table 2, "Diesel Vehicle Test Roster," along with vehicle and test information including dynamometer road-load, date, odometer, and vehicle test identification number. Table 3 gives a summary of the 19 vehicles and 34 tests by manufacturer. This categorization does not separate vehicles by model year, drive train or engine calibration differences that may affect individual vehicle emissions. Fig. 2 shows the distribution of vehicle tests by odometer reading at each test point to indicate the range of vehicle mileage for tests on sample group vehicles. Only two vehicles have accumulated sufficient mileage and have been tested enough times to provide a measure of mileage accumulation effects. Several different dynamometer road-load values were used to certify different vehicles produced by a given manufacturer employing a common engine. These variations in road-load may have impact on the emissions parameters grouped by "vehicle manufacturer" in subsequent tabulations.

Privately owned vehicles were solicited by letter to registrants of diesel vehicles in the Albany, N.Y. area. In consideration for the use of their vehicle for testing, vehicle owners receive the use of a loan car during the vehicle test, their vehicle is returned with two car wash tickets, a full fuel tank and an oil and oil filter change. Two project vehicles (Cars #1 and #5) serve as the loan vehicles, and also are used for special test purposes, such as the large particulate sample collection work for chemical characterization mentioned previously. Two Oldsmobiles (cars #2 and #3) belong to the NYS Thruway Authority and

Table 2-Diesel Vehicle Test Roster

Vehicle Data							(Test #) / (Month/Day/Year) / (Odometer)				
Project I. D. #	Make Year	Model	Engine Displace. (liters)	Trans-mission	Dynamometer Road Load	Dynamometer Inertia Weight	1	2	3	4	5
1	VW 79	Rabbit	I4 1.5 l	M4	7.3	2250	3 10/23/79 3,575	24 6/10/80 10,500			
2	Olds 79	Cutlass Cruiser	V8 5.7 l	A3	12.5	4000	1 9/27/79 2,991	7 1/22/80 11,679	18 4/21/80 20,069	30 7/28/80 28,000	
3	Olds 79	Cutlass Cruiser	V8 5.7 l	A3	12.5	4000	2 10/5/79 6,159	6 1/14/80 14,138	20 5/15/80 24,000	31 8/4/80 33,400	
4	Olds 79	98 Regency	V8 5.7 l	A3	12.8	4500	5 1/7/79 6,728	21 5/20/80 11,500			
5	Olds 79	Cutlass Cruiser	V8 5.7 l	A3	12.5	4000	4 10/30/79 3,617	23 6/2/80 15,300			
6	VW 80	Rabbit	I4 1.5 l	M5	6.8	2250	8 1/28/80 4,279	33 8/18/80 13,700			
7	Cadillac 79	Eldorado	V8 5.7 l	A3	10.6	4500	9 2/4/80 15,115	29 7/22/80 25,200			
8	VW 78	Rabbit	I4 1.5 l	M4	8.8	2250	10 2/25/80 48,055	26 6/23/80 56,000			
9	VW 79	Rabbit	I4 1.5 l	M4	7.3	2250	11 3/3/80 16,340	27 6/30/80 24,600			
10	Dodge 78	D-10 Mitsu-bishi	I6 4.0 l	A3	14.4	5500(+)	12 3/10/80 34,165				
11	VW 77	Rabbit	I4 1.5 l	M4	8.8	2250	13 3/17/80 119,190	28 7/14/80 134,000			
12	M-B 77	240-D	I4 2.4 l	M4	11.2	3500	14 3/24/80 43,444				
13	M-B 78	300 CD	I5 3.0 l	A4	12.0	4000	15 3/31/80 26,006				
14	M-B 79	240 D	I4 2.4 l	M4	12.6	3500	16 4/7/80 22,317				
15	Audi 79	5000	I5 2.0 l	M5	11.8	3000	17 4/14/80 22,754	34 9/8/80 37,600			
16	Olds 79	Delta 88	V8 5.7 l	A3	13.3	4500	19 4/28/80 31,800				
17	Puegeot 79	504	I4 2.3 l	M4	10.7	3500	22 5/27/80 34,700				
18	Olds 80	Cutlass Cruiser	V8 5.7 l	A3	12.6	4000	25 6/17/80 4,300				
19	M-B 1979	300-SD	I-5 3.0 l	A4	13.0	4000	32 8/11/80 23,000				

TABLE 3

Sample Group Vehicles by Manufacturer and Tests

Manufacturer	No. of Vehicles	Vehicle Tests
General Motors	8	16
Volkswagen	4	10
Mercedes-Benz	4	4
Audi	1	2
Puegeot	1	1
Mitsubishi	1	1
TOTALS	19	34

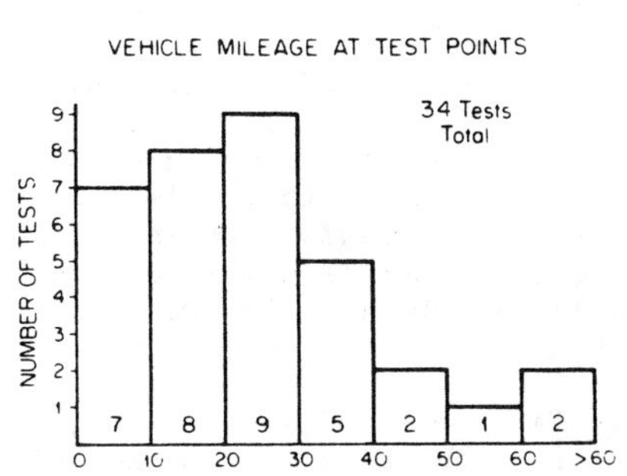

Fig. 2 - Vehicle Mileage at Test Points.

accumulate approximately 30,000 miles per year in normal administrative service, and provide the first project data on mileage accumulation effects.

Fuel and Oil - Vehicle owners are now encouraged to call to arrange for a test on their vehicle when it is due for an oil change, although for a few early tests the "as received" oil was not necessarily due for change. Vehicles delivered with 1/4 to 1/2 tank-full fuel are acceptable for testing but if delivered with less than this amount they may run out of fuel during the first test phase, and thus are filled with commercial (non-AEL reference) fuel prior to the beginning of the vehicle-test.

When ready for testing, three fuel samples are taken from the tank. One sample is analyzed for: API gravity, specific gravity, heating value, carbon, hydrogen, nitrogen, sulfur, oxygen, ash, complete distillation, and cetane index by Fuel Engineering, Thornwood, N.Y. A second sample is analyzed for saturates, olefins, and aromatics using ASTM D-1319 by the Source, Fuels, and Molecular Chemistry branch of EPA/RTP. A third portion is kept at AEL for project reference. A sample of "as received" crankcase oil from each vehicle is also collected and kept at AEL for project reference. Fuel analyses data for the reference fuel stored at AEL for vehicle testing are given in Appendix A.

SAMPLE HANDLING PROCEDURES

TABLE 4

Emission and Fuel Economy Results - Average All Phases (g/mi) and (mi/gal)

Car No.	Make Year	Model	FTP HC	FTP CO	FTP NO_x	FTP Particulate	FTP Mileage	HFET HC	HFET CO	HFET NO_x	HFET Particulate	HFET Mileage
1	VW 79	Rabbit	0.37	1.18	0.84	0.40	41.8	0.25	0.93	0.77	0.34	55.0
2	Olds 79	Cutlass Cruiser	0.60	1.64	1.61	0.76	20.2	0.30	0.92	1.33	0.38	30.3
3	Olds 79	Cutlass Cruiser	0.47	1.46	1.48	0.74	20.6	0.24	0.85	1.22	0.30	30.5
4	Olds 79	98 Regency	0.46	1.84	1.62	1.07	20.3	0.21	0.98	1.39	0.59	30.5
5	Olds 79	Cutlass Cruiser	0.51	1.58	1.68	0.81	21.2	0.24	0.87	1.29	0.37	32.0
6	VW 80	Rabbit	0.24	1.03	0.79	0.37	41.8	0.21	0.63	0.52	0.20	60.5
7	Cadillac 79	Eldorado	0.53	1.62	1.59	0.84	18.5	0.27	0.87	1.14	0.34	30.5
8	VW 78	Rabbit	0.31	1.19	0.80	0.44	44.0	0.18	0.80	0.76	0.32	54.3
9	VW 79	Rabbit	0.22	0.90	0.99	0.32	43.6	0.20	0.75	0.89	0.29	57.4
10	Dodge 78	D-10 Mitsubishi	0.33	1.62	2.70	0.66	25.5	0.24	1.10	3.17	0.45	27.4
11	VW 77	Rabbit	0.33	1.04	0.75	0.32	43.2	0.17	0.64	0.66	0.27	56.6
12	M-B 77	240-D	0.39	1.37	1.29	0.55	27.3	0.21	0.90	1.03	0.36	37.9
13	M-B 78	300 CD	0.21	1.41	1.87	0.67	21.9	0.10	0.79	1.49	0.43	30.2
14	M-B 79	240 D	0.22	1.29	1.43	0.48	26.0	0.11	0.88	1.25	0.32	33.4
15	Audi 79	5000	0.30	1.39	1.62	0.49	26.4	0.20	0.88	1.16	0.36	36.8
16	Olds 79	Delta 88	0.42	1.55	1.67	0.83	18.9	0.20	0.86	1.15	0.40	29.7
17	Puegeot 79	504	0.65	1.43	0.79	0.45	26.8	0.25	0.72	0.63	0.24	35.8
18	Olds 80	Cutlass Cruiser	2.83	2.83	0.73	0.83	20.0	0.52	1.11	0.55	0.21	32.1
19	M-B 1979	300-SD	0.13	1.03	1.40	0.50	23.0	0.07	0.70	1.27	0.39	29.8

Weighing of Filters - The 47 mm diameter Pallflex filters were placed in a constant humidity chamber for at least 24 hours before each weighing. The temperature was maintained at 18 to 21°C with a relative humidity range of 55 to 60%. The equilibrated filters were weighed on a Mettler ME 30 microbalance in micrograms. 50 cm x 50 cm filters were not equilibrated to any standards but were kept in areas where the temperature was usually about 20°C and the relative humidity in the 30 to 50% range. Tare weights were taken by carefully rolling the filters into approximately 15 cm cylinders and placing them on a semicircular weighing pan in a Mettler H10T balance. The balance doors were kept open to accommodate the length of the rolled filters. The balance itself was in a large five-sided box to reduce air currents and the weighings were performed in a balance room without air circulation. Gross weights were taken after the filters had been folded into quarters and were performed in a normal manner. Weights were taken to the nearest milligram with the uncertainty in the net weight estimated to be about ± 5 mg.

Extraction of Filters - The 50 cm x 50 cm filters used for bulk particle collection were extracted with dichloromethane to obtain the SOF as a percentage of the particulate mass on the 50 cm x 50 cm filters. The filters were folded and placed in 50 mm Soxhlet extractors without thimbles. Extraction was carried out with approximately 300 ml of dichloromethane (Burdick and Jackson Laboratories, Distilled in Glass™ grade) for 24 hr at three-four cycles per hour.

The extracts were filtered into 500 ml flasks by vacuum through an 0.2 μm Fluoropore filter (Millipore FGLP) to remove any particles which may have carried over. The solution was heated to boiling before filtering and the filtering apparatus was kept hot during filtration to prevent precipitation of any sparingly soluble material on the filter. The 500 ml flasks were rotary evaporated to about 75 ml volume and the sample transferred to acid and solvent washed, dried and tared 125 ml borosilicate glass bottles with Teflon lined caps. These bottles were then partially immersed in a warm water bath while attached to the rotary evaporator to remove the remaining solvent to yield the dried particulate extract. Final weighing provides by difference the mass of extract from which the SOF percentage was computed directly. Bottles with dried extract are then stored at -80°C until needed for bioassay.

Each vehicle test results in 21 50 cm x 50 cm filters. 15 of these have been selected for extraction, the remainder kept in -80°C storage. The six unextracted filters are the CFDS and HFET filters from the afternoon portion of each phase. Four vehicle tests for which gaseous and particulate emissions (g/mi) are available at this writing are yet to be extracted (tests 29, 32, 33, 34 - see roster Table 2). This leaves about 450 discrete 50 cm x 50 cm filters that have been extracted as the SOF data base in this paper.

GASEOUS AND PARTICULATE EMISSIONS

Gaseous Emissions and Fuel Economy - Average emissions (gaseous and particulate) and fuel economy results by individual vehicles are given in Table 4 for the FTP and HFET cycles. The results are averaged over all three test-phases and with repeat tests (if any) to give a single value for each sample group vehicle. Although no detailed partition of FTP particulate emissions by individual bags is presented in this paper, results from 99 FTP runs have averaged 44% greater particulate emission in bag 1 (cold start at 72°F) than for bag 3 (hot start at 72°F).

Car #10 is a Dodge pickup equipped with a Mitsubishi diesel engine, which was certified by the manufacturer as a heavy duty engine. This limited-production vehicle is in New York State Thruway service and has been included in this study to increase the variety of engines in the project data base. One 1980 General Motors 5.7 liter diesel vehicle has been tested to date (Car #18, tested at 4,000 miles) and the results are probably quite abnormal. This engine has been significantly changed for 1980, including new hardware (EGR, etc.) and new calibrations. The effect of these changes on in-use emission is unknown at this time. The vehicle ran very poorly throughout testing as characterized by excessive engine rocking at all but acceleration modes. The cause of this was not determined, but AEL technicians did recommend to the owner that the vehicle be returned to the dealer for check-out. The owner did have the dealer examine the vehicle and new injector lines were installed; however, at 12,000 miles the vehicle runs roughly and will be returned for further service.

With the exception of the two above mentioned vehicles, the HC/CO/NO_x emissions for the sample group are below the certification standards of 1.5/15/2. g/mi applicable to the 1979 vehicles. The inherent low HC/CO emissions characteristics of the diesel are evident in the results on Car #11, with tests at 120,000 and 133,000 miles averaging 0.33/1.04/0.75 g/mi HC/CO/NO_x in FTP tests.

Particulate Backup Filters - Twelve vehicle tests have been conducted with backup filters for every 47 mm filter collected to investigate their importance for the various vehicles, fuels, and

TABLE 5 Particulate Emissions Reproducibility

Average Particulate Emissions (g/mi) Differences Between Duplicate Cycles

[% difference ± 1 σ]

	Obs. in each Average	Phase I		Phase II		Phase III		Three Phase Average	
		CFDS	HFET	CFDS	HFET	CFDS	HFET	CFDS	HFET
All Tests	34	1.0%± 8.8	1.2%±10.6	0.1%±10.2	1.8%±8.1	1.2%± 9.2	0.8%± 8.0	0.7%	1.3%
General Motors	16	0.5%± 7.2	2.7%± 9.0	-1.2%± 7.5	0.0%±6.3	0.6%± 5.8	1.3%± 5.5	0.8%	1.7%
Volkswagen	10	1.0%± 12.3	1.8%±14.9	3.5%±15.7	0.6%±7.2	2.9%±13.4	3.7%± 9.2	2.5%	2.0%
Mercedes-Benz	4	1.8%± 8.3	-5.7%± 7.9	-0.3%± 5.6	-3.6%±3.4	3.1%± 8.8	0.2%± 2.0	1.7%	-3.2%

Note: % > 0 indicates first cycle > second cycle
% < 0 indicates first cycle < second cycle

driving schedules. Average backup filter contribution to the combined mass (primary plus backup) was 1.65% (σ = 0.95) for 295 individual test segments (i.e. 295 separate filters). The backup contribution for FTP testing was 2.66% (σ = 1.6) for 96 observations. The group of filters with the highest backup contribution came from the idle test mode, where 35 tests gave backup contributions that averaged 5.68% (σ = 5.75). Prior to the time of this finding for the idle test, the CVS flow setting had been left at a constant value for all cycles for a given vehicle. This resulted in the idle sampling being conducted at a significantly higher exhaust dilution ratio than the other cycles used. Backup filters from idles under these conditions frequently gave higher absolute mass loadings than for those from other (higher speed) driving cycles. The collection efficiency for filter media is known to be increased by previously deposited particulate on the filter, and in the dilute idle exhaust the primary filter never received sufficient mass loading for this efficiency enhancement to occur. A subsequent decision to decrease the dilution ratio for idles increased the primary filter loadings and decreased the backup filter loading.

In contrast to the general finding that backup filters collected a small percentage of the total particulate, tests on a Peugeot (Car #17) exhibited distinctly different results. Particulates from this vehicle did not mat on the face of the filter media but penetrated to give a grey appearance on both the front and back of the filter. Backup filters on 30 Peugeot tests averaged 9.4% of the total particulate mass. In 7 Peugeot tests, two 50 cm x 50 cm sheets were used in series in the same holder, and the second 50 cm x 50 cm filter contained an average of 9.5% of the total. This observation parallels that of the USEPA for a Peugeot vehicle (36, 37). Backup filters have been discontinued for all vehicles except the Peugeot, but variations in sampling conditions and vehicle type can impact the need for backup filters.

Particulate Emissions Reproducibility - Within each test phase the duplicate CFDS and HFET cycles performed in the afternoon and again the following morning can be compared to determine the stability of particulate emissions during the given phase. Table 5 shows the reproducibility for these two driving cycles for each vehicle test phase for all 34 tests and also by vehicle make subcategory. On the average, the reproducibility is in the 1-3% range, but the standard deviations for the sample are in the 5-15% range. The Volkswagens show a larger standard deviation than other groups, possibly due to their manual transmissions in contrast to the automatic transmissions on most other sample group vehicles.

Filter Pair Correspondence - The two 47 mm diameter filter collection systems used for the gravimetric measurements by the 1981 FTP particulate standards are routinely operated simultaneously during all CFDS cycles as an ongoing check of this system's correspondence and data integrity. For all 111 CFDS tests, the mean absolute difference of these simultaneous measurements was 0.015 ± 0.013 g/mi which is 3.3% ± 2.8% of the overall average CFDS particulate emission rate of 0.457 g/mi, indicating acceptable performance of these systems.

The 50 cm x 50 cm bulk particulate collection system was designed to obtain significant quantities of particulate for chemical and biological analysis and as such is different in design and was not optimized to insure strict gravimetric correspondence with the 47 mm diameter filter systems discussed above. The mean absolute difference of the 50 cm x 50 cm particulate emission rate referenced to the 47 mm system was -0.04 ± 0.02 g/mi, indicating a small but systematic lower collection rate by the bulk particulate system.

In consideration of the standard deviations, the small percentage differences indicates no systematic change in emissions between the duplicate cycles, while individual tests can be expected to exhibit 5 - 15% positive and negative deviations for duplicate tests.

Particulate Emissions Results - Average particulate emissions, g/mi, are presented in Fig. 3 for FTP, HFET, CFDS, and 50 mph cruise driving cycles for General Motors, Volkswagen, and Mercedes-Benz groupings. Fig. 3 depicts the mean value as the top of the bar, with the ± 1 σ bounds as shown by the error flag on the bar. This nomenclature is used for all bar graphs throughout the paper. Values in Fig. 3 include results from all three test phases. The Volkswagen results are relatively constant for all

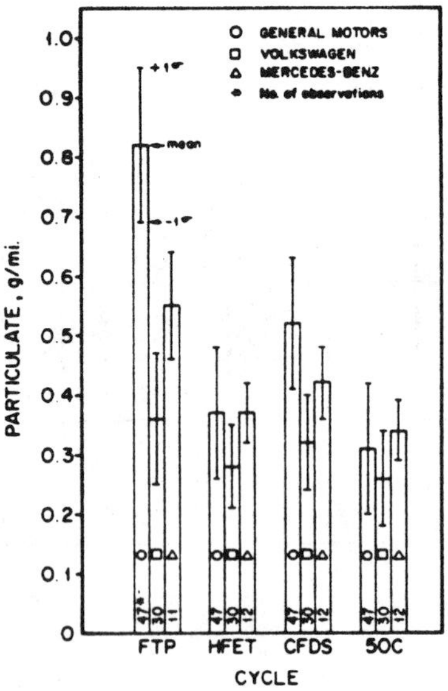

Fig. 3 - Particulate Emissions, g/mi, by Test Cycle and Vehicle Group.

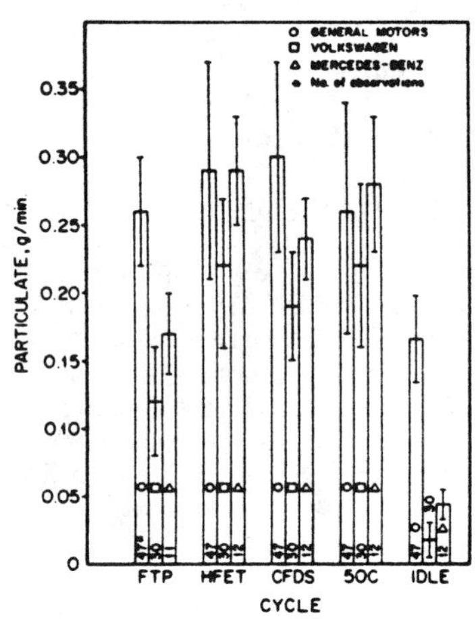

Fig. 4 - Particulate Emissions, g/min, by Test Cycle and Vehicle Group.

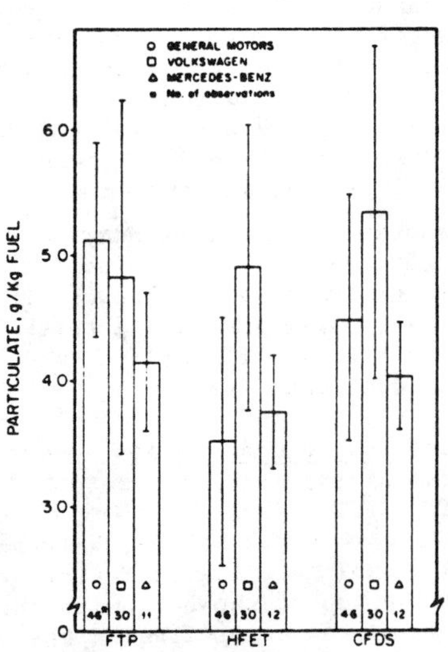

Fig. 5 - Particulate Emissions, g/Kg Fuel, by Test Cycle and Vehicle Group.

four driving cycles, whereas the General Motors and Mercedes-Benz groups show higher emissions during the FTP than for the other cycles. For each of the three vehicle groups the CFDS particulate emission levels are higher than the value for the respective groups in HFET and 50 mph cruise. This may be due to the more numerous acceleration/deceleration modes of the CFDS compared to the HFET and 50 cruise cycles.

Particulate emissions for each driving cycle have been computed on an emission per unit time basis to facilitate comparison of idle emissions with those from other driving cycles. Fig. 4 shows the results of the three vehicle groupings in grams particulate/minute for the various vehicle operating cycles (all test phases included). Idle emission rates for Volkswagen and Mercedes-Benz vehicles are significantly lower than for General Motors vehicles; the time emission rate of particulate for the General Motors vehicles is only a factor of 1.7 lower for the Idle mode than the average of other modes tested, whereas the Volkswagen and Mercedes-Benz vehicles have comparable ratios of 9.4 and 5.4 respectively.

The Volkswagen and Mercedes-Benz follow almost identical patterns with Mercedes-Benz consistently higher by about 0.06 g/min except for the Idle where the difference is much smaller. The ratios of Mercedes-Benz to Volkswagen are 1.46, 1.30, 1.31, 1.25 and 2.46 for the FTP, HFET, CFDS, 50C and Idle, respectively. The Volkswagens and Mercedes-Benz also differ from the General Motors vehicles in that there is an increase in particulate emissions per minute as the average speed of the driving cycle increases, not observed with the General Motors vehicles.

Particulate emissions have been calculated on a fuel specific basis for those driving cycles in which gaseous emissions are measured (FTP, HFET, and CFDS). Fig. 5 gives these results as particulate grams per kilogram fuel consumed for the three vehicle groupings used previously (all test phases included), Fuel specific particulate ranges average from 3.5 to 5.5 g/Kg fuel for the FTP, HFET and CFDS cycles with the three vehicle groupings. Volkswagen vehicles are relatively constant for the three cycles (4.8 to 5.4 g/Kg fuel), as are the results on Mercedes-Benz vehicles (3.7 to 4.2 g/Kg fuel). General Motors vehicles exhibit larger differences among cycles (3.5 to 5.1 g/Kg fuel) with the lower value of the range found on HFET driving.

Soluble Organic Fraction (SOF) Results - The term soluble organic fraction (SOF) is applied to that portion of the particulate which is extracted by dichloromethane. Fig. 6, 7, 8, show the SOF data in terms of %, g/mi and g/Kg fuel respectively. In Fig. 6 it can be seen that the Mercedes-Benz shows very little variation in SOF percentage among driving cycles. The Volkswagen rate is almost constant for the driven cycles but increases very sharply for the Idle. The General Motors vehicles show a definite cycle effect and have higher values for all cycles except the Idle.

Fig. 7 shows this data in g SOF/mi. In this comparison no vehicle group shows any consistent value but there is a common trend in the relative values for each cycle. In each case the g SOF/mi decrease in the order FTP > CFDS > HFET

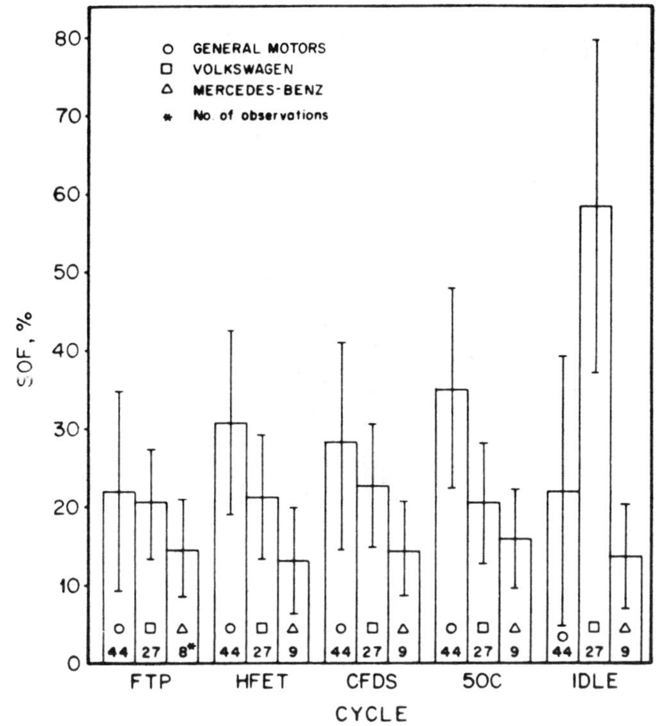

Fig. 6 - Soluble Organic Fraction, %, by Test Cycle and Vehicle Group.

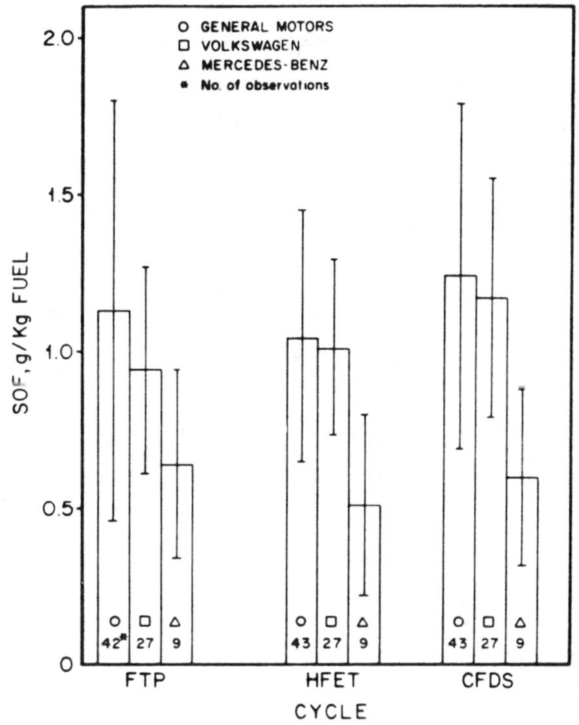

Fig. 8 - Soluble Organic Fraction, g/Kg Fuel, by Test Cycle and Vehicle Group.

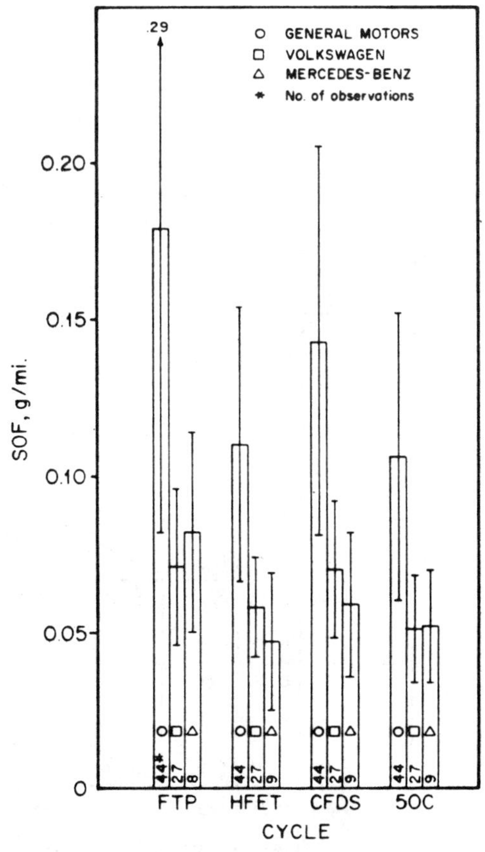

Fig. 7 - Soluble Organic Fraction, g/mi, by Test Cycle and Vehicle Group.

> 50 C. The General Motors vehicles are again the highest and are even more greatly separated from the other vehicles than in Fig. 6. This is caused by the higher g/mi particulate emission rate for these vehicles. The Idle value for Volkswagen is now comparable to that for Mercedes-Benz due to the very low g/mi particulate emission rate.

Fig. 8 shows the SOF in fuel specific terms, i.e. g SOF/Kg fuel for the three cycles for which gaseous emission data were collected. The trend here is similar to that for SOF in g/mi and %. In each case the General Motors vehicles have the highest values followed by Volkswagen and Mercedes-Benz. In this case, however, the General Motors vehicles have values close to those for the Volkswagen; Mercedes-Benz results are considerably lower. No consistent cycle-to-cycle variation is evident.

SOF (%) average results are given in Fig. 9 for all tests on each individual car in the sample group. Both FTP and HFET results are shown with the standard deviation indicated. Vehicles with several tests at various mileage test points show larger deviations since the SOF (%) has changed from test to test. The unique SOF % results for the single 1980 Oldsmobile (Car #18) are notable. The Peugeot tests (Car # 17) also show high SOF (%) values.

Particulate and SOF Emissions Variability - The three fuel/oil test conditions can be compared in terms of particulate emissions and particulate extract (SOF) emissions to see if any systematic

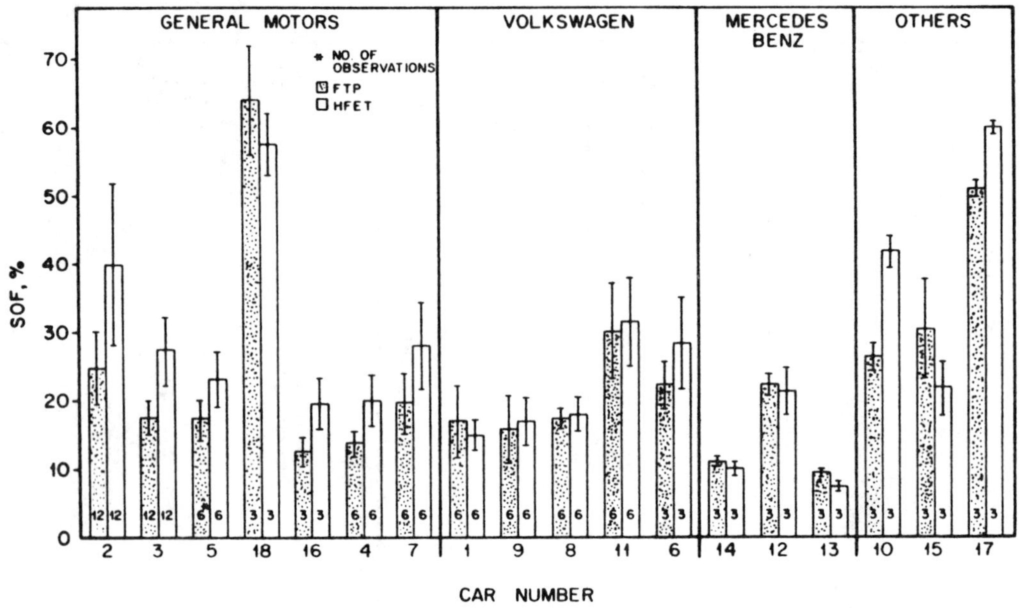

Fig. 9 - Soluble Organic Fraction, %, for FTP and HFET for Individual Vehicles.

Table 6 - Particulate and SOF Emissions: Reproducibility Within Test-Phases and Variability Among Test-Phases

$$R = \left[\frac{(grams/mile)_A}{(grams/mile)_B}\right]$$

A : First Test
B = Second Test

			CFDS						HFET					
			Particulate (g/mi)			SOF (g/mi)			Particulate (g/mi)			SOF (g/mi)		
			$\bar{R}$	σ	obs<1 / obs>1	$\bar{R}$	σ	obs<1 / obs>1	$\bar{R}$	σ	obs<1 / obs>1	$\bar{R}$	σ	obs<1 / obs>1
ALL	Reproducibility in each vehicle test phase	I	1.01	0.09	14/20				1.01	0.11	19/15			
		II	1.00	0.10	15/18				0.98	0.08	17/16			
		III	1.01	0.09	18/16				1.01	0.08	13/21			
	Phase-to-Phase Variability	I/III	0.99	0.11	19/15	1.02	0.21	13/17	1.04	0.10	14/20	1.06	0.24	18/12
		II/III	1.01	0.13	15/18	1.03	0.20	17/12	1.02	0:10	12/21	1.01	0.21	16/13
GENERAL MOTORS	Reproducibility in each vehicle test phase	I	1.00	0.07	8/8				1.03	0.09	7/9			
		II	0.99	0.08	6/9				1.00	0.06	6/9			
		III	1.01	0.06	8/8				1.01	0.05	6/10			
	Phase-to-Phase Variability	I/III	1.00	0.08	9/7	0.97	0.24	10/5	1.02	0.08	9/7	0.96	0.23	11/4
		II/III	1.00	0.07	7/8	0.95	0.19	10/4	1.01	0.09	6/9	0.96	0.20	9/5
VOLKSWAGEN	Reproducibility in each vehicle test phase	I	1.01	0.12	4/6				1.02	0.15	6/4			
		II	1.04	0.16	4/6				1.01	0.07	4/6			
		III	1.03	0.13	6/4				1.04	0.09	2/8			
	Phase-to-Phase Variability	I/II	0.99	0.13	7/3	1.05	0.17	2/7	1.08	0.13	2/8	1.16	0.22	4/5
		II/III	1.00	0.16	5/5	0.98	0.20	6/2	0.99	0.09	5/5	0.98	0.20	6/3
MERCEDES-BENZ	Reproducibility in each vehicle test phase	I	1.02	0.08	1/3				0.96	0.03	3/1			
		II	1.00	0.06	2/2				0.94	0.08	3/1			
		III	0.97	0.09	2/2				1.00	0.02	2/2			
	Phase-to-Phase Variability	I/III	0.96	0.17	1/3	1.14	0.07	0/3	1.10	0.10	2/2	1.13	0.14	1/2
		II/III	0.98	0.03	3/1	1.16	0.17	1/2	1.00	0.09	3/1	1.15	0.24	0/3

variations occur in the three test-phases. First, however, a measure of the repeatability for the emissions within a given test condition must be established. Since two CFDS and HFET tests are performed for each of the three fuel/oil combinations, comparisons of particulate emissions (g/mi) for these repeated cycles are easily obtained. For each pair of repeated cycles, the ratio of particulate (g/mi) is computed. Table 6 shows average values, standard deviations and number of observations greater than and less than unity for CFDS and HFET cycle ratios. These results are presented for all tests as one group, and also by vehicle type. Results are given in each case for all three vehicle test phases. For all 34 vehicle tests combined the ratio for repeated CFDS and HFET tests is very nearly unity, but the standard deviation is about 10% (.10 in Table 6). This variation in repeated measurements at one fuel/oil condition impacts the magnitude of differences among phases that can be resolved.

Examination of the phase-to-phase variability for all tests in one vehicle group shown in Table 6 shows these ratios and standard deviations to be comparable to the repeated measurements within a single phase, suggesting little influence on total particulate emissions of the fuel/oil combinations. This is not totally unexpected in that the reference fuel is merely a lot of commercially available fuel.

Repeatability checks within a test-phase for SOF (g/mi) determinations are not available since only one of the two repeated filters has been extracted at this time. Phase-to-phase comparisons of SOF (g/mi) emissions are shown, and for both CFDS and HFET tests show slightly higher SOF emissions in phases I and II than for phase III (~ 4%). The standard deviation for SOF is considerably larger than for the total particulate (~ 20%) so the systematic change among test phases is unlikely to be a definite trend.

Vehicle sub-groups in Table 6 can be examined to seek more detailed phase-to-phase and repeatability information. The number of observations less than and greater than unity provide a rough indication of the shape of the distribution from which each entry was obtained, and, of course, the number of observations available from which the results were computed.

Phase-to-phase variabilities for the sample group as a whole are given in Table 7 for a variety of specific driving schedules. Particulate and SOF data are both shown as ratios using the results from phase III as a common denominator in each given vehicle test. Little systematic difference is evident, but the standard deviation in FTP tests is seen to be larger than for other cycles (20% vs. 11%) with the exception of the idle mode. Phase-to-phase SOF variations result in roughly the same standard deviations for all cycles.

For the range of test conditions encountered up to this point the data do not indicate any significant systematic variation in either particulate or SOF emissions, although the magnitude of standard deviations clearly show that a single test on a particular vehicle may result in significant variations in even a repeated test cycle, as well as for a new fuel/oil combination.

Mileage Accumulation Effects - As an aid in assessing the effects of mileage accumulation on vehicle emissions two 1979 Oldsmobile Cutlass Crusiers were purchased for use by the New York State Thruway Authority. These two vehicles are operated under essentially the same highway conditions in administrative service; and each accumulates about 2,500 miles/month (30,000 miles/year). These vehicles are being tested at 8,000 to 10,000 mile intervals and have now been each tested four times. Results for particulate emissions and SOF are shown in Fig. 10. The values shown in Fig. 10 are averages of all applicable test cycles and phases at each mileage test point. Car #5, another Cutlass Crusier stationed at AEL as a control test vehicle and loan car for access to privately owned vehicles, has also

Table 7 - Particulate and SOF Emissions: Variability Among Test-Phases by Driving Cycle.

All vehicle tests taken as one group of measurements, Emission ratios computed:

$$R = \left[\frac{(g/mi)\ phase\ I}{(g/mi)\ phase\ III}\right] \text{ and } R = \left[\frac{(g/mi)\ phase\ II}{(g/mi)\ phase\ III}\right]$$

Ratios for all tests compared to obtained mean ($\bar{R}$) and standard deviation (σ)

Vehicle Test Cycle =		FTP		HFET-1st		HFET-2nd		CFDS-1st		CFDS-2nd		50C		Idle	
Ratio of Phases =		I/III	II/III	I/III	II/III	I/III	II/III	I/III	II/III	I/III	II/III	I/III	II/III	I/III	II/III
Total Particulate (g/mi)	$\bar{R}$	1.001	0.991	1.046	0.998	1.043	1.024	0.994	0.989	0.995	1.006	1.037	1.047	1.047	1.091
	σ	0.24	0.17	0.12	0.07	0.10	0.10	0.12	0.09	0.11	0.13	0.12	0.12	0.34	0.21
	obs<1/obs>1	12/15	20/12	16/18	16/17	14/20	12/21	21/13	16/17	19/15	15/18	15/19	12/21	11/21	6/26
Particulate SOF (g/mi)	$\bar{R}$	1.025	0.961	-	-	1.057	1.013	-	-	1.024	1.029	1.058	1.069	1.04	1.074
	σ	0.35	0.23	-	-	0.24	0.21	-	-	0.21	0.20	0.35	0.29	0.54	0.29
	obs<1/obs>1	19/11	17/11	-	-	18/12	16/13	-	-	13/17	17/12	16/14	11/18	12/17	12/15

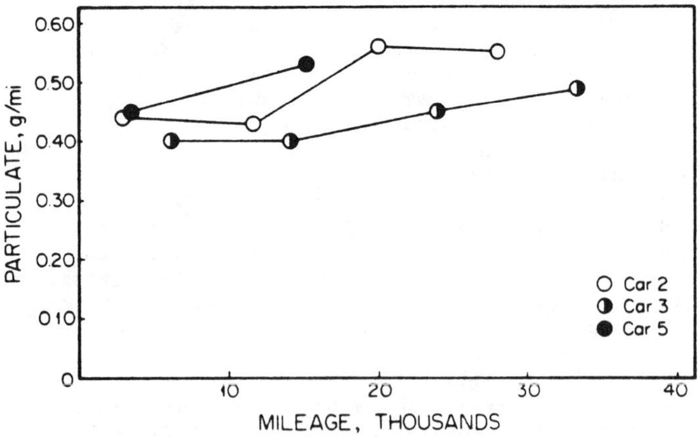

Fig. 10 - Mileage Accumulation Effects on Particulate Emissions, g/mi, for Oldsmobile Cutlass Cruisers.

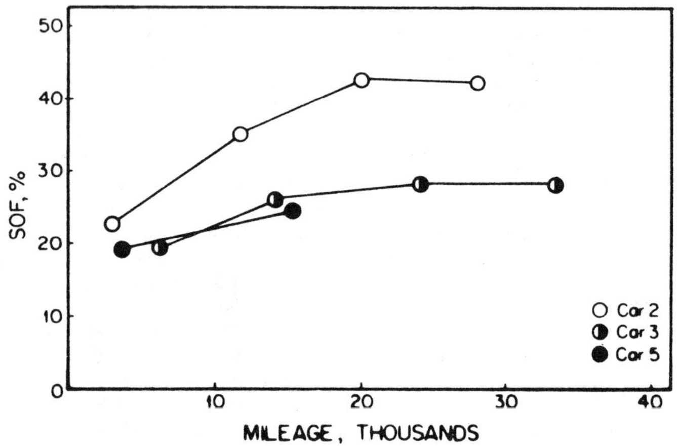

Fig. 11 - Mileage Accumulation Effects on Soluble Organic Fraction, %, for Oldsmobile Cutlass Cruisers.

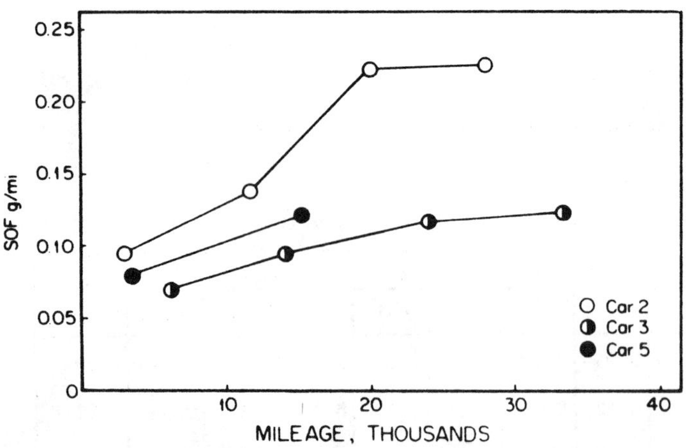

Fig. 12 - Mileage Accumulation Effects on Soluble Organic Fraction, g/mi, for Oldsmobile Cutlass Cruisers.

been included in these figures for comparison. Fig. 10 shows a general, although nonsystematic, increase in particulate emissions with increasing mileage which averages about 0.05 g/mi increase per 10,000 miles. It is uncertain whether this trend will continue in further tests.

Fig. 11 shows that the % SOF also increases with accumulated mileage but is more systematic than the particulate data. The % SOF generally increases rapidly for the first approximately 20,000 miles and then levels off for an as yet undetermined period. Vehicle-to-vehicle differences are very apparent in this Figure where Car #2 shows consistently higher and more rapidly increasing % SOF than the other two vehicles. Fig. 12 shows g/mi SOF and demonstrates a similar trend for all three vehicles with Car #2 again higher than the other two vehicles. The combination of increased particulate emissions and increased % SOF leads to an average increase in SOF, g/mi, of 80% for each 10,000 miles accumu-

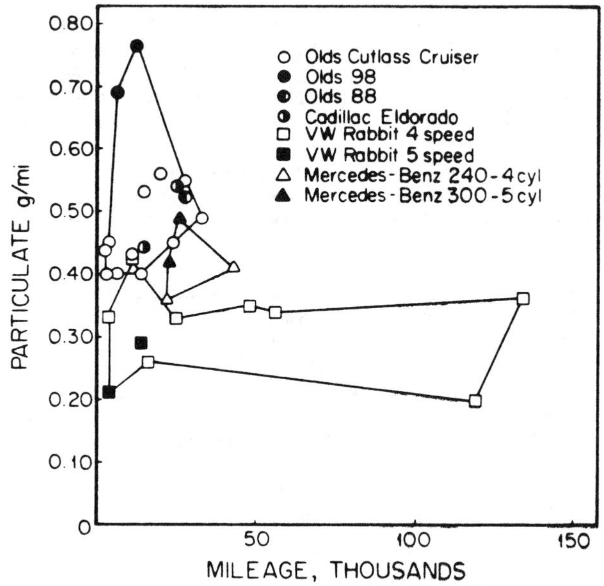

Fig. 13 - Mileage Accumulation Effects on Particulate Emissions, g/mi, for General Motors, Volkswagen and Mercedes-Benz vehicles.

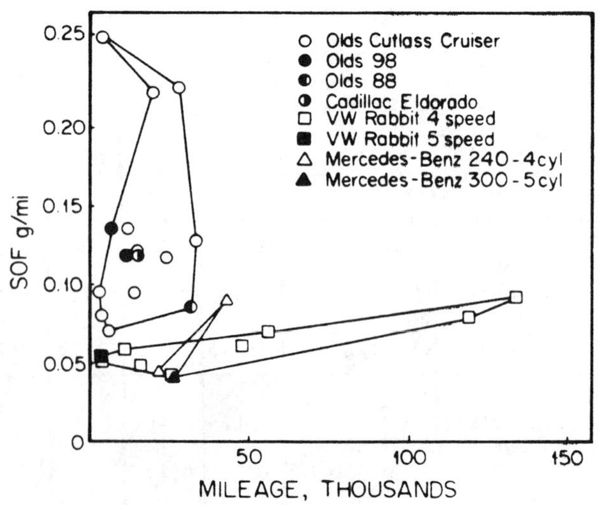

Fig. 14 - Mileage Accumulation Effects on Soluble Organic Fraction, g/mi, for General Motors, Volkswagen and Mercedes-Benz vehicles.

lated by Car #2 up to 20,000 miles. In general the three vehicles are much more similar to each other at low mileage than at higher mileage.

In Figs. 13 and 14 each major vehicle class i.e.General Motors, Volkswagen and Mercedes-Benz is represented by a different type of symbol and a solid line delineates the envelope which contains all the points for a given vehicle class. The vehicle classes show some overlap but are generally well resolved with General Motors showing the highest particulate emissions followed by Mercedes-Benz and then Volkswagen. More detailed comparisons are difficult as there is a lack of high mileage data except for two tests on one Volkswagen at this time. All the Cutlass Cruisers and the General Motors 5.7 liter V8 diesel vehicles in general fall into a relatively discrete envelope of values as compared to Volkswagen and Mercedes-Benz vehicles.

Figs. 15, 16, and 17 show particulate and SOF data for Volkswagen Rabbits only. Fig. 15 shows particulate emissions for five vehicles each of which was tested twice. In all but one case the particulate emissions increased markedly (0.07 g/mi/10 K mi) between tests although further tests are needed to determine if this is a mileage related change or random variation due to other unknown factors. Fig. 16 which shows % SOF indicates that particulate emissions and % SOF change in opposite directions. This trend would tend to keep g/mi SOF fairly constant but as is apparent in Fig. 17 the increase in particulate emissions more than compensates for the decrease in % SOF and there is an increase in absolute amount of material extracted. The apparent decrease in SOF g/mi for Car #9 may have been caused by the correction of a significant injector timing error performed between tests.

No Mercedes-Benz vehicles have been tested more than once and the four tests had similar mileages. Therefore, no mileage accumulation data will be presented.

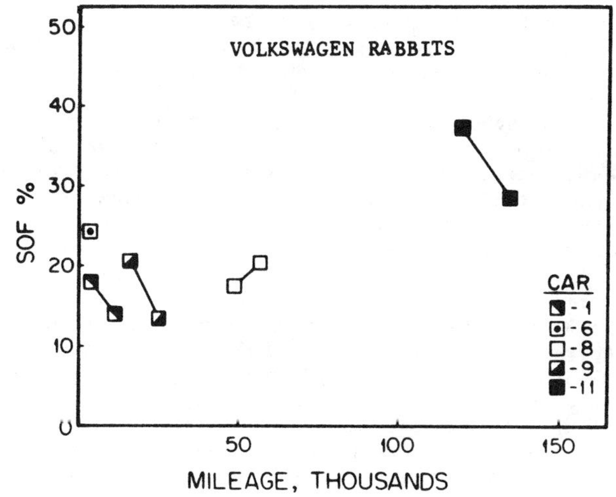

Fig. 16 - Mileage Accumulation Effects on Soluble Organic Fraction, %, for Volkswagen Rabbits.

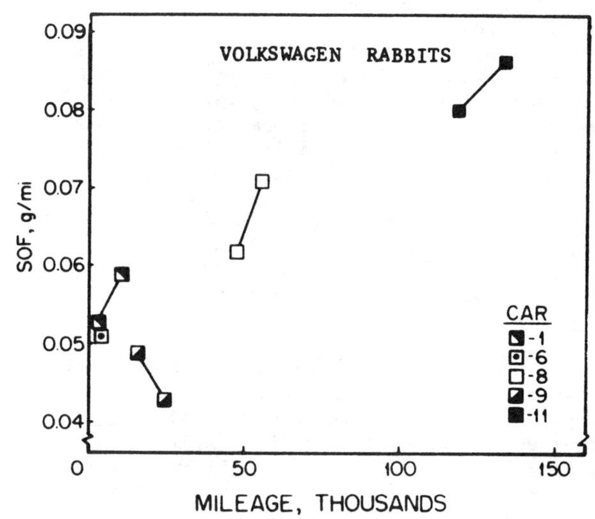

Fig. 17 - Mileage Accumulation Effects on Soluble Organic Fraction, g/mi, for Volkswagen Rabbits.

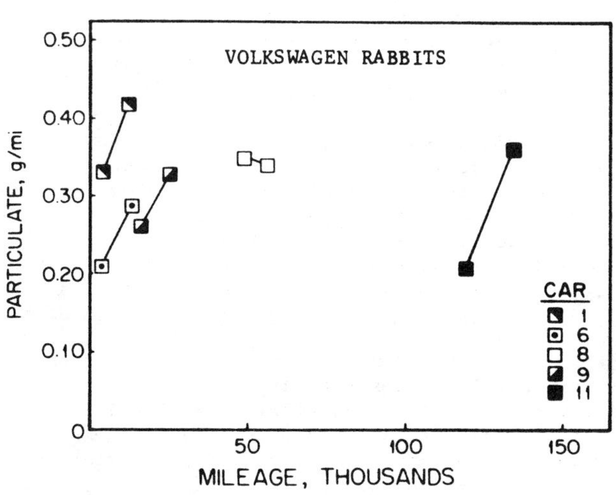

Fig. 15 - Mileage Accumulation Effects on Particulate Emissions, g/mi, for Volkswagen Rabbits.

AMES BIOASSAY OF SOF SAMPLES

Ames Test Procedures - The bacterial mutagenesis plate incorporation assay according to the method of Ames (38) has given positive indications of mutagenic activity for all SOF samples tested. Since the chemical compounds responsible for the observed Ames activity have not been fully identified at this time, the Ames assay itself is presently the only available tool for routine detection of their presence. The potential for mutagen artifact from sampling in the dilution tunnel cannot be presently ruled out, as evidence of nitro-PAH mutagens in SOF is advanced (28, 39, 40). Until the chemical understanding is more fully developed, interactions of vehicle emission products and sampling and handling artifacts cannot be separated, and therefore the sampling procedures are not able to quantify mass emissions of the critical species. Thus, the SOF sample mutagenic potency values are reported here to provide a relative measure of the project variables for internal comparison purposes only, and are not intended to represent absolute emission values.

Every effort has been made to perform all steps of sampling and SOF handling in a consistent manner so as not to introduce random artifact errors.

SOF samples from the in-use vehicle tests in this project are analyzed for dose/response mutagenic potency by the Ames tester strain TA98 without metabolic activation. The use of other strains with and without metabolic activation have been undertaken on project samples but only TA98 (-) data are included here. The freezer stored SOF extracts are redissolved in dichlormethane, split in two portions, one of which is returned to cold storage and the other dried and redissolved in DMSO for the bioassay work. Duplicate plates are run at doses of 10, 20, 30, 40, 50, 75, 100, and 200 μg. Straight-line regressions are obtained to dose/revertant response data sets utilizing first all doses, and subsequently with the highest dose eliminated from the regression, and successively with the two highest doses eliminated, etc. to look at the dose/response data for toxic effects that cause a plateau or falloff of revertant counts with increasing dose. Appendix B contains three typical dose/response and data regression plots. The "best" slope of the dose/response curve is selected by considering the R^2 value, intercept of the fitted line, total number of revertants at the highest dose selected, and number of data points available. This judgement is by no means absolute or unequivocal but has been applied consistently to obtain a slope (revertants/μg dose SOF) in the 0-200 μg dose range with an R^2 value of typically 0.90 and above using 5-8 doses with duplicates at each dose entered as separate data. From 10 to 16 points are used in each regression.

In addition to chemical standards to check response, freezer stored aliquots of SOF obtained from an Oldsmobile operated on project reference fuel are routinely used. For the TA98(-) tests reported here, the coefficient of variance of the response to this diesel SOF standard has been ± 23% of the mean. To increase the comparability of dose/response results for all samples within a given vehicle test, the 15 SOF samples from each vehicle test are run as a group on a single day. Two sets of SOF samples were analyzed in April of 1980, and reanalyzed in September. Table 8 shows the ratio of slopes obtained for each SOF sample for which duplicate runs were available, and provides a guide for interpreting the relative changes. The specific activity for all samples was about 25% reduced in the September analyses. Degradation of mutagens cannot be completely ruled out as contributing to these observed decreases, but bacterial response variations are considered more likely in view of the variable response to standards. The mutagenic potency of SOF samples are best considered as relative indicators and not absolute values, and large variations in specific activity (revertants/μg) are easily accumulated between the preparation of the SOF samples and the choosing of a slope from a SOF dose - revertant colony response curve.

Table 8-Change in Ames Response with Five Months Between Assays.

SOF Samples Rerun on TA98(-)

Table Entries are Ratio of (Revertants/μg) April / (Revertants/μg) Sept.

	Test Phase	50 mph Cruise	FTP	CFDS	HFET
Vehicle/Test	1	1.4	0.8	1.0	1.1
Car #3	2	1.6	0.8	1.5	1.5
Project Test #2	3	1.6	0.9	1.6	1.4
Average		1.5	0.8	1.4	1.3
Vehicle/Test	1	1.2	0.9	2.0	1.5
Car #3	2	1.3	1.1	1.8	1.3
Project Test #2	3	1.0	1.7	1.2	1.8
Average		1.2	1.2	1.7	1.5

Average = 1.3, or Activity in September assay is 77% of April

Ames Test Results - The quantitative results of the Ames test are usually expressed simply as revertants/μg sample. In this case, where the composition of the sample (i.e. the SOF) is unknown, the amount of bio-active material in the sample is unknown. Therefore, the discussions below will generally consider mutagenic activity in terms of experimentally measurable parameters, i.e. grams of particulate, miles and Kg fuel consumed (Some data in terms of revertants/μg SOF are found in Fig. 24). The twelve sets of data below represent results of four tests each on two Oldsmobile Cutlass crusiers and one test each on an Oldsmobile 98, a Cadillac Eldorado, a Volkswagen Rabbit and a Mercedes-Benz 300 CD; and are all the data available at this writing. The data for the results of the Ames test in terms of the measured slope of the dose-response curve and the derived parameters are presented in Appendix C. Except for the results on the Volkswagen Rabbit and Mercedes-Benz, all data are those of the September 1980 assay.

The amount of mutagenic activity may be expressed as revertants per gram particulate so as to remove, at least partially, the effects of changes in % SOF which may be caused by fuel and/or lubricant dilution of the extract. Revertants/g particulate (reported in units of 10^6 revertants/g which is equivalent to 1 revertant/μg) are equal to the slope of the dose response curve (rev/μg) times the % SOF divided by 100. These data are presented in Figures 18 and 19 for Car #2 and Car #3 respectively and for Cars #5, 8, 9 and 15 in Fig. 20. These twelve sets of data are the only results available at this time. In these figures, each data point is the average of three vehicle test phases, with the exception of Test 1 which had only two complete test-phases.

Car #2 and Car #3 are essentially identical Oldsmobile Cutlass cruisers which were each tested four times over an interval of 25-30,000 miles. As is evident from a comparison of Figs. 18 and 19, the two vehicle's emissions exhibit different mutagenic responses both in magnitude and in cycle-to-cycle differences. The values for Car #2 and Car #3 are almost mutually exclusive of each other and cover a range with an approximately five-fold change. The two sets of data are similar in that in each the FTP shows the largest test-to-test variations, the Idle generally shows the lowest mutagenic activity and the overall activity decreases with accumulated mileage (higher test number in each case except for Test 30). Except for the lower Idle values there are no apparent cycle-to-cycle differences. Fig. 20 shows data analogous to that in Figs. 18 and 19 but for an Oldsmobile 98 (Test 5), a Volkswagen Rabbit (Test 8), a Mercedes-

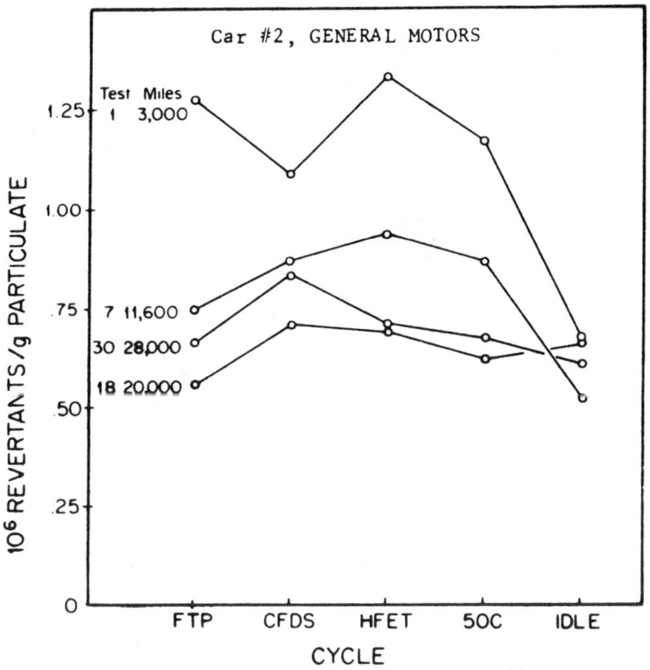

Fig. 18 - Ames Activity for Car #2 by Test and Cycle in Revertants /g Particulate.

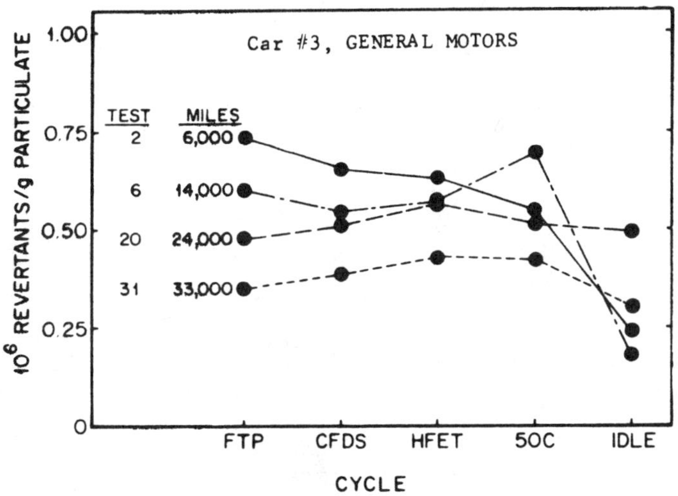

Fig. 19 - Ames Activity for Car #3 by Test and Cycle in Revertants /g Particulate.

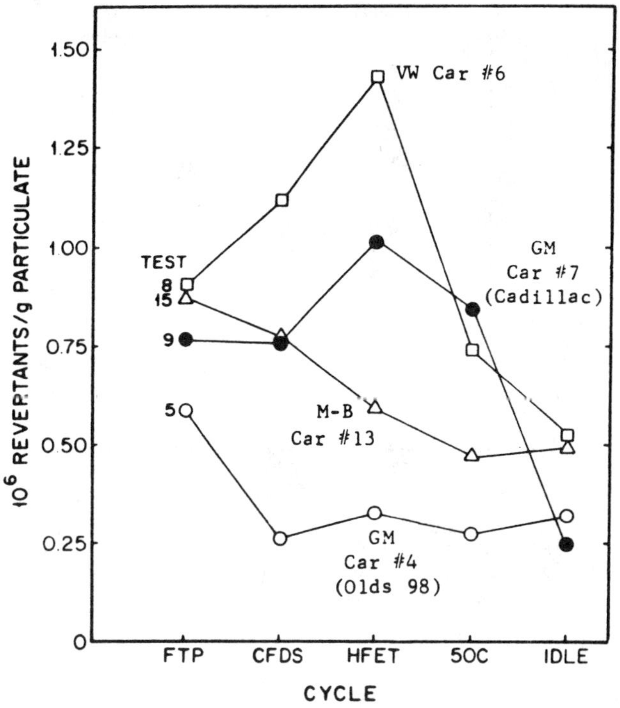

Fig. 20 - Ames Activity for Car #6, Car #13, Car #7 and Car #4 by Cycle in Revertants /g Particulate.

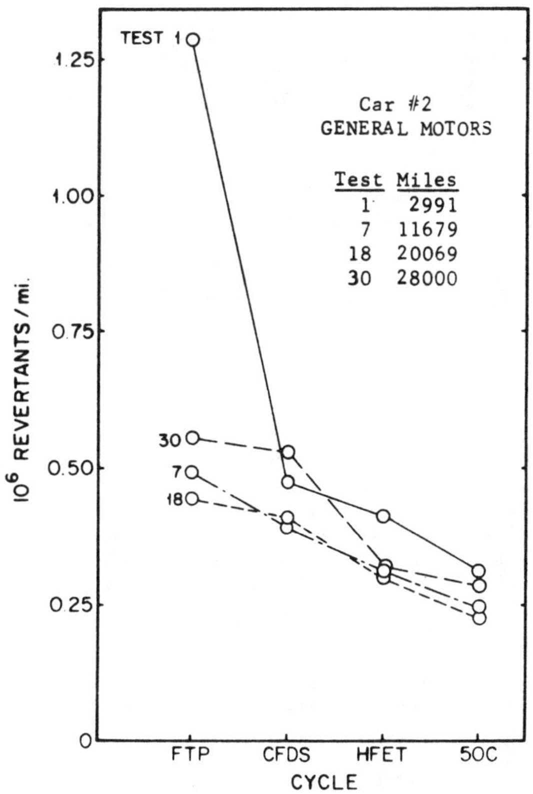

Fig. 21 - Ames Activity for Car #2 by Test and Cycle in Revertants/mi.

Benz 300 CD (Test 15) and a Cadillac Eldorado (Test 9). These data are generally in the range of values exhibited the Oldsmobile Cutlass Cruisers. These vehicles also display a wide divergence of absolute values and cycle-to-cycle variations.

Figs. 21, 22 and 23 display the data for each test in terms of revertants/mi and driving cycle. In this representation the test-to-test variations for a single vehicle are relatively small and cycle-to-cycle trends are more pronounced and reproducible. In each case, the number of revertants decreases in the order FTP>CFDS>HFET>50C (Test 8 on a Volkswagen has HFET slightly greater than CFDS); and the FTP shows the largest test-to-test differences. Except for Test 30, there is generally a reduction in revertants/mi as the vehicle

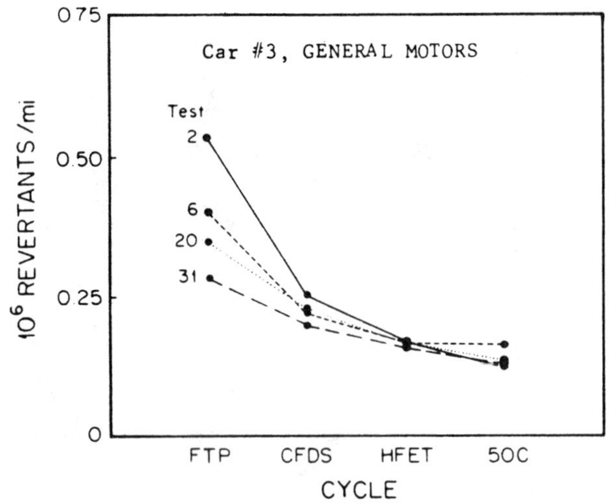

Fig. 22 - Ames Activity for Car #3 by Test and Cycle in Revertants/mi.

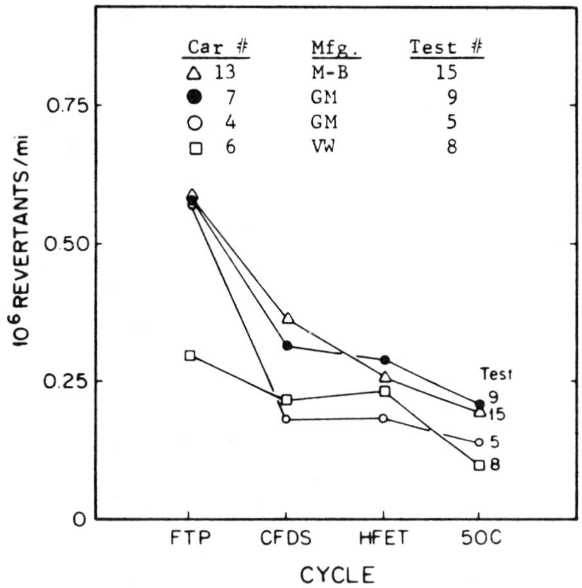

Fig. 23 - Ames Activity for Car #6, Car #13, Car #7 and Car #4 by Cycle in Revertants/mi.

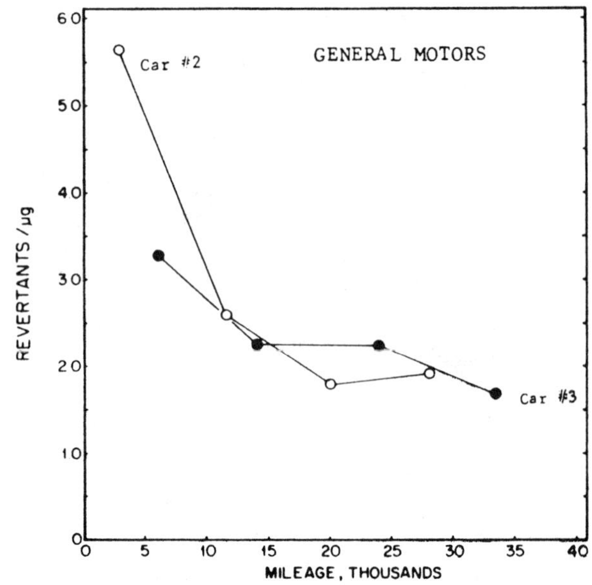

Fig. 24 - Mileage Accumulation Effects on Ames Activity for Car #2 and Car #3 in Revertants/μg SOF.

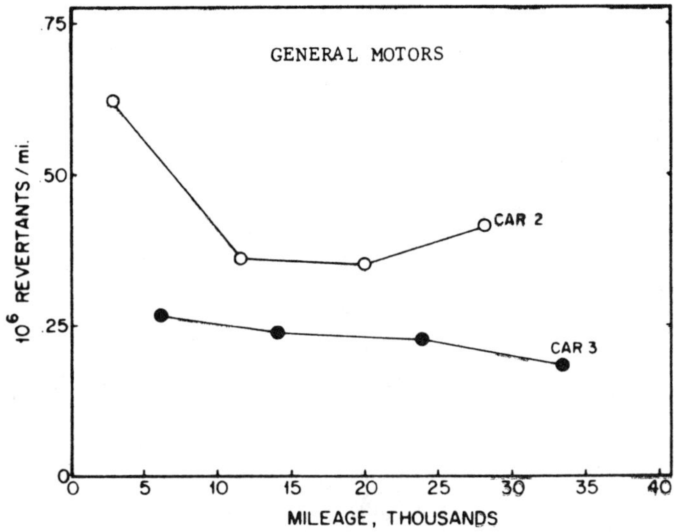

Fig. 25 - Mileage Accumulation Effects on Ames Activity for Car #2 and Car #3 in Revertants/mi.

mileage increases. As with revertants/g particulate data (Figs. 18 and 19), the two Oldsmobile Cutlass Cruisers have widely divergent values of revertants/mi. especially when compared to each other on a mileage basis (cf. Figs. 24 and 25 below). The data in Fig. 23 are for the same set of vehicles as in Fig. 20 and shows that other General Motors models and other manufacturers vehicles exhibit similar trends.

Mileage Accumulation Effect - Fig. 24 and Fig. 25 show the Ames activity in terms of revertants/μg extract and revertants/mi as a function of accumulated mileage. Both show an overall decrease in Ames activity with increasing mileage. Fig. 24 shows that the two Oldsmobile Cutlass Cruisers have extracts of comparable activity at similar mileages. However, because Car #2 has both higher particulate emissions and % SOF it has considerably higher revertants/mile. Figures for revertants/Kg fuel and revertants/g particulate vs. mileage are essentially the same as Fig. 25 and, therefore, have not been included graphically but the tabular data are given in Appendix C. The large divergence of the lowest mileage point for Car #2 is possibly related to the fact that the vehicle had only 3,000 miles of usage at the time of the test.

Effect of Vehicle Test Phase - Table 9 contains average values and standard deviations for the vehicle test-phase ratios of specific activity (revertants/μg SOF) using the results from test-phase III as the comparison base. The sample number is considerably smaller than for similar comparisons presented above for particulate and SOF emissions, and combined with the inherent

Table 9 - Variability of Ames Response by Vehicle Test-Phase.

Specific Activity (revertants/μg SOF)
TA98(-)

$$R = \left[\frac{\text{Phase I or II SOF activity (rev/μg)}}{\text{Phase III SOF activity (rev/μg)}}\right]$$

$\bar{R}$ = average of all R values from individual vehicle tests

		FTP	CFDS	HFET	50C	Idle
[Test Phase I / Test Phase III]	$\bar{R}$	1.16	1.14	1.17	0.98	1.14
	σ	0.80	0.59	0.28	0.28	0.43
	N	12	12	12	12	12
[Test Phase II / Test Phase III]	$\bar{R}$	1.31	1.14	1.24	0.96	1.01
	σ	0.50	0.30	0.31	0.23	0.33
	N	11	11	11	11	11

variability in the assay the average values presented must be considered as tentative. The standard deviations are large for all driving cycles, but the FTP appears to be significantly larger than other cycles. Ratios of activity between phases are fairly uniform for the driving cycle, and would suggest a 0-20% increase in specific activity for both phases I and II relative to phase III results, but the high standard deviation and limited sample size make this a preliminary estimate. Since phase III involves a fresh engine oil test condition whereas phases I and II are only a fuel switch, the uniformity of the results for comparisons of I/III and II/III are not completely surprising. The oil change effect on specific activity of SOF as measured by TA98(-) appears in these results not to be a massive effect.

CONCLUSIONS

The project to obtain emissions data from in-use diesel cars is not complete, and this paper is somewhat of a status report midway through the work. Conclusions at this point are preliminary and contingent upon data collection of vehicular emission factors over much larger spans of mileage accumulation, the inclusion of more engine configurations in the vehicle sample group, and a more thorough understanding of the chemistry of diesel particulate emissions as related to emission products, sampling procedures, and bioactivity.

Summarizing the data presented here, the following conclusions are drawn:

(1) Particulate Emissions Variability - Duplicate vehicle tests within a given fuel/oil condition gave a coefficient of variation of approximately 10% of the mean emission. Little differences were observed among vehicle makes or driving cycles tested.

The coefficient of variation for particulate emissions between driving cycles repeated at different fuel/oil conditions was approximately 10-15% of the mean emission rate, with little driving cycle or vehicle make dependence.

The coefficient of variation for SOF emissions (g/mi) between driving cycles repeated at different fuel/oil conditions was approximately 25% of the mean emission rate with little driving cycle or vehicle make dependence.

On the average, no systematic variations in either particulate or SOF emissions were found for the fuel/oil conditions tested although individual vehicle tests did exhibit significant changes, especially in terms of SOF. A more detailed analysis of the collected data (cf. 5, 25) may reveal correlating parameters not accessed by overall averaging.

(2) Particulate Emissions - Particulate emissions (g/mi) generally were found to increase by driving cycle in the following order: 50C < HFET < CFDS < FTP. For the vehicles tested, particulate emissions increased in the order: Mercedes-Benz < Volkswagen < General Motors.

Mileage accumulation effects for two General Motors vehicles tested up to 30,000 mi exhibited marked increases in particulate and SOF emissions up to 20,000 mi, and have subsequently stabilized for an undetermined period. Preliminary mileage accumulation effects on particulate for sample group vehicles with multiple tests show approximately 0.06 g/mi increase per 10,000 miles.

In general, particulate emissions expressed in time-based units increased in the order: Idle < FTP < 50C < CFDS < HFET, and Volkswagen < Mercedes-Benz < General Motors.

Particulate emissions expressed on a time basis show little variation among driving cycles for General Motors vehicles, with the exception of the idle mode where emissions averaged 60% of the average for driven cycles. Particulate emissions in the Idle mode for Volkswagen and Mercedes-Benz vehicles averaged 10-20% of the time-based emissions from driven cycles.

Particulate emissions of Volkswagen and Mercedes-Benz vehicles exhibited a stronger driving cycle dependence when expressed in g/min than as g/mi.

(3) Fuel Specific Emissions - Fuel specific particulate emissions increased in the order: HFET<CFDS<FTP for General Motors vehicles, but exhibited no definite trend for Volkswagen or Mercedes-Benz types, and increase in the order Mercedes-Benz < General Motors < Volkswagen.

Fuel specific particulate emission averages were in the range 3.5-5.5 g/Kg fuel consumed for all vehicles and FTP, CFDS, and HFET driving cycles.

Fuel specific SOF emissions averages were in the range 0.5-1.2 g SOF/Kg fuel consumed for all vehicles tested on the FTP, CFDS and HFET driving cycles. Fuel specific SOF (g/mi) showed no definitive driving cycle dependence but increased in the order Mercedes-Benz < Volkswagen < General Motors.

(4) SOF Emissions - Average SOF expressed as a percentage of collected particulate increased in the order Idle ≈ FTP < CFDS < HFET < 50C for General Motors vehicles. SOF (%) was relatively constant

for all cycles for Mercedes-Benz vehicles at 13% and averaged about 20% for Volkswagen, except in the Idle mode where Volkswagen SOF (%) averaged 60%.

Mass emission of SOF (g/mi) increased in the general order: 50C < HFET < CFDS < FTP and Mercedes-Benz < Volkswagen < General Motors.

(5) Ames Test Activity of SOF Samples - All SOF samples which have been tested exhibit a positive mutagenic response in the Ames test. One set of SOF samples kept in cold storage and assayed at a 5-month interval showed an average 25% reduction in specific activity.

For multiple tests on two General Motors vehicles, mutagenic activity of SOF samples exhibited a general decrease with accumulated mileage (0-30,000 mi) whether expressed as activity/gram of particulate or activity of SOF emissions/mi.

No definitive effect of driving cycle on mutagenic activity was identified, except that SOF samples from the idle mode frequently showed a decreased response.

At this time it is not possible to partition the observed mutagenic activity of SOF samples into emission products and sampling artifacts. The decreased activity noted at idle could be related to engine variables or exhaust dilution variations, for example.

ACKNOWLEDGEMENTS

The vehicle testing, filter handling and extraction and data processing were by Robert Johnson, Paul Werner, Ben Hill and Tom Hoffman. Bioassay tests were performed by Mary Franke under the direction of Dr. Doudney at the New York State Dept. of Health, Division of Laboratories and Research. Dr. Delip Choudhury assisted in the preparation of extract samples for bioassay, and his constructive discussions and assistance are acknowledged. The diverse contributions of these people to the results presented here are sincerely appreciated. Peter Gabele, Roy Zweidinger, and several other members of the EPA Mobile Sources Research Branch are acknowledged for their helpful discussions in the course of this work. The work was partially supported by USEPA Grant R805934010.

Special thanks are expressed by the authors to Carol Clas and Garry Lanphear for their preparation of the graphics, and Kay Cassidy and Kathy Blackwell for preparation of the typewritten manuscript.

REFERENCES

1. K. Springer and R. Stahman, "Emissions and Economy of Four Diesel Cars." SAE Paper 750332, Presented February 24-28, 1975.

2. K. Springer and T. Baines, "Emissions From Diesel Versions of Production Passenger Cars." SAE Paper 770818, Presented September 26-September 30, 1977.

3. K. Springer and R. Stahman, "Diesel Car Emissions - Emphasis on Particulate and Sulfate." SAE Paper 770254, Presented February 28-March 4, 1977.

4. K. Springer and R. Stahman, "Unregulated Emissions from Diesels Used in Trucks and Buses." SAE Paper 770258, Presented February 28-March 4 , 1977.

5. C. Hare and T. Baines, "Characterization of Particulate and Gaseous Emissions from Two Diesel Automobiles as Functions of Fuel and Driving Cycle." SAE Paper 790424, Presented February 26-March 2, 1979.

6. K. Springer, "Characterization of Sulfates, Odor, Smoke, POM and Particulates from Light and Heavy Duty Engines - Part IX." Final Report, EPA Contract 68-03-2417, NTIS Document PB80-121551, June, 1979.

7. R. M. Schreck, J. J. McGrath, S. J. Swarin, W. E. Hering, P. J. Groblicki, and J. S. MacDonald, "Characterization of Diesel Exhaust Particulate under Different Engine Load Conditions." Paper 78-33.5, Presented at 71st Annual Meeting of the Air Pollution Association, Houston, June 25-30, 1978.

8. W. R. Pierson and W. W. Brachczek, "Particulate Matter Associated with Vehicles on the Road." Paper 760039, Presented at SAE Automotive Engineering Congress and Exposition, Detroit, February 23-27, 1976.

9. W. Lipkea, J. Johnson, C. Vuk, "The Physical and Chemical Character of Diesel Particulate Emissions - Measurement Techniques and Fundamental Considerations." SAE Special Publication 430, 1978.

10. E. Funkenbusch, D. Leddy, and J. Johnson, "The Characterization of the Soluble Organic Fraction of Diesel Particulate Matter." SAE Paper 790418, Presented February 26-March 2, 1979.

11. D. B. Kittelson, D. F. Dolan, and J. A. Verrant, "Investigation of a Diesel Exhaust Aerosol." Paper 780109, Presented at SAE Congress and Exposition, Detroit, February 27-March 3, 1978.

12. W. F. Marshall, D. E. Seizinger, and R. W. Freedman, "Effects of Catalytic Reactors on Diesel Exhaust Composition." Bureau of Mines Technical Progress Report 105, U.S. Department of the Interior, April 1978.

13. C. Hare, K. Springer, and R. Bradow, "Fuel and Additive Effects on Diesel Particulate Development and Demonstration of Methodology." SAE Paper 760130, Presented February 23-27, 1976.

14. H. Dietzmann and F. Black, "Unregulated Emissions Measurement Methodology." SAE Paper 790816, Presented September 10-13, 1979.

15. W. Mayer, D. Hilden, and D. Lechman, "The Contribution of Engine Oil to Diesel Exhaust Particulate Emissions." SAE Paper 800256, Presented February 25-29, 1980.

16. F. Black and L. High, "Methodology for Determining Particulate and Gaseous Diesel Hydrocarbon Emissions." SAE Paper 790422,

Presented February 26-March 2, 1979.

17. L. Frisch, J. Johnson, and D. Leddy, "Effect of Fuels and Dilution Ratio on Diesel Particulate Emissions." SAE Paper 790417, Presented February 26-March 2, 1979.

18. J. MacDonald, S. Plee, J. D'Arcy, and R. Schreck, "Experimental Measurement of the Independent Effects of Dilution Ratio and Filter Temperature on Diesel Exhaust Particulate Samples." SAE Paper 800185, Presented February 25-29, 1980.

19. S. Plee and J. MacDonald, "Some Mechanisms Affecting the Mass of Diesel Exhaust Particulate Collected Following a Dilution Process." SAE Paper 800186, Presented February 25-29, 1980.

20. K. Duleep, "Analysis of Automotive Particulate Sampling Techniques." SAE Paper 800184, Presented February 25-29, 1980.

21. Federal Register, 40CFR86, page 14496, March 5, 1980.

22. J. Shelton, "Exhaust Emissions from Twenty Light Duty Diesel Trucks." EPA Report EPA-AA-TEB-80-20, August, 1980.

23. J. White III and G. Jones, "A Study of Exhaust Emissions from Twenty High Mileage Oldsmobile Diesel Passenger Cars." EPA Report EPA-AA-TEB-80-11, March, 1980.

24. J. Huisingh, R. Bradow, R. Jungers, L. Claxton, R. Zweidinger, S. Tejeda, J. Bumgarner, R. Duffield, M. Waters, V. Simmon, C. Hare, C. Rodriguez, and L. Snow, "Application of Bioassay to the Characterization of Diesel Particulate Emissions: Parts I and II." Presented at Symposium on Application of Short Term Bioassays in the Fractionation and Analysis of Complex Environmental Mixtures, Williamsburg, February, 1978.

25. C. Hare, "Characterization of Gaseous and Particulate Emissions from Light Duty Diesels Operated on Various Fuels." Final Report, EPA Contract 68-03-2440, NTIS Document PB80-122443, July, 1979.

26. E. Wei, Y. Wang and S. Rappaport, "Diesel Emissions and the Ames Test: A Commentary." Journal of the Air Pollution Control Association, Vol. 30, No. 3, pg. 267, March, 1980.

27. Proceedings of the International Symposium on Health Effects of Diesel Engine Emissions, USEPA, December 3-5, 1979, Cincinnati, Ohio. (in press).

28. R. Bradow, "Diesel Particle Emissions." New York Academy of Medicine, Symposium on Health Aspects of Automotive Emissions, May 15-16, 1980.

29. D. Choudhury, "Mutagenic Activity of Diesel Emissions Particulate Extracts and Isolation of the Mutagenic Fractions." In Proceeding of the International Symposium on Health Effects of Diesel Engine Emissions, Cincinnati, Ohio, December, 1979, (in press).

30. D. Choudhury and B. Bush, "Polynuclear Aromatic Hydrocarbons in Diesel Emissions Particulate." Proceedings of the International Symposium on Health Effects of Diesel Engine Emissions, Cincinnati, Ohio, December, 1979, (in press).

31. D. Choudhury and C. O. Doudney, "Mutagenic Activity of the Diesel Exhaust Extract and an Approach to Characterization of the Active Fraction." Presented at the Pittsburgh Conference on Analytical Chemistry and Applied Spectroscopy, Atlantic City, New Jersey, March, 1980, Abstract of Papers p. 522.

32. D. Choudhury, "Detection of Polycyclic Ketones and Quinones in Diesel Particulate Emissions." Presented at 2nd Chemical Congress of the North American Continent, Las Vegas, Nevada, September 1980.

33. D. Choudhury and B. Bush, "Gas Chromatography-Mass Spectrometric Characterization of Polynuclear Aromatic Hydrocarbons in Particulate Diesel Emissions," in "Measurement and Control of Chemical Hazards in the Workplace Environment," American Chemical Society Symposium Series, in press.

34. R. Gibbs, G. Wotzak, S. Byer and J. Hyde, "Emissions From Diesel Vehicles in Consumer Use." Proceedings of the International Symposium on Health Effects of Diesel Engine Emissions, Cincinatti, Ohio, December, 1979, in press.

35. T. Penninga, "Evaluation of Exhaust Collection Configuration on Diesel Particulate Measurement." NTIS Document PB80-104995, May 1979.

36. K. Stulp, "Particulate Measurement-Efficiency of Pallflex T60A20 Filter Media." EPA Report SDSB-79-22, NTIS PB80-115975, July 1979.

37. J. Braddock and P. Gabele, "Emission Patterns of Diesel-Powered Passenger Cars - Part II." SAE Paper 770168, presented February 28-March 4, 1977.

38. B. Ames, J. McCann, and E. Yamasaki, Mutation Research, Vol. 31, pp. 347-364, 1975.

39. D. Schuetzle, F. Lee, T. Prater, and S. Tejada, "The Identification of Polynuclear Aromatic Hydrocarbon Derivatives in Mutagenic Fractions of Diesel Particulate Extracts." Presented at 10th Annual Symposium on The Analytical Chemistry of Pollutants, Dortmund, Germany, May 28-30, 1980.

40. T. Pederson and J. Siak, "Characterization of Direct Acting Mutagens in Diesel Exhaust Particulates by Thin Layer Chromatography." 180th National ACS Meeting, Las Vegas, August 24-29, 1980.

Appendix A

AEL Reference Fuel Specification Data

Distillation		
	Initial Boiling Point	336°F
	10% Recovery	398°F
	20% "	423°F
	30% "	442°F
	40% "	461°F
	50% "	483°F
	60% "	501°F
	70% "	522°F
	80% "	546°F
	90% "	585°F
	95% "	600°F
	End Point	635°F
	% Recovery	98.8
	% Residue	1.1
	% Loss	0.1
Elemental Analyses		
	Carbon	86.33%
	Hydrogen	13.06%
	Nitrogen	<0.05%
	Oxygen	0.47%
	Sulfur	0.14%
Ash		<0.001%
API Gravity @ 60°F		36.2
Specific Gravity @ 60°F		0.8438
Heating valve (BTU/lb)		19,458
Cetane Index		45.5
% Saturates		67.1
% Olefins		0.3
% Aromatics		32.6

Appendix B-Typical Dose Response Curves for TA98(-) Ames Tests

STRAIN	TEST	CAR	PHASE	CYCLE	FILTER	BOTTLE	RUN
TA98 -	7	2	2	7	561	261	910

DOSE	COLONIES	
BACKGROUND	46	46
10	113	124
20	126	156
30	162	193
40	173	213
50	200	231
75	246	294
100	350	456
200	601	612

HIGH DOSE	SLOPE	INTERCEPT	R-SQUARE	REV/100μgms
200	2.72	82.42	0.96	355
100	3.13	68.29	0.91	382
75	2.76	77.24	0.89	353
50	3.16	69.48	0.88	386
40	3.53	64.60	0.87	418
30	4.17	58.20	0.89	475

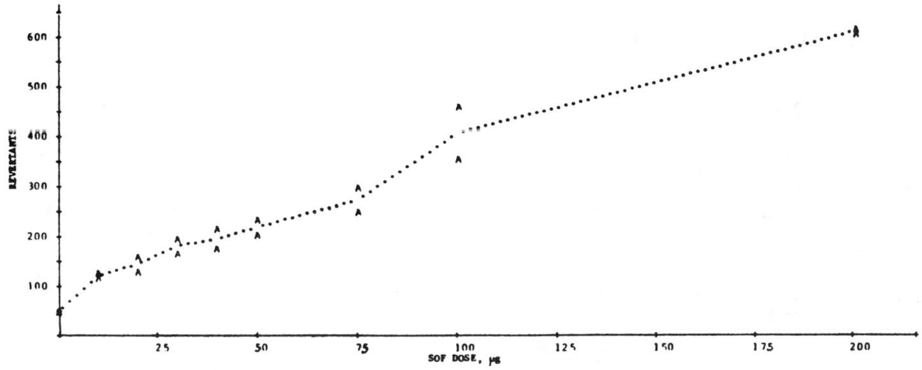

STRAIN	TEST	CAR	PHASE	CYCLE	FILTER	BOTTLE	RUN
TA98 -	8	6	3	123	586	273	924

DOSE	COLONIES	
BACKGROUND	37	37
10	72	88
20	136	121
30	135	174
40	174	185
50	192	197
75	281	320
100	363	382
200	601	612

HIGH DOSE	SLOPE	INTERCEPT	R-SQUARE	REV/100μgms
200	2.87	61.05	0.98	348
100	3.27	47.60	0.98	374
75	3.29	46.93	0.96	376
50	3.11	50.52	0.92	361
40	3.59	44.00	0.94	403
30	4.01	39.85	0.93	441

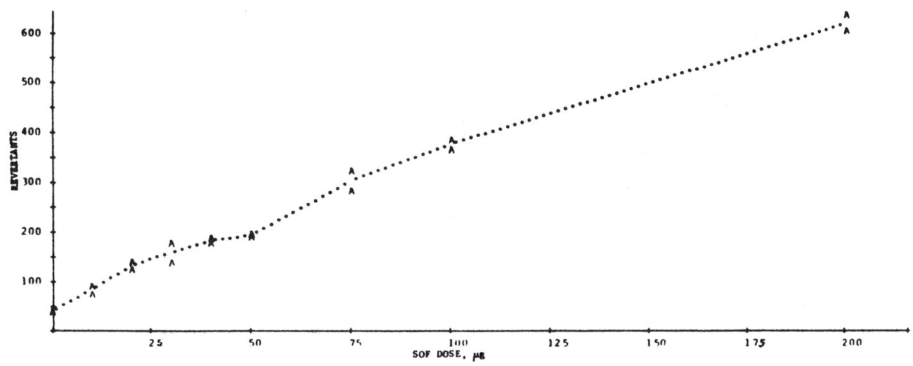

STRAIN	TEST	CAR	PHASE	CYCLE	FILTER	BOTTLE	RUN
TA98 -	30	2	3	123	1082	518	930

DOSE	COLONIES	
BACKGROUND	37	37
10	34	47
20	74	70
30	71	101
40	121	112
50	143	112
75	191	215
100	231	256
200	345	365

HIGH DOSE	SLOPE	INTERCEPT	R-SQUARE	REV/100ugms
200	1.68	44.20	0.95	212
100	2.20	26.44	0.97	246
75	2.25	25.27	0.95	250
50	1.98	30.31	0.90	229
40	2.05	29.50	0.90	234
30	1.79	32.10	0.80	211

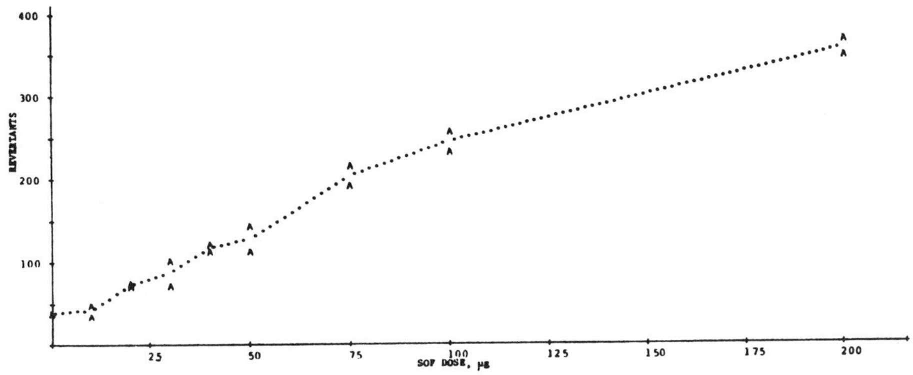

Appendix C- Ames Test Activity Data Base

			Slope (revertants / µg SOF)					Revertants*/ gram Particulate					Revertants*/ KG Fuel			Revertants*/ Mile			
TEST	CAR	PHASE	FTP	CFDS	HFET	50C	IDLE	FTP	CFDS	HFET	50C	IDLE	FTP	CFDS	HFET	FTP	CFDS	HFET	50C
1	2	1	8.00	4.40	5.90	4.10	5.00	1.63	1.09	1.57	0.83	0.87	8.54	4.47	4.94	1.22	0.51	0.51	0.27
1	2	3	10.40	5.40	5.00	5.20	2.90	1.76	1.07	1.09	1.49	0.47	9.31	3.74	2.85	1.34	0.44	0.30	0.34
2	3	1	5.20	3.90	3.60	2.20	2.40	0.74	0.70	0.71	0.50	0.26	3.77	2.37	1.85	0.56	0.28	0.20	0.12
2	3	2	5.00	2.80	2.90	2.00	2.10	0.74	0.51	0.62	0.55	0.22	3.51	1.60	1.51	0.51	0.19	0.16	0.12
2	3	3	5.20	4.70	2.80	2.20	2.00	0.72	0.75	0.55	0.58	0.23	3.41	2.35	1.22	0.53	0.28	0.14	0.12
5	4	1	4.10	2.20	1.70	1.50	1.20	0.59	0.42	0.36	0.34	0.28	4.06	2.49	1.90	0.59	0.29	0.20	0.17
5	4	2	5.30	0.80	1.60	0.70	1.40	0.75	0.15	0.36	0.16	0.31	4.90	0.84	1.79	0.72	0.10	0.19	0.08
5	4	3	2.50	0.80	1.00	1.20	1.50	0.41	0.20	0.24	0.30	0.34	2.78	1.18	1.41	0.41	0.14	0.15	0.16
6	3	1	5.60	2.00	2.20	2.20	1.40	0.85	0.46	0.57	0.67	0.18	4.06	1.58	1.52	0.59	0.19	0.16	0.16
6	3	2	3.60	2.50	2.60	2.10	1.60	0.59	0.58	0.68	0.62	0.19	2.73	1.96	1.84	0.39	0.24	0.20	0.14
6	3	3	1.60	1.90	1.40	1.90	1.00	0.34	0.57	0.43	0.77	0.15	1.58	1.99	1.24	0.22	0.24	0.13	0.20
7	2	1	3.20	2.30	2.60	1.90	2.50	0.66	0.76	0.96	0.88	0.52	2.78	2.98	3.01	0.41	0.33	0.32	0.23
7	2	2	4.30	2.80	2.90	1.80	2.70	0.84	0.90	1.02	0.84	0.54	3.96	3.39	3.10	0.56	0.40	0.32	0.24
7	2	3	2.90	2.60	2.00	1.80	2.20	0.74	0.95	0.81	0.88	0.49	3.37	3.75	2.68	0.50	0.44	0.29	0.25
8	6	1	4.20	4.00	4.40	3.10	1.40	0.93	1.03	1.17	0.59	0.51	3.50	3.45	3.68	0.24	0.21	0.21	0.09
8	6	2	4.80	5.20	5.90	4.00	1.20	0.89	1.15	1.38	0.72	0.45	3.08	4.06	3.87	0.22	0.25	0.21	0.10
8	6	3	3.30	3.90	4.80	3.70	1.10	0.86	1.14	1.71	0.87	0.58	6.17	3.39	5.49	0.43	0.19	0.28	0.11
9	7	1	2.40	2.10	3.20	1.70	1.20	0.55	0.63	0.99	0.78	0.22	3.64	2.10	2.75	0.57	0.25	0.28	0.21
9	7	2	4.00	2.70	3.00	2.60	1.60	0.85	0.80	0.87	0.96	0.26	4.14	2.63	2.45	0.64	0.32	0.25	0.25
9	7	3	3.70	3.00	3.10	2.20	1.40	0.88	0.83	1.17	0.77	0.25				0.58	0.39	0.34	0.19
15	13	1	7.70	7.40	8.90	5.70	7.80	0.80	0.68	0.72	0.58	0.75	4.00	2.82	2.82	0.54	0.31	0.29	0.21
15	13	2	10.40	10.70	7.00	4.60	3.10	0.97	0.95	0.55	0.40	0.30	4.59	3.91	2.31	0.63	0.45	0.25	0.19
15	13	3	9.40	8.40	7.10	5.20	4.80	0.83	0.67	0.49	0.42	0.42	3.94	2.64	1.85	0.56	0.30	0.21	0.16
18	2	1	1.50	1.10	1.10	1.10	2.50	0.44	0.47	0.54	0.57	0.57	2.34	2.25	2.09	0.35	0.27	0.23	0.21
18	2	2	1.80	1.80	1.80	1.00	2.70	0.54	0.77	0.82	0.57	0.66	2.99	4.02	3.38	0.44	0.48	0.37	0.22
18	2	3	2.30	2.20	1.40	1.30	2.90	0.65	0.85	0.68	0.67	0.70	3.74	4.10	2.74	0.55	0.49	0.30	0.25
20	3	1	1.90	1.90	1.60	0.81	1.70	0.37	0.47	0.45	0.29	0.26	1.92	1.92	1.38	0.27	0.22	0.14	0.08
20	3	2	3.00	2.30	1.90	1.90	2.40	0.59	0.63	0.58	0.66	0.33	3.19	2.62	1.70	0.44	0.29	0.18	0.18
20	3	3	2.20	1.50	2.00	1.50	6.70	0.46	0.41	0.65	0.58	0.89	2.47	1.74	1.98	0.35	0.20	0.21	0.16
30	2	1	2.20	2.20	1.40	1.10	2.00	0.60	0.94	0.66	0.59	0.52	3.21	5.09	2.68	0.48	0.58	0.29	0.22
30	2	2	2.30	2.30	1.90	1.50	2.80	0.66	0.96	0.91	0.78	0.75	3.48	5.31	3.58	0.51	0.62	0.39	0.32
30	2	3	2.20	1.50	1.40	1.30	2.20	0.73	0.61	0.57	0.66	0.56	4.29	3.10	2.17	0.65	0.36	0.24	0.28
31	3	1	1.40	1.80	1.60	1.60	3.60	0.25	0.45	0.48	0.63	0.38	1.37	2.06	1.59	0.19	0.23	0.17	0.19
31	3	2	1.70	1.40	1.20	0.60	2.00	0.30	0.35	0.38	0.21	0.24	1.89	1.54	1.16	0.27	0.17	0.13	0.06
31	3	3	2.30	1.20	1.30	1.00	2.20	0.46	0.32	0.41	0.39	0.25	2.55	1.45	1.35	0.37	0.16	0.14	0.11

*In units of 10^6

Automotive Diesel Engines–Fuel Composition vs Particulates**

Harvey A. Burley and Theodore L. Rosebrock
Advance Product Engineering
Engineering Staff
General Motors Corporation

As our society becomes more energy conscious, the automotive diesel engine becomes more attractive. The superior fuel economy of this engine is expected to lead to more and more sales of diesel vehicles of all kinds. The superior fuel economy is increasingly outweighing the traditional disadvantages of the diesel: cost, size and weight, odor, noise, and cold start difficulties. Moreover, these disadvantages are being mitigated significantly as they come into the focus of intensive engineering efforts.

At this point the main dark cloud on the horizon of the automotive diesel engine (and this is meant both literally and figuratively) is particulate emissions. It is true that there are other major concerns: for instance the question of adequate fuel supplies and also the control of hydrocarbons and oxides of nitrogen emissions. However, none of these concerns seems so difficult to resolve as that of particulate emissions.

Low exhaust smoke has been a development objective of all diesel passenger cars in current production. Smoke is the visible corollary of particulate emissions. It follows that much effort has, in effect, been expended to achieve low particulate emissions. (1-8)*

More than four years ago General Motors recognized a desirability to reduce diesel particulate emissions below then current levels. A program with that objective was activated at the Engineering Staff. The program has included a study of particulate measurement methods and an evaluation of three particulate reduction techniques; engine modification, fuel modification and particulate trapping.

The authors wish to make clear it has not been established that the fuel blends or additives that were evaluated in this study are practical from the standpoints of bulk producibility, cost, overall acceptability (cold starting, noise, lubricity, etc.) or even an emission benefit in all diesel engines. Answers to these questions can be obtained only through analyses by fuel and additive suppliers to define potentially producible fuels plus extensive evaluation of candidate fuels in a variety of diesel engines. Maximum cooperation among petroleum suppliers, engine manufacturers, and regulatory agencies will be required for execution of any fuel change found desirable in such an effort.

*Numbers in parentheses designate references at end of text.

**Paper 790923 presented at the Fuels and Lubricants Meeting, Houston, Texas, October, 1979.

ABSTRACT

FORTY-SIX EXPERIMENTAL DIESEL FUELS have been tested in a modern automotive diesel engine, and the effect of fuel characteristics on particulate emissions, gaseous emissions, and fuel economy has been studied. Regression analysis of results established a relationship between particulate emissions and two fuel characteristics: (1) aromatic content and (2) volatility, especially as described by the 90% boiling point. Particulates could be reduced by lowering either aromatic content or 90% boiling point. The foregoing conclusions were based on a weighted five-load dynomometer schedule but appear to be confirmed by limited vehicle testing on the Federal Test Procedure.

PRELIMINARY FUEL MODIFICATION TESTS

The study of fuel modification techniques has included investigations in four main areas; fuel blends, fuel additives, emulsions, and heated fuels. Table 1 summarizes our preliminary findings with each of these four approaches. Dynamometer tests were generally run at 25, 45, 55 mph road loads and at some heavier loads. Particulate reductions were found to vary greatly with speed and load.

The following amplifies somewhat the summary of information in Table 1.

Table 1 - Observed particulate reduction utilizing various fuel modification techniques. A 5.7 liter V-8 engine was used in most tests. These preliminary tests led to the fuel blend study which is the subject of this paper.

	PARTICULATE REDUCTIONS DYNAMOMETER TESTS	VEHICLE FTP	OTHER EFFECTS
FUEL BLENDS	16 – 70%	27%	CITY MPG REDUCED 2.5%
FUEL ADDITIVES	4 – 30%	17%	
EMULSIONS – WATER	34%	7%	HC INCREASED 100%, NOx REDUCED 40%
– METHANOL	31%	–	NOx REDUCED 15%
HEATED FUEL	11%	–	INJECTOR DAMAGE

FUEL BLENDS: The results of tests of a few light hydrocarbon blends are shown here. Of the various fuel modification schemes tried, the fuel-blending approach showed the greatest promise as a means of particulate reduction. The main disadvantage was the increase in fuel consumption observed in some but not all cases. Because of this apparent particulate reduction potential it was decided to investigate a large number of test fuels in an effort to learn which fuel characteristics appear related to particulates. The study of the forty-six fuels chosen for testing is the subject of this paper. (9-18)

FUEL ADDITIVES: Three types of additives were investigated - ignition improvers, those that catalyze carbon and spray dispersion improvers. One finding was that additive effectiveness differed greatly between the 2.1 liter and the 5.7 liter engines used in this study. Both catalysts or ignition improvers were effective in reducing particulates in the 2.1 liter engine. The 5.7 liter engine required a blend of these two types of additives for significant particulate reduction. So far no additive has been found that significantly reduces particulates in vehicle tests on the Federal Test Procedure (FTP). (19-21)

EMULSIONS: Water-in-fuel emulsions have been shown to reduce particulate emissions. The mechanism is believed to involve the shattering of the oil droplet by the water when it becomes steam upon injection. This break-up of the oil droplet is said to improve air-fuel mixing. The role of the OH radical is not known at this time but may also be significant. Microemulsions and macroemulsions with and without surfactants have been briefly studied in this investigation. We have tested emulsions of up to 30% water. The use of water-in-fuel emulsions usually resulted in lower particulates and NOx. Unfortunately, a way to overcome the increased HC has not been found. Methanol emulsions of up to 20% have also been briefly tested and found worthy of further study because they do not increase hydrocarbon emissions. (22-27)

HEATED FUEL: The objective in heating the fuel prior to injection was to promote burning that was less droplet-like and more vapor-like and thereby achieve a more complete combustion. The fuel lines were electrically heated in this experiment. Injector malfunctions above 260°F prevented proper evaluation of the potential for heated fuel. (28-30)

CARBON-14 FUEL TRACER STUDY

Early in these fuel studies it was decided to learn which constituents of the fuel (i.e. paraffins, aromatics or olefins), if any, were most responsible for particulates. (31-34) The method selected was to (1) blend radioactive carbon-14 tagged fuel constituents one at a time with our standard baseline D-2 fuel, (2) run these fuels in the engine and collect all particulates and (3) analyze the particulates for the presence of carbon-14.

Seven carbon-14 tagged compounds were used. The three paraffins - decane, hexadecane and octadecane - were selected to give an indication of the effect of boiling point on particulates. The olefin, decene, was selected because of its low boiling point. One aromatic, xylene, was selected because of its low boiling point. The other two aromatics were two forms of 1-methylnaphthalene chosen to gain an insight into particulate formation. In one of the naphthalene compounds a carbon in the ring was labeled. In the other compound a carbon in the methyl radical was labeled. Table 2 gives specific information on the compounds used.

Table 2 - Petroleum compounds used in carbon-14 fuel tracer studies.

COMPOUND	CARBON NUMBER	FORMULA & STRUCTURE
DECANE	C-10	$CH_3-(CH_2)_8-{}^{14}CH_3$
HEXADECANE	C-16	$CH_3-(CH_2)_{14}-{}^{14}CH_3$
OCTADECANE	C-18	$CH_3-(CH_2)_{16}-{}^{14}CH_3$
DECENE	C-10	$CH_2{=}CH-(CH_2)_7-{}^{14}CH_3$
ORTHO METHYL XYLENE	C-8	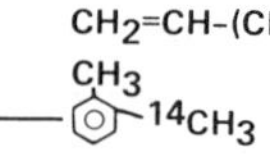 CH_3, ${}^{14}CH_3$
1-METHYLNAPHTHALENE (RING LABEL)	C-11	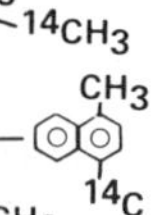 CH_3, ${}^{14}C$
1-METHYLNAPHTHALENE (METHYL LABEL)	C-11	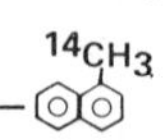 ${}^{14}CH_3$

When one of the tracer compounds was blended with the test fuel, the quantity added was so small that the general physical properties of the fuel were unchanged. The level of radioactivity in the fuel was kept below the maximum permissible exempt concentration of 8 x 10^{-3} microcuries per ml. When this fuel was burned in the engine, part of the tracer became carbon dioxide and was diluted to below the maximum permissible level of 5×10^{-9} microcuries/cc by the test cell exhaust system. The portion that became particulate was caught on an 8 x 10 inch filter mounted in the tunnel to capture all particulates. Approximately 4 minutes sample collection time was required per test. Particulate samples were sent to a commercial laboratory for analysis of carbon-14 content.

The engine used in this series of tests was a 2.1 liter indirect injection Opel. This is a four-cylinder engine but it was modified so that only cylinder number one was operating. Tests were at the equivalent of 25 mph road load and 55 mph road load for each fuel. The results as seen in Figure 1 show clearly that: (1) each of the compounds that we chose made a contribution to particulates, (2) particulate emissions from the paraffins were independent of boiling point over the range we tested, 174°C to 317°C, (3) as a group, aromatics were the greatest contributors to particulates with ring-labeled 1-methyl naphthalene contributing the most of those tested.

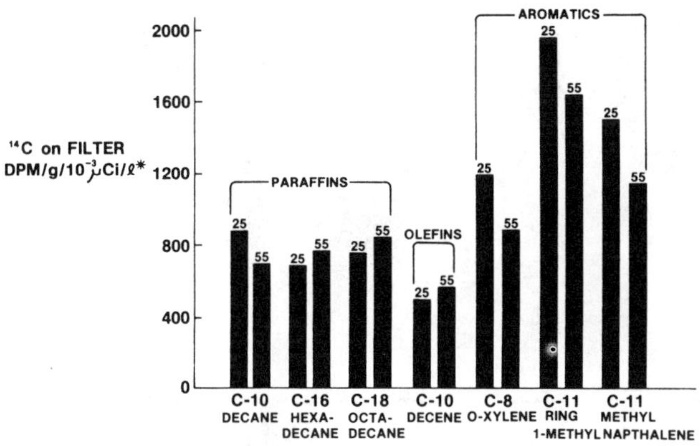

Fig. 1 - Contribution of various components of diesel fuel to particulate emissions. Results are adjusted to a constant quantity of carbon-14 in the fuel for each fuel component. Therefore they indicate the relative probability of each fuel component contributing to particulate emissions. *Disintegrations per minute carbon-14, per gram particulates, per 10^{-3} microcuries carbon-14, per liter of fuel.

Soxhlet extractions were performed on the particulates from these tests. Both the insoluble carbonaceous fractions and the soluble fractions were checked for radioactivity. The radioactivity levels were too low for exact determinations but it was clear that in all cases the major portion of the carbon-14 went to the insoluble fraction. The remainder, probably less than 20%, went to the soluble fraction.

FUEL BLEND TEST - PROCEDURE AND EQUIPMENT

Our early experiments in fuel modification and its effect on particulate showed that of the four techniques tried - additives, fuel blends, emulsions and heated fuels - the greatest particulate reductions occurred during the fuel blend experiments. Accordingly, it was decided to greatly expand the scope of these experiments. Many of the oil companies cooperated by supplying us with a number of petroleum products to test in our engine. The characteristics of the 46 fuels included in this test are listed in Table A-1 found in Appendix A. Figure 2 shows the range of distillation characteristics these fuels represented.

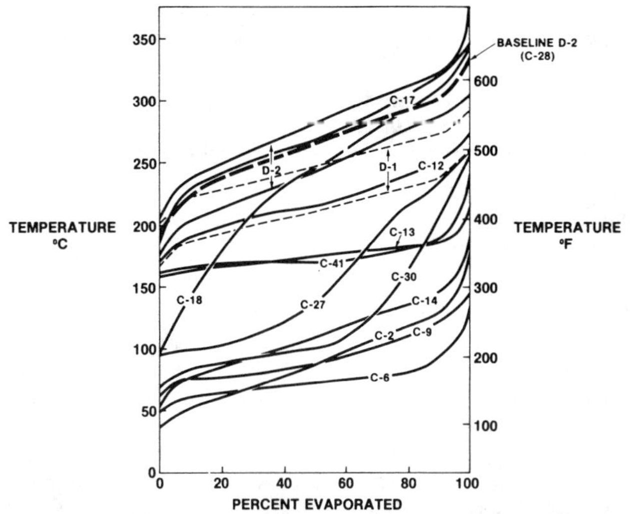

Fig. 2 - Distillation range of some of the fuels tested this diesel fuel study. Range of Federal specification D-1 and D-2 fuels are also shown.

The objective of these tests, of course, was to learn the relationship between fuel characteristics and particulate emissions. Because of the importance of the Federal Test Procedure (FTP), dynamometer loads were tailored to give an indication of how each fuel might perform in a vehicle on the FTP. This was done by analyzing the loads and speeds of the engine during the FTP, selecting a number of representative speeds and loads and applying an appropriate weighting factor to each of these operating modes. The five modes selected are shown in Table 3 which also shows the weighting factors used.

Table 3 - Engine loads and weighting factors used in analyzing test fuels.

TEST POINT (MPH)	SPEED RPM	LOAD FT.LBS.	LOAD Nm	WEIGHTING FACTOR
40 ROAD LOAD	1340	71.0	96.2	.19
IDLE	600	24.5	35.0	.43
45 LIGHT LOAD	1500	40.0	55.0	.22
25 ROAD LOAD	920	57.0	77.2	.10
55 HEAVY LOAD	1780	129.0	174.8	.06

The engine used in this series was an Oldsmobile 5.7-liter, indirect injection eight-cylinder, four-cycle, light-duty diesel. It was similar to a '78 and '79 production engine with an important exception: Poppet-type injectors were substituted for the two-hole pencil-type injectors usually found in this engine. This change necessitated the use of a different injection pump and revised fuel lines. The injector change made the engine similar to, but not identical to, engines that will be marketed in the early 1980s. Additional details are given in Table 4. It can be seen that by tailoring the dynamometer tests toward the FTP and by choosing an engine configuration of advanced design, we were attempting to obtain test results that were as meaningful as possible with respect to future particulate emission legislation.

Table 4 - Description of engine used in study of various diesel fuels.

TYPE	1978 OLDS V-8 MODIFIED
DISPLACEMENT	350 CID, 5.7 LITER
COMPRESSION RATIO	22.5:1
COMBUSTION SYSTEM	INDIRECT INJECTION
ROOSA MASTER DISTRIBUTOR PUMP	DB 2825 PC
INJECTORS	C.A.V. POPPET
INJECTION TIMING	STANDARD, +5°, -5°

Injection timing requirements vary from fuel to fuel. To partially overcome this problem, at each load each fuel was tested at three injection timings: standard, 5 engine degrees advance and 5 engine degrees retard. All together each fuel was tested at fifteen points - five loads times three points per load. Particulates were collected for 20 minutes at each point and, with the time required to take other measurements, flush fuel, etc., each fuel required about eight hours for performance evaluation.

In this series of fuel tests the exhaust from one cylinder was fed to a 300 CFM constant volume tunnel. Particulate and other emission readings are all on the basis of this one cylinder. Fuel consumption, however, was based on input to all eight cylinders. The equipment used in these tests is itemized in Appendix B. To separate the soluble portion of the particulate from the sooty or carbon portion, Soxhlet extractions were performed according to the procedure outlined in Appendix C. (35-57)

As a means of monitoring possible changes in the engine condition, baseline tests were run after each set of five or six fuels. Before tests were begun several hundred gallons of baseline D-2 were set aside as a control fuel. This fuel was designated as C-28 and its characteristics are found in Table A-1 in Appendix A. These baseline values were later used in normalizing all particulate and emission data before analysis.

It should be noted that some fuels were tested at two cetane levels. Fuels labeled C-6 and C-7 in Appendix Table A-1 are an example of this. Fuel C-6, a benzene raffinate, as received was found to have a cetane number of 27.5 and the fuel was tested in the engine at this cetane number. An outside laboratory performed cetane improver sensitivity tests which permitted us to determine the amount of cetane improver needed to increase the cetane number to 47. Fuel C-6 then with 2.6% cetane improver became test fuel C-7 which had a cetane number that was closer to the current national average of diesel fuels.

FUEL BLEND TEST - RESULTS

The test engine was run on a three shift basis for over two months in the evaluation of the test fuels. Figure 3 is an overview of this test series. It can be seen that the variation in particulates with the baseline fuel C-28 amounted to about ± 11% over the duration of the test. This was felt to be quite good stability when one considers the wide variety of petroleum products that went into the engine. In this graph the particulates are in percent of the baseline particulates for Data Set Number 18. Each data point represents the weighted average of the five loads listed in Table 3.

Note in Figure 3 that a wide variation in particulate emissions was observed between the

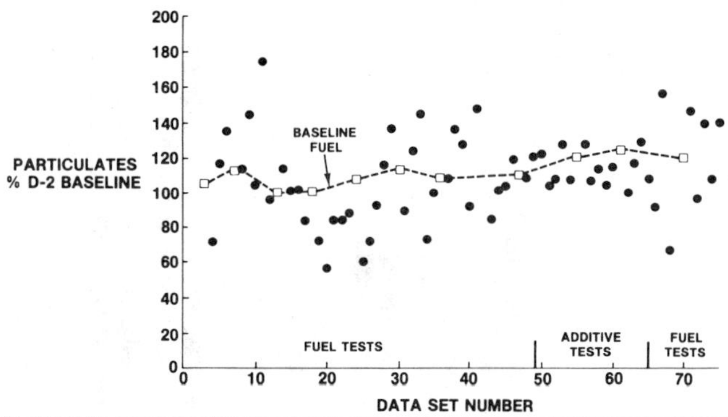

Fig. 3 - Overview of fuel test series showing variation of baseline fuel particulate measurement with time. Each data point is the weighted average of the five load points given in Table 3. Some diesel additives were tested during this test series but were not very effective in reducing particulates.

test fuels: The cleanest fuels had a particulate level about half that of the baseline D-2; some fuels had up to 80% higher particulates. This fuel test series was interrupted briefly for a short series of fuel additive tests. Note that, as pointed out in the previous section, the additives were not as effective in reducing particulates as the fuel blends.

The test results were next normalized in an attempt to remove the effect of slight changes in the engine over the duration of the test. In the curves that follow, all results are given in arbitrary units. These units are proportional to the weight of the emission divided by work done or distance traveled. Low numbers are the best numbers. In the case of economy, the arbitrary units are proportional to work done or distance traveled divided by volume of fuel. Here a high number is best. Data to be analyzed was chosen on the basis of the basic engine timing that gave the lowest weighted particulates.

At this point in the study, test data had been obtained on a large number of fuels, and the physical characteristics of these fuels were known (Table A-1). The next step was to employ multiple linear stepwise regression analysis to learn which fuel characteristics appeared related to the various emissions. In this analysis the dependent variables were the particulates, the other emissions and fuel consumption. The independent variables were the fuel characteristics - specific gravity, 90% boiling point, viscosity, etc. These fuel variables are of course not independent of each other; in fact they are closely related in some cases. Table 5 lists the dependent and independent variables and also gives the range of each fuel characteristic.

Table 5 - Dependent and independent variables used in multiple stepwise linear regression analysis of fuel test results and also range of fuel physical characteristics.

FUEL TEST RESULTS: DEPENDENT VARIABLES	FUEL CHARACTERISTICS: Physical Characteristic	INDEPENDENT VARIABLES Range of Test Fuels
PARTICULATES	SPECIFIC GRAVITY	.678 - .879
	VISCOSITY	.50–3.5 cs @ 40°C
HC	10% BOILING POINT	48–233°C
	50% BOILING POINT	71–267°C
NOx	90% BOILING POINT	92–365°C
	CETANE NUMBER	21–80
ECONOMY	AROMATIC CONTENT	0–95%
	API GRAVITY	27.13–77.20
SOLUBLES	SULFUR CONTENT	.002–1.25%

The regression analysis showed that aromatic content and 90% boiling point were the fuel characteristics most closely related to particulates. Figures 4 and 5 show this relationship. Here particulate index is proportional to the weighted sum of the particulates collected on our five load schedule. It should be pointed out that other fuel volatility factors such as the 10% and 50% boiling points also showed a relationship to particulates. These had a weaker correlation to particulates than the 90% boiling point, however, and it is a characteristic of the stepwise regression analy-

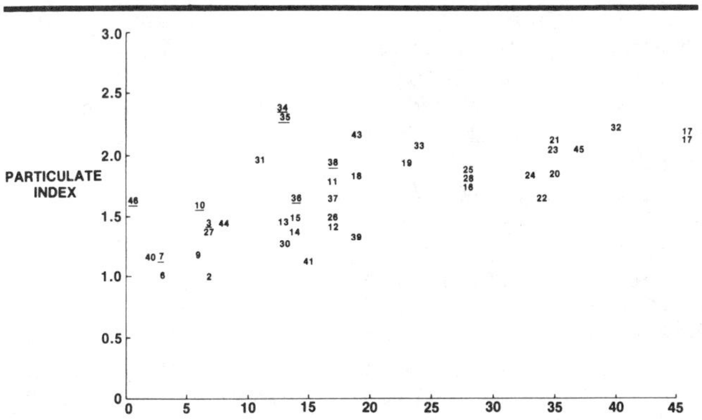

Fig. 4 - Particulate index in arbitrary units versus fuel aromatic content. Underlined numbers indicate fuels to which cetane improver had been added.

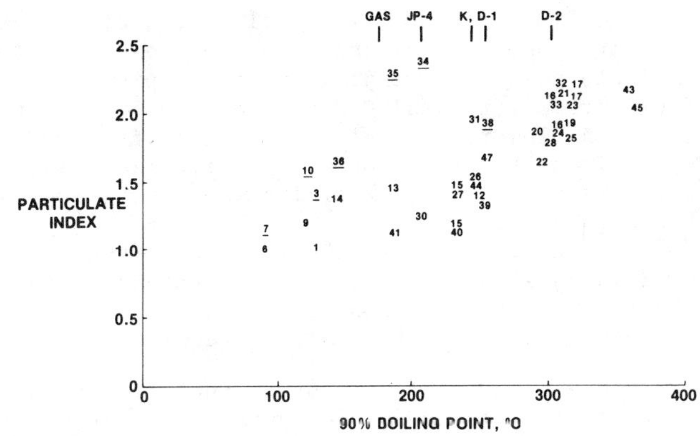

Fig. 5 - Particulate index in arbitrary units versus fuel 90% boiling point. Underlined numbers indicate fuels to which cetane improver had been added. At the top of the chart are some representative 90% boiling points.

sis to choose the independent variable with the strongest correlation and then exclude similar but weaker independent variables.

Figures 4 and 5 in conjunction with Table A-1 also show that there is another factor that is related to particulate output. In every case where a cetane improver was used to increase the cetane number of one of the light fuels, particulate emissions increased.

In using the regression analysis the first step was to remove all tests in which a cetane improver was used. This improved the quality of the analysis significantly. The resulting mathematical expression for particulate index is as follows:

$$\text{Particulate Index} = 0.57 + .00343\ (90\%\text{B.P. }^{\circ}\text{C}) + .0102\ (\%\ \text{Aromatics}) \qquad (1)$$

This expression is shown graphically in Figure 6.

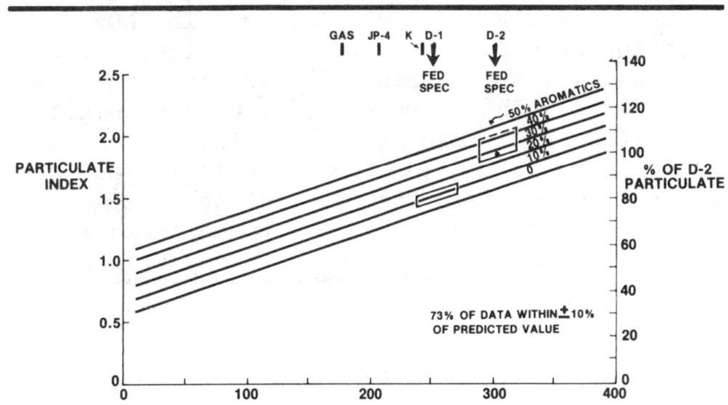

Fig. 6 - Graph of Equation 1 relating particulate index to fuel aromatic content and fuel 90% boiling point. Boxes indicate position of Federally specified D-1 and D-2 fuels.

In determining how well the above expression can predict the particulate level of a fuel, a comparison was made between predicted particulates and measured particulates for each of the test fuels. The result was that 73% of the measured data was found to be within ± 10% of the predicted value. The boxes on this figure indicate the characteristics of the Federally specified D-1 and D-2 diesel test fuels. The relative positions of the boxes on these curves indicate that in a comparison of a typical D-1 fuel versus a typical D-2 fuel one would expect about 25% lower particulate emissions from the D-1. This is the approximate particulate reduction observed at this facility when comparing these two fuels.

Let us turn now to the results of the regression analysis on the other dependent variables: HC, NOx, fuel economy and solubles. Table 6 summarizes the analysis results. The

Table 6 - Results of regression analysis of fuel characteristics (independent variables) versus emissions, economy and solubles (dependent variables). The R^2 column indicates the fraction of variation accounted for in the analysis. The other columns indicate the degree of correlation with the dependent variables. In all cases 1.0 would indicate perfect correlation.

	R^2	SPECIFIC GRAVITY	VISCOSITY	90% B.P.	CETANE NUMBER	AROMATIC %	SULFUR %
PARTICULATES INDEX	.814			.88		.81	
HC INDEX	.722	-.55	-.35	-.53	-.37		
NOx INDEX	.502	.40		.40		.44	
ECONOMY INDEX	.747	.65	.49	.61	.52	.45	
SOLUBLES %	.309				-.56		

first column, R^2, is the coefficient of determination. This is a statistical term meaning the fraction of the variation that has been accounted for by the analysis. An R^2 of 1.0 indicates all variability accounted for. The .814 R^2 value for particulates indicates that the fuel characteristics of aromatics and 90% boiling point account for 81% of the difference in particulate readings between the fuels, and is high enough to be meaningful. The R^2 values 0.722 and 0.747 for hydrocarbons and economy indicate fairly good correlation. The NOx correlation of 0.502 indicates only slight correlation and is useful mainly for identifying possible trends. The R^2 value of 0.309 for solubles indicates that there was no significant correlation between the fuel characteristics and the percent solubles in the particulates.

The slight correlation between aromatic content and NOx index is illustrated in Figure 7.

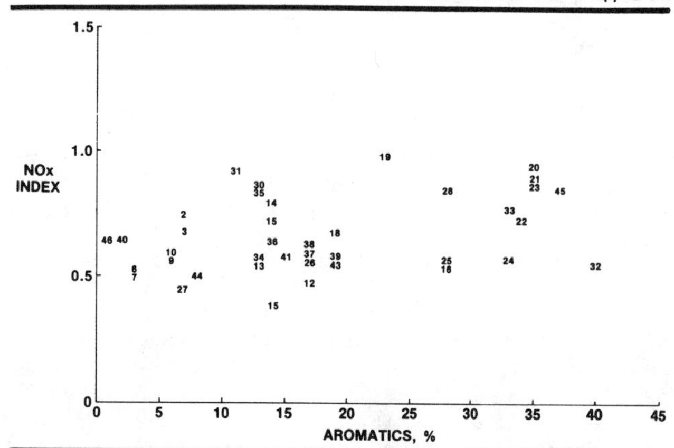

Fig. 7 - NOx index versus fuel aromatic content showing relatively poor correlation.

A better correlation is found between specific gravity and economy index as shown in Figure 8. This is expected since our economy index is an indication of work done per unit volume (not weight). It should be pointed out, however, that some of the lighter fuels do extremely well. To illustrate this, compare fuel C-44, a blend of benzene raffinate and kerosene, with the baseline fuel C-28. Their densities were .740 and .833 respectively and yet work done per unit volume was approximately the same.

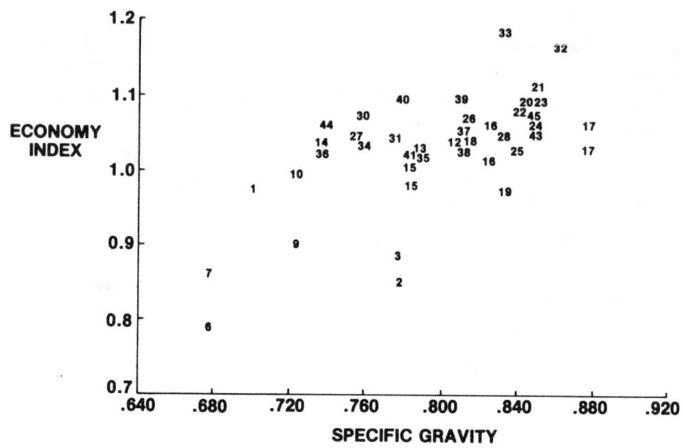

Fig. 8 - Correlation between specific gravity and economy index (work done per unit volume).

VEHICLE TESTS AND SUPPLEMENTAL DYNAMOMETER TESTS

VEHICLE TESTS: For comparison purposes fuel C-40 (a light, narrow distillation range, low aromatic product) was tried in two vehicles on the Federal Test Procedure. The vehicles were equipped with 5.7 liter engines similar to the one used in the fuel test program. The engines in the vehicles, however, had conventional two-hole pencil injectors rather than the poppet-type injectors previously used.

When tested with fuel C-40, particulate emissions averaged 0.474 grams per mile in three FTP tests per vehicle. Using the baseline D-2 fuel, C-28, the particulate emissions in three tests per vehicle averaged 0.826 grams per mile or 74% higher. This tended to substantiate our dynamometer tests that predicted lower particulates from fuel C-40. However, there is no certainty that this reduction would carry over the same degree to low particulate engine configurations.

EFFECT OF AROMATICS: Two series of fuel blend tests were run on a two-hole pencil injector-equipped 5.7 liter engine on the dynamometer. In one series the baseline D-2 fuel, C-28, was blended with various amounts of the 95% aromatic product, C-42. In this way blends of 28%, 35%, 45%, 55% and 65% aromaticity were formulated. Each formulation was tested at the five loads shown in Table 3 and the results weighted as described previously. The results are shown in Figure 9. Note that there was little change in particulates until the aromatic level exceeded 55%; then the increase was dramatic. Hydrocarbons behaved similarly but NOx emissions climbed steadily as aromatics increased.

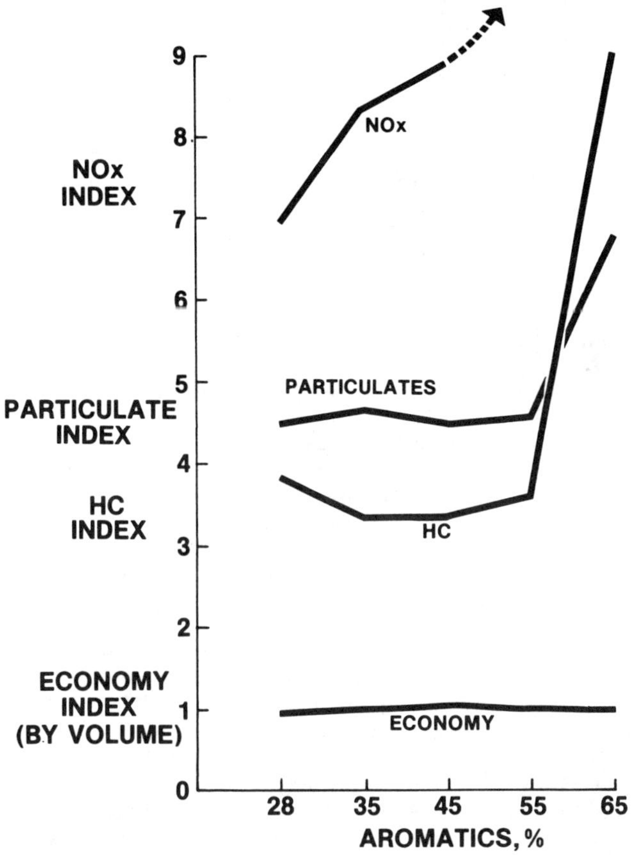

Fig. 9 - Emission and economy variation with fuel blends of increasing aromaticity and specific gravity starting with D-2 fuel.

In the second series C-42 (the 95% aromatic product) was blended with C-40 (a 2% aromatic product). Blends of 2%, 10%, 20%, 30% 40%, 50% and 60% aromatics were formulated and tested. The five standard weighted load points were used as before, and the results are presented in Figure 10. In this test series NOx, HC and particulates all increased steadily with increasing aromaticity while the fuel consumption remained relatively constant on a volume basis in spite of the density change. A 60% aromatic blend was tried, but engine operation was too unstable to obtain meaningful data.

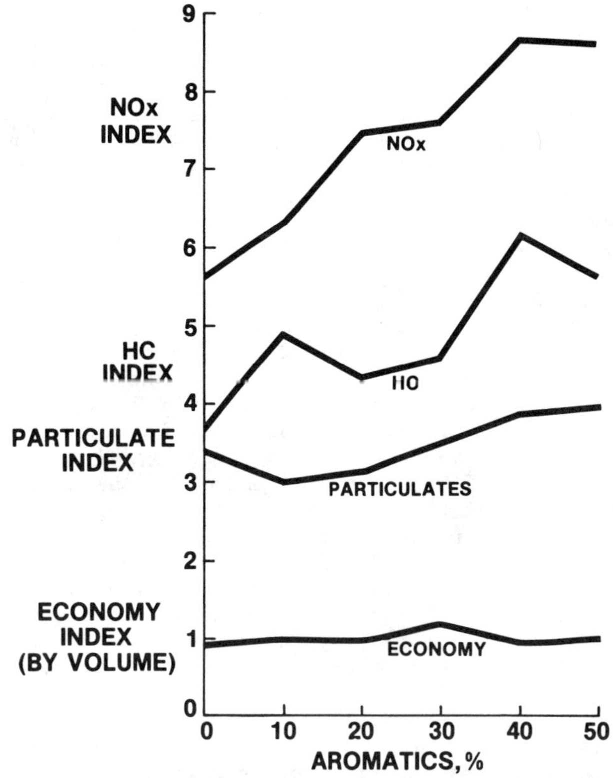

Fig. 10 - Emission and economy variation with fuel blends of increasing aromaticity and specific gravity starting with low-aromatic narrow-cut product.

CONCLUSIONS AND SUMMARY

In attempting to reduce particulate emissions through fuel modifications, four approaches were investigated: additives, water-in-fuel emulsions, heated fuels, and experimental fuel blends. Greatest particulate reductions were observed by fuel blending, and this led to a study of forty-six conventional and unconventional diesel fuels. The objective was to learn the relationship, if any, between fuel characteristics and diesel engine emission output.

We found that there is a relationship between fuel characteristics and the quantity and quality of emissions of a diesel engine. The results of the experiments can be summarized as follows:

1. As a prelude to these fuel studies, seven carbon-14 tagged fuel constituents were traced through a diesel engine. The tests showed that paraffins, olefins and aromatics all contributed to particulates but that ring carbons in aromatics made the greatest contribution.

2. Particulates were found to be related mainly to two fuel characteristics: (1) aromatic content and (2) volatility, especially as described by the 90% boiling point. Low aromatic content and/or low 90% boiling point was accompanied by low particulates.

3. When an ignition improver was blended with certain light petroleum products in order to raise the cetane number to about 46, there

was a significant increase in particulates that was independent of engine injection timing.

4. One of the fuels was tested in two vehicles on the Federal Test Procedure. The particulate reduction observed was similar to that from the dynamometer schedule. However, there was no certainty that this reduction would carry over to low particulate engine configurations.

5. Two further experiments were conducted to confirm the role of aromatics in diesel emissions. A high aromatic product was mixed with D-2 and also with a very low aromatic product to various levels of aromaticity. As aromatic content increased, so did particulates, HC and NOx.

6. In the analysis of forty-six fuels, hydrocarbons were found to be related mainly to four fuel characteristics: specific gravity, viscosity, 90% boiling point and cetane number. High values of these characteristics were accompanied by low hydrocarbons.

7. Oxides of nitrogen in the forty-six fuel analysis did not correlate well with fuel characteristics but appeared related to aromatic content, specific gravity and 90% boiling point. Low values of these characteristics were accompanied by low NOx.

8. Fuel economy on a work done per unit volume basis was found to be related to the following fuel characteristics: specific gravity, 90% boiling point, cetane number, and viscosity. Raising these values improved fuel economy.

9. A further observation on fuel economy was that most lighter fuels burned with relatively high brake specific fuel consumption and this was enough to overcome in part their lower specific gravity when comparing fuels on a work done per unit volume basis.

10. Soluble content of the particulate did not correlate well with fuel characteristics. Our analysis showed only a very tenuous relationship with cetane number; soluble content decreased as cetane number increased.

ACKNOWLEDGMENTS

The authors acknowledge with sincere thanks the careful work of the three dynamometer operators who tested these fuels - Donald Veres, James Earle, and Michael Kaurich. They appreciate the help of William Mayer and Robert Hill of the GM Research Laboratories Analytical Chemistry Department in procuring, blending and analyzing carbon-14 tracer fuels and particulates. Thanks are due Lewis Allison and Levi Woods, Jr. for their help in computer programming and data analysis respectively. Gilliam Clark and the Ethyl Corporation were very cooperative in providing diesel fuel additives and in making diesel fuel cetane number determinations of most of the fuels tested. A number of oil companies were very helpful in providing non-standard experimental fuels and these include Texaco, Gulf, Mobil, Amoco, Exxon and Shell. And finally the authors extend their thanks to the project sponsor, David Dimick, Executive Engineer, Advance Product Engineering, for his helpful suggestions and encouragement.

REFERENCES

1. M.L. Brubacher, "Reduction of Diesel Smoke in California", SAE Paper No. 660548, August 8-11, 1966.

2. A.W. Carey, Jr., "Smoke Reduction in Diesel Engines", SAE Paper No. 670224, February 3, 1967.

3. H. Hiroyasu and T. Kadota, "Models for Combustion and Formation of Nitric Oxide and Soot in Direct Injection Diesel Engines", SAE Paper No. 760129, February 23-27, 1976.

4. J.A. Breslin, A.J. Strazisar and R.L. Stein, "Size Distribution and Mass Output of Particulates From Diesel Engine Exhausts", [Washington] U.S. Bureau of Mines, 1976. Report No. 8141r

5. W.Bryzik and C.O. Smith, "Relationships Between Exhaust Smoke Emissions and Operating Variables in Diesel Engines", SAE Paper No. 770718.

6. G.L. Scofield, "Diesel Engines and Their Particle Signatures", SAE Paper No. 780426, June 15, 1977.

7. C.T. Hare and T.M. Baines, "Characterization of Particulate and Gaseous Emissions from Two Diesel Automobiles as Functions of Fuel and Driving Cycle", SAE Paper No. 790424, February 26 - March 2, 1979.

8. D.L. Stivender and C.A. Amann, "The Formation and Oxidation of Diesel Soot -- A Rudimentary Study", Comb. Inst., April, 1979.

9. D. Lyon, A.H. Howland, W.L. Lom, "Controlling Exhaust Emissions from a Diesel Engine by LPG Dual Fuelling", Instn Mech Engrs, July 7, 1971. C126171.

10. S. Fukazawa, T. Murayama and Y. Fujiwara, "An Experimental Study on Exhaust Smoke in Diesel Engines", SAE of Japan, Bulletin no. 6,22-32, April, 1974.

11. E.E. Milkins, H.C.Watson, M. Considine and G. Sprigg, "Emissions Control by Dual-Fuelled Operation of a Diesel Engine", Univ. of Melbourne.'75 Int. Clean Air Conf.

12. H. Tesarek, "Investigations Concerning the Employment Possibilities of the Diesel-Gas Process for Reducing Exhaust Emissions, Especially Soot (Particulate Matters), SAE Paper No. 750158, February 24-28, 1975.

13. C.T. Hare, K.J. Springer and R.L. Bradow, "Fuel and Additive Effects on Diesel Particulate Development and Demonstration of Methodology", SAE Paper No. 760130, February 23-27, 1976.

14. R.L. Bechtold and S.S. Lestz, "Combustion Characteristics of Diesel Fuel Blends Containing Used Lubricating Oil", SAE Paper No. 760132, February 23-27, 1976.

15. F.J. Hills, C.G. Schleyerbach, "Diesel Fuel Properties and Engine Performance", SAE Paper No. 770316, February 28 - March 4, 1977.

16. G.P. Gross and K.E. Murphy, Jr., "The Effects of Diesel Fuel Properties on Performance, Smoke, and Emissions", ASME Paper No. 78-DGP-26, July 21, 1978.

17. L.E. Frisch, J.H. Johnson and D.G. Leddy, "Effect of Fuels and Dilution Ratio

on Diesel Particulate Emissions", SAE Paper No. 790417, February 26-March 2, 1979.

18. J. Moon, "Fuel-Tolerant Engine Developments", Diesel & Gas Turbine Progress Worldwide, April, 1979.

19. C.G. McCreath, "The Effect of Fuel Additives on the Exhaust Emissions from Diesel Engines", Combustion and Flame, 17, 359-366, 1971.

20. K.C. Tessier and H.E. Bachman, "Fuel Additives for the Suppression of Diesel Exhaust Odor and Smoke, Part I: Proposed Mechanism for Smoke Suppression", ASME, December 1-5, 1968.

21. K.C. Tessier and H.E. Bachman, "Fuel Additives for The Suppression of Diesel Exhaust Odor and Smoke, Part II: Field Trials, ASME No. 68-WA/DGP-5, December 1-5, 1968.

22. K. Muller-Dethlefs and A.F. Schlader, "The Effect of Steam on Flame Temperature, Burning Velocity and Carbon Formation in Hydrocarbon Flames", Comb. & Flame, 27, 1976.

23. C.K. Law, "A Model for the Combustion of Oil/Water Emulsion Droplets", Combustion Science and Technology, Vol. 17, pp. 29-38, 1977.

24. T. Murayama, M. Tsukahara, Y. Morishima, N. Miyamoto, "Experimental Reduction of NOx, Smoke, and BSFC in a Diesel Engine Using Uniquely Produced Water (0-80%) to Fuel Emulsion", SAE Paper No. 780224, February 27-March 3, 1978.

25. R.A. Walter, "The Emissions and Fuel Economy of a Detroit Diesel 6-71 Engine Burning a 10-Percent Water-In-Fuel Emulsion", Defense Documentation Center, Alexandria VA, Report No. CG-D-10-78, July, 1978.

26. J.O. Storment and C.W. Coon, "Single-Cylinder Diesel Engine Tests with Unstabilized Water-In-Fuel Emulsions", Defense Documentation Center, Alexandria, VA, Report No. CG-D-13-78, August, 1978.

27. H.S. Fogler, S.R. Reddy, M.L. Storbeck, J.B. Cross and S.G. Siegmund, "Physical Properties and Stability of Water In Fuel Oil Emulsions", National Technical Info Center, Springfield, VA, Report No. DOT/RSPA/DPB-50/78/17, March, 1979.

28. H.C. Gerrish and B.E.Ayer, "Influence of Fuel-Oil Temperature on the Combustion in a Prechamber Compression-Ignition Engine", National Advisory Committee for Aeronautics, April 1936. No. 565

29. J.H. Morgenthaler, "Surface Vaporization Studies with No. 2 Fuel Oil", API Research Conference on Distillate Fuel Combustion, June 19-20, 1962. CP62-8

30. R.W. Temple-Pediani, "Effect of Pre-Injection Fuel Temperature Upon Diesel Engine Ignition Delay and Soot Emission", Instn Mech Engrs, 1973. Vo. 187 32/73

31. W.H. Lipkea, J.H. Johnson and C.T. Vuk, "The Physical and Chemical Character of Diesel Particulate Emissions". SAE SP-430, 1978.

32. F.Black and L. High, "Diesel Hydrocarbon Emissions, Particulate and Gas Phase". EPA Symposium on Diesel Particulate Emissions Measurement and Characterization, May 17-18, 1978, Ann Arbor, MI.

33. M.L. Studebaker, "Carbon Black and Reinforcement of Rubber, Part 1. Carbon Black and its Properties". Phillips Petroleum Company R and D. Center, Bartlesville, OK, 1976.

34. P.H. Johnson and C.R. Eberline, "Carbon Black, Furnace Black" Encyclopedia of Chemical Processing and Design, Vol. 6, J.J. McRetta, Editor, Marcel Dekker, Inc. Publ. 1977.

35. W. D. Davis, "Continuous Mass Spectrometric Analysis of Particulates by Use of Surface Ionization", Environmental Science & Technology Vol. 11, June 1977.

36. E.F. Funkenbusch, D.G. Leddy and J.H. Johnson, "The Characterization of the Soluble Organic Fraction of Diesel Particulate Matter", SAE Paper No. 790418, February 26-March 2, 1979.

37. F. Black and L. High, "Methodology for Determining Particulate and Gaseous Diesel Hydrocarbon Emissions", SAE Paper 790422, February 26-March 2, 1979.

Appendix A

A-1

Table A-1-Physical Characteristics of Test Fuels

Fuel Sample Number	C-1	C-2	C-3	C-4	C-5	C-6	C-7	C-8	C-9	C-10
Name of Fuel	Normal Heptane	Straight Run Gasoline	Straight Run Gasoline	Stoddard Solvent	Stoddard Solvent	Benzene Raffinate	Benzene Raffinate	Gasoline Component	Reformer Feed	Reformer Feed
Distillation $^{\circ}$C										
IBP		34.5	34.5	158.5	158.5	48.0	48.0	32.0	63.0	63.0
10%		53.5	53.5	166.0	166.0	62.5	62.5	48.0	78.5	78.5
50%		86.0	86.0	173.0	173.0	72.0	72.0	70.5	90.5	90.5
90%		130.0	130.0	184.0	184.0	92.0	92.0	96.0	121.5	121.5
EP	209	179.5	179.5	217.5	217.5	145.0	145.0	144.0	174.0	174.0
Gravity, $^{\circ}$API	74.1	68.64	68.64	48.07	48.07	77.20	77.20	74.77	64.21	64.21
Cetane Number	54	27.0	45.0	38.8	45	27.5	47.0	21.0	27.5	45.0
Sulfur, WT%	.005	.02	.02	.005	.005	.005	.005	.01	.005	.005
Flash Point, $^{\circ}$C	-2	21.1	21.1	48.9	48.9	20	20	21.1	20	20
Viscosity at 40.0°C		.56@23°C	.56@23°C	1.24@23°C	1.24@23°C	.50	.50	.53@23°C	.64	.64
Gum, mg/100ml	.5	2.8	2.8	3.0	3.0	1.8	1.8		1.4	1.4
Paraffins, %	100	89	89	80	80	94	94	96	93	93
Olefins, %	-	4	4	1	1	3	3	1	1	1
Aromatics %		7	7	19	19	3	3	3	6	6
H/C	2.286	2.12	2.12	1.94	1.94	2.24	2.24	2.18	2.09	2.09
Specific Gravity	.6880	.707	.707	.7880	.7880	.678	.678	.686	.723	.723
Energy, MJ/Kg	44.513	47.61	47.61	46.47	46.47	48.81	48.81	-----	47.62	47.62
Cetane Improver, %V	-----	-----	2.2	-----			2.6	-----	-----	1.71

A-2

Table A-1-Physical Characteristics of Test Fuels

Fuel Sample Number	C-11	C-12	C-13	C-14	C-15	C-16	C-17	C-18	C-19	C-20
Name of Fuel	Diesel No. 1	Kerosene A	Oleum	Naptha	ISR-LFE	SRA-ALFE	SRA-AHTE	SRA-LFE	2D SRA	2D AT
Distillation $^{\circ}$C										
BIP	170.5	169.0	158.0	49.5	83.0	64.5	175.0	96.0	173.5	176.0
10%	201.0	189.5	164.5	78.5	125.0	133.0	266.5	143.0	219.0	198.0
50%	226.0	215.5	171.5	107.5	171.5	242.0	267.0	246.5	266.5	243.5
90%	254.5	251.0	185.5	145.0	235.0	308.0	319.5	308.5	313.5	296.0
EP	276.0	272.0	213.0	192.0	268.0	333.0	371.5	339.0	336.5	326.5
Gravity, $^{\circ}$API	41.69	43.41	47.84	60.49	48.53	39.81	29.48	41.91	38.16	35.96
Cetane Number	49.0	44.0	38.8	27.0	40.5	42.0	40.0	45.5	52.0	46.5
Sulfur, WT%	.23	.07	.005	.09	.02	.18	.14	.20	.24	.26
Flash Point, $^{\circ}$C	76.7	73.9	57.2	22	29.4	20	107.2	32.2	96.1	90.6
Viscosity at 40.0°C	1.75	1.56	1.27@22°C	.57	1.06	1.59	2.56	1.75	2.68	1.99
Gum, mg/100ml	-----	-----	94.6	2.0	-----	-----	-----	-----	-----	-----
Paraffins, %	79	80	85	85	85	70	53	79	73	62
Olefins, %	4	3	2	1	1	2	1	2	4	3
Aromatics, %	17	17	13	14	14	28	46	19	23	35
H/C	1.90	1.92	1.95	1.98	1.94	1.80	1.60	1.88	1.87	1.77
Specific Gravity	.817	.809	.789	.737	.786	.826	.879	.816	.834	.845
Energy, MJ/Kg	46.28	46.00	46.36	47.61	46.43	45.62	45.15	46.30	45.95	45.75
Cetane Improver, %V	-----	-----	-----	-----	-----	-----	-----	-----	-----	-----

A-3

Table A-1-Physical Characteristics of Test Fuels

Fuel Sample Number	C-21	C-22	C-23	C-24	C-25	C-26	C-27	C-28	C-29	C-30
Name of Fuel	2D BT	2D A	2D B	SRA 50C	SRA 25C	1D SR	50% Heptane 50% Kerosene	Baseline Diesel No.2		JP-4
Distillation $^{\circ}$C										
IBP	184.0	176.5	189.5	181.5	147.0	170.5	95.5	179.0		68.0
10%	221.5	203.0	221.0	213.0	213.5	190.0	100.5	221.5		84.5
50%	268.5	246.0	286.0	256.5	256.5	214.5	136.0	256.0		101.5
90%	313.5	297.0	313.5	307.0	310.5	246.5	233.0	303.0		208.0
EP	337.5	329.0	339.0	329.5	337.5	276.5	261.0	333.5		255.0
Gravity, $^{\circ}$API	34.77	35.96	34.77	35.36	36.95	42.55	56.17	36.35		54.93
Cetane Number	49.5	42.5	46.5	43.0	46.5	47.0	54.0	44.5		30.5
Sulfur, WT%	.30	.27	.29	.31	.28	.11	.01	.31		.005
Flash Point, $^{\circ}$C	96.1	79.4	98.9	87.8	85.0	73.9	21.1	93.3		20
Viscosity at 40.0°C	2.64	2.00	2.65	2.31	2.40	1.48	1.02@23°C	2.40		.69
Gum, mg/100ml	-----	-----	-----	-----	-----	-----	1.4	-----		-----
Paraffins, %	63	63	61	64	69	80	92	68		86
Olefins, %	2	3	5	3	3	3	1	4		1
Aromatics, %	35	34	35	33	28	17	7	28		13
H/C	1.78	1.76	1.77	1.77	1.82	1.90	2.06	1.79		1.97
Specific Gravity	.851	.845	.851	.848	.840	.813	.754	.843		.759
Energy, MJ/Kg	45.64	45.75	45.64	45.69	45.84	46.36	48.06	46.33		47.02
Cetane Improver, %V	-----	-----	-----	-----	-----	-----	-----	-----		-----

A-4 Table A-1-Physical Characteristics of Test Fuels

Fuel Sample Number	C-31	C-32	C-33	C-34	C-35	C-36	C-37	C-38	C-39	C-40
Name of Fuel	25% Heptane 75% Kerosene	No.42	No.52	JP-4	Oleum	Naptha	JP-5	JP-5	Kerosene B	Odorless Level Oil
Distillation °C										
IBP	100.0	199.0	165.0	68.0	158.0	49.5	169.0	169.0	174.0	209.0
10%	116.5	232.5	204.0	84.5	164.5	78.5	193.5	193.5	187.5	214.5
50%	202.0	261.0	253.5	101.5	171.5	107.5	226.5	226.5	214.0	221.5
90%	246.0	311.0	309.0	208.0	185.5	145.0	255.5	255.5	251.5	232.5
EP	264.0	339.0	339.5	255.0	213.0	192.0	268.5	268.5	268.5	244.0
Gravity, °API	50.85	32.84	38.57	54.93	47.84	60.49	42.98	42.98	43.19	49.91
Cetane Number	54	44	50	47	47	47	30	47	45	36
Sulfur, WT%	.06	.36	.26	.005	.005	.09	.06	.06	.03	.005
Flash Point, °C	22	102	85	20	57.2	22	77	77	77	85
Viscosity at 40.0°C	1.03	2.47	2.25	.69	1.27@22°C	.57	1.67	1.67	1.46	2.15
Gum, mg/100ml	-----	-----	-----	-----	94.6	2.0	-----	-----	-----	-----
Paraffins, %	88	57	73	86	85	85	79	79	78	96
Olefins, %	1	3	3	1	2	1	4	4	3	2
Aromatics, %	11	40	24	13	13	14	17	17	19	2
H/C	2.01	2.07	1.86	1.97	1.95	1.98	1.91	1.91	1.89	2.11
Specific Gravity	.776	.861	.832	.759	.789	.737	.811	.811	.810	.780
Energy, MJ/Kg	45.59	45.62	45.78	47.07	46.36	47.61	45.82	45.82	45.92	46.16
Cetane Improver	-----	-----	-----	1.13	.2	2.21	-----	1.18	-----	-----

A-5 Table A-1-Physical Characteristics of Test Fuels

Fuel Sample Number	C-41	C-42	C-43	C-44	C-45	C-46
Name of Fuel	Mineral Spirits #37	Solvent #150	SRA-PHTE	50% Kerosene 50% Benzene Raffinate	Very High End Pt. Fuel	Synthetic Diesel Fuel
Distillation °C						
IBP	161.5	181.0				
10%	165.5	187.0	221	68	215	190
50%	170.5	191.5	280	145	287	260
90%	186.5	201.0	358	250	365	320
EP	225.5	233.0	384	267	397	
Gravity, °API	49.22	27.13	35.20	58.69	35.80	54.20
Cetane Number	39	21	52	36	61	80
Sulfur, WT%	.02	.005	.03	.03	1.25	.002
Flash Point, °C	52	74	20	20	60	60
Viscosity at 40.0°C	.98	1.07	3.36	1.00	3.50	2.20
Gum, mg/100ml						
Paraffins, %	84	2				
Olefins, %	1	3				
Aromatics, %	15	95	19	8	37	0.1
H/C	1.96	1.36	1.84	2.07	1.84	2.1
Specific Gravity	.783	.892	.849	.744	.846	.840
Energy, MJ/Kg	46.40	46.66	45.68	47.37	45.31	48.8
Cetane Improver, %V						

Appendix B

Table B-1 - Emission Equipment Used In Evaluation Of Diesel Fuels

Measurement	Manufacturer	Model No.
HC, Outside Cell	Beckman	400
Inside Cell	"	402
CO	"	864
O_2	"	778
CO_2	"	864
NOx	"	951
Pump, (Particulate Sample Line)	Doerr	1022-V103-G272X
Blower, Tunnel	Fuller	5LV-B
Scale, Filter Weighing	Sartorius	2405

APPENDIX C

SOXHLET EXTRACTION PROCEDURE

The following laboratory apparatus and procedure are recommended for use in the extraction and measurement of organic solubles from diesel particulate filters.

APPARATUS

- o Laboratory microbalance in chamber controlled to approximately 20°C and 45% relative humidity.
- o Soxhlet extraction apparatus - Sargent Welch Catalogue No. S-31265A or equivalent.
- o Low pressure steam supply.
- o Filtered low pressure air or nitrogen supply.
- o Predistilled dichloromethane solvent.
- o Pyrex rods: 3/4 inch dia. x 75 mm (approx) length.
- o 2-3 dram vials.

PROCEDURE

Filter Sample

Particulate samples will be on 142 mm diameter Gelman Type AE fibreglas filters. Numbered filters will be delivered in individual containers marked with the initial and final masses and difference (ie. particulate mass). The particulate mass will range from about 20-60 mg (approx. 1.5-5% of the total filter mass). Teflon coated filters may also be used.

Sample Weighing

Filters and vials shall be stabilized for at least twelve hours at approximately 20°C and 45% relative humidity before each weighing to avoid humidity effects.

Soxhlet Extraction

The recommended Soxhlet extraction setup is shown in Figure C-1. Steam heating is recommended to avoid local overheating. A 3/4 inch diameter pyrex rod inserted into the apparatus with the filter will reduce chamber net volume and increase the siphon cycle rate. A few teflon chips may be used as boiling stones if required to prevent superheating. An insulating wrap of Filterdown or equivalent will minimize heat loss. As diagramed, this setup should operate at about one siphon cycle per three minutes. A total of about 60 cycles is required.

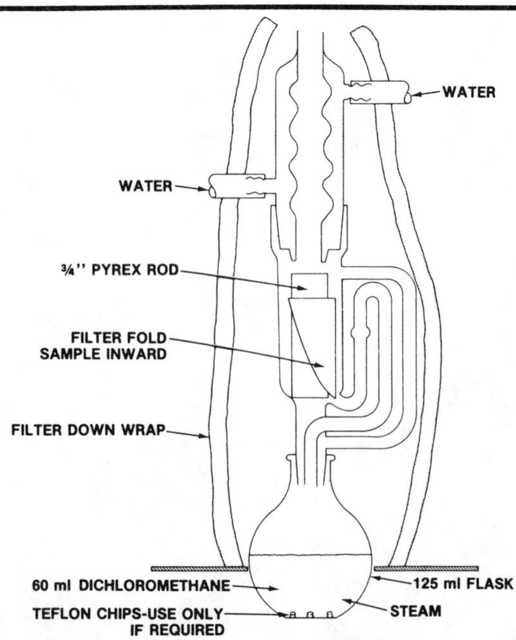

Fig. C-1 - Soxhlet extraction apparatus.

Proceed as follows:

o Clean Soxhlet apparatus (and teflon boiling chips if used) by precycling several times with dichloromethane. Discard solvent.

o Fold preweighed filter with particulates inside. Insert filter and pyrex rod into extraction chamber with filter edges up.

o Add 60 ml of predistilled dichloromethane to 125 ml flask of Soxhlet apparatus. Add a few clean teflon chips as boiling stones if required.

o Adjust steam and cooling water. Cycle for 3 hours.

o Remove filter and lay on foil to dry. Restabilize at 20°C and 45% relative humidity after drying. Reweigh to determine mass loss. Return filter in original container.

o Evaporate the solutions from the flask without heat to approximately 5 ml. (a mild jet of filtered air or nitrogen may be used as shown in Figure C-2 to increase the evaporation rate.)

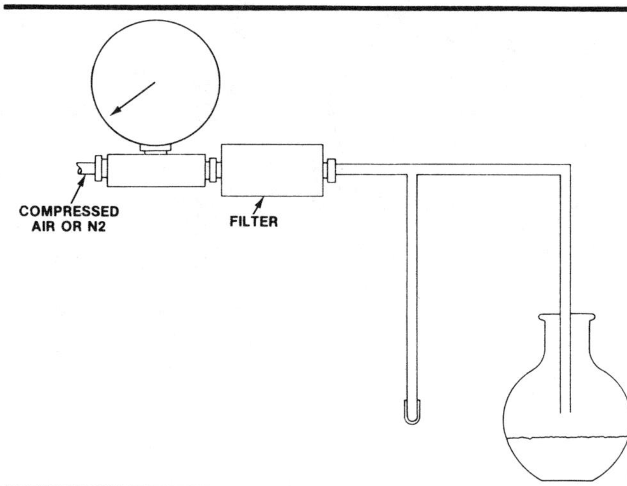

Fig. C-2 - Evaporation apparatus.

Transfer remaining solution to prestabilized and weighed 2 or 3 dram vial. Rinse the flask with 2 ml of predistilled dichloromethane and transfer to same vial. Repeat rinse.

o Evaporate solution in vial to dryness. Stabilize vial for 12 hours at 20°C and 45% relative humidity and reweigh. Cap vial with clean cap and return in upright position.

o Determine total solubles and fraction solubles from vial and filter weighings and total particulate mass.

EXHAUST AFTERTREATMENT

Assessment of Diesel Particulate Control – Direct and Catalytic Oxidation*

M. J. Murphy, L. J. Hillenbrand, and D. A. Trayser
Battelle, Columbus Laboratories
Columbus, OH

J. H. Wasser
Industrial Environmental Research Laboratory
U. S. Environmental Protection Agency
Research Triangle Park, NC

DIESEL ENGINES ARE BECOMING AVAILABLE in increasing numbers in passenger car service because of their good fuel economy in comparison to conventional gasoline engines. Although diesel exhaust is relatively clean with respect to unburned hydrocarbons (HC) and carbon monoxide, it contains particulate emissions that are 30 to 50 times greater than those produced by the catalyst-equipped gasoline engine. Not only will these diesel particulate emissions contribute to already high levels of total suspended particulate (TSP) in urban areas, but certain components of the particulates have been identified as carcinogenic, thereby creating a potentially greater health hazard.

Several approaches to the control of diesel emissions are being pursued by the automotive industry, EPA, and others. These include operating mode modifications, engine design and component modifications, fuel modifications, and exhaust devices.

This paper discusses the technology of and potential devices for: thermal and catalytic oxida- of particulate while suspended in the exhaust stream, thermal oxidation of filter-trapped particulate, catalytic oxidation of particulate, oxidation of particulate trapped by other techniques, and catalytic oxidation of the organic vapors before deposition on the particulate.

The following discussions will center primarily on the application of "add-on" devices in the engine's exhaust system. However, the successful application of a thermal or catalytic oxidation system for particulate or organic vapor control may also depend on operating mode, engine design, or fuel modification for maximum effectiveness. These aspects will also be covered in the discussion.

DIESEL PARTICULATE

The particulate matter emitted by the engine

*Paper 810112 presented at the International Congress and Exposition, Detroit, Michigan, February, 1981.

ABSTRACT

The technology for disposal of diesel particulate by oxidation is discussed. Properties of typical diesel particulate are given, including the size (on the order of 0.1 μm diameter) and the presence of a portion extractable with an organic solvent. Available reaction rate data are used to derive particle lifetimes at various temperatures; these exceed likely exhaust system residence times. The use of catalysts to increase oxidation rates and lower ignition temperatures is discussed. Small amounts of many metals increase the rate of oxidation and lower the ignition temperature. Chemical reactor theory is used to derive ignition and operational characteristics of trap/oxidizers. Special note is taken of the tendency of these devices to go rapidly from a cold unignited state to an ignited state close to the adiabatic flame temperature of the fuel. Design techniques to ameliorate undesirable temperature excursions are presented.

Table 1. Characteristics of Loose Particulate Samples

Speed/Load Condition, rpm/hp	Soluble Organic Content, wt %	H wt %	C wt %	Ash wt %	Exh Temp, C [a]	Filter [a] Soluble Organic Content, wt %
1170/5, 10, 20	6.7	--	--	--	213	15
830, 1170, 1830/10	9.0	--	--	--	170	18
1170/20*	10.5	--	--	--	240	16
1170/10	7.7	--	--	--	162	16
1170/5	7.2	2.0	77.1	2.0	120	20
1830/40	6.5	0.4	77.8	1.0	350	3
830/variable	2.7	0.4	70.0	0.8	182	12
1830/variable	5.1	1.0	89.6	2.0	240	14
1170,1830/variable	1.9	0.5	82.0	0.9	220	12
1170/10, 20	1.8	0.5	71.0	0.6	200	12
1170, 1830/20	4.8	0.7	71.9	1.5	225	7

(a) At same or similar conditions.

consists of solid particles with liquid and adsorbed vapor. The particles can be considered either as single entities or as aggregates of subunits. Because of this complexity, exact specification is difficult; one must measure properties that seem to be meaningful in the design of control strategies. Moreover, each engine, fuel, and set of operating conditions causes the properties of the particulate to vary. The following section gives representative values for some of the properties that are important in the design of control strategies.

PROPERTIES OF DIESEL PARTICULATE

Physical Form: Small, solid, irregularly shaped particles, agglomerates of roughly spherical subunits. High molecular weight HC adsorbed on surface. Also may have liquid coating.

Size: Mass median diameter 0.1 to 0.3 μm; subunits about 10 nm

Size variation: See Fig. 1

Composition: Mostly carbonaceous material 10-50 percent extractable with organic solvent. See Table 1.

Bulk Density: ρ bulk $\cong 0.075$ g/cm^3

Surface Area: 100 m^2/g

Chemical Composition: See Table 2.

Effect of Operating Parameters-A 4.3 liter General Motors passenger-car diesel engine coupled to a 175-hp eddy-current dynamometer was used to investigate the effect of various operating and exhaust conditions on particulate properties.

The exhaust system consisted of standard vehicle exhaust pipe, muffler, and tailpipe components. A dilution tunnel-CVS system combination was used consistent with present practice for collecting and studying diesel exhaust particulates.

The site distribution measurements by Electrical Aerosol Analyzer showed a shift in the mass median diameter as engine load was changed. Fig. 1 illustrates this shift. As engine load (represented by exhaust temperature) was increased at constant speed, the mass median diameter increased also. This phenomenon may also be contributing to the observed increase in mass concentration with load increase. Particles too small to be captured on a filter at lower loads agglomerate to larger particles as load increases, thereby increasing the filter-collected mass. Others have also observed this effect.(1)* Fig. 1 also shows that higher speed operation results in smaller mass median diameter, a phenomenon probably related to residence time; i.e., less time available for agglomeration at higher engine speeds.

Fig. 3 shows the effect of engine speed and load on the particulate soluble organic content.

Table 2. Spectrographic Analysis of Particulate Samples

Element	Sample [a] 830 rpm/2 hp [b]	1830 rpm/40 hp
Li	10	10
B	10	3
C	Major	Major
F	3	<3
Na	500	500
Mg	150	150
Al	1500	1500
Si	3000	1500
P	∿1%	∿2%
S	∿4%	∿1%
Cl	50	50
K	100	50
Ca	3000	3000
Ti	300	3
Cr	500	300
Mn	500	300
Fe	∿30%	∿10%
Co	5	5
Ni	150	150
Cu	1500	1000
Zn	2000	1500

(a) Values given in ppm unless otherwise noted.
(b) Data shown in Tables 1 and 2 were obtained from a 4.3 liter Oldsmobile diesel engine.

* Numbers in parentheses designate references at the end of the paper.

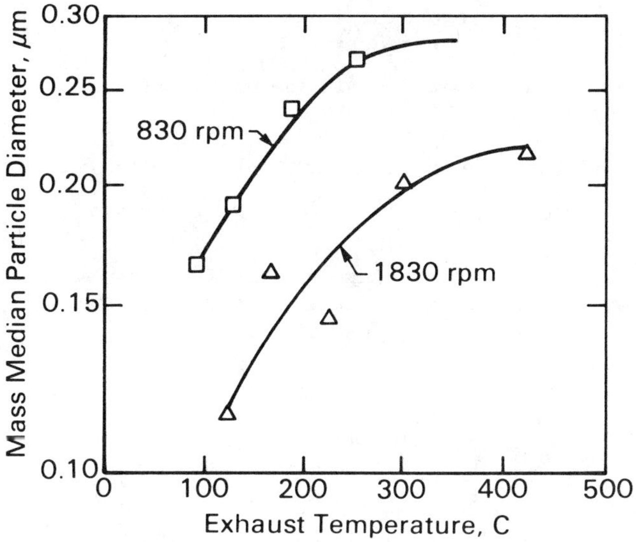

Fig. 1-Influence of exhaust temperature and engine speed on particulate mass median diameter

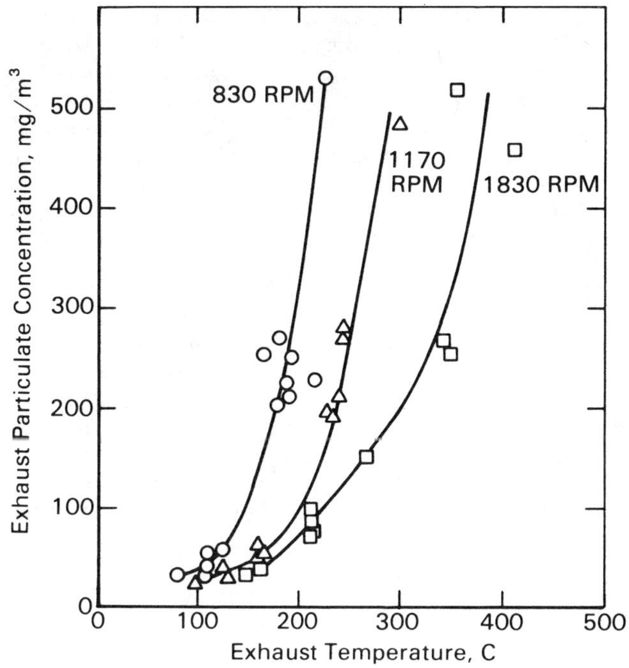

Fig. 2-Influence of exhaust temperature and engine speed on particulate emissions concentration

As with the particulate mass, the soluble organic content is plotted against temperature rather than engine horsepower. Engine speed does not seem to have as significant an influence on the soluble organic content as it does on the mass concentration. The temperature surely affects volatiles condensation. Figs. 1-3 show the range in exhaust properties a potential trap/oxidizer must handle.

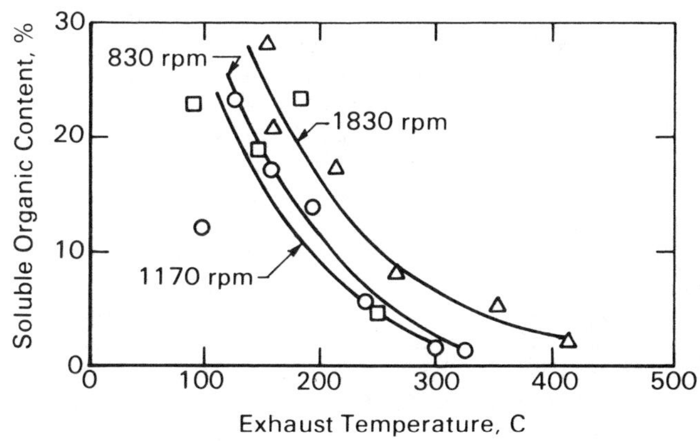

Fig. 3-Influence of exhaust temperature and engine speed on soluble organic content

PARTICULATE EMISSIONS RATE CHARACTERISTICS OF DIESEL PASSENGER CARS

Particulate emissions from passenger-car diesels range from 0.2 to 0.7 gm/km on the 1975 cycle, and from 0.1 to 4.0 gm/km on steady speed/load operation. Expressed in terms of concentration in the exhaust gases, the particulates range from about 20 to 130 mg/m^3.

In general, the particulate emission rate is a function of vehicle weight; the heavier vehicles produce more particulates in the same manner in which they produce more gaseous emissions. EPA certification vehicle emissions data for the 1979 model year are presented in Table 3.

Table 3. Light-Duty Vehicle Particulate Emission Rates As A Function of Vehicle Type and Weight*

Vehicle (Diesel)	Approximate Curb Weight, kg	FTP Particulate Emissions Rate, gm/km
VW Rabbit	900	0.14
Peugeot 504	300	0.18
Mercedes 300 D	1600	0.52
Oldsmobile 4.3 l	1600	0.45-0.63
Oldsmobile 5.7 l	1700	0.53

* Federal Register, Vol. 44, No. 23, February 1, 1979

IN-STREAM OXIDATION OF PARTICULATE

One of the simplest control techniques would be an exhaust system that completes the combustion of the carbon particles. Some aspects of this method are explored below.

THERMAL IN-STREAM OXIDATION-One strategy for the control of diesel particulates is the use of in-stream oxidation. Since the exhaust always contains some oxygen, the diesel particulate can be eliminated by oxidation if the temperature is high enough and the time for burning long enough. However, raising the particle temperature by heating the bulk of the exhaust gas requires a large amount of energy. Since such very small particles will reach temperature equilibrium with the surrounding fluid very rapidly, the bulk of the gas must be kept at the temperature desired for the particles. Unless almost all of the energy used to heat the bulk gas could be recovered, this method of control would be quite impractical. Design of a compact heat exchanger for the temperatures and flow rates involved would be difficult.

The analysis below explores some aspects of in-stream oxidation of particulate. The following exhaust conditions are assumed:

exhaust temperature:	200C
exhaust flow rate:	0.05 kg/s
particle size:	1 μm, 0.1 μm or 0.0287 μm*
particle concentration:	100 mg/m^3

(*Particles of this size are estimated to have a surface area of 112 m^{2}g. Lee, Thring, and Beer[2] report that soot from a hydrocarbon flame has a BET surface area of 112 m^2/g. Battelle surface area measurements are in this range.)

These conditions might represent a 5.7 liter diesel engine at medium speed and high load.

To gain perspective on the time available for oxidation, particle residence time was calculated using the above conditions. An exhaust velocity of 21 m/s was calculated based on a 50-mm ID exhaust pipe. Assuming a 1 m long path at high temperature, the time available for oxidation will be 50 ms.

The residence time at high temperature could be increased by increasing the volume of the afterburner device. However, since the engine may supply a flammable mixture to the exhaust system, a large confined volume would present a safety hazard, thus limiting this approach to increasing residence time.

For particles in this size range the rate of oxidation should be kinetically controlled at all reasonable temperatures.[3] Thus, by examining the chemical kinetics, the temperature necessary for oxidation in this time interval can be determined. Good data for the oxidation of diesel particulate is not available, so this calculation was carried out using data for graphite. It was also assumed that the particulate consisted of small spheres. The kinetic expression of Nagle and Strickland-Constable[4] is used for these calculations. This expression gives the mass of carbon oxidized per unit surface area per unit time at a given temperature and partial pressure of oxygen. For this type of rate expression

$$\frac{dm}{dt} = RS \text{ or } dm = RS\,dt$$

where

m = mass of carbon burned
t = time
R = reaction rate
S = surface area .

Since for a sphere $S = \pi D^2$ where D = sphere diameter

$$dm = R\,\pi D^2\,dt\ .$$

To find the change in sphere volume we note that:

$$\frac{dV}{dD} = \frac{D^2}{2}\ .$$

By considering the change in surface area with diameter, the burning times of the particles can be found. The following expression results:

$$t = \frac{\rho D_c}{2R} \qquad \text{(I)}$$

where

ρ = density of carbon
D = particle diameter
R = reaction rate

Using Equation (1) and the kinetic expression of Nagle and Strickland-Constable, the time required to oxidize completely a particle of a given size can be found. Some results of this calculation are shown below:

Particle Size, μm	Temp, K(C)	Burning time, approximately
1	1000 (727)	20 s
1	1500 (1227)	1 s
1	2000 (1727)	50 ms
0.1	1000 (727)	2 s
0.1	1500 (1227)	100 ms
0.1	2000 (1727)	5 ms
0.0287	1000 (727)	575 ms
0.0287	1500 (1227)	30 ms
0.0287	2000 (1727)	1 ms

The above results show that the particle temperature must be quite high to complete oxidation in a reasonable time. It is of interest to estimate the heat required to bring the exhaust to this temperature.

To heat a kg of exhaust from 200 C to 1200 C requires about 1120 kJ of heat. At an exhaust flow of 0.08 kg/s, 90 kJ/s of heat must be supplied.

At this speed and load the engine uses about 13 kg/hr of fuel or 0.0036 kg/s. If the heat content of the fuel is taken to be 45,000 kJ/kg, then the engine is using 162 kJ/s of energy. It can be seen that the energy required to heat the exhaust to the particulate burning temperature is over half of that used by the engine. This is obviously unacceptable.

It might be argued that heat will be released when the particulate burns. The heat released will be given by

$$\dot{Q} = \dot{m}_p H$$

where $\dot{Q}$ = heat released

$\dot{m}_p$ = mass flow of particulates

$$= \left(\frac{100\ \text{mg}}{\text{m}^3}\right)\left(\frac{\text{m}^3\text{ex}}{0.737\ \text{kg ex}}\right)$$

$$\frac{0.08\ \text{kg ex}}{\text{s}} = 11\ \text{mg/s}$$

H = heat of combustion for particulate.

Assuming that H is 45,000 kJ/kg, we find 0.5 kJ/s released by the burning particulate. This is negligible compared with the 90 kJ/s required to heat the exhaust.

If it were possible to recover most of the heat in the hot exhaust, this afterburning concept might be feasible. A heat exchanger operating at a temperature of about 1200 C would be required for this purpose.

CATALYTIC IN-STREAM OXIDATION-The burning rate of the diesel particulate appears not to be rapid enough at typical exhaust temperatures to eliminate a significant amount of particulate. One way to speed up a reaction rate is to use a suitable catalyst.

Such a catalyst would, of course, have to be incorporated in the particle as it is formed. This suggests that the catalyst be an additive to the diesel fuel. Although the exact mechanism that would cause incorporation is not known, several could be suggested involving chemical reaction of the soot with the additive, agglomeration of soot and additive particles, additive particles acting as nuclei for soot formation, or adsorption of the gaseous additive by the soot. If such incorporation can be achieved and a suitable catalyst found, the burning rate at exhaust temperatures could increase dramatically.

A number of metals show catalytic activity for the oxidation of graphite. Some likely candidates are discussed in detail in connection with the oxidation of material collected on traps. As the following tabulation[5,6] shows, the reaction rate can increase dramatically. And we can expect that the presence of the catalyst will be most noticeable at lower temperatures.

Metal	Rate of Reaction Relative to Pure Graphite
Ba	100
Na	230
Au	240
V	340
Cu	500
Ag	1,340
Cs	64,000
Mn	86,000
Pb	470,000

Some of these metals would have toxicity problems, but others may be acceptable. Even lead may be acceptable since the amount required may be very small--in tests with graphite, a lead concentration of 1.3 ppm was enough to increase the rate of oxidation at 430 C by a factor of 120.

Exactly how large an increase in the rate of oxidation will occur is uncertain. One reason is that the impurities and trace elements in the diesel fuel may already be acting to catalyze the oxidation reaction. It has been suggested that this is the case for most coal combustion where the measured energy of activation is about 170 kJ/mol although the energy of activation of pure graphite is reported to be 250 kJ/mol. Certainly diesel fuel does not have as many metallic impurities as coal and probably lacks the optimum amount of metallic catalyst.

Such a catalytic approach could make in-stream oxidation of the diesel particulate feasible. This would have the advantage of not requiring any add-on devices. Or, if a trap were found necessary, the particulate loading would be reduced, and the light-off characteristics would be improved.

OXIDATION OF FILTER-TRAPPED PARTICULATE

A basic problem with the particulate in the diesel exhaust stream is that it is too dilute to burn. This can be remedied by collecting the material on a filter. Although a filter is usually thought of as a way to remove particles from the flowstream, it can also be thought of as a device for concentrating the particulate material. In this concentrated form the average heat loss is reduced and conditions for combustion are more favorable. In fact they may be so favorable that damaging temperature excursions may occur.

TRAP/OXIDIZER STRATEGIES-There are a number of design concepts for trapping and oxidizing the particulate emissions from diesel engine operation. Although the requirements for successful application of each of these concepts can be described qualitatively, the character of the particulate, the service requirements, and the limitations imposed by automobile application are not well enough described to permit firm choice among these concepts or the complete rejection of any. Some concepts, such as the catalytic oxidation of a solid carbon particle that has been casually deposited on a solid catalytic surface, immediately imply such severe limitations as to appear improbable. This section of the paper assumes that the particulate has been trapped in a fashion suitable to each concept. Four broadly based strategies must be considered:

Strategy 1-Trapped particulate is brought to ignition by heat derived from the engine operation, usually from the hot exhaust gases themselves.

Strategy 2-Trapped particulate is brought to ignition by an auxiliary heater used to trigger ignition at programmed times in the engine operation cycle.

Strategy 3-Particulate is brought to ignition at substantially lower temperatures than either of the above strategies by catalytic materials that promote particulate oxidation.

Strategy 4-The condensible portion of the material forming the ambient particulate is first oxidized (catalytically), then the remaining particulate is burned in a trap-oxidizer according to any of the above three strategies.

These four strategies are used to discuss the problems and critical features associated with the burning of trapped particulate emitted by automotive diesel engines. Any of these strategies might be applied successfully in the laboratory; the primary source of difficulty is the automobile application with its adjunct limits on space, cost, and energy consumption.

In any of these choices the trapped particulate must be ignited and burned, and the implications of this are basic to all strategies. The next section discusses a kinetic model for ignition.

The Ignition Concepts-The trap/oxidizer used to control particulate is basically a chemical reactor in which oxidation occurs. The rate of oxidation varies with the temperature--as the temperature increases, the rate increases rapidly due to the exponential nature of the kinetic rate expression. Eventually, the rate of the reaction is limited by mass transfer (diffusion of oxygen to the reacting surface and reaction products away from the surface), and no longer increases with temperature. As the temperature rises the heat loss from the device also rises. Thus, the operating temperature will be that temperature for which the heat released by the reaction is equal to the heat loss. Because the reaction rate curve flattens at high temperatures, the heat loss curve intersects in two places and the device tends to be in one of two conditions--unignited or ignited. Onceignition occurs the temperature very rapidly reaches a value near the combustion temperature very rapidly reaches a value near the combustion temperature of the fuel. This is because the exhaust flow through the filter bed increases mass transfer to the carbon on the filter. Thus, the mass transfer limitation is not severe. Also, the heat loss occurs at the surface of the device and conditions at the center will be near adiabatic. So the two factors that tend to limit the temperature are not very effective. The ignited conditions therefore may destroy the control device.

Application of chemical reactor theory allows a more detailed picture to be drawn.

Salient features of any exothermic reaction (e.g., combustion or oxidation) in a flow-through reactor are the transient nature of both the temperature of any volume element with time, and the large gradients of temperature throughout the reactor. Afterburners, whether catalytic or not, exhibit ignition characteristics similar to flame processes and the performance of the afterburner is directly related to the achieving and stabilizing of a suitable "ignited" state.

With suitable sample geometry, only the ignition need be catalytic and this is the effect being proposed here. The situation for achieving ignition in some volume element of a reactor can be described with the help of the theory employed by Wagner(7) and Frank-Kamenetskii(8). Usually the ignition temperature lies well above the temperature at which the first significant heat effect due to oxidation is detected, and at ignition the excess heat generated over that lost by the volume element causes that element to suddenly rise in temperature well above the environment temperature. In so doing it passes from activity-limited oxidation to transport-limited conditions and usually exceeds the kinetics that can be expected for the catalytic surface reaction processes.

Strategy 1-Thermal Ignition of Trapped Particulate-Experimental trials at Battelle indicate that accumulations of diesel particulate ignite at temperatures of about 600 C. Below this temperature oxidation rates are very slow. Such temperatures are unrealistically high for any reliable mechanism whereby the particulate is regularly ignited by waste heat from the engine without some assistance. No further consideration is given to this strategy, but the temperature cycle for the exhaust gases can serve as the basis for more frequent and/or scheduled ignition in the options discussed for the other strategies.

Strategy 2-Use of Ignition Devices-Since the heat from the engine alone is insufficient to ignite the deposit of particulate material on the trap, some type of ignition device must be used. One possible device is a glow-plug type of electric resistance heater. The basic problem with this strategy is that the automotive electrical system is not capable of delivering an amount of energy that is significant compared with the energy that might be lost to the relatively cool exhaust flowing through the device. For example, the electrical system is hardly able to supply more than 1 kW or 1 kJ/s. Something like 50 kJ/s would be required to heat the exhaust stream to the deposit ignition temperature. Obviously this is not the purpose of the ignition device. The point is that if one tried to heat the entire deposit on the trap to the ignition temperature, the heat losses to the exhaust would be too large. The only alternative is to provide a few hot spots in the deposit and let the combustion propagate through the remainder. This could present a problem, since the deposit must be sufficiently evenly distributed for this propagation to occur. If it is not, fuel may accumulate in part of the trap, only to ignite in an unscheduled or uncontrolled fashion.

Alternatively some diesel fuel could be burned to raise the trap/oxidizer device to its ignition temperature. The diesel fuel would, in turn, be ignited by an ignition source. This approach has a number of advantages. The amount of heat that can be put into the system is large. The fuel is already at hand. An ignition system should pose few problems. Since the additional fuel needs to be supplied for only a few seconds, the effect on overall fuel economy will be small.

There will be design problems, however, The fuel spray must ignite smoothly--a controlled addition of heat is required, not a backfire. The fuel would probably be added with a small spray nozzle; this nozzle must not foul when not in use. The heat from the burning ignition spray must not harm the device--the additional heat may make burn-off temperatures even higher. The presence of another fuel and ignition system adds cost and complexity.

Either system would need some kind of control to ensure that ignition occurred at necessary intervals. This could be done on the basis of engine hours, pressure drop across the device, distance traveled, or fuel used. Because of the wide variation in ambient conditions, it is anticipated that the control system will have to be set for worst case conditions with extra cycles at other times.

It seems then, that there is no barrier to

making an auxiliary ignition device for the trap/oxidizer. Of course it would be more desirable that the device stand alone. If the ignition temperatures were lower, additional ignition would not be necessary. It is possible that a catalytic approach may help here.

Strategy 3-Catalytic Ignition of Particulate-For the present application, the catalytic effect must be obtained at the surface of carbon particulate at temperatures less than that at which the unassisted thermal oxidation is detectable. The heat release rate due to this catalytic oxidation must be sufficient to achieve the ignition temperature described earlier. Both the design of the catalytic chamber and the activity of the catalyst influence the ability of the chamber to achieve this ignition. Once the most favorable form of catalysis is identified, a study of chamber design must be undertaken to maximize the repeatability of the catalytic ignition at acceptable performance levels.

The oxidation of a solid carbon particle casually implanted on a solid catalytic surface is most difficult to catalyze since the contact between the particle and the catalyst is uncontrolled and its area is anticipated to be small. The presence of hydrocarbon condensate in that carbon particle (as in diesel particulate) may alleviate this difficulty somewhat since that condensate may transfer to the condensate may transfer to the catalytic surface and the higher H/C ratio, expected for the condensate as compared to the solid carbon portion, augments both the rate and heat release of the catalytic oxidation. This anticipated influence of the level of condensate accumulation on either the thermal or catalytic ignition of diesel particulate does not appear to have been examined systematically.

The only form of catalysis that has been described in any detail for oxidation of carbon particulate is that developed years ago for removal of soot accumulation from flues and furnaces by the use of inorganic salts deposited on the carbon particles. U.S. Bulletin of Mines Bulletin 360[9] provides an understanding of the level of development of this form of carbon particulate oxidation catalysis. More recently, salt additives have been studied as catalysts for the oxidation of high purity graphites.(5,6)

In all of these tests the volatility of the salt used was an important characteristic. Initially, the salt was applied by vaporizing it and allowing it to deposit onto soot that had been accumulated on a test screen. The action by which the salt catalyzed burning was not specified, but some relatively contemporary publications suggest mechanisms that do not seem to have been further investigated.

For instance, Cassel (10) noted that soot deposited on the etched or ground surfaces of Jena glass ignited much more easily and burned more rapidly than soot on adjacent smooth surfaces of the glass. From evidence for oriented crystal growth of soot particles on surfaces, he suggested that the ground surfaces interrupted crystal growth during soot deposition because of surface irregularity and also helped to prevent secondary crystal growth during oxidation (burning). He reasoned that the salt deposits condensed on soot similarly promote ignition and burning by maintaining the highly dispersed soot structure prior to and after ignition.

Alternatively, Taylor and Neville (11) and later Day, et al. (12) noted large increases in soot burning rate following deposition of soot onto surfaces previously coated with salts by evaporation. Both of these investigations concluded that the salts probably functioned by hastening the decomposition of carbon-oxygen surface complexes. With this barrier removed the carbon surface was believed to be more rapidly attacked by oxidizing gases. A brief review of the succeeding years' publications has failed to disclose any further discussion of these or other hypotheses for the action of the salts. It is of interest to note that the promotion effect is claimed either for salts vaporized onto the soot or for soot condensed (deposited) onto the previously deposited salt.

In more recent work of Duval and coworkers (5,6) inorganic cations were impregnated into graphite carbon from water solution and then the rates of oxidation were noted. The results showed that the introduction of cations at 120 ppm concentration produced rates of oxidation up to 470,000 times that of the untreated graphite. In this study the effective cations were identified as representing those elements that have variable oxidation states, and can exist in defect states of oxidation. By this means they serve as oxygen carriers to the graphite interface.

The differences in relative effectiveness of the various metal salts in these two studies seem to underline differences in mechanisms of action that have occurred, either because of the difference in method of application or because of the differences in reactivity and structural stability of soot and graphite. The variable valence postulate for the catalytic action on graphite is the same as that widely used to explain the catalytic action of high-surface-area oxide catalysts in catalytic hydrocarbon oxidizers. The large increases in oxidation rate illustrated by these catalytic applications indicate that considerable catalytic assistance is available for the burning of diesel-exhaust particulate.

For the present application, the catalytic effect should start at temperatures less than that at which the unassisted thermal oxidation has been detectable and the heat release rate due to this catalytic oxidation must be sufficient to achieve the ignition temperature. In the U.S. Bureau of Mines study (9), trials of uncatalyzed ignition of furnace soot established the ignition temperature at about 615 C, and it was observed that insignificant oxidation rates were obtained below 480 C. The reduction of inlet oxygen concentration from normal air to 8 percent had little effect on the ignition temperature for untreated samples.

Experimental Study of Soot Ignition-A reactor used to study soot ignition is shown in Fig. 4. A sample of diesel soot, 0.15 g, is placed in a

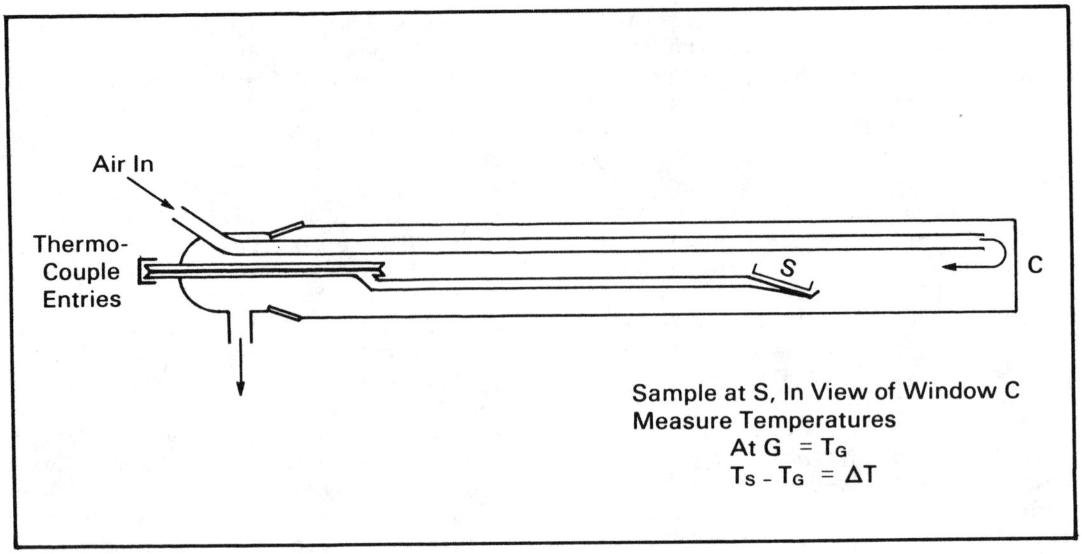

Fig. 4-Arrangement of Quartz Reactor for Measurement of Ignition Temperatures

stainless steel pan that is about 10-mm wide, 47-mm long, and about 3-mm deep. One thermocouple junction, chromel-alumel 26 gauge, is buried in the sample; another is in the gas stream just upstream of the sample. The entire quartz reactor up to the standard taper joint is placed in a tube furnace that is heated gradually from room temperature to 600 C in about 1.5 hr. During most of the heating the sample lags behind the indicated gas temperature by about 15 to 20 C. On ignition the sample temperature rises abruptly at a rate of 100 to 200 C per minute and the occurrence of ignition on some portion of the sample can be confirmed by observa- of the orange-yellow glow produced as the combustion site moves through the sample. Since the gas temperature typically is rising about 5 to 10 C per minute, the ignition temperature can be identified with an uncertainty of only a few degrees. Typically, the indicated temperature rise, T_S-T_G, reaches a maximum of about 120 to 200 C, and this lags behind the true combustion-produced temperature rise by an amount depending on the actual location of the hot zones in the sample with respect to the thermocouple. The apparatus is sensitive to a temperature change of less than 1 C so that the occurrence of ignition somewhere in the sample is detected readily. Sample traces of the temperature plots obtained during experiments are shown in Fig. 5.

Catalytic salts were added to the diesel soot as methanol solutions; the solvent was removed before trial by heating the sample 16 to 24 hours at 115 to 130 C. In each case the salt solution was made up 1 x 10^{-5} molar so that a few milliliters of these solutions sufficed to impregnate 0.20 g of soot; after drying, 0.15 g of the sample was used for the experiment.

The soot samples used in this study were brushed from the walls of the exhaust pipe after various experiments. They were identified by the date of collection and by the engine operation data for that date. It was difficult to get good thermocouple contact with these extremely fluffy loose samples. It was found that samples dispersed in methanol had a more suitable density after removal of that solvent at 110 to 130 C and so the uncatalyzed samples were examined in this form. Since all catalyst applications were made from methanol solution in the same way, the uncatalyzed performance should represent a correct blank for comparison with the catalyzed samples.

The primary intent of these experiments was to survey the parameters that control catalytic ignition; most of the data listed in Table 4 were obtained using soot from a single test run, identified by an asterisk in Table 1.

Results-The activities of a few metal chlorides for promoting ignition of the diesel particulate are compared in Table 4. The superiority of copper chloride is surprising in view of the relatively high oxidation activity that has been demonstrated by other investigators for catalysts based on Co, Mn, and Pt at lower temperatures than used here. In each case the question arises concerning the temperature required before a catalytic form of the metal can be obtained under the conditions used here where no preactivation treatment is employed. Persistence of some part of the water of crystallization in a few cases may be retarding the decomposition of the catalytic salt during the heating period prior to ignition.

Preformed catalyst surfaces, even if effective, may accumulate ash residues from lubricating oil and/or fuel components of the particulate and the active life of such surfaces may be materially shortened by such accumulations. On the other hand, catalysts incorporated in the particulate by inorganic residues from fuel or oil additives, by spray application or by vaporization onto the accumulating particulate, are self-renewing and should occupy very little residual volume once burning is complete. The effectiveness of such inorganic catalyst applications for burning of

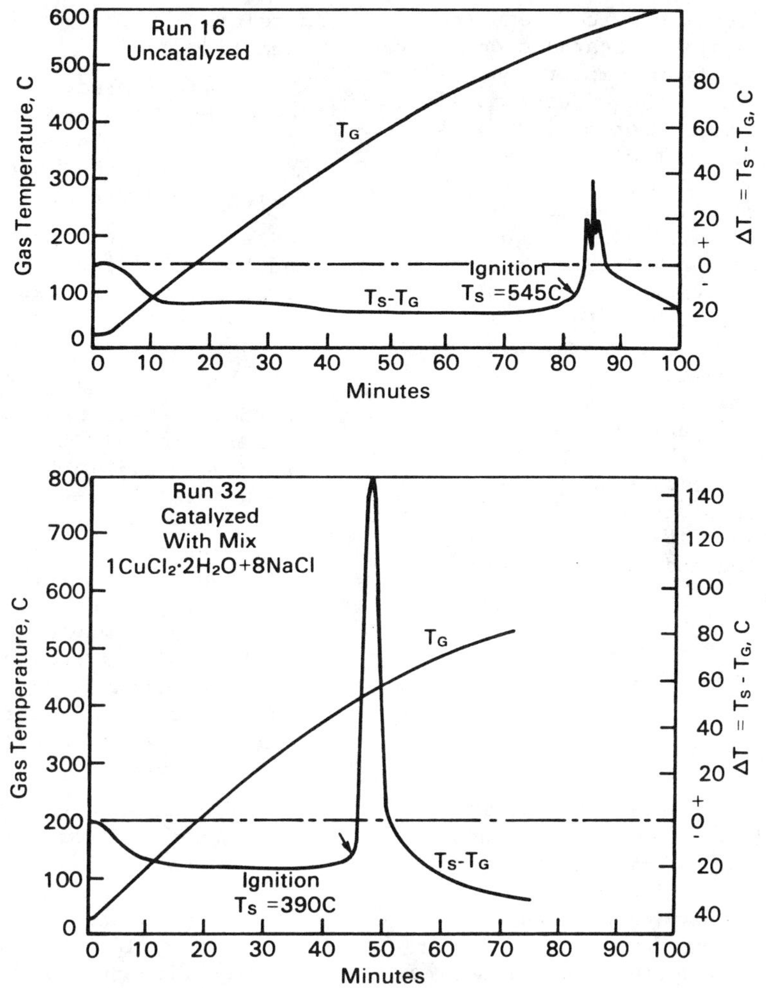

Fig. 5-Sample Temperature Traces During Ignition Trials

Table 4. Comparison of Ignition Temperatures For Various Metal Chlorides

Catalytic Salt Used[(a)]	Ignition Temperature, C	Maximum Temperature Rise,[(d)] ΔT C
None[(b)]	504-545	50-121
$MnCl_2.4H_2O$	456	125
$CoCl.6H_2O$	461	122
H_2PtCl_6 [(c)]	455	237
$CuCl_2.2H_2O$	421	139

(a) The salt used in each case was equivalent to 1.18×10^{-4} moles of the metal per gram of particulate. Since each was added in methanol and then dried at 110 to 130 C, the water of crystallization probably was lost on drying.

(b) Diesel particulate collected 12/12/79, slurried in methanol, and dried 110 to 130 C.

(c) In methanol-ethanol, not previously activated to decompose the salt other than the 110 to 130 C drying conditions.

(d) This is not the combustion temperature, but the overall maximum rise sensed by thermocouple reflecting the relative rate and completeness of oxidation.

diesel exhaust particulate has been demonstrated. The means for application of such catalytic methods to automobile service remains to be examined.

Strategy 4-Combined Use of Catalytic Hydrocarbon Oxidizer and Particulate Trap/Oxidizer-The condensate portion of diesel particulate is a potential problem in operation of the trap/oxidizer because of its volatility. Balanced against the usefulness of the condensate for promotion of ignition is the possibility that partially oxidized portions of this condensate may be re-evaporated from the advancing combustion front and emitted with the exhaust gases. Such re-emitted material may be toxic. Elemental carbon can be expected to burn largely to CO, CO_2, and H_2O and the troublesome re-emissions should derive from the condensate portion of the trapped particulate.

Separate oxidation of the condensibles would reduce the mass load on the trap/oxidizer by 5 to 40 percent depending on the engine operation conditions. The question arises then of the degree to which separate oxidation is feasible and the effect of that oxidation on the particulate trap. Published experience (13) with catalytic hydrocarbon oxidizers for stationary diesel service seems to imply that the condensation of hydrocarbons onto the particulate has been far from complete when the oxidizer is located as close to the exhaust manifold as possible. These units, built for minimum back pressure, do not function as particulate traps so that the hydrocarbon portion apparently is being oxidized at a catalyst surface without any effective retention of the solid particulate. Thus, Sercombe (13) has described a platinum catalyst system that is claimed to provide overall control of hydrocarbon, nitrogen oxide, and CO emissions from diesel engines, but which apparently has no effect on the particulate. Similarly, another hydrocarbon oxidizer system was described (14) as able to remove 70 percent of the hydrocarbon emissions from diesel engines that, without the oxidizer, produced particulate with 35 percent condensibles. Catalytic hydrocarbon oxidizers of this sort perform well only if "ignition" is attained and maintained, and failure to achieve this may have contributed to the comparatively poor performance of the oxidizer tried by Seizinger, et al. (15) in their trials of two derated diesel engines operating at low emission levels.

APPROACHES TO TRAP/OXIDIZER DESIGN

The trap/oxidizer is a chemical reactor in which oxidation occurs. As outlined above, such a reactor tends to be in either an ignited or an unignited condition. When the device is in the ignited condition, the temperature very rapidly reaches the combustion temperature. The ignited condition therefore may destroy the control device.

Application of chemical reactor theory, along with data on the heat transfer, mass transfer, and chemical kinetics of the device would allow a quantitative analysis of the above phenomena.

But even in the absence of such an analysis, it is clear from experience that is available for trap/oxidizers that the qualitative picture presented above is correct and that ignition of the particulate often destroys the device.

In view of the above, a number of approaches to trap/oxidizer design present themselves:

- If the flow through the device were stopped when ignition occurred, the mass transfer limitation would occur at a lower temperature. This could be accomplished by temporarily bypassing the device, or by switching to an alternate. A possible disadvantage of this approach is that the combustion reaction will be oxygen-starved and may produce noxious products.
- The heat transfer can be increased. The critical area is in the center of the reactor where conditions are adiabatic rather than at the edge, The solution is to make the reactor all "edge"; that is, to increase the surface-to-volume ratio. This could be done by making a flat pancake-shaped device or by including cooling passages in the internal structure. It would be desirable that each filter fiber "see" a relatively cool wall, and thus be cooled by radiation.
- The chemical kinetics could be changed by an inhibitor. The rate of reaction at a given temperature would be less and there would be more time for the heat to dissipate, thus making the device temperature lower.
- The situation can be accepted as unchangeable and the filter designed to withstand the temperatures expected during burn-off. It may well be that ceramic fibers would have the necessary ability to keept their integrity when very hot.
- The device will proceed to the adiabatic flame temperature only if a sufficient supply of fuel is available. If the supply is limited, the highest temperature would never be reached. Also, the reactor does not consist entirely of fuel, but contains the filter as well. The filter has a small thermal mass that requires heat energy to be raised in temperature. Thus, this approach would ensure that the trap/oxidizer always runs out of fuel before reaching a temperature that would be too high for the filter. The oxidation would never reach steady-state. This might be accomplished in a number of ways.
  - If ignition occurs frequently enough, the amount of fuel accumulated will necessarily be small.
  - Frequent ignition would be encouraged by the presence of a catalyst
  - Fuel could be added to the exhaust and ignition supplied regularly.
  - The thermal mass can be deliberately increased. Ceramic "marbles" might be "mixed" with the filter to ensure that the fuel was gone before the ceramic and filter were overheated.

A successful device might well use several of these approaches. However, each should receive a detailed evaluation to determine its effectiveness.

COMBUSTION OF COLLECTED MATERIAL

A discussion of the combustion of collected material must be prefaced with the warning that the physical configuration of the device will be a major factor in its operation. If particulate material is removed (say, by rapping) from a filter, it will be in some kind of container, and it will be in a thick layer (many particles thick). Thus, when this material burns the situation will be similar to the combustion of a piece of solid fuel. Just as a thin piece of wook will not support a flame if the heat loss is too great, a layer of particulate will not burn if too much heat is lost to the container. Striking a balance between maintaining combustion and preventing excessive temperatures should be difficult.

CONCLUSION

This paper lists a number of conceptually different ways to attack the diesel particulate problem. Each concept could become the basis for different ingenious applications of creative design. The acquisition of engineering data relevant to the various concepts will make that design effort easier and will suggest the concepts that have the greatest potential for success. The specific design will ensure that success; it is difficult to consider this factor in a general evaluation.

With this caveat in mind, it appears that the catalytic oxidation of the particulate on some type of filter-trap holds the greatest promise for the removal and destruction of the diesel particulate. The key to the successful operation of such a device would appear to be frequent, if not continuous, ignition combined with filter fibers that maintain their integrity at high temperatures.

The work in progress, at Battelle and elsewhere, will help fill the need for a store of experience with these devices. This experience is needed both to identify the concepts worthy of future development and to identify the specific materials and dimensions of the final choice.

REFERENCES

1. J.A. Verrant and D.B. Kittelson, "Sampling and Physical Characterization of Diesel Exhaust Aerosols", SAE Publication PT-79-17, 1979.
2. K.B. Lee, NW Thring, and JM Beér, Combustion and Flame, V 6, 137, 1962.
3. M.F.R. Mulcahy and IW Smith, Reviews of Pure and Applied Chemistry, V 19, 81, 1969.
4. J. Nagle and RF Strickland-Constable, Proceedings of the Fifth Conference on Carbon, V 1, 154, 1962.
5. C. Heuchamps and X Duval, Carbon, V 4, 243-253, 1966.
6. H. Amoriglio and X. Duval, Carbon, V 4, 323-332, 1966.
7. C. Wagner, Chem. Tech. (Leipzig) V 18, 28-34, 1945.
8. D.A. Frank-Kamenetskii, Diffusion and Heat Transfer in Chemical Kinetics, Plenum Press, New York, 1969.
9. P. Nicholls and CW Staples, US Dept of Mines Bulletin 360, 1932.
10. H.M. Cassel, J. Am. Chem. Soc. V 43, 2055, 1921.
11. H.S. Taylor and H.A. Neville, J. Am. Chem. Soc., V 43, 2055, 1921.
12. J.E. Day, R.F. Robey, and H.J. Dauben, J. Am. Chem. Soc. V 57, 2725-6, 1935.
13. E.J. Sercombe, Platinum Metals Review, Johnson Matthey, V 19, 2-11, 1975.
14. J. Mooney, Englehard Industries, personal communication.
15. D.E. Seizinger, B.H. Eccleston, and R.W. Hurn, SAE 790420, Detroit, 1979.

Mathematical Modeling of Fibrous Filters for Diesel Particulates – Theory and Experiment**

Se H. Oh, J. Scott MacDonald,
Gerald L. Vaneman, and L. Louis Hegedus
General Motors Research Labs.

IN VIEW OF INCREASINGLY STRINGENT fuel economy standards for the coming years, diesel engines appear to be attractive alternative powerplants for passenger vehicles. However, despite its superior fuel economy, a diesel engine generally presents a challenging emission control problem. For example, emission data of an Oldsmobile 5.7-L V-8 diesel engine [1]* indicate that the simple manipulation of engine operating conditions, such as the degree of exhaust gas recirculation (EGR), is not adequate to meet the future emission standards (Table 1), primarily due to the associated tradeoffs between NO_x and particulate emissions. As illustrated in Figure 1, an EGR rate chosen on the basis of the NO_x standard may lead to a particulate emission level above its emission standard, and thus some additional scheme for controlling particulate emission is necessary.

TABLE 1 - Emission Standards (g/mile) for Light-Duty Diesel Vehicles

	1983	1985
NO_x	1.0	1.0
HC	0.41	0.41
PARTICULATES	0.6	0.2
CO	3.4	3.4

*Numbers in brackets designate references at end of paper.

**Paper 810113 presented at the International Congress and Exposition, Detroit, Michigan, February, 1981.

In general, there are several ways of controlling diesel particulates [1,2]: engine modifications, fuel modifications, fuel additives, and exhaust aftertreatment. The first three approaches will not be discussed here. One of the most promising aftertreatment schemes is filtration of diesel exhaust particulates. In this report, we will limit our attention to the problem of particulate collection using fibrous filters. Disposal of the particulates collected in filters is beyond the scope of this work.

Fibrous filters have been frequently used for removing fine particles (especially those in the submicron range) from gaseous streams because of their high collection efficiency. Recently, the application of fibrous filters to the diesel particulate removal problem has been considered, the work by Springer and Stahman [3], General Motors [1], and Tessier et al. [4] representing major efforts along this line.

Rational design and optimization of diesel particulate filtration systems require a quantitative understanding of particle deposition phenomena on fibers. Problems of particle deposition on clean (i.e., particle-free) fibers are, in general, reasonably well understood and constitute the main content of the classical theory of aerosol filtration. (A comprehensive review on this subject can be found in a book by Davies [5] and in a review article by Chen [6].)

ABSTRACT

A mathematical model has been developed to describe the initial performance of fibrous filters for diesel exhaust particulates. In addition to the basic mechanisms of particle deposition on fibers, the size distributions of both diesel particulates and fibers were incorporated into the model in order to account for the polydispersity of particles and fibers encountered in realistic diesel particulate filters. Filtration experiments with a single-cylinder diesel engine were conducted to test the validity of the model, and the calculated filter efficiencies agreed very well with the experimental data.

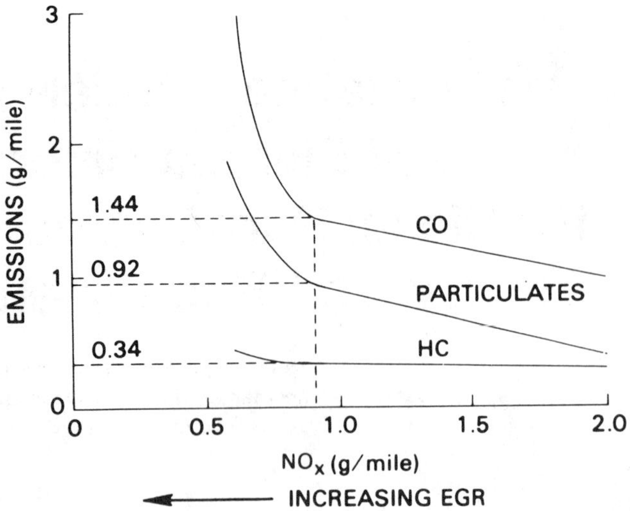

Fig. 1 - Emission tradeoffs of an Oldsmobile 5.7-L V-8 diesel engine (FTP measurements)

The theory of the time-dependent behavior of fibrous filters (i.e., the change in filtration efficiency with particle accumulation), on the other hand, is still in the development stage; it has received considerable attention only recently [7,8,9], although its importance has been recognized for some time [10,11].

This paper documents our modeling studies of the performance of fibrous diesel particulate filters. In the present work it is not intended to develop an entirely new theory of aerosol filtration. Instead, we concentrate on how the well-developed filtration theories can be applied to the diesel particulate problem to obtain a quantitative model of fibrous diesel particulate filters. To be specific, we will first briefly review the theory of aerosol filtration and then extend the theory to incorporate particle and fiber size distributions. Finally, the predictions of the model will be compared with the results of engine experiments to test the validity of the model.

It should be noted that our discussion will be limited to the initial performance of fibrous filters, that is, the behavior of sufficiently clean filters. Although this covers only the period during which the deposition of particles is not influenced by those already held on the fiber surface, the initial filter efficiency is still an important criterion for selection of appropriate filters. The time-dependent behavior of a fibrous filter is the subject of ongoing work in our laboratories.

THEORETICAL

PARTICLE COLLECTION MECHANISMS - During aerosol filtration with fibrous filters, particles are deposited on the surface of a fiber from the gas layer adjacent to the fiber; therefore, it is important to understand possible pathways of particles near the fiber surface in a moving gas stream.

Consider an aerosol stream moving towards a cylindrical fiber which is perpendicular to the direction of flow and effectively infinite in length (see Figure 2). When a particle has negligible inertia (Particle 1), its trajectory will follow the gas streamlines shown by solid lines in Figure 2. In this case the particle, because of its finite size, will touch the fiber when its center approaches within a distance of one particle radius from the fiber surface. This effect is called interception, and this collection mechanism is characterized by the dimensionless number (interception parameter)

$$R = \frac{d_p}{d_f} \tag{1}$$

The symbols used in the equations are defined in the Nomenclature section.

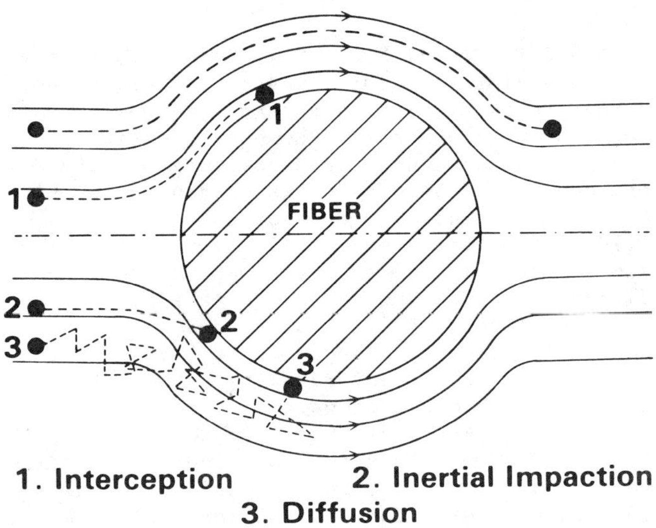

Fig. 2 - Particle collection mechanisms by a single fiber

The trajectory of particles having significantly large inertia, on the other hand, deviates from the gas streamlines when the streamlines of flow spread around the cylindrical fiber. This leads to particle impaction on the fiber surface, as illustrated by Particle 2 in Figure 2. The characteristic parameter for such an inertial impaction mechanism is the Stokes number, which is defined as

$$Stk = \frac{C \rho_p d_p^2 v}{9 \mu_g d_f} \tag{2}$$

The Cunningham factor, C, which is introduced to account for the effects of gas slip past the particle, can be calculated from the expressions given in [e.g., 5,12].

As a result of their collisions with gas molecules, small particles show considerable random diffusive motion called Brownian movement and consequently tend to deviate from the gas streamlines, leading to particle deposition on the fiber surface (Particle 3 in Figure 2). The dimensionless number characterizing this diffusion mechanism is the Peclet number

$$Pe = \frac{v\, d_f}{\mathcal{D}} \tag{3}$$

where $\mathcal{D}$ is the diffusion coefficient or diffusivity of the particles. For Brownian movement, the particle diffusivity can be calculated from

$$\mathcal{D} = \frac{C\, k_b\, T}{3\pi\, \mu_g\, d_p} \tag{4}$$

Particle collection due to the diffusion mechanism becomes increasingly significant as the particle size is decreased, because of the associated increase in the particle diffusivity (see Equation 4). This mechanism also plays a predominant role when the gas velocity is low, since the particles remain longer in the neighborhood of the fiber.

In addition to the three basic mechanisms of particle collection discussed above, the particles may be under the influence of external forces, such as gravitational or electrical. Such effects will not be considered in the present study.

SINGLE FIBER EFFICIENCY - The most commonly used approach in the modeling of fibrous filters is to focus on a single fiber in the filter and examine the flow field around it. Most of the early theoretical studies [e.g, 13,14] were based on the well-known flow field around an isolated fiber. However, such a simplified flow field is inadequate for filtration studies, since the gas flow around individual fibers in a fibrous filter occurs under the condition of mutual hydrodynamic interferences of neighboring fibers. In 1959, Happel [15] and Kuwabara [16] considered the problem of creeping flow (i.e., low Reynolds number regime) in an array of parallel cylinders transverse to the flow direction and obtained expressions for the flow field around a single fiber. This stimulated further development of the theory of aerosol filtration.

A recent review article by Kirsch and Stechkina [17] describes how the Kuwabara flow field can be applied to what they term a "fan model" to obtain expressions for the single fiber efficiencies due to various particle collection mechanisms. In the fan model, a fibrous filter is visualized as a system of parallel but randomly oriented rows, each row being composed of parallel cylinders. In view of its structural similarity to real filters, the fan model seems to be a good representation of real filters and therefore will be adopted here.

The semiempirical expressions for the single fiber efficiencies for the fan model are given by [17]

$$\eta = \eta_D + \eta_R + \eta_{DR} \tag{5}$$

where

$$\eta_D = 2.7\, Pe^{-2/3} \{1 + 0.39\, k^{-1/3}\, Pe^{1/3}\, Kn\} + 0.624\, Pe^{-1} \tag{6}$$

$$\eta_R = (2k)^{-1} \{(1+R)^{-1} - (1+R) + 2(1+R)\ln(1+R) + 2.86\, Kn\, (2+R)\, R(1+R)^{-1}\} \tag{7}$$

$$\eta_{DR} = 1.24\, k^{1/2}\, Pe^{1/2}\, R^{2/3} \tag{8}$$

and

$$k = -0.5\ln\alpha - 0.52 + 0.64\,\alpha + 1.43(1-\alpha)Kn \tag{9}$$

Here the Knudsen number, Kn, is introduced to allow for the gas slip effects near the fiber surface. In the above equations, η_D, η_R, and η_{DR} represent the single fiber collection efficiencies due to diffusion, interception, and their combination effect, respectively. The combination term η_{DR} is typically small compared to the main terms η_D and η_R. The single fiber efficiency is defined as the ratio of the cross-sectional area of the original stream from which the particles are removed (i.e., particle-cleared area) to the projected area of the fiber in the direction of the flow.

The simplifying assumptions invoked in the derivation of Equations 6 to 9 include: (1) spherical equivalent particles, (2) cylindrical fibers perpendicular to the flow, (3) uniform packing in the filter, (4) sticking probability of particles equal to one, and (5) a flow field which is not influenced by the particles. Also, the expressions for the single fiber efficiences given above are valid only for $Re < 1$ because of the nature of the flow field employed in the derivation. This condition, however, does not significantly limit the utility of our model because, as our calculations will show, the optimum filter configurations fall into this low Reynolds number regime.

It should be noted that the inertial impaction mechanism is not included in our calculations of the single fiber collection efficiency (see Equation 5). This simplification follows directly from the fact that for an aerosol stream approaching a cylinder, particle collection due to inertial impaction is negligible when $Stk < 0.3$ [12]. (The inertial impaction efficiency for cylindrical fibers can be approximated by $\eta_I = 3.2 \times 10^{-4} \exp(10.5\, Stk)$ for $Stk \leq 0.7$ and $\eta_I = 0.094\, (9.72)^{Stk}$ for $0.7 \leq Stk < 1.04$.) This condition implies that for the particle size range encountered in diesel exhaust filtration (typically $d_p < 1\ \mu m$ [18]), inertial impaction does not contribute significantly to particle collection unless the linear gas velocity is high and the fiber size is small (e.g., $v > 100$ cm/s and $d_f < 10$ μm). Such a combination of high linear velocity and small fiber size, however, is rarely used in filtration practice because of the resulting excessive pressure drop across the filter.

It may be instructive to plot the single fiber collection efficiency as a function of the particle size. As the solid curve of Figure 3 shows, the single fiber efficiency goes through a minimum as the particle size is varied. This interesting phenomenon can be readily explained by examining the contributions of the individual particle collection mechanisms (see the dotted curves of Figure 3). There are two points worthy of mention here: (1) diffusion and interception are primarily responsible for particle collection for diesel exhaust applications, and (2) minimum efficiency occurs in the particle size range where both of these two collection mechanisms are weak. In general, the particle size corresponding to a minimum efficiency is a function of the linear gas velocity and the fiber size. It is interesting to note that the minimum efficiency of Figure 3 occurs at a particle size close to the mass median diameter of diesel exhaust particulates.

FILTERS OF FINITE DEPTH - Having established the expressions for the single fiber efficiency, the efficiency of a filter of finite depth can

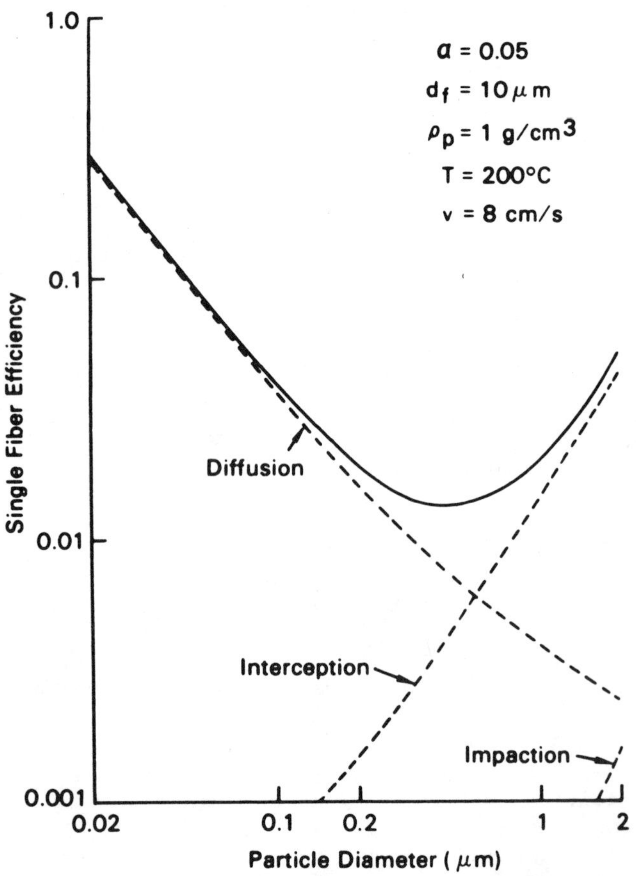

Fig. 3 - Single fiber collection efficiency as a function of particle size

be readily calculated by integrating the following particle mass balance equation:

$$\frac{dm}{dx} = \frac{-4\,\eta\,\alpha}{\pi\, d_f}\, m \,, \qquad 0 < x < H \tag{10}$$

with a known aerosol concentration at the filter inlet. That is, the mass collection efficiency of the whole filter, E, is related to the single fiber efficiency by the simple formula

$$E = 1 - \frac{m(H)}{m(0)} = 1 - \exp\left[-\frac{4\,\eta\,\alpha\,H}{\pi\, d_f}\right] \tag{11}$$

The pressure drop across the filter is given by [17]

$$\Delta P = \frac{4\,F\,\mu_g\, v\,\alpha\, H}{\pi\, d_f^2} \tag{12}$$

where

$$F = 4\pi/[-0.5\ln\alpha - 0.52 + 0.64\,\alpha + 1.43(1-\alpha)Kn] \tag{13}$$

Equations 6 through 9, 11, and 12 show that the key parameters influencing filter performance (i.e., collection efficiency and pressure drop) are (1) linear superficial gas velocity (= gas volumetric flow rate/filter face area), (2) filter thickness, (3) fiber size, (4) packing density, (5) particle size, (6) particle density, and (7) temperature. For a given set of operating conditions, then, there are four filter design parameters which we can vary, namely filter face area, filter thickness, fiber size, and packing density. These design parameters can be divided into two groups; the first two parameters are related to the size and shape of a filter, while the last two refer to the properties of a filter medium. This leads to two important areas of model application: filter size determination, and sensitivity analysis with respect to the properties of the filter medium. Each of these will be briefly discussed in the following paragraphs.

Figure 4 illustrates how our filter model can be used for filter sizing calculations. The operating conditions considered (see Table 2) correspond to approximately 64 km/h (40 mph) road-load of an Oldsmobile 5.7-L diesel engine.

TABLE 2 - Operating Conditions Considered

Exhaust flow rate	= 2832 L/min (at 20°C, 1 atm)
Exhaust temperature	= 200°C
Particle diameter	= 0.2 μm (mass median)
Particle density	= 1 g/cm³

The filter design criteria were chosen to be $E \geq 90\%$ and $\Delta P \leq 3.3$ kPa (1 inch Hg). Curves A and B in Figure 4 represent the conditions for which the filter efficiency and pressure drop constraints are marginally satisfied, respectively. Thus the feasible operating region is given by

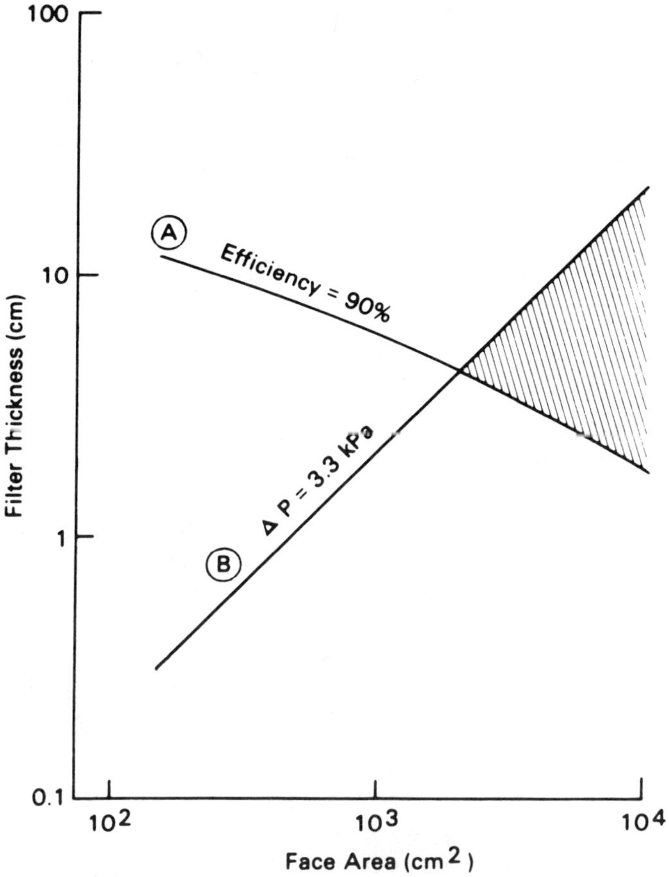

Fig. 4 - Determination of feasible operating region (cross-hatched) for α = 0.05 and d_f = 10 μm

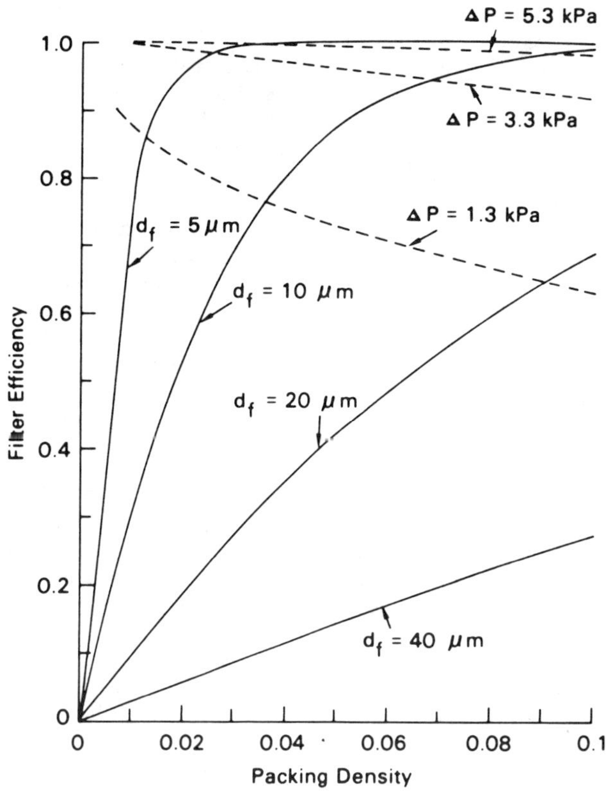

Fig. 5 - Effects of filter-medium properties on filter performance (A = 2500 cm^2, H = 3.4 cm)

the cross-hatched area (that is, the area above Curve A and below Curve B). Of course, each point in the cross-hatched area represents a possible combination of face area and filter thickness which satisfies the specified design criteria. Also, it is interesting to note that the intersection point of Curves A and B corresponds to a minimum filter volume required to satisfy the design criteria.

Other items of practical importance are the effects of the filter-medium properties, such as fiber size and packing density, on the performance of a filter. Figure 5 summarizes the results of such parametric calculations for a given filter configuration (face area = 2500 cm^2, thickness = 3.4 cm). The operating conditions were taken to be the same as those given in Table 2. It can be seen from Figure 5 that increasing the packing density and decreasing the fiber size show a similar trend; that is, the filter collection efficiency improves, but at the expense of increasing pressure drop. Figure 5 also provides useful information about the system's sensitivity to variations in the filter-medium properties. For example, the filter efficiency for d_f = 5 μm is a very strong function of the packing density in the range of small packing densities, while the curve for d_f = 40 μm exhibits a more gradual increase in efficiency over the entire packing density range of practical interest.

EXTENSION TO POLYDISPERSE PARTICLES AND FIBERS - The model discussed above simulates the filter behavior for the case of uniform particle and fiber sizes. Although this model provides useful insight, it is inadequate for predicting the performance of fibrous diesel particulate filters of practical significance, since they are usually composed of polydisperse fibers and operate under the conditions of particle size distributions.

In order to account for the effect of polydispersity of particles and fibers on the filter efficiency, we need to integrate Equation 11 over all particle sizes and fiber sizes of interest. Using an approach similar to that used in [19], and defining the normalized particle mass distribution function as $f(\log d_p)$ and the fiber volume distribution function as $g(d_f)$, Equation 11 becomes:

$$E = 1 - \int_{-\infty}^{\infty} \exp\left(-\frac{4\,\bar{\eta}\,\alpha\,H}{\pi}\right) f(\log d_p)\; d\log d_p \qquad (14)$$

where

$$\bar{\eta} = \int_0^{\infty} \frac{\eta}{d_f}\, g(d_f)\; d\,d_f \qquad (15)$$

It should be noted that the log-normal function is used here to describe the particle size distribution, while the fiber size distribution is taken to be normal. The log-normal function was chosen for the particle size distribution because experimental data [e.g., 20] indicate that aerosol size distribution can usually be cast into this functional form. A detailed review on the use of a log-normal function in particle size analysis is given by Raabe [21]. Also, the normal distribution function seems to be a good representation of fiber size distributions; as will be shown later, the size distributions for the two different fibers used in our experiments can be well approximated by normal functions.

The integrations with respect to the particle size as well as the fiber size distributions in Equations 14 and 15 were carried out numerically using a 10-point Hermite quadrature formula. The abscissas and weight factors for this formula are given by Abramowitz and Stegun [22].

Similarly, the pressure drop in polydisperse fibrous filters can be calculated by integrating Equation 12 with respect to the fiber size distribution. That is,

$$\Delta P = \frac{4 \bar{F} \mu_g v \alpha H}{\pi} \qquad (16)$$

where

$$\bar{F} = \int_0^\infty \frac{F}{d_f^2} g(d_f)\, d\, d_f \qquad (17)$$

It should be noted that integration over the particle size range is not necessary for the pressure drop calculation because of the assumption that the flow field is not influenced by the presence of particles.

EXPERIMENTAL

TEST APPARATUS - A schematic of the filtering test setup and its physical arrangement are shown in Figures 6 and 7, respectively. The essential elements of the test system are a test engine, a filter test fixture, and two raw-gas sampling systems. Each of these elements will be briefly discussed here.

The diesel particulate aerosol was generated using a 0.5-L single-cylinder test engine which was derived from a production 2.1-L passenger car (Opel) diesel engine. Fuel and air flow rates to the engine were measured with a burette system and a sonic-nozzle metering system, respectively. Exhaust gas analyzers were used to measure the concentrations of CO, CO_2, O_2, NO_x, and HC for each data point. In addition to indicating the repeatability of the combustion process from day to day, these concentration data permitted a check on the measured air and fuel flow rates based on the calculated exhaust air-fuel ratio.

The filter test fixture was designed to house filters of different sizes with diameters up to 20 cm and thicknesses up to 7.5 cm. This design allowed us to vary the face area and thickness of the filter easily for a series of parametric studies. As shown in Figure 6, the transition from the exhaust pipe to the filter was very gradual (7° angle) in an attempt to ensure a uniform flow distribution at the filter face. As shown in the photograph of Figure 7, the entire exhaust system was insulated to maintain realistic exhaust gas temperatures at the filter location. The exhaust gas temperature was measured with a chromel-alumel thermocouple mounted 5 cm upstream of the test filter. Heater tape was also wrapped on the exhaust circuit of the filter test fixture to preheat the metal surface prior to the start of each test.

The two raw-gas sampling systems shown in Figures 6 and 7 were used to make simultaneous measurements of particulate mass concentrations at the inlet and outlet of the test filter for the purpose of determining its mass collection

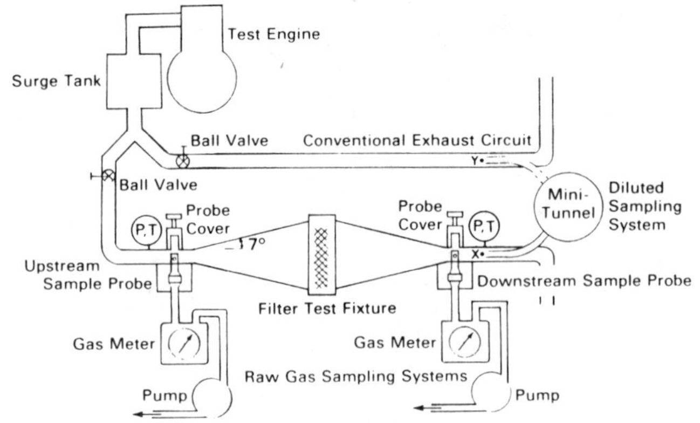

Fig. 6 - Particulate filtering test setup

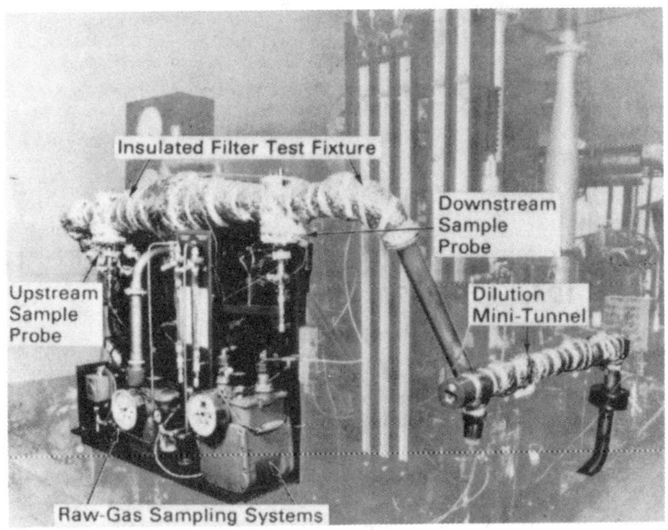

Fig. 7 - Photograph of particulate filtering test setup

efficiency. The sample probes were closed-end, multi-hole probes which were welded directly to 4.7-cm Gelman filter holders. This design eliminated the need for any sample line upstream of the sample filter which frequently results in loss of part of the particulate sample. The particulate samples were collected on Dexiglass No. 225 glass fiber filters. Flow rates through the sample filters were measured with Singer dry-gas meters and checked with Fischer & Porter rotameters. Ice baths were used to remove the combustion-generated water between the sample filter and the dry-gas meter. This prevented unwanted condensation in the sample lines and gas measuring instruments. Gast vane pumps were used to draw the aerosol samples through the sampling networks.

In our experiments, considerable caution was exercised to maintain the sample temperature close to the exhaust temperature in order to ensure that the particulate aerosol collected in the sample filter was truly representative of the aerosol flowing through the test filter element. In an attempt to minimize the temperature transients in the sampling system, two sample-probe covers (see Figure 6) were used to shield the sampling probes from the exhaust stream prior to the start of sampling. These probe covers allowed the sampling system to be heated to the desired exhaust temperature without contaminating the sample filters. The sample temperature was monitored just behind the sample filter in the filter housing.

TEST CONDITIONS - For all of the data presented in this paper, the test engine was run at a constant load and speed of 620 kPa IMEP and 1500 r/min, respectively. The air-fuel ratio at this condition was nominally 26:1. A constant engine back pressure of 3.3 kPa (1 inch Hg) was maintained for all of the test conditions, independently of the pressure drop across the filter. These operating conditions yielded an exhaust temperature at the filter of approximately 200°C and an exhaust flow rate of 300 L/min (at 20°C, 1 atm).

TEST PROCEDURE - Prior to each test, the exhaust gas was routed through the conventional exhaust circuit shown in Figure 6 until the engine stabilized at the desired test condition. The stability of the engine operation was determined by monitoring the engine fuel rate, oil temperature, coolant temperature, and exhaust emission levels. This warmup process usually required about one hour of engine operation. During the engine warmup process, power was supplied to the heater tapes in order to heat the filter exhaust system to the normal exhaust temperature of 200°C.

After the stability of the engine operation was achieved, the exhaust flow was diverted to the exhaust system containing the filter test fixture and a test was started. The sample-probe covers were then lifted, and the sampling systems upstream and downstream of the filter were started simultaneously. Sampling was continued until the pressure drop across one of the sample filters reached 33 kPa (10 inches Hg). The sample flow rates were chosen so that both sampling systems reached the 33 kPa pressure drop criterion after approximately ten minutes of sampling. This procedure resulted in nominal flow rates of 16 L/min and 23 L/min (at 20°C, 1 atm) upstream and downstream of the filter, respectively. The resulting particulate mass samples ranged between 2.5 and 7.5 mg.

The mass of particulate collected on the sample filters was measured in a temperature- and humidity-controlled room after 12 hours of stabilization. A Sartorius microbalance was used to make these mass measurements.

The particulate mass concentration was determined by dividing the particulate mass by the volume of sample flow and then correcting the sample flow to the standard temperature and pressure. The mass collection efficiency of the filter was then calculated from the formula

$$E = 1 - \frac{m(H)}{m(0)}$$

where m(0) and m(H) are the particulate mass concentrations upstream and downstream of the filter, respectively.

PARTICLE SIZE-DISTRIBUTION MEASUREMENTS - The particle size distribution of the exhaust aerosol was measured using a Model 3030 Electrical Aerosol Analyzer (EAA) manufactured by Thermo Systems, Inc. The exhaust gas sample to the analyzer was diluted with the dilution mini-tunnel shown in Figures 6 and 7. This dilution system is described in detail by MacDonald et al. [23]. The exhaust gas was diluted 10:1 in the main dilution nozzle and 30:1 in the secondary dilution nozzle to give a total dilution of 300:1 for the EAA sample. Particle size distributions upstream and downstream of the test filter were obtained from the conventional exhaust circuit (sampling point labeled Y in Figure 6) and from the filter test-fixture exhaust circuit (sampling point labeled X in Figure 6), respectively.

Figure 8 shows the particle size distribution data (circles) of the exhaust aerosol upstream of the filter as well as the log-normal function (solid curve) used to fit the data. Each point represents the average of six separate measurements. A detailed description of the principle of EAA measurements and the data reduction method used here can be found elsewhere [24]. It should be noted that the particle <u>volume</u> distribution shown in Figure 8 was directly used in this work to calculate the mass efficiency of the filter, assuming that the particle density remains constant over the particle size range of interest.

CHARACTERIZATION OF FIBERS - In our experiments, two different grades of commercial steel wool fibers were used: medium (Grade 1) and medium-fine (Grade 0). Because of its production method, steel wool is a very nonhomogeneous material that consists of continuous fibers or filaments (Figure 9) having a wide variety of cross-sectional shapes and sizes (Figure 10). Such irregularity in shape and size makes it

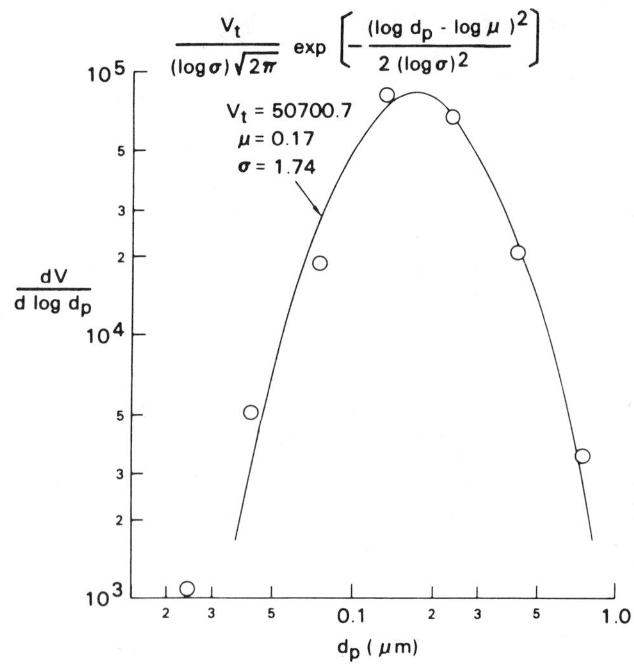

Fig. 8 - Particle size distribution of the exhaust aerosol upstream of the filter

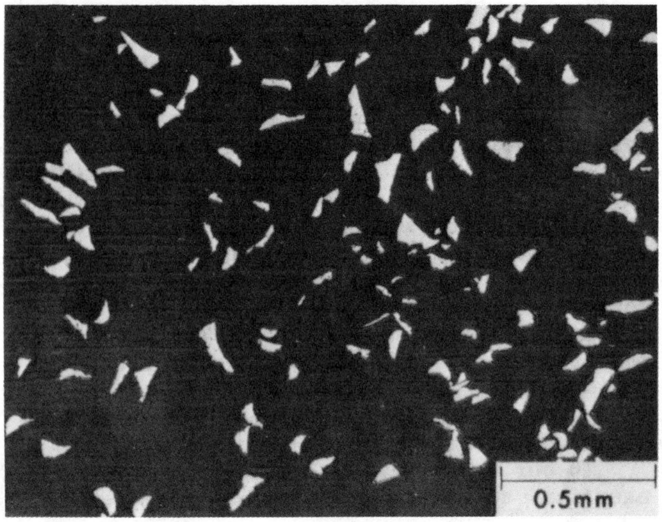

Fig. 10 - Cross-section of the steel wool fiber sample (medium-grade, 4% packing density)

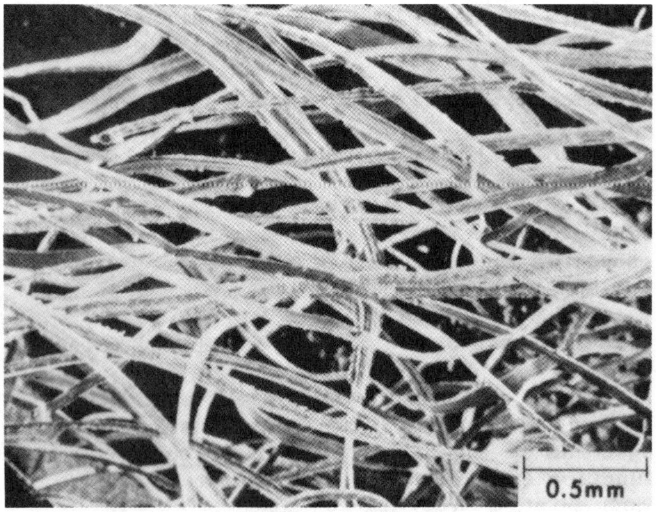

Fig. 9 - Photograph of the steel wool fiber sample (medium-grade, 4% packing density)

necessary to define and determine an "effective" fiber diameter for the purpose of efficiency predictions. The accurate determination of the fiber size distribution is important since the filter efficiency is a very sensitive function of the fiber size (see Figure 5).

The steel wool fibers used in the tests were characterized by examining their cross-sections using an image analyzer. The instrument used was a Model 720 Quantimet unit produced by the Imanco Division of Metals Research, Ltd. It consists essentially of a microscope, a TV-type Vidicon tube, a CRT for display purposes, and a variety of electronic controls and packages designed to quantify selected microscopic features displayed on the Vidicon tube. Based primarily on the number of scanning lines interrupted by certain displayed microscopic features and the number of picture points involved in those features, various parameters characterizing the selected microscopic features, such as their number, cross-sectional area, perimeter, projected length, and spacing, can be quickly determined.

For the purpose of fiber characterization, samples of the two commercial steel wool fibers were mounted and metallographically prepared such that the cross-sections of the fibers were exposed, as shown in Figure 10. Individual fibers in such a sample (containing 50 to 120 fibers) were examined on the Quantimet to determine the cross-sectional area and perimeter of each fiber. Hydraulic diameters (defined as four times the cross-sectional area divided by the perimeter) were then calculated to characterize the sizes of the individual fibers of irregular shape (see Figure 10), and a histogram showing the volume fraction of fibers vs. the hydraulic diameter was constructed. Figures 11 and 12 show such histograms for the medium- and medium-fine-grade steel wool fibers, respectively. These two fiber-volume distributions were found to be reasonably well approximated by normal distribution functions, as shown in Figures 11 and 12. The median and standard deviation of the normal functions were determined by plotting the distribution data on probability paper. Note that the medium-fine-grade steel wool exhibits a smaller median diameter, as well as a smaller deviation around the median, than the medium-grade steel wool.

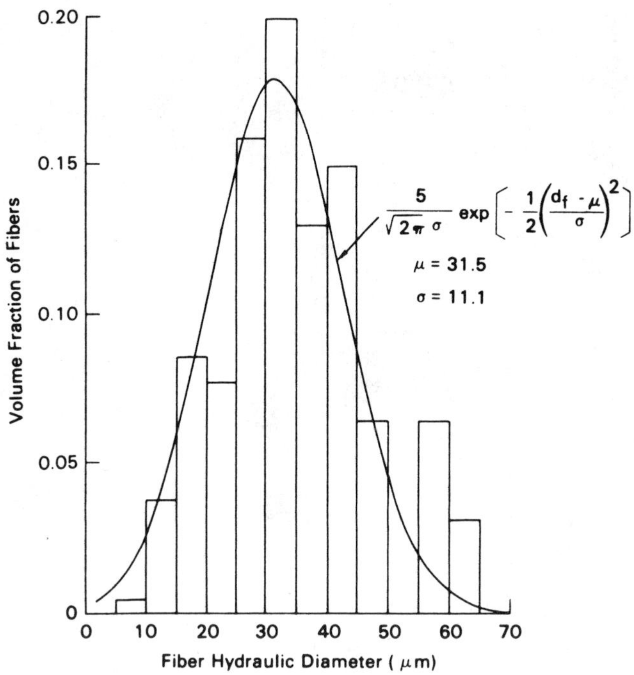

Fig. 11 - Fiber size distribution of the medium-grade steel wool

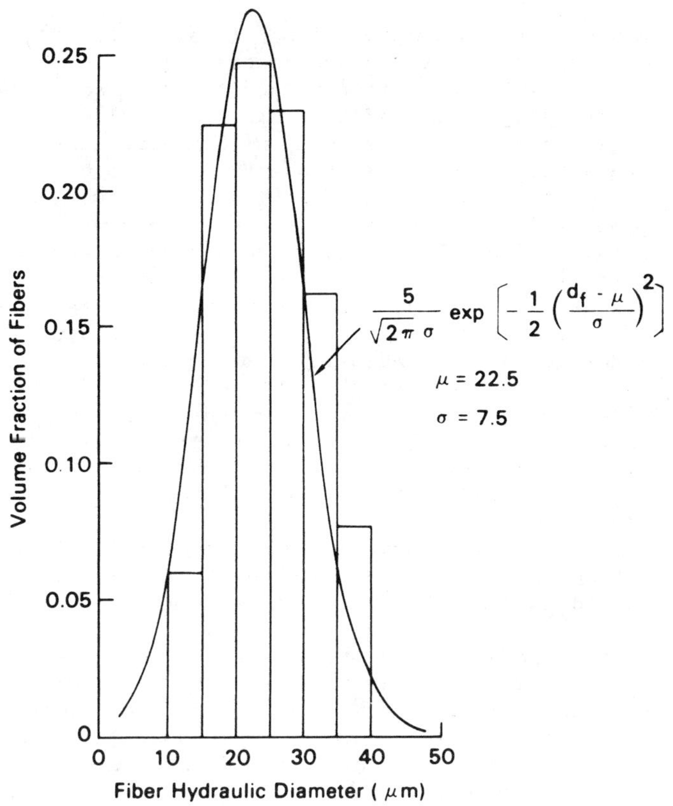

Fig. 12 - Fiber size distribution of the medium-fine-grade steel wool

COMPARISON OF THEORY AND EXPERIMENTS

As discussed in the Theoretical section, our modeling studies led to the identification of four important filter design parameters: filter face area, filter thickness, packing density, and fiber size. In our experiments we systematically varied these key design parameters in order to investigate the parametric sensitivity of the system and to test the validity of the model.

Table 3 compares measured and predicted initial collection efficiencies for five different test conditions involving two different fibers. For the experiments with the

TABLE 3 - Comparisons of Initial Filter Performance

	Efficiency (measured)	Efficiency (predicted)	Pressure Drop (predicted)
Medium-grade steel wool:			
(I) α = 4%, H = 5.1 cm, D = 20.0 cm	61%*	61%	0.5 kPa
(II) α = 4%, H = 5.1 cm, D = 14.1 cm	50%	50%	1.0 kPa
(III) α = 4%, H = 2.5 cm, D ≐ 20.0 cm	45%	39%	0.25 kPa
(IV) α = 3%, H = 5.1 cm, D = 20.0 cm	48%**	51%	0.34 kPa
Medium-fine-grade steel wool:			
(V) α = 3%, H = 5.1 cm, D = 20.0 cm	66%	63%	0.5 kPa

* average of four runs

** average of two runs

medium-grade steel wool fiber (Cases I to IV), we took Case I as a standard case and perturbed one parameter at a time around the standard case to investigate the effects of the individual parameters. The experiment with medium-fine-grade steel wool fiber (Case V) was also conducted to study the effect of fiber size. The predicted efficiencies shown in Table 3 were obtained from the model, which accounted for the particle size and fiber size distributions discussed previously in the Experimental section of this report. It should be noted that our model contains no adjustable parameters; that is, all the parameters of the model were determined from independent measurements. As can be seen from Table 3, very good agreement was obtained between measured and predicted values of the filter efficiency for all the cases considered.

It should be mentioned that measured values of the pressure drop across the filter are not listed in Table 3 because it was difficult to make reliable pressure drop measurements. At least part of this difficulty with the pressure drop measurements can be attributed to the limited resolution of the standard flush-mounted pressure taps used in the experiments. These taps were not designed to provide accuracy in the relatively low pressure range encountered in our experiments. (For Case I in Table 3, for example, the pressure drop data from four repeated runs show a mean value of 0.10 kPa and a relatively large standard deviation of 0.06 kPa.) In addition, the pressure drop measurements were complicated by cyclic fluctuations in the manometer readings (about $\pm$ 0.05 kPa), which were caused by engine generated pressure pulsations in the exhaust system. Although direct comparison between theory and experiment cannot be made for the pressure drop for the reasons discussed above, predicted values of the pressure drop (obtained using Equations 16 and 17) are shown in Table 3 in order to illustrate the effects of the various parameters on the pressure drop across the filter.

In order to test the validity of the model further, we measured the particle size distribution downstream of the filter for the standard case (Case I in Table 3) and compared it with the prediction of the model. As shown in Figure 13, the model again correctly predicts the particle size distribution downstream of the filter, confirming the validity of the model developed here. It is worth mentioning that the particle size distribution upstream of the filter was also measured in this experiment and used as a basis for calculating the theoretical curve shown in Figure 13, so that a more meaningful comparison between theory and experiment can be made. The particle size distribution data obtained by this measurement (average of three runs: μ = 0.196, σ = 1.70) showed no significant deviation from those of Figure 8 (μ = 0.170, σ = 1.74).

As a result of these model-verification studies, we conclude that our model describes the essential features of the system well enough

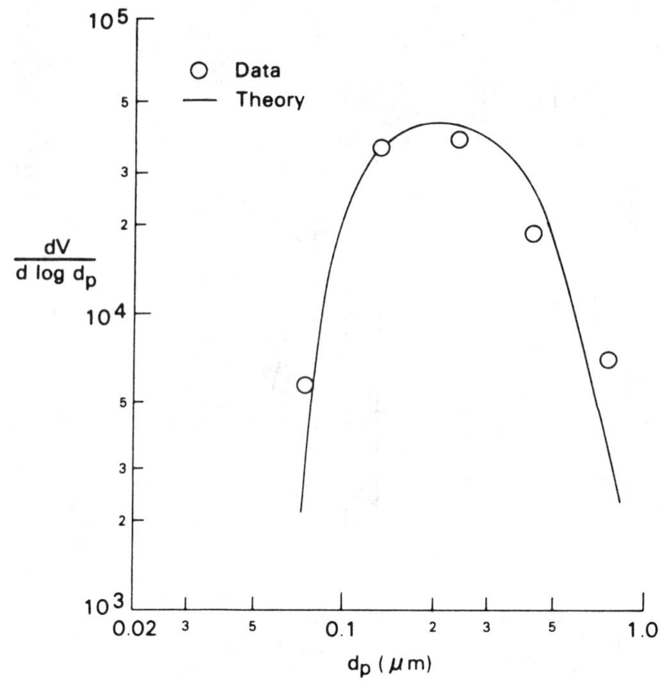

Fig. 13 - Particle size distributions downstream of the filter (α = 0.04, H = 5.1 cm, D = 20.0 cm, medium-grade steel wool)

to permit the quantitative prediction of the initial collection efficiency of fibrous diesel particulate filters.

ACKNOWLEDGMENT

The authors gratefully acknowledge S. R. McEwen and L. S. Jerzy of the Engine Research Department for their efforts in building the test apparatus and in collecting the experimental data for this study. R. C. Lints, J. M. Gulish, and W. W. Bach of the Metallurgy Department provided help both in the preparation of the test filters and in the characterization of the steel wool samples. Thanks also go to P. J. Groblicki of the Environmental Science Department for use of his data reduction program in reducing the Electrical Aerosol Analyzer data. Discussions with Professors B. Y. H. Liu (Univ. of Minnesota) and A. C. Payatakes (Univ. of Houston) were also valuable.

NOMENCLATURE

- A = filter face area, cm^2
- C = Cunningham correction factor
- D = diameter of filter face, cm
- $\mathcal{D}$ = diffusion coefficient of particle, cm^2/s
- d_f = fiber diameter, cm
- d_p = particle diameter, cm
- E = mass efficiency of total filter
- F = dimensionless drag force given by Equation 13

$\overline{F}$ = quantity given by Equation 17
$f(\log d_p)$ = normalized particle size distribution function
$g(d_f)$ = normalized fiber size distribution function
H = filter thickness, cm
k_b = Boltzmann constant, g $cm^2/(s^2 K)$
k = hydrodynamic factor given by Equation 9
Kn = Knudsen number, $2\lambda/d_f$
m = mass concentration of aerosol, g/cm^3
ΔP = pressure drop across the filter, $dyne/cm^2$
Pe = Peclet number defined by Equation 3
R = interception parameter defined by Equation 1
Re = Reynolds number, $d_f\ v\ \rho_g/\mu_g$
Stk = Stokes number defined by Equation 2
T = temperature, K
v = linear gas velocity, cm/s
V = particle volume per unit volume of gas, $\mu m^3/cm^3$
V_t = total particle volume per unit volume of gas, $\mu m^3/cm^3$
x = length measured from the filter inlet, cm

Greek Letters

α = packing density (volume fraction of fibers in filter)
η = single fiber efficiency
η_D = single fiber efficiency due to diffusion
η_I = single fiber efficiency due to inertial impaction
η_R = single fiber efficiency due to interception
η_{DR} = combination term in the single fiber efficiency due to diffusion and interception of particles
$\overline{\eta}$ = quantity given by Equation 15
λ = mean free path of gas molecules, cm
μ = median fiber or particle diameter, μm
μ_g = gas viscosity, g/(cm s)
ρ_g = gas density, g/cm^3
ρ_p = particle density, g/cm^3
σ = standard deviation of fiber or particle size distributions, μm

REFERENCES

1. General Motors Response to EPA Notice of Proposed Rulemaking on Particulate Regulation for Light-Duty Diesel Vehicles, submitted to Environmental Protection Agency, April 19, 1979.
2. Public Hearing on EPA's Proposed Particulate Emission Standards for Light-Duty Diesel Vehicles, Washington, D.C., March 19 and 20, 1979.
3. K. J. Springer and R. C. Stahman, "Removal of Exhaust Particulate from a Mercedes 300D Diesel Car," SAE Paper 770716, presented at Off-Highway Vehicle Meeting and Exhibition, Milwaukee, September 12-15, 1977.
4. L. P. Tessier, H. F. Sullivan, G. M. Bragg, and C. E. Hermance, "The Development of a High Efficiency Diesel Exhaust Particulate Filter," SAE Paper 800338, presented at SAE Congress and Exposition, Detroit, February 25-29, 1980.
5. C. N. Davies, "Air Filtration," Academic Press, New York, 1973.
6. C. Y. Chen, "Filtration of Aerosols by Fibrous Media," Chem. Rev. 55, 595 (1955)
7. A. C. Payatakes, "Model of Transient Aerosol Particle Deposition in Fibrous Media with Dendritic Pattern," AIChE J. 23, 192 (1977).
8. A. C. Payatakes, "Model of the Dynamic Behavior of a Fibrous Filter. Application to Case of Pure Interception During Period of Unhindered Growth," Powder Technology 14, 267 (1976).
9. A. C. Payatakes and L. Gradon, "Dendritic Deposition of Aerosols by Convective Brownian Diffusion for Small, Intermediate and High Particle Knudsen Numbers," AIChE J. 26, 443 (1980).
10. R. Leers, "Die Abscheidung von Schwebstoffen in Fasernfiltern," Staub 50, 402 (1957).
11. C. E. Billings, "Effect of Particle Accumulation in Aerosol Filtration," Ph.D. Dissertation, Calif. Inst. Technol., Pasadena (1966).
12. H. E. Hesketh, "Fine Particles in Gaseous Media," Ann Arbor Science, Ann Arbor, Michigan, 1977.
13. A. Kaufmann, "Die Faserstoffe für Atemschutzfilter," Z. Verein Deutscher Ing. 80, 593 (1936).
14. I. Langmuir, "Report on Smokes and Filters," U.S. Office of Scientific Research and Development Report No. 865, Part IV (1942).
15. J. Happel, "Viscous Flow Relative to Arrays of Cylinders," AIChE J. 5, 174 (1959).
16. S. Kuwabara, "The Forces Experienced by Randomly Distributed Parallel Circular Cylinders or Spheres in Viscous Flow at Small Reynolds Numbers," J. Phys. Soc. Japan 14, 527 (1959).
17. A. A. Kirsch and I. B. Stechkina, "The Theory of Aerosol Filtration with Fibrous Filters," in "Fundamentals of Aerosol Science," Ed. D. T. Shaw, John Wiley & Sons, New York, 1978.
18. P. J. Groblicki and C. R. Begeman, "Particle Size Variation in Diesel Car Exhaust," SAE Paper 790421, presented at SAE Congress and Exposition, Detroit, February 26 - March 2, 1979.
19. F. Löffler, "Collection of Particles by Fiber Filters," in "Air Pollution Control," Part I, Ed. W. Strauss, Wiley-Interscience, New York, 1971.
20. N. A. Fuchs, "The Mechanics of Aerosols," Pergamon Press, Oxford, 1964.
21. O. G. Raabe, "Particle Size Analysis Utilizing Grouped Data and the Log-Normal Distribution," Aerosol Science 2, 289 (1971).
22. M. Abramowitz and I. A. Stegun, "Handbook of Mathematical Functions," Dover Publications, New York, 1965.
23. J. S. MacDonald, S. L. Plee, J. B. D'Arcy, and R. M. Schreck, "Experimental Measurements of the Independent Effects of

Dilution Ratio and Filter Temperature on Diesel Exhaust Particulate Samples," SAE Paper 800185, presented at SAE Congress and Exposition, Detroit, February 25-29, 1980.

24. B. Y. H. Liu, D. Y. H. Pui, and A. Kapadia, "Electrical Aerosol Analyzer: Principle and Data Reduction," presented at Aerosol Measurement Workshop, University of Florida, Gainesville, Florida, March 1976.

Cellular Ceramic Diesel Particulate Filter**

John S. Howitt and
Max R. Montierth
Corning Glass Works

ABSTRACT

An innovative design of ceramic filter for diesel engine exhaust particulate matter is described. It is a cellular ceramic honeycomb with porous walls which act as the filter medium. Test results relating the effects of wall thickness, cell density, porosity, and filter volume on collection efficiency, pressure drop and filter capacity are described.

Regeneration of the filter by ignition of the accumulated soot has been accomplished with conclusions drawn about the influence of exhaust temperature, oxygen content, and the effect of the filter design.

INTRODUCTION

Diesel powered passenger cars and light duty trucks are steadily increasing their market share in the U.S. It is expected that the majority of automakers will offer diesel options by 1984. Diesels power about .5% of all vehicles on the road in 1980 and that is expected to increase to approximately 25% of the new vehicles and to 5% of the total of all vehicles in 1985 and perhaps to 15% by 1990.

The reason for this explosive growth is mainly economic. The diesel provides an approximate 25% gain in average fuel economy over gasoline engines with equivalent displacement * (1) and additionally the price of diesel fuel is at the present time approximately 10% cheaper than unleaded gasoline. This conversion to diesel is certainly one of the major changes in the automotive scene in the past few decades.

The diesel has a substantial emissions problem which is quite different than that of gasoline engines. It is relatively clean with respect to unburned hydrocarbons and carbon monoxide but it produces large amounts of particulate comprised of soot and adsorbed organics.

The particulate matter is a complex mixture of pure carbon plus hundreds of organic compounds, not all of which have been identified and understood. The particulate is extremely fine and light with a flour-like consistency. More than 50% of it is less than .5 microns in size.

If totally unregulated, with the expected market penetration several years from now, diesels could add over 200,000 tons of particulate matter to the air each year. The effect of these emissions on air quality and public health is the subject of a large amount of research. Chemical and biological tests have shown that certain compounds associated with diesel particles are potentially carcinogenic. At this point, it is questionable if long term exposure to concentrated levels of diesel exhaust will increase the incidence of cancer in exposed populations. (2)

The amount of particulate matter emitted by a diesel is generally related to vehicle weight with heavy vehicles producing greater amounts. It is 60-100 times greater than comparable gasoline powered vehicles.

The EPA has imposed future limits on the amount of particulate that may be emitted by light duty diesel vehicles. The standards based on the Federal Testing Procedures are .6 g/mi for all light duty vehicles beginning in 1982 Model Year. In the 1985 Model Year the limits will be reduced to .2 g/mi for passenger vehicles, .26 g/mi for light duty trucks. (3)

The EPA estimates that these standards will force the reduction of the total particulate emissions from light duty diesels by 74% by 1990 with respect to what could be expected without these regulations. The automotive industry has approached the control of particulate in a number of ways: fuel modifications, engine modifications, and after-treatment separation devices. Complicating matters is the 1.0 ppm NO_X standard because the methods typically used to reduce NO_X (such as EGR) have the effect of increasing particulates.

It is generally agreed that all manufacturers can meet the .6 g/mi specification with engine modifications which do not necessitate the use of exhaust filters. However, it is most likely that all diesel vehicles will require a regenerative type filter as part of their emissions control system to meet the .2 g/mi specification in 1985.

*Numbers in parentheses designate References at end of paper

**Paper 810114 presented at the International Congress and Exposition, Detroit, Michigan, February, 1981.

This report describes the design and performance of a device to filter the majority of the particulate matter in the exhaust.

CONCEPT

The Diesel Particulate Filter, (DPF), is an outgrowth of the extruded cellular ceramic technology successfully utilized in the production of monolithic catalyst supports. It is a square cell shaped honeycomb with parallel channels running the length of the unit. The body of the structure is basically a porous cordierite (2 MgO - $2AL_2O_3$ - $5SiO_2$) which has the necessary mechanical strength, chemical resistance, thermal fracture resistance, and melt resistance to survive effectively in the hostile environment of a diesel engine exhaust. (4)

The filter concept involves blocking alternate cell channel openings on the monolith face in a checkerboard type fashion as seen in Figure 1. The opposite end is similarly blocked but one cell displaced so that the gas cannot flow directly through a given channel. This manifolded cellular DPF is mounted in the exhaust system of the diesel engine. The exhaust gas enters the upstream open end of cells. Since the downstream end of the cell is blocked with a ceramic plug, the exhaust gas is forced through the porous wall to exit through an adjacent cell.

This is shown in Figure 2. The exhaust gas is filtered by the porous wall. Obviously the filtration pressure drop and total filtrate loading characteristics are determined by the wall properties and cell geometry. This monolithic honeycomb structure offers the advantage of a large potential filter area in a reasonably compact volume.

The nature of the material used to block the cell openings at the face is similar to the ceramic body in thermal expansion and composition.

It is estimated that a diesel engine will typically emit a gallon of particulate in every thousand miles. Therefore, it is acknowledged that the accumulated particulate matter must be periodically removed, presumably by thermal oxidation. (5) This involves raising the temperature of the exhaust exiting the engine and entering the filter sufficiently to cause the soot to ignite (450°C or greater). The ceramic body provides a thermally stable base for this process.

DESIGN CRITERIA

The successful application of wall flow type ceramic filters will depend upon meeting the following criteria without penalty to vehicle economy or performance.

1. COLLECTION EFFICIENCY - Typically the higher the efficiency, the more effective the filter, but in this case, it must not be accomplished at the expense of excessive contributed back pressure.

Some automakers will need a reasonably high (75%-90%) collection efficiency to allow their vehicles to meet the 1985 standard (.2 g/mi). Others with engines which emit lower levels of particulate may require only 50%-60%. This must be accomplished by a compromise of properties to maintain the back pressure at as low a level as possible.

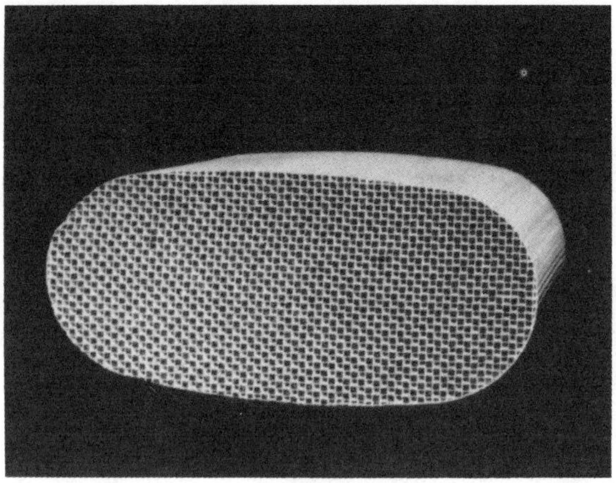

Fig. 1

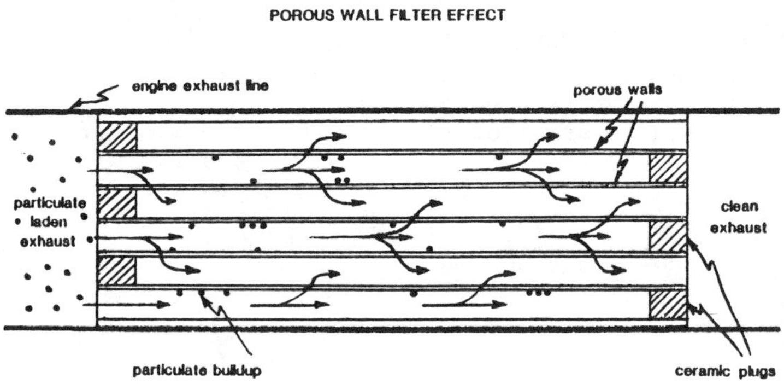

Fig. 2

2. PRESSURE DROP - When exhaust flows through the filter, the particulate becomes trapped in the internal pores of the body of the cell wall and then begins to deposit as a layer on the surface of the channel. As the accumulated matter collects, the thickness of this layer increases, thereby increasing the pressure drop across the unit. The soot accumulation can be seen in Figure 3. Back pressure in an exhaust system is important because it has a direct and negative effect on engine operation and vehicle fuel economy.

Pressure drop must be controlled to as low a level as possible and the maximum tolerated amount will vary due to engine size and style. It is a function of cell geometry, wall properties, and volume of the ceramic unit. These are controllable and, as previously stated, they must be balanced against maintaining the required filter efficiency.

3. REGENERATION - Due to the increasing pressure drop across the filter, resulting from rapidly accumulating material, periodic regeneration by ignition of the accumulated soot is required. Filter configuration can affect the regeneration time and peak pressure. These properties must

Fig. 3

be controlled by filter design to the desired ranges for engine or environmental requirements.

4. DURABILITY - The ceramic DPF must demonstrate structural integrity for the useful lifetime for the vehicle. This can be thought of in terms of the number of regenerations rather than the accumulated miles because it is the regeneration cycle which produces the most physical demand on the filter.

Ignition of the accumulated soot in the regeneration operation releases a large amount of energy and generates a rapid temperature rise. It is not necessarily evenly distributed throughout the body, resulting in thermal gradients in both radial and axial directions. It is an important design requirement that the ceramic body be resistant to fracture from thermal stresses induced by these gradients. A low expansion ceramic composition is necessary to maintain structural integrity over the lifetime of the filter. An excessive build up of particulate material can result in a large amount of released energy upon ignition, and temperatures exceeding the melting point of the cordierite may occur. Acceptable operating characteristics and filter life, therefore, will require the proper balance among the operating variables of filtration time between regeneration cycles, filter pressure drop, and degree of particulate loading.

TEST PROGRAM - As part of the program to develop this device, a laboratory testing program was used to evaluate the effects of the honeycombs parameters on the above criteria.

In order to generate representative particulate for filter testing, it was judged necessary and most expedient to use actual diesel engine exhaust.

EXPERIMENTAL PROCEDURES

1. SAMPLES - All DPF samples consisted of variations of monolithic honeycombs with manifolded ends. The basic ceramic composition consists of a modified cordierite ($2MgO - 2Al_2O_3 - 5SiO_2$) encompassing ranges in porosity from 35% to 55%, mean pore diameter from 10 to 35μm, and thermal expansion coefficient (RT to 1000°C) of $9.5 - 12.5 \times 10^{-7}/°C$.

The manifolding cement consists of a modified cordierite composition similar to the DPF body in thermal expansion characteristics.

DPF shapes tested included 6" and/or 12" lengths of 3.66" rounds, 5.66" rounds and 3.18" x 6.68" ovals. Cell densities of 50, 100, 200 and 300 Sq's/in^2, and wall thickness of .012", .017", and .025" were included in the study.

2. FILTRATION - Filtration tests and regeneration tests were both conducted using stationary 1980 Oldsmobile 350 CID diesel V-8 engines coupled to water brake dynamometers. Baseline emissions from each of the two engines were measured periodically to insure test consistency. Filtration studies were conducted at engine conditions of 1000 rpm and 100 ft.-lbs. torque. Filter cleaning between filtration tests, when conducted, was accomplished by furnace heating to 1100°F for 1 hour. Filtration tests were conducted on one exhaust bank from the engine.

Baseline particulate data was obtained through weight gain, using a 47 mm Millipore filter to collect particulate from a portion of the exhaust stream. Flow rate and time were carefully controlled. Sample ports were located downstream from filter cans, and the collection system consisted of a small tube positioned longitudinally in the exhaust stream with open end directed into the gas flow, with flow rate controlled through a metered pump. The diesel particulate filter can was replaced with a valve during particulate baseline measurement, and the valve ΔP adjusted to approximate the DPF ΔP. Baseline data was monitored over a range of exhaust back pressure.

Filtration data was obtained by first achieving steady state engine operations, then installing a clean filter in the filter can, and then monitoring filter pressure drop and weight gain over the test period. Particulate baseline conditions were periodically sampled. Filtration tests were normally terminated at a filter ΔP of 140 cm H_2O. This valve was suggested by the engine manufacturer as the maximum tolerable DPF pressure drop. Weight loading for the filter was calculated from initial and final weight measurements, and filter efficiency calculated from the loading and baseline data. These valves were checked with particulate levels downstream of the filter compared with baseline particulate level for the appropriate back pressure. Filter efficiencies calculated from filter weight gain agreed closely with the calculations based on downstream untrapped particulate levels.

3. REGENERATION - Regeneration tests were conducted on the engine, with full engine exhaust through the filter. Exhaust temperature increase to initiate regeneration was accomplished by combinations of increased engine speed, throttling, and retardation. Engine conditions for each regeneration temperature and oxygen concentration were determined with the exhaust gas channeled through a bypass line of pressure drop equivalent to the filter loaded with particulate to test conditions. After achieving steady state condi-

tions the exhaust stream was rapidly switched to the line containing the preloaded filter, and the regeneration parameters recorded. A schematic of the regeneration test system can be seen in Figure 4. Temperature, oxygen concentration, HC, CO, CO_2, and pressure valves were continuously measured immediately upstream and downstream of the filter throughout the regeneration test. Regeneration cycle conditions are described in Table 1 for all types of regeneration systems considered.

Table 1 - Regeneration Cycle Conditions

1. Particulate Collection

Flow = 70-75 CFM
Temp = 450°F, ΔT/in = 10-20°F
Collection at .25g/mile (one exhaust bank)
Collection period ~ 200 miles
ΔP = 10 cm H_2O - 140 cm H_2O

2. Regeneration Initiation

Flow = 5-200 CFM (engine intake)
Temp = 1000° - 1700°F
ΔP = 140 cm H_2O - 550 cm H_2O
ΔT/in = 10-1500°F

3. Filter Regeneration

Flow = 5-200 CFM
Temp = 1000°F-2000°F
ΔP = 550 cm H_2O - 10 cm H_2O
Time = 2 seconds - 2 hours
ΔT/in = 10-1500°F

EXPERIMENTAL DESIGN

Modified 2^N factorial experiments were used in both filtration and regeneration studies, encompassing the range of filter design variables, and those variables consistent with operation and regeneration on the engine. The range in sample design variables was described in the sample description section.

The range in filtration test variables and regeneration test variables can be seen in Table 2.

Table 2 - Regeneration Study - Variable Range

Variable	Min.	Max.
Cells/in^2	50	300
Wall Thickness (in.)	.012	.025
Porosity (%)	35	55
Mean Pore Size/μm)*	10	35
Regeneration Temp. (°F)	900	1200
Oxygen Concentration (%)	3	9
Particulate Loading Pressure drop (cm H_2O)**	60	120

* Regeneration test conducted primarily with samples of 50% porosity and 12.5μm mean pore size, as determined by results of the filtration study.

** DPF loaded pressure drop in one cylinder bank.

REGENERATION TEST STAND SCHEMATIC

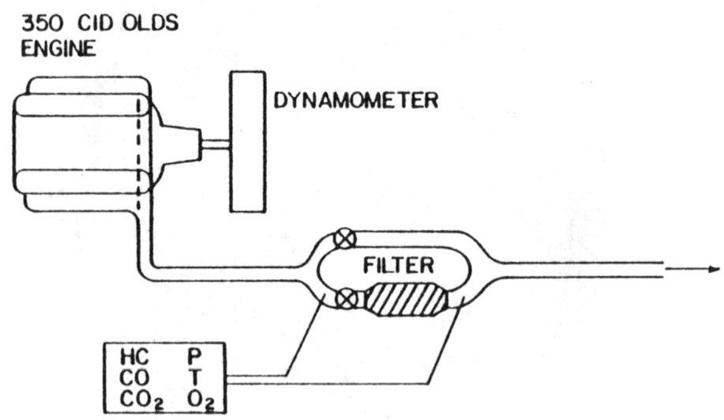

Fig. 4

EXPERIMENTAL RESULTS AND DISCUSSION

Filtration

A typical filtration curve can be seen in Figure 5. The filtration behavior consists of three basic regions, (1) rapid initial ΔP rise with negative d^2P/dt^2, (2) a relatively long linear ΔP/ΔT region, and (3) a rapidly increasing ΔP region of positive d^2P/dt^2. It is felt the first region corresponds to particulate plugging of the wall pores and formation of the particulate blanket. The steady state region of relatively constant dP/dt represents the particulate blanket buildup. The third region, represented by the increasing dP/dt behavior is probably caused by increasing restricted exhaust gas flow as the cells become plugged with particulate.

Calculations of the particulate packed density, based upon loading weight and measured particulate film thickness were found to be relatively constant at 0.05 to 0.06 g/cm^3. There were no significant radial variations in amount of particulate collected within the filter. There were significant axial variations in particulate loading, with the upstream end (~5/8") loaded 43% in excess of the remainder of the cell. These results can be seen in Figure 6.

The filtration behavior of filters was observed to describe two basic categories; that typical of the small pore filters (MPS $\leq$ 15μm), and that represented by the larger pore size materials. Filter efficiency curves for each can be seen in Figure 7. Low initial filtration efficiency was obtained with the large mean pore size material (35μm MPS), presumably resulting from a large fraction of the particulates passing through the wall. With gradual plugging of the wall pores during continued operation, the filtration efficiency was observed to increase significantly to the value characteristic for that material. This contrasts sharply with filters with mean pore size $\leq$ 15μm, which exhibited filtration efficiencies $\geq$ 80% for all wall thicknesses over the entire filtration test.

The effects of wall thickness, mean pore size, and filter size upon filtration time can be

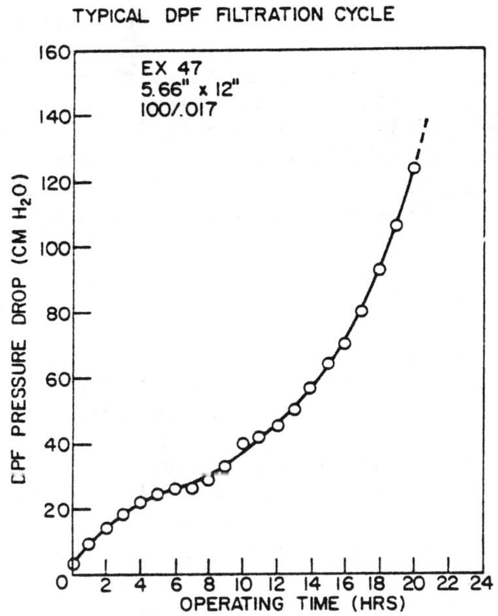

Fig. 5

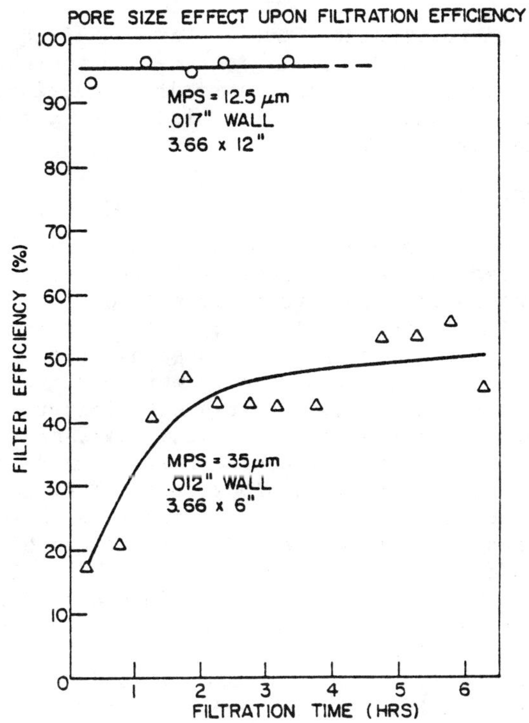

Fig. 7

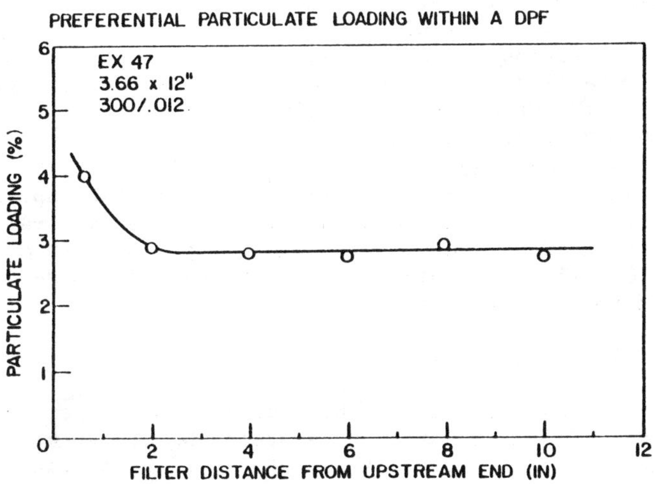

Fig. 6

seen in Figure 8. A doubling of the wall thickness effectively halves filtration time to achieve ΔP_L = 140 cm H_2O for the 12.5µm mean pore size filter. The same wall thickness increase reduces the filtration time by more than a factor of 5-10x, depending upon filter volume, for a 35µm mean pore size filter. The relative changes in filtration time (to ΔP_L = 140 cm H_2O pressure drop) with changes in wall thickness and filter area are summarized in Table 3. The results indicate doubling the large pore size filter wall thickness to be approximately 3x more influencial on filtration time than doubling the wall thickness of the small pore size material. In fact, the influence of wall thickness on filtration time for this sample appeared to be more important than filter size. This probably results from a "tunneling" effect for the large mean pore size thin-walled samples, particularly in the early portions of the filtration cycle. This can be significantly reduced by increasing filter wall thickness. However, in some cases, shorter operating times to a given back pressure were observed for thick-walled large-pore filters than for smaller pore filters. This is probably a result of plugging within the pores of the large pore materials causing a more rapid rise in pressure drop than in the small pore filter where most of the particulate remains on the filter surface.

Table 3 - Relative Filtration Time Factors As A Function of Wall Thickness and Filter Size

Filter Mean Pore Size (µm)		Relative Change in Filtration Time to Obtain ΔP_L=140 cm H_2O 12.5µm	35µm
Double Wall Thickness .012" .025"	@ L 6"	0.5	0.1
	12"	0.6	0.2
Double Filtration Area 6"L 12"L	@ wall .012	4.9	1.8
	.025	5.4	3.7

Operating time was found to be proportional to filter area for filters with mean pore size ≤15µm. The operating times for these filters were also found to be linearly proportional to the g/mi particulate level. For the large pore size filter, a fourfold increase in the particulate level resulted in a twofold reduction in operating time, as can be explained by the lower filtration efficiency.

FILTER CAPACITY

Particulate weight measurements proved to be an accurate measurement of filter capacity when normalized to a back pressure of 140 cm H_2O. In Figure 9, filter capacity as a function of filter surface area can be seen. Both filtration time and filter capacity were observed to exhibit a linear relationship with filter area at low filter loadings. At heavy loading for filter area $\geq$ 16 ft^2, however, the cell density was found to significantly affect filter capacity.

The divergence of capacities at high loading as a function of cell density in filters of approximately the same surface area but differing volumes suggests that filter volume and/or cell density may be more important than surface area at high loadings.

A possible explanation of this divergence is the nonuniform axial loading within the filter. Weight loss measurements indicated a 43% heavier particulate film in the leading 1" of filter relative to the relatively uniformly loaded remainder of the filter. Such excessive front end loading could rapidly block off the partially filled cells to further loading in the small celled filters, with decreasing effect for decreasing cell density.

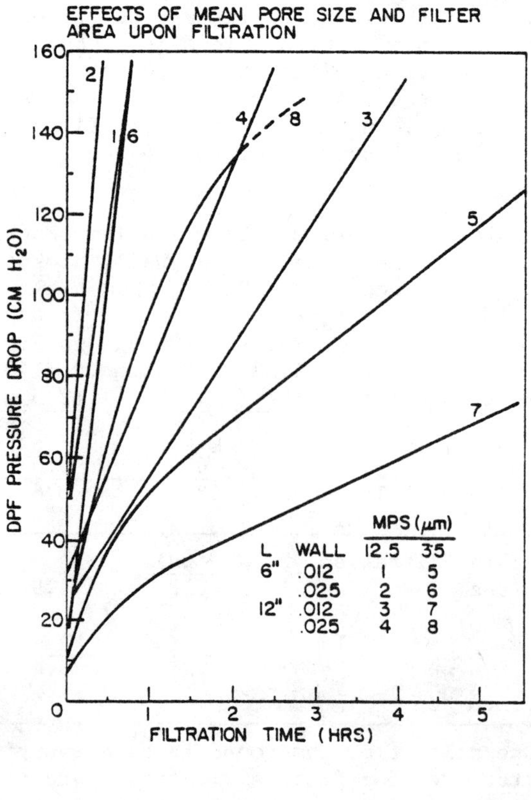

Fig. 8

It would appear there is a separate capacity vs. surface area relationship for each cell density, with all curves coinciding at low surface areas. For a given filter volume there is a cell size yielding maximum capacity, as indicated in Figure 10. This cell size increases as the filter size increases. The scatter at 100 cells/in^2 represents samples of differing wall thickness.

FILTRATION MODEL

The models for both filter efficiency and operating time were determined by multiple regression analysis describing 83% of the variance of the data.

The mean pore diameter was found to be the most important influence upon filter efficiency. Porosity and wall thickness also influence efficiency, but this effect is primarily confined to the large pore size (> 15μ) filters. Operating time was also found to be strongly influenced by average pore diameter as shown in Figure 11.

When operating time and efficiency plots are superposed, as shown in Figure 11, one can determine the desired porosity and pore size characteristics needed to yield the desired combination of efficiency and operating time. Filter volume could be used to compensate for differences in engine particulate levels, and to allow longer operating times, or equivalent times at lower average back pressure levels. Figure 11 data is specific for the 1980 Oldsmobile 350 CID engine used for filtration tests at steady state conditions.

On the basis of a filter efficiency of ~80% required to attain the 1985 EPA particulate specification of 0.2g/mi, a material designated EX-47 with a 12μm pore diameter and 50% porosity in a 100 cell/in^2, 0.017" wall configuration was selected for the bulk of regeneration testing.

FILTER REGENERATION

It is not within the scope of this paper to investigate regeneration systems, but to examine and define those characteristics and phenomena of the DPF in a regeneration mode.

Filter regeneration occurs when the particulate temperature is sufficient to initiate combustion with the excess oxygen present. This oxidation reaction is temperature dependent according to the relation

$$R = Ae^{-\Delta E/RT}$$

Where R is the oxidation rate, A the reaction rate constant, ΔE the activation energy in K cal/mole, and T the reaction temperature in °K yielding a rate increasing exponentially with temperature.

The regeneration process will ultimately be a compromise among many conflicting requirements. Among these will be the drive to minimum regeneration time, minimum filter size/volume, maximum loading capacity for given filter pressure drop, minimum fuel penalty to initiate burnoff, and maximum temperature/thermal stress consistent with

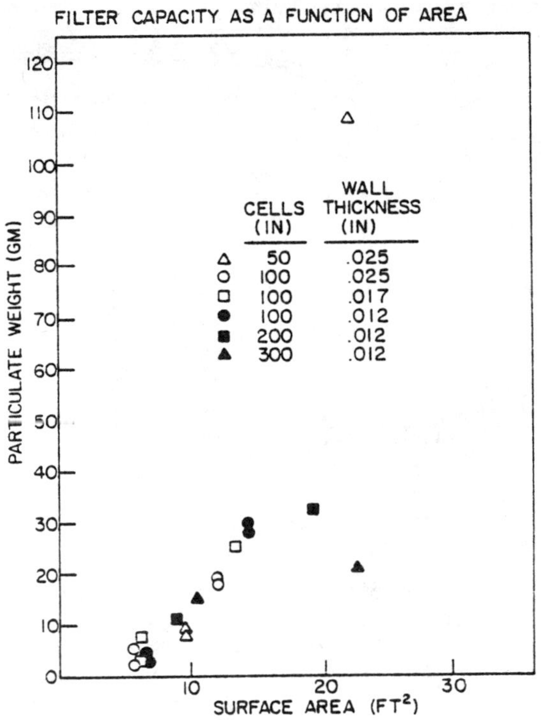

Fig. 9

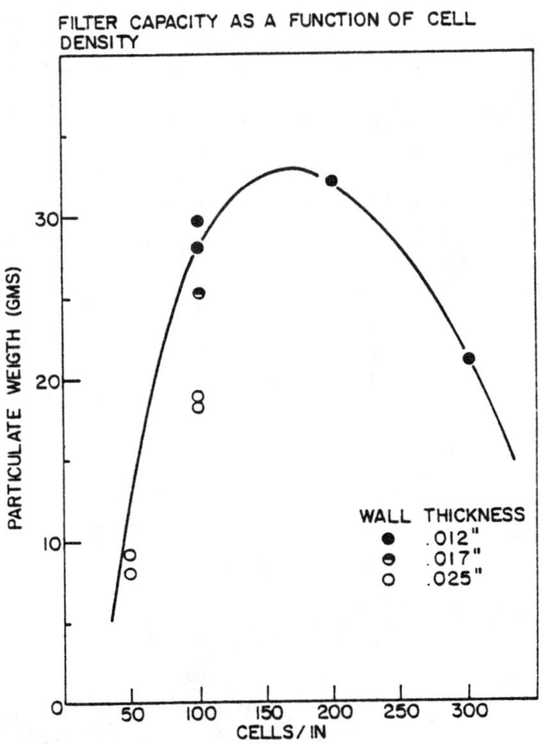

Fig. 10

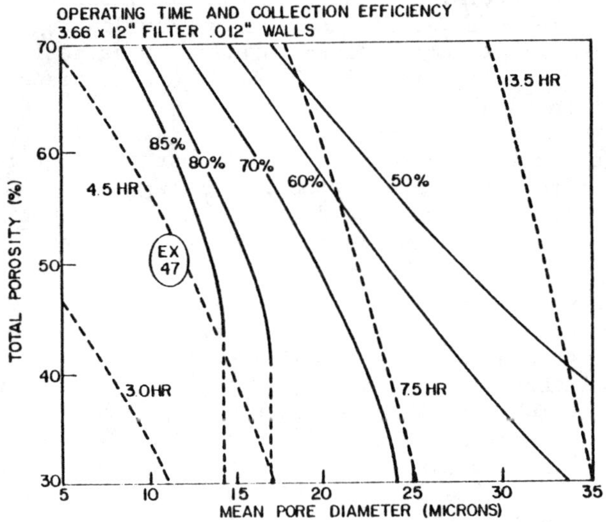

Fig. 11

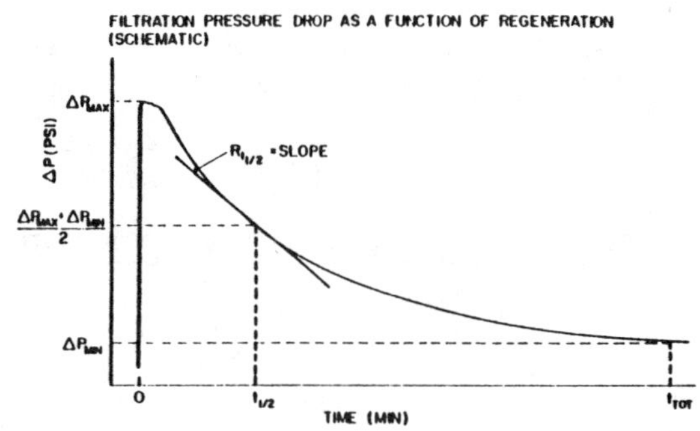

Fig. 12

filter durability over the expected filter lifetime.

Several methods have been considered for initiating filter regeneration. However, for this study engine throttling and rpm were used to obtain all regeneration test conditions under full exhaust flow.

Available data indicate greatly increased axial thermal gradients within the DPF as the flow rate is decreased, as is the case for alternate regeneration techniques. For this study, the DPF axial gradients during regeneration were ~10°F/in.

A schematic of the filter regeneration pressure drop can be seen in Figure 12. Upon achieving steady state engine conditions with the bypass line, the exhaust gas is valved over to the exhaust line containing the preloaded filter at time zero. This results in an immediate increase in pressure drop across the filter as it heats under full exhaust flow. Upon ignition, there is a resultant drop in filter ΔP, as shown schematically in Figure 12. It is sometimes very difficult to determine total reaction time, particularly at low particulate loadings and/or low regeneration temperatures. Therefore, $t_{1/2}$, as defined in $(\Delta P\ max + \Delta P\ min)/2$ has been a useful measurement to use.

The most influential variable affecting regeneration was found to be exhaust temperature. The effect of regeneration temperature upon regeneration time at a filter loading of 120 cm H_2O can be seen in Figure 13. Actual filter temperature can greatly exceed the exhaust gas temperature depending upon filter loading and gas flow rate. For engine regeneration at 70-240 CFM exhaust flow rate and loading to 120 cm H_2O ΔP the axial thermal gradient was observed to be approximately 10°F/in, with increasing temperature in the flow direction. There was no distinct "flame front" moving through the filter, and the thermal gradient decreased to near zero within a few minutes, as indicated in Figure 14.

The radial thermal gradient was observed to be approximately 100°F for the previous sample, primarily from OD cooling rather than from reaction thermal effects. The radial gradient did not change with time during regeneration.

It was also possible to initiate regeneration at $T \geq 1000°F$, and then sustain combustion at a reduced rate even though exhaust temperature was reduced to ~700°F. Combustion was monitored by CO ratio, before and after the filter, which

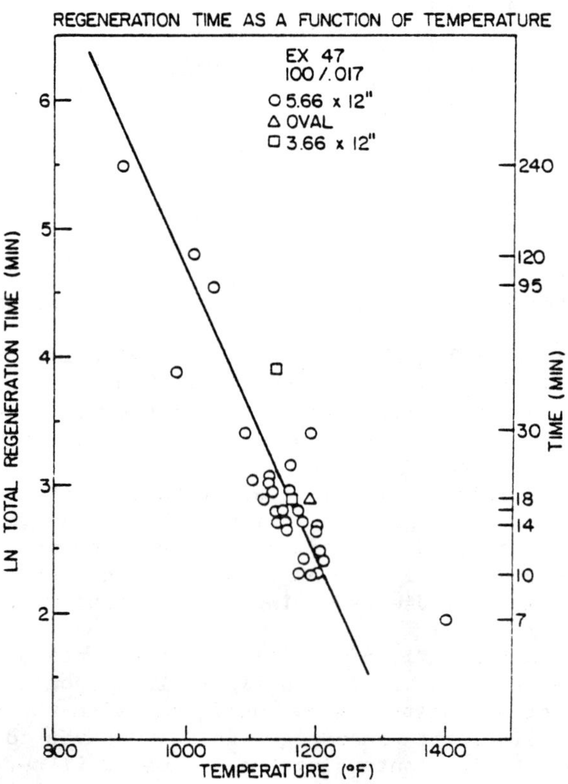

Fig. 13

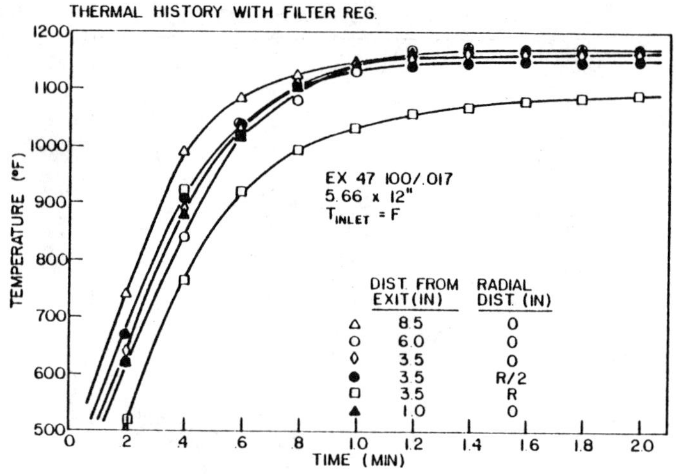

Fig. 14

proved to be a very sensitive measure of this chemical reaction.

A strong relationship between oxygen concentration and temperature was observed for particulate ignition, as indicated in Figure 15. These data represent filters of all configurations tested in the 2^N factorial regeneration experiment. This ignition point curve appears to be independent of filter wall thickness, cell density, size, and particulate loading within the limits of the test conditions. It is not known if this curve is valid for low flow conditions.

The sample starting at 905°F and 8.5% O_2 may have ignited during the test, as results were not conclusive for this particular test. Additionally, the 2-3% O_2 tests conducted at 1100°F-1300°F appeared to be borderline conditions as there appeared to be a time factor influencing ignition. All temperatures indicated in both Figures 13 and 15 represent the inlet gas temperature values

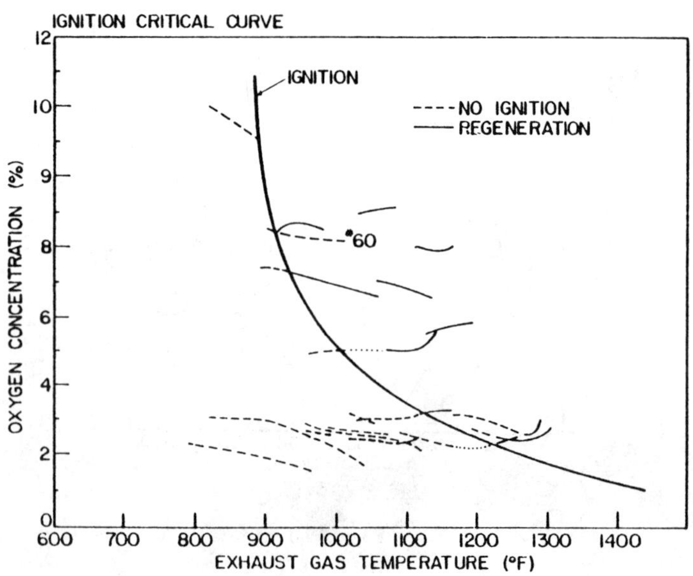

Fig. 15

approximately 10 cm ahead of the filter. Actual filter temperatures may have exceeded these values significantly. The thermocouple data would suggest filter temperature to be as much as 130°F higher than the inlet gas temperature under some conditions.

Insufficient data is available at this time to determine significant effects upon filtration as a function of cell density, wall thickness, loading, or filter size. However, it does appear that regeneration is relatively independent of filter geometry. Future experiments will, hopefully, quantify the effects of these variables upon the regeneration cycle.

CONCLUSIONS

Preliminary diesel engine testing has shown the manifolded porous monolithic ceramic filter to be a promising candidate for downstream control of diesel exhaust particulates.

In the ceramic body the average pore size of the filter wall was determined to be the major factor influencing filtration efficiency. Wall thickness and total porosity affected filtration to a lesser extent, and primarily with the large pore (>15μm) filter materials.

Filter capacity was found to be linearly related to filter volume at low loadings, and dependent upon cell density at high particulate loadings. This results, presumably, from excessive particulate layer thickness at the cell entrance with substantially reduced flow and excessive pressure drop in small dimension cells. The filtration optimum cell density appears to be in the 100-200 cell/in^2 range. Based upon a 2-3 hour operating time and $\geq$ 80% filtration efficiency requirements a candidate filter material, EX-47, was selected for further testing. This material exhibits 50% porosity with a mean pore size of 12-13μm. Cell geometry, based upon efficiency and capacity requirements were selected at 100 cells/ in^2, a 0.017" wall, and a filter dimension of 5.66" Ø and 12" long. This geometry was selected specifically for the 1980 Oldsmobile 350 CID diesel engine.

Steady state engine tests have shown the DPF is capable of efficiencies $\geq$ 80%, operating times $\geq$ 2 hours with maximum filter pressure drop $\leq$ 140 cm H_2O.

Filter regeneration was found to be strongly influenced by exhaust gas temperature. Regeneration times for particulate loading of 120 cm H_2O were from ~10 minutes at 1200°F to 4-5 hrs. at 900°F.

Both the log of the total regeneration time and the log of the $t_{1/2}$ time were found to be linearly related to the inlet gas temperature. It appears possible to initiate regeneration at T $\geq$ 1000°F and then sustain regeneration to temperatures as low as 700°F. It may be possible to achieve a steady state condition during vehicle operation with particulate filtration rate equal to regeneration rate if exhaust gas temperature reaches a high enough value to initiate ignition.

A critical ignition curve was found relating oxygen concentration and inlet gas temperature. The curve appears to be independent of filter geometry and level of particulate loading. It does not appear possible to initiate particulate ignition and filter regeneration until the minimum temperature - oxygen concentration conditions defined by the critical curve are met. The curve was defined within engine generated conditions. It is not known if the curve is valid or displaced for low flow rate regeneration, or for particulate from alternate diesel engines.

Filter wall thickness, cell density, and filter size variations were not observed to significantly affect the regeneration characteristics.

ACKNOWLEDGEMENT

The authors greatfully acknowledge the contribution of their colleagues at Corning to this report, notably Dr. W. Pitcher for his initial filtration studies, Dave Smith for experimental analysis in modeling, F. E. Noll, R. G. Reid, Jr., J. Belmonte, and Art Hillman for preparing the samples. Thanks are also due to Kim Lind of EDI for his test work and William Dreyer of Walker Manufacturing for his contributions of supplying containers.

REFERENCES

1. Issues Concerning the Light Duty Diesel-DOE/EV-0050/1, Sept. 1979.

2. The Diesel Emissions Research Program-EPA-625/9-79 004, December 1979.

3. Federal Register - Volume 45, Number 45-Wednesday, March 5, 1980.

4. Howitt, J.S. - Thin Wall Ceramics as Monolithic Catalyst Supports-SAE Paper 800082-February 1980.

5. Murphey, M. J., Hillenbrand, L. J., Trayzer, D. A. - Assessment of Diesel Particulate Control: Direct and Catalytic Oxidation. EPA-600/7-79 232b - October 1979.

Diesel Particulate Trap Regeneration Techniques**

W. R. Wade, J. E. White, and J. J. Florek
Engineering and Research Staff
Ford Motor Co.

ALTHOUGH INDIRECT INJECTION (I.D.I.) DIESEL ENGINES in passenger cars and light trucks can provide 20 to 35% improvement in the fuel economy relative to conventional gasoline engines, more stringent exhaust emission regulations in the future may severely restrict the availability of Diesel powered passenger cars and light trucks in the United States (1, 2)*. Reduction of the Federal NO_x emission requirement to 1.0 gm/mi in 1981 required many light duty Diesel engine manufacturers to seek NO_x waivers up to the 1.5 gm/mi limit for the 1981-1984 time period. However, by 1985, when the provision for NOx waivers expires, control systems required to achieve 1.0 gm/mi NO_x emission levels could possibly be developed for moderate inertia weight vehicles. Several studies (3, 4) have shown that controlling NO_x emissions to the 1.0 gm/mi level at vehicle inertia weights up to 4250 lbs. is possible in experimental vehicles at low mileage in the laboratory.

The capability of Diesel engines to meet the 1985 .2 gm/mi Federal particulate emission regulation is one of the major open issues regarding the future use of Diesel engines in light duty vehicles (5). The general trends of CVS-CH particulate emissions vs. NO_x emissions for indirect injection Diesel powered vehicles with inertia weight classes ranging from 2250 lbs. to 4500 lbs. are shown in Figure 1. The NO_x and particulate emissions of production vehicles, represented by the right-hand termination points of the curves in this figure, increase significantly as the vehicle inertia weight increases. Reductions in NO_x emissions with the use of EGR and modified injection timing increase particulate emissions. The assumed low mileage engineering objective levels for the 1985 1.0 gm/mi NO_x and .2 gm/mi particulate emission regulations are shown on this figure. The low mileage engineering objective levels were assumed to be equal to 80% of the emission standards. Only vehicles below 2000 lbs. inertia weight can approach both of these low mileage emission objectives simultaneously.

* Numbers in parentheses designate References at the end of paper.

**Paper 810118 presented at the International Congress and Exposition, Detroit, Michigan, February, 1981.

ABSTRACT

Diesel engine particulates collected on a trap cause the exhaust back pressure to increase and adversely affect fuel economy and vehicle performance. Therefore, a trap must be periodically regenerated by oxidizing the collected particulates. Several techniques for regenerating a Diesel particulate trap are discussed. Regeneration was achieved with high speed and high load engine operation. Lead, added to the Diesel fuel, acted as a catalyst and reduced the ignition temperature of particulates collected on a trap by about 300°F. Throttling the intake air flow increased exhaust temperature to facilitate regeneration at moderate vehicle speeds. An externally fueled burner provided regeneration over the widest range of engine operating conditions, including idle.

For vehicles above 2000 lbs. inertia weight, currently available indirect injection Diesel engine combustion systems do not appear to be capable of meeting the engineering objective levels for the .2 gm/mi particulate regulation at a 1.0 gm/mi NO_x level. Significant efforts are underway to develop combustion system modifications for lower NO_x and particulate emissions (3). However, limited published results have indicated that these efforts have not been successful, to date, in modifying the HC/NO_x/Particulate emission relationship of current indirect injection Diesel combustion systems.

Fuel modifications have also been investigated to determine their potential for lowering Diesel particulate emissions. An experimental fuel, in which the aromatic content was reduced from about 30%, which is characteristic of current No. 2 Diesel fuel, to nearly zero and the 50% boiling point temperature was reduced by about 100°F, yielded a 40% reduction in particulate emissions from a Diesel vehicle tested over the CVS cycle. However, fuels of this type are not currently available for large scale use. In contrast, tests with currently available fuels with less drastic modifications of fuel properties have not shown significant changes in the particulate emission levels. Therefore, until combustion system technology or fuel blends are developed which may yield lower in-cylinder particulate emissions, the particulate emission trends for current technology light duty Diesel powered vehicles, shown in Figure 1, indicate that some form of exhaust treatment will probably be required to meet the 1985 .2 gm/mi particulate emission regulation.

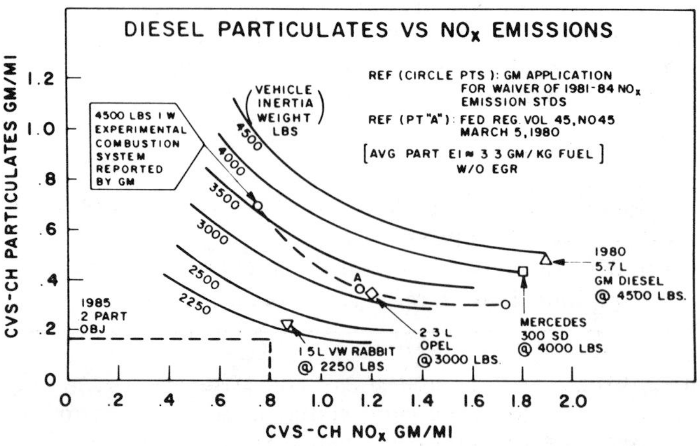

Fig. 1 - CVS-CH Particulate emissions vs. NO_x emissions for light-duty Diesel vehicles with inertia weights ranging from 2250 lbs. to 4500 lbs.

This paper will discuss the application of regenerative particulate traps to light-duty Diesel powered vehicles. After a discussion of trap requirements and candidate trap systems, the following trap regeneration techniques will be discussed: high speed/high load regeneration, catalysts for the reduction of particulate ignition temperature, air throttling and an externally fueled burner.

TRAP REQUIREMENTS

Many requirements must be met with a Diesel particulate trap. The most important requirement which dictates the design and size of the particulate trap is the attainment of adequate particulate collection efficiency over the CVS-CH emission test cycle. Collection efficiency of a particulate trap is defined as follows:

$$\%\ \text{Collection Efficiency} = \frac{W_T}{W_E} = \frac{W_E - W_L}{W_E}$$

where:

W_T = Weight of particulates collected on the trap per mile or unit time

W_E = Weight of particulates entering the trap per mile or unit time

W_L = Weight of particulates leaving the trap per mile or unit time

Particulate trap collection efficiency requirements, computed from the vehicle particulate emission trends illustrated in Figure 1, are shown in Figure 2 as a function of vehicle inertia weight. A Diesel vehicle at an inertia weight of 3500 lbs. is projected to require a particulate trap with a collection efficiency of approximately 80%, whereas a 2500 lb. inertia weight vehicle would require a trap with a collection efficiency of only 35%.

Other particulate trap requirements are shown in Figure 3. In addition to providing adequate collection efficiency, the particulate trap must have a low pressure drop to ensure minimum loss in fuel economy and performance. Since a low pressure drop must be maintained during mileage accumulation, a low pressure drop increase, or rise rate, must be provided as the particulate material is deposited on the trap. Minimum particulate trap volume is required for packaging considerations. However, the requirement for small trap volumes conflicts with the requirement for low pressure drop and pressure rise rate. A trap must be fabricated from a material which can withstand the high temperatures encountered during the regeneration process when the collected particulate material is oxidized. Specific regeneration requirements will be defined in a later section of the paper. The particulate trap must be capable of extended durability. During extended mileage accumulation, all of the above particulate trap requirements must be maintained.

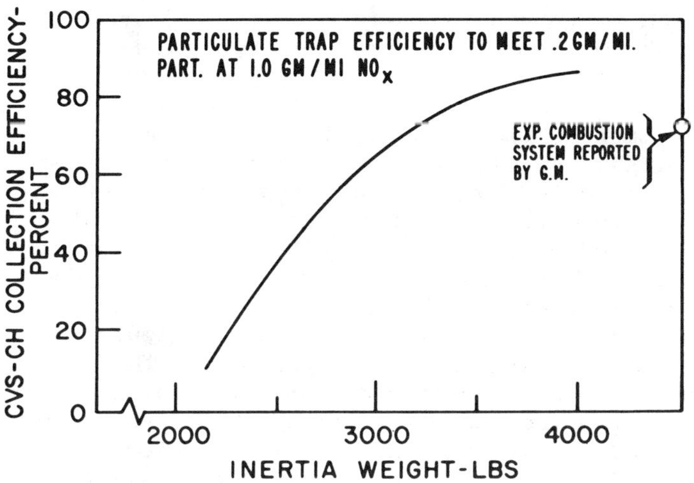

Fig. 2 - Particulate trap collection efficiency required for .2 gm/mi particulates at 1.0 gm/mi NO_x over the CVS-CH emission test cycle as a function of vehicle inertia weight

DIESEL PARTICULATE TRAP REQUIREMENTS

- COLLECTION EFFICIENCY ADEQUATE FOR .2 GM/MI (80% FOR A 3500 LB. INERTIA WEIGHT VEHICLE)
- LOW PRESSURE DROP TO ENSURE MINIMUM LOSS IN FUEL ECONOMY AND PERFORMANCE
- LOW PRESSURE RISE RATE
- MINIMUM VOLUME FOR PACKAGING
- HIGH TEMPERATURE CAPABILITY TO WITHSTAND REGENERATION (PARTICULATE OXIDATION)
- EXTENDED DURABILITY

Fig. 3 - Diesel particulate trap requirements

TRAP SYSTEMS

Only mechanical particulate trapping systems will be discussed in this paper, although other systems, such as electrostatic precipitation, thermal precipitation and cyclone separation, have also been investigated as possible trapping systems for particulate emissions from light-duty Diesel powered vehicles. Mechanical trapping mechanisms, shown in Figure 4, consist of impaction, interception and diffusion.

Impaction and diffusion are the primary trapping mechanisms of a wire mesh or ceramic fiber trap (6, 7). The larger sizes of particulates impact on filaments of the mesh and adhere to the surface of the filaments or particulate material previously collected on the filaments. Some of the smaller sizes of particulates migrate to the surfaces of the filament or previously collected particulate material by diffusion and are retained. This type of trap is sometimes called a non-blockable trap, because an exhaust flow path will usually exist which cannot be blocked by the accumulation of particulate matter. Although these traps tend to have relatively low pressure drops, disadvantages that have been observed include moderately low collection efficiency and blow-off of collected particulates.

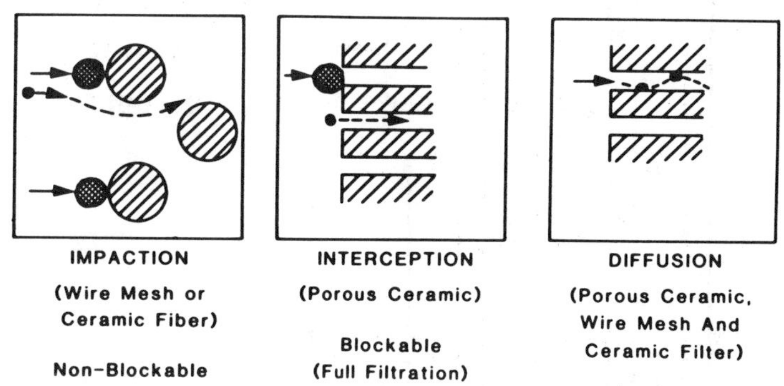

Fig. 4 - Mechanical trapping mechanisms

Interception is the primary trapping mechanism of a porous material trap, although diffusion may also enhance the collection efficiency of this type of trap. Particulates larger than approximately the mean pore size of the material are intercepted and prevented from passing through the material. As additional particulate material is accumulated on the surface of the trapping material, the effective pore size may be reduced, thereby causing efficiency to increase with the collection of smaller sizes of particulates. Traps using the interception method of collection are sometimes called blockable, or full filtration traps. Although blow-off of collected particulates is not a problem with this type of trap, back pressure and rate of back pressure increase tend to be somewhat higher than with non-blockable traps.

A schematic of a candidate ceramic honeycomb particulate trap using the interception method of mechanical filtration is shown in Figure 5. This experimental trap, supplied by Corning Glass Works, Inc., consists of a porous ceramic honeycomb substrate similar to that used for monolithic catalysts on gasoline engines (8). The open channels of the honeycomb are alternately blocked with high temperature ceramic cement at the top and bottom so that all of the inlet flow must pass through the porous ceramic walls before exiting from the trap. This honeycomb trap provides very high filtration surface area per unit volume. A 119 in^3 trap of this configuration with 100 cells/in^2 and 0.017 inch wall thickness has approximately 1,970 in^2 of filtering surface area. The filtering surface area per unit volume for this trap is 16.6 in^2/in^3.

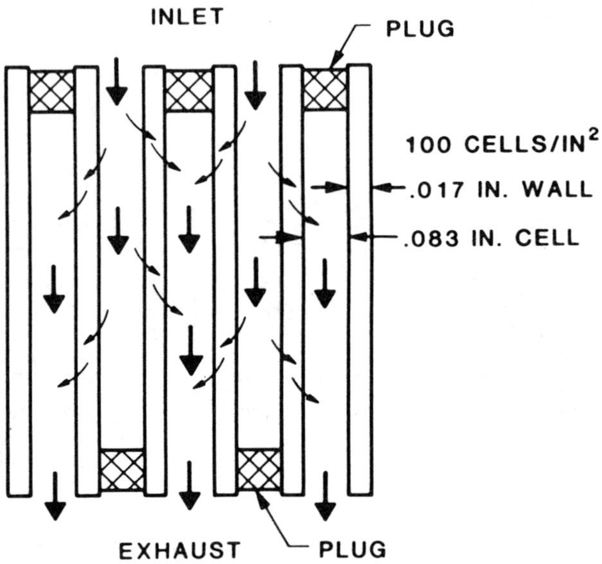

Fig. 5 - Schematic of a ceramic honeycomb particulate trap

A photograph of a honeycomb trap partially loaded with particulates is shown in Figure 6. The large filtering surface area per unit volume is apparent. The initial layer of particulates on the surface of the porous ceramic walls have been collected by the interception method. Successive layers of collected particulates provide an additional filtration medium. The total mass of deposited particulates on the trap has an influence on the regeneration characteristics of the trap as will be discussed later in the paper.

An example of the packaging of a 119 in^3 honeycomb trap on a 4-cylinder Diesel engine is shown in Figure 7. The trap is mounted on a modified exhaust manifold as close as possible to the exhaust ports of the engine to preserve exhaust gas temperatures to assist in the regeneration process. Most of the test results shown in this paper are with a trap of this configuration mounted on a 2.3L, 4-cylinder, indirect injection Diesel engine. The 119 in^3 trap volume was approximately equal to the engine displacement.

The cold flow pressure drop characteristics of the 119 in^3 honeycomb trap are shown in Figure 8. By applying appropriate correction factors, the cold flow pressure drop data can be used to estimate the pressure drop across the trap for an engine application. Engine exhaust pressure and temperature were used to calculate the actual volumetric flow through the trap on the engine. An additional correction was applied to the cold flow pressure drop data to estimate the effects of the higher viscosity of the flow at the actual engine exhaust temperature. The measured pressure drop at a 40 mph cruise condition obtained with this trap installed on a vehicle is shown in Figure 8 to be in relatively good agreement with the pressure drop estimated from cold flow data.

CVS test results with a 119 in^3 honeycomb trap installed on a 2.3L indirect injection Diesel engine in a 3000 lb. inertia weight vehicle are shown in Table 1. A CVS-CH collection efficiency of 81% was obtained for this trap at low mileage. At the .27 gm/mi feedgas particulate emission level of the test vehicle, low mileage tailpipe emissions of less than .2 gm/mi were achieved with a clean trap. However, NO_x emissions were over 1.0 gm/mi with this vehicle that did not have an EGR system. The application of EGR to a Diesel engine is not expected to significantly affect the particulate collection efficiency of the trap, although the feedgas particulate emission level would increase as shown in Figure 1. Therefore, the 81% collection efficiency of this clean trap is projected to provide particulate emissions below the low mileage engineering objective level for the .2 gm/mi requirement when NO_x emissions are controlled to the 1.0 gm/mi level. A clean trap pressure drop of 1.0 in.Hg. was measured at a steady 40 mph cruise condition prior to the start of the CVS test. An average trap pressure rise rate of 3.3 in. Hg/100 miles of operation was observed for this trap at the cruise 40 mph condition. The trap pressure rise rate is used as the primary parameter which determines the frequency of trap regeneration required to minimize losses in fuel economy and performance.

Trap collection efficiency, pressure drop and the pressure rise rate are interrelated and are strong functions of trap volume, cell density, porosity and mean pore size for the honeycomb trap. An example of the effect of trap volume on the trap operational characteristics is shown in Figure 9 for a steady state engine dynamometer operating condition. Collection efficiency appears to be relatively insensitive to changes in the trap volume, or effective filtering surface area, although the larger volume trap was slightly more efficient. Since the traps shown in Figure 9 were fabricated from similar material with the same mean pore size, the interception method of filtration removes from the exhaust gas approximately the same number of particles which are larger than the mean pore size of the trapping material.

Trap volume has a significant effect on the clean trap pressure drop. Doubling the trap volume to reduce the face velocity of the exhaust gas through the trap material resulted in a 50% reduction in clean trap pressure drop. The pressure rise rate is also significantly affected by the trap volume. As shown in Figure 9, doubling the trap volume resulted in an eight-fold reduction in the pressure rise rate.

For a specific trap design, the pressure drop can be used as an approximate indication of the particulate loading on the trap. The mass of particulates collected on a trap is approximately proportional to the increase in pressure drop over the pressure drop of the clean trap. Deviations from this proportionality occur in a clean trap, where the pressure drop increases more rapidly than the average value for a moderately loaded trap.

A particulate loading parameter, defined as grams of particulates collected on a trap per in.Hg. of pressure drop increase, can be calculated in the following manner. The feedgas particulate emission rate and the collection efficiency can be used to calculate the mass of particulates collected per unit time

Fig. 6 - A honeycomb trap partially loaded with particulates

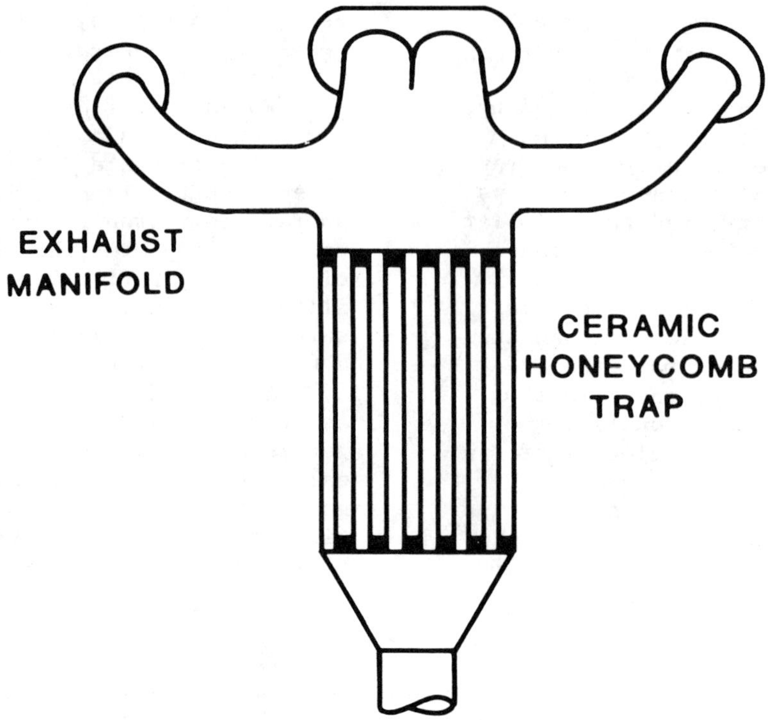

Fig. 7 - An example of the packaging of a 119 in^3 honeycomb trap on a 4-cylinder Diesel engine

or distance. Dividing the particulate collection rate by the pressure rise rate yields an approximate trap loading parameter expressed as gms. of particulates per in. Hg. Approximate loading parameters for traps with two different volumes are shown in Figure 9. The larger volume trap can contain a significantly larger mass of particulates for the same increase in pressure drop. For a moderately loaded trap, the trap loading parameter provides a cursory assessment of the total particulate loading on a trap by observing the increase in pressure drop over the pressure drop of the clean trap. The total particulate loading is an important parameter for controlling trap regeneration frequency and temperatures during regeneration.

REGENERATION FUNDAMENTALS

The ceramic honeycomb trap provides for the following two functions: 1) filtration of a portion of the particulates from the exhaust stream, and 2) storage of the filtered or collected particulates. The storage function is limited by two criteria, which are interrelated: 1) the void volume within the trap available to store the collected particulates, and 2) the increase in filter media pressure drop due to the deposition of particulates directly on the filtration material.

The limitations of trap void volume for storing the collected particulates are described in Table 2. A trap with a volume of 119 in^3 and a cell density of 100 cells/in^2 has a void volume of 43.4 in^3 available for storage of particulates. The void volume available for storage of particulates is only one half of the total void volume in the honeycomb trap. The low void volume occurs because the flow enters the trap through the alternately open channels and filtration occurs only on one side of the porous ceramic walls of the trap, as shown in Figure 6. If the density of the particulates collected in a honeycomb trap is .056 gm/cm^3, then only 40 gms. of particulates could be stored within the trap void volume. This quantity of particulates would be collected in approximately 125 miles of operation for an assumed feedgas particulate emission level of .4 gm/mi and a trap collection efficiency of 80%. Larger amounts of particulate material may be contained if it is compacted by the dynamic pressure of the exhaust flowing through the trap. The limited void volume available for storage of the collected particulate material dictates the requirement for periodic regeneration to remove the collected particulates.

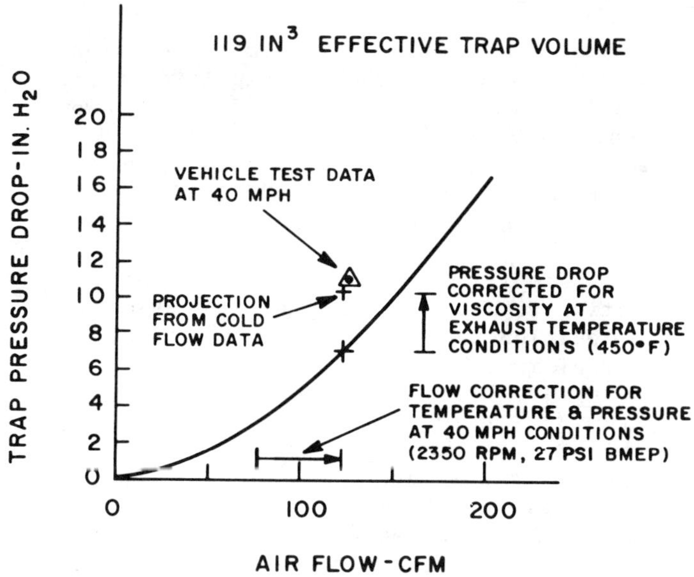

Fig. 8 - Cold flow pressure drop characteristics of a 119 in^3 honeycomb trap

TABLE 1

LOW MILEAGE PERFORMANCE OF A CERAMIC HONEYCOMB PARTICULATE TRAP WITH EFFECTIVE VOLUME OF 119 IN^3

2.3L I.D.I. Diesel
3000 lb. I.W.
10.3 hp PAU
3.89 Axle Ratio
Manual Transmission

	HC GM/MI	CO GM/MI	NO_x GM/MI	Particulates GM/MI
Baseline	.50	1.50	1.25	.27
With Trap	.54	1.49	1.12	.05

Particulate Collection Efficiency (%)	Cruise 40 MPH Clean Trap Pressure Drop (In Hg)	Pressure Rise Rate (In Hg/100 Mi)
81	1.0	3.3

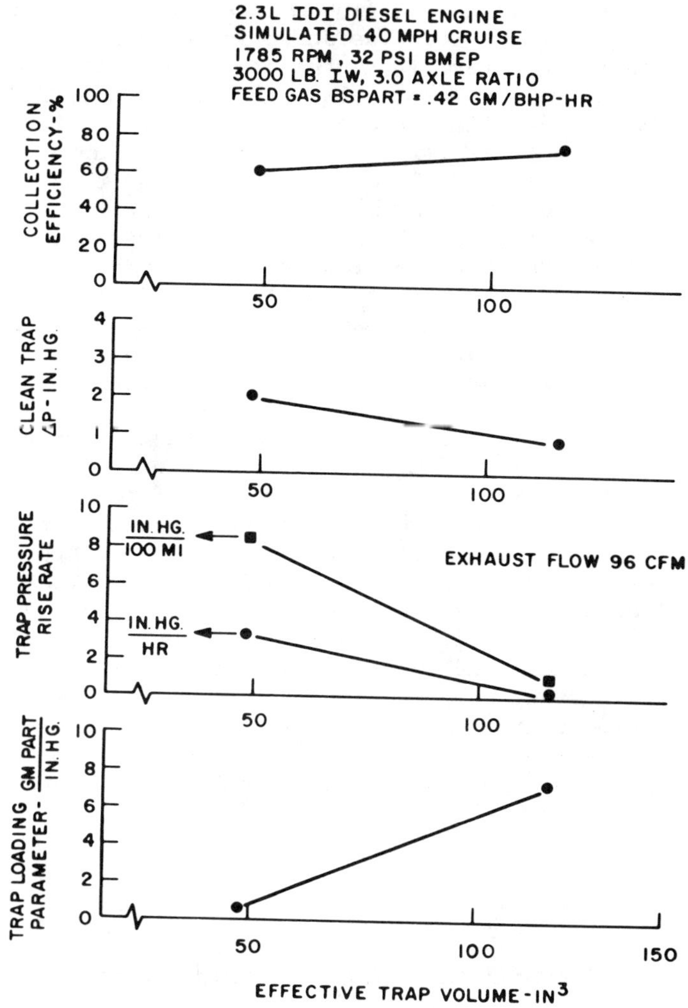

Fig. 9 - Trap collection efficiency, pressure drop, pressure rise rate and trap loading parameter as a function of ceramic honeycomb trap volume

The accumulation of particulates on the surface of the ceramic filtration material causes the pressure drop through the trap to increase. The resulting increase in exhaust back pressure can have an adverse effect on maximum power levels of the engine and on vehicle fuel economy. The effect of back pressure, measured at a cruise of 40 mph condition, on CVS-H fuel economy for a 3000-lb. inertia weight vehicle with a 2.3L Diesel engine is shown in Figure 10. Approximately 1% loss in fuel economy occurs for each 1 in. Hg. increase in engine back pressure. Therefore, to minimize the loss in fuel economy due to trap back pressure, the particulate trap must be regenerated periodically.

The relationship between fuel economy loss and miles of operation can be estimated from the data available. As shown in Figure 10, a 3% loss in fuel economy occurs with a back pressure of 3 in. Hg. A pressure rise rate of 3.3 in. Hg./100 mi. is characteristic of a 119 in^3 trap, as shown in Table 1. Using this parameter, it can be concluded that a 3% fuel economy loss would result after approximately 90 miles of operation. Either a prescribed limit for the vehicle fuel economy loss or the total particulate storage capability of the trap may be used as the criteria for determining the frequency of trap regeneration. However, the requirement to minimize fuel economy loss is usually the criteria that results in the shorter intervals between trap regenerations.

Numerous techniques can be applied to dispose of the collected particulates on a trap. However, in this paper, the only trap regeneration concept that will be discussed is the oxidation, or combustion, of the collected particulate material within the trap. The collected particulate material consists of carbon and adsorbed organic material that was not oxidized within the engine. The carbon material can be oxidized according to the simplified reaction shown in Figure 11. This reaction produces an adiabatic combustion temperature rise of approximately 4700°F at a chemically correct air/carbon mass ratio of 11.4:1 (9). At moderate exhaust gas temperatures, adiabatic flame temperatures of carbon can exceed 5000°F. Significant air or exhaust diluent is required to achieve leaner air/carbon ratios necessary to reduce the reaction temperature to a level so that the melting temperature of 2500-2700°F of the ceramic trap material is not exceeded.

In addition to controlling the maximum reaction temperatures during regeneration, the ignition temperature of the particulate material must be exceeded to initiate the oxidation process. Particulate ignition temperature data (10), shown in Figure 12, indicate that temperatures over 900°F are required to initiate combustion of the collected particulate material; however, some reduction in the mass of collected particulates occurs at lower temperatures as the volatile organic material evaporates. If the particulate material is reheated after a short term exposure to temperatures slightly above the ignition temperature, the reheat ignition temperature is increased as shown in Figure 12. The data shown were obtained at an oxygen concentration of 21% which is higher than the level normally available in the exhaust of a Diesel engine.

Although adequate temperature and oxygen are required to initiate regeneration, adequate time must also be available to complete the particulate oxidation process. The time required for regeneration is strongly dependent upon the trap inlet gas temperature. The time required to regenerate a trap loaded

TABLE 2

THEORETICAL MAXIMUM COLLECTED PARTICULATE VOLUME VS. TRAP VOLUME

Trap Effective Volume	118.7	in^3
Cell Density	100	cpsi
Wall Thickness	.017	in
Void Volume for Particulate Storage	43.4	in^3
Particulate Density	.056	gm/cm^3
Total Particulate in Trap Void Volume	40.0	gm
Feedgas Particulate Emissions - 2.3L Diesel @ 3000 lb. I.W.	.40	gm/mi
Collection Efficiency	80	%
Particulates Collected	.32	gm/mi
Mileage to Fill Trap - (Total Available Void Volume)	125	miles

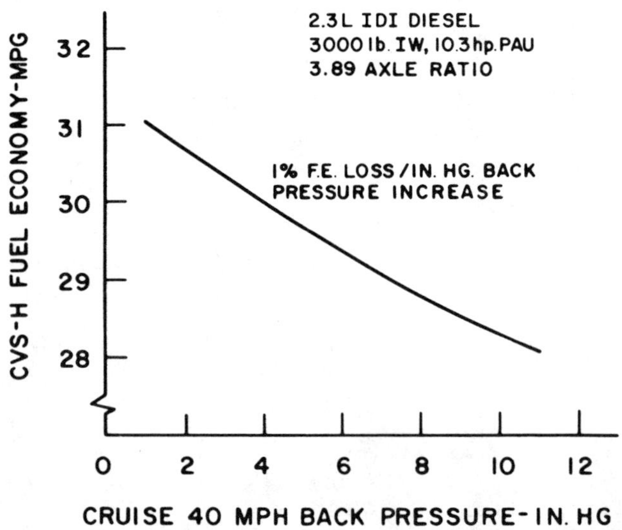

Fig. 10 - Effect of back pressure measured at a cruise 40 mph condition on CVS-H fuel economy for a 3000 lb. inertia weight vehicle

with particulates will also depend upon the following factors:

- Total accumulated particulate quantity
- Particulate packing density and distribution on the trap
- Reactivity of the collected soot
- Exhaust flow rate
- Heat losses from the trap

Data is presented later to illustrate the effect of some of these factors on trap regeneration time.

In summary, trap regeneration can be achieved if the following three major requirements are met:

- Trap inlet temperature exceeds about 900°F
- Oxygen concentration exceeds about 2%
- Adequate time is provided for complete combustion of the particulate material.

In the following sections of this paper, techniques for achieving these regeneration conditions for a Diesel engine will be discussed.

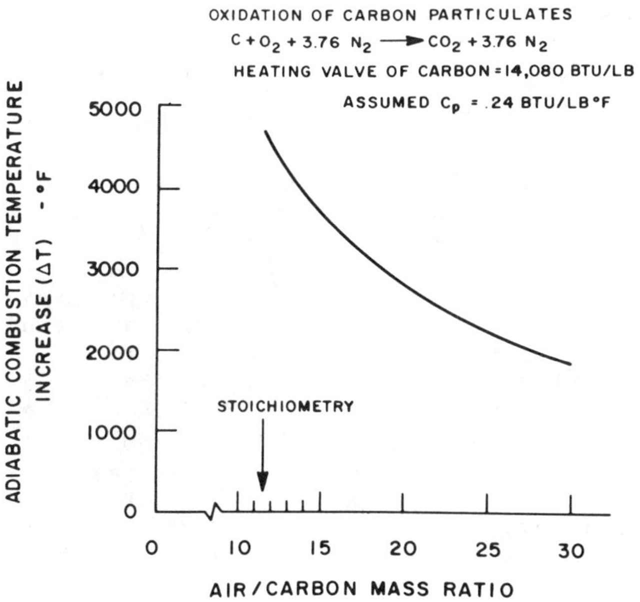

Fig. 11 - Adiabatic combustion temperature increase as a function of air/carbon mass ratio

HIGH SPEED/HIGH LOAD REGENERATION

One technique to initiate trap regeneration is to use the Diesel engine exhaust gas to produce the required temperature and oxygen concentration. However, since light-duty, indirect injection Diesel engines typically have high compression/expansion ratios of over 20:1 and operate unthrottled with lean air-fuel ratios, the exhaust temperatures of the Diesel engine are inadequate to initiate regeneration at most speed and load operating conditions encountered in normal driving.

An exhaust temperature map for a 2.3L Diesel engine operated at steady state conditions is shown in Figure 13. Road load cruise conditions from 30 mph to 60 mph, which are shown superimposed on this temperature map, produce exhaust temperatures significantly below the 900°F level required to initiate regeneration. In addition, most of the CVS cycle region, also superimposed on this temperature map, is below the 900°F level required for regeneration. Although a portion of the CVS cycle extends to temperatures exceeding the 900°F level, inadequate time is spent at these conditions to achieve and sustain regeneration during the actual transient CVS cycle. During a transient acceleration, the mass of the trap and associated container and piping absorb heat supplied in the exhaust gas and prevent the accumulated particulate material from rapidly reaching its ignition temperature.

The temperature map, shown in Figure 13, indicates that, at a limited range of moderate to high speed and high load conditions, exhaust temperatures are above 900°F, which is the level required to initiate trap regeneration. In addition, more than 2% oxygen is present in the exhaust wherever temperatures above 900°F are reached. Actual exhaust temperatures required for regeneration are usually higher than the nominal 900°F level to compensate for heat loss from the exhaust system and the trap.

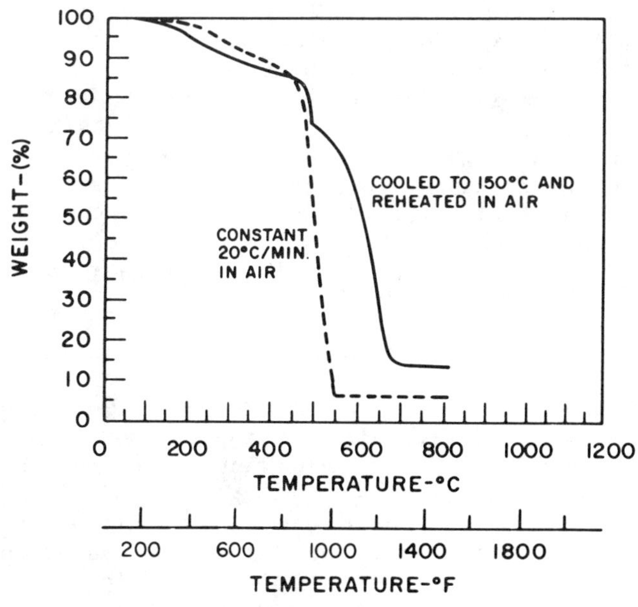

Fig. 12 - Diesel particulate ignition temperatures as determined from thermogravimetric analysis (Ref. 10)

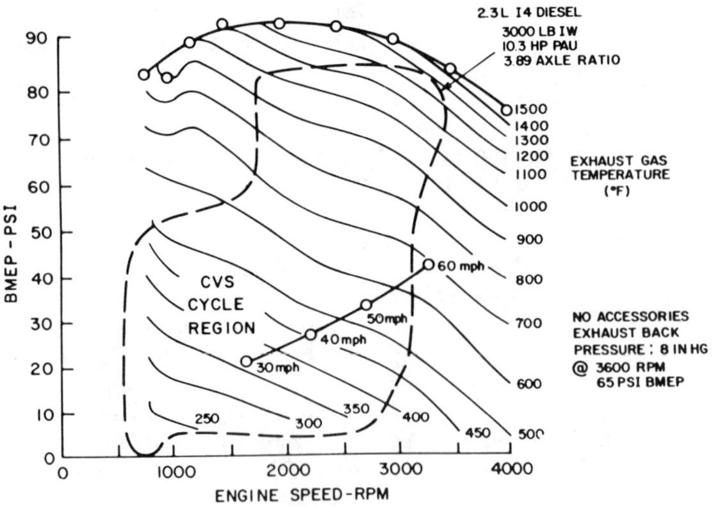

Fig. 13 - An exhaust temperature map for a 2.3L Diesel engine at steady state conditions

To evaluate trap regeneration at high speed/high load conditions, a simulated steady state representation of the AMA durability cycle was developed. The steady state operating conditions of this cycle are shown in Figure 14. The entire cycle is approximately two hours in duration with maximum power operation occurring only in the last few minutes of the cycle.

Two honeycomb traps with 47 and 119 in^3 effective volumes were evaluated on this dynamometer cycle. The increases in exhaust back pressure resulting from the particulate traps as a function of the operating time over the simulated AMA cycle are shown in Figure 14. As indicated by the changes in pressure rise rates, particulates were collected on a trap at different rates for the various operating conditions. For each trap, the pressure rise rates depend on the particulate concentrations, collection efficiency and exhaust flow rates at the various operating conditions. The pressure rise rates for the 47 in^3 trap were significantly higher than for the 119 in^3 trap.

At the first maximum power operating condition where the exhaust temperature exceeded 1200°F, the collected particulates were rapidly oxidized. Trap regeneration is shown by the sharp reduction in the exhaust back pressure after the beginning of the first maximum power operating condition. Complete regeneration of the trap was demonstrated by the restoration of the initial value of the moderate load pressure drop at the beginning of the next simulated AMA cycle. Trap regeneration could not be achieved when a similar trap was installed in a vehicle and the vehicle was operated over the actual transient AMA cycle. In this cycle, maximum power operation required to accelerate the vehicle from idle to 55 mph lasted only about 20 seconds. Due to thermal inertia effects, this time was inadequate to increase the temperature of the relatively cool trap and particulate material to the ignition temperature required for regeneration.

During the high speed/high load regeneration process, the carbon particulate material can theoretically be oxidized completely to yield CO_2. However, imperfect distribution of the available oxygen around the collected particulate material on the trap results in the partial oxidation of a portion of the carbon to produce CO. The emissions produced from combustion of the collected particulates during a high speed/high load regeneration process on a 2.3L Diesel engine operated on a dynamometer are shown in Figure 15. The emissions produced from the regeneration process were determined from the difference in measured emissions and particulates at the maximum power condition for a loaded, regenerating trap and a clean, non-regenerating trap. Approximately 88% of the collected particulates were converted to CO_2 and about 12% of the particulates were converted to CO. The regeneration conditions for these tests were not optimized to minimize the CO emissions during regeneration.

The temperature and time requirements for the trap regeneration process were defined from a series of engine dynamometer tests conducted with a 119 in^3 honeycomb trap installed on a 2.3L Diesel engine. For these tests, trap loadings of 2.25 times the clean trap pressure drop were used. The trap was loaded with particulates by operating the engine at 2000 RPM, 64 psi BMEP. The engine was subsequently operated at 3500 RPM and the load was selected to achieve the specified engine exhaust/trap inlet temperature, which was maintained until trap regeneration was complete or until the regeneration process was terminated. The results are shown in Figure 16 where the trap pressure drop is plotted as a function of time for several trap inlet temperatures. As illustrated, temperatures below 1000°F did not result in complete regeneration of the trap, as shown by the leveling out of the 900°F pressure drop vs. time curve above the clean trap value. Above 1000°F, complete trap regeneration could be achieved. The time required to complete the regeneration process varied from over ten mimutes at 1000°F to less than two minutes at temperatures above 1100°F.

Heat loss from the trap can affect the results shown in Figure 16. These test results were obtained with an insulated trap, which minimized the extent of radial or axial heat loss. Significant radial or axial heat loss would require higher exhaust temperatures or longer times for regeneration, because local regions of the collected particulate material will be at temperatures below the average trap inlet temperature.

The loading of particulates on the trap also affects the time required to complete the regeneration process. The effect of particulate loadings on trap regeneration times at a constant trap inlet temperature of 1100°F is shown in Figure 17. The reduction in trap pressure drop as regeneration proceeds is shown as a function of time for several values of trap loadings, ranging from 1.8 to 3.25 times the clean trap pressure drop. As illustrated, the time required to complete the regeneration process increased as the loading of particulates on the trap was increased. However, even with a trap loading of 3.25 times the clean trap pressure drop, regeneration was completed in less than two minutes for the 1100°F trap inlet temperature used in these tests.

Trap regeneration can be achieved with high speed/high load engine operation in a vehicle by sustained high speed operation. A 119 in^3 honeycomb trap was installed on a vehicle with a 2.3L Diesel engine. The trap was close-coupled to the engine as shown in Figure 7. After loading the trap with particulates to an exhaust back pressure of about 8 in. Hg. at a cruise 40 mph condition, the vehicle was operated on the test track over a range of high vehicle speeds. The remaining fraction of the loaded trap pressure drop increase (pressure drop of the loaded trap - pressure drop of the clean trap) was used to provide a cursory estimate of the fraction of the total collected particulates remaining within the trap as the regeneration process progressed.

The results from high speed/high load regeneration tests in a vehicle are shown in Figure 18. Below a trap inlet temperature of 800°F, corresponding to a vehicle speed of about 63 mph, no regeneration was observed. For temperatures above 800°F, a 2.5 mile trip around the test track at constant speed produced varying degrees of regeneration. Higher temperatures significantly accelerated the regeneration process so that at a trap inlet temperature of 1300°F, corresponding to 90 mph, complete regeneration of the trap was achieved.

High speed/high load regeneration is not practical for normal vehicle operation. However, some of the fundamental characteristics of the trap regeneration process can be useful in developing other regeneration techniques, which will be described in the following sections of this paper.

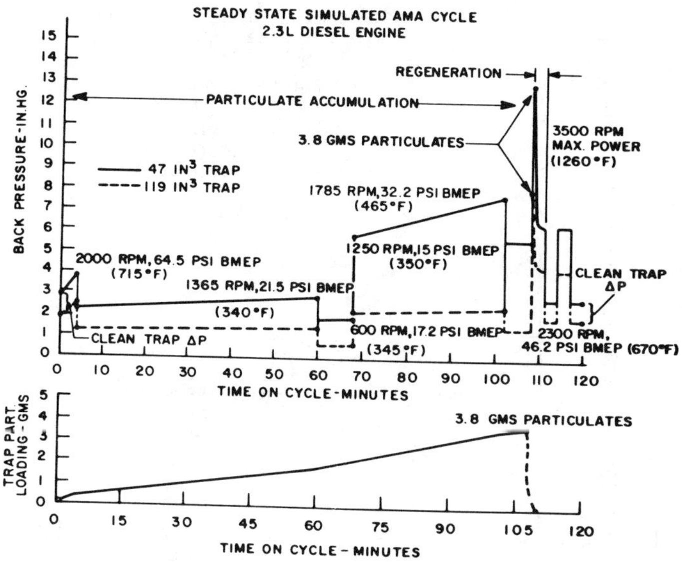

Fig. 14 - Exhaust back pressure with Diesel particulate traps of 47 and 119 in^3 effective volumes as a function of time on the steady state simulated AMA cycle with regeneration occurring during the first maximum power acceleration.

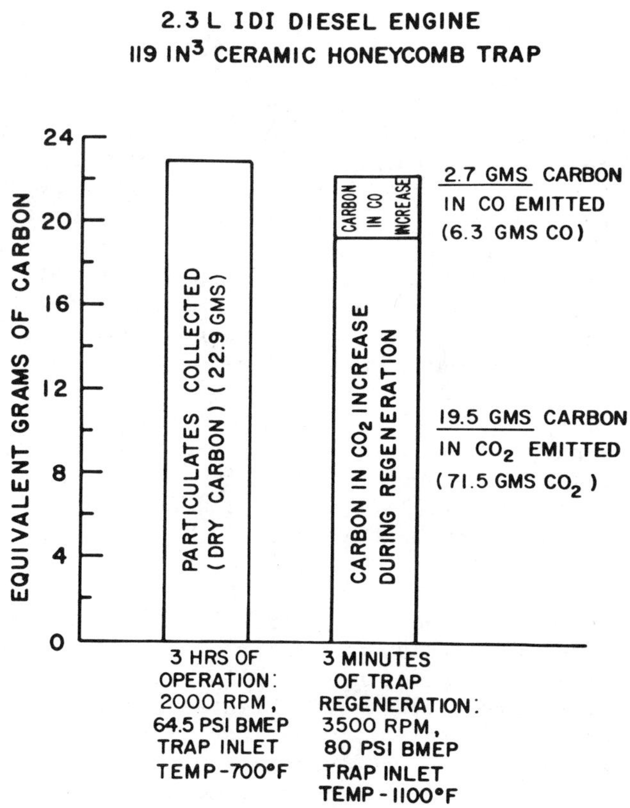

Fig. 15 - Emissions produced from collected particulates in a trap during a high speed/ high load regeneration

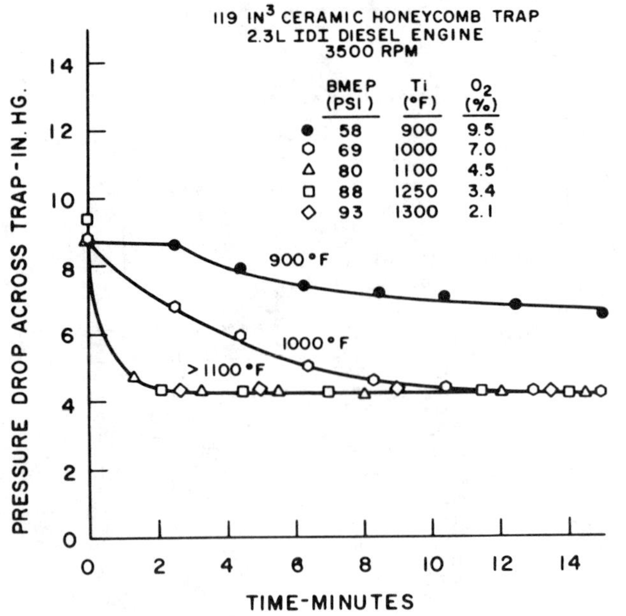

Fig. 16 - Effect of trap inlet temperature on regeneration time as indicated by the pressure drop across the trap vs. time

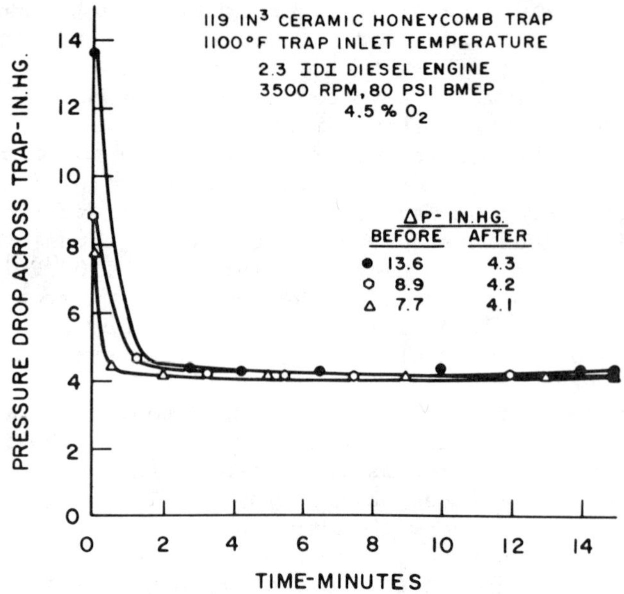

Fig. 17 - Effect of trap loading on regeneration time at a constant trap inlet temperature of 1100°F as indicated by the pressure drop across the trap vs. time

CATALYZED TRAP

Tests were conducted to determine if catalysts impregnated in the trap could reduce the ignition temperature of the collected particulates in the trap and accelerate the particulate oxidation process. Platinum and lead were investigated as candidate catalysts. Lead was selected because of its ability to promote high carbon oxidation rates and to reduce the ignition temperature of carbon as shown in Table 3 (11). The platinum catalyst and the lead catalyst were impregnated into 119 in^3 honeycomb traps, without washcoats, using conventional catalyst technology.

Test results from high speed/high load vehicle regeneration tests with the catalyzed traps are shown in Figure 18. These data indicate that the catalysts did not lower the regeneration temperature of the trap nor did they accelerate the particulate oxidation process. Since the catalyst was impregnated into the ceramic trap material, only the first layer of particulate material was in contact with the catalyst as shown in Figure 6. Since successive layers of collected particulate material were not in contact with the catalyst, singificant advantages from an impregnated catalyst would not be expected. Methods to improve contact between the catalyst and a larger amount of the particulate material may have the potential to reduce the particulate ignition temperature.

CATALYTIC FUEL ADDITIVES

The inorganic elements such as lead and copper have the capability to reduce the ignition temperature of carbon by more than 300°F as shown in Table 3 (11). As discussed in the previous section, catalysts impregnated into the trap structure did not reduce the trap regeneration temperature due to the lack of intimate contact between the catalyst and the particulate material. However, the addition of the catalytic element to the fuel can provide the intimate contact required to reduce the ignition temperature of the carbon.

Tests were conducted to determine the effect of adding neat tetraethyl lead to the Diesel fuel to reduce the regeneration temperature of a loaded trap. The results are shown in Figure 19, where exhaust back pressure is plotted as a function of trap inlet temperature. Exhaust back pressures of the loaded trap are shown as the horizontal portion of the curves at the low trap inlet temperatures. Regeneration is indicated by the rapid reduction in back pressure resulting from the oxidation of particulates collected on the trap. With 1 gm/gal of lead (as tetraethyl lead) added to the fuel, the regeneration temperature was reduced by over 300°F. The lead oxide, which exists in the combustion chamber as a vapor, is believed to condense on the carbon particles to form a nearly ideal distribution at the 1 gm/gal concentration to produce the maximum reduction in carbon ignition temperature. The magnitude of the effect is nearly identical to that obtained using vaporized lead chloride as shown in Table 3. The large loss in the catalytic effect experienced by reducing the lead level to 0.1 gm/gal, shown in Figure 19, indicates the importance of the catalyst distribution and population density. Although these cursory tests at steady state engine conditions have shown significant results, the effects of an additive on the engine, fuel injection system and the trap have not been evaluated.

THROTTLING REGENERATION

As shown on the exhaust temperature map for a Diesel engine in Figure 13, normal road load conditions do not produce adequate exhaust temperatures required for regenerating a particulate trap. However, the exhaust temperatures of a Diesel engine can be increased by throttling the airflow supplied to the engine. By reducing the airflow to the engine, the overall air-fuel ratio of the engine is decreased from the normal unthrottled level. Reducing the air-fuel ratio causes higher average combustion gas temperatures and subsequent higher exhaust temperatures.

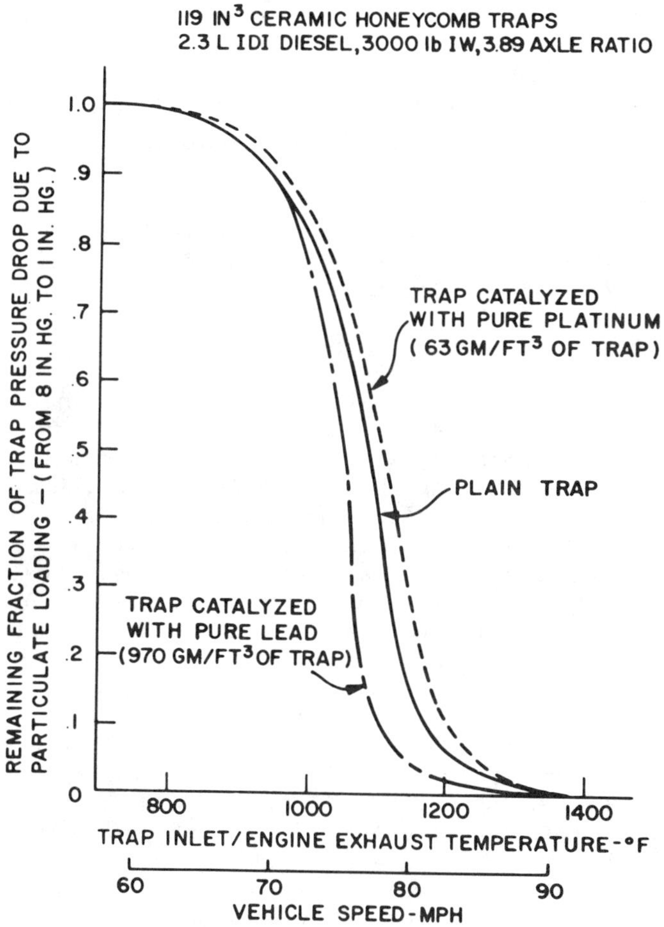

Fig. 18 - Remaining fraction of trap pressure drop due to the particulate loading as a function of trap inlet temperature during high speed vehicle regeneration for plain and catalyst impregnated traps

TABLE 3

RELATIVE ACTIVITY ESTIMATES FOR CATALYTIC INORGANIC COMPOUNDS ADDED TO CARBON (from Ref. 11)

	Lowering of Ignition Temperature °C (1)	Relative Oxidation Rate (2)
Na	92	230
Ca	124	4
Zn	130	--
Mn	130	86,000 (3)
Fe	131	--
K	137	--
NH_4	137	--
Sn+2	153	--
Ni	162	32
Pb	180	470,000 (3)
Cu	284	500
Ba	--	100
Au	--	240
V	--	340
Ag	--	1,340 (3)
Cs	--	64,000 (3)

(1) Compound added to soot particulate by vaporization of the corresponding chloride.

(2) Compound added to graphite powder by impregnation of a soluble salt. Oxidation rate compared to untreated graphite.

(3) For most of these elements the normal oxide structure melts below 500 C; in the case of Mn a series of defect oxides are known with uncertain melting points. All of these elements exhibit thermal decomposition of normal oxide structure at or below 500 C.

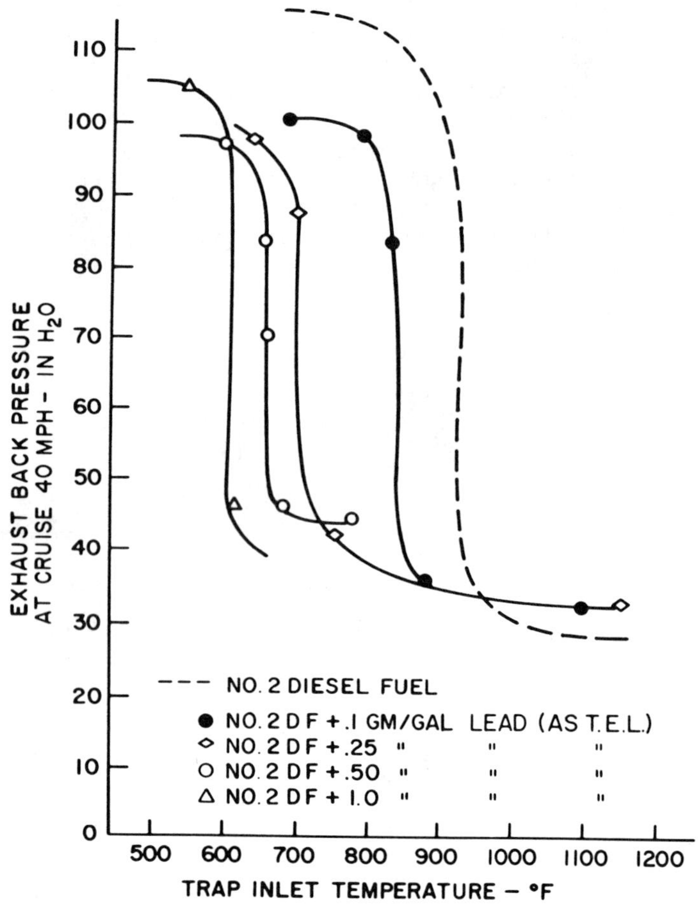

Fig. 19 - Effect of tetraethyl lead in Diesel fuel on regeneration temperatures as shown by trap back pressure vs. trap inlet temperature characteristics

The temperature increases obtained by throttling a 2.3L Diesel engine at several loads are shown in Figure 20 where exhaust temperature is plotted as a function of intake manifold vacuum for an engine speed of 1800 RPM. Lines of constant air-fuel ratio, which correspond to lines on constant exhaust oxygen level, are also shown in the figure. The maximum amount of throttling for a Diesel engine is limited to an exhaust oxygen concentration in the range of 2 to 5%. Below this value, the smoke and particulate production are excessive. Throttling produces a greater rise in the exhaust temperature at the lighter loads. However, by throttling to the same oxygen level, a higher exhaust temperature can be produced at the heavier loads. Therefore, at light loads, throttling will not produce adequate exhaust temperatures required to initiate regeneration. At heavier loads, where the normal unthrottled exhaust temperature exceeds approximately 600°F, throttling can be applied to achieve adequate temperatures for regeneration. As shown in Figure 13, this requirement restricts successful throttling regeneration to loads above 40 psi BMEP for moderately high engine speeds and 65 psi BMEP for low engine speeds. These results indicate that throttling regeneration cannot be applied at steady state vehicle speeds below approximately 55 mph for the test vehicle employed for this study where the drivetrain N/V ratio was about 55 RPM/mph.

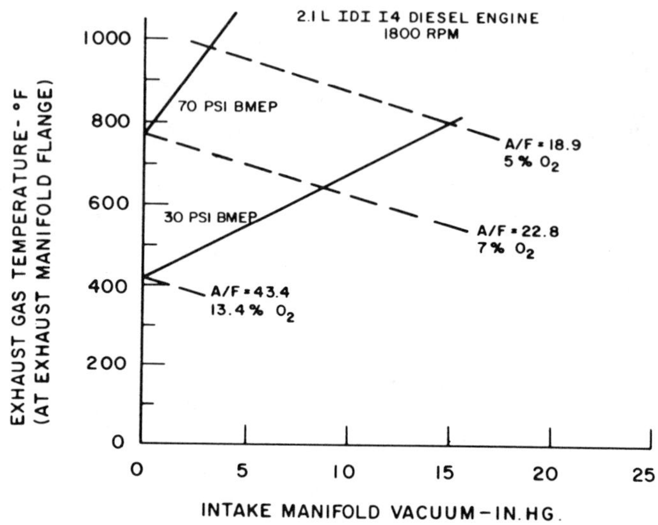

Fig. 20 - Exhaust temperature increases obtained by throttling a 2.3L Diesel engine over a range of intake manifold vacuum levels

Throttling generally has an adverse effect on the engine exhaust emissions and fuel consumption, as illustrated by the data in Figure 21. As throttling is applied, HC, CO and smoke emissions increase substantially while the NO_x emissions decrease due to the richer combustion zone air-fuel ratios. Fuel consumption increases with throttling due to the richer air-fuel ratios, higher heat losses and negative pumping work.

Throttling was used to regenerate a ceramic honeycomb trap on a 2.3L Diesel engine installed on a dynamometer and operated at a simulated 55 mph cruise condition (3200 RPM, 43 psi BMEP). A trap was initially loaded with particulates to a back pressure of 16 in. Hg. at a simulated 55 mph cruise condition. Operating characteristics of the engine and the trap during the throttling regeneration process are plotted as a function of time in Figure 22. Throttling was applied to enrichen the air-fuel ratio from 35:1 at the base unthrottled condition to 18:1. Since throttling reduced the air flow to the engine, the exhaust flow was likewise decreased, which resulted in an initial reduction in the pressure drop across the trap. The exhaust temperature, or trap inlet temperature, at the unthrottled condition was about 750°F. After a 45 second delay due to the thermal inertia of the exhaust system and the trap, the trap inlet temperature increased to about 1200°F. At this temperature, the regeneration process was initiated and an exothermic temperature rise was produced through the trap. The oxygen concentration at the trap outlet was reduced from 3.25% to about 1% by the rapid particulate oxidation process.

The effects of throttling and the regeneration process on tailpipe emissions are shown in Figure 23. HC emissions decreased slightly as throttling was initiated. In contrast to the increase in HC emissions shown in Figure 21 for the base engine when throttling was applied, some HC oxidation apparently occurred within the trap during throttling due to the increased residence time at the higher temperatures. During the regeneration period where the high exothermic temperature rise through the trap was observed, HC oxidation was increased which resulted in significant reductions in HC emissions below the unthrottled base engine level. Conversely, CO emissions increased during the regeneration period with the high exothermic temperature rise through the trap. Exhaust smoke levels at the tailpipe during the throttling regeneration process were very low and were equal to the levels during normal operating conditions.

Regeneration cannot be achieved if the engine is throttled to produce exhaust oxygen concentration levels below about 2%. Although adequate trap inlet temperatures can be achieved, inadequate oxygen levels are available for the particulate oxidation process.

A loaded particulate trap can fail catastrophically as a result of overthrottling the engine. An example of such a failure occurred during a throttling regeneration test on a vehicle. During this test, the oxygen level was accidentally allowed to drop below 2% by overthrottling so that temperatures adequate for regeneration were achieved

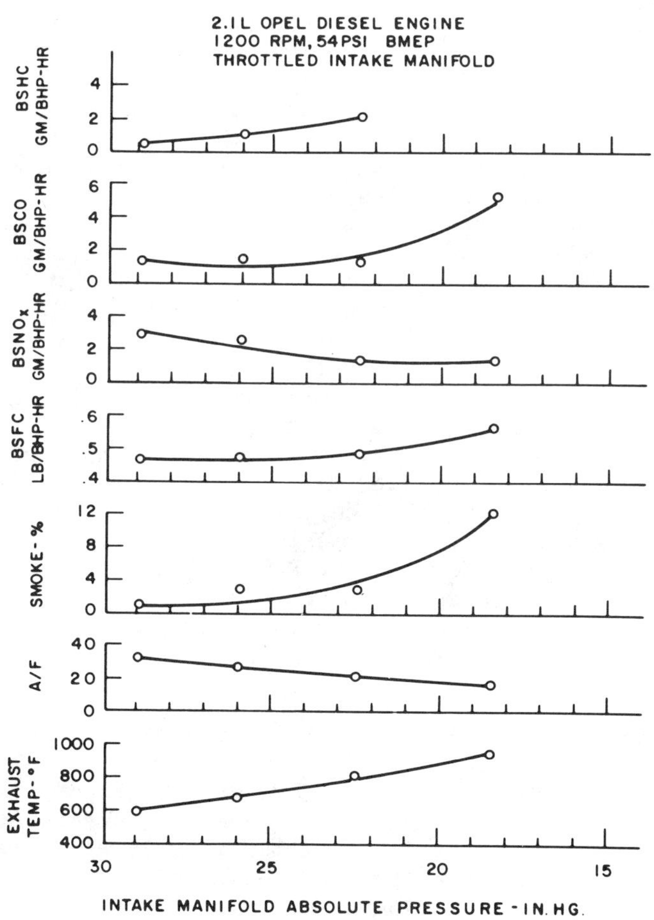

Fig. 21 - Effect of throttling on emissions, fuel consumption and smoke and operating characteristics of a 2.3L Diesel engine

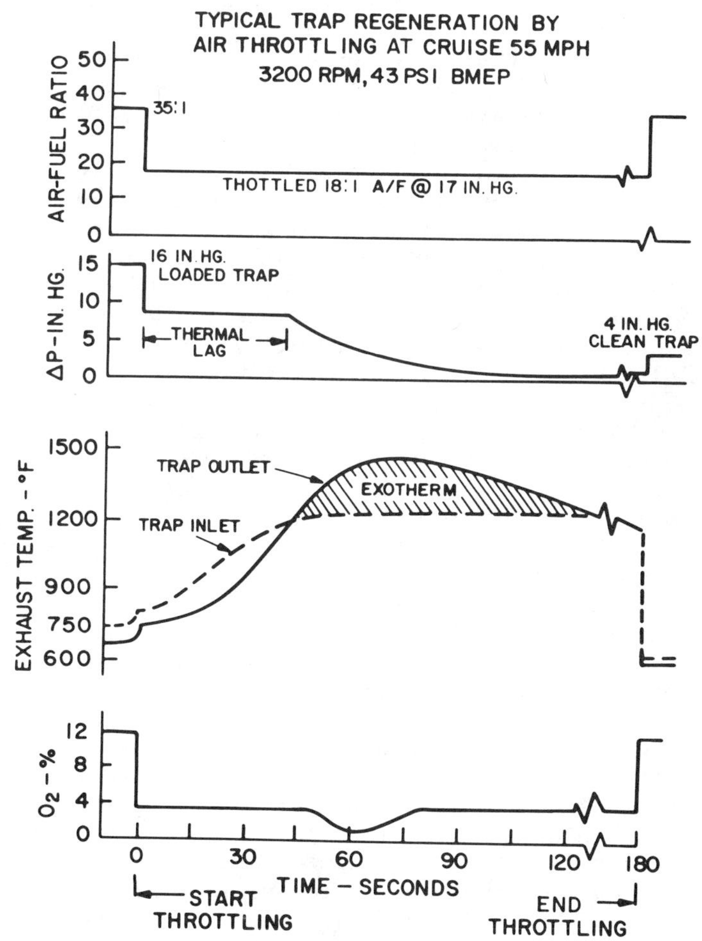

Fig. 22 - Operating characteristics of a 2.3L Diesel engine equipped with a trap during a throttling regeneration process at cruise 55 mph as a function of time

without sufficient oxygen to support combustion. When the engine was returned to the unthrottled condition and the engine speed was reduced, adequate oxygen was supplied to the superheated particulates on the trap. Without sufficient exhaust flow to remove the heat, a "run-away" regeneration condition occurred. This rapid oxidation of the particulates caused the ceramic honeycomb trap to melt. A photograph of the trap, which failed in this manner, is shown in Figure 24. Failures of this nature can be prevented by providing adequate exhaust flow rates without excessive particulate loadings at the high temperature throttled conditions.

AIR-FED BURNER REGENERATION

Idle exhaust temperatures of a Diesel engine are in the range of 250-300°F, as illustrated in the exhaust temperature map shown in Figure 13. Throttling the engine intake air is not adequate to produce exhaust temperatures required for regeneration below about a 55 mph cruise condition. For low speed, light load conditions and for extended idling, alternate regeneration methods must be considered.

An externally fueled burner is one system which can be considered for producing trap inlet temperatures required for regeneration at low exhaust temperature conditions. A schematic of an air-fed burner regeneration system is shown in Figure 25. In this system, the burner is supplied with air from an external pump, although shop air was used for the dynamometer tests described in this paper. The burner is also supplied with fuel through an atomizing nozzle. Either a glow plug or spark plug was used to ignite the burner when regeneration was required.

The burner air flow and fuel flow must be controlled to provide an adequate regeneration temperature at the inlet to the trap where the burner exhaust and engine exhaust have been mixed. The burner flow conditions required to achieve adequate trap regeneration temperatures over 1000°F are shown in Figure 26 as a function of engine speed and corresponding vehicle cruise speeds. The engine exhaust temperature and flow rate increase as the vehicle speed is increased. However, as the speed is increased to approximately 50 mph, the fuel flow of the burner must be increased to maintain the required trap inlet temperature. Above 50 mph, the burner fuel flow requirement decreases as the engine exhaust temperature approaches the 1000°F required for regeneration. The burner supplies a smaller fraction of the total enthalpy of the trap inlet flow as the vehicle speed is increased.

Engine dynamometer tests were conducted to evaluate the capability of an air-fed burner to regenerate a loaded particulate trap at low speed, light load conditions. A small burner was installed near the inlet to the ceramic honeycomb trap on the 2.3L Diesel engine. A moderate volume below the burner was provided to ensure mixing of the engine exhaust and burner exhaust before the inlet to the trap.

The test results obtained from using the air-fed burner for regenerating a loaded particulate trap at an engine idle condition are shown in Figures 27 and 28. Figure 27 illustrates the trap pressure drop, trap inlet and outlet temperature and oxygen concentration as a function of time during the regeneration process. When the burner was ignited, the trap inlet temperature increased from the normal idle exhaust temperature of 300°F to over 1100°F. The pressure drop across the trap increased as the temperature and flow passing through the trap were increased by the addition of the burner exhaust gas to the engine exhaust gas. As the trap inlet temperature increased, particulate oxidation was initiated as shown by the decrease in oxygen concentration and the decrease in back pressure. After about 80 seconds, regeneration was completed and the burner was turned off. The trap pressure drop decreased as the flow of burner exhaust gases through the trap was eliminated and as the trap inlet temperature decreased to 300°F, typical of the normal engine idle condition.

The exhaust emissions in the total flow through the trap during the air-fed burner regeneration process at an engine idle condition are shown in Figure 28. As illustrated in this figure, HC and CO emissions peaked during the initial light-off period of the air-fed burner. Lower than steady state air and fuel flow settings were used during burner light-off. After light-off, the burner air and fuel flow rates were increased to obtain the desired trap inlet temperature of about 1100°F. Since this was an initial screening test, low emissions during light-off were not developed. However, after about 15 seconds of burner operation, the emissions decreased to relatively low, stabilized levels.

EXHAUST-FED BURNER REGENERATION

Although the air-fed burner described in the previous section had adequate stability from an engine idle condition to a 55 mph cruise condition, the complexity of an engine driven air pump for this type of burner is undesirable. Since the Diesel engine operates at lean air-fuel ratios, excess oxygen in varying quantities, depending upon engine speed and load, is always available in the exhaust gas. Therefore, the exhaust gas of the Diesel engine could be used as the combustion oxygen supply for an exhaust-fed trap regeneration burner.

A map of oxygen concentrations as a function of engine speed and load is shown in Figure 29. At idle, the exhaust oxygen concentration is 15-17%, whereas at full load, the oxygen concentration decreases to less than 2%. The exhaust temperature map shown in Figure 13 indicates that the largest temperature increase through an exhaust-fed burner to reach regeneration temperatures of 1000° to 1200°F must be provided at the light load conditions with the highest available oxygen concentrations.

The oxygen concentrations consumed by an exhaust-fed burner in producing a range of temperature increases are shown in Figure 30. The depleted oxygen concentrations and the temperature increases required to reach a trap regeneration temperature of 1100°F are shown in this figure as a function of load (BMEP) at 2000 RPM. The calculations in this figure indicate that oxygen concentrations are adequate to produce trap regeneration temperatures with an exhaust-fed burner at all engine loads. The achievement of adequate combustion stability at all of these conditions is another important consideration. Application of the combustion stability criteria described in Reference 11, to these conditions indicated that a burner should be able to operate at all of the conditions shown in Figure 30.

A schematic of an exhaust-fed burner for trap regeneration in which all of the engine exhaust is supplied to the burner is shown in Figure 31. A comparison of operating parameters and the resulting emissions for an air-fed and an exhaust-fed burner operated to produce trap regeneration temperatures over 1100°F at a steady 40 mph cruise condition are shown in Table 4. The exhaust-fed burner required lower fuel flow than the air-fed burner since the total gas flow rate supplied through the trap was lower. However, the data in Table 4 indicate that the air-fed burner produced lower HC emissions than the exhaust-fed burner at steady state conditions. On the other hand, the exhaust-fed burner produced lower CO emissions than the air-fed burner.

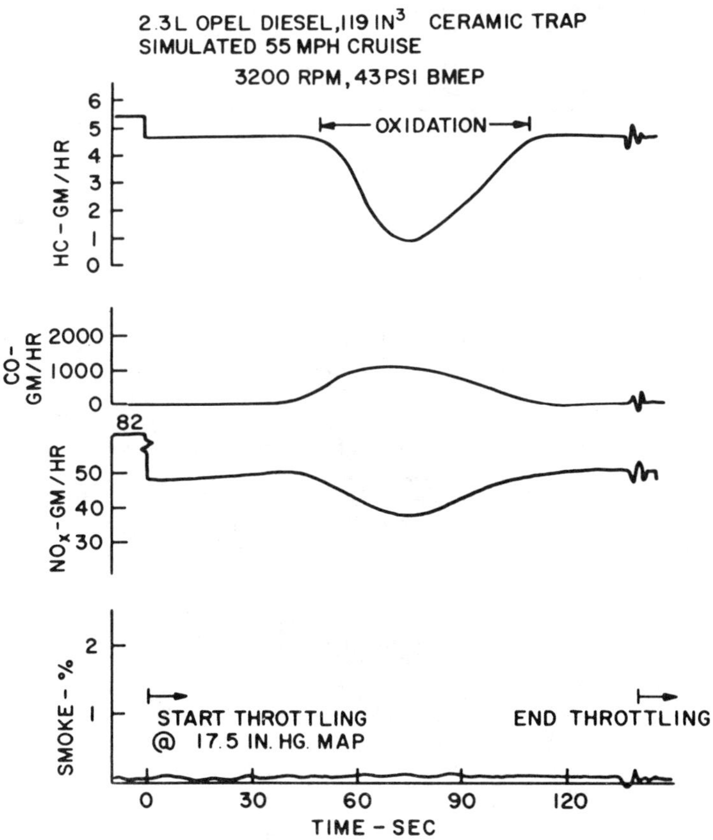

Fig. 23 - Effect of throttling and the regeneration process on tailpipe emissions as a function of time

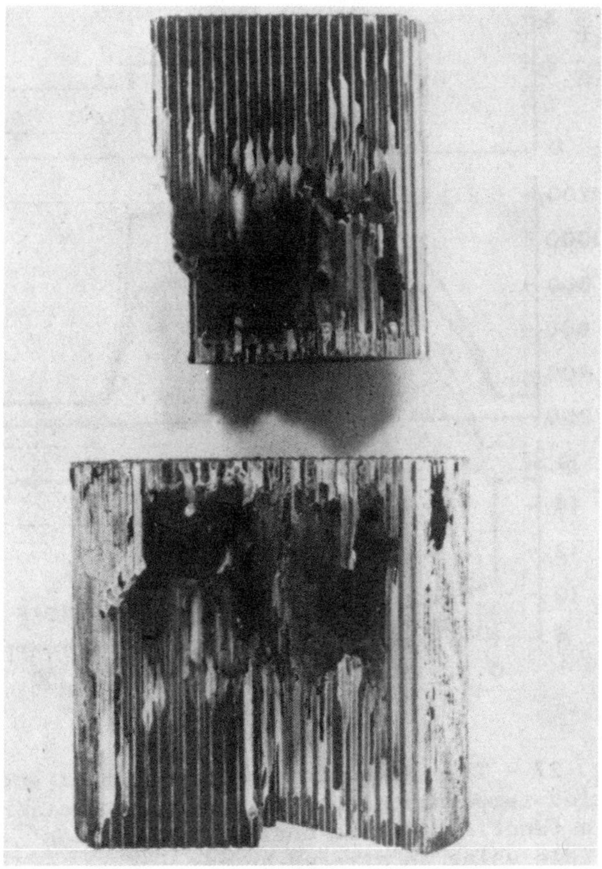

Fig. 24 - A particulate honeycomb trap that failed due to overthrottling followed by low flow, high oxygen concentration conditions

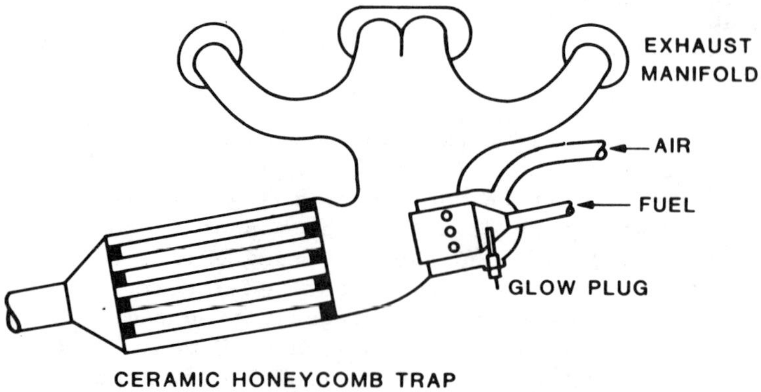

Fig. 25 - Schematic of an air-fed burner regeneration system

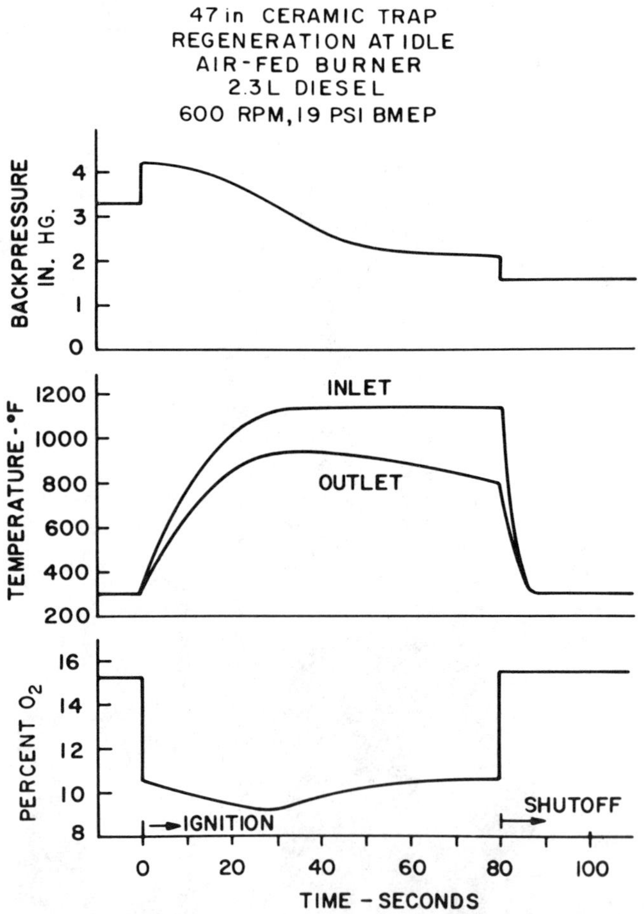

Fig. 27 - Trap pressure drop, trap inlet and outlet temperatures and oxygen concentration as a function of time during regeneration at idle using an air-fed burner

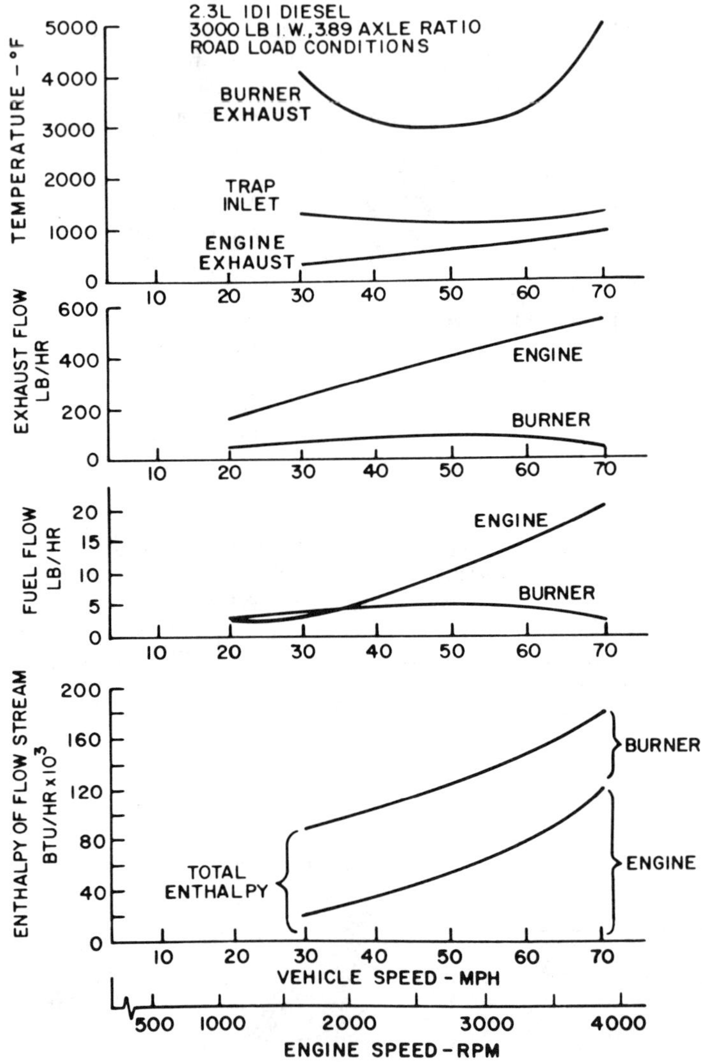

Fig. 26 - Burner flow conditions required to achieve a trap regeneration temperature of over $1000^{\circ}F$ as a function of engine and and vehicle speeds

REGENERATION EMISSIONS AND FUEL CONSUMPTION

All of the trap regeneration systems discussed in this paper produced changes in emissions and fuel consumption during the regeneration process. A comparison of the changes in emissions and fuel consumption during regeneration is shown in Figure 32 for the following regeneration systems: throttling at a steady 55 mph cruise speed, an air-fed burner at a steady 55 mph cruise speed and an air-fed burner at idle. The changes in emissions and fuel consumption are shown in Figure 32 as absolute values and as a percent of the CVS values for a 2.3L Diesel engine in a 3000 lb. inertia weight vehicle.

At a 55 mph cruise condition, both the burner and throttling result in about a 5% increase in fuel consumption over the fuel consumption of the vehicle operated on the CVS cycle. The burner produced about a 25% increase in HC emissions over the HC emissions produced by the vehicle during the CVS cycle. In contrast, the throttling system showed a minor decrease in HC emissions. Both systems resulted in significant increases in CO emissions. The throttling system produced approximately a 100% increase in CO emissions, whereas the burner produced approximately a 50% increase in CO emissions over the CO emissions produced by the vehicle during the CVS cycle.

Regeneration at idle with an air-fed burner consumed about 1/2 of the fuel required for regeneration at 55 mph. Although the increase in HC emissions was slightly less than the increase with the burner at a 55 mph cruise condition, the increase in CO emissions was significantly lower for the idle regeneration condition. A portion of the HC emissions from the burner regeneration system was due to the light-off, which was not optimized for low emissions in these tests.

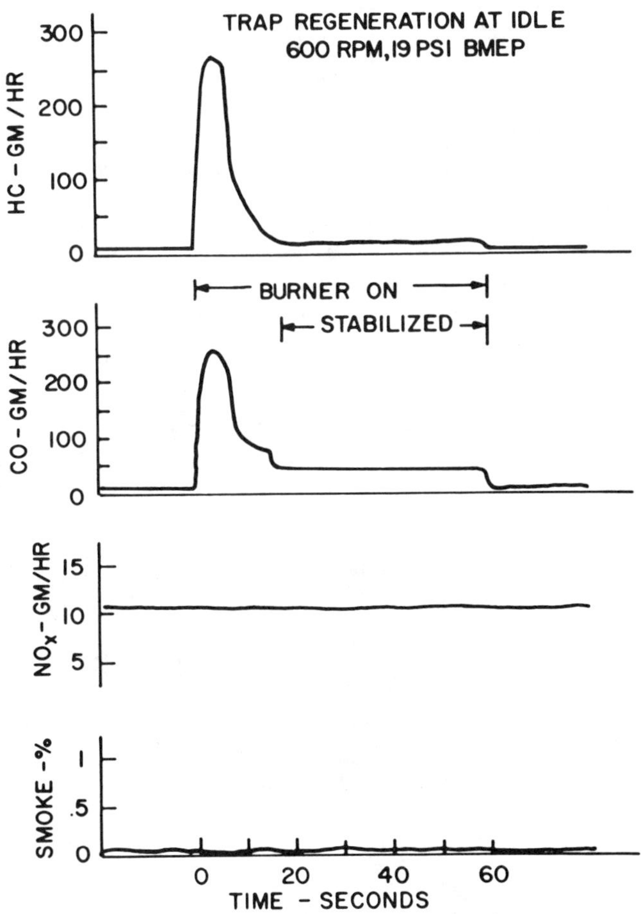

Fig. 28 - Exhaust emissions during regeneration at idle using an air-fed burner

REGENERATION RANGE

The regeneration systems described in this paper operate over a limited range of engine operating conditions. The current operating ranges that have been developed for high speed/high load regeneration, throttling regeneration and burner regeneration are shown on the engine BMEP vs. RPM map in Figure 33. Road load cruise conditions are also shown on this map. This figure shows that throttling is limited to a range of engine operation that may not be encountered with normal low speed urban driving. In contrast, the exhaust-fed burner covers a range of engine operation frequently encountered by most drivers. Furthermore, with additional burner development, the operating range of the burner regeneration system can probably be expanded.

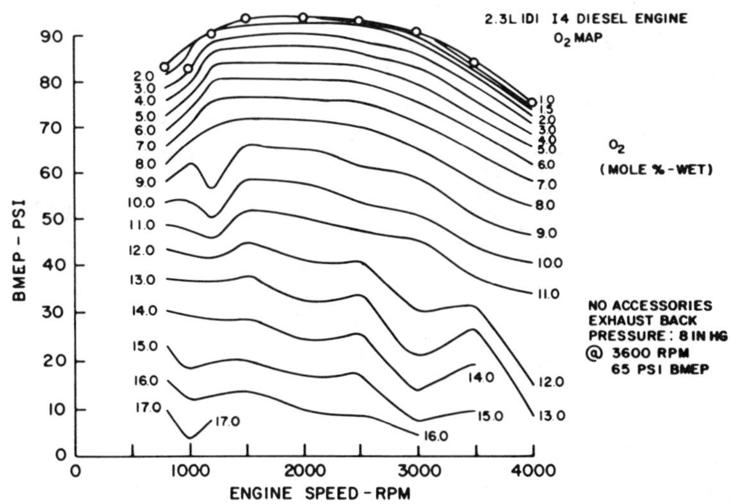

Fig. 29 - An oxygen concentration map for a 2.3L Diesel engine at steady state conditions

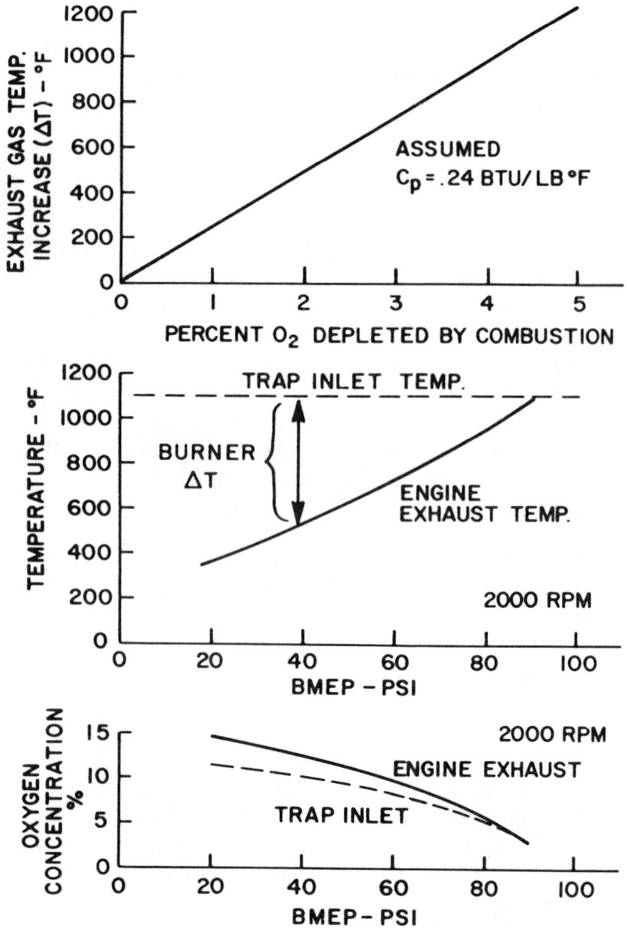

Fig. 30 - Effect of engine load on the required temperature rise through an exhaust-fed regeneration burner and the depletion in exhaust oxygen concentrations to produce an 1100°F trap inlet temperature

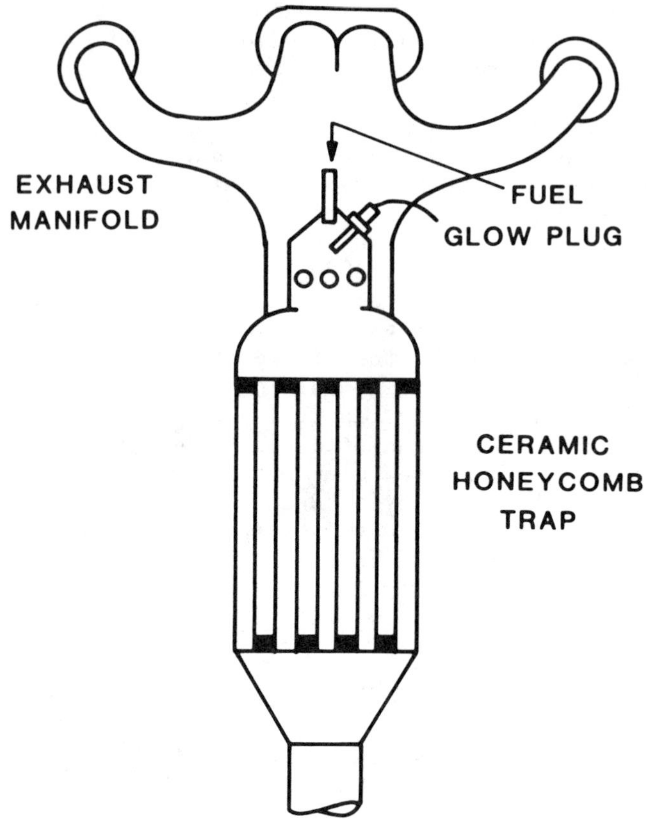

Fig. 31 - Schematic of exhaust-fed burner regeneration system

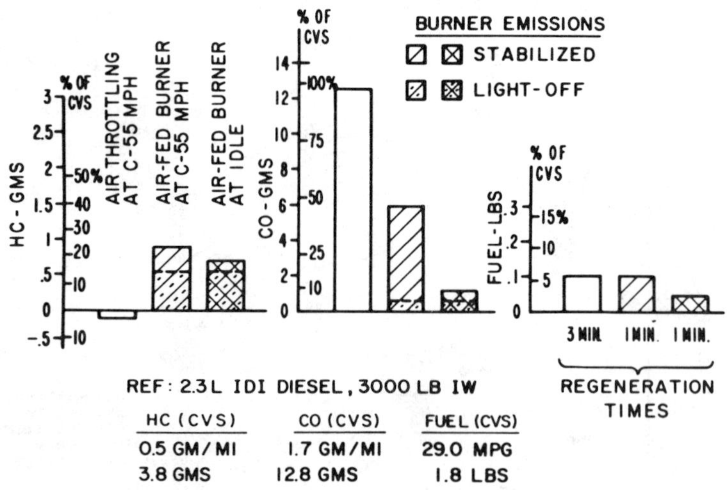

Fig. 32 - A comparison of emissions and fuel consumption during regeneration for air-throttling at 55 mph, air-fed burner at 55 mph, and an air-fed burner at idle

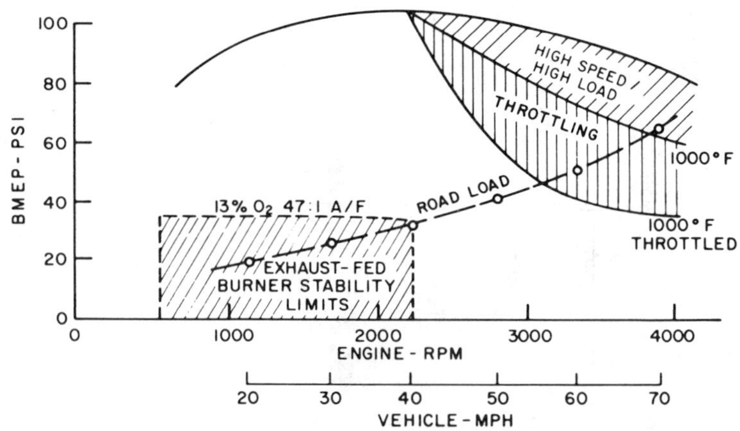

Fig. 33 - Operating ranges for high speed/high load regeneration, throttling regeneration and burner regeneration on an engine BMEP vs. RPM map

All of the regeneration systems in this study were evaluated at steady state operating conditions of the engine, vehicle and regeneration system. An investigation of the capability of these systems to operate during transients experienced during normal driving was beyond the scope of the present study.

TABLE 4

COMPARISON OF AIR-FED AND EXHAUST-FED BURNER REGENERATION OPERATING PARAMETERS AT STEADY STATE CRUISE 40 MPH

(Trap Inlet Temperature 1150-1250°F)

Operating Parameters:	Air-Fed Burner	Exhaust-Fed Burner
Fuel Flow - lb/hr	4.0	3.5
Inlet Air Flow - lb/hr	60	0
Total Trap Flow - lb/hr	385	320
Emissions: (Produced from Burner)		
HC - gm/hr	8.4	62.4
CO - gm/hr	177	103

CONCLUSIONS

1. Diesel particulate traps with collection efficiencies up to 80% are required to approach the 1985 .2 gm/mi particulate emission regulation in vehicles up to 3500 lb. inertia weight. Collection efficiencies over 80% have been obtained at low mileage with clean experimental traps.

2. The particulates collected on a Diesel trap must be periodically oxidized, or burned off the trap, to prevent high engine back pressure and the associated fuel economy and performance loss.

3. By increasing the volume of a ceramic honeycomb trap, moderate reductions in clean trap pressure drop and significant reductions in trap pressure rise rate were obtained and the frequency between regenerations was extended.

4. High speed/high load engine operation was used to regenerate a Diesel particulate trap; however, this technique is impractical because these conditions are not normally encountered during vehicle operation.

5. Impregnating a trap with a catalyst did not produce the potential catalyst effect of reducing the trap regeneration temperature due to insufficient contact between the catalyst and the particulate material.

6. Lead added to the Diesel fuel acted as a catalyst for carbon combustion and reduced the ignition temperature of particulates collected on a trap by about 300°F.

7. Throttling the engine intake air extended the range of trap regeneration from maximum engine speeds and loads down to a steady 55 mph cruise condition.

8. An externally fueled burner was used to achieve trap regeneration at idle through 55 mph cruise conditions. Current burner stability limits prevent operation at other conditions.

9. Trap regeneration at a 55 mph condition with either throttling or a burner consumed about 5% additional fuel over the fuel required to operate a Diesel powered vehicle on the CVS cycle.

10. Trap regeneration at an idle condition with an air-fed burner consumed about 2.5% additional fuel over the fuel required to operate a Diesel powered vehicle on the CVS cycle.

ACKNOWLEDGEMENT

The authors acknowledge the many people who have contributed to the preparation of this paper. Messrs. H. Cikanek, J. Monty and L. Gruich made especially valuable contributions by conducting many of the tests reported in this paper.

REFERENCES

1. J. D. Collins, "Powertrains - Now to the Year 2000", presented to the Institute of Mechanical Engineers in London, England, October, 1979.

2. H. A. Nickol, "Automotive Powertrains - Now and Into the 1990's," SAE Paper No. 801340, October, 1980.

3. W. R. Wade, "Light Duty Diesel NO_x - HC - Particulate Trade-Off Studies," SAE Paper No. 800335, February, 1980.

4. G. Stumpp and W. Banghaf, "An Exhaust Gas Recirculation System for Diesel Engines," SAE Paper No. 780222, February, 1978.

5. Federal Register, March 5, 1980, Part III, Environmental Protection Agency, "Standard for Emission of Particulate Regulation for Diesel - Fueled Light-Duty Vehicles and Light-Duty Trucks."

6. S. H. Oh, et al, "Mathematical Modeling of Fibrous Filters for Diesel Particulates - Theory and Experiment," SAE Paper No. 810113, February 1981.

7. M. G. Faulkner, et al, "Assessment of Diesel Particulate Control: Filters, Scrubbers, and Precipitators," EPA Report No. 600/7-79-232a prepared by Southern Research Institute, October, 1979.

8. J. S. Howitt and M. R. Montierth, "Cellular Ceramic Diesel Particulate Filter," SAE Paper No. 810114, February 1981.

9. G. J. Van Wylen, "Thermodynamics," John Wiley and Sons, Inc., New York, 1961, p. 460.

10. K. Otto, et al, "The Oxidation of Soot Deposits from Diesel Engines," SAE Paper No. 800336, February 1980.

11. M. J. Murphy, et al, "Assessment of Diesel Particulate Control: Direct and Catalytic Oxidation," EPA Report No. 600/7-79 - 232b prepared by Battelle Columbus Laboratories, October 1979.

12. M. G. Zabetakis, "Flammability Characteristics of Combustible Gases and Vapors," Bureau of Mines Bulletin 627, 1965.

The Effect of an Oxidation Catalyst on the Physical, Chemical, and Biological Character of Diesel Particulate Emissions**

G. Hunter, J. Scholl, F. Hibbler, S. Bagley, D. Leddy, D. Abata, and J. Johnson
Michigan Technological Univ.

THE DIESEL ENGINE, by virtue of its combustion system, offers two primary advantages over spark ignition engines. The diesel, by various estimates, offers 25% to 45% more fuel economy than a comparable spark ignition engine. Corporate average fuel economy (CAFE) standards and consumer demand for vehicles with better fuel economy are causing manufacturers in most vehicle classes to evaluate the diesel engine as a viable alternative to the spark ignition engine. Industry predictions estimate that by 1985 about 25% of the new automobiles sold in the U.S. could be diesel powered [1].* The diesel engine is also inherently low in carbon monoxide (CO) and unburned total hydrocarbons (HC), while maintaining reasonable levels of nitrogen oxide (NO_x) emissions. Current technology light duty diesel engines were able to meet the 1980 Environmental Protection Agency (EPA) standards for these pollutants without major emission control modifications.

Diesel engine manufacturers, users, and the general public are becoming keenly aware of the problems associated with increased use of diesel engines. Problems such as higher noise levels, odor and smoke have always been aesthetically objectionable. The physiological effects of diesel smoke and odorous pollutants have recently come under close scrutiny. The 1977 Clean Air Act amendments required the EPA to study the health and welfare effects of particulate emissions and to set particulate matter emission standards for the 1981 model year [2]. Due to certification leadtime difficulties, light-duty standards will become effective in 1982. The complex problems presented by the nature of particulates resulted in the 1981 heavy-duty standards being based on smoke (related to particulate) measurements while the setting of heavy-duty particulate mass emission standards was deferred to the 1985 model year. The interim period is to be used to intensively study the character, effects, and possible

*Numbers in parentheses designate Reference at end of paper.

**Paper 810263 presented at the International Congress and Exposition, Detroit, Michigan, February, 1981.

ABSTRACT

A diesel oxidation catalyst (Engelhard PTX Series) was evaluated on a medium-duty diesel engine (Caterpillar 3208, naturally aspirated, direct injection). Tests were conducted at six modes of the EPA 13 mode heavy-duty cycle to measure the total particulate, soluble organic fraction (SOF), sulfates, NO, NO_2, NO_x and hydrocarbons emitted by the engine with and without the oxidation catalysts. Chemical analysis of the SOF collected was carried out to determine the effects of the catalysts on each of the subfractions composing the SOF. The Ames Salmonella/microsome bioassay was employed to quantify the mutagenic properties of the particulate SOF.

Test results show large increases in the amounts of total particulate and sulfate emissions due to the catalyst while the amounts of SOF are reduced by the catalyst. The amounts of NO_x produced with and without the catalyst are similar, but the equivalent NO_2 emitted with the catalyst installed is increased at most modes. The Ames bioassay results show that the mutagenicity/ g of the SOF increased with the use of the catalyst, but the effect of this increase on revertant emission rates is partially offset by the reduction in mass emission rates of the SOF.

control of diesel particulate emissions, leading to the establishment of acceptable standards.

Transportation related particulate, while not a significant source on a mass basis, present several special problems. They are usually fairly highly concentrated in areas of high population, are emitted at or near ground level, and because of their small size they have relatively long lifespans in the atmosphere [3]. Due to their small size and optical characteristics, they are also very visible.

The health effects of a particle are dependent on its extent and mode of entry into the body, and on the body's ability to remove or neutralize the compounds present on the particle. Diesel particulate matter is small enough to result in deposition of the particle in the respiratory tract. It has been determined that 90% by mass of the particulate produced by a diesel is less than 1 micrometer in diameter [3], and that up to 70% by mass of the particulate is less than 0.3 micrometers in diameter [4]. The bulk of the particulate greater than 0.3 micrometers in diameter will be removed by the upper respiratory tract, while the remaining small particles can enter the lungs. It should be noted that while in the respiratory system, the particles may or may not be deposited. Due to their small size, the particulate are subject to Brownian motion and must rely on diffusion to the tissue to be retained [4].

Due to the above facts and because diesel exhaust particulate SOF has been shown to be mutagenic, it is important to develop a better understanding of the effects of engine operating parameters (load and speed) and control technology on the physical, chemical, and biological character of particulate emissions. It is to this end that our paper is directed relative to determining the effect that an oxidation catalyst has on emissions as a function of load and speed. From this data base we have also shown for this engine that the specific activity (revertants/μg) of the non-paraffin portion of the SOF is directly proportional to the concentration of equivalent NO_2 in the exhaust. This conclusion could indicate that the biologically active compounds in the SOF are nitro-organic compounds and are formed on the expansion stroke, in the exhaust system, in the dilution tunnel or on the filter. Further basic experiments and calculations need to be conducted to better understand these conclusions since the control of NO_2 or other nitrogen compounds in the exhaust might be a more effective approach to reducing the biological activity of the SOF than reducing either the total particulate mass or the mass of SOF.

This paper is divided into three main sections. The first gives the BACKGROUND literature supporting the overall paper. The EXPERIMENTAL section describes the experimental methods and the test matrix used in the study. This is followed by the RESULTS section which presents the physical, chemical and biological emission data.

BACKGROUND

The particulate emissions from a diesel engine are composed of both solid and liquid compounds and are very complex in character. The solid portion of the particulate contains primarily agglomerations of small carbonaceous subparticles [5]. The liquid portion may be in the form of small droplets or may be adsorbed onto the surface of the solid particles. The liquid portion consists primarily of various organic compounds, termed the soluble organic fraction (SOF), and sulfates. The formation and destruction of particulate matter in a diesel engine is the result of many physical and chemical transformations occurring before, during, and after combustion.

The SOF of diesel exhaust particulate matter contains a wide variety of chemical compounds, many of which arise from the combustion process [6]. Table 1 gives the major compound classes present in diesel SOF, as well as typical compounds for each class. Because of the correlation between compounds that are mutagenic in bacterial test systems and those that are carcinogenic in mammalian test systems [11,12,13], microbial assays are used to screen particulate SOF samples for mutagenicity as a first step to assess potential adverse health effects [10,14,15,16,17,18, 19]. The aromatic and transitional classes have shown the most potential for being harmful to health, however, many of the oxygenated compounds as yet unidentified in diesel exhaust may be responsible for much of the mutagenic activity observed [10]. Wei et al. [18] demonstrated mutagenicity for cyclopeteno<c,d>pyrene anhydride, a compound isolated from diesel exhaust. The mutagenicity of this compound was too low to account for the total mutagenic activity of the sample. Schuetzle et al. [10] identified four nitro-containing polynuclear aromatic (PNA) compounds in diesel exhaust, one of which (1-nitropyrene) is mutagenic [10]. This compound may be responsible for a significant portion of the mutagenic activity of a diesel particulate extract despite its relatively low concentration [20]. The great variety of compounds present in diesel SOF suggests that no single compound or class of compounds should account for the total observed mutagenic activity. The resolution and identification of the mutagenic compounds would require extensive analytical separation work and biological characterization.

Diesel fuel, on the average, contains about

Table 1. Selected Organic Compounds Identified from Diesel Particulate Matter SOF [7, 8, 9, and 10]

Paraffinic Compounds	Aromatic Compounds	Acidic Compounds
Octane	Ethyl naphthalene	Cresol
n-Tetradecane	Phenanthrene	Phenylphenols
n-Hexadecane	Fluoranthene*	Naphthols
n-Eicosane	Benzo(a)pyrene*	Naphthoic acids
Cyclohexane	Methyl phenanthrene*	Benzoic acids
Methyl cyclohexane	Dibenzanthracenes	
Basic Compounds	**Transitional Compounds**	**Oxygenated Compounds**
Benzacridine	1-Nitropyrene*	9,10-Anthraquinone
Methyl aniline	9-Fluorenone	Hydroquinone
o-Toluidene	2-Methyl-9-anthrone	Benzanthrone
Dibenzo(c,g)carbazole	1,8-Naphthalic acid anhydride	Dinitro-o-cresol

*known mutagens

10 times the sulfur present in motor gasoline [21,22]. The conversion of fuel sulfur to exhaust sulfate in a diesel engine without a catalytic exhaust treatment device is usually about 2-4% by mass and varies with engine load and speed. This conversion factor decreases slightly with increasing fuel sulfur content [23,24]. As with catalyst equipped spark ignition engines, the primary sulfate species thought to be present is sulfuric acid (H_2SO_4) [24], with the remainder present as ammonium sulfate and metal sulfates. The balance of the sulfur in diesel fuel that is not released as sulfate is primarily SO_2, although a portion of the sulfur appears as elemental sulfur in the particulate.

Of the regulated gaseous emissions found in diesel exhaust, CO and total HC are generally at low levels. The HC level is of additional interest because the SOF is formed from a portion of the hydrocarbons that are measured in the raw (undiluted) exhaust. The fraction of the total HC emissions that are measured as particulate SOF is a function of the engine operating conditions, fuel properties, and dilution parameters such as dilution ratio and dilution mixture temperature. A study of the independent effects of dilution ratio and temperature are discussed in a series of papers from General Motors [25,26,27].

The NO_x present in diesel exhaust are primarily NO and NO_2. The photochemical atmospheric reactions of NO_x have been studied, as have the health and welfare effects [28]. The ability of NO_2 to convert a pro-mutagen such as benzo<a>-pyrene (B<a>P) to a direct mutagen in the Ames bioassay has been shown [29]. The nitration of B<a>P as well as other PNA's was reported at ppm levels of NO_2 in the presence of ppb levels of nitric acid, which purportedly catalyzed the nitration [29]. Activation due to chemical reactions with other oxidant constituents of photochemical smog was also observed [29].

Many modifications and devices have been proposed for the reduction of diesel pollutants. There are three broad approaches to control technology; fuel modifications, engine modifications, and exhaust aftertreatment.

Fuel modifications may include variations in the feedstocks used in blending the fuel to increase volatility, which increases the mixing rate during injection. Increased mixing rate results in more complete combustion, reducing particulate emissions, while keeping NO_x at acceptable levels by reducing high temperature zones in the combustion regions. Fuel additives may be developed to reduce particulate emissions. Barium additives currently used have been found to reduce visible smoke, but their effect on particulate mass emission is minimal [30]. The fuel sulfur content may also be reduced. Although sulfate emissions account for only a few percent of the total particulate emissions of untreated diesel exhaust, the possible use of an oxidation catalyst aftertreatment device may make low fuel sulfur content desirable or possibly a requirement. Fuel sulfur content may also affect the complex combustion reactions resulting in the formation of particulate.

The effects of fuel injection parameters on diesel emissions are also the subject of much research. The control of particulate emissions by turbocharging, higher swirl ratios, increased compression, and changes in combustion chamber shapes is also possible. Electronic control of the fuel injection system is also being investigated [31]. In general there is a trade-off between the control of particulate and NO_x emissions by changes in the fuel injection parameters and/ or engine modifications [32].

Various types of aftertreatment devices are also being developed to control particulate emissions. There are two main types of aftertreatment devices proposed; those that physically collect the particulate matter as the exhaust stream passes, generally referred to as traps; and those that oxidize the particulate as it passes through the device without collection. Once collected, the handling and disposal of the particulate becomes a problem. Diesel particulate has a low bulk density, $\rho = 0.075$ g/cm^3 [33], requiring the collection device to be fairly large to retain a reasonable amount of material. A collection device that has the ability to regenerate itself by oxidizing the collected particulate would seem advantageous to one that required periodic manual cleaning. Oxidation of the particulate may occur continuously or cyclically. Cyclic oxidation may be obtained by periodic throttling of the engine intake air or by the use of other ignition methods as described in Murphy et al. [33]. Addition of catalytic agent to reduce the required light-off temperature is also possible. Most collection devices result in higher back pressure and mesh type traps suffer from blowoff problems under some operating conditions such as heavy accelerations [33].

Oxidation catalysts have also been proposed for use with diesel engines as a possible means of controlling particulate emissions. Although oxidation catalysts have found widespread application in the control of spark ignition engine emissions, the scope of their use on diesel engines has been limited primarily to the reduction of CO levels when the engine is operated in an enclosed environment, such as underground mining vehicles. Very little research has been reported on the effects of oxidation catalysts on diesel particulate emissions. Research done on oxidation catalyst equipped spark ignition engines has shown that particulate emissions are increased by the use of an oxidation catalyst and that the increase is primarily due to the catalytic oxidation of SO_2 to SO_3 and eventual release as sulfates [34].

An oxidation catalyst has the capability to drastically alter the physical, chemical, and biological characteristics of the treated exhaust gas. Along with the desired reduction of the particulate and gas phase hydrocarbon emissions, a catalytic converter may also increase the amounts of other pollutants. Due to the higher sulfur content of most diesel fuel as compared to gasoline, sulfate emissions may become excessive when an oxidation catalyst is used on a diesel engine. Changes in the amounts of NO and NO_2

emissions may also occur. In addition, the use of the catalyst may result in changes in the mutagenicity of the organics that are not completely converted. The evaluation of any emission control modifications should be based not only on the desired reduction of a specified pollutant, but also on the total effect the modification has on the entire exhaust.

EPA has supported research on the control of heavy-duty diesel particulate emissions at Michigan Technological University (MTU). The scope of this work includes evaluations of changes in fuel injection parameters, alternate fuels, and exhaust aftertreatment devices. This paper deals with the evaluation of an oxidation catalyst.

EXPERIMENTAL

ENGINE TEST SET-UP - The oxidation catalyst chosen for evaluation was an Engelhard Minerals and Chemical Corporation model PTX-D-616-300NKG platinum exhaust gas purifier. The specifications of the catalyst are given in Table 2. This catalytic converter is considered typical of oxidation catalysts that could possibly be used on diesel powered vehicles in the near future.

Table 2. Oxidation Catalyst Specifications [35]

Manufacturer	Engelhard Industries Div. Engelhard Minerals & Chemicals Corp.
Model	PTX-D 616-300NKG
Catalytic Agent	Platinum
Catalyst Loading	<3500 gm/m^3
Wash Coat	Alumina
Substrate Material	Cordierite
Substrate Type	Laminar Flow Monolith
Substrate Cell Size	46.5 cells/cm^2
Substrate Size	14.6 cm dia x 15.24 cm long

The catalytic converters were installed on a 1979 Caterpillar Tractor Co. 3208 diesel engine. The specifications of this engine are given in Table 3. Since this engine is a V-8, a separate catalytic converter was installed for each bank of cylinders.

The Engelhard PTX catalysts were installed as close as possible to the exhaust manifolds of the engine. This was done to allow the exhaust to enter the catalyst at the highest possible temperature. Blank tubes were fabricated to replace the catalytic converters for baseline testing. Uncatalyzed monoliths had been furnished by Engelhard to be used for baseline testing, but it was felt that their use would not result in an accurate evaluation of the catalyst and substrate as a system. The addition of an uncatalyzed monolith to the exhaust system would have resulted in a large increase in surface area that the exhaust components would have been subjected to, and may have resulted in the catalysis or quenching of reactions occurring in the exhaust stream. The installation of the catalysts and blank tubes are shown in Figures 1a and b, respectively. Approximately 80 hours of use were accumulated on the catalysts before actual testing was begun, as recommended by the catalyst manufacturer [35].

Table 3. Engine Specifications

Engine Manufacturer and Model: Caterpillar 3208
Engine Type: V-8, Direct Injection, Natural Aspirated
Bore X Stroke: 114mm x 127mm (4.5 in. x 5.0 in.)
Displacement: 10.4 liter (636 cu.in.)
Rated Power: 157 KW (210 hp) at 2800 rpm
Rated Torque: 654 N M (485 FT LBF) at 1400 rpm

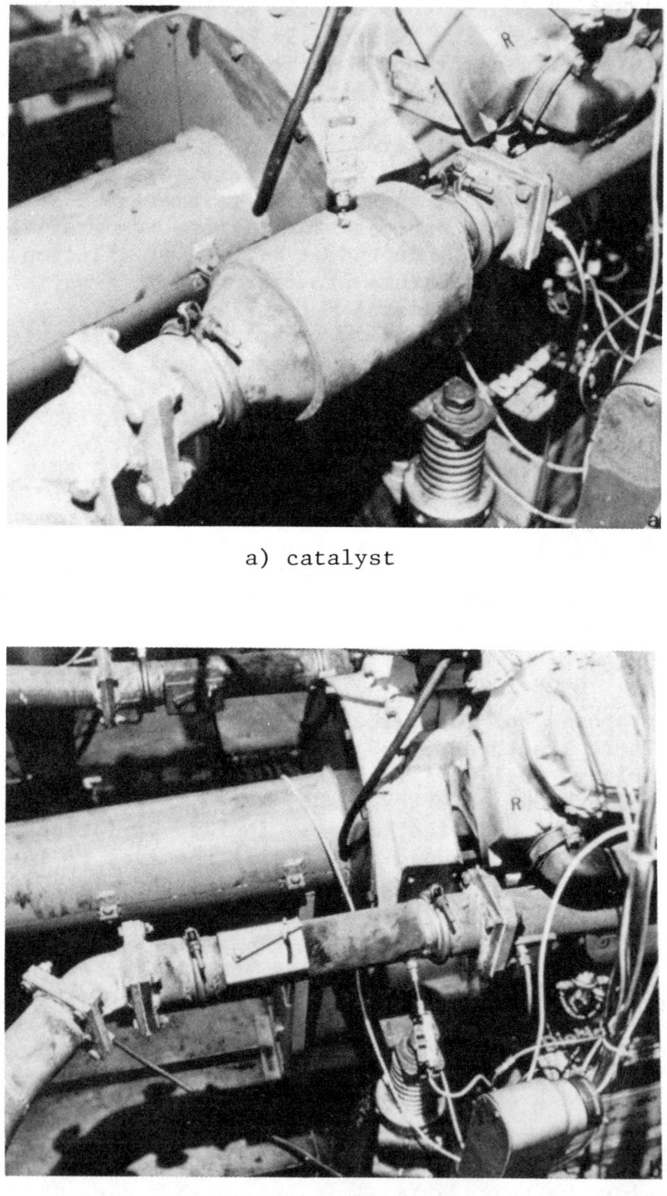

a) catalyst

b) blank tube

Fig. 1 - Installation of oxidation catalysts and blank tubes for baseline tests

SAMPLE COLLECTION PROCEDURE - All particulate samples were collected using the MTU dilution tunnel. The specifications of this tunnel are given in Table 4. A complete discussion of this system can be found in Carpenter [36], and is discussed briefly in references 3, 23, and 37.

Table 4. Dilution Tunnel Specifications

Material	304 Stainless Steel
Diameter	.305 m
Distance from Injection to Sample	4.57 m
Flow	15.29 m^3/min
Transient Time (s)	1.34 s
Reynolds Number	47,000
Injection	Counter Current
Exhaust Utilized	Variable
Intended Use	Diesel Emissions

It was discovered during preliminary testing that the exhaust system can act as a large storage place for diesel particulates. Of interest was the amount of sulfates stored when operating with the catalyst installed. Large variations in total mass and sulfate concentrations were observed during early testing when tests without the oxidation catalysts were conducted soon after tests with the catalyst had been completed. It is believed that these variations were caused by the re-entrainment of particulate matter (probably sulfates [38,39,40]) that had been deposited on the exhaust system walls from prior tests with the catalysts installed. Exhaust system conditioning times were then increased when tests were to be conducted at different conditions than those run immediately prior to allow for system equilibrium to be obtained at the new test conditions. The conditioning time was increased to at least 2 hours when a mode change was made and to at least 8 hours when the catalysts were either removed or installed.

Oxides of nitrogen were measured in both the raw exhaust and the dilution tunnel using a Beckman Instruments model 951H chemiluminescence analyzer. Samples from the raw exhaust were drawn through a heated (121°C) sample line, while dilution tunnel samples were drawn through an unheated line. Both sample streams were passed through a heated filter to remove any particulate matter before introduction into the analyzer. The volume dilution ratio (DR) was determined by the equation:

$$DR = \text{raw } NO_x/\text{dilute } NO_x \qquad (1)$$

The recorded values for NO_x and NO concentration were those that were directly measured by the instrument used. The value for equivalent NO_2 concentration was found by the equation

$$NO_2 \text{ (ppm)} = NO_x \text{ (ppm)} - NO \text{ (ppm)} \qquad (2)$$

The values reported should be construed as NO_x, NO, and equivalent NO_2 as detected by the analyzer. No corrections for NO_2 readings due to undetermined types and amounts of other forms of nitrogen oxides or other compounds which may affect the instrument response were attempted [41].

The brake specific NO_x emissions were calculated by assuming the molecular weight of NO_x is that of NO_2, i.e. 46 kg/kmol. The brake specific NO_2 and NO emissions were calculated using the appropriate molecular weights of each species, 46 kg/kmol and 30 kg/kmol, respectively. Thus the brake specific NO_x emissions may be calculated as

$$bsNO_x \text{ (g } NO_2\text{/kW*hr)} = bsNO_2 \text{ (g } NO_2\text{/kW*hr)}$$
$$+ 46/30 \text{ bsNO (g NO/kW*hr)} \qquad (3)$$

Measurements of gaseous HC emissions were made using a Beckman model 402 heated flame ionization detector (HFID) hydrocarbon analyzer. The sample line and pre-filter of this instrument were both operated at 191°C. The brake specific HC emissions were calculated by assuming an average atomic hydrogen/carbon ratio of 1/1.85 for the exhaust HC [42].

Engine exhaust opacity was measured by a Celesco model 107 green light, water cooled opacity meter.

Pallflex Corporation 47 mm EMFAB TX40HI20WW filters were used to determine total particulate mass (TPM), SOF, and sulfate emissions. The Pallflex EMFAB filter consists of a Teflon* coating applied to a woven glass fiber strand backing. The woven backing allows this filter to be extracted in a Soxhlet extractor for SOF determination and then re-extracted in water/isopropyl alcohol for sulfate analysis with no deterioration of the filter. The sampling procedure and equipment used for the 47 mm filter tests are described in Frisch et al. [23]. Although for the particulate size range found in diesel exhaust isokinetic sampling is not critical [3,23], an effort was made to sample within 20% of the isokinetic rate.

It was determined during preliminary testing that ideally 0.4 mg of SOF should be collected on each 47 mm filter for accurate mass analysis of SOF emission rates. This value is probably valid only for the sampling techniques and precautions used at MTU. Many times a much smaller amount of SOF was collected per filter and still yielded repeatable results. The exact minimum required mass of SOF per filter that results in repeatable SOF measurements depends on the experience of the personnel and the care taken during sample collection and analysis. The approximate time required to collect the desired amount of SOF with a 47 mm filter was determined from the 47 mm sample time nomograph shown in Figure 2. During this sample time the engine and filter data were recorded. The time period between recording these data was dependent on the total sample period length, but was usually 5 to 10 minutes.

*Registered Trademark

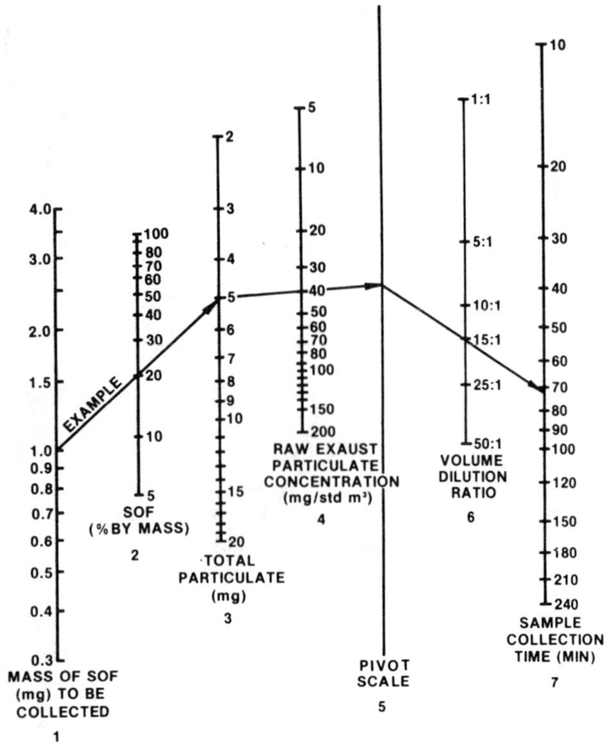

Fig. 2 - 47 mm sample time nomograph. Select on on scale 1 the desired amount of SOF to be collected. Draw line from point on scale 1 through approximate %SOF (scale 2) to determine amount of total particulate required (scale 3). Draw line from point on scale 3 through approximate raw exhaust particulate concentration to scale 5. From scale 5, a line is drawn through the desired dilution ratio (scale 6) to determine the required sample time (scale 7)

Filter plugging was common during sampling at most of the operating conditions tested. A 47 mm filter was arbitrarily considered to be plugged when the inlet vacuum to the sample pump reached 67.7 kPa gage. The test was stopped when either the desired SOF sample mass was collected or when the filter became plugged. The approximate sample time available before filter plugging occurred was determined to be a function of exhaust solid particulate matter concentration and dilution ratio. At least 1.5 mg of solid particulate matter can usually be collected on a 47 mm Palliflex EMFAB filter before plugging occurs. The minimum available sample time can therefore be calculated from the equation:

$$\text{time} = 1.5\ \left(\frac{1}{C_m}\right)(DR)\left(\frac{1}{FR}\right) \qquad (4)$$

Where: time = time to collect 1.5 mg of solid particulate (min).,
C_m = mass concentration of solid in raw exhaust (mg/Act. m^3)
DR = volume dilution ratio, and
FR = sample flowrate (Act. m^3/min.).

C_m can be approximately determined from data by using the opacity-mass concentration correlation [5]:

$$C_M = \frac{-1}{\left(\frac{AQL}{PV}\right)}\ \ln\left(1 - \frac{\%\ \text{Opacity}}{100}\right) \qquad (5)$$

Where $\frac{AQ}{PV} = 8.2\ m^2/g$, and L = path length (m)

At MTU a sample flowrate of 0.028317 Act.m^3/min. and a path length of 0.152 m are used. The minimum available sample time can be found from exhaust opacity and dilution ratio data by combining equations (4) and (5):

$$\text{time} = \frac{-0.066\ DR}{\ln\left(1 - \frac{\%\ \text{Opacity}}{100}\right)} \qquad (6)$$

After testing, the filters were placed in a desiccator with the Petri dish tops ajar and exposed to an ammonia atmosphere for one hour. This ammoniation converted all of the sulfates collected on the filter (primarily sulfuric acid [34]) to ammonium sulfate, which resulted in less interference in the EPA barium chloranilate sulfate analysis procedure. Ammoniation also changed all of the extremely hygroscopic sulfuric acid to the less hygroscopic ammonium sulfate, which allowed for more repeatable weighing by eliminating changes in filter mass due to water pick-up by the sulfuric acid.

The ammoniated filters were placed back into the humidity chamber and allowed to soak for 24 hours before being weighed for total particulate mass analysis.

The filter was then extracted in a glass Soxhlet micro-extractor (50 ml reflux volume) using dichloromethane (DCM) as the solvent for 4 hours. The extracted filters were then allowed to soak an additional 24 hours in the humidity chamber before being weighed to determine the amount of SOF that was present on the filter.

The filter was then re-extracted in the 60/40 water/isopropyl alcohol solution used for the EPA barium chloranilate sulfate analysis. The total mass of sulfate present was then determined following the procedure outlined in Khatri et al. [37]. Since ammonium sulfate has a higher molecular weight than the sulfate ion, the mass of sulfate determined by the barium chloranilate analysis and also the total particulate mass collected must be corrected to account for this difference. The data collected were then analyzed to determine the mass emission rates and other operating parameters of the engine and catalysts.

All samples for chemical and biological characterization were collected on Pallflex Corporation 20 in. x 20 in. (20x20) type T60A20 filters. These filters consist of a non-woven glass fiber with a coating of Teflon applied over it. The Teflon coating provides a relatively inert surface for sample collection and therefore minimizes any artifact formation due to surface reactions catalyzed by the collection media [43]. All of the handling of the 20x20 filters was done under yellow light (Kodak type 908 filters) to eliminate the effects of ultra-violet radiation on the SOF.

The ultra-high volume (UHV) sampler shown in Figure 3 was used in conjunction with these filters to collect large mass samples for chemical and biological analysis. The sampling rate of this unit was 4.25 Act.m^3/min. which drastically reduced the sample time required to collect the needed amounts of SOF. The plans for the filter holder portion of this sampler were obtained from EPA [44], while the sample pump system was designed and constructed at MTU. The sample system consists of a Roots blower type positive displacement pump (PDP) driven at a constant speed by a 2.24 kW electric motor. The exhaust from the blower is fed back into the dilution tunnel downstream of the sample section in order to maintain a constant flowrate in the dilution tunnel whether or not the UHV is in use. The sample volume/blower revolution was determined to be 1.863 x 10^{-3} Act.m^3/rev using a laminar flow element and a digital revolution counter mounted on the blower shaft. The total volume of diluted exhaust sampled may be calculated by multiplying the number of blower revolutions by this factor.

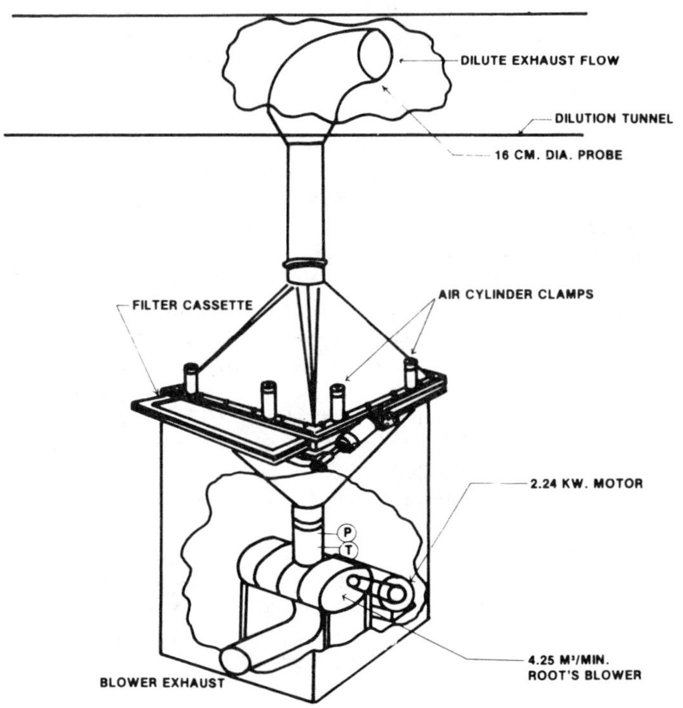

Fig. 3 - Schematic of ultra-high volume sampler used to collect greater than 50 mg amounts of particulate for chemical and biological characterization

Ideally, 100 mg of SOF were required for each individual chemical fractionation, and 60 mg were required for a 5 strain Ames mutagenicity bioassay. The total sample time required to collect enough mass of SOF for chemical and biological characterization was determined from the 20x20 test time nomograph shown in Figure 4. Because the time required for sampling with the 20x20 filters was usually longer than the sample time of the 47 mm filters, some of the 20x20 filters were run without sampling simultaneously with the 47 mm filters. Sampling with the 47 mm filters was done during the mid-point of the 20x20 sample period. All total particulate, SOF, and sulfate emission rates reported were determined from samples collected on 47 mm filters. The assumption made was that the particulate mass emissions did not greatly change throughout the 20x20 sample time period and that the results obtained using the 47 mm filters were representative of the entire sampling period.

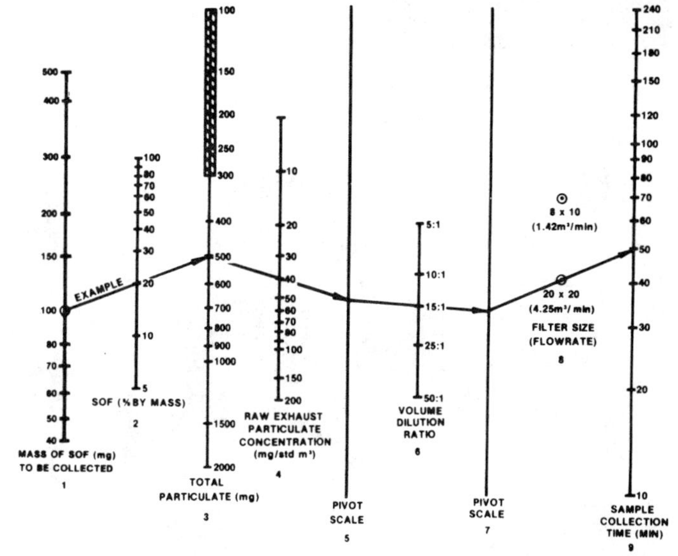

Fig. 4 - 20x20 sample time nomograph. Select on scale 1 the desired amount of SOF to be collected. Draw line from point on scale 1 through approximate %SOF (scale 2) to determine amount of total particulate required (scale 3). Shaded area depicts insufficient particulate/filter and should be avoided. Draw line from point on scale 3 through approximate raw exhaust particulate concentration to scale 5. From scale 5, a line is drawn through the desired dilution ratio (scale 6) to scale 7. Draw a line from scale 7 through the appropriate filter size to determine the required sample time (scale 9)

To begin sampling with the 20x20 filter, the filter was first loaded into the filter cassette. This was done by placing the 20x20 filter face up on the filter cassette back-up screen and then placing the cassette lid on top of the filter. The filter cassette was then lodged into the UHV sampler through the air actuated access door and the cylinder cassette clamps were actuated at the same time the Roots blower PDP was started. Figure 5a and b show the cassette lid being placed over the filter and the cassette being pushed

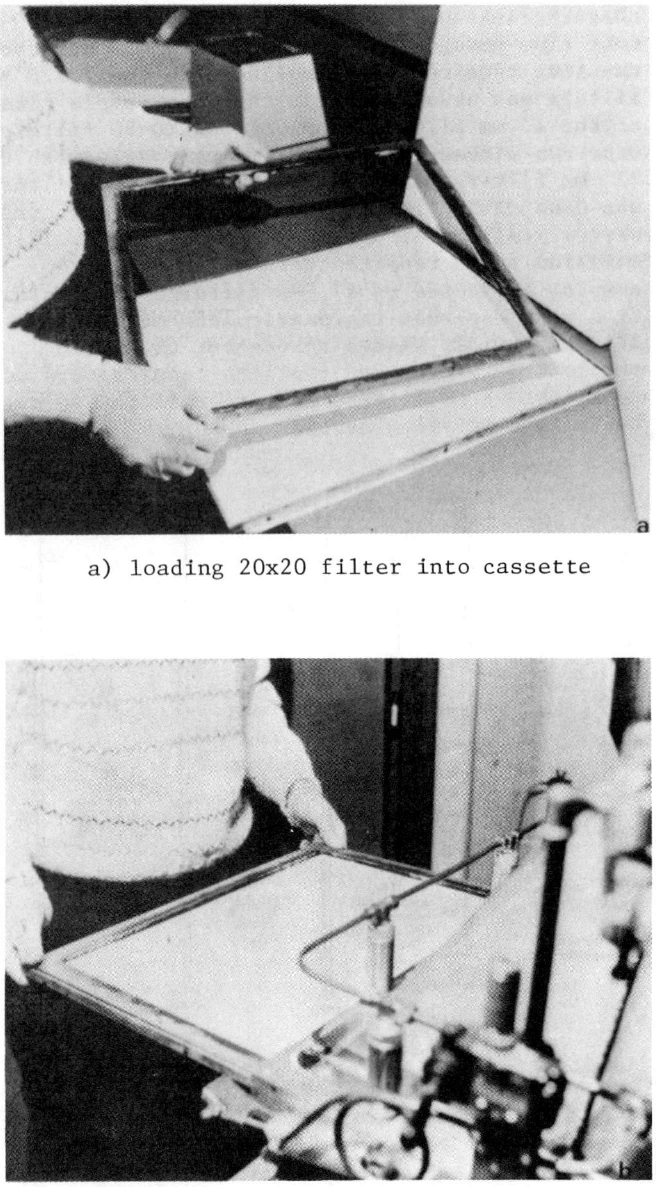

a) loading 20x20 filter into cassette

b) sliding filter into UHV

Fig. 5 - Pictures of ultra-high volume (UHV) sampler

into the UHV sampler. It was important to start the sample pump as quickly as possible when the cassette was in place to prevent the slight vacuum in the dilution tunnel (approximately 0.25-0.75 kPa) from pulling the filter out of the cassette and thus ruining the filter.

Filter plugging was also a problem at most modes with the 20x20 filters. A 20x20 filter was arbitrarily considered plugged when a vacuum of approximately 14 kPa gage was created across the filter. The pressure drop at plugging was different for the 20x20 filters and 47 mm filters because of the different types of sample pumps used with each filter.

At the end of the sample period or when the filter became plugged, the access door and air cylinder clamps were released and the cassette removed. The filter was then removed from the cassette and folded in sixths. The filter was placed in a folder and then sealed in a Teflon autoclave bag. The bag was then placed in a freezer until the filter was to be extracted.

The 20x20 filters were extracted in a 200 ml Soxhlet extractor using DCM for 24 hours. The extended extraction time was required because of the longer cycling time of the larger extractor. The extractors were covered with aluminum foil as insulation in order to decrease siphoning time, as well as to protect the compounds from photo-decomposition. After extraction the solvent was removed from the SOF using a rotary evaporator to a volume of approximately 10 ml. The samples were then taken to dryness under a stream of nitrogen at room temperature and stored at -5°C until further use.

CHEMICAL CHARACTERIZATION - Compound class characterization of the SOF was performed using a fractionation method developed by Huisingh et al. [15], as modified by Funkenbusch et al. [9]. Pooled SOF from a number of filters corresponding to the same engine test mode and exhaust after-treatment was divided into at least three portions after removal of 60-120 mg for biological testing. Each portion was then re-dissolved in DCM and filtered through a nitrocellulose membrane filter just prior to fractionation, in order to eliminate any insoluble degradation products which may have accumulated during storage. The SOF sample was then placed in a tared 50 ml round bottom flask and the DCM removed until the flask obtained constant weight. The SOF mass used as the basis for fractionation was typically 100 to 150 mg.

The fractionation procedure used to obtain SOF compound classes is described in Funkenbusch et al. [9]. Basic and acidic subfractions were obtained using liquid-liquid extraction with strong acid and strong base, respectively. Paraffins, aromatics, transitionals, and oxygenates were eluted from a 1 cm x 10 cm silica gel chromatagraphic column using hexane, 1% ether in hexane, and acetone:methanol (1:1), sequentially. Ether insoluble and hexane insoluble subfractions were also obtained.

Detailed analysis of an aromatic subfraction was performed using a Hewlett-Packard model 5840 gas chromatograph/FID using a 30 m x 0.25 mm ID open tubular glass capillary column, wall coated with SP-2100 methyl silicone fluid. Compounds were identified by combined gas chromatography-mass spectrometry with a Hewlett-Packard model 5985 GC-MS data system. The same capillary column was employed in gas chromatography. Programming conditions were 3 C/min from 100-250 C, after an initial 10 minute hold. The final temperature was held for 20 minutes.

The Hewlett-Packard probability based search program was used to select matches for spectra generated during the chromatographic run. The National Bureau of Standards library of nearly 30,000 spectra was used as the data base, and the three most probable spectral matches for each chromatographic peak were obtained for comparison with the sample spectra. Identification of the molecular weight was deemed necessary information for tentative assignment of compound identities.

The SOF or subfractions to be used for the Ames mutagenicity bioassay were prepared by dissolving a filtered, preweighed amount of the sample in DCM and adding dimethyl sulfoxide (DMSO) to give a known concentration. DMSO and DCM do not form an azeotrope [45]. The DCM was removed with a rotary evaporator, assuming negligible loss of the much less volatile DMSO (DMSO BP @ 189°C, DCM BP @ 40°C). Solvent removal was complete after 30 minutes using a water bath no warmer than 35°C. The SOF or subfraction was transferred to a 15 ml Teflon vial and stored at -10°C until use.

BIOLOGICAL CHARACTERIZATION - The SOF and subfractions were tested for biological activity using the *Salmonella*/microsome mutagenicity bioassay developed by Ames et al. [11]. This test was chosen because it is the most commonly used bacterial bioassay for determining any type of potential mutagenicity [12,19]. It is also possible to quickly screen SOF and subfraction samples for biological activity (e.g., mutagenicity or toxicity). By use of the Ames bioassay, a quantitative dose-response curve can be obtained depicting the overall biological activity of the test samples.

The Ames bioassay utilizes several mutant strains of *Salmonella typhimurium* for defining the mutation produced by a sample or compound. Mutagenicity, as defined by the Ames bioassay, is the ability of a sample compound to revert the *Salmonella typhimurium* histidine-dependent mutants to their histidine-independent wild types. *Salmonella typhimurium* mutant strains TA98, TA100, TA1535, TA1537, and TA1538 were used as the indicator strains in the tests. All strains were obtained directly from Dr. B. N. Ames, University of California, Berkeley. Strains TA98, TA1537, and TA1538 respond to frameshift mutagens; strain TA1535 responds to base-pair substitution mutagens; and strain TA100 will respond to both base-pair substitution and frameshift mutagens. Base-pair substitution mutations in the Ames tester strains occur when a nucleotide is substituted by a different nucleotide thus causing a change in the reading of the codon. A frameshift mutation in the Ames tester strains occurs when one or two nuclotides are deleted from or added to the DNA, thus causing a change in the reading of subsequent codons.

Plate incorporation tests were performed on all samples under the laminar flow hood shown in Figure 6. The protocol followed that established by Ames et al. [11] except that the Petri plates containing the tester strain and the test samples were incubated for 63 hours instead of 48 hours. In order to determine if any of the samples' mutagenic response was direct-acting or required metabolic activation [13], some of the SOF samples were tested in conjunction with a mamalian-microsome (S-9) activation mixture.

A variety of measures were employed to ensure the accuracy and repeatability of the Ames test results, including the following:

Fig. 6 - Laminar flow hood used for preparation of samples for biological characterization

1) Use of forced air incubators with internal temperature measured to be 37±0.2°C;
2) use of an automatic colony counter, (shown in Figure 7) with routine checks on counter accuracy using a calibration slide;
3) routine performance of strain identification checks and top agar sterility checks;
4) use of tester strain density of 1-2x10^8 cells/ml monitored by routine viability counts;
5) holding of cells and top agar mixture at 43°C for no more than 2 minutes before pouring;
6) use of a vortex to thoroughly mix the top agar suspension;
7) dilution of positive controls and sample not more than 2 hours before each test;
8) conducting all test procedures at the same time of day for each series of tests;
9) testing several concentrations of each appropriate positive control to ensure that each tester strain was responding properly;
10) running all plates in triplicate; and
11) testing the samples on the same day to eliminate any interexperimental differences when comparing SOF or subfractions from the same mode but with and without the catalyst installed.

For an initial screening test, all five *Salmonella typhimurium* strains were tested using at least five sample concentrations over a thousand-fold range, in order to establish the linear portion of the dose-response curve (plotted in linear-log form). The most responsive strains were then tested using at least five sample concentrations within the linear dose-response range.

To determine the mutagenic activity on both a mass basis and in terms of actual engine operating conditions, test data were plotted using a dose-response format and also as brake specific revertants. Analysis of the dose-response curve

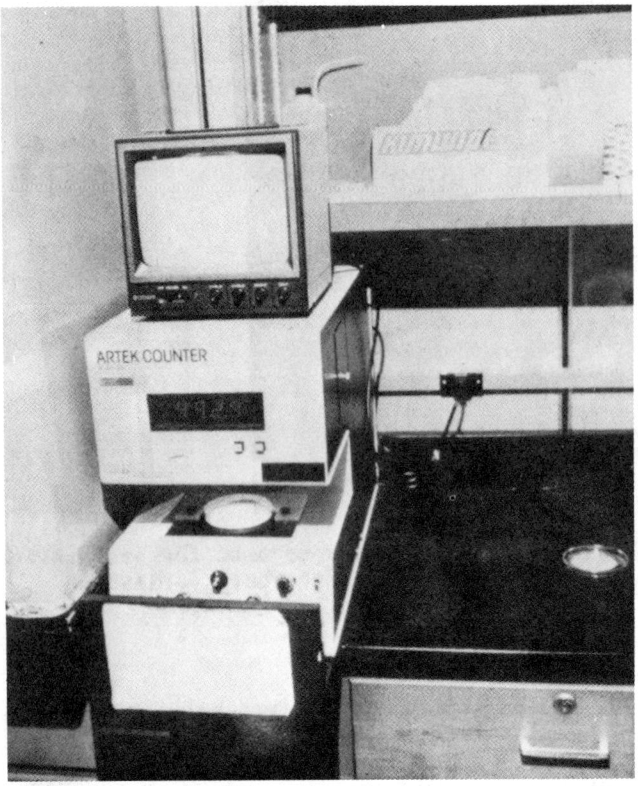

Fig. 7 - Colony counter used to determine the number of revertant colonies

was done as follows [46]:

1) Statistically determine the curve fit for the dose-response data by least squares;

$$R = a + b\ (\log_{10} C) \qquad (7)$$

where R = Number of Revertants/plate;
C = Dose concentration (µg), and
a and b are determined statistically;

2) the slope at any C is given by:

$$\frac{dR}{dC} = \frac{1}{\ln 10}\left(\frac{b}{C}\right) \qquad (8)$$

3) the average slope of the dose-response curve in Cartesian coordinates is found from;

$$\overline{\frac{dR}{dC}} = \frac{\int_{C_1}^{C_2} \frac{dR}{dC}\, dC}{C_2 - C_1} \qquad (9)$$

$$= \frac{b}{C_2 - C_1} \log_{10} \frac{C_2}{C_1} \qquad (10)$$

where C_1 = minimum dose (µg) observed on linear portion of dose-response curve and
C_2 = maximum dose (µg) observed on linear portion of dose-response curve;

4) this slope is then used to calculate the value for brake specific revertant emissions from the following equation:

$$bsREV = \overline{\frac{dR}{dC}} \times 10^6 \times bsSOF \qquad (11)$$

where bsREV = brake specific revertant emissions (REV/kW*hr), (rate of revertant emissions per rate of work done by engine);
dR/dC = specific activity (REV/µg) of SOF, (average slope of dose-response curve in Cartesian coordinates)
bsSOF = brake specific SOF emissions (g/KW*hr) (Rate of SOF emissions per rate of work done by engine).

TEST MATRIX AND SAMPLING PARAMETERS - A suitable test sequence consisting of six modes of the EPA 13 mode heavy-duty cycle was devised to parametrically evaluate the effect of the PTX oxidation catalysts on the emissions measured. The cycle was selected to subject the catalysts to a wide variety of exhaust gas compositions and operating conditions (temperature and space velocity) that heavy-duty vehicles produce under typical over the road operating conditions. Modes 3,4,5,9,10, and 11 were chosen for these reasons. The selected steady state cycle did not represent transient operation. A steady state cycle was necessary to understand the basic processes that occur in the catalysts without the influence of exhaust system storage and re-entrainment that may occur during a transient cycle.

The emissions to be measured both with and without the oxidation catalysts included total particulate, soluble organic fraction, sulfate, NO_x, NO, NO_2 and HC emission rates, and opacity. Rates of CO emissions were not studied because of the diesel engine's already low CO levels, and also because data have already been published to document the effects of oxidation catalysts on CO emissions with diesel engines [47].

Chemical and biological studies of the collected SOF were performed at each of the six test modes both with and without the oxidation catalysts installed to determine the effects of the catalysts on the chemical and biological character of the organics present in the particulate matter. In addition, the acidic, ether insoluble, and oxygenated subfractions from mode 11 with and without the catalysts installed were assayed over a broad range of concentrations. The mode 11 uncatalyzed aromatic subfraction was also subjected to analysis using the chromatographic technique described above. This work was done in an attempt to identify the most biologically active subfractions present in the diesel SOF.

All tests were conducted at 15:1 volume dilution ratio, while the dilution mixture temperature was allowed to vary. At least three repetitions of each test were made to verify the results obtained.

The properties of the fuel used for this study are given in Table 5. The AMOCO Premier fuel tested is representative of a currently available, high quality no. 2 diesel fuel. This fuel contains approximately the national average fuel sulfur level of 0.23% by weight for no. 2

Table 5. Fuel Properties

Calculated Cetane Index	52.0
API Gravity	39.1
Sulfur (% by mass)	.26

Distillation Profile

% Recovered	Boiling Point °C
0	146
5	191
50	248
90	307
95	323
98 (max. recovery)	338 (end point)

Elemental Analysis (% by Mass)

C	86.20
H	13.52
S	0.26
N	0.02
ash	0.00

Compound Types by FIA Method (% by Volume)

Saturates	72.3
Olefins	2.9
Aromatics	24.8

fuel [21].

The test matrix of the CAT 3208/oxidation catalyst evaluation is given in Table 6. The dilution and filter sampling parameters observed during this study are listed in Tables 7 to 9. The divergence of the mode 9, without catalyst test from the EPA specified maximum sampling temperature of 52°C was due to unusually high dilution tunnel inlet temperature. Since the sample temperature was not too far above 52°C, the results obtained are acceptable for research purposes. It should be noted that only the 47 mm filters used for particulate mass analysis were collected that day, to be used to replace damaged filters that had been collected earlier. Samples at mode 9 without the catalyst to be used for chemical and biological analysis had been collected earlier at 52°C sample temperature. This was the only condition where data from simultaneous sampling with both the 20x20 and 47 mm filters were not obtained. All of the detailed measured and calculated data are given in Appendix A.

RESULTS

EFFECTS OF CATALYST ON MASS EMISSIONS - As shown in Table 10 and in Figures 8 and 9, the use of the catalyst had little effect on the operating parameters of the engine. The greatest effects the catalyst had were a slight increase in exhaust backpressure and an increase in exhaust opacity at modes 4 and 10 (50% rated load at intermediate and rated speeds) because of increased particulate emissions. The use of the catalyst had no effect on the brake specific fuel consumption or on the air/fuel (A/F) ratio of the engine.

The results of NO_x, NO, and equivalent NO_2 (as NO_x-NO) emissions measured by chemiluminescence analyzer are shown in Figures 10a and b. The brake specific NO_x and NO_x concentration by volume (ppm) were unchanged by the catalyst. The levels of NO_2 emissions are, with the exception of mode 3, increased by the catalyst, depending on the engine operating condition. This increase in NO_2 is at the expense of NO emission rates and concentrations, which are reduced. NO_2 is generally considered the least desirable of the oxides of nitrogen and is suspected of converting many pro-mutagens to direct-acting mutagens in the Ames bioassay [29].

The results of gas phase hydrocarbon analysis using a 191°C HFID with and without the oxidation catalysts installed are given in Figure 11. The use of the catalyst resulted in a decrease in the HC emissions of the engine. This decrease was expected due to the known performance of platinum catalysts.

The effects of the oxidation catalyst on the mass emission rates of particulate matter are given in Figures 12a and b. As discussed earlier, the total particulate emissions are composed of the sum of the SOF, sulfate, and solids emitted by the engine.

The use of the catalyst reduces the amount of SOF emissions of the engine at all of the test modes. Since the amount of SOF found in the particulate at a given engine speed and load is a function of the dilution parameters and raw exhaust HC concentration, the reduction in particulate SOF is related to the reduction in HFID HC emissions by the catalyst.

As shown in Figures 12a and b, the amounts of sulfate emissions are increased with the catalyst installed. This phenomenon has been observed with both catalyst equipped spark ignition and diesel engines. The percent conversion of fuel sulfur to exhaust sulfate on a mole basis is shown in Figure 13.

The use of the catalyst also increases the amounts of total particulate emissions at 5 of the 6 modes tested, as shown in Figures 12a and b. The only mode that showed a decrease in total particulate emissions is mode 3. This mode is characterized by a low exhaust temperature, which reduces the catalyst's activation of the oxidation of SO_2 to SO_3, and also by a relatively high baseline SOF emission rate, which causes even a moderate reduction in SOF (61%) to result in a significant decrease in total particulate emissions.

Close examination of Figures 12a and b reveals that the increase in total particulate emissions is not due entirely to the observed increase in sulfate emissions alone. An increase in the unextractable residue (largely solid) of the particulate at most of the modes of the test cycle used was observed. The solid portion of

Table 6. Test Matrix for Catalyst Evaluation

EPA MODE	3		4		5		9		10		11	
Speed (rpm)	1680		1680		1680		2800		2800		2800	
Load (N*M)	160		320		480		399		266		133	
BMEP (kPa)	192		383		575		485		322		161	
Catalyst	w/o	with	w/o	with	w/o	with	w/o	with	w/o	with	w/o	with
Total Particulate, SOF, and SO_4= Mass Emissions	X	X	X	X	X	X	X	X	X	X	X	X
Chemical Characterization of SOF	X	X	X	X	X	X	X	X	X	X	X	X
Ames Mutagenicity Bioassay on SOF	X	X	X	X	X	X	X	X	X	X	X	X
Ames Mutagenicity Bioassay on Subfractions											X	X
HC, NO, and NO_x Mass Emissions	X	X	X	X	X	X	X	X	X	X	X	X

Table 7. Ambient Test Conditions

MODE	Ambient Temp. (°C) unc.	cat.	Barometric Pressure (kPa) unc.	cat.	Specific Humidity (g H_2O/kg air) unc.	cat.
3	29	30	98.9	99.9	4.5	3.0
4	31	32	99.2	99.6	2.0	2.7
5	31	31	99.0	100.3	3.1	2.4
9	37	22	99.4	98.6	6.2	7.4
10	25	27	101.5	98.9	2.1	4.3
11	32	27	99.4	98.8	9.8	2.3

Table 8. Dilution Tunnel Conditions

MODE	Volume Dilution Ratio unc.	cat.	Dilution Temp. (°C) unc.	cat.	NO_x @Filter (ppm) unc.	cat.	NO_2 @Filter (ppm) unc.	cat.
3	15.0	15.1	39	40	32.7	33.4	2.0	1.4
4	14.7	14.7	48	47	64.2	60.6	1.3	19.5
5	14.9	14.4	51	50	84.3	80.9	1.3	8.4
9	15.3	15.0	60	52	63.5	73.7	0.5	1.1
10	14.9	15.0	50	51	48.1	48.4	1.0	1.7
11	15.0	14.2	49	48	29.8	28.7	0.9	3.9

Table 9. 47mm Filter Operating Conditions

MODE	Sample Time (min.) unc.	Sample Time (min.) cat.	Sample Rate (% Isokinetic) unc.	Sample Rate (% Isokinetic) cat.	Face Velocity (m/s) unc.	Face Velocity (m/s) cat.	Filter Loading $\left(\frac{mg\ TPM^*}{cm^2\ stain}\right)$ unc.	Filter Loading $\left(\frac{mg\ TPM^*}{cm^2\ stain}\right)$ cat.	Total Mass (mg) unc.	Total Mass (mg) cat.
3	32	30	73	84	0.28	0.33	0.175	0.122	2.54	1.77
4	42	53	86	84	0.33	0.32	0.201	1.299	2.92	18.86
5	40	32	83	83	0.32	0.31	0.216	1.324	3.14	19.22
9	11	15	81	61	0.31	0.24	0.218	0.493	3.17	7.16
10	39	10	79	98	0.30	0.35	0.424	0.268	6.16	3.89
11	21	16	81	59	0.31	0.23	0.318	0.341	4.62	4.95

*TPM = Total Particulate Matter

Table 10. Engine & Catalyst Operating Conditions

MODE	A/F unc.	A/F cat.	Exh. Temp.[a] (°C) unc.	Exh. Temp.[a] (°C) cat.	Back Pressure (kPa) unc.	Back Pressure (kPa) cat.	Exh. O_2[b] (%)	Cat. Temp.[c] (°C)	Space Vel. x 10^{-3} (hr^{-1})
3	66	67	252	258	101	104	16.1	269	167
4	41	41	365	372	101	104	13.1	383	200
5	29	31	499	481	102	105	10.8	498	242
9	24	25	642	666	104	108	8.4	695	450
10	36	35	474	490	105	106	11.9	507	349
11	51	51	377	374	104	106	14.7	391	295

[a] Measured at exhaust manifold (inlet to catalyst)
[b] Calculated assuming ideal combustion
[c] Measured at center of catalyst monolith

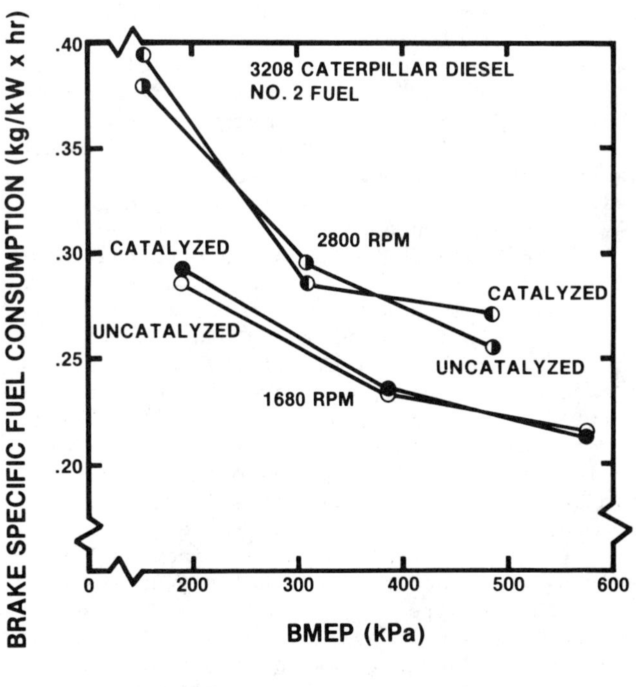

Fig. 8 - Brake specific fuel consumption

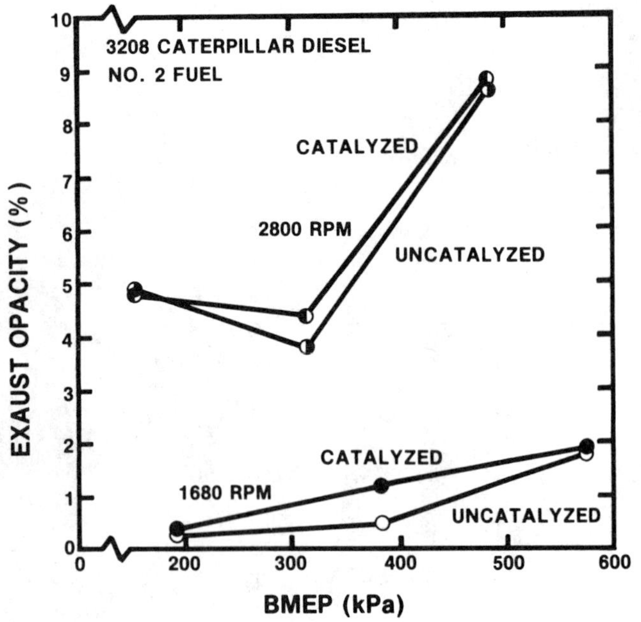

Fig. 9 - Effect of the oxidation catalyst on the exhaust opacity

the particulate is defined as

$$\text{Solid} = \text{TPM} - \text{SOF} - SO_4 \qquad (12)$$

The solid portion may contain species such as graphitic carbon, high molecular weight hydrocarbons, and other compounds that are insoluble in DCM, and that may or may not be soluble in the water/isoproyl alcohol solution used for the sulfate extraction. Some of these compounds might be liquids, but the bulk of them are expected to be solid. The increase in opacity observed at modes 4 and 10 with the catalyst installed would suggest that the measured increase in "solid" emissions could be due to an increase in these types types of compounds.

The increase in backpressure with the catalyst installed (Table 10) might cause the increase in solids. It is not felt that the backpressure increase was large enough to account for the measured change in solids. If the backpressure increase was acting to increase the solid portion, the effect should be similar to that of exhaust gas recirculation (EGR). The effects of EGR on a diesel engine include a decrease in the air/fuel ratio and a decrease in NO_x emissions, these phenomena were not observed in these tests. Therefore we feel that the slight increase in backpressure caused by the catalyst was insignificant to our results.

The increase in solids caused by the catalyst may be due to dehydrogenation of the organic compounds present in the exhaust resulting in the formation of solid carbon and low H/C ratio hydrocarbons. The definition of oxidation of an organic compound includes both the addition of oxygen to the molecule (oxygenation) or the removal of hydrogen from the molecule (dehydrogenation) [48]. An order of magnitude study was

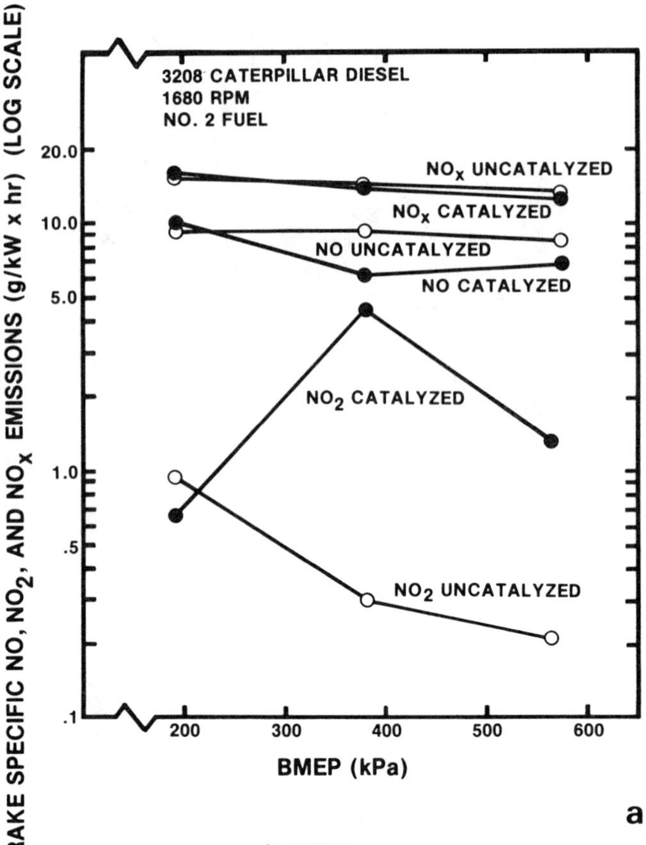

a) 1680 rpm

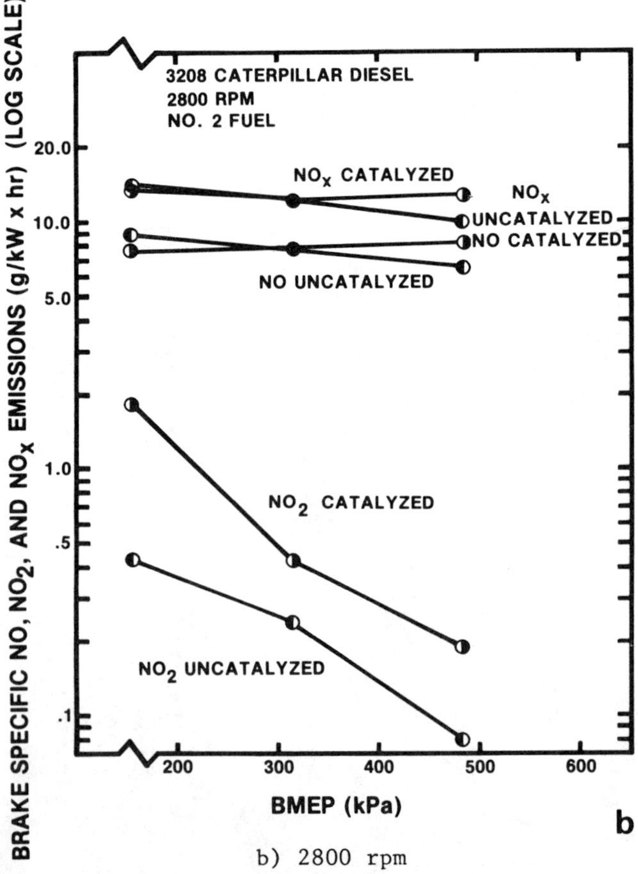

b) 2800 rpm

Fig. 10 - Brake specific NO, NO_2 and NO_x emissions with and without oxidation catalyst

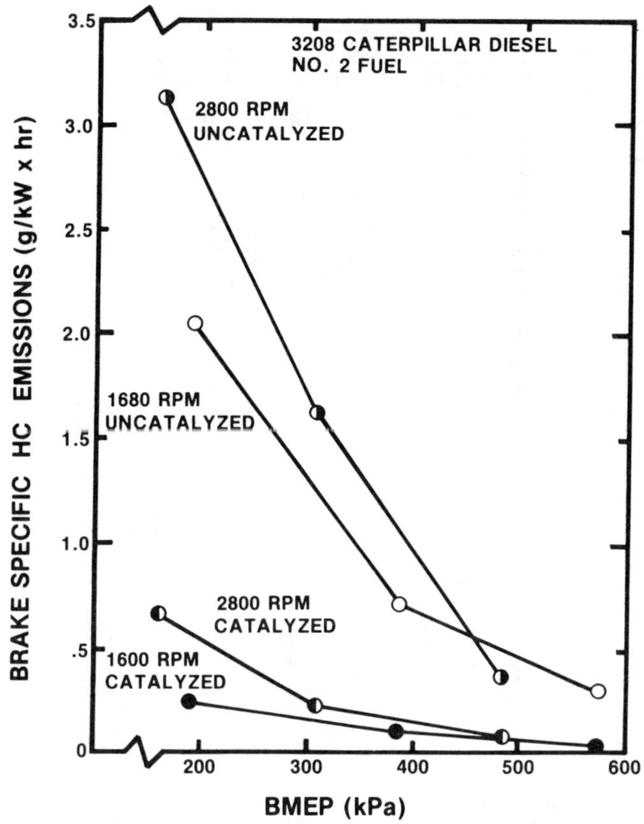

Fig. 11 - Effect of the oxidation catalyst on brake specific hydrocarbon emissions

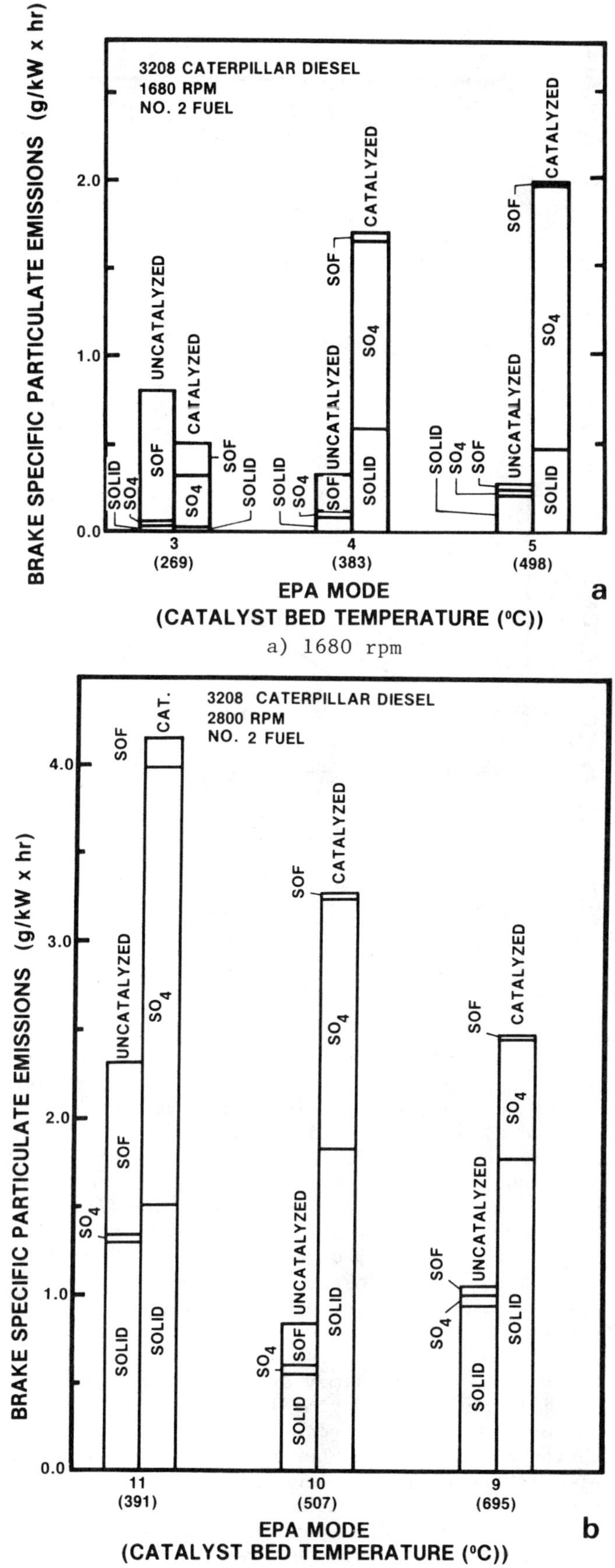

a) 1680 rpm

b) 2800 rpm

Fig. 12 - Effect of the oxidation catalyst on brake specific particulate emissions

performed to determine if the change due to the installation of the catalyst in the amounts of HC measured in the raw exhaust was sufficient to account for the increased amount of solids measured at modes 4,5,9,10, and 11. The HC measurements done with the HFID both with and without the catalyst at each of these modes were used for this study. Since the response of a HFID is in ppm carbon present in the gas phase HC, the difference in HFID response from uncatalyzed to catalyzed exhaust should be related and not more than the measured increase in solids due to the catalyst. The solid portion of the particulate is probably composed of very low H/C ratio compounds, therefore a H/C ratio of 0.014 by weight [5] was assigned to the compounds measured by the change in HFID response. This available "carbonaceous material" which would be measured as increased solid particulate matter due to the catalyst was then calculated from the change in HFID response and compared to the actual measured increase in solids (as defined by Equation (12)), as shown in Figure 14. With the exception of mode 9, there was enough available gaseous phase HC's in the exhaust to account for the increase in solids due to dehydrogenation in the catalytic convertors.

The formation and subsequent condensation of aromatic free radicals above the catalyst sites would result in the formation of low H/C ratio compounds [49] that would be detected as

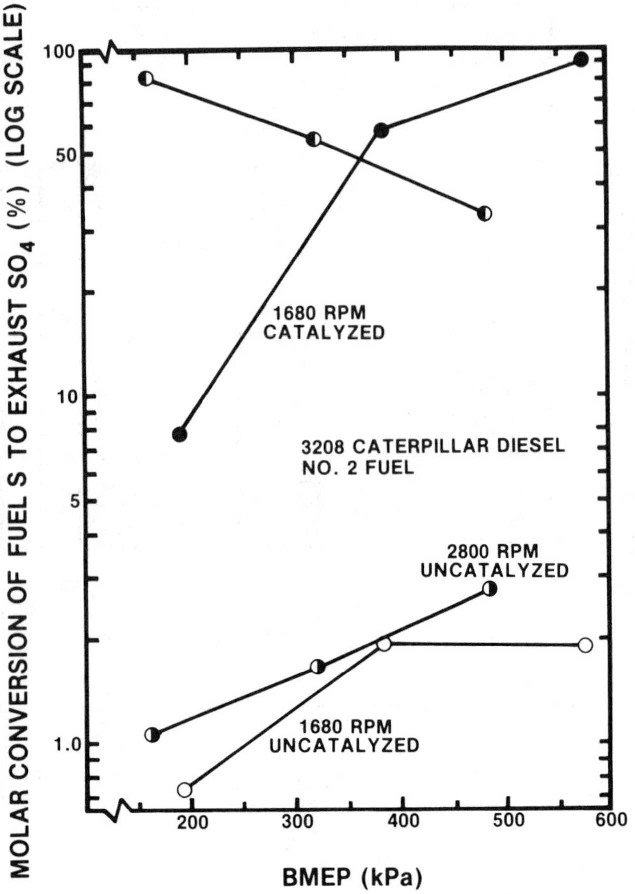

Fig. 13 - % conversion of fuel sulfur to exhaust sulfate (on a mole basis) with and without the oxidation catalyst

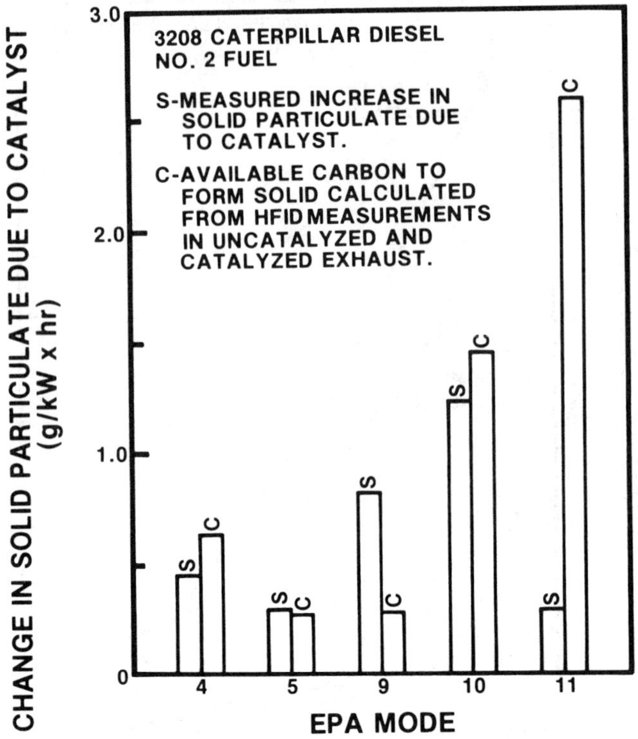

Fig. 14 - Comparison of measured increase in unextractable residue (solids) with the catalyst to the calculated increase due to HFID HC decrease assuming .014 H/C ratio compounds

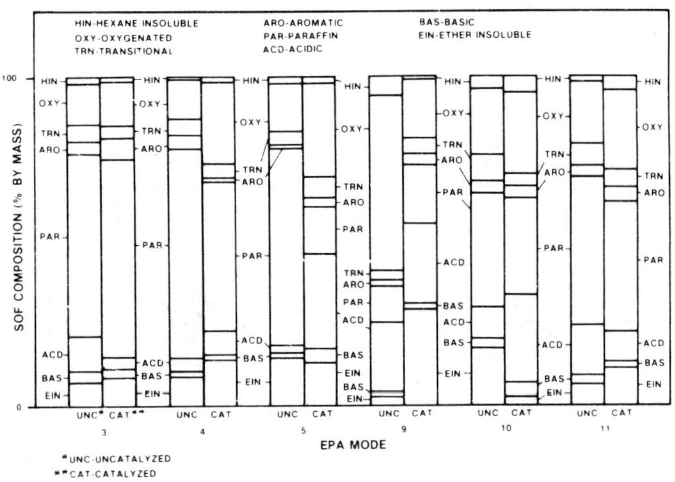

Fig. 15 - Fractional composition of SOF with and without oxidation catalyst

solids. The formation of these radicals is an endothermic process [50], with the required heat being supplied by exothermic reactions occurring on the catalyst sites. The oxidations of NO, SO_2, CO, and HC's on the catalyst sites are all highly exothermic. The formation of benzene radicals by hydrogen dissociation and subsequent condensation to diphenyl has been studied and found to occur both with and without the presence of oxygen over platinum catalysts [51]. There is a chemically and kinetically complex system of reactions occurring at each catalyst site. Variables such as fuel sulfur which would result in changes in the concentration of SO_2 in the exhaust to be oxidized may affect the amount of solid formation occurring in the catalyst by reducing the amount of heat available for the formation of radicals. When the factors of residence time in the catalyst, temperature, exhaust gas composition, and catalyst activity are considered, any thermodynamically possible compound of carbon, hydrogen, oxygen, nitrogen, and sulfur may be formed to account for the measured increase in solids.

DETAILED CHEMICAL CHARACTERIZATION - Figure 15 shows the SOF composition for the modes investigated. Use of the oxidation catalyst caused an increase in the percentage oxygenates for 5 of the 6 modes tested. The only mode for which oxygenated compounds were a smaller percentage of the total in catalyzed exhaust was mode 9. This may be due to a long storage time (greater than 3 months) for a portion of the mode 9 filters. Even though the samples were stored frozen, they were thawed and opened repeatedly for the addition of SOF as it became available. This was necessary due to the small amount of SOF collectable on each filter before plugging, requiring addition of SOF from a large number of filters. Otherwise, no reasonable explanation is available for this discrepancy.

Another effect of the oxidation catalyst on the SOF was the general decrease in the percentage of paraffins and increase in the percentage of aromatic compounds with its use. The only modes for which this was not observed was the 50% load modes (modes 4 and 10).

With the exception of the 25% load modes (modes 3 and 11), an increase in the percentage of the acidic compounds present in the SOF was observed with the catalyst in use. Since acidic compounds are also partially oxygenated, one might expect the high load modes (higher exhaust and catalyst bed temperature) to produce the greatest total increase in oxygenates and acidics. This was not observed, but the differences in residence time in the catalysts from mode to mode (higher load results in decreased residence time; see Table 10), and the difficulty encountered with the mode 9 sample may have obscured this trend.

Figure 16 is a gas chromatogram of the mode 11 uncatalyzed aromatic subfraction. Numbered peaks correspond to compounds identified by mass spectrometry, as listed in Table 11. Compounds no. 1 and no. 17 are possibly not derived from engine particulate samples. Compound 1 is a preservative in the diethy ether used for fractionation, while compound no. 17 is a component of many plastic materials. The compounds listed have been previously identified by Hanson et al. [7] and Schuetzle et al. [10]. Compounds 4-8 and 13 are known mutagens [7], but the mutagenic activity of this aromatic subfraction has not yet been fully determined.

AMES BIOASSAY - The statistical analysis of the effect of different assay dates upon the Ames assay data for the same extract is presented in Table 12. Mode 5 catalyzed SOF and mode 11 catalyzed oxygenated subfraction, both without the S-9 activation mixture, were each assayed on two different dates using identical concentrations within each sample and the Ames tester strain types TA98, TA100, TA1537 and TA1538.

For each strain/mode/date combination a dose-response curve was established and the slope of its linear portion (in Cartesian form) calculated. The two slopes from the different dates within a strain and mode were then compared using the two sample t-statistics. The null hypothesis (Ho) of no difference between the two slopes from the two assay dates within each strain/mode was tested against the alternate hypothesis (Ha), which states that a difference does exist between the slopes from the different assay dates with each strain/mode. For Table 12 (and Table 13), linear model $R = a + b \cdot C$ was used for the testing of differences between slopes. The analyses involved a t-statistic that was calculated based on the differences of the two slopes divided by their pooled residual mean square.

In addition to each date's slope and the calculated t-statistic for each pair of dates within a strain and mode, Table 12 gives the probability that the differences which exist between the slopes being compared are due to chance. Also shown in Table 12 is the value of the 5% two-tailed t-statistic with 14 degrees of freedom. The Ha was rejected at the 5% level for all comparisons except for the TA100 mode 11 catalyzed oxygenated subfraction.

Table 13 shows the effect of varying all phases (collection, extraction, and assaying) of diesel exhaust particulate testing.

Samples 3-1 and 3-2 were collected at mode 3 while samples 11-1 and 11-2 were collected at mode 11. The distinction between samples 3-1 and

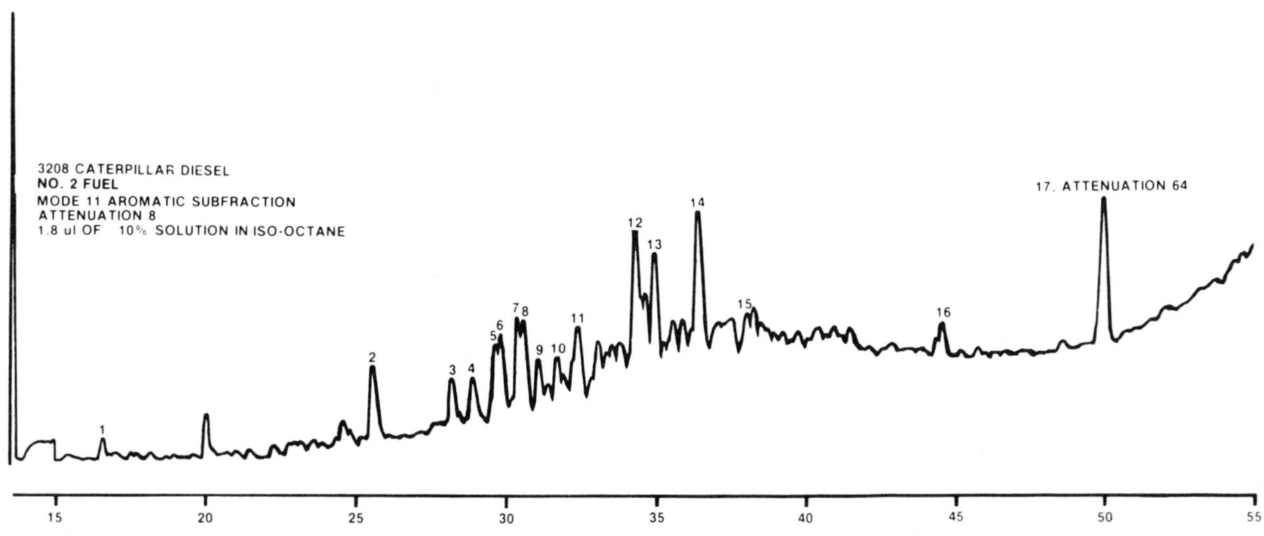

Fig. 16 - Gas chromatogram of mode 11 uncatalyzed aromatic subfraction. Numbered peaks correspond to compounds listed in table 11.

Table 11. Tentative Identities of compounds in Mode 11 Aromatic Subfraction as Determined by Gas Chromatography-Mass Spectrometry

Peak Number (figure)	Retention Time (min.)	Compound	Molecular Weight
1	19.2	2,6-di(*t*-butyl)-4-methyl phenol	220
2	28.6	Phenanthrene	178
3	31.2	Monomethyl dibenzothiophene isomer	198
4[a]	32.0	Monomethyl dibenzothiophene isomer	198
5[a]	32.8	Monomethyl phenanthrene isomer	192
6[a]	33.0	Monomethyl phenanthreme isomer	192
7[a]	33.6	Monomethyl phenanthrene isomer	192
8[a]	33.8	Monomethyl phenanthrene isomer	192
9	34.8	Dimethyl naphthothiophene isomer	212
10	35.6	Dimethyl naphthothiophene isomer	212
11	36.3	Dimethyl naphthothiophene isomer	212
12	37.6	Dimethyl phenanthrene isomer	206
13[a,b]	38.3	Fluoranthene	202
14[b]	39.9	Pyrene	202
15	41.2	Trimethyl phenanthrene isomer	220
16	48.0	Benzo[ghi]fluoranthene	226
17	53.0	Phthalate ester (insignificant mass ion)	--

[a]Known mutagens [7]

[b]Despite nearly identical mass spectra, identity of these compounds was confirmed by retention times

Table 12. Effect of Assay Date on Ames Test Results

Hypotheses Tested:
H_o = No difference between experimental dates within sample and strain.
H_a = There is a difference between experiment dates within sample and strain.

Ames Strain[a]	TA98				TA100				TA1537				TA1538			
Mode[b]	5		11		5		11		5		11		5		11	
Extract	SOF		OXY[c]		SOF		OXY		SOF		OXY		SOF		OXY	
Date of Experiment	072980	081480	072380	080780	072080	081480	072380	080780	072980	081480	072380	080780	072980	081480	072380	080780
Slope[d] $\left(\frac{REV}{\mu g}\right)$	4.45	4.64	1.64	1.39	4.61	4.01	2.03	1.66	2.90	2.59	.807	.749	4.44	4.41	1.97	1.68
t-statistic	.1295		.9470		.9640		2.9105		.6800		.7266		.0333		.9288	
Probability	.50<P		.20<P<.50		.20<P<.50		.01<P<.02		.50<P		.20<P<.50		.50<P		.20<P<.50	
$t^e_{.05(2),14}$	2.404		2.404		2.404		2.404		2.404		2.404		2.404		2.404	

[a]Without S-9 activation system.
[b]All samples collected with catalyst installed.
[c]OXY - Oxygenated subfraction.
[d]Colony counter sensitivity was increased for this study which increased relative slope.
[e]Two-tailed t-statistic for 14 degrees of freedom.

Table 13. Effect of Experimental System Variability on Ames Test Results

Hypotheses Tested:
H_o = No difference between samples collected and extracted on different days within strains.
H_a = There is a difference between samples collected and extracted on different days within strains.

Ames Strain[a]	TA98				TA100				TA1537				TA1538			
Mode	3		11		3		11		3		11		3		11	
Sample[b]	1	2	1	2	1	2	1	2	1	2	1	2	1	2	1	2
Slope[c] $\left(\frac{REV}{\mu g}\right)$	.469	.541	.506	.605	.809	1.036	.954	.936	.115	.102	.033	.082	.337	.338	.321	.191
t-statistic	1.385		1.251		.1010		.1610		1.258		3.155		.001		2.404	
Probability	.10<P<.20		.20<P<.50		.20<P<.50		.10<P<.20		.20<P<.50		.005<P<.01		.50<P		.02<P<.05	
$t^d_{.05(2),14}$	2.404		2.404		2.404		2.404		2.404		2.404		2.404		2.404	

[a]Without S-9 activation system.
[b]Total SOF collected and extracted from engine on different days, without catalyst.
[c]Colony counter sensitivity was increased for this study which increased relative slope.
[d]Two-tailed t-statistic for 14 degrees of freeedom.

3-2 (also samples 11-1 and 11-2) is that they were collected on different days, extracted on different days, and assayed in separate experiments.

For each strain and mode combination (without the S-9 activation system), a dose-response curve was obtained and the slope of its linear portion (in Cartesian form) was calculated. For each strain, sample 3-1 was compared with 3-2 and sample 11-1 with 11-2 for significant differences between their slopes. The two sample t-statistic is used for testing the null hypothesis of no difference against the alternate hypothesis that a difference exists.

In addition to the calculated t-statistic, Table 13 gives the probability that the differences which exist between the slopes are due to chance. Also shown is the table-t value of the 5% two-tailed t-statistic with 14 degrees of freedom. The Ha was rejected at the 5% level for all cases except for strain TA1537 with samples 11-1 and 11-2.

The statistical analysis therefore indicate that the Ames bioassay data were reproducible in 14 of 16 comparisons, whether the same SOF samples were repeated or whether SOF samples from different engine runs were tested. The difference found with strain TA1537 tested against two mode 11 uncatalyzed SOF samples (Table 13) may be due to the fact that this strain has relatively low-level response (compared to strains TA98, TA100, and TA1538). The slope differences were therefore likely due to natural variation in the Ames bioassay. Strain TA100 provides high-level response with usually good reproducibility. The slopes presented in Table 12 for repeats of mode 11 catalyzed SOF oxygenated subfraction were not significant at the $\underline{\alpha} = 0.1$ significance level and therefore the differences probably also represent natural variation in the test results.

Figure 17 shows the response of all five Salmonella typhimurium tester strain genotypes to the catalyzed mode 11 SOF. Dose-response data indicate the mutagenic components of the diesel SOF are not base-pair substitution mutagens as no mutagenic response was demonstrated by strain TA1535, which is the only strain used responding solely to base-pair substitution mutagens. The greatest response was observed with strains TA100 and TA98, followed by stains TA1538 and TA1537. All of these strains are frameshift mutagen detectors. Similar results have been reported by other investigators working with a variety of diesel and gasoline engines [14,15,16, 17]. For purpose of further discussion, only the TA100 tester strain data are presented in graphical form. All test results are listed in Appendix A.

Some of the SOF samples were tested for mutagenic activity in the presence of the S-9 activation mixture. As demonstrated by strain TA100 in Figure 18, the results from mode 11 and mode 11 catalyzed, with and without the S-9 mixture, indicate that the mutagenic activity is direct-acting (i.e., metabolic activation is not required) and that the addition of the S-9 mixture causes a general decrease in mutagenic response.

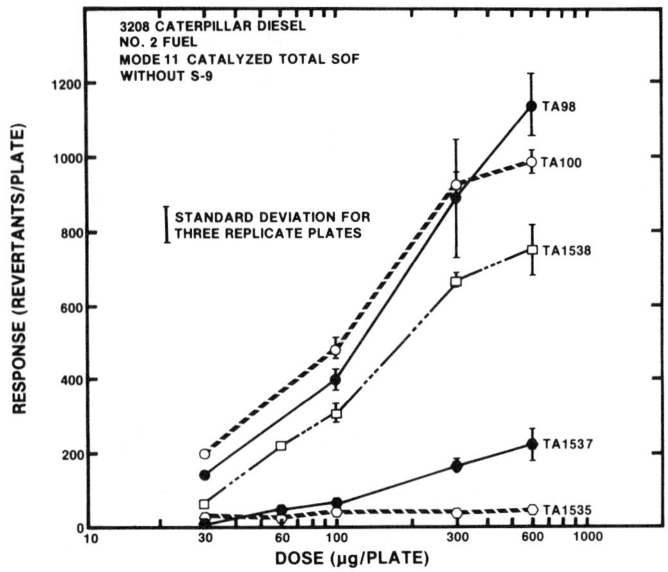

Fig. 17 - Response of all five Ames tester strains to mode 11 catalyzed SOF

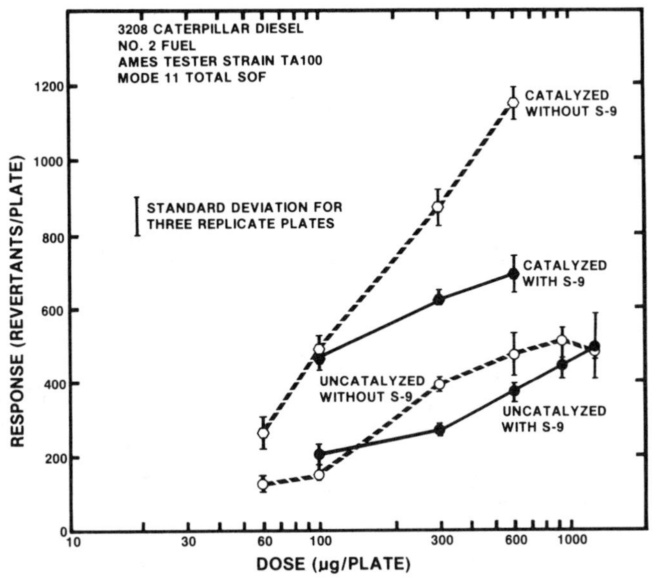

Fig. 18 - Effects of catalyst and S-9 mammalian activation system on mode 11 SOF

These responses have also been noted for SOF from other diesel and fuel systems [15,16,17,18]. Tests with other samples and tester strains showed similar results (Appendix A).

The dose-response curves presented in Figure 18 for strain TA100 and mode 11 SOF (without S-9 activation) also indicate that the catalyzed mode 11 SOF (on a revertant/μg of SOF basis) resulted in a greater mutagenic response than the uncatalyzed mode 11 SOF sample. This effect was generally observed with other engine modes, as demonstrated by the dose-response results from modes 4 and 10 catalyzed and uncatalyzed (Figure 19).

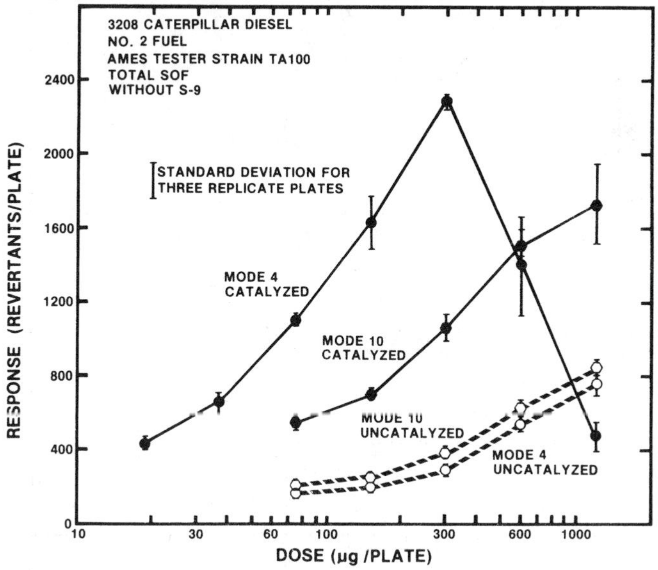

Fig. 19 - Response of Ames tester strain TA100 to SOF from modes 4 and 11 catalyzed and uncatalyzed

The only exception occurred with mode 3, where the catalyzed SOF produced fewer revertants/µg of SOF than the uncatalyzed SOF. Additionally, the SOF from the catalyzed engine modes often displayed a toxic effect which was not found with the uncatalyzed modes (Figure 19, modes 4 uncatalyzed and 4 catalyzed).

As discussed earlier, NO_2 is believed to be capable of activation of PNA compounds [29]. The specific activity (average slope of the dose-response curve) of the non-paraffin portion of the SOF was empirically correlated to the concentration of NO_2 in the raw exhaust using a linear regression least squares curve fit. The paraffin subfraction is believed to be inactive when tested using the Ames bioassay and therefore acts to dilute the more active compounds. The specific activity of the total SOF may then be found by accounting for the dilution of the active compounds by the paraffins by the following equation

$$SA = \left(\frac{100 - \%\ PAR}{100}\right) SA^* \qquad (13)$$

where % PAR = Percent of Paraffin in SOF, SA = Specific Activity of total SOF, and SA^* = Specific Activity of non-paraffin portion of the SOF. For strain TA98 SA^* was found to be:

$$SA^* = 0.0287(NO_2) + 0.7107 \qquad r = 0.952 \qquad (14)$$

and for strain TA100 SA^* was found to be:

$$SA^* = 0.0295(NO_2) + 1.371 \qquad r = 0.917 \qquad (15)$$

It has not yet been determined whether the activation of the SOF occurs in the raw exhaust, in the dilute exhaust before or after collection, or in some combination of these regions. It is also unclear what are the effects of HC and NO_x concentrations, time in the exhaust system and dilution tunnel, and exhaust or dilution tunnel mixture temperatures. It is also possible that the activating species may not actually be the equivalent NO_2, but some compound whose concentration is related to the concentration of equivalent NO_2.

The brake specific revertants and specific activity calculated from data taken from strain TA100 are shown in Figures 20a and b. Although the brake specific SOF emissions are reduced by the oxidation catalyst, the increased mutagenicity of the SOF with the catalyst at least partially negates the reduction in the SOF mass.

Initial mutagenicity bioassays on some of the mode 11 uncatalyzed and 11 catalyzed subfractions (i.e., acidic, ether insoluble, and oxygenated) indicate that their mutagenic activity is both direct-acting and frameshift in nature (Figures 21a, b, and c). These results are consistent with the results of other investigations [15,16]. Maximum mutagenic activity occurred in the acidic and oxygenated subfractions. As with the unfractionated SOF, the catalyzed SOF subfractions displayed a greater mutagenic activity than their uncatalyzed counterparts (Figures 21a, b, and c).

CONCLUSIONS

Based on the data and results obtained for the Caterpillar 3208 with the Engelhard PTX catalysts and the given test conditions using no. 2 fuel, the following conclusions can be made:

1) Exhaust opacity is increased slightly with the catalyst at modes 4 and 10, the 50% load modes. This increase is probably due to increased particulate emissions.
2) Brake specific NO_x emissions remain unchanged by the catalysts. The brake specific equivalent NO_2 emissions are increased by the catalysts at most modes, with a similar decrease in NO emissions. The increase in NO_2 with the catalysts installed is greatest at modes 4 and 5 (50% and 75% of rated load at intermediate speed).
3) Brake specific HC and brake specific SOF emissions are both decreased by the use of the catalysts. The percent reduction of these species by the catalyst was relatively unaffected by changes in engine speed or load.
4) Brake specific sulfate emissions are greatly increased by the catalysts. This trend was observed at all test modes. The increase in sulfate emissions with the catalyst installed does not follow any trends with engine load or speed. The increase is due to the oxidation of SO_2 and SO_3 and eventual release as sulfates.
5) Brake specific particulate emissions are generally increased by the catalysts. The increase is due to both the increase in sulfate emissions and to an increase in the solid (unextractable) portion of the particulate.
6) The percentages of oxygenated, aromatic, and acidic compounds in the SOF are generally

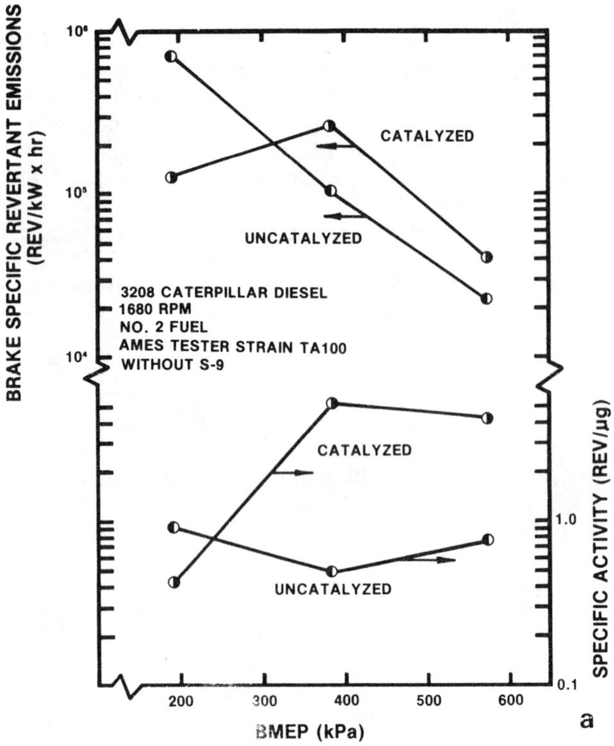

a) 1680 rpm

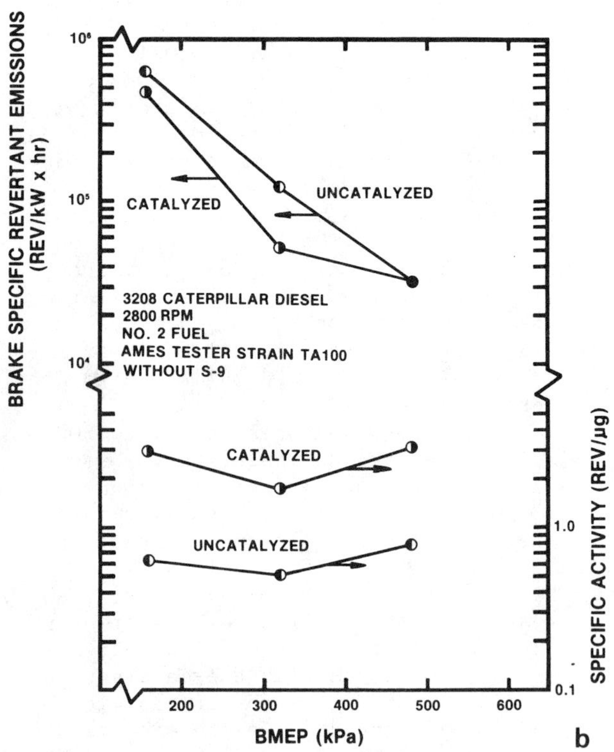

b) 2800 rpm

Fig. 20 - Effect of catalyst on brake specific revertant emissions and specific activity

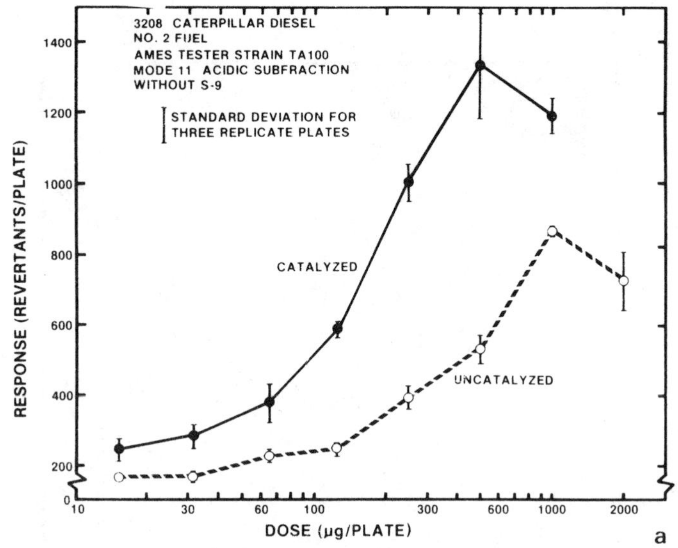

a) acidic subfraction

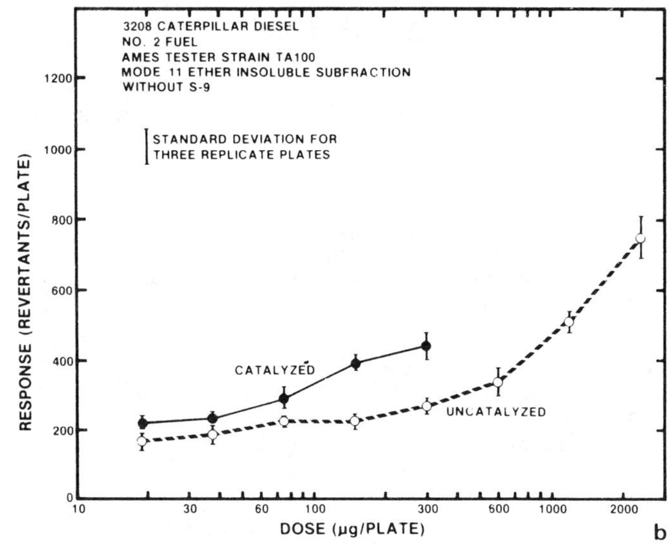

b) ether insoluble subfraction

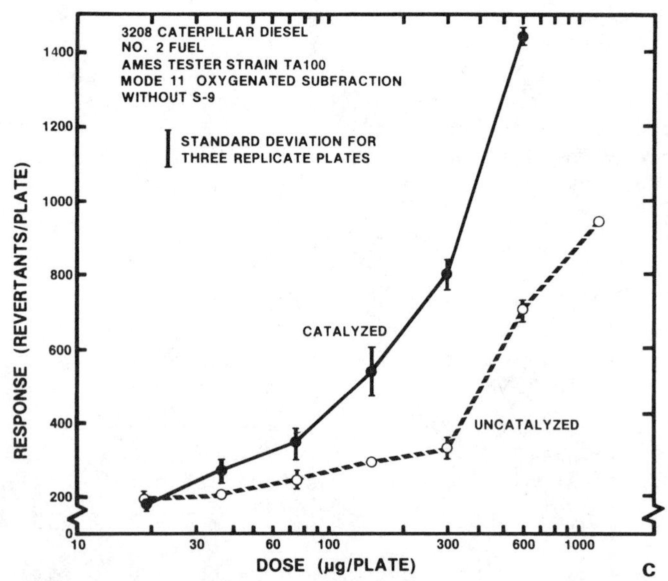

c) oxygenated subfraction

Fig. 21 - Effect of catalyst on the response of Ames tester strain TA100 to the mode 11 acidic, ether insoluble, and oxygenated subfractions increased by the catalysts. The percentage of paraffinic compounds in the SOF is generally decreased by the catalysts.

7) The use of the catalysts generally results in greater mutagenic specific activity (revertants/μg SOF) of the SOF. At some modes the increase in the biological activity of the SOF is sufficient to offset the reduction in mass emissions, resulting in higher brake specific revertant emissions. Using the Ames bioassay, the SOF mutagenic activity was found to be direct-acting and frameshift in nature. Greatest mutagenic response was found when using frameshift mutation detecting strains TA100, TA98, TA1538, and TA1537. Strain TA1535, which detects only base-pair substitution mutagens had no response to the SOF tested.
8) The addition of the S-9 activation mixture resulted in a slight decrease in the activity of both the uncatalyzed and catalyzed mode 11 SOF samples, indicating that the SOF mutagenic activity was direct-acting in nature.
9) Results of Ames mutagenicity bioassays on the oxygenated, acidic, and ether insoluble SOF subfractions showed that these compounds are also direct-acting, frameshift mutagens. These fractions show a higher specific activity than the total SOF.
10) The Ames bioassay was found to be a repeatable method for the analysis of the biological activity of diesel SOF.
11) The specific activity of the non-paraffin subfractions was found to correlate with the concentration of NO_2 present in the exhaust and the diluted exhaust in the tunnel.

From this research it is clear that more and better measurements of the nitrogen compounds in the exhaust are needed. Additional data and understanding about the kinetics of NO_2 and the hydrocarbons in the combustion process and the exhaust system-dilution tunnel are needed. It is also clear that research effort is needed to identify the biologically active compounds in the SOF. This information will provide the technical approach for control of these compounds if its deemed necessary from a health viewpoint.

ACKNOWLEDGEMENTS

The authors would like to express their appreciation to EPA for sponsoring this research program. We also appreciate the grant from Ford Motor Co. that provided the graduate research assistantship for Gary Hunter. In addition, the Ford Grant, the Michigan Tech. Development Fund, EPA, the Upjohn Company, the Center for Waste Management Programs, and the State of Michigan provided funds for developing our Environmental Microbiology Laboratory for carrying out the Ames mutagenicity bioassay.

The helpful advice of Larry Claxton (EPA) and Dr. Jim Pitts and his staff (University of California, Riverside) in helping us set up the protocol for accurate and repeatable Ames testing on diesel SOF samples is appreciated.

The advice received on the overall program and this phase of the study from Andy Kaupert, Jack McFadden, and Dr. Karl Hellman of EPA is greatly appreciated.

Finally, the assistance and gifts of the engine from Caterpillar Tractor Company and the catalysts from Engelhard Minerals Corporation are sincerely acknowledged.

This combined support has provided the basis for a modern real world engineering and scientific educational opportunity on a multidiscipline project.

REFERENCES

1. D. J. Barth and S. M. Blacker, "The EPA Program to Assess the Public Health Significance of Diesel Emissions", J. Air Pollution Control Assoc., Vol. 28, p. 769, 1978.

2. "Clean Air Act Amendments of 1977", Public Law 95-95, 95th Congress, August 7, 1977.

3. W. H. Lipkea, J. H. Johnson, and C. T. Vuk, "The Physical and Chemical Character of Diesel Particulate Emissions-Measurement Techniques and Fundamental Considerations", SAE Paper No. 780108 (SP-430), February, 1978.

4. "GMR Studies Diesel Health Effects", Search, General Motors Research Lab., Vol. 14, No. 5, 1980.

5. C. T. Vuk, M. A. Jones, and J. H. Johnson, "The Measurement and Analysis of the Physical Character of Diesel Particulate Emissions", SAE Paper No. 760131, February, 1976.

6. A. E. Aaronson and R. A. Matula, "Diesel Odor and the Formation of Aromatic Hydrocarbons During the Heterogeneous Combustion of Pure Cetane in a Single-Cylinder Diesel Engine", Thirteenth International Symposium on Combustion, pp. 421-481, 1971.

7. R. L. Hanson, R. F. Royer, and A. L. Brooks, "Identification of Organic Compounds in Diesel Exhaust Particulate", Inhalation Toxicology Research Institute Annual Report, Lovelace Biomedical and Environmental Research Institute, pp. 203-206, 1979.

8. M. Mentser and A. G. Sharkey, Jr., "Chemical Characterization of Diesel Exhaust Particulate", Pittsburgh Energy Research Center Publication No. RI-77/5, 1977.

9. E. F. Funkenbusch, D. G. Leddy, and J. H. Johnson, "The Characterization of the Soluble Organic Fraction of Diesel Particulate Matter", SAE Paper No. 790418, February, 1979.

10. D. Schuetzle, F. S.-C. Lee, T. J. Prater, and S. B. Tejada, "The Identification of Polynuclear Aromatic Hydrocarbons in Mutagenic Fractions of Diesel Particulate Extract", Presented at the 10th Annual Symposium on the Analytical Chemistry of Pollutants, Dortmund, Germany, 1980.

11. B. N. Ames, J. McCann, and E. Yamasaki, "Methods for Detecting Carcinogens and Mutagens with the Salmonella/Mammalian-Microsome Mutagenicity Test", Mutat. Res., Vol. 31, pp. 374, 1975.

12. M. Hollstein, J. McCann, F. A. Angelgsanto, and W. W. Nichols, "Short-Term Tests for Carcinogens and Mutagens", Mutat. Res., Vol. 65,

pp. 133, 1979.

13. A. S. Wright, "The Role of Metabolism in Chemical Mutagenesis and Chemical Carcinogenesis", Mutat. Res., Vol. 75, pp. 215, 1980.

14. A. L. Brookes, C. R. Clark, and R. E. Royer, "Mutagenic Characterization of Particulate Materials in Diesel Exhaust", Inhalation Toxicology Research Institute Annual Report, Lovelace Biomedical and Environmental Research Institute, 1979-1980.

15. J. Huisingh, R. Bradow, R. Jungers, L. Claxton, R. Zweidinger, S. Tejada, J. Bumgarner, F. Duffield, M. Waters, V. F. Simmon, C. Hare, C. Rodriguez, and L. Snow, "Application of a Mutagenicity Bioassay to Monitoring Light-Duty Diesel Particle Emissions", Presented at Symposium on Application of Short-Term Bioassays in the Fractionation and Analysis of Complex Environmental Mixtures, Williamsburg, February, 1978.

16. Y. Y. Wang, M.-S. Lee, C. M. King, and P. O. Warner, "Evidence for Nitroaromatics as Direct-acting Mutagens of Airborne Particulates", Chemosphere, Vol. 9, p. 83, 1980.

17. Y. Y. Wang, S. M. Rappaport, R. F. Sawyer, F. E. Talcott, and F. T. Wei, "Direct-acting Mutagens in Automobile Exhaust", Cancer Letters, Vol. 5, p. 39, 1978.

18. E. T. Wei, Y. Y. Wang and S. M. Rappaport, "Diesel Emissions and the Ames Test: A Commentary", J. Air Poll. Control Assoc., Vol. 30, p. 267, 1980.

19. B. N. Ames, "Identifying Environmental Chemicals Causing Mutations and Cancer", Science, Vol. 204, p. 587, 1979.

20. D. Schuetzle, T. J. Prater, T. L. Riley, A. M. Durisin, and I. Salmeen, "Analysis of Nitrated Derivatives of PAH and Determination of Their Contribution to Ames Assay Mutagenicity for Diesel Particulate Extracts", Fifth International Symposium on Polynuclear Aromatic Hydrocarbons, Columbus, Ohio, October 28-30, 1980.

21. E. M. Shelton, "Diesel Fuel Oils, 1977, Petroleum Products Survey", BERC/pps-77/5, Nov. 1977.

22. E. M. Shelton, "Motor Gasolines, Summer 1977", BERC/pps-78/1, Jan. 1978.

23. L. E. Frisch, J. H. Johnson, and D. G. Leddy, "Effect of Fuels and Dilution Ratio on Diesel Particulate Emissions", SAE Paper No. 790417, February, 1979.

24. T. J. Truex, W. R. Pierson, and D. E. Mckee, "Sulfate in Diesel Exhaust", Environmental Science and Tech., Vol. 14, p. 1118, Sept. 1980.

25. J. S. MacDonald, S. L. Plee, J. B. D'Arcy and R. M. Schreck, "Experimental Measurements of the Independent Effects of Dilution Ratio and Filter Temperature on Diesel Particulate Samples", SAE Paper No. 800185, February, 1980.

26. S. L. Plee and J. S. MacDonald, "Some Mechanisms Affecting the Mass of Diesel Exhaust Particulate Collected Following a Dilution Process", SAE Paper No. 800186, February, 1980.

27. C. A. Amann, D. L. Stivender, S. L. Plee, and J. S. MacDonald, "Some Rudiments of Diesel Particulate Emissions", SAE Paper No. 800251, February, 1980.

28. K. Wark and C. F. Warner, "Air Pollution", Dun-Donnelly, New York, N.Y., 1976.

29. J. N. Pitts, Jr., K. A. Van Cauwenberghe, D. Grosjean, J. P. Schmid, D. R. Fitz, W. L. Belser, Jr., G. B. Knudson, and P. M. Hynds, "Atmospheric Reactions of Polycyclic Aromatic Hydrocarbons: Facile Formation of Mutagenic Nitro Derivatives", Science, Vol. 202, p. 515, 1979.

30. T. J. Truex, W. R. Pierson, D. V. Mckee, M. Shelef, and R. E. Baker, "Effects of Barium Fuel Additive and Fuel Sulfur Level on Diesel Particulate Emissions", Environmental Science and Tech., Vol. 14, p. 1121, Sept. 1980.

31. J. A. Kimberly and R. A. DiDomenico, "UFIS-A New Diesel Injection System", SAE Paper No. 770084, February, 1977.

32. W. R. Wade, "Light-Duty Diesel NO_x-HC-Particulate Trade-Off Studies", SAE Paper No. 800335, February, 1980.

33. M. J. Murphy, L. J. Hillenbrand, and D. A. Trayser, "Oxidation of Diesel Particulate", Report to U.S. EPA by Battelle, 1979.

34. W. R. Pierson, R. H. Hammerle, and J. T. Kummer, "Sulfuric Acid Aerosol Emissions from Catalyst-Equipped Engines", SAE Paper No. 740287, February, 1974; SAE Transactions, Vol. 83, 1974.

35. Personal Communication, J. Hansel, Engelhard Industries Div., Edison, N. J., 1979.

36. K. A. Carpenter, "The Design and Development of a Dilution Tunnel for the Physical and Chemical Characterization of Diesel Particulate Material", M.S. Thesis, MTU, 1979.

37. N. J. Khatri, J. H. Johnson, and D. G. Leddy, "The Characterization of the Hydrocarbon and Sulfate Fractions of Diesel Particulate Matter", SAE Paper No. 780111, February, 1978.

38. D. A. Trayser, E. R. Blosser, F. A. Creswick, and W. R. Pierson, "Sulfuric Acid and Nitrate Emissions from Oxidation Catalysts", SAE Paper No. 750091, February, 1975; SAE Transactions, Vol. 84, 1975.

39. F. A. Creswick, E. R. Blosser, D. A. Trayser, and J. F. Foster, "Sulfuric Acid Emissions from an Oxidation Catalyst-Equipped Vehicle", SAE Paper No. 750411, February, 1975.

40. D. A. Trayser, F. A. Creswick, E. R. Blosser, W. R. Pierson, and R. F. Bauer, "Effect of Catalyst Operating History on Sulfate Emissions", SAE Paper No. 760036, February, 1976.

41. R. D. Matthews, R. F. Sawyer, and R. W. Schefer, "Interferences in Chemiluminescent Measurement of NO and NO_2 Emissions from Combustion Systems", Environmental Science and Tech., Vol. 11, p. 1092, November, 1977.

42. EPA Draft Recommended Practive for Measurement of Gaseous and Particulate Emissions from Heavy-Duty Diesel Engines Under Transient Conditions, April, 1979.

43. Personal Communication, F. Black, U.S. EPA, 1979.

44. P. Killough, J. Watson, and R. Bradow, "Filter Type, High Volume Particulate Sampler for Automotive Diesel Emission Studies", Report on EPA Contract No. 68-02-2566, 1979.

45. R. Philippe, C. Jambon, and R. Clechet, "Thermodynamic Properties of Dimethylsulfoxide + Halomethane Mixtures II Vapor Pressures and Excess Thermodynamic Functions", Journal of Chemical Thermodynamics, Vol. 5, No. 3, pp. 431-444, May, 1973.

46. Personal Communication, J. Campbell, MTU, 1980.

47. W. F. Marshall, "Emission Control for Diesels Operated Underground: Catalytic Converters", BERC/RI-75/8, 1975.

48. H. Zimmerman and I. Zimmerman, "Elements of Organic Chemistry", Glencoe Press, Encino, California, 1977.

49. Personal Communication, M. Skaates, MTU, 1980.

50. E. M. Goodger, "Hydrocarbon Fuels", Halsted Press, New York, N.Y., 1975.

51. L. F. Marek and D. A. Hahn, "The Catalytic Oxidation of Organic Compounds in the Vapor Phase", The Chemical Catalog Comp., Inc., New York, N.Y., 1932.

APPENDIX

Table A-1. Engine Performance and Emission Data

MODE		3U*	3C**	4U	4C	5U	5C	9U	9C	10U	10C	11U	11C
bsfc (kg/kw*hr)	Ave	0.286	0.293	0.233	0.236	0.217	0.214	0.255	0.271	0.296	0.285	0.379	0.342
	S.D.	0.004	0.003	0.001	0.002	0.002	0.002	0.001	0.007	0.006	0.002	0.003	0.003
Opacity(%)	Ave	0.3	0.4	0.5	1.2	1.8	1.9	8.6	8.8	3.8	4.4	4.9	4.8
	S.D.	0.0	0.0	0.0	0.0	0.0	0.0	0.1	0.5	0.1	0.1	0.1	0.0
NO_x Conc.(ppm)	Ave	490	503	943	891	1261	1174	967	1101	715	726	445	409
	S.D.	5	5	10	6	23	34	27	28	8	17	1	2
$bsNO_x$ (g/kw*hr)	Ave	15.5	16.2	14.9	14.2	13.6	12.9	10.2	13.0	12.4	12.3	14.2	13.8
	S.D.	0.2	0.2	0.2	0.1	0.5	0.3	0.3	0.3	0.1	0.3	0.2	0.1
NO_2 Conc.(ppm)	Ave	30	21	19	286	19	120	8	16	14	26	14	56
	S.D.	3	2	2	4	6	25	6	11	2	3	3	4
$bsNO_2$ (g/kw*hr)	Ave	1.0	0.7	0.3	4.6	0.2	1.3	0.1	0.2	0.3	0.4	0.4	1.9
	S.D.	0.1	0.1	0.0	0.1	0.1	0.3	0.1	0.1	0.0	0.0	0.1	0.1
NO Conc.(ppm)	Ave	460	484	924	605	1242	1044	959	1085	701	700	431	353
	S.D.	8	4	9	6	28	47	26	22	7	17	2	6
bsNO (g/kw*hr)	Ave	9.5	10.2	9.5	6.3	8.7	7.5	6.6	8.3	7.9	7.7	9.0	7.8
	S.D.	0.2	0.1	0.1	0.1	0.4	0.4	0.2	0.1	0.1	0.2	0.2	0.1
HC Conc.(ppm)	Ave	215	26	210	23	95	15	120	25	310	41	325	30
	S.D.	15	5	25	3	15	3	12	5	18	7	25	10
bsHc (g/kw*hr)	Ave	2.05	0.25	0.73	0.11	0.31	0.05	0.38	0.09	1.61	0.21	3.13	0.67
	S.D.	0.20	0.05	0.10	0.02	0.05	0.01	0.04	0.02	0.10	0.04	0.30	0.20
bs Total Particulate	Ave	0.89	0.57	0.33	1.71	0.27	2.10	1.04	2.46	0.84	3.27	2.32	4.15
(g/kw*hr)	S.D.	0.04	0.00	0.01	0.03	0.01	0.27	0.05	0.30	0.02	0.14	0.12	0.10
Particulate Emission	Ave	2.13	1.35	1.00	5.03	0.83	6.78	2.83	6.27	1.95	7.96	4.23	7.35
Factor (g/l)	S.D.	0.06	0.07	0.00	0.12	0.06	0.83	0.15	0.67	0.07	0.29	0.21	0.17
$bsSO_4^=$ (g/kw*hr)	Ave	0.02	0.04	0.03	1.06	0.03	1.51	0.06	0.68	0.04	1.41	0.03	2.48
	S.D	0.01	0.02	0.01	0.04	0.02	0.10	0.04	0.16	0.02	0.24	0.02	0.15
$SO_4^=$ Emission Factor	Ave	0.04	0.09	0.10	3.11	0.10	4.87	0.15	1.73	0.09	3.43	0.06	4.40
(g/l)	S.D.	0.01	0.05	0.02	0.10	0.04	0.30	0.12	0.46	0.05	0.59	0.03	0.26
bsSOF (g/kw*hr)	Ave	0.76	0.30	0.21	0.05	0.03	0.01	0.04	0.01	0.24	0.03	0.98	0.16
	S.D.	0.03	0.04	0.01	0.00	0.01	0.00	0.01	0.01	0.06	0.02	0.06	0.05
SOF Emission Factor	Ave	1.53	0.59	0.72	0.16	0.13	0.04	0.14	0.03	0.61	0.09	1.58	0.25
(g/l)	S.D.	0.06	0.07	0.01	0.01	0.02	0.02	0.04	0.03	0.15	0.05	0.06	0.07
ACD(%SOF)	Ave	10.5	3.9	4.5	7.1	12.6	28.5	21.1	24.2	9.6	26.9	15.5	8.9
	S.D.	1.3	0.4	0.8	0.6	1.4	5.7	0.3	2.5	0.0	3.7	3.1	3.2
bsACD (mg/kw*hr)	Ave	79.8	11.7	9.5	3.6	3.8	2.9	8.4	2.4	23.0	8.1	152.0	14.2
	S.D.	10.4	2.0	1.7	0.3	1.3	0.6	2.1	2.4	5.8	5.5	31.8	6.8
ARO(% of SOF)	Ave	3.7	6.3	4.2	1.2	0.6	2.4	1.6	3.6	3.6	3.2	3.3	4.2
	S.D.	3.4	0.6	0.6	0.1	0.1	0.9	0.2	0.2	0.6	0.5	0.8	0.1
bsARO (mg/kw*hr)	Ave	28.1	18.9	8.8	0.6	0.2	0.2	0.6	0.4	8.6	1.0	32.3	6.7
	S.D.	25.9	2.5	1.3	0.1	0.1	0.1	0.2	0.4	2.6	0.7	8.1	2.1
BAS(% of SOF)	Ave	3.3	2.1	1.0	1.7	1.5	4.5	1.8	1.7	2.6	4.6	2.7	1.7
	S.D.	1.2	0.2	0.3	0.4	0.3	3.8	0.3	0.2	0.1	1.1	0.4	0.4
bsBAS (mg/kw*hr)	Ave	25.1	6.3	2.1	0.9	0.5	0.5	0.7	0.2	6.2	1.4	26.5	2.7
	S.D.	9.2	0.8	0.6	0.2	0.2	0.4	0.2	0.2	1.6	0.3	4.2	1.1
EIN(% of SOF)	Ave	7.5	9.0	9.4	14.2	14.7	13.3	3.0	29.7	17.8	2.3	6.6	11.8
	S.D.	4.9	0.7	1.3	0.3	1.3	3.4	1.9	1.1	0.1	0.3	4.1	0.4
bsEIN (mg/kw*hr)	Ave	57.0	27.0	10.7	7.1	4.4	1.3	1.2	3.0	42.7	0.7	64.7	18.9
	S.D.	37.3	4.2	2.9	0.2	0.1	0.3	0.8	3.0	10.7	0.5	40.4	5.9
HIN(% of SOF)	Ave	2.3	1.8	0.6	1.4	1.7	1.8	5.4	0.9	3.7	4.8	1.7	4.6
	S.D.	1.6	0.4	0.3	0.2	0.3	0.8	4.3	1.2	0.8	0.5	0.6	2.8
bsHIN (mg/kw*hr)	Ave	17.5	5.4	1.3	0.7	0.5	0.2	2.2	0.1	8.9	1.4	16.7	7.4
	S.D.	12.2	1.4	0.7	0.1	0.2	0.1	1.8	0.2	2.9	0.9	6.0	5.1
OXY(% of SOF)	Ave	11.8	13.0	12.2	24.8	14.4	28.7	33.5	17.9	20.1	23.8	19.2	24.1
	S.D.	1.4	0.1	0.3	0.1	0.6	7.7	3.0	3.1	0.1	9.4	0.4	0.5
bsOXY (mg/kw*hr)	Ave	89.7	39.0	25.6	12.4	4.3	2.9	21.4	1.8	48.2	7.1	188.2	38.6
	S.D.	11.2	5.2	1.4	0.1	1.4	0.8	5.5	1.8	12.1	5.5	12.2	12.1
PAR % of SOF	Ave	55.3	60.0	63.3	45.2	49.7	14.5	10.6	17.6	34.6	29.7	44.6	39.6
	S.D.	2.0	0.6	4.8	3.5	5.7	2.4	0.2	0.0	0.6	0.9	1.9	2.5
bsPAR (mg/kw*hr)	Ave	420.3	180.0	132.9	22.6	14.9	1.5	4.2	1.8	83.0	8.9	437.1	63.4
	S.D.	22.5	24.1	11.9	1.8	5.3	0.2	1.1	1.8	20.8	5.9	32.6	20.2
TRN % of SOF	Ave	5.6	4.0	4.9	4.4	4.7	6.3	3.1	17.9	8.0	3.2	6.4	5.3
	S.D.	1.5	0.0	0.0	0.6	1.0	0.8	1.2	3.1	0.0	0.5	0.1	0.8
bsTRN (mg/kw*hr)	Ave	42.6	1.2	10.3	2.2	1.4	0.6	1.2	1.8	19.2	1.0	62.7	8.5
	S.D.	11.5	0.2	0.5	0.3	0.6	0.1	0.6	1.8	4.8	0.7	4.0	2.9

* U = uncatalyzed

** C = catalyst installed

Table A-2. Ames Bioassay Dose-Response Statistics

Mode	Cat	S-9		Ames Tester Strain TA98	TA100	TA1535	TA1537	TA1538
3	w/o	with	SA*	0.444	0.804	0.003	0.062	0.389
			C_1	150.000	150.000	150.000	150.000	150.000
			C_2	1200.000	1200.000	1200.000	1200.000	1200.000
			a	-1043.415	-1691.931	19.918	-143.661	-912.434
			b	516.560	935.044	3.955	71.817	451.861
			r	0.978	0.990	0.239	0.934	0.970
3	w/o	w/o	SA	0.480	0.927	0.031	0.115	0.346
			C_1	300.000	300.000	150.000	300.000	300.000
			C_2	1200.000	1200.000	600.000	1200.000	1200.000
			a	-1492.963	-2778.271	-17.13	-373.027	-1088.637
			b	718.090	1386.351	28.253	172.187	516.560
			r	0.992	0.998	0.746	0.976	0.982
3	with	w/o	SA	0.443	0.423	0.016	0.077	0.373
			C_1	150.000	300.000	150.000	300.000	300.000
			C_2	600.000	1200.000	600.000	1200.000	1200.000
			a	-569.199	-1237.757	14.755	-245.266	-1181.587
			b	323.888	632.827	11.627	115.160	556.977
			r	0.983	0.983	0.443	0.913	0.971
4	w/o	w/o	SA	0.339	0.502	0.013	0.088	0.453
			C_1	300.000	300.000	300.000	300.000	300.000
			C_2	1200.000	1200.000	1200.000	1200.000	1200.000
			a	-1007.968	-1500.508	-20.630	-285.854	-1456.714
			b	506.040	750.202	18.824	131.770	676.566
			r	0.994	0.978	0.953	0.961	0.949
4	with	w/o	SA	4.492	5.206	0.074	2.821	4.841
			C_1	75.000	75.000	75.000	75.000	75.000
			C_2	300.000	300.000	300.000	300.000	300.000
			a	-1664.074	-2344.234	-1.573	-1632.723	-2163.185
			b	1678.681	1945.543	27.630	1054.159	1809.344
			r	0.963	0.986	0.566	0.997	0.965
5	w/o	w/o	SA	0.914	0.771	0.011	0.198	0.641
			C_1	150.000	300.000	150.000	150.000	150.000
			C_2	600.000	1200.000	300.000	300.000	300.000
			a	-1256.616	-2451.751	9.317	-278.405	-787.988
			b	683.210	1152.155	8.305	147.826	478.911
			r	0.988	0.990	0.592	0.982	0.988
5	with	w/o	SA	5.139	4.201	0.033	2.680	4.201
			C_1	75.000	75.000	75.000	75.000	75.000
			C_2	300.000	300.000	300.000	300.000	300.000
			a	-2029.462	-1941.822	10.605	-1666.60	-1941.822
			b	1920.628	1570.165	12.180	1001.561	1570.165
			r	0.936	0.984	0.359	0.937	0.984

Table A-2. (Continued)

				Ames Tester Strain				
Mode	Cat	S-9		TA98	TA100	TA1535	TA1537	TA1538
9	w/o	w/o	SA	0.542	0.804	0.003	0.062	0.389
			C_1	100.000	60.000	100.000	100.000	100.000
			C_2	600.000	300.000	600.000	600.000	600.000
			a	-582.657	-1691.931	19.918	-143.661	-912.434
			b	348.386	935.044	3.955	71.817	451.861
			r	0.848	0.990	0.239	0.934	0.970
9	with	w/o	SA	1.044	3.146	0.058	0.154	2.783
			C_1	60.000	30.000	60.000	100.000	30.000
			C_2	600.000	100.000	300.000	600.000	100.000
			a	-53.985	-296.428	-6.952	-136.766	-363.534
			b	188.260	421.155	19.800	98.728	372.629
			r	0.958	0.951	0.420	0.665	0.986
10	w/o	w/o	SA	0.487	0.570	0.009	0.089	0.272
			C_1	150.000	300.000	150.000	150.000	300.000
			C_2	600.000	1200.000	600.000	600.000	1200.000
			a	-602.833	-1481.025	10.320	-112.354	-815.548
			b	363.751	758.507	6.644	66.439	406.383
			r	0.977	0.990	0.448	0.930	0.978
10	with	w/o	SA	1.309	1.756	0.023	0.364	1.018
			C_1	150.000	150.000	150.000	150.000	150.000
			C_2	600.000	600.000	600.000	600.000	600.000
			a	-1650.832	-2162.494	-13.849	-524.836	-1281.288
			b	978.308	1312.162	17.163	271.844	760.722
			r	0.965	0.988	0.581	0.980	0.986
11	w/o	w/o	SA	0.260	0.650	-0.004	0.023	0.211
			C_1	100.000	100.000	100.000	300.000	300.000
			C_2	600.000	600.000	600.000	1280.000	940.000
			a	-263.035	-664.439	32.857	-61.659	-577.426
			b	166.927	417.441	-2.789	35.194	272.019
			r	0.990	0.965	-0 189	0.874	0.964
11	with	w/o	SA	1.294	2.962	0.009	0.504	0.863
			C_1	100.000	60.000	100.000	100.000	100.000
			C_2	600.000	300.000	600.000	600.000	600.000
			a	-1380.734	-1542.725	33.802	-581.110	-931.501
			b	831.653	1017.036	5.595	323.672	554.609
			r	0.988	0.992	0.304	0.960	0.984
11	w/o	with	SA	0.167	0.319	0.012	0.024	0.099
			C_1	600.000	300.000	100.000	300.000	300.000
			C_2	1280.000	940.000	600.000	1280.000	940.000
			a	-782.713	-751.512	5.960	-58.781	-207.170
			b	346.012	411.162	7.645	37.317	127.669
			r	0.901	0.982	0.573	0.886	0.860

Table A-2. (Continued)

Mode	Cat	S-9		TA98	TA100	TA1535	TA1537	TA1538
				Ames Tester Strain				
11	with	with	SA	0.422	0.440	0.021	0.199	0.378
			C_1	100.000	100.000	100.000	100.000	100.000
			C_2	600.000	600.000	600.000	600.000	600.000
			a	-384.270	-88.145	0.824	-215.011	-297.733
			b	270.949	282.613	13.351	127.773	242.656
			r	0.957	0.966	0.512	0.935	0.909
11 Oxygen-ated	w/o	w/o	SA	0.338	0.572	0.021	0.067	0.318
			C_1	150.000	300.000	150.000	150.000	150.000
			C_2	600.000	1200.000	600.000	600.000	600.000
			a	-388.724	-1685.438	-9.217	-73.099	-406.472
			b	252.467	854.843	16.056	49.829	237.518
			r	0.952	0.993	0.820	0.857	0.981
11 Oxygen-ated	w/o	with	SA	1.672	2.014	0.033	0.813	2.043
			C_1	150.000	150.000	150.000	75.000	75.000
			C_2	600.000	600.000	600.000	300.000	300.000
			a	-2131.452	-2803.582	-23.718	-510.104	-1056.312
			b	1253.747	1505.387	24.914	303.956	763.490
			r	0.972	0.968	0.479	0.967	0.951
11 Ether Insol-uble	w/o	w/o	SA	0.124	0.229	0.005	0.043	0.202
			C_1	600.000	600.000	600.000	600.000	300.000
			C_2	2400.000	2400.000	2400.000	2400.000	1200.000
			a	-848.072	1577.918	-12.840	-334.368	-672.395
			b	372.056	685.978	13.841	129.555	301.742
			r	0.918	0.970	0.556	0.925	0.954
11 Ether Insol-uble	with	w/o	SA	N.A.	1.366	N.A.	N.A.	N.A.
			C_1		37.500			
			C_2		150.000			
			a		-169.803			
			b		255.235			
			r		0.935			
11 Acidic	w/o	w/o	SA	0.176	0.741	-0.002	0.092	0.233
			C_1	500.000	125.000	500.000	250.000	500.000
			C_2	2000.000	500.000	2000.000	1000.000	2000.000
			a	-848.589	-714.466	52.493	-221.764	-1256.468
			b	439.048	461.748	-6.090	114.607	580.230
			r	0.966	0.979	-0.272	0.891	0.970
11 Acidic	with	w/o	SA	1.918	1.978	N.A.	0.538	1.447
			C_1	125.000	125.000		125.000	125.000
			C_2	500.000	500.000		500.000	500.000
			a	-2040.249	-1978.201		-616.328	-1480.494
			b	1194.787	1231.828		334.961	901.350
			r	0.946	0.970		0.960	0.914

Table A-2. (Continued)

Mode	Cat	S-9		Ames Tester Strains TA98	TA100	TA1535	TA1537	TA1538
11 Aromatic	w/o	w/o	SA	0.117	0.204	N.A.	N.A.	N.A.
			C_1	150.000	150.000			
			C_2	600.000	600.000			
			a	-140.248	-151.710			
			b	87.477	152.255			
			r	0.907	0.898			
11 Aromatic	with	w/o	SA	0.160	0.083	N.A.	N.A.	N.A.
			C_1	18.750	75.000			
			C_2	75.000	300.000			
			a	0.248	122.864			
			b	14.949	31.005			
			r	0.613	0.568			

*SA = specific activity (REV/μg) (Average slope of dose-response curve).
C_1 = minimum dose in linear portion of dose-response curve.
C_2 = maximum dose on linear portion of dose-response curve.
a = response intercept of dose-response curve, determined statistically.
b = slope of dose-response curve in linear-log format, determined statistically.
r = coefficient of correlation.
N.A. = data not available.

BIBLIOGRAPHY

The literature in the Bibliography references additional material on the measurement and control of diesel particulate emissions. Copies of SAE papers are available in original or photocopy format from SAE. For ordering information, contact the Publications Division, SAE, 400 Commonwealth Drive, Warrendale, PA 15096.

This Bibliography is divided into six sections:

- Formation Studies
- Measurement Techniques and Fundamental Considerations
- Characterization, Engine Controls and Fuel Effects
- Exhaust Aftertreatment
- Combustion and Cycle Analysis
- Other Related Material

Formation Studies

T. Shiozaki, T. Suzuki, and M. Shimoda, Observation of Combustion Process in D.I. Diesel Engine via High Speed Direct and Schlieren Photography, SAE Paper 800025, 1980.

Direct and Schlieren Photographing of the combustion process in an actual D.I. diesel engine have been investigated, and by comparing both direct and schlieren methods with cylinder pressure measurement, the characteristics of mixing, ignition and burning process were analyzed. The different combustion characteristics induced by various in-cylinder air motions were also observed. The injected fuel spray jet pattern, which impinged on the combustion chamber wall, seems to affect the mixing process due to combination with in-cylinder air motion. The development of fuel jet spray was observed and compared with some theoretical treatments. The contribution of turbulence on combustion both in pre-mixed and diffusion type burning period was recognized and the effect of swirl especially on diffusion burning period was also confirmed.

Kei Miwa, Makoto Ikegami, and Ryoji Nakano, Combustion and Pollutant Formation in an Indirect Injection Diesel Engine, SAE Paper 800026, 1980.

A high-speed gas-sampling technique has been applied to reveal the time development of spatial distributions of several chemical species produced during combustion in the swirl and the main chambers. To enable traversed gas-sampling in each chamber, experiments have been made by using a double-scavenged two-stroke cycle engine with a simulated two-dimensional chamber configuration. By these experiments the following matters have been examined in detail; the extent of fuel-rich zones, their decay with time, the action of swirling air motion, the formation of nitric oxide, the formation of hydrocarbons and of carboneous substance, the state of the outflow of gas from the swirl chamber into the main chamber, and the flame spread within the main chamber. Besides, the influences of some operating conditions and of design parameters, such as overall fuel-air ratio, injection timing, dimension of connecting passage, and direction of the fuel spray, upon these items have been investigated.

T. J. Williams and M. J. Tindal, Gas Flow Studies in Direct Injection Diesel Engines with Re-Entrant Combustion Chambers, SAE Paper 800027, 1980.

The paper describes an investigation of the gas flow inside the cylinders of motored direct injection diesel engines using fixed and moving hot wire probes. Particular attention has been paid to the transfer of swirl into the bowl as the piston approaches the end of the compression stroke and the flow pattern in the bowl around T.D.C. The investigation covered two types of bowl in current use, one a toroidal design and the other a re-entrant shape which had been found to give low emissions. Measurements were also made in a re-entrant chamber of simplified geometry with the object of providing data for use in the theoretical modelling of gas flow and combustion processes.

N. D. Whitehouse, E. Clough, and S. O. Uhunmwangho, The Development of Some Gaseous Products During Diesel Engine Combustion, SAE Paper 800028, 1980.

An extensive experimental program of in-cylinder sampling is described and the integrity of the results discussed. The sampling data of the gaseous products NO, CO and CO_2 have been used to produce concentration contours on various planes around the fuel spray at different times during the combustion process. Some calculations of spray penetration and NO production have been compared with experimental data.

R. C. Yu, V. W. Wong, and S. M. Shahed, Sources of Hydrocarbon Emissions from Direct Injection Diesel Engines, SAE Paper 800048, 1980.

Some results of a systematic study on sources of unburned hydrocarbons from direct injection diesel engines are presented. The following possible sources are considered and investigated experimentally and/or analytically: local over-mixing, local under-mixing, bulk quenching, cyclic misfire, cyclic variation, and wall effects. The significance of each source under a variety of operating conditions including simulated deceleration, light loads, high loads, and simulated acceleration are discussed.

The results show that the formation of unburned hydrocarbons is mainly controlled by transient fuel-air mixing and bulk quenching processes. The fraction of fuel appearing as unburned hydrocarbons in the exhaust is greatest at light loads and retarded conditions.

J. C. Dent, Turbulent Mixing Rate—Its Effect on Smoke and Hydrocarbon Emissions from Diesel Engines, SAE Paper 800092, 1980.

The importance of turbulent energy dissipation rate on smoke and unburned hydrocarbon emission from Diesel engines has been identified quantitatively, the latter through a quenching mechanism which is predominant at idling and light load operation of the engine.

A parameter has been derived to assess the contribution to unburned hydrocarbon emissions from various design and operational characteristics of an engine. Further experimental data are required to support the general validity of the parameter.

Otto A. Uyehara, Effect of Burning Zone A/F, Fuel H/C on Soot Formation and Thermal Efficiency, SAE Paper 800093, 1980.

In engine combustion chambers, where the overall fuel-air mixture is stoichiometric or leaner, the effect of localized pockets or zones of stoichiometric or richer mixture are presented. Possible explanations for the existence of CO, with soot in the exhaust are offered. Also the effect of localized zones of richer than stoichiometric have on the thermal efficiency are also presented. Experimental data from literature are used to support the ideas.

Takeyuki Kamimoto, Yuzo Aoyagi, Yukio Matsui, and Shin Matsuoka, The Effects of Some Engine Variables on Measured Rates of Air Entrainment and Heat Release in a DI Diesel Engine, SAE Paper 800253, 1980.

The rate of air entrainment into the flame and the rate of heat release are thermodynamically calculated in a DI diesel engine: A two-zone model is proposed which uses as input data three measured values of cylinder pressure, flame temperature, and injection rate. The correlations between both rates under various conditions make it clear that the combustion during early and main periods of diffusion combustion is mainly controlled by air entrainment into the flame. The effects of injection pressure, piston configuration, and swirl intensity on the air entrainment are also studied. And the extent of mixing in the flame is evaluated by the equivalence ratio in the flame which is also obtained by the same model. The trends of exhausted NO and soot concentrations well correlate with the equivalence ratios in the flame and measured flame temperatures under all conditions studied.

Otto A. Uyehara, Diesel Combustion Temperature on Soot, SAE Paper 800969, 1980.

From recently published papers, mostly from Japanese authors, which deals with soot containing flames, the author chose certain experimental results contained in these papers to support the idea that 2000 to 2400°K is the temperature range in diffusion flames where the flame contains soot. The maximum soot concentration in diffusion flames occurs at a temperature of about 2150°K. From this peak, soot concentration containing flame at 2150, the soot concentration drops off on both sides until at about 1950 and at 2400 the flame contains negligible soot. Supportive arguments are presented and backed by published experimental data.

Measurement Techniques and Fundamental Considerations

Steven H. Cadle, George J. Nebel, and Ronald L. Williams, Measurements of Unregulated Emissions from General Motors' Light-Duty Vehicles, SAE Paper 790694, 1979.

Sampling and analysis methods for unregulated exhaust constituents are discussed. Emission results for more than fifteen exhaust consituents from both gasoline- and diesel-powered automobiles are presented. It is shown that the catalytic converter substantially lowers the emission rates of aldehydes, benzene, benzo(a)-pyrene, hydrogen cyanide, and nitrogen dioxide. However, under certain rich-malfunction conditions, small increases in hydrogen sulfide, carbonyl sulfide, hydrogen cyanide, and ammonia occur. Particulate emissions are the primary concern for diesels since other unregulated emissions occur at the same low levels as from gasoline-powered vehicles. It is concluded that although steady improvements in chemical analysis technology have led to the detection of more and more minor impurities in exhaust, none of these substances are emitted at concentrations that can be considered dangerous.

K. G. Duleep, Analysis of Automotive Particulate Sampling Techniques, SAE Paper 800184, 1980.

The accurate sampling and characterization of automotive particulate emissions is important to quantify any potential threat to human health. EPA has issued a draft recommended practice to measure the total weight of particulates emitted over the FTP, but there are many parameters that are uncontrolled or can be varied within the scope of the draft recommended practice. The parameters were separated into two categories: those affecting particulate formation and those affecting particulate sampling. The variables under the former category were identified as tailpipe conditioning and heat transfer, dilution ratio, and mixing rate. Test data on the effects of each of these variables on measured particulate emissions are presented and analyzed. The variables under the latter category were identified as the tunnel configuration, sample probe and sample line, and the type of filter used for sampling. Only the filter type was identified as having a significant effect on measured particulate emissions, and test data on filters is presented and analyzed.

D. F. Dolan, D. B. Kittelson, and D. Y. H. Pui, Diesel Exhaust Particle Size Distribution Measurement Techniques, SAE Paper 800187, 1980.

Diesel exhaust aerosol particle size measurements have been made by three techniques and compared. The techniques used were based on electrical mobility, diffusion, and inertial separation. The instruments used were a Thermo Systems Incorporated (TSI) Model 3030 Electrical Aerosol Analyzer, TSI Model 3040 diffusion battery in conjunction with a TSI Model 3020 Condensation Nucleus Counter, and a University of Minnesota micro-orifice impactor (0.1 μm cut size). Particle samples were also examined by electron microscopy and volatile fractions were determined by vacuum sublimation. Measurements were made on two engines, one direct injection and one indirect injection swirl chamber. Engine operating conditions were chosen to give a wide range of exhaust particle characteristics. The size measurements agreed well in the regions of instrument overlap. Measured number mean diameters varied from 0.01 μm to 0.1 μm. Total particle number concentrations measured with the electrical aerosol analyzer and the condensation nucleus counter were also in good agreement.

Yukio Matsui, Takeyuki Kamimoto, and Shin Matsuoka, A Study on the Application of the Two-Color Method to the Measurement of Flame Temperature and Soot Concentration in Diesel Engines, SAE Paper 800970, 1980.

Flame temperature and KL factor in a DI diesel engine are measured optically by the two-color method. Some differences are observed between the measure values at visible and infrared wavelengths. These differences are caused by: (1) effect of change of index α in time at infrared wavelength during combustion period; (2) effect of distributions of temperature and soot concentration along optical path; and (3) effect of reflection at the walls.

The optical characteristics and some other problems on the instrumentation of the two-color method at both wavelengths are also discussed.

Yasuhiro Fujiwara and Shoichi Fukazawa, Growth and Combustion of Soot Particulates in the Exhaust of Diesel Engines, SAE Paper 800984, 1980.

The size and aggregation of soot particulates emitted from a diesel engine have been investigated using a transmission electron microscope, and the chemical composition in soot particulates were analyzed by a gas chromatograph in order to clarify the mechanisms of growth and combustion in the exhaust pipe.

In the present report, the following results have been obtained: (1) When the gas temperature was lowered below 400°C in the exhaust pipe, the soot particle distribution shifted to a larger size, from 50 Å to 100 Å. This tendency was remarkable in the case of lower gas temperatures, (2) In this case, polynuclear aromatic hydrocarbons in the soot particulates exhibited a tendency to increase in the downstream of the exhaust pipe. (3) When hydrocarbons, for instance C_2H_2, C_3H_8, etc., were added to the exhaust gas, the rate of growth of soot particles increased compared with no addition. (4) When the exhaust gas was heated up to above 600°C, it was observed that soot particulates burnt from their surface, and they became smaller or burnt out. In this case, smoke density decreased and soot particle distributions shifted to a smaller size.

Gary L. Green and Don Wallace, Correlation Studies of an In-Line, Full-Flow Opacimeter, SAE Paper 801373, 1980.

In-line, full-flow diesel smoke opacity meters have been developed to avoid the many installation and operational problems of using the EPA required end-of-line (open stack) PHS smokemeter. Extensive correlation tests were run to establish relationships between the in-line and PHS meters. These correlation studies included determining the effect of emitted light wave length on particle absorption, effective optical path length measurements and correlation studies with other (Bosch and Hartridge) meters.

E. Ulku Goff, James R. Coombs, David H. Fine, and Thomas M. Baines, Nitrosamine Emissions from Diesel Engine Crankcases, SAE Paper 801374, 1980.

Crankcase emissions from four different Diesel engines, using four different types of lubricating oils, were analyzed for volatile nitrosamines. The results showed large differences in nitrosamine output between different crankcase oils tested in the same engine. Composite emission rates of NDMA from the heavy duty diesel engines tested ranged from 4.4 to 136 μg/hr. Large differences were also observed when the same brand of oil was used in different engines. The controlling factors appear to be the nitrosamine producing capacity of the oils, the NO_x levels in the crankcase, and the emission flow rate from the crankcase.

Characterization, Engine Controls, and Fuel Effects

Hideyuki Tsunemoto and Hiromi Ishitani, The Role of Oxygen in Intake and Exhaust on NO Emission, Smoke and BMEP of a Diesel Engine with EGR System, SAE Paper 800030, 1980.

EGR reduces NO emission, but increases smoke and decreases BMEP in diesel engines. This paper describes the relationships between these behaviors and the effect of decreased oxygen with EGR in direct injection, prechamber, and turbocharged diesel engines.

The results indicate that the reduction of NO depends on decreasing the rate of the incoming oxygen. The increase in smoke and the decrease in BMEP is due to a reduced rate of exhaust oxygen. Also the reduction of NO is due to increased ambient humidity which can be explained by the decreased oxygen in the incoming charge.

With these results, it becomes possible to predict the ratio of the reduction of NO emission, the increase in smoke and the decrease in BMEP.

R. J. Crookes, M. A. A. Nazha, M. S. Janota, and T. Storey, Investigation into the Combustion of Water/Diesel Fuel Emulsions, SAE Paper 800094, 1980.

The results of steady-flow combustion tests for emulsions of 5, 10 and 20% by volume of water at two pressures are compared with those obtained with diesel oil. A reduction in CO, H/C and soot concentrations are obtained with increase in water content. Results from a single cylinder 4-stroke diesel engine show a significant reduction in NO_x emissions.

Photographic examination of the burning spray for different water in diesel emulsions at atmospheric pressure indicates that an increase in water content increases the rate of combustion and reduces the luminosity of the flame up to a critical value of water content.

R. J. Slone, R. W. Scheffel, S. M. Shahed, and B. Petersen, Nitrosamine Emissions from Diesels, SAE Paper 801375, 1980.

Diesel engine exhaust and blowby gases were sampled and analyzed for nitrosamines. Tenax-GC traps were used as collection media. Three different engines were sampled. Tenax-GC material used contained significant quantities of dimethyl nitrosamine as impurities. A cleaning procedure was developed to precondition the traps. No nitrosamines were detected in either exhaust or blowby down to levels of 90 pptr and 10 pptr respectively.

J. Kraft and K.-H. Lies, Polycyclic Aromatic Hydrocarbons in the Exhaust of Gasoline and Diesel Vehicles, SAE Paper 810082, 1981.

The purpose was to study the precision and the separation efficiency of the analytical methods which are currently discussed for the sampling and the analysis of polycyclic aromatic hydrocarbons (PAH) in the exhaust of automobiles. The sampling of the PAH is done by condensing the total exhaust gas flow. The PAH collected are concentrated by known liquid-liquid extractions, purified by silica gel and separated into fractions using Sephadex.

Two quantitative analytical methods are used, the advanced technique of glass capillary gas chromatography and thin-layer chromatography coupled with in situ fluorescence spectrometry. The results of these chromatographic methods correlate within an error of 10% for both gasoline and diesel exhaust samples. The PAH contents in the exhaust gas of several diesel and gasoline cars are compared for different driving cycles. The corresponding PAH mass emissions from diesel and gasoline cars are of the same order of magnitude.

A. Postulka and K.-H. Lies, Chemical Characterization of Particulates from Diesel-Powered Passenger Cars, SAE Paper 810083, 1981.

Particulate emissions from VW and Audi diesel vehicles ranging from 230 to 1000 mg/mi were analysed by different analytical techniques. The results from the classical elemental analysis shows that the particulate matter consists up to 90% carbon. For the determination of metals the X-ray fluorescence method was used, but only traces were found. Methods for the measurement of particle bound organics based on organic extraction and thermogravimetry. Results so far obtained indicate an organic fraction between 20 and 70%. Sulfates were also determined in a portion of 2 to 4% using the BCA method.

Exhaust Aftertreatment

Lynn P. Tessier, Harry F. Sullivan, Gordon M. Bragg, and Clark E. Hermance, The Development of a High Efficiency Diesel Exhaust Particulate Filter, SAE Paper 800338, 1980.

Several materials have been evaluated for their potential as a filter medium in a high-efficiency diesel exhaust particulate filter. One material, a graphite treated woven fiberglass cloth, was chosen for further development work. The filtration and backpressure characteristics of this material were examined over a wide range of face velocities and exhaust particle loadings. From these results an equation to estimate the required filter area for a given engine application, desired service interval and maximum backpressure was developed.

Combustion and Cycle Analysis

N. Watson, A. D. Pilley, and M. Marzouk, A Combustion Correlation for Diesel Engine Simulation, SAE Paper 800029, 1980.

The use of cycle simulation computer programs has become an established part of turbocharged diesel engine research and development. However, the utility of these programs has in the past been limited by the need for combustion information at the operating point. This difficulty has been overcome by the development of an empirical correlation simulating the combustion process (heat release) via an analytical expression whose governing parameters are linked to in-cylinder conditions.

A method of deriving the governing parameters from only a minimum of experimental test data is presented, but enabling performance to be predicted over a wide range of operating conditions.

The use of the combustion correlation enables the effects of changing ambient conditions, turbocharger match, valve timing and other engine design parameters to be predicted automatically, and includes their influence on combustion as well as the turbocharging process.

A. D. Gosman and R. J. R. Johns, Computer Analysis of Fuel-Air Mixing in Direct-Injection Engines, SAE Paper 800091, 1980.

The extension is described of an existing multidimensional method of calculating in-cylinder air motion to the representation of the injection of a liquid fuel spray. Sample calculations are presented of the droplet and gas motion and fuel-air mixing in an axisymmetric representation of an open-chamber direct-injection engine, in the absence of combustion, and are believed to be the first in which a realistic representation of the gas-phase turbulence behaviour is employed. One of the more important findings is that the spray induces velocities and turbulence levels in the gas which are comparible to, and sometimes greater than, those produced by other mechanisms such as swirl and squish. It is concluded however that considerable further work is required to make such models truly predictive and detailed experimental data is urgently required to assist this task.

Other Related Material

Tanvir Ahmad, Effect of Particulates on the Measurement of Oxides of Nitrogen in Diesel Exhaust, SAE Paper 800189, 1980.

Presence of soot in gas analysis systems is believed to cause errors in measurement of the concentration of oxides of nitrogen (NOx) because of adsorption of NOx by soot. The work reported herein involves a preliminary experimental and theoretical analysis to determine the extent of the errors caused by the interaction of NOx with soot suspended in diesel exhaust measurement systems. The experimental results, which are in qualitative agreement with the theory, suggest that at room temperature the percentage of NOx adsorbed by soot suspended in the exhaust stream is a function of the residence time of the gas sample in the sample line. Therefore, a short sample line and a high flow rate would minimize the error in the measurement of NOx. Application of the model to a production passenger-car diesel engine tested on the Federal Test Procedure and using the standard Constant Volume Sampling system indicates that the error in NOx measurement attributable to adsorption by soot is negligible.

Michael P. Walsh and George D. Kittredge, Future Trends in the Control of Emissions from Motor Vehicles, SAE Paper 801359, 1980.

Progress of the Federal program for control of pollutant emissions from motor vehicles is summarized, through the 1968-1980 period. The standards mandated by the Clean Air Act have now been achieved. The future will see greater emphasis on maintaining the control of emissions from passenger cars in service, as opposed to more stringent emissions standards for new cars. More emphasis will also be placed on standards for highway vehicles other than passenger cars and on control of added pollutant types, such as particulate emissions from diesel powered vehicles.

Timothy P. Cox, Heavy-Duty Gasoline Engine Emission Sensitivity to Variations in the 1984 Federal Test Cycle, SAE Paper 801370, 1980.

An attempt at quantification of emission sensitivity of heavy-duty gasoline engines to variations in the 1984 Federal Certification transient test cycle was made. A brief overview of the Federal Test Procedure is presented, followed by emission test results from several dynamometer controller miscalibration modes. In general, the results indicated that certain modes of controller calibration can produce significant emission variations well within the limits of the 1984 FTP validation criteria. The data indicated that some validation criteria require tightening or respecification to assure repeatable test results.

ADDITIONAL REFERENCES REGARDING THE MEASUREMENT AND CONTROL OF DIESEL PARTICULATE EMISSIONS

The topic of the measurement and control of diesel particulate emissions has been addressed by many other organizations in addition to the Society of Automotive Engineers. The bibliography included with this special program lists only papers presented at SAE meetings. To broaden the reader's perspectives, listed below are other publications regarding the measurement and control of diesel particulate emissions.

These Additional References are divided into six sections:

- Formation Studies
- Measurement Techniques and Fundamental Considerations
- Characterization, Engine Controls and Fuel Effects
- Exhaust Aftertreatment
- Health Effects
- Other Related Material

Formation Studies

S. M. Shahed, P. F. Flynn, and W. T. Lyn, A Model for the Formation of Emissions in a Direct-Injection Diesel Engine, Combust. Model. Reciprocating Engines, [Proc. Symp.], 1978 (Pub. 1980), 345–68 (Eng). Edited by James N. Mattavi and Charles A. Amann, Plenum: New York, NY.

A math. model of diesel spray combustion in direct-injection diesel engines is described. The model is based on concepts of quasi-homogeneous multiple-zone combustion zones evolve progressively as a result of the fuel-air mixing process. The thermodn. and chem. kinetics model is flexibly designed without geometrical constraints to accept any fuel-air mixing model. Calcn. of mixing rates is based on transient spray-plume-growth equations together with concn. profiles within the spray plume. Unlike concepts of apparent heat-release rates, the model combines, in a consistent set of equations, the fuel-air mixing process, local chem. energy release, pressure evolution, and pollutant-formation kinetics. The paper illustrates the use of themodel in a fundamental study of the effect of high injection pressure on the combustion and pollutant-formation process.

Measurement Techniques and Fundamental Considerations

Barry R. Belinky, Improved resolution in high performance liquid chromatography analysis of polynuclear aromatic hydrocarbons using ternary solvent systems, ACS Symp. Ser. 1980, 120 (Anal. Tech. Occup. Health Chem.), 149–68 (Eng).

The anl. of polynuclear arom. hydrocarbons (PAH) from engine exhausts, tobacco smoke, coking operations, and other occupations is a problem due to the need to sep. large nos. of structurally similar compds. and simultaneously detect them at extremely low levels. An acetonitrile–MeOH–water element was used to improve resoln. of PAH chromatog. on a reverse phase column. Statistical design techniques were used to improve high performance liq. chromatog. resoln. by optimization of solvent selectivity effects in a ternary solvent system.

Frank Black and Lisa Doberstein, Filter Media for Collecting Diesel Particulate Matter, U.S. Environmental Protection Agency EPA-600/S2-81-071, June, 1981.

Certification of particulate emissions from diesel motor vehicles involves filtration of measured aliquots of the total air diluted exhaust. Seven commercially available filter media were examined for this purpose. The media included a variety of PTFE membrane filters, glass fiber filters, and PTFE coated glass fiber filters. Relative flow resistance (pressure drop), collection efficiency, and gas phase adsorption were examined. Filter structural differences, which influence particulate collection mechanisms, sample flow rates and pressure drops, were studied microscopically. Two media, a membrane and a fiber filter, were also examined microscopically with varying levels of particulate load to determine the role of collected particles in the filtration of subsequent particles.

The results obtained indicate that under the defined test conditions the membrane filters yield low gram per mile emissions rates due to difficulties with collection of Federal Test Procedure phase 1 emissions, and the Gelman A-E glass fiber filter high rates due to adsorption of gas phase emissions. The mechanics of particle collection are similar for both membrane and fiber filters as applied. Diffusion deposition is important with fiber filters for about 5 percent of the collected particulate matter, the remaining 95 percent collected primarily by direct interception. With the membrane filters direct interception is the dominate process.

Charles R. Clark, Timothy J. Truex, Frank S. C. Lee, and Irving T. Salmeen, Influence of Sampling Filter Type on the Mutagenicity of Diesel Exhaust Particulate Extracts.

The effects of filter types on the mutagenicity and chemical characteristics of organic extracts of diesel engine particulate exhaust were studied by collecting exhaust particles in a dilution tube simultaneously on three different types of filters: Teflon membrane (Zefluor), Teflon impregnated glass fiber (Pallflex T60A20), and a quartz fiber (Pallflex 2500QAO). The particles were extracted with dichloromethane and subsequently with acetonitrile. The dichloromethane extracts were evaluated in the Salmonella reversion (Ames) assay using strains TA 98, TA 100 and TA 1538 and analyzed by high performance liquid chromatography (HPLC) with fluorescence detection. The filter loadings ranged from 0.3 to 0.7 mg cm^2 typical of loadings in studies of diesel engine particulate exhaust. No major differences in relative concentrations were observed in the polycyclic aromatic hydrocarbon, oxygenated of transition fractions for the three filter types. Furthermore, no differences in the mutagenicity of the samples could be detected.

Chemical Characterization Panel of the CRC-Program Group on Composition of Diesel Exhaust, Informational Report of the Measurement and Characterization of Diesel Exhaust Emissions (CRC-APRAC Project No. CAPI-1-64), Edited by: J. M. Perez, F. J. Hills, D. Schuetzle, and R. L. Williams, Coordinating Research Council, Inc., December 1980.

This Informational Report on the Measurement and Characterization of Diesel Exhaust Emissions was prepared by the Chemical Characterization Panel of the CRC-APRAC Program Group on Composition of Diesel Exhaust (CRC-APRAC Project No. CAPI-1-64). The report provides guidance on selection of methods for collection and analysis of the constituents in diesel exhaust to enable a complete chemical characterization of the exhaust. It was prepared by a thorough review of the state-of-the-art for measuring diesel emissions in 1980. An attempt was made to compile validated, or at least tested, methods and to evaluate their engineering applicability. To establish the state-of-the-art, available methods were reviewed, and an attempt was made to establish whether the following data were available:

- Procedure details and experimental data
- Limits of detection
- Precision and accuracy
- Problem areas, interferences, etc.
- Applicability to raw and/or diluted exhaust
- Applicability to light- and heavy-duty diesel applications
- Resources required
- Time constraints, if any

Eugene Danielson, Diesel Hydrocarbon Measurement – Series Filter Test, Report 1978, LDTP-78/04; Order No. PB-301410, 13 pp. (Eng). Avail. NTIS. From Gov. Rep. Announce. Index (U.S.) 1979, 79(26), 270.

Filters were loaded by driving various cycles (FTP, LA-4 and 50 mph SS) and collecting the particulate materials. Wet filters are those on which particulate matter was collected and no drying steps were taken. Dry filters were prepd. by baking wet filters at 380°F for one hour. Two different particulate matter collection systems were used for loading the second filter. One was an isokinetic particulate matter collection device designed to collect a large vol. of diesel exhaust and is called a high vol. sampler. The other was the existing Scott filter holder and heated probe set-up which is used to sample exhaust from the diln. tunnel.

Reginald Stanley Emerson, Apparatus for Measuring Smoke Density, Brit. 1,549,193 (Cl. G01N21/26), 25 Jul 1979, Appl. 75/17,848, 29 Apr 1975; 15 pp.

Opacimeters for measuring the d. of smoke emission from diesel vehicle exhausts by the free acceleration method include a probe for insertion in the exhaust pipe, a tube contg. a plug which forms an annular venturi, a light source which passes a beam transversely through the venturi via holes in the tube, reflectors for transmitting the beam back through the venturi 4 times along different paths, a photoelec. cell for receiving the reflected light, and a device for displaying the d. reading.

Teruyuki Fujisawa, Problems with Exhaust Emission Measurement Methods for Heavy-Duty Diesel Trucks, Jidosha Gijutsu 1977, 31(5), 356–62 (Japan).

A review with 9 refs.

Joseph A. Gardella, Jr.; David M. Hercules, Surface Spectroscopic Examination of Diesel Particulates – A Preliminary Study, Int. J. Environ. Anal. Chem. 1979, 7(2), 121–36 (Eng).

Preliminary studies on the use of combined surface anal. techniques such as XPS, Auger electron spectroscopy, secondary ion mass spectrometry and ion scattering spectrometry to probe the surfaces of 2 different diesel particulate systems are reported. The first study shows common chem. forms assocd. with the surface of particulate emission. O, S, and N in various chem. states are seen on the surface of a carbonaceous-graphitic (pyrolytic) center. These elements are postulated as organically bound surface segregated species. Observation of particles extd. for routine adsorbed org. anal. by gas chromatog. show that N and S compds. are left behind, along with residual orgs. Trace metals were not seen on any sample. The second study involves samples collected under different running speed conditions of a diesel engine. Typical anal. procedures for surface examn. by the above techniques were used to monitor the forms and amts. of elements present. Data are shown to support the applicability of these techniques to a study of source particulate emission, and the unique results obtainable.

R. A. Gorse, Jr., I. T. Salmeen, and C. R. Clark, Effects of Filter Loading and Filter Type on the Mutagenicity and Composition of Diesel Exhaust Particulate Extracts, Atmos. Env., 1981.

No Abstract.

C. V. Hampton, W. R. Pierson, T. M. Harvey. W. S. Updegrove, R. S. Marano, and D. Schuetzle, Characterization of Organic Emissions from Vehicles on the Road, Proceedings of The 29th Annual Conference on Mass Spectrometry and Allied Topics, Minneapolis, Minnesota, May 25, 1981.

A survey of gas-phase organic compounds generated by motor vehicles in highway operation was conducted in the Allegheny Mountain Tunnel of the Pennsylvania Turnpike. Concurrent sampling, using Tenax GC polymer adsorbent, of the tunnel air and of the tunnel intake ventilation air permitted discrimination between vehicle-derived and ambient-background components. Strong diurnal and weekly traffic patterns in the tunnel permitted sampling in periods dominated by gasoline-powered vehicles as well as periods dominated by Diesel vehicles. The samples were analyzed by computer-assisted glass capillary gas chromatography/mass spectrometry. Over three hundred of the four hundred compounds found in the tunnel air have been identified. The major components are homologous series of n-alkanes (C_5–C_{26}), branched alkanes, alkenes, and various alkyl series based on cyclopentane, cyclohexane, benzene, styrene, indene, naphthalene and decalin. The samples were found to be too complex for adequate species identification by mass spectral analysis alone. Additional criteria needed to assist in identifications will be presented. Advantage was taken of the thermodynamic properties inherent in gas chromatographic data to devise a new version of the classical carbon number and boiling point versus retention time relationships. Aspects of specific compound quantitation methodology and resolution of pertinent compounds into Diesel and gasoline-powered vehicle contributions are discussed.

David B. Kittelson and Daniel F. Dolan, Dynamics of Sampling and Measurement of Diesel Engine Exhaust Aerosols, Lawrence Berkeley Lab., [Rep.] LBL 1979, LBL–9037, Carbonaceous Part. Atmos., 212–15 (Eng).

The dynamics of sampling and measurement of diesel exhaust aerosols and the exhaust of a single-cylinder prechamber diesel engine were studied. Diesel engine exhaust aerosols consist primarily of 0.02 μ-diam. carbonaceous nuclei which coagulate into chain aggregates of ~0.1–0.2 μ diam. These aggregates may also contain condensed hydrocarbons, sulfates, and other materials. The form of these particles in the atm. depends upon processes occurring inside the engine and exhaust systems, the dynamics of the process of diln. in the exhaust plume, and coagulation and condensation processes which occur after emission into the atm. Because of the complex nature of these interactions, great care must be taken in attempting to relate lab. measurements of diesel exhaust aerosols with their potential atm. impact. In this study an elec. aerosol analyzer, transmission electron microscope, and filter sampler were used in conjunction with a rapid diln. sampling system for measurements of particle size distribution, particle morphol., total particle mass concn. Particle size distributions were bimodal with modes at ~0.02 and 0.10 μ diam. Total aerosol mass concns. in the exhaust were 10–110 mg/m^3 and depended on engine load and fuel.

F. S.-C. Lee, T. M. Harvey, T. J. Prater, M. C. Paputa, and D. Schuetzle, "Chemical Analysis of Diesel Particulate Matter and an Evaluation of Artifact Formation," Sampling and Analysis of Toxic Organics in the Atmosphere, ASTM STP 721, American Society for Testing and Materials, 1980, pp. 92–110.

Methods for the sampling and analysis of diesel particulates have been evaluated. A number of different solvents were utilized for the extraction of diesel particulates. Beno(a)pyrene (BaP) standards spiked on glass-fiber or quartz-fiber filters were found to be oxidized during the sampling of ambient air or exhaust gases. Such degradation was largely reduced by using filters with more inert surfaces, such as Teflon membranes. The results suggest that sample degradation due to gas-BaP reactions requires the presence of a reactive or catalytic medium, such as the glass-fiber filter surface. Such reactions can be minimized with the proper choice of a more inert filter surface such as Teflon. Oxidation of BaP was also observed in diesel extracts during analytical procedures or prolonged storage for several weeks in the dark. The implication of the sampling and analytical artifact problem is a question not only of analytical recovery but also of whether significant biologically active conversion products are produced. Analytical methods to study the extraction efficiency for BaP from diesel particulates are presented. High-performance liquid chromatography (HPLC), high-resolution mass spectrometry (HRMS), and gas chromatography-mass spectrometry (GC-MS) are being investigated for the identification and analysis of these species and preliminary results indicate the presence of a number of oxygenated polynuclear aromatic hydrocarbon (PAH) compounds, which potentially could be produced by oxidation during sampling and analysis.

Frank S.-C. Lee and Dennis Schuetzle, Sampling, Extraction and Analysis of PAH from Internal Combustion Engines, Chapter 1.3 in Handbook of Polycyclic Aromatic Hydrocarbons, A Bjorseth, editor, Marcel Dekker Publishing Company, 1981, 70 pages.

No Abstract.

Thomas Penninga, Evaluation of Exhaust Collection Configuration on Diesel Particulate Measurement, May 1979.

On February 1, 1979, the EPA published proposed rules regulating particulate emissions for light-duty diesel vehicles. Questions both internal and external to EPA concerning the practicality and necessity for these requirements resulted in the initiation of a testing program. The test data was run according to the proposed Federal Register requirements for both FTP and HFET cycles. The two test vehicles were a 1975 Mercedes Benz 300D and a 1979 Oldsmobile 350 CID Diesel. Five different configurations of exhaust collection hardware were tested. The two vehicles were tested twice in each configuration. Previous testing programs had established a baseline particulate emission number for each vehicle. All of the individual FTP and HFET tests results are tabulated. The average FTP and HFET results for each configuration were calculated and compared to both the baseline data average, if available, and to the average result of this study. No baseline HFET data was available. In conclusion, there appears to be direct correlation between collection configurations and mass particulate emissions.

T. Prater, T. Riley, and D. Schuetzle, Capillary GC/MS Characterization of Diesel Exhaust Particulates, Proceedings of The 29th Annual Conference on Mass Spectrometry and Allied Topics, Minneapolis, Minnesota, May 25, 1981.

The chemical characterization of methylene chloride extracts of diesel particulate is described. Normal phase semi-preparative high performance liquid chromatography was used to prefractionate these extracts. Individual fractions were then characterized by fused silica capillary gas chromatography-mass spectrometry (GC-MS). Qualitative and quantitative results for diesel particulates generated from three different engines are presented. A comparison of results from the GC-MS analysis using electron impact and positive and negative ion chemical ionization techniques will be discussed.

T. Riley, T. Prater, D. Schuetzle, T. M. Harvey, and D. Hunt, The Analysis of Nitrated Polynuclear Aromatic Hydrocarbons in Diesel Exhaust Particulates by MS/MS Techniques, Proceedings of the 29th Annual Conference on Mass Spectrometry and Allied Topics, Minneapolis, Minnesota, May 25, 1981.

The analysis of nitrated-PAH (nitro-PAH) in diesel particulate extracts is described using mass analyzed ion kinetic energy spectrometry (MIKES) and triple stage quadrupole (TSQ) mass spectrometry/mass spectrometry (MS/MS) techniques. Qualitative and quantitative results for diesel particulates generated from four different engines are presented. A comparison of results from the MIKES and TSQ techniques are described and factors influencing quantitative results are discussed.

Robert Bosch, G.m.b.H., Detector for Ensuring the Absence of Soot in Exhaust Gas, Fr. Demande 2,433,745 (Cl. G01N27/16), 14 Mar 1980, Ger. Appl. 2,836,002, 17 Aug 1978; 8 pp.

The sensor for detecting soot in the exhaust gases of diesel engines consists of a ceramic noncond. shell tipped with Al_2O_3 contg. 35% Pt. Two electrodes 0.05–0.2 m apart just protrude through the Al_2O_3–Pt layer. Any deposition of soot between the electrodes produces elec. current between them. The resulting temp. increases cause an immediate oxidn. of the soot and evolution of CO_x. The oxidn. was catalyzed by the Al_2O_3–Pt layer. The sensor could be used to monitor the operation of diesel engines and to prevent excessive air pollution by soot in the engine exhaust gases.

R. E. Royer, J. C. Sturm, and R. A. Walter, Extraction and Fractionation of Organics from Diesel Exhaust Particles, Annu. Rep. Inhalation Toxicol. Res. Inst.

(Lovelace Biomed. Environ. Res. Inst.) 1978 (Pub. 1979), 207–10 (Eng).

The methods suited for extg. and fractionating org. compds. from diesel exhaust particles were untrasonic extn. with $CHCl_2$ and fractionation by gel chromatog.

Dennis Schuetzle and Joseph M. Perez, A CRC Cooperative Comparison of Extraction and HPLC Techniques for Diesel Particulate Emissions, Paper #81-56.4, Proceedings of the 74th Annual Meeting of The Air Pollution Control Association, Philadelphia, PA, June 21, 1981.

The projected increase in the number of light-duty diesel-powered U.S. passenger cars has prompted concern about adverse human health effects that may be caused by an increased exposure to submicron respirable exhaust particulates, since diesel engines emit about 25–100 times more particulate mass than that from catalyst-equipped gasoline-powered vehicles.

Issuance of the EPA Advisory Circular #76(AC-76) on Unregulated Diesel Emissions in June 1978 clearly required the development of improved chemical characterization methods. The Coordinating Research Council's Chemical Characterization Panel of the Composition of Diesel Exhaust Project No. CAPI-1-64 was formed in an attempt to accelerate the validation of the chemical analysis methods.

The goal of the Panel was "to conduct a technical evaluation of available methods to measure trace constituents in diesel exhaust."

The objectives to meet the goal included:

1. Reviewing current analytical methods, their precision and accuracy, limits of detection and applicability to both light- and heavy-duty diesels;
2. Establishing a priority of the constituents in the exhaust to be studied;
3. Obtaining consensus of the Panel and conducting round-robin cooperative tests to establish engineering applicability;
4. Identifying problem areas for further research.

The Panel chose as their first objective to undertake studies of extraction and fractionation procedures for diesel particulates.

A number of studies have indicated that the extractable organic components contain certain compounds of environmental concern. This soluble organic fraction (SOF) of the diesel particulate is often used for chemical and biological studies. It is important, therefore, that the organic compounds present in the SOF are extracted with quantitative accuracy while preserving their chemical integrity.

Several techniques have been described in the literature for the extraction of air particulates, including ultrasonic agitation and Soxhlet extraction procedures. However, these procedures required further evaluation for application to the extraction of the SOF for diesel particulates before a suggested procedure could be used in a round-robin by the Panel.

The separation of the SOF into individual compounds is very difficult since a large number of compounds are present. Open column liquid chromatography and liquid-liquid separation procedures have been the most widely used method for the fractionation of complex environmental particulate samples according to chemical functionalities. The application of HPLC as a means of direct fractionation of the particulate extracts has recently begun to gain popularity due to its reproducibility and speed.

The most comprehensive study of an HPLC procedure for the fractionation of diesel particulate extracts to date involves a modification of an EPA-RTP separation scheme which was adopted for further study as a round-robin procedure as described herein. Data used for the development and validation of the round-robin techniques, as well as results from the two round-robins, are reported.

Dennis Schuetzle, Loretta M. Skewes, Gerald E. Fisher, Steven P. Levine, and Robert A. Gorse, Jr., The Determination of Sulfates in Diesel Particulates, Anal. Chem., May 1981.

Analytical techniques for the determination of total sulfates in diesel particulates are described utilizing the techniques of total combustion/non-dispersive infrared (NDIR) detection, and modified extraction/barium perchlorate-thorin titration (BPT). The combustion technique utilizes a commercially available sulfur analyzer (LECO – IR32) that requires less than fifteen minutes for triplicate analysis and has a precision and accuracy of better than ±5% for filter samples containing from 100–10,000 μg of sulfate. The combustion technique measures soluble and insoluble sulfate such as $BaSO_4$ and is not affected by the presence of large quantities of carbon in the particulate sample. The extraction efficiency for sulfates in diesel particulates using standard procedures (11) were found to average 67% for samples containing only trace quantities of anions and 59% for samples containing stoichiometric levels of barium. These procedures have been modified to include ultrasonication with isoproponal/water in order to overcome the problem of incomplete sulfate extraction. Extracted sulfate samples are then analyzed using the BPT method. The extraction/BPT method showed an average recovery of 1.02 ± 0.12 (1σ) when compared to the combustion technique. ESCA studies showed that the chemical state of the surface sulfur species for the diesel particulates averaged 93 ± 3% sulfate and 7 ± 3% elemental sulfur.

Lawrence Smith, Mary Ann Parness, E. Robert Fanick, Harry E. Dietzmann, and Ronald L. Bradow, Analytical Procedures for Characterizing Unregulated Emissions from Vehicles Using Middle-Distillate Fuels, U.S. Environmental Protection Agency, EPA-600/2-80-068, April, 1980.

This research program was initiated with the objective of developing, codifying and testing a group of chemical analytical methods for measuring toxic compounds in the exhaust of distillate-fueled engines (i.e. diesel, gas turbine, Stirling, or Rankin cycle powerplants). It is a part of a larger effort to characterize these components from a number of prototype powerplants and, thus, represents a logical first step in the process.

Methods of collection and analysis for aldehydes and ketones, for hydrogen cyanide and cyanogen, for hydrogen sulfide, carbonyl sulfide and organic sulfides, for ammonia and amines, for nitrous oxide, sulfur dioxide, individual hydrocarbons, for soluble sulfate and N-nitrosodi-methylamine, benzo-a-pyrene, and phenols were studied in detail. Ten analytical procedures were developed and codified. Interference studies and proof-tests in diesel engine exhaust were conducted with every procedure and the results of these experiments are reported in detail.

All of the procedures were found to be suitable for use in exhaust emissions characterization studies. The sampling parameters were found to be adequate for the collection of trace levels of exhaust components using standard CVS sampling techniques. Interferences were, in general, minimal although there were two significant problem areas. Phthalate ester interferes with crotonaldehyde determinations and this contaminant must be avoided in the procedure. In the hydrogen sulfide method, S02 decreases the apparent sulfide, and its presence must be corrected for. While other interferences were noted, all could be avoided with the appropriate precautions noted in the final procedure.

Qualification tests were conducted by introducing known quantities of these pollutants into the exhaust of a diesel engine operating on a standard emissions test CVS tunnel. The results of these experiments indicated completely quantitative recovery for aldehydes and ketones, S02, nitrous oxide, total cyanide and phenols. Hydrogen sulfide is lost to the extent of ten percent at normal exhaust levels. Amines, ammonia and organic sulfides can be lost in sampling in significant amounts in the CVS apparatus. These losses must be taken into account when calculating exhaust contributions.

J. Stauff, I. Fleischhauer-Simo, and W. L. Tsai, Determination of Epoxides in Automotive Emissions by Chemiluminescence, Proc. – Int. Symp. Anal. Appl. Biolumin. Chemilumin. 1978 (Pub. 1979), 278–93 (Eng). Edited by Eric Schram and Philip Stanley, Westlake Village, Calif.

Epoxides in the 10^{-8} g range are detd. in exhaust gases by chemiluminescence; the substances detd. enhance the mutation rate of bacteria in the Ames test without microsomal activation. The chemiluminescent response of the exhaust exts. are correlated with mutagenicity.

Ulf Stenberg, Thomas Alsberg, Lars Blomberg, and Thomas Waennman, Gas Chromatographic Separation of High-Molecular Polynuclear Aromatic Hydrocarbons in Samples from Different Sources, Using Temperature-Stable Glass Capillary Columns. Polynucl. Aromat. Hydrocarbons, Int. Symp. Chem. Biol. – Carcinog. Mutagen., 3rd 1978 (Pub. 1979), 313–26 (Eng). Edited by Peter W. Jones and Philip Leber.

Glass capillary gas chromatog. columns coated with Me silicone gums were used in the identification of polynuclear arom. hydrocarbons (PAH) in ambient air and in emissions from mobile and stationary sources. SiO_2 columns and gel filtration chromatog. on Sephadex LH-20 were used for cleaning the gas or air samples prior to chromatog. detns. Benzo [g, h, i] perylene (I) [191–24–2] and coronene [191–07-1] amts. in exhaust gas from automobiles fueled with MeOH-gasoline were rather higher, as compared to benzo[a]pyrene [50–32–8]. The PAH profile exhaust gases from diesel engine differed from that of the Otto engine. The combustion of peat gave more PAH in the high-mol. wt. region (>270) than the combustion of wood. Ambient air contained relatively high amts. of I and coronene.

Kevin D. Stulp, Particulate Measurement – Efficiency of Pallflex T60A20 Filter Media, U.S. Environmental Protection Agency, July, 1979.

An experiment designed to measure the efficiences of different lots of Pallflex T60A20 teflon-coated glass fiber filters has been completed. Results of these tests indicated that: (1) batch-to-batch variation of filters is generally less than 2 percent, and (2) this media will not satisfy the 98 percent filter efficiency requirement on all vehicles. The test procedure specified in the light-duty diesel particulate NPRM requires: (1) a filter efficiency of 98 percent, or greater efficiency and (2) that flurocarbon-coated glass fiber filters must be used for testing. Some manufacturers have expressed concern that the 98 percent filter media efficiency requirement will be difficult to meet, and that there is significant batch-to-batch variability with this media. Hence, a brief test program was conducted to evaluate the significance of these concerns. Two different production vehicles were used to evaluate batch-to-batch variation of the above filters, and also to investigate difficulties in meeting the 98 per cent filter efficiency requirement.

Characterization, Engine Controls, and Fuel Effects

Thomas Baines and Joseph H. Somers, Heavy-Duty Diesel Particulate Emission Factors, Proc., Annu. Meet. – Air Pollut. Control Assoc. 1978, 71st. Vol. 4, Paper No. 54.3, 15 pp. (Eng).

The 2-stroke diesel engines emitted almost twice the particulates (4.74 g/kg fuel) as the 4-stroke engines (2.64 g/kg fuel). The particulate emission factor range for urban areas which permit only local buses is 1.8–2.7 g/km. Diesel trucks emit, on a nationwide scale, 80,000 metric tons particulates/yr. A formula is given for calcg. the particulate emission factors, based on particulate emissions/kg fuel and fuel consumption/100 km.

Thomas M. Baines, Joseph H. Somers, and Craig A. Harvey, Heavy Duty Diesel Particulate Emission Factors, J. Air Pollut. Control Assoc. 1979, 29(6), 616–21 (Eng).

For the past several years, the Environmental Protection Agency has been measuring particulate emissions from a variety of heavy-duty diesel engines. Particulate emissions samples have been collected by passing the exhaust into a std. diln. tunnel. A small fraction of the dil. exhaust from the tunnel is pulled through a filter from which particulate mass and org. content is detd. A discussion is given of the sampling system and particulate emission factors that have been computed from truck and bus fuel consumption data as well as av. truck and bus speed data from New York and Los Angeles (freeway and nonfreeway usage). Av. particulate emission test results (steady state tests) for 2-stroke engines were 4.74 g/kg fuel and for 4-stroke engines were 2.64 g/kg fuel. Using av. particulate emissions results, a particulate emission factor range of 0.8–1.3 g/km was computed. Nationwide diesel particulate emissions were calcd. to be 88,000 metric tons/yr.

Ronald L. Bradow, Ph.D., Diesel Particle Emissions, Bull. N.Y. Acad. Med. Vol. 56, No. 9, November–December 1980.

Diesel engines can substantially improve passenger car fuel economy in any size vehicle, and it is virtually certain that a significant fraction of the new cars sold 10 years from now will have diesel engines. This technology may produce environmental effects that counterbalance the social advantage of fuel savings. The size and shape of some of these disadvantages and some new information that has progressively narrowed our attention to a small fraction of the exhaust components is presented.

Diane M. Cavagnaro, Diesel Exhaust Emission Control Motor Vehicles (Citations from the Engineering Index Data Base), Report 1979, Order No. NTIS/PS-79/0038, 266 pp. (Eng). Avail. NTIS. From Gov. Rep. Announce. Index (U.S.) 1979, 79(8), 95.

A bibliog. review, with 261 refs., on the control of exhaust gases from diesel engines.

Diane M. Cavagnaro, Diesel Exhaust Emissions (Citations from the American Petroleum Institute Data Base). Report 1979, Order No. NTIS/PS-79/0039, 160 pp. (Eng). Avail. NTIS. From Gov. Rep. Announce. Index (U.S.) 1979, 79(8), 95.

A bibliog. review, with 155 refs., on control of exhaust emissions from diesel motor vehicles with emphasis on engine design, techniques of measuring the gases, different fuel additives, and odor control.

Diane M. Cavagnaro, Diesel Exhaust Emissions. Volume 1. 1964–1976 (citations from the NTIS data base). Report 1979, Order No. NTIS/PS-79/0036, 205 pp. (Eng). Avail. NTIS. From Gov. Rep. Announce. Index (U.S.) 1979, 79(8), 94.

A bibliog. review with 199 refs. of Federally-funded research on exhaust gases from stationary and vehicular diesel engines. The citations include their pollution potential, compn., control, and formation in combustion reactions. Types of exhaust gases are covered. Research concerned with health effects of these exhaust gases is included.

Diane M. Cavagnaro, Diesel Exhaust Emission Control for Motor Vehicles. Volume 1. 1970–1977 (Citations from the Engineering Index Data Base). Report 1980, Order No. PB80-805393, 213 pp. (Eng). Avail. NTIS. From Gov. Rep. Announce. Index (U.S.) 1080, 80(9), 1513.

A review with 206 refs.

H. E. Dietzmann, M. A. Parness, and R. L. Bradow, Emissions from Gasoline and Diesel Delivery Trucks by Chassis Transient Cycle, Contributed by the Diesel & Gas Energy Power Division of The American Society of Mechanical Engineers for presentation at the Energy-Sources Technology Conference and Exhibition, January 18–22, 1981, Houston, Texas.

Gaseous, particulate and unregulated emissions are reported from two gasoline and one diesel powered delivery trucks operated on the chassis version of the 1983 transient heavy-duty engine test procedure. Total exhaust dilution was employed to collect particulate and unregulated gaseous emissions for chemical characterization. Analyses conducted were: Ames bioassay, sequential extractions, cycle segment particulate analysis, metals, N_2O, NO_2, individual hydrocarbons, benzo(a)pyrene, HCN and fuel consumption. Comparisons of gasoline and diesel emission rates for regulated and unregulated emissions are included.

Mitchell D. Erickson, David L. Newton, Edo D. Pellizzari, Kenneth B. Tomer, and David Dropkin, Identification of Alkyl-9-Fluorenones in Diesel Exhaust Particulate, J. Chromatogr. Sci. 1979, 17(8), 449–54 (Eng).

A homologous series of 9-fluorenones ranging from the parent compd. through C_4-alkyl-9-fluorenone were identified in a fraction of diesel exhaust particulate exts. A variety of anal. techniques were employed to achieve identification, including high resoln. gas chromatog./mass spectrometry/computer using both electron impact and chem. ionization, gas chromatog. with flame ionization detection, high resoln. mass spectrometry, and high resoln. gas chromatog./Fourier transform IR spectrometry. The identification of the 9-fluorenone derivs. was made on the basis of precise mass detns. using high resoln. mass spectrometry, IR spectra of one of the gas chromatog./Fourier transform IR spectrometry peaks, comparison of retention times with authentic compds., and interpretation of the mass spectra. Any single anal. would have been insufficient for confident identification, and indeed, the mass spectra strongly correlated with std. spectra of alkyl benzo[c] cinnolines. The results of this study clearly illustrate the need for complementary anal. techniques for analyses of complex unknowns.

Evaluation of Two Turbocharged Diesel Volkswagen Rabbits, U.S. Environmental Protection Agency, Ann Arbor, MI, October, 1979.

The EPA was requested to test two Volkswagen Rabbits with turbocharged Diesel engines by the U.S. Department of Transportation. These were research vehicles which coupled turbochargers to small displacement Diesel engines with the objective of maintaining the fuel efficiency of the Diesel with a power output similiar to that of equal displacement gasoline engines. The DOT requested testing by the EPA as the EPA has the capability to assess the vehicle performance and emission levels. There has been limited experience in testing turbocharged Diesel engines. A small displacement Diesel with a matched turbocharger were to be evaluated by the EPA in terms of performance, fuel economy, particulate emissions, and gaseous emissions. These vehicles offered the opportunity of evaluating an automotive engine which may indicate the direction the auto industry will take in the future in an effort to minimize harmful emissions, maximize fuel economy and maintain current vehicle performance. This test program was run during the period of February 1977 to May 1978.

Thomas L. Gibson, Alice I. Ricci, and Ronald L. Williams, Measurement of Polynuclear Aromatic Hydrocarbons, Their Derivatives, and Their Reactivity In Diesel Automobile Exhaust, General Motors Research Laboratories, GMR – 3478, ENV #95, November, 1980.

Samples of diesel exhaust particulate are mutagenic when tested with histidine-dependent Salmonella typhimurium, the Ames test (3, 5, 7, 9). The mutagenic response is considered direct activity because it is observed without addition of mammalian enzymes to activate the compounds as is required by indirect mutagens such as unsubstituted polynuclear aromatic hydrocarbons (PAH). The direct mutagenicity is, however, of the frameshift type associated with derivatives of PAH (4). Recently, considerable attention has been given to the nitro derivatives of PAH (nitro-PAH), and estimates based on GC-MS (7) and TLC-UV spectroscopic (5) data have suggested that these compounds make a significant contribution to the mutagenicity of diesel particulate.

By measuring the levels of PAH and nitro-PAH in diesel particulate samples, we can study the effects of nitro-PAH on the observed mutagenicity and observe whether nitro-PAH are formed in particulate matter on a filter when exhaust gases pass through it. To measure these compounds, high-performance liquid chromatography (HPLC) with fluorescence detection was chosen for its ability to analyze complex mixtures.

G. P. Gross and K. E. Murphy, Jr., The Effects of Diesel Fuel Properties on Performance, Smoke, and Emissions, J. Eng. Power 1979, 101(4), 524–32 (Eng).

Diesel fuels were blended from selected components to provide aroms. contents 10–57% and viscosities 2.21–6.95 cSt at 100°F in a 14 fuel set which included a com. diesel fuel as a ref. fuel. Tests of the fuels were conducted under full load at several speeds and in the Federal 13-mode and smoke-cycle procedures, using a 2-stroke naturally aspirated engine and a 4-stroke turbocharged engine. Fuel properties such as viscosity, aromaticity, cetane no., gravity, distn. points, and heat of combustion, some of which were partially correlated, were examd. individually and in combinations as predictors of the engine performance data. The 2 test engines responded similarly to fuel variables, but with some differences in sensitivity. Power output and fuel economy were correlated with the heats of combustion on vol. and wt. bases, resp. Smoke increased with the amt. of fuel boiling 640°F and was not apparently affected by fuel arom. content. Emissions of NO_x and of NO_x + hydrocarbons increased with increasing fuel aroms. by itself or with increasing fuel sp. gr. and decreasing fuel 50% distn. temp. Hydrocarbon emissions decreased with increasing viscosity or cetane no. CO emissions increased with increasing 90%-distn. temp. and with decreases in cetane no.

Yoshisuke Hamamoto, Tomoyuki Wakisaka, and Seiichi Tanabe, Effects of Air Swirl on Diesel Combustion, Investigation by Exhaust Gas Analysis, Nippon Hakuyo Kikan Gakkaishi 1980, 15(11), 880–8 (Japan).

The effects of air swirl on the burning of diesel fuel air mixts. were studied by analyzing the soot and NO_X emissions and the combustion efficiency of a direct-injection diesel engine equipped with a single-cylinder (1.35 L) and a disk-type combustion chamber. The effects were influenced by the no. of fuel-injection holes (H) on the nozzle and the air velocity-engine speed ratio (S). The magnitudes of the pollutant emissions and the combustion efficiency were proportional to the HS values. At air-fuel ratios (A/E) slightly above 1:1, an undesirable over-swirl phenomenon occurred at SH> ~15. The effect became increasingly significant when the magnitudes of the A/F ratio and the O concn. in the intake air (R) decreased. Equations were formulated to relate the soot concn. and the combustion efficiency with R.

R. L. Hanson, R. E. Royer, and A. L. Brooks, Identification of Organic Compounds in Diesel Exhaust Particles, Annu. Rep. Inhalation Toxicol. Res. Inst. (Lovelace Biomed. Environ. Res. Inst.) 1978 (Pub. 1979), 203–6 (Eng).

Diesel exhaust particles were analyzed by gas chromarog./mass spectrometry. Tentative identification of components indicated that several mutagens were present in some of these fractions. The aliph. hydrocarbons isolated in the 1st of 2 fractions (of $CHCl_2$ exts.) had no mutagenic activity, but the polynuclear arom. hydrocarbons found in the remaining 4 fractions did have mutagenic activity.

Charles T. Hare and T. M. Baines, Characterization of Gaseous and Particulate Emissions from Light Duty Diesels Operated on Various Fuels, U.S. Environmental Protection Agency, EPA-460/3-79-008, July, 1979.

Gaseous and particulate emissions of a non-routine nature were measured in the exhausts of two light-duty Diesel-powered automobiles. These vehicles were a Mercedes 240D and a Volkswagen Rabbit Diesel. Visible exhaust smoke, regulated gaseous pollutants, and exhaust odor were also measured. Five fuels were used in this investigation, representing broad ranges in sulfur content, hydrocarbon-type composition, density, cetane index, and a number of other properties.

Vehicle operating procedures used for test purposes included both those specified in Federal Regulations (FTP, FET) and several others simulating different situations (CFDS, NYCC, steady-state, odor test conditions). Gas samples were acquired from both direct and dilute exhaust streams. Particulate samples were taken using an exhaust dilution tunnel operating on the entire exhaust stream of each engine. Filter-collected particulate weights provided the basis for particulate mass emission calculations. Most of the sampling and analytical procedures used were developed during earlier EPA Contracts 68-02-1230, 68-03-2196 Task Order 4, and 68-02-1777.

A statistical analysis of the particulate emissions data was conducted, using some of the methods developed under Contract 68-02-1777. Analysis of gaseous emissions data and particulate size data was also conducted.

Charles T. Hare and Frank M. Black, Motor Vehicle Particulate Emission Factors, APCA 81–56.5, 74th Meeting & Exposition, June 21–26, 1981, Philadelphia, PA.

Development of emission factors for motor vehicles requires not only emissions measurement, but also consideration of vehicle population makeup and vehicle usage. Experimental studies with the specific purpose of determining comprehensive vehicle particulate emission factors have not been performed, so the results of numerous programs conducted for a variety of purposes must be brought together to provide broadly applicable information. This publication presents the results of such a data-gathering and analytical effort, providing input to national, regional, and local (microscale) impact assessments.

The end results of this paper are particulate emission factors for a range of vehicle types and sizes. Different factors have been developed for diesel and gasoline-fueled light-duty vehicles, motorcycles, diesel and gasoline-fueled heavy-duty vehicles, and several vehicle subgroups. Engine-related particulate matter is considered separately from vehicle-related particulate emissions (clutches, brakes, and tires). Reentrained dust has not been considered. Factors are given for highway vehicles only.

David B. Kittelson and Daniel F. Dolan, Diesel Exhaust Aerosols, Gener. Aerosols Facil. Exposure Exp., [Tech. Pap. Symp.] 1979 (Pub. 1980), 337–59 (Eng). Edited by Klaus Willeke.

Diesel exhaust particles are primarily submicron agglomerates of carbonaceous nuclei. From 5 to 50% of the mass of these agglomerates is solvent extractable. The exts. are mainly hydrocarbons, but also contain O, N, and S. Calcns. of aerosol growth by coagulation show that markedly larger particles should be formed under typical diln. tunnel/exposure chamber conditions than under roadway diln. conditions. Adsorption/condensation processes and chem. reactions lead to chem. differences between diln. tunnel and roadway aerosols. The exhaust particle-size distributions produced by a small diesel engine depend upon fuel and operation conditions.

R. V. Malov, Underground Transport Diesel Engines Producing Exhaust Gases of Low Toxicity, Zashch. Vozdushn. Basseina Zagryaz. Toksichnymi Vybrosami Transp. Sredstv, Dokl., Vses. Nauchn. Konf. 1977, 2, 142–68 (Russ). Edited by I. L. Varshavskii and E. A. Pogosyan.

Theor. anal. and evaluation of expts. identified these factors as lowering toxicity of exhausts: (a) partition of combustion chamber; (b) use of 4 cycle engines; (c) avoidance of pressure charging; (d) increased rpm of the crankshaft; (e) preference to heavy fuels; (f) injection of water to cool and humidify the fresh charge; (g) partial recirculation of exhausts; and (h) decreasing the advantage angle of fuel injection.

W. F. Marshall, Emissions Characteristics of Six High-Speed Diesel Engines, Report 1978, BETC/TPR-78/7, 13 pp. (Eng). Avail. NTIS. From Energy Res. Abstr. 1979, 4(5), Abstr. No. 11289.

Low NO_X emissions, CO and smoke emissions that can be reduced to acceptable levels, and relatively high power densities of the tested diesel engines indicate good potential for their use in underground mines.

Joseph D. McCain, M. Gregory Faulkner, and Dennis C. Drehmel, Assessment of Diesel Particulate Control: Particle Size Measurements, U.S. Environmental Protection Agency, EPA-600/7-79-232c, December 1979.

The report gives results of tests on an Oldsmobile 98 diesel automobile to obtain data on parameters typical of a light duty vehicle. Data are presented for operations under four simulated driving conditions. (The increasing appearance of the diesel engine on the passenger car market has led to concern over particulate emissions (smoke) from these engines. Possible control strategies involving particulate collection in the hot exhaust stream requires knowledge of the concentrations and size distributions of the particulate matter at exhaust conditions.) Typical particulate concentrations at exhaust line conditions for the 350 CID diesel engine were found to be about 50 mg/Ncm. Aerodynamic mass median diameters were about 0.3 to 0.5 micrometers with the larger median diameters from higher engine speed/load conditions under steady state operation. Test results are qualitatively similar in size distribution to those found by other investigators in measurements of emissions from heavy duty diesel engines insofar as the impactor data are concerned. Additional particle size and composition data are required under conditions where the engine speed/load can be maintained at a constant value while the exhaust temperature is varied. This data would determine the suitability of particle collection at exhaust gas temperatures and would help define collection device cooling needs.

D. A. Pike, Modulated Exhaust Gas Recirculation System for Light Duty Diesel Engines, Report 1978, EPA/460/3-78/007; Order No. PB-288939, 221 pp. (Eng). Avail. NTIS. From Gov. Rep. Announce. Index (U.S.) 1979, 79(7), 198.

A system is described for exhaust gas recirculation, modulated to reduce the NO_X emission of a Mercedes 300D light duty diesel vehicle, and to limit other parameters, such as hydrocarbon and CO emissions, and performance loss, etc. To offset losses in performance due to the application of exhaust recycle, a turbo charger was fitted to the vehicle to provide excess air. No extra fuel was supplied.

E. Plassmann and F. Dursbeck, Exhaust-Emission Situation of Automobiles in Traffic. Means for Detection and Reduction of Pollutant Emissions, Mater. – Umweltbundesamt (Ger.) 1980, 4/80, KOLLOQ. FRAGEN AUSBREITUNGSRECHN ABGASEMISSIONEN DEM KRAFTFAHRZEUGVERKEHR, 1–33 (Ger).

The review, with refs., covers exhaust gas control legislation in the US, Japan, and Europe; methods of exhaust anal.; exhaust emission factors; exhaust emission control methods; field measurement; soot detn.; carcinogens; and odor emissions.

S. K. Rastogi and G. N. Pandey, Control of Diesel Engine Exhaust, Chem. Era 1979, 15(3), 17–27 (Eng).

A review with 19 ref.

Dennis Schuetzle, Frank S.-C. Lee, Thomas J. Prater, and Silvestre B. Tejada, The Identification of Polynuclear Aromatic Hydrocarbon Derivatives in Mutagenic Fractions of Diesel Particulate Extractions. International Journal of Environmental Analytical Chemistry. 93-144, 1981.

The soluble organic fraction (SOF) of particulate matter (from point sources, ambient air, etc.) contains hundreds of organic constituents. Normal-phase high pressure liquid chromatography (HPLC) has been used to separate the SOF into subfractions suitable for subsequent chemical analysis and bioassays. These fractions consist of non-polar, moderately polar (transition) and highly polar constituents. The non-polar fractions have been well characterized and consist of polynuclear aromatic hydrocarbons (PAH) and aliphatic hydrocarbons. The specific compounds present in the transition and polar fractions are for the most part unknown. This analytical information has been difficult to obtain since these compounds are highly labile, polar, of low volatility and in very low concentrations when compared to the bulk of material found in the SOF. Mutagenicity tests using the Ames Salmonella typhimurium assay indicate that the transition fraction accounts for most of the mutagenicity when compared to the non-polar and polar fractions.

A variety of chromatographic and mass spectrometric techniques are described that have been used to determine the composition of the HPLC fractions. More than one hundred species have been identified in the transition fraction of diesel particulate matter using high resolution gas chromatography (HRGC)/high resolution mass spectrometry (HRMS), HPLC and direct-probe high resolution mass spectrometry. It has been found that the transition fraction contains mostly PAH derivatives consisting of hydroxy, ketone, quinone, carboxaldehyde acid anhydride and dihydroxy derivatives of PAH. Three nitro-PAH species have been tentatively identified and 1-nitropyrene positively identified in the transition fraction. The problem of PAH oxidation during the analytical procedures and possible effect on bioassay results are discussed.

D. Schuetzle, T. Riley, T. J. Prater, T. M. Harvey, and D. Hunt, The identification of Nitrated Derivatives of PAH in Diesel Particulates, Submitted to Anal. Chem. March, 1981.

The analysis of nitrated polynuclear aromatic hydrocarbons (nitro-PAH) in diesel particulate extracts is described using a combination of high performance liquid chromatography, gas chromatography/mass spectrometry, high resolution mass spectrometry and mass spectrometry/mass spectrometry (MS/MS) techniques. Qualitative and quantitative results for diesel particulates generated from four different engines are presented using both triple stage quadrupole (TSQ) and mass analyzed ion kinetic energy spectrometry (MIKES) MS/MS techniques. Twenty-three different groups of nitro-PAH were identified using the TSQ instrument. The compound 1-nitropyrene (1-NP) was found to be one of the most abundant nitro-PAH species in these extracts. The 1-NP was quantified by standard addition techniques and found to represent from 55 ± 11 to 2080 ± 220 ppm in the four diesel extracts. A comparison of results from the MIKES and TSQ techniques are described and factors influencing quantitative results are discussed.

Karl J. Springer and T. M. Baines, Characterization of Sulfates, Odor, Smoke, POM and Particulates from Light and Heavy Duty Engines – Part IX, U.S. Environmental Protection Agency, EPA-460/3-79-007, June 1979.

This report expands EPA's data base on regulated and unregulated emissions from Diesel-powered cars and trucks. To the extent possible, comparisons were made to similar vehicles powered by engines that use gasoline as a fuel. For example, one part of the report is a comparison of the emissions, fuel economy, noise and acceleration characteristics of a pair of Volkswagen Rabbit and a pair of Oldsmobile Cutlass passenger cars. Each pair included a vehicle powered by a Diesel engine and a vehicle powered by a gasoline engine. Emissions characterization of heavy-duty engines included a Caterpillar 3208 with exhaust gas recirculation (EGR), and a Mack ETAY (B) 673A (both Diesels) and a Chevrolet 366 Heavy-Duty Gasoline (HDG) engine. The effect of several engine parameters on particulate and sulfate emissions is also discussed. A Caterpillar 3406 Heavy-Duty Diesel (HDD) engine was operated at several timings, with EGR, and in an open chamber (direct injected) as well as prechamber configuration. A pair of Daimler-Benz OM-352 HDD engines, one turbocharged and one not, were used to evaluate the effect of turbocharging. An American Bosch APS high pressure fuel injection system was compared to the standard system using the Mack ETAY (B) 673A HDD engine. The passenger car studies were based on recognized transient chassis dynamometer driving cycles for city, congested freeway, and highway type operation. The HDD engines were operated by mainly the 13-mode Federal cycle, or short versions thereof, on a stationary dynamometer. Emissions of major interest were unburned hydrocarbons, carbon monoxide, carbon dioxide, oxides of nitrogen, sulfate, particulate, smoke, odor, various aldehydes and specific hydrocarbons, benzo(a)pyrene, and particle size distribution. In all cases, vehicle fuel economy or engine fuel efficiency was obtained and reported.

V. I. Smailis and V. M. Kurov, Development of Modifications to Lower the Toxicity of Exhaust Gases from Turbocharged Diesel Engines, Zashch. Vozdushn. Basseina Zagryaz. Toksichnymi Vybrosami Transp. Sredstv, Dokl., Vses. Nauchn. Konf. 1977, 2, 280–98 (Russ). Edited by I. L. Varshavskii and E. A. Pogosyan.

Turbocharged diesel engines produced 10–25% less CO and NO per unit of fuel consumed than regular diesel engines but showed higher particulate (smoke) emission when equipped with a catalytic converter, thus requiring antismoke fuel additives. Turbocharged diesels have better performance but in general offer only marginal advantages in reducing pollution. Recirculation of the exhaust gases could reduce NO_X emission.

A. Toth, E. Pudmer, S. Borzsonyi, P. Bencze, and A. Zalai, Decrease of the Smoke Emission of Diesel Engines by the Use of Additives, (Hung.). Magy. Asvanyolaj-Foldgazkiserl. Intez. Kozl. 1980, 20, 203–24 (Hung).

Ba sulfonate (M-255) and Ca sulfonate (M-256) combustion improver-detergent 0.25 wt. % concn. in diesel fuel decreased the soot formation to 0.2–0.4 g/m^3 exhaust gas which is well below the EPA std., i.e., 0.5 g/m^3.

Timothy J. Truex, William R. Pierson, and Douglas E. McKee, Sulfate in Diesel Exhaust, American Chemical Society, Environmental Science & Technology, 1980, 14, 1118.

A study of sulfate emissions from diesel-powered light-duty vehicles has been conducted. The applicability of a controlled-condensation sampling procedure for determining the chemical speciation of vehicle sulfate emissions has been demonstrated and used to determine that sulfuric acid is the major sulfate species present in diesel exhaust. Diesel sulfate emissions were found not to be proportional to fuel sulfur level, the percent $S \rightarrow SO_4^{2-}$ conversion increasing as fuel sulfur level decreases. An Opel diesel vehicle was equipped with an experimental oxidation catalyst to determine the effect on diesel sulfate emissions.

Timothy J. Truex, William R. Pierson, Douglas E. McKee, Mordecai Shelef, and Richard E. Baker, Effects of Barium Fuel Additive and Fuel Sulfur Level on Diesel Particulate Emissions, American Chemical Society, Environmental Science & Technology, 1980, 14, 1121.

The effects of a barium fuel additive and of fuel sulfur level on diesel particulate emissions have been investigated. The Ba fuel additive results in a 30–40% decrease in exhaust opacity as measured by a smokemeter. The opacity reduction does not result from a reduction in total particulate mass emission rate, which is relatively unaffected by the Ba additive. There is, however, an ~30% reduction in the mass emission rate of carbonaceous material. More than 90% of the Ba introduced with the fuel can be accounted for in the vehicle exhaust. In the presence of the Ba additive, the amount of emitted SO_4^{2-} is approximately equimolar with the Ba, and electron spectroscopy shows that both the Ba and the SO_4^{2-} are wholly in the form of $BaSO_4$–provided that the amount of sulfur in the fuel is sufficient to react with all of the Ba. This $BaSO_4$ represents a substantial increase in diesel particulate SO_4^{2-} emission rate compared to that observed without the Ba additive. The level of fuel sulfur does not have a significant effect on total particulate mass emission rate. In the absence of the Ba additive, the diesel SO_4^{2-} emission rate is related to fuel sulfur level, but not linearly; the percent of sulfur converted to sulfate decreases as the fuel sulfur level is increased.

Thomas A. Tupaj, Testing of Five Diesel-Powered Passenger Cars at High and Low Altitude, U.S. Environmental Protection Agency, EPA-AA-TEB-81-6, October 1980.

A group of five in-use diesel-powered passenger cars was tested for emissions and fuel economy at both high altitude and low altitude locations. The test sequence consisted of the current Federal Test Procedure (FTP) and the Highway Fuel Economy Test (HFET). Particulate emissions were also measured over both cycles at both locations. On the average, a substantial increase was noted at high altitude for hydrocarbons, carbon monoxide and particulates as compared to the low altitude measurements. As measured over the FTP, the average increases for these three pollutants were 130%, 75% and 45% respectively. Results of the HFET were in the same direction but were of a slightly smaller magnitude. The emission levels of oxides of nitrogen were approximately the same at each location. The average fuel economy values at high altitude were approximately 14% lower than those at low altitude.

Robert A. Walter, John O. Storment, and Carlos W. Coon, Single-Cylinder Diesel Engine Tests with Unstabilized Water-in-Fuel Emulsions, Proc., Annu. Meet. – Air Pollut. Control Assoc. 1978, 71st. Vol. 4, Paper No. 54.1, 24 pp. (Eng).

Tests were conducted on unstabilized water-in-fuel emulsions in diesel engines using 2 emulsification devices to define favorable and adverse consequences. The benefits obtained were an overall redn. in exhaust smoke d., NO_x, and specific fuel consumption at some operating conditions. Adverse effects obsd. were sometimes substantial increases in concn. of unburned hydrocarbons and CO.

Exhaust Aftertreatment

M. G. Faulkner, E. B. Dismukes, J. R. McDonald, D. H. Pontius, A. H. Dean, and Dennis C. Drehmel, Assessment of Diesel Particulate Control: Filters, Scrubbers, and Precipitators, U.S. Environmental Protection Agency, EPA-600/7-79-232a, October 1979.

The report discusses an investigation of three types of devices that might be used for the aftertreatment of diesel exhaust to lower particulate emissions from light-duty vehicles. The devices are filters, electrostatic precipitators (ESPs), and wet scrubbers. The conclusions reached are that filters and ESPs merit further consideration, but wet scrubbers do not. Wet scrubbers were eliminated from further consideration on the basis of excessive size, low efficiency, and excessive fluid loss by evaporation (assuming that water is the fluid of choice). Filters and ESPs, although appearing to offer significant potential, have possible disadvantages that can only be assessed experimentally. Prototype filters and electrostatic devices that appear worthy of experimental study are described.

M. J. Murphy, L. J. Hillenbrand, D. A. Trayser, and John H. Wasser, Assessment of Diesel Particulate Control: Direct and Catalytic Oxidation, U.S. Environmental Pro-

tection Agency, EPA-600/7-79-232b, October 1979.

The report discusses the technology and potential for disposal of diesel particulate by oxidation. Relevant properties of typical diesel particulate are given; note is taken of the small size (on the order of 0.1 micrometer diameter) and the presence of a portion extractable with an organic solvent. Available reaction rate data is used to derive particle lifetimes at various temperatures; these exceed likely exhaust system residence times. The use of catalysts to increase oxidation rates and lower ignition temperatures is discussed. Small amounts of many metals are effective in increasing the rate of oxidation by 2 to 5 orders of magnitude. Chemical reactor theory is used to derive ignition and operational characteristics of trap/oxidizers. Note is also taken of the tendency of these devices to go rapidly from a cold unignited state to an ignited state close to the adiabatic flame temperature of the fuel. Design techniques to ameliorate undesirable temperature excursions are presented.

John B. Strachan, Emission Control Apparatus for Diesel Engines, U.S. 4,190,629 (Cl. 422–169; F01N3/02), 26 Feb 1980, Appl. 855,433, 28 Nov 1977; 4 pp.

Exhaust gas from the waters-jacketed exhaust manifold of a diesel engine passes through a water-jacketed adapter to a catalytic purifier which lowers hydrocarbon and CO emissions. The water jacket maintains a low temp. on the outside of the catalytic purifier, although the inside is at a high temp. for optimum performance. The exhaust gas then passes through a scrubber consisting of 3 concentric tubes, the 2 outer tubes being perforated, placed in a water reservoir. The scrubber lowers the temp. of the exhaust gas, removes combustion products, and quenches exhaust flames.

Toshinobu Suzuki, Kiyoshi Chiba, Tomoya Tokuhiro, Yoshiyuki Iwasawa, Yasuo Kajioka, and Tomio Komine, Device for Purifying Exhaust Gas of Diesel Engine, PCT Int. Appl. 80 00, 362 (Cl. F01N3/15), 06 Mar 1980, Japan. Appl. 78/92,132, 27 Jul 1978; 21 pp.

The title device comprises heat-resistant filters upstream of a cylindrical catalytic converter contg. laminated layers of filtering and catalytic materials. Soot entrained in the exhaust gas is completely removed, and the soot-contaminated filters are regenerated by burning the soot. The converter is equipped with a high-pressure burner or a neg.-pressure generator that blows hot gas into the filtering layers to remove the retained soot.

Yoshinobu Suzuki, Kiyoshi Chiba, Tomoya Tokuhiro, Yoshiyuki Iwasawa, Yasuo Kajioka, and Tomio Komine, Device for Diesel Exhaust Gas Treatment, PCT Int. Appl. 80 00, 168 (Cl. F01N3/02), 07 Feb 1980, Japan. Appl. 78/77,794, 27 Jun 1978; 16 pp.

The title device comprises a heat-resistant filter in a gas-purifying conduit, a heater at the upstream side of the filter, a catalytic converter, and a sedimentation tank. The exhaust gas contg. water vapor, soot, CO, and hydrocarbons is heated and passed through the filter to remove the dried soot and through the converter to remove the toxic components. The soot is removed by washing the filter with water. The wastewater formed is drained to the sedimentation tank for sepn. of the soot.

V. A. Zvonov, Z. T. Zvonova, P. P. Fesenko, and I. P. Vasil'ev, Study of the Catalytic Neutralization of Diesel Exhaust Gases, Zashch, Vozdushn. Basseina Zagryaz. Toksichnymi Vybrosami Transp. Sredstv, Dokl., Vses. Nauchn. Konf. 1977, 1, 348–62 (Russ). Edited by I. L. Varshavskii and E. A. Pogosyan.

Removal of >80% of NO_x from diesel exhaust gases was achieved with addn. of 100% excess of NH_3 (calcd. on NO_x) on a Cu–Cr oxide catalyst at 350–450°. The optimum height of the catalyst layer was 90 mm. Changes of concn. of NO_x (12–13% decrease at 200–400° and 50% increase at 500°) were obsd. in front of the reactor in presence of soot in the NH_3 contg. exhaust gases.

Health Effects

Health Effects of Diesel Engine Emissions, Proceedings of an International Symposium, December 3-5, 1979, Sponsored by the Health Effects Research Laboratory, Edited by W. E. Pepelko, R. M. Danner, N. A. Clarke, U.S. Environmental Protection Agency, EPA-600/9-80-057a, November 1980.

The Health Effects Research Laboratory of the U.S. Environmental Protection Agency sponsored an International Symposium on the Health Effects of Diesel Emissions. It was held in Cincinnati, Ohio on December 3-5, 1979.

The Symposium brought together scientists, engineers, and Federal, state, and local public health officials for the purpose of determining the state-of-knowledge regarding the physical nd chemical characteristics of diesel emissions, in vitro carcinogenic and mutagenic effects of diesel emissions and their components, their biochemical and metabolic effects, the toxicological effects of inhaled diesel emissions, the mutagenic and carcinogenic potency of extracts of diesel and related emissions, the mutagenicity of inhaled diesel emissions, the carcinogenic effects of exposure to diesel emissions, and the results of epidemiological studies involving human exposure to diesel emissions. The Symposium culminated in a panel discussion on the health risk assessment of diesel emissions.

The Proceedings consists of 77 manuscripts and associated discussions, as well as the panel discussion.

Diane M. Cavagnaro, Diesel Exhaust Emissions, Volume 2, 1977–January, 1979 (citations from the NTIS data base). Report 1979. Order No. NTIS/PS-79/0037, 113 pp. (Eng). Avail. NTIS. From Gov. Rep. Announce, Index (U.S.) 1979, 79(8), 95.

Exhaust gases from stationary and vehicular diesel engines are reviewed with a bibliog. of 107 refs. with emphasis on control from a health standpoint.

J. Huisingh, R. Bradow, R. Jungers, L. Claxton, R. Zweidinger, S. Tejada, J. Bumgarner, F. Duffield, M. Waters, et al. Environ. Sci. Res. 1979, 15(Appl. Short-Term Bioassays Fractionation Anal. Complex Environ. Mixtures), 381-418 (Eng).

Org. exts. of diesel particle emissions were mutagenic in Salmonella typhimurium and the mutagenicity was dependent on the org. solvent used in the extn. The particle emissions from light-duty passenger cars had a wide range of benzo [a] pyrene content and mutagenic activity. Ext. sample storage also affected mutagenicity. Emission particle mutagenicity was affected by the fuel and vehicle. Min. quality fuel had the highest I content and mutagenic activity. However, the fractions contg. the highest I levels were not the most mutagenic fraction and a min. of 4 mutagenic components were present.

I. P. Lee, K. Suzuky, S. D. Lee, and R. L. Dixon, Aryl Hydrocarbon Hydroxylase Induction in Rat Lung, Liver, and Male Reproductive Organs Following Inhalation Exposure to Diesel Emissions. Toxicol. Appl. Pharmacol. 1980, 52(1), 181-4 (Eng).

The effects of inhalation exposure of diesel emission on the specific activities of aryl hydrocarbon hydroxylase (AHH) [9037-52-9] and epoxide hydrase (EH) [9048-63-9] in rat lung, liver, testis, and prostate gland were described.

Adult male rats (8–10 wk old) were exposed to diesel emission (1:13 diln.) contg. 14.2 ppm (vol./vol.) total hydrocarbon, 20 h/day for 42 days. Although liver exhibited the greatest overall AHH activity, the percentage increase over control values was highest in the prostate gland. On day 15 of exposure, AHH activity had increased to 4.5-fold the control and remained elevated throughout the entire exposure period. In contrast, AHH induction in liver and lung occurred on day 33 of exposure, and the max. AHH induction was obsd. on day 42 (1.4- and 1.3-fold increase in lung and liver, resp.). EH activity did not changed during the exptl. period.

Irving Salmeen and Anna Marie Durisin, Some Effects of Bacteria Population on Quantitation of Ames Salmonella-Histidine Reversion Mutagenesis Assays, Mutation Research, 00 (1961) 000-000 © Elsevier/North Holland Biomedical Press, Mutation Research, 1981.

We show from simple theory that the number of mutants observed in the Ames assay should be approximately proportional to αCn_oP where α is the mutation rate per concentration of mutagen, $C{:}n_o$ is the initial inoculum; and P is the average number of bacteria per colony in the background lawn. We tested this theory by carrying out Ames assays for the directly acting mutagen 2-nitrofluorene on TA98 as a function of initial inoculum over the range 10^4–10^8 bacteria/plate. P was assumed to be proportional to the average volume of the background colonies and was estimated from 100X photomicrographs of the background lawn. The initial inoculum was determined by counting background colonies in the photomicrographs, by dilution-plating, and by electronic particle counting. We found that the number of mutants depended weakly on n_o, but the dependence could be quantitatively accounted for by the simple theory. These data and the theory explain at least some of the reported variations in quantitation of the Ames assays show that the slopes of dose-response curves depend on n_o; explain the range observed for spontaneous revertants; suggest some limitations to quantifying the Ames assay and, within these limits, suggest a method for normalizing independently obtained Ames assay data.

Irving Salmeen, Anna Marie Durisin, Thomas J. Prater, Timothy Riley, and Dennis Schuetzle, Contribution of 1-Nitropyrene to Direct Acting Ames Assay Mutagenicities of Diesel Particulate Extracts, Science, 1981.

We have determined that 1-nitropyrene (1-NP) accounts for the following percentages of the direct acting mutagenicities (Ames assays with Salmonella strains TA 1538, 98, and 100) of exhaust particulate extracts obtained from three different diesel-powered passenger vehicles: 12, 16, and 24% (TA98); 7, 13, and 17% (TA1538), and 1, 4, and 9% (TA100). 1-NP was quantified using a recently developed mass spectrometer/mass spectrometer technique. We discuss requirements necessary to achieve accurate determinations of the Ames assay mutagenicity of 1-NP relative to that of particulate extract.

Dennis Schuetzle, Frank S.-C. Lee, Thomas J. Prater, an and Silvestre B. Tejada, The Identification of Polynuclear Aromatic Hydrocarbon (PAH) Derivatives in Mutagenic Fractions of Diesel Particulate Extracts, Intern. J. Environ. Anal. Chem. 1981, Vol 9(2), page 93. Gordon and Breach Science Publishers Inc., Great Britain.

The soluble organic fraction (SOF) of particulate matter from diesel exhaust (from point sources, ambient air, etc.) contains hundreds of organic constituents. Normal-phase high pressure liquid chromatography (HPLC) has been used to separate the SOF into subfractions suitable for subsequent chemical analysis and bioassays. These fractions consist of non-polar(PAH), moderately polar (transition) and highly polar constituents. The non-polar fractions have been well characterized and consist of PAH and aliphatic hydrocarbons. The specific compounds present in the transition and polar fractions are for the most part unknown. This analytical information has been difficult to obtain since these compounds are highly labile, polar, of low volatility and in very low concentrations when compared to the bulk of material found in the SOF. Mutagenicity tests using the Ames Salmonella typhimurium assay indicate that the transition fraction accounts for most of the mutagenicity when compared to the non-polar (PAH) and polar fractions.

A variety of chromatographic and mass spectrometric techniques are described that have been used to determine the composition of the HPLC fractions. More than one hundred species have been identified in the transition fraction of diesel particulate matter using high resolution gas chromatography (HRGC), high resolution mass spectrometry (HRMS), HPLC and direct-probe high resolution mass spectrometry. It has been found that the transition fraction contains mostly PAH derivatives consisting of hydroxy, ketone, quinone carboxaldehyde, acid anhydride and dihydroxy derivatives of PAH. Three nitro-PAH species have been tentatively identified and 1-nitropyrene positively identified in the transition fraction. The 1-nitropyrene was found to account for approximately 45",, and 30",, of the direct-acting mutagenicity observed for the transition fraction and total extract, respectively. The HPLC separation procedure was shown to give better than 95",, recovery of the mass and mutagenic activity. The problem of PAH oxidation during the analytical procedures and possible effect on bioassay results are discussed.

Dennis Schuetzle, Thomas Prater, Timothy Riley, Ann Durisin, and Irving Salmeen, Analysis of Nitrated Derivatives of PAH and Determination of Their Contribution to Ames Assay Mutagenicity for Diesel Particulate Extracts. Fifth International Symposium on Polynuclear Aromatic Hydrocarbons, Columbus, Ohio – October 28–30, 1980.

We have recently reported the identification and semi-quantitative analysis of 1-nitropyrene (1-NP) and other nitro-PAH in diesel particulate extracts(1). We now describe further studies to quantify 1-NP in particulate extracts using mass analyzed ion kinetic energy spectroscopy (MIKES).

Particulates were collected on filters in a dilution tube from two diesel vehicles run on a chassis dynamometer. One sample was collected in-house (OP-1), the other provided by the EPA(NI-1). The 1-NP was quantified by external and internal standardization (standard addition). The 1-NP was measured with ±10% precision (standard error of replicate runs) at 20 ng levels in 10μg of whole diesel extract. At the 2 ng level interferences reduced the precision to ±50%. These interfering compounds were removed using wide cut liquid chromatography resulting in precision of ±20% at the 2 ng level. Concentrations of 1-NP in the total particulate extracts from the two vehicles were found to be 55 ± 11 and 2030 ± 120 ppm (wgt/wgt).

We carried out Ames assays (TA 98 without activation) of unfractionated particulate extracts and found 1.2×10^3 revertants/mg/(9.6×10^7 bacteria/plate) and 18.6×10^3 revertants/mg/(4×10^7 bacteria/plate) for samples OP-1 and NI-1, respectively.

We determined the Ames assay mutagenicity of 1-NP because direct injection mass spectrometry of our reference sample showed that it contained 1.4% of an unknown isomer of dinitropyrene which may be 75 times more mutagenic than 1-NP(2). Furthermore literature values of mutagenicity for 1-NP are uncertain because the papers do not include analysis for dinitropyrenes. In addition no account is taken of the dependence of mutagenicity on the initial bacterial inoculum, which we have shown(3) can cause a 2-3 fold variation in apparent mutagenicities over the range of initial inocula acceptable in the "standard" Ames assay. We found that the mutagenicity of 1-NP was 2.3×10^6 revertants/mg/9.2×10^7 bacteria/plate and 1.9×10^6 revertants/mg/4.6×10^7 bacteria/plate. These different mutagenicities at different initial inocula are consistent with our previous observations(2). Therefore, the 1-NP in samples OP-1 and NI-1 account for 11 ± 2 and 21 ± 4% of the total extract mutagenicities, respectively. We did not analyze for dinitropyrene in the diesel extracts. Previous work(1), however, has shown that the HPLC fraction in which dinitropyrene is expected to elute accounts for only about 20% of the direct acting mutagenicity of the total extract, whereas the fractions eluting with 1-NP account for more than 50% of the total extract mutagenicity. Further work will be needed in order to determine if 1-NP is a combustion product or a sample collection, handling, and storage artifact.

1. D. Schuetzle; F.S.C.-Lee, T. J. Prater, and Sylvester B. Tejada, "The Identification of Polynyuclear Aromatic Hydrocarbon Derivatives in Mutagenic Fractions of Diesel Particulate Extracts, Presented at the 10th Annual Symposium on the Analytical Chemistry of Pollutants, Dortmund, Germany, May 28–30, 1980. International Journal of Environmental Analytical Chemistry, In Press 9(2), 1981.

2. H. S. Rosenkranz, E. C. McCoy, D. R. Sanders, M. Butler, D. Kiriazides, R. Mermelstein, "Nitropyrenes; Isolation, Identification, and Reduction of Mutagenic Impurities in Carbon Black and Toners", Science, 209, 1039(1980).

3. I. Salmeen, A. M. Durisin, "Some Effects of Bacteria Population on Quantitation of Ames Salmonella Histidine Reversion Mutagenesis Assays," Mutation Research, In Press (1981).

A. P. Wehner, B. O. Stuart, and C. L. Sanders, Inhalation Studies with Syrian Golden Hamsters. Prog. Exp. Tumor Res. 1977 (Pub. 1979). 24(Syrian Hamster Toxicol. Carcinog. Res.), 177-98 (Eng).

The Syrian golden hamster was used for chronic inhalation studies to investigate the biol. effects of cigaret smoke, chrysotile asbestos, NiO, CoO, the latter 3 in combination with cigaret smoke, talc baby powder, Rn daughters, U ore dust, diesel engine exhaust, and Rn daughters in combination with U ore dust and(or) diesel engine exhaust, asbestos cement, fly ash, and plutonium dioxides. Exposure to asbestos (23 μg/L respirable aerosol concn., 7 h/day, 5 days/wk) for 11 mo resulted in asbestosis in all animals, in an increased incidence of lung adenomas, indicating an early neoplastic response, and in a shorter life-span. Life-span exposures to NiO (53 μg/L) and to CoO (10 μg/L) resulted in pneumoconiosis. Life-span expsoure to cigaret smoke (3 x /day, 5 days/wk, in Hamburg II smoking machines) decreased body wt. and increased the incidence of tumors and epithelial lesions of the larynx. The increased incidences in the smoke-exposed hamsters were not necessarily caused by the smoke exposures, but might be explained by the longer life-span of the smoke-exposed hamsters. Exposure to talc baby powder (8 μg/L $\leq$150 min/day, 5 days/wk, $\leq$300 days) caused no changes. Exposure to 1200 working levels of Rn daughters 6 h/day, 5 days/wk for 15 mo produced severe epithelial proliferation culminating in pulmonary squamous carcinoma; the same exposure in combination with 15 μg/L U ore dust also produced invasive carcinoma. Exposure to 15 μg/L U ore dust alone caused pneumoconiosis and epithelial proliferation. Daily (6 h/day, 5 days/wk) life-span exposures to diesel engine exhaust (50 ppm CO) caused accumulation of soot in alveolar macrophages, with emphysema, fibrosis and cuboidal metaplasia; the same exposure plus Rn daughters and U ore dust caused acceleration of these effects plus pulmonary squamous metaplasia. Exposure to $^{238}PuO_2$ and $^{239}PuO_2$ particles in male and female Syrian golden hamsters produced a shortened life-span only in male hamsters with the highest deposition. Radiation pneumonitis and fibrosis and epithelial metaplasia of the lung were increased with increasing radiation dose. Only 3/448 life-span hamsters developed malignant lung tumors, all adenocarcinomas and all at initial alveolar depositions >160 nCi. Thus, the hamster was resistant to lung tumor formation following inhalation of α-emitting radionuclides. Results for asbestos cement and fly ash are not yet available.

M. J. Wiester, R. Iltis, and W. Moore, Altered Function and Histology in Guinea Pigs After Inhalation of Diesel Exhaust. Environ. Res. 1980, 22(2), 285-97 (Eng).

Health effects of inhaled diesel engine exhaust were evaluated in infant guinea pigs following 4 and 8 wk of exposure. Animals were exposed to 1 part exhaust dild. by 13 parts clean air for 20 h/day, 7 days/wk. Lung function, ECG, growth rate, and histopathol. were assessed following exposure. After 4 wk, animals exposed to irradiated exhaust, showed a 35% increase in pulmonary air flow resistance, and a decrease in heart rate. Necropsy, after 8 wk, showed exhaust exposed animals to have black discoloration of the lungs and increased lung to body wt. ratios. Microscopic examn. of the lungs from these animals revealed black particulate material in the cytoplasm of alveolar macrophages and in draining lymph nodes.

Other Related Material

R. Dwight Atkinson, Localized Air Quality Impacts of Diesel Particulate Emissions, U.S. Environmental Protection Agency, EPA-AA SDSB 79-31, November 1979.

A number of studies are available which attempts to predict the level of diesel exhaust particulate along the roadway resulting from the increased use of diesel engines as power plants for light- and heavy-duty vehicles. Among these are reports prepared by EPA, the Southwest Research Institute, General Motors Corp., Toyota Motor Co., and the Aerospace Corp. When trying to evaluate the results of these studies, the rate of diesel penetration into the market, vehicle emission factors, traffic density and meteorological conditions are among the variables encountered. This report attempts to establish a broader base of comparison among these studies than presently exists. This report consists of three basic sections, the first provides a brief description of the various studies, the second incorporates the standardizing assumptions in the modification procedure, and the third compares the studies based on the changes made in section two.

Axel Berner and Christian Luerzer, Mass Size Distributions of Traffic Aerosols at Vienna. J. Phys. Chem. 1980, 84(16), 2079-83 (Eng).

The mass size distributions of aerosols were measured directly with low-pressure cascade impactors. Because of the wide span from 0.06 to 16μ of aerodynamic equiv. diam., the accumulation mode and part of the nucleation and coarse mode fall into the measuring range of the impactor. Urban aerosols were sampled near a road by 3 identical impactors at different heights above street level, and the structure of the mass size distributions and their change with increasing height were investigated.

S. S. Filatov, Problem of Protecting the Air Basin of Deep Open-Cut Mines in the Development of Heavy-Freight Trucking. Zashch. Vozdushn. Basseina Zagryaz. Toksichnymi Vybrosami Transp. Sredstv, Dokl., Vses. Nauchn. Konf. 1977, 2, 364-82 (Russ). Edited by I. L. Varshavskii and E. A. Pogosyan. Akad. Nauk Ukr. SSR,

Inst. Probl. Mashinostr.: Kharkov, USSR.

The possibilities of the application of gas-turbine engines and catalytic and liq. exhaust treatment systems for heavy trucks, as well as of ventilation systems, were described and discussed.

Melvin N. Ingalls and Ronald L. Bradow, Particulate Trends with Increasing Dieselization 1977 to 2000, APCA 81-56.2, 74th Meeting & Exposition, June 21-26, 1981, Philadelphia, PA.

Diesel engines are being used in increasing numbers for automotive power to meet the demand for improved fuel economy. From an environmental impact viewpoint this increase merits study since diesel exhaust contains more particulate matter than gasoline engine exhaust. This paper presents the results of a study to evaluate how the increase in diesel vehicles will affect the mass of particulates emitted from highway vehicles on a nationwide basis. The total particulate mass from vehicles includes both engine exhaust and vehicle generated (from tires, brakes, etc.) particulate. It does not include reentrained dust.

These national particulate totals in metric tons per year are presented for three assumed rates of dieselization, both with diesel particulate regulations and assuming no diesel particulate regulations. Because specific compounds in the particulate emissions may affect biological systems differently, emission rates are also presented for the fractions of lead, sulfate, soluble organics and carbonaceous material in the total particulate mass. The highway vehicle particulate mass for total particulate and each of the particulate fractions is estimated for the baseline year 1977 and for three future years, 1985, 1990 and 2000.

Jean L. Muhlbaier and Ronald L. Williams, Fireplaces, Furnaces, and Vehicles as Emission Sources of Particulate Carbon. Presented at International Symposium on "Particulate Carbon: Atmospheric Life Cycle." Sponsored by Environmental Science Department, General Motors Research Laboratories, Warren, Michigan 48090, October 12–14, 1980.

Particulate carbon in the atmosphere results from the incomplete combustion of fossil and contemporary fuels. Fuel-specific emission rates of elemental and organic carbon are presented for residential wood-burning fireplaces and gas furnaces as well as precatalyst, catalyst, and diesel automobiles. The fireplace emissions from several types of wood were collected from residential chimneys using standard EPA source-sampling procedures. Furnace emissions were collected with a laboratory setup and the effects of variations in stoichiometry and cycling rate were evaluated. Automotive emissions were collected from a dilution tunnel using two driving cycles both at sea level and high altitude. Carbon analysis was performed by several methods, including temperature programming, combustion analysis, and solvent extraction. Organic carbon associated with the particulate emissions was also tested for benzo(a)pyrene content. The emission rates and annual fuel consumption data were used to estimate the relative strength of these particulate carbon sources on a national basis.

Roy A. Paul and Richard Atherton, The Impact of Future Diesel Emissions on the Air Quality of Large Cities. U.S. Environmental Protection Agency, EPA-450/5-79-005, May 1979.

The impact of diesel-powered vehicles in the central business districts of New York, Chicago and Los Angeles was studied. Annual particulate matter emissions from automobiles, taxis, buses and light-duty trucks were projected to the year 2000. The emissions were correlated with air monitoring data in order to calculate the impact on air quality. Significant increases were predicted in atmospheric particulate matter because of increased diesel engine use.

James N. Pitts, Jr., Karel A. Van Cauwenberghe, Daniel Grosjean, Joachim P. Schmid, and Dennis R. Fitz, Chemical and Biological Aspects of Organic Particulates in Real and Simulated Atmospheres. [Rep.] LBL 1979, LBL–9037, Carbonaceous Part. Atmos., 169-76 (Eng).

The reactions of benzo[a]pyrene (I) [50-32-8] deposited on glass fiber filters with ambient photochem. smog, as well as with O_3, NO_2, and peroxyacetyl nitrate [2278-22-0] in simulated atms. were studied. A variety of products were readily formed, including phenols, diphenols, dihydrodiols, quinones, etc. Directly mutagenic nitro derivs. were formed upon exposure of I and of perylene [198-55-0] to 1 ppm of NO_2 in air contg. traces of HNO_3 (~10 ppb). If such reactions occur in urban atms., they may account in part for the excess carcinogenicity (over I and certain other polycyclics) obsd. in org. particulates collected from smog and exhausts of gasoline and diesel engines. However, historically, such gas-solid interface processes have occurred on the filters used in particulate sampling. Thus there is a possibility of filter artifacts.